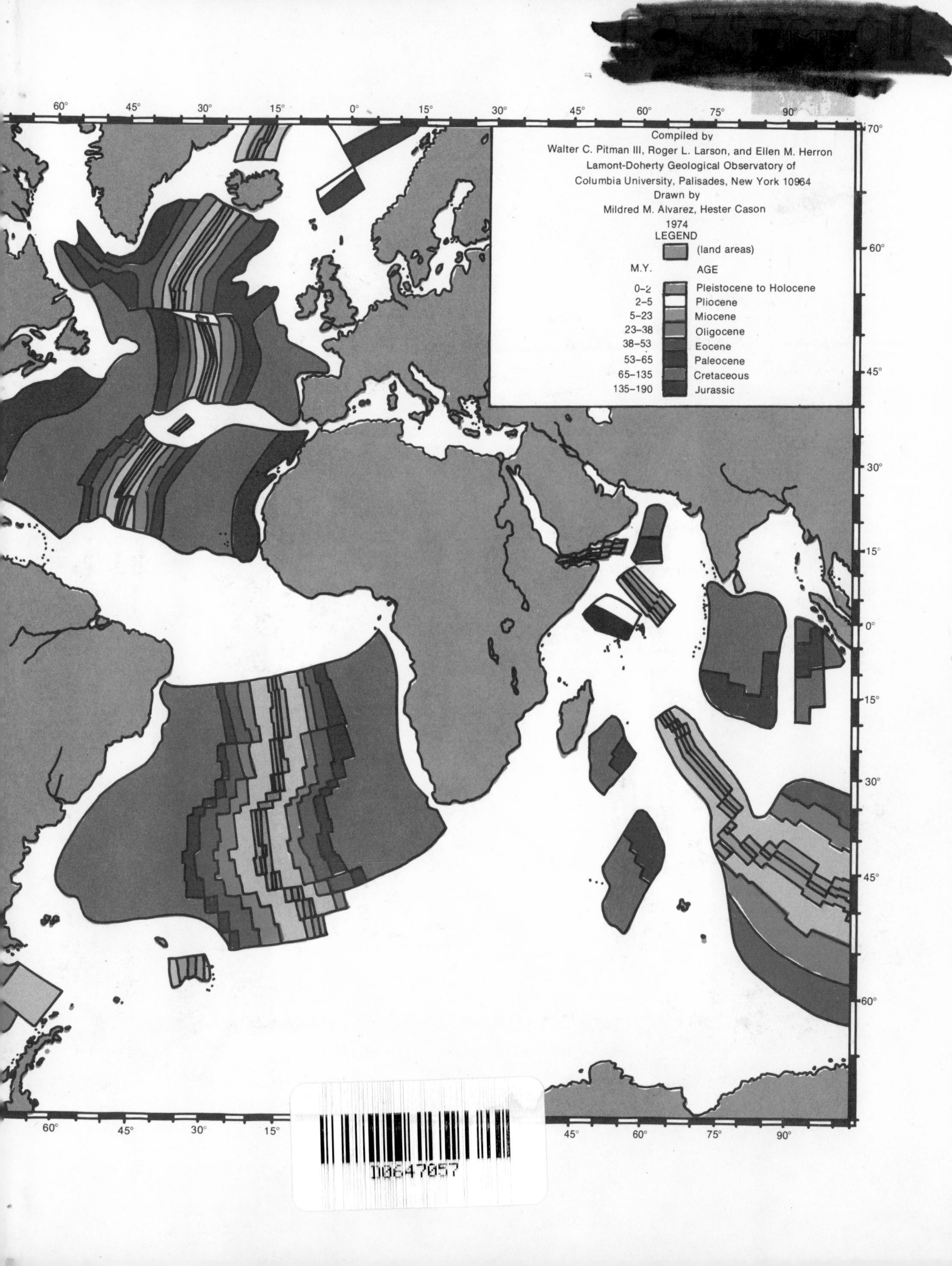
Compiled by
Walter C. Pitman III, Roger L. Larson, and Ellen M. Herron
Lamont-Doherty Geological Observatory of
Columbia University, Palisades, New York 10964
Drawn by
Mildred M. Alvarez, Hester Cason
1974
LEGEND
(land areas)
M.Y.
AGE
0–2 Pleistocene to Holocene
2–5 Pliocene
5–23 Miocene
23–38 Oligocene
38–53 Eocene
53–65 Paleocene
65–135 Cretaceous
135–190 Jurassic
60°
45°
30°
15°
0°
15°
30°
45°
60°
75°
90°
70°
60°
45°
30°
15°
0°
15°
30°
45°
60°

essentials of earth science

HAROLD L. LEVIN

Professor of Geology, Washington University, St. Louis

SAUNDERS COLLEGE PUBLISHING

*Philadelphia New York Chicago San Francisco
Montreal Toronto London Sidney Tokyo
Mexico City Rio de Janeiro Madrid*

Address orders to:
383 Madison Avenue
New York, NY 10017

Address editorial correspondence to:
West Washington Square
Philadelphia, PA 19105

Text Typeface: Goudy Old Style
Compositor: Progressive Typographers Inc.
Acquisitions Editor: John Vondeling
Developmental Editor: Lloyd Black
Project Editor: Joanne Fraser, Patrice L. Smith
Copyeditor: Sally Atwater
Art Director: Carol C. Bleistine
Art/Design Assistant: Virginia A. Bollard
Text Design: Adrianne Onderdonk Dudden
Cover Design: Lawrence R. Didona
New Text Artwork: Larry Ward and Linda Maugeri
Production Manager: Tim Frelick
Assistant Production Manager: Maureen Iannuzzi

Cover credit: Photo by Ruska Hadian, Volcanology Survey of Indonesia.
Courtesy of the U.S. Geological Survey.

Library of Congress Cataloging in Publication Data

Levin, Harold L. (Harold Leonard), 1929–
Essentials of earth science.

Includes bibliographies and index.
1. Earth Sciences. I. Title.
QE26.2.L48 1985 550 84-14168
ISBN 0-03-062411-8

ESSENTIALS OF EARTH SCIENCE ISBN 0-03-62411-8

3456 032 987654321

CBS COLLEGE PUBLISHING
Saunders College Publishing
Holt, Rinehart and Winston
The Dryden Press

preface

This is a book about the earth and the earth's neighbors in space. It is written for all those with an interest in their planetary home and its relationship to the rest of the universe. Students today are fortunate, for they live at a time when the frontiers of earth and planetary sciences are advancing with exceptional speed. The advance is catalyzed by new methods for examining the abyssal depths of the oceans, by spacecraft that explore formerly unreachable planets, and by artificial satellites that report the endlessly varied activity of the atmosphere. With a virtual flood of new information for their use, today's scientists have been able to recognize the earth as a truly vibrant planet on which ocean floors are in constant motion, continents break apart to admit new seas, and even the magnetic poles do an occasional turnabout. These events, together with the more traditional aspects of the earth sciences, are discussed in this text.

Essentials of Earth Science is particularly directed toward first- and second-year college students taking their initial course in earth science. It is also useful as a beginning text for those preparing to teach earth sciences in the secondary schools. The text has an essentially nonmathematical approach, emphasizing basic principles and providing broad coverage in geology, oceanography, meteorology, and astronomy. An effort has been made to provide flexibility and sufficient background in each chapter to permit adjustment of the topic sequence to the instructor's preferred course design. For the student, there are ample illustrations, a comprehensive glossary, and questions for use in study and classroom discussions. The appendices include conversion factors for measurement systems; quantitative information about the earth, moon, and sun; a table of properties of minerals; and a discussion of topographic maps.

The plan of the book is to begin with an examination of the solid earth. Because rocks and minerals are the basis for an understanding of many geologic processes, these earth materials are introduced early. Similarly, an understanding of the evidence for sea floor spreading is enhanced if the student has already acquired an introduction to the methods for determining the age of rocks and understands how earthquake waves are used for deciphering the earth's internal structure. These topics, therefore, precede the chapter on plate tectonics. A chapter on geochronology provides the perspective of time and helps the student understand correlation and the endless operation of geologic processes in shaping the landscapes of the continents.

From events on land, the discussion shifts to the oceans, with emphasis on oceanic circulation, the ocean floors, and shoreline processes. This is followed by discussion of the "ocean of air" we call the atmosphere. For humans, the atmosphere is the most immediate and vital of all the major environmental components of the planet. This thin, gaseous envelope of the earth is irradiated by the sun, and this provides the energy that heats the air, drives the winds, and influences climate and weather.

The final chapters of *Essentials of Earth Science* view the earth as a planet, and consider its relation to the solar system and the universe. Our planet is not unique. Its development, and the development of the sun on which it depends, came about by processes that operate in all galaxies. As we have progressed in our study from solid earth to sea and thence to air, it is fitting to close with a look at the firmament and the earth's neighbors in the universe.

Contemporary Physical Geology owes much to the guidance and wisdom of the book's reviewers. Their painstaking work and incisive comments resulted in many useful revisions and corrections. I deeply appreciate the help of Stephen Berman (University of Northern Iowa), John C. Cleveland (Indiana State University), Charles W. Dimmick (Central Connecticut State University), Thomas H. Dunham (Old Dominion University), Benjamin C. Friedrich (Jersey City State College), Jay Hackett (University of Northern Colorado), Darrell B. Hoff (University of Northern Iowa), Jim L. Jackson (University of Akron), Roberta Landau (St. Louis Community College at Forest Park), Kenneth W. Landon (Contra Costa Community College), David W. Marczely (Southern Connecticut State University), James P. McGuirk (Texas A & M University), William S. McLoda (Mountain View College), Paul D. Nelson (St. Louis Community College at Forest Park), James N. Pierce (Mankato State College), Robert Schwob (Itasca Community College), Kenneth E. Sheppard (East Texas State University), James D. Stewart (Vincennes University), Louis Unfer, Jr. (Southeast Missouri State University), Chester H. Wilson (Charles Steward Mott Community College), and Robert Zitter (Southern Illinois University). Even with the diligent work of all of these reviewers, undetected errors may still exist, and for these I am solely responsible.

I am also indebted to professors Jay M. Pasachoff (Williams College), John G. Navarra (Jersey City State College), and the late George O. Abell (University of California, Los Angeles) for their kindness in allowing me to use illustrations from their own Saunders College Publishing textbooks.

The support and assistance I received from my editors at Saunders College Publishing was indispensible. Don Jackson's persuasion initiated the project, Lloyd Black served as a particularly devilish devil's advocate, and Joanne Fraser efficiently accomplished a mind-boggling multitude of tasks. The job of typing the manuscript and test bank was accomplished with her customary efficiency by my friend Mrs. Margaret Bewig.

HAROLD L. LEVIN

To Kay and our students Janet, Stephen, and Linda

Supplements

Several teaching supplements accompany ESSENTIALS OF EARTH SCIENCE. Available upon adoption of the text are the following items:

TEST BANK: This contains 570 multiple choice, true or false, and association questions and answers for each of the chapters.

OVERHEAD TRANSPARENCIES: Over a hundred of the illustrations from the text are reproduced on 100 two-color acetate sheets.

MINERALS OF THE WORLD POSTER: Full-color photographs of over 200 mineral specimens from around the world are presented on this large poster.

contents overview

contents

introduction

The light in the world comes principally from two sources—the sun, and the student's lamp.
BOVEE, 1842

Prelude We humans are questioning creatures. We are not content to accept our surroundings without asking how things came to be. Very likely, such questioning began over a million years ago when an ancestral member of the human family hefted two rocks and wondered which would make a better tool. Somewhat later, a questioning human found a way to kindle a flame, and to utter sounds that became speech. Now the accumulated knowledge of each generation could be passed on to the next. You and I are the beneficiaries of this progressive accumulation of learning. It includes the discoveries of Galileo, Newton, Einstein, and others, all of whom have helped us answer questions about our planetary home.

One way we can appreciate at least a small part of the knowledge that has been handed down to us is through the modern discipline of earth science. *Earth science* is the study of the earth and its neighbors in space. Geology, oceanography, and meteorology are components of earth science that deal with the earth itself. It is also customary, however, to include some aspects of astronomy under the rubric of the earth sciences. The earth, after all, is just one body in a boundless universe and is subject to the same physical laws that govern other bodies.

The earth's crust is a constantly changing battleground where forces operating from the depths raise mountains, while other forces work constantly to wear the land away. Mt. Stephen, British Columbia. (British Columbia Government photograph.)

THE THREE SPHERES

The photographs of the earth taken from space by the Apollo astronauts (Fig. I-1) remind us of the three major components of the earth. Whispy patterns of white clouds tell us of the presence of an atmosphere. The azure blue of the ocean reveals the existence of a hydrosphere. Here and there one discerns patterns of tan and green along with outlines of continents, which are part of the solid earth, or lithosphere.

The atmosphere

We live beneath a thin but vital envelope of gases called the atmosphere. We refer to these gases as air. Nitrogen forms 78.03 percent of the total volume of dry air, and oxygen, 20.99 percent. The remaining 0.98 percent of air is made of argon, carbon dioxide, and minute quantities of other gases. One of these other components found mostly in the upper atmosphere is a form of oxygen called ozone (O_3). Ozone absorbs much of the sun's lethal ultraviolet radiation and is thus of critical importance to organisms on the earth. Air also contains from 0.1 percent to 5.0 percent water vapor. However, because this moisture content is so variable, it is not usually included in lists of atmospheric components.

Each day, the atmosphere receives radiation from the sun. This solar radiation, reflected off the earth's surface, provides the energy that heats the atmosphere and drives the winds. Distribution of solar radiation is one of the most important factors in determining the various kinds of climate we experience on the earth.

Figure I-1 A view of the earth photographed from the Apollo 17 spacecraft. Near the top center, Saudi Arabia, the Red Sea, and coastline of eastern Africa are visible. The Antarctic polar ice cap can be seen clearly. (Courtesy of National Aeronautics and Space Administration.)

The hydrosphere

The discontinuous envelope of water that covers 71 percent of the earth's surface is called the **hydrosphere.** It includes the ocean, as well as water vapor in air, water in streams and lakes, water frozen in glaciers, and water that occurs underground in the pores and cavities of rocks. If such surface irregularities as continents and deep oceanic basins and trenches were smoothed out, water would completely cover the earth to a depth of more than 2 kilometers.

Water is an exceedingly important geologic agent. Glaciers composed of water in its solid form alter the shape of the land by scouring, transporting, and depositing rock debris. Because water dissolves many natural compounds, it contributes significantly to the development of soils and thus the plants on which we depend for food. Water moving relentlessly downhill as sheetwash, in rills, and in streams loosens and carries away the particles of rock to lower elevations where they are deposited as layers of sediment. Clearly, the process of sculpturing our landscapes is primarily dependent upon water.

By far the greatest part of the hydrosphere is in the ocean basins. These basins are of enormous interest to geologists, who have discovered that they are not permanent and immobile, as was once believed, but rather are dynamic and ever changing. There is ample evidence that the sea floors move and that these movements have a direct relation to the formation of mountains, chains of volcanoes, deep sea trenches, and midoceanic ridges. On the ocean bottom are the layers of sediment from which geologists decipher earth history. Here one also finds mineral resources and clues to the location of ore deposits elsewhere on the planet. The ocean provides part of our food supply and has a pervasive influence on climate.

The lithosphere and the "spheres" beneath

Somewhat like the concentric shells of an onion, the solid earth is composed of a series of layers (Fig. I-2). As will be described in a later chapter, the existence of these layers has been deduced from the study of earthquake waves that have passed through the earth. At the surface is the thin outer shell known as the **crust.** The base of the crust is defined by a plane below which the velocity of certain earthquake waves is significantly greater than in the rocks above. A plane of this kind is called a seismic discontinuity, and the seismic discontinuity at the bottom of the crust has been named the **Mohorovičić discontinuity** after its discoverer. The term is often shortened to simply **Moho.**

The crust of the earth really consists of two kinds of rock, each with its distinctive general composition, thickness, and density. The continental crust has a composition somewhat similar to granite, has a relatively low density, and ranges in thickness from about 35 to 60 kilometers. The crust beneath the ocean basins is somewhat denser, rarely exceeds 5 km in thickness, and is composed of blackish rocks similar to those that form the Hawaiian Islands.

The layer beneath the earth's crust is called the **mantle.** The mantle has not yet been penetrated by drilling, but earthquake data indicate that it extends from the base of the crust to a depth of about 2900 km. It accounts for nearly 83 percent of the earth's volume. At the base of the mantle is yet another discontinuity that serves as the boundary between the mantle and the core. Geophysicists have recognized two parts of the core—a liquid outer zone and a solid inner core. Both parts are believed to be composed mainly of iron and some nickel. As we have noted earlier, the density of the earth as a whole is greater than that of the common rocks making up its outer parts. Thus, we may conclude that the material of the core is indeed very dense.

Two additional terms for upper zones of the earth's interior have come into wide usage because of their relationship to the movement and evolution of the crust. There exists a layer of the upper mantle, beginning at depths from about 60 to 120 km and extending to 650 to 700 km, in which the velocity of earthquake (seismic) waves is distinctly lower. Geophysicists believe that seismic waves are slowed in this layer because it is composed of relatively weak material, which exhibits plastic flow. The zone has been named the **asthenosphere.** Above the asthenosphere is a more rigid layer that includes not only the

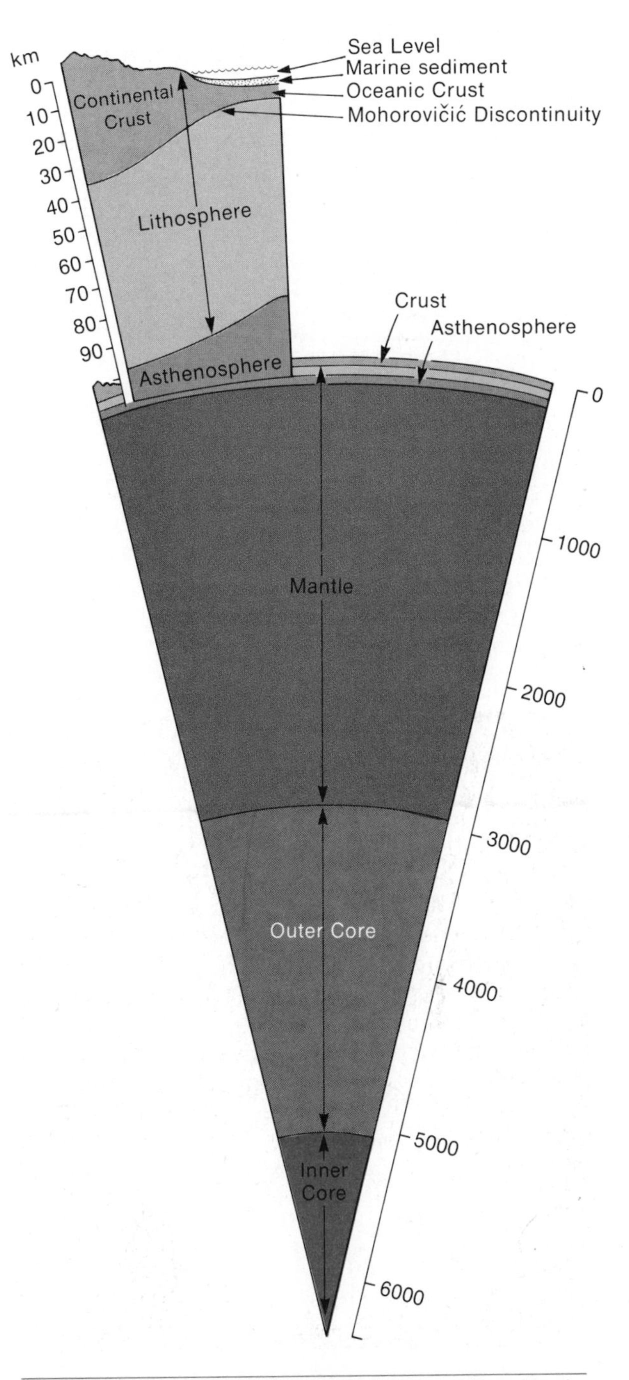

Figure I-2 *Internal structure of the earth. Notice that the mantle extends from the base of the crust to the top of the core. The lithosphere includes both the crust and the thin layer of mantle that overlies the asthenosphere.*

crust but an important part of the mantle as well. This cooler and more solid layer is called the **lithosphere.**

MAJOR FEATURES OF THE EARTH'S SURFACE

The most conspicuous elements of the earth's surface, continents and ocean basins, contain distinctive geologic features that have developed in response to particular geologic processes. For example, the major features of continents are **stable regions** and **orogenic belts.** As is suggested by the name, stable regions are parts of the continents that are no longer disturbed by the kind of geological forces that tend to distort rock layers and raise mountains. Plains and plateaus are characteristic of stable regions. **Orogeny** means mountain building, and thus orogenic belts are zones in which great thicknesses of rocks have been strongly compressed, altered, and raised into lofty mountain chains.

The ocean floors were once considered to be rather featureless plains. Actually, they exhibit a variety of major features. Around the edges of the oceans are the submerged margins of the continents, called **continental shelves.** The shelves are of enormous importance because they contain many offshore oil traps, as well as deposits of sand, gravel, oyster shells, and even diamonds. They are bounded seaward by the steeper **continental slopes,** which in turn drop off into less steep continental rises, and eventually into the **abyssal plains** (Fig. I–3). Rising above the floors of the abyssal plains in the Atlantic, Pacific, and Indian oceans are perhaps the most impressive features of the ocean basins. These features, called **midoceanic ridges** (Fig. I–4), tower more than 3500 meters above the sea floor. The ocean floor is also broken by long, narrow, earthquake-prone **deep-sea trenches** (Fig. I–5). The Marianas Trench in the western Pacific descends to the awesome depth of 11,034 meters below sea level.

The midoceanic ridges, the deep-sea trenches, and even our great mountain ranges are neither haphazardly formed nor randomly located. These features are manifestations of a dynamic process called **plate tectonics.** Plate tectonics helps explain the formation of new oceanic crustal material along the midoceanic ridge, the migration of the newly produced ocean floor away from the ridge, and the ultimate descent of that material into the mantle along zones marked by deep-sea trenches. It is an exciting concept that cannot be fully appreciated without knowing more about earth materials. For this reason, it will be more fully examined in Chapter 5.

A delicate balance

In speculating about the origin of such features as deep-sea trenches, midoceanic ridges, and mountain ranges, one becomes quickly aware of the mobility of the lithosphere. The old notion embodied in the phrase "good old terra firma" is no longer valid. Crustal movements are continuous and may vary from imperceptibly slow to rapid and violently destructive. They may affect only small areas or disturb a large region of a continent or ocean basin. The earth's instability is apparent to anyone who has been jolted by an earthquake or who has witnessed an erupting volcano. To the geologist, the continuum of unrest can be traced through folded strata, layers of lava, and great displacements of crustal rocks. These features are manifestations of powerful inner forces capable of uplifting entire continents and changing the dimensions of the sea floors. However, just as there are powerful construction forces on the earth, there are less dramatic but equally significant destructive forces.

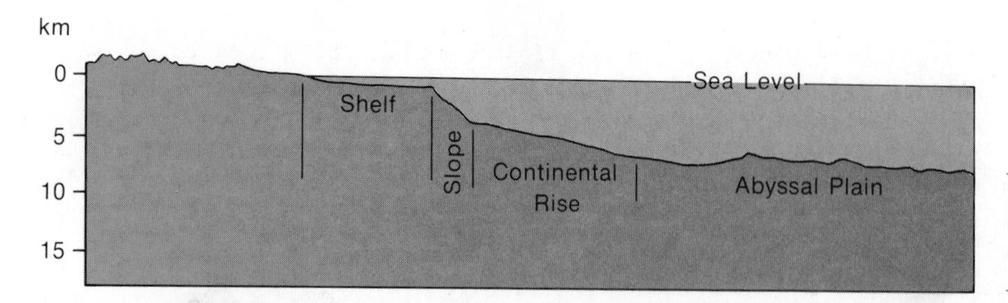

Figure I–3 Profile showing the relationships of the continental shelf, continental slope, and continental rise.

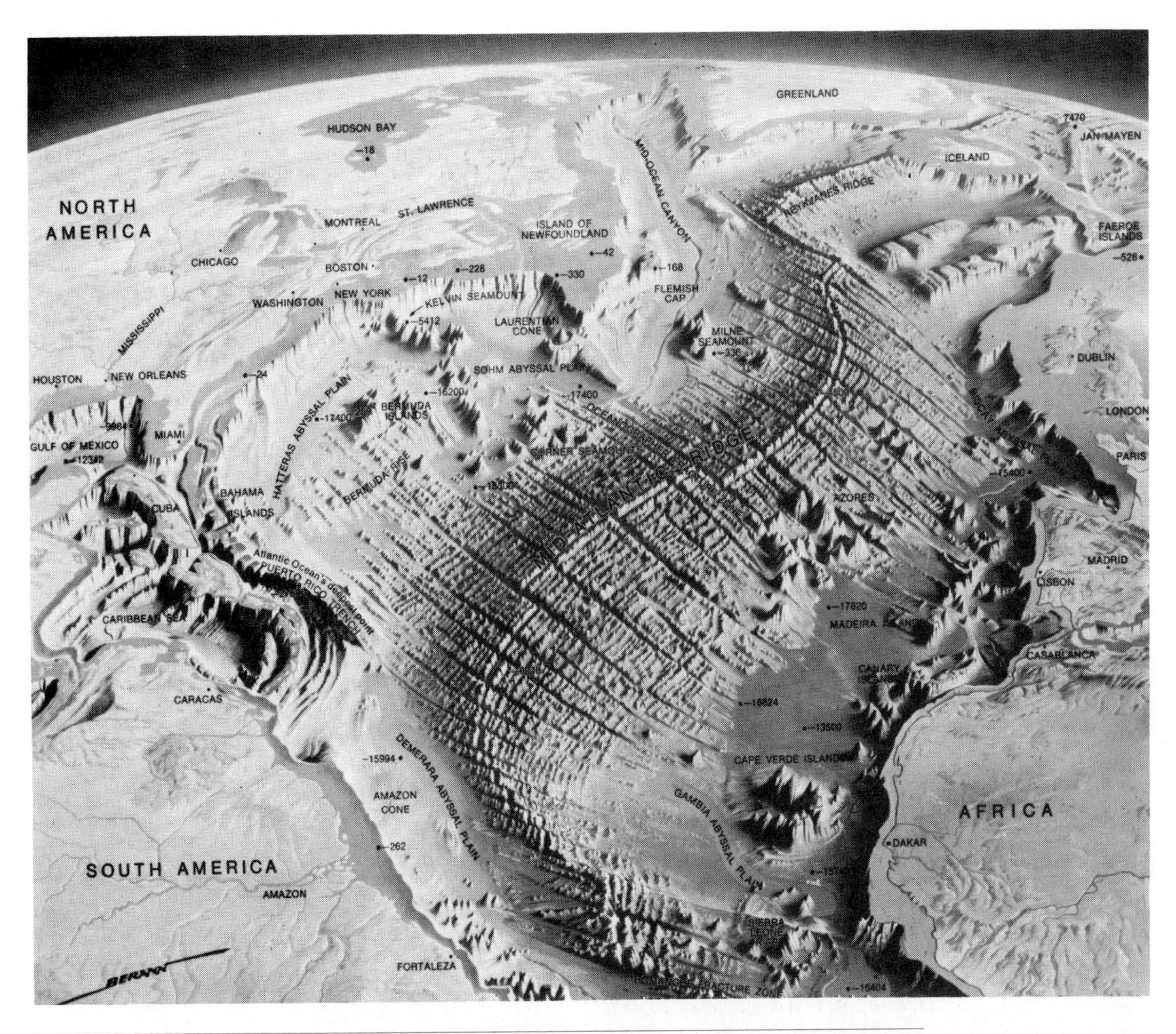

Figure 1–4 *Artist's conception of the floor of the North Atlantic Ocean. (From a painting by Heinrich Berann, Courtesy of Alcoa; from McCormick, J.M., and Thiruvathukal, J.V.,* Elements of Oceanography, *2nd ed. Philadelphia, Saunders College Publishing, 1981.)*

Such geologic agents as streams, glaciers, and wind erode the lands and carry the products of erosion to the ocean. One might surmise that the floors of the ocean would be the final resting place for these land derived sediments, but this is not likely. As suggested earlier, the sea floors are not static but move slowly like gargantuan conveyor belts. Sediment accumulating in the ocean basins is carried along on the moving floor to subduction zones, and there it may be pulled downward into the mantle and melted. Thus, the earth's surface is rather like a constantly changing battleground where internal forces raise and produce new land areas, where gravity and geologic agents operating near the surface wear the lands away, and where earth materials are continuously being changed and recycled.

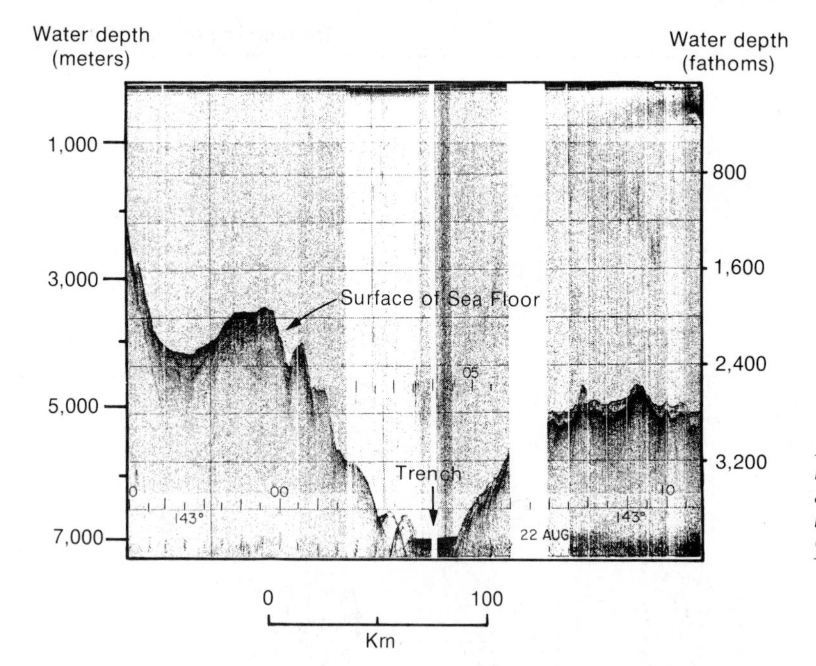

Figure I–5 *A continuous seismic profile across the Aleutian Trench. (Courtesy of Lamont-Doherty Geological Observatory, R.V. Conrad Cruise #12, 1969.)*

Figure I–6 *Crystals of the mineral quartz. (Courtesy of the Institute of Geological Sciences, London.)*

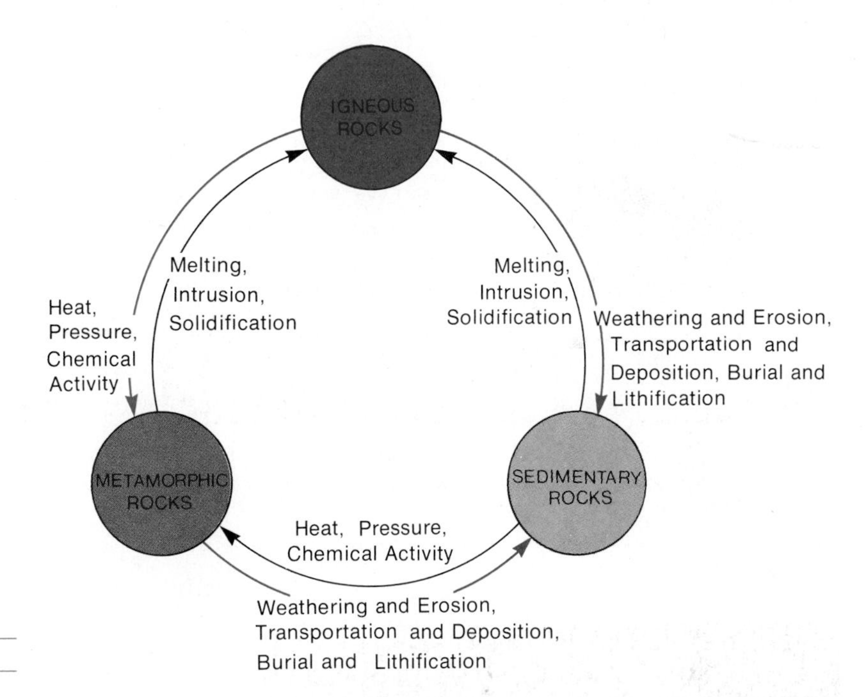

Figure I–7 *The rock cycle.*

THE RECYCLING OF EARTH MATERIALS

We have only to look about us to see that the crust of the earth is composed of rocks and minerals. **Minerals** (see Color plates 1 to 3) are naturally formed chemical elements or compounds having a definite range in chemical composition and usually a characteristic crystal form (Fig. I–6). **Rocks** are aggregates of one or more minerals that constitute an appreciable part of the earth's crust. They can be identified according to their physical properties and mineralogic composition.

The properties that provide clues to the origin of rocks are of greatest importance. Geologists have grouped rocks into three great families according to their origin. **Igneous rocks** are those that have cooled from a molten state. **Sedimentary rocks** are formed by the accumulation and consolidation of sediment. **Metamorphic rocks** are those that have formed in the solid state from preexisting rocks of any kind in response to heat, pressure, and associated chemical activity.

In thinking about the three major groupings of rocks, it is important to remember that they are not immune to change. The earth's crust is dynamic and everchanging. Any sedimentary or metamorphic rock may be melted and become igneous rock, and an existing rock of any category can be affected by the pressure and heat accompanying mountain building and become metamorphic rock. The weathered and eroded residue of any kind of rock can be observed being transported by rivers to the sea for deposition and eventual conversion into sedimentary rock. These changes can be incorporated into a schematic diagram that is designated the **rock cycle** (Fig. I–7).

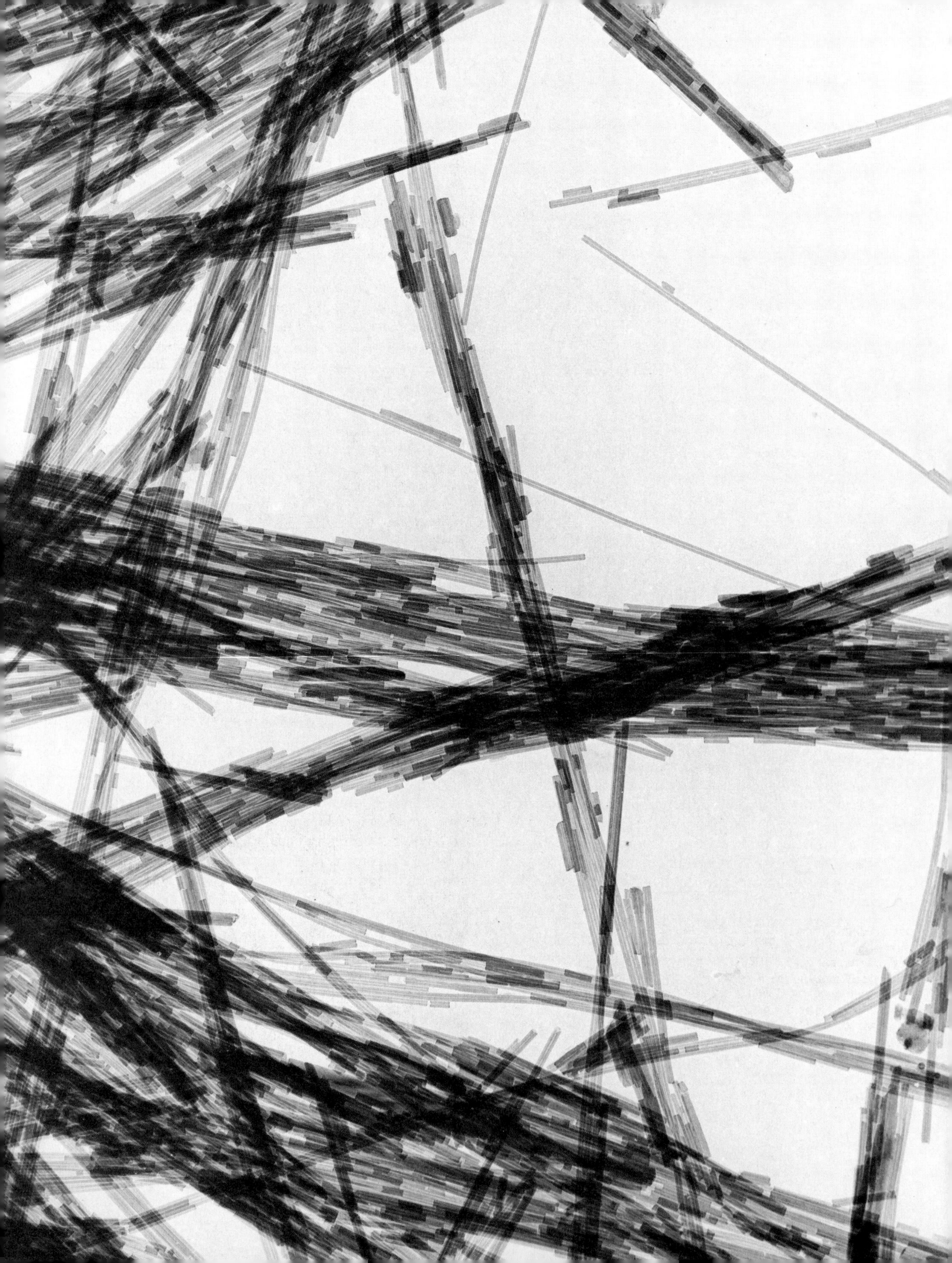

1 minerals

A casual glance at crystals may lead to the idea that they were pure sports of nature, but this is simply an elegant way of declaring one's ignorance. With a thoughtful examination of them, we discover laws of arrangement.
ABBÉ RENÉ JUST HAÜY, 1801

Prelude The quotation beneath the title of this chapter is a translation from the French of an article by René Just Haüy. It was published in the first year of the nineteenth century. Haüy, who had originally trained for the church and entered the priesthood, became interested in minerals when he noticed how large pieces of calcareous spar (Fig. 1 – 1) can be broken into smaller pieces all having a shape similar to that of the original piece. From his measurements of these and other solids displaying crystal form, he surmised that crystals are composed of small structural units that are of identical shape in all crystals of the same form. Today, mineralogists have demonstrated that Haüy's building blocks are actually groupings of still smaller particles called atoms. In this chapter, we will discuss the characteristics of atoms, the manner in which atoms are systematically

Photographs of crystals of chrysotile asbestos taken with the aid of the electron microscope. Total magnification ×42,000. One might guess that chrysotile was a chain structure silicate, but the electron microscope shows the fibers have a sheet structure, with the sheet curled around an axis rather like a roll of paper. (Courtesy of William J. Murphy and P.A. Ross, Lakehead University Thunder Bay, Ontario. Micrograph by A.J. McKenzie.)

Figure 1 – 1 *Calcareous spar or calcite showing the characteristic rhombohedral shapes formed when large crystals of the mineral are broken. (Courtesy of Wards Natural Science Establishment, Inc., Rochester, N.Y.)*

located and repeated in three dimensions within minerals, and the way in which combinations of atoms affect the properties of all earth materials. Before beginning, however, it will be useful to recall that *minerals* are naturally formed solids having three-dimensional, orderly internal arrangements of atoms and a definite chemical composition or range of compositions. The phrase "range of composition" is necessary because some atoms are similar in size and properties to one another and, to a limited degree, can replace one another in the internal atomic structure of minerals.

ATOMS

The architecture of the atom

If one were to disjoin the constituents of a mineral, one might then be left with certain components that could not be separated further by chemical and physical methods. These basic ingredients, of which no further separation is possible, are called **elements.** The fundamental unit of an element is an **atom.** There are 92 naturally occurring chemical elements, as well as 13 additional elements that are man-made. Geology students, however, need not be overly concerned about the rather large number of elements, since only eight occur abundantly in the earth's surficial rocks. These eight abundant elements are oxygen, silicon, aluminum, iron, calcium, sodium, potassium, and magnesium. They constitute at least 98 percent of the earth's crust and over 95 percent of the elements in the entire earth from core to crust (Table 1-1).

The idea that earth materials are composed of the atoms of elements was conceived long before the present atomic age. More than 25 centuries ago, the Hindu philosopher Kanadu proposed that matter was made of tiny, invisible "eternal particles." This same concept was proposed in classical Greek time by Democritus, who taught that atoms were small, indivisible particles capable of producing the properties of the substances they formed. Democritus, however, formulated his concept of the atom purely on speculation and intuition. He had no idea about the structure of the atom, or that it might contain even smaller constituents.

Knowledge of some of the smaller constituents of atoms and their characteristics began to evolve during the late 1800s and early 1900s, mostly as a result of experiments carried out with radioactive substances. The atom was conceived as having an extremely small but heavy **nucleus** surrounded by a cloud of rapidly moving particles with negative electric charge, called **electrons.** Electrons whirl around the nucleus at

TABLE 1-1 **ABUNDANCES OF CHEMICAL ELEMENTS IN THE EARTH'S CRUST***

Element and symbol	*Percentage by weight*	*Percentage by number of atoms*	*Percentage by volume*
Oxygen (O)	46.6	62.6	93.8
Silicon (Si)	27.7	21.2	0.9
Aluminum (Al)	8.1	6.5	0.5
Iron (Fe)	5.0	1.9	0.4
Calcium (Ca)	3.6	1.9	1.0
Sodium (Na)	2.8	2.6	1.3
Potassium (K)	2.6	1.4	1.8
Magnesium (Mg)	2.1	1.9	0.3
All other elements	1.5		
	100.0	100.0†	100.0†

* Based on B. Mason, 1966. Principles of Geochemistry. New York, John Wiley & Sons, Inc.
† Includes only the first eight elements.

speeds so great that if permitted to do so, they would encircle the earth in less than one second. They do not revolve in a set path, as do planets around the sun, but swirl around the nucleus rather like a swarm of tiny insects around a light. Thus, it would be incorrect to think of an atom as a solid sphere, because in reality, atoms consist mostly of empty space. The electrons move so rapidly that they effectively fill that space and thereby give size to the atom.

Located in the nucleus of the atom are closely compacted subatomic particles called **protons,** each of which carries a unit charge of positive electricity equal to the unit charge of negative electricity carried by the electron. Protons contain an amount of matter almost identical to that of a hydrogen atom, which has only one proton and an electron of negligible weight.

Associated with the protons in the nucleus are electrically neutral particles having the same mass as protons. These particles are called **neutrons.** Modern atomic physics has provided evidence for the existence of other subatomic particles, but their contribution to the chemical behavior of elements is minimal.

Atomic number and atomic mass

The number of protons within a nucleus determines the type of element to which an atom belongs. For example, if 6 protons occur in the nucleus of an atom, the element is carbon; if the number is 8, the element is oxygen; 14, silicon; and so on. Thus, all atoms of any one element have the same number of protons. The number of protons in the atom of an element defines that element's **atomic number.** In an electrically neutral atom, there are equal numbers of protons and electrons. The number of protons in naturally occurring elements ranges from 1 in hydrogen (Table 1–2) to 92 in uranium.

The **atomic mass** of an atom is approximately equal to the sum of the masses of its protons and neutrons. The mass of an electron is so small that it

TABLE 1–2 **STRUCTURE OF SOME GEOLOGICALLY IMPORTANT ELEMENTS**

Element and symbol	*Atomic number (number of protons in nucleus)*	*Number of neutrons in nucleus*	*Atomic mass*	*Electrons in various levels*	*Total number of electrons*
Hydrogen (H)	1	0	1	1	(1)
Helium (He)	2	2	4	2	(2)
Carbon 12 (C)*	6	6	12	2–4	(6)
Carbon 14 (C)	6	8	14	2–4	(6)
Oxygen (O)	8	8	16	2–6	(8)
Sodium (Na)	11	12	23	2–8–1	(11)
Magnesium (Mg)	12	13	25	2–8–2	(12)
Aluminum (Al)	13	14	27	2–8–3	(13)
Silicon (Si)	14	14	28	2–8–4	(14)
Chlorine 35 (Cl)*	17	18	35	2–8–7	(17)
Chlorine 37 (Cl)	17	20	37	2–8–7	(17)
Potassium (K)	19	20	39	2–8–8–1	(19)
Calcium (Ca)	20	20	40	2–8–8–2	(20)
Iron (Fe)	26	30	56	2–8–14–2	(26)
Barium (Ba)	56	82	138	2–8–18–18–8–2	(56)
Lead 208 (Pb)*	82	126	208	2–8–18–32–18–4	(82)
Lead 206 (Pb)	82	124	206	2–8–18–32–18–4	(82)
Radium (Ra)	88	138	226	2–8–18–32–18–8–2	(88)
Uranium 238 (U)	92	146	238	2–8–18–32–18–12–2	(92)

* When two isotopes of an element are given, the most abundant is starred; for other elements, only the most abundant isotope is given. Note carefully that ordinary chemical atomic weights are not given; these are mixtures of isotopes and are therefore not whole numbers.

need not be considered in determining the atomic mass. Indeed, it would require 1837 electrons to equal the atomic mass of a single proton. By convention, atomic mass is noted as a superscript preceding the chemical symbol of an element, and the atomic number is placed beneath it as a subscript. Thus, $^{40}_{20}Ca$ is translated as the element calcium having an atomic number of 20 and an atomic mass of 40.

Isotopes

Although the atoms of any specific element are essentially identical in the way they behave chemically, they are not always precisely identical in the number of neutrons they contain. Atoms of a particular element that differ in the number of neutrons they contain, and therefore also differ in atomic mass, are called **isotopes.** As an example, there are at least three isotopes of oxygen (Table 1-3). All three have the same atomic number (8) but have mass numbers of 16, 17, and 18, respectively. Of these, oxygen-16 is the most abundant in nature.

As has been suggested, the chemical properties of isotopes of the same element are nearly identical. This is because the chemical behavior of atoms is determined not by protons or neutrons but by the electrons. The number of electrons does not vary among the isotopes of an element.

Most isotopes have a high degree of stability and tend to remain unchanged over long spans of time. Others, however, are called unstable isotopes because they tend to break down into other isotopes. Later we will discuss how these changes occur and explain how unstable isotopes can be used to determine the ages of rocks and to ascertain the temperature of the ocean hundreds of millions of years ago.

The electronic structure of atoms

Because of their negative charge, electrons are attracted to the positively charged nucleus of the atom but are prevented from collapsing inward to the nucleus by the centrifugal force associated with their rapid orbital spin. Although electrons may spin along any path, *on the average* a certain quantity will remain within a particular zone or **shell.** The shells are really areas in which the probability of finding electrons is good (Fig. 1-2).

TABLE 1-3 **STABLE ISOTOPES OF OXYGEN***

Isotope	*Number of protons*	*Number of neutrons*	*Atomic mass*
^{16}O	8	8	16
^{17}O	8	9	17
^{18}O	8	10	18

* Isotopes of oxygen (oxygen 16, oxygen 17, and oxygen 18) are written ^{16}O, ^{17}O, and ^{18}O. The superscript designates the sum of the protons and neutrons.

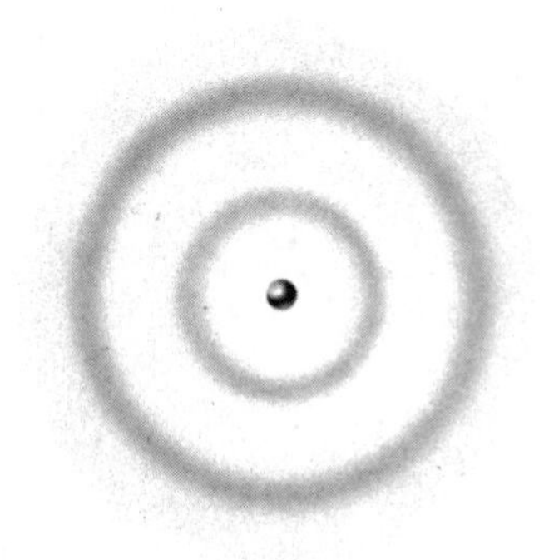

Figure 1-2 An atom is not a solid sphere, but is composed mostly of space occupied by negatively charged electrons which whirl around the positively charged nucleus at such enormous speeds as to effectively "fill" the space. Within the cloud of rapidly orbiting electrons, there is a high probability of a specific number of electrons occupying particular zones surrounding the nucleus. Such a zone is called an electron shell.

It is not possible to know the position of a particular electron at any designated moment. However, scientists have deduced the probable arrangement of the shells from x-ray studies, from atomic spectra, and from the atom's behavior. Such deductions provide a working hypothesis for the number of electron shells and the probable number of electrons in each shell (see Table 1-2). As the simplest of all the elements, hydrogen has only one proton and one planetary electron. The helium atom has two electrons close to the nucleus—a condition also present in the innermost shell of all elements heavier than helium.

The greatest number of electron shells in even the most complex of atoms is seven. Each of the first four shells may contain no more than—*but may contain less than*—the following number of electrons: 2 in the first shell, 8 in the second, 18 in the third, and 32 in the fourth. Beyond the fourth shell, electrons are distributed in such a way that the *outermost shell always contains eight or fewer electrons.* Indeed, once an atom has acquired the eight outer shell electrons, it becomes chemically quite stable. The chemical properties of atoms thus depend on their tendency to obtain eight

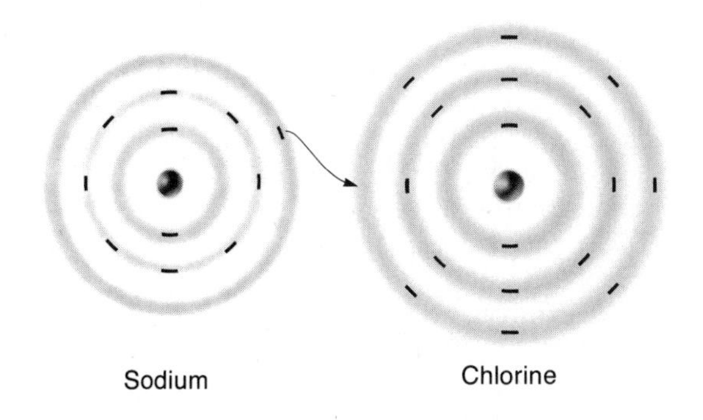

Figure 1–3 Diagrammatic representation of the formation of sodium chloride from sodium and chlorine.

electrons in this all-important outermost shell. For example, if the outer shell contains exactly eight electrons (or two in helium, which has only two electrons), the atom will show little tendency to chemically interact. Thus, atoms like neon, argon, and krypton are called **noble** or **inert gases** because their outer shells contain the maximum quota of eight electrons. They are, in a symbolic sense, "satisfied."

For those elements whose atoms have an outer ring containing fewer than eight electrons, there is a tendency to share, gain, or lose outer-shell electrons so as to assume the stable configuration in which eight electrons occur in the outermost shell. For example, the conceptual two-dimensional diagram of the chlorine atom shown in Figure 1–3 exhibits only seven electrons in the outer shell. Chlorine is "inclined" to borrow one electron, often from sodium, thereby giving itself and its chemical partner a stable configuration.

COMBINING ATOMS

As described above, the guiding rule in chemical combination is that when two or more atoms react with one another, they do so in a manner that will promote the attainment of noble gas electron configurations. The process is accomplished by accepting and donating electrons, or by sharing electrons. Furthermore, atoms tend to achieve a complete outer shell by undergoing the smallest possible change. Thus, an element having fewer than four electrons in its outer shell will tend to donate those electrons, but an element having more than four will tend to borrow electrons to complete its outer ring. Once an atom has either borrowed or donated an electron, it is no longer electrically neutral. It is now called an **ion.**

Examples of chemical bonding

Ionic bonding

The reaction between atoms of sodium (Na) and chlorine (Cl) can again be used to illustrate one mechanism by which atoms may bond together. In the combination of these two elements, it is easier for sodium to lend the single electron in its outer ring (see Fig. 1–3) than to borrow seven. Upon lending this electron, the sodium atom becomes positively charged. A chlorine atom carries 17 electrons: 2 in the first shell, 8 in the second shell, and the remaining 7 in the third, or outermost shell. Thus, the chlorine's outer shell is lacking one electron. For chlorine, it is far easier to acquire one electron than to donate seven. In the reaction between the two elements, each sodium atom gives its one outer-shell electron to one chlorine atom with the result that each has the complete outer shell structure of a noble gas. The acquisition of the extra electron by the chlorine atom converts it into a negative ion. Because unlike charges of electricity attract, the negatively charged chlorine ion is drawn tightly to the positively charged sodium ion to form sodium chloride. Studies of halite, the mineral form of sodium chloride, reveal that each sodium ion is surrounded by exactly six of the larger chlorine ions, and similarly, each chlorine atom is in contact with six sodium ions. The cubic crystal form (Fig 1–4) is the result of the manner in which the ions are packed, their electrical properties, and the sizes (atomic radii) of the participating ions (Fig. 1–5).

Figure 1–4 Cubic crystals of the mineral halite (NaCl). (Courtesy of the Institute of Geological Sciences, London.)

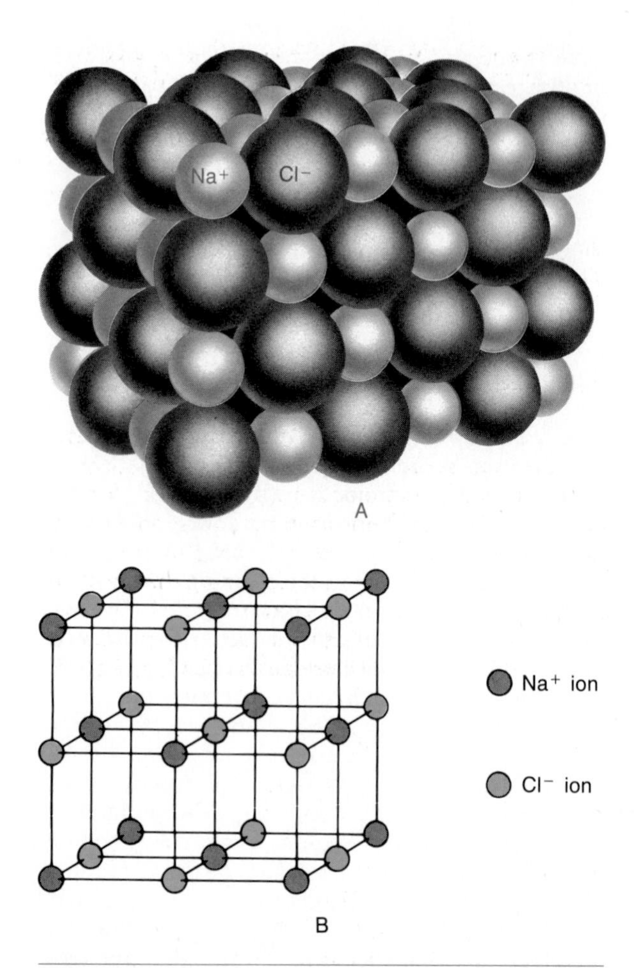

Figure 1–5 Structure of sodium chloride crystal. (A) Model showing relative sizes of the ions. (B) Ball-and-stick model showing cubic symmetry. (From Jones, M.M., Johnston, D.O., Netterville, J.T., Wood, J.L., Chemistry, Man and Society. *Philadelphia, Saunders College Publishing, 1983.)*

Covalent bonding, a sharing situation

The completion of the outer shell of atoms by *sharing* pairs of electrons is another way atoms achieve the noble gas configuration. This manner of combination is termed **covalent bonding.** In covalent bonding, the electrons of each atom orient themselves so that their influence is extended equally between the participating atoms. The shared electrons become indistinguishable as to the atom of origin and serve both nuclei simultaneously.

Examples of covalent bonding are the combining of two atoms of hydrogen to form the H_2 molecule, two oxygen atoms to form O_2, or the combination of four hydrogens and one carbon to form methane. Diamonds are also bonded covalently. In a diamond, each carbon atom is surrounded by four other carbon atoms to form a tetrahedron. Within the tetrahedra, every atom shares an electron pair with each of its neighboring atoms (Fig. 1–6).

Water provides another example of bonding by sharing of electrons. In a molecule of water (H_2O), the single electrons from each of two hydrogen ions are shared by an oxygen ion that needs them to complete the eight-electron outer-shell configuration. The ions, however, do not arrange themselves in a straight line; rather, the two positively charged hydrogen ions become fixed at one end of the molecule, whereas two nonbonding or unshared electrons occur at the opposite or oxygen side (Fig. 1–7). Thus, the molecule be-

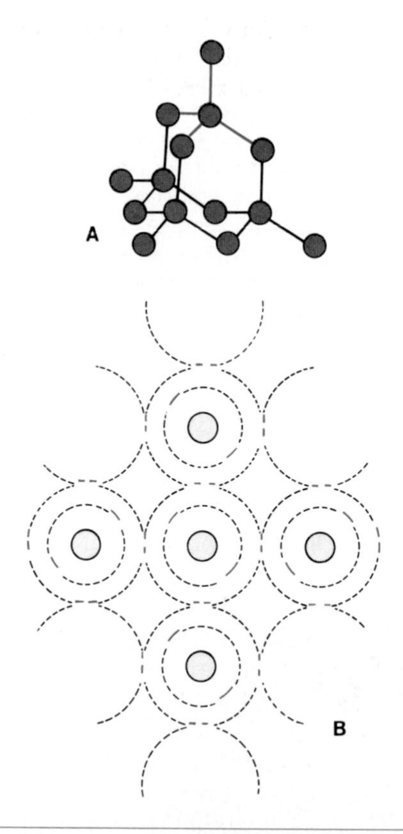

Figure 1–6 (A) The carbon atoms in a diamond are arranged in tetrahedrons consisting of four carbon atoms surrounding a central carbon atom. (B) A plan view of the atoms in a diamond. Note that each carbon atom, which individually has four electrons in its outer shell, shares one outer electron with each of its four neighbors. A stable outer shell of eight electrons is achieved by sharing pairs of electrons, and this type of bonding is termed covalent.

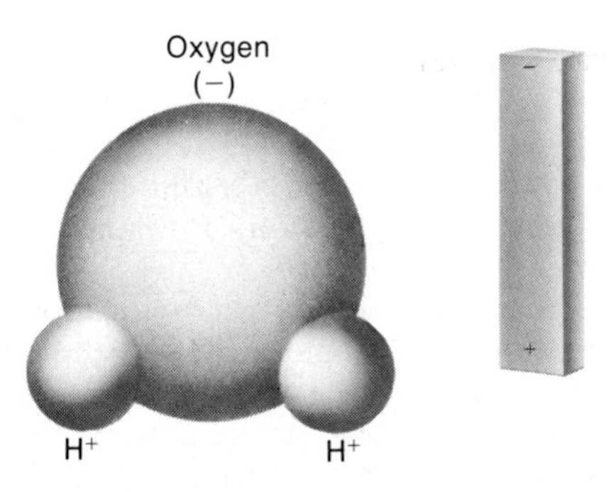

Figure 1–7 *Model of a water molecule, H_2O. The negative charges are crowded close to the oxygen end of the molecule, while the side with the two hydrogen nuclei contains more positive charge. Thus, the molecule has polarity—somewhat like a bar magnet.*

haves as if it were positively charged on one end and negatively charged on the other. This characteristic gives water molecules a weak electrical attraction for other ions. If these other ions are at the surface of mineral grains with which water comes into contact, the attraction may pull ions away from the mineral (Fig. 1–8). In fact, the mineral is being dissolved. Such solution work by water molecules is an important aspect of the weathering of rocks at the earth's surface.

Ionic and covalent bonding are the most common kinds of bonds found in minerals. They are not, however, mutually exclusive. Both types of bonding may occur within certain minerals, and in many instances bonding may be intermediate or mixed in character.

Metallic bonding

Metals can be visualized as closely packed aggregates of positive ions afloat in a "sea" of electrons. The outer electrons roam about independently of their atoms of origin. Rather than being associated with only one or two atoms, these electrons are shared by the entire metal ion aggregate. The atoms forming the metal become positive ions by giving their outer shell electrons to be shared among all other ions. The freedom of those electrons in being able to move about and not be

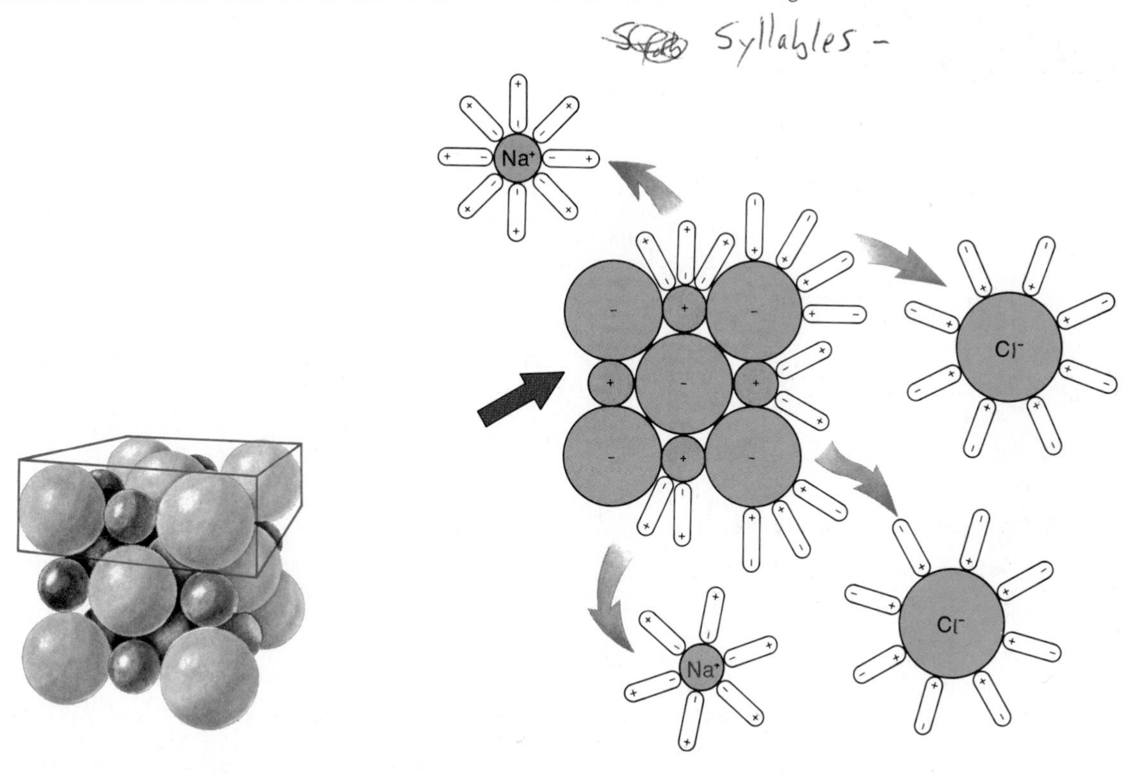

Figure 1–8 *The mechanism by which water dissolves halite. At lower left is a packing diagram of a segment of a halite crystal. Shown in plan view in the center are the ions in the topmost layer of the packing diagram. Dipolar water molecules attach themselves to the sodium and chlorine ions, overcome their ionic bonds, and convey them into the liquid.*

retained by a particular atom accounts for such metallic properties as electrical conductivity, thermal conductivity, softness, and malleability.

Atomic and ionic sizes

Another factor that influences which elements will combine to form minerals is the size of the ions and atoms (Fig. 1-9). The ions in minerals tend to be efficiently packed into tight geometric packages that leave little opportunity for any given ion to have so much surplus space that it can wobble about (Fig. 1-10). Combinations of ions of certain sizes can produce only specific kinds of atomic structures. The size of an ion depends on the number of electrons surrounding its nucleus and on the charge on the nucleus. If an atom loses an electron, the excess positive charge on the nucleus pulls the electron shells inward. Thus, positive ions are much smaller than the neutral atoms from which they were formed. Negative ions are larger than the neutral atoms of the same element.

The space-lattice

Crystals are solid bodies bounded by natural plane surfaces. The ions and atoms in crystals are arranged in three-dimensional geometric patterns that we call **space-lattices.** The concept of an orderly internal arrangement of the constituents of crystals was accepted by scientists long before the formulation of atomic theory. In the seventeenth century, Nicolaus Steno observed that similar crystal faces found on different crystals of the same kind of mineral always met at the same angle, regardless of the size and distortions of shape in the crystal (Fig. 1-11). Steno's observation, which has since been called the **Law of Constancy of Interfacial Angles,** indicates that the internal arrangement of constituent units (later termed ions and molecules) must be orderly, and the components must be present in definite proportions. If this were not the case, then the relationships of the crystal faces would not be consistent.

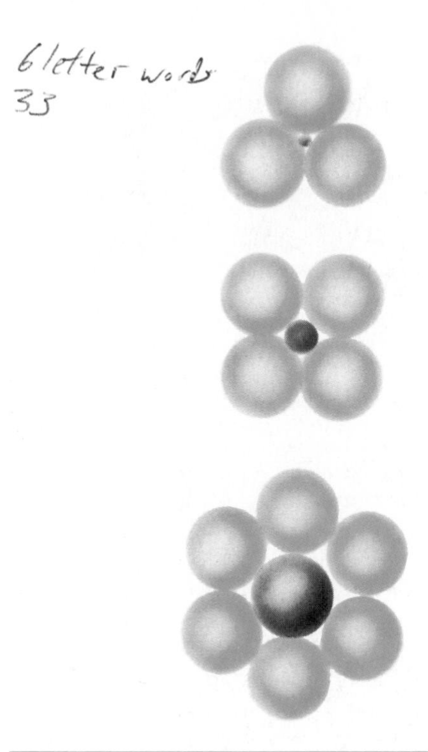

Figure 1-10 Diagram illustrating how the relative sizes of two ions can influence the packing arrangement.

Relative Sizes of Ions	Ions	Radius (in Å)	Coordination Number
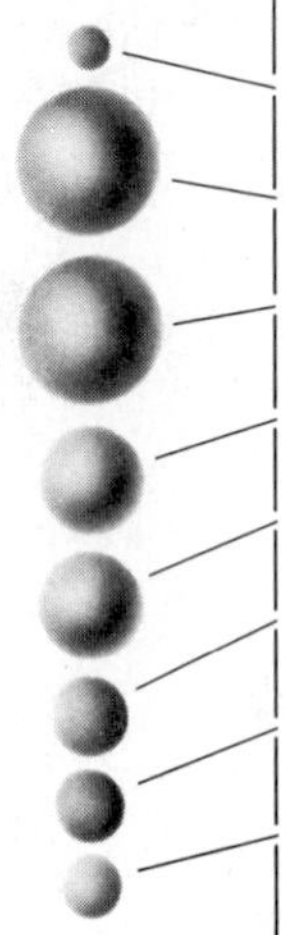	Silicon	0.40	4
	Oxygen	1.40	—
	Potassium	1.42	10
	Calcium	0.99	6
	Sodium	0.97	6
	Iron(+2)	0.74	6
	Magnesium	0.66	6
	Aluminum	0.49	4

Figure 1-9 Sizes of some geologically important ions (one angstrom unit equals a length of 10^{-10} meter).

Unfortunately, Nicolaus Steno (and later, Abbé René Just Haüy) could only infer the existence of internal ordering from exterior form. More direct evidence eluded scientists until 1912. At that time, the German physicist Max von Laue and his associates provided proof of the internal ordering of atoms in a crystal by the use of the mysterious x-rays discovered 17 years earlier by William Roentgen. Von Laue con-

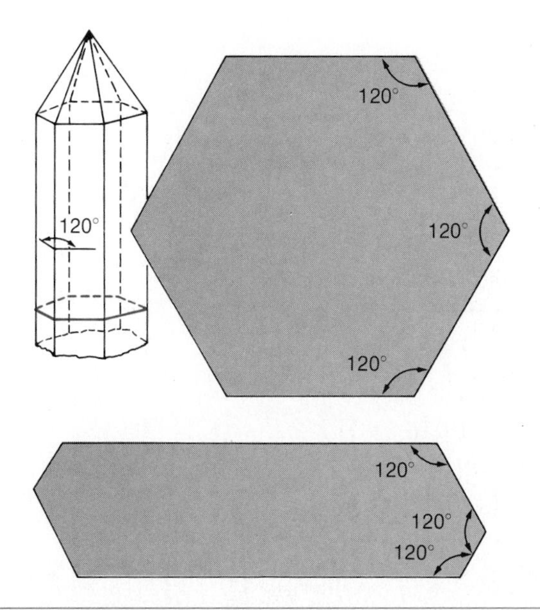

Figure 1 – 11 Cross-sections of crystals of the same mineral having six-sided (hexagonal) symmetry. Although the lower crystal is misshaped, the angles between comparable surfaces (faces) of both crystals are 120°. This is because the internal atomic structure is identical in both cyrstals and there is no internal distortion of that structure in the misshapen crystal.

jectured that x-rays, like light rays, might be wavelike in character, but x-ray wavelengths would be smaller than those in light rays. Assuming atoms in a crystal might be systematically arranged, von Laue speculated that the space between atoms would be about equal to the calculated wavelengths of x-rays, and thus, some of the rays would be reflected by striking layers of atoms. A simple apparatus was constructed so that x-rays could be passed through a crystal and onto a photographic plate (Fig. 1 – 12). Von Laue was rewarded with a photograph showing a uniform geometrical pattern of dots (now called von Laue spots) that reflected a uniform geometric arrangement of atoms. Steno's belief that the constituent units (ions) that build a crystal are packed into a symmetrical three-dimensional array had been amply confirmed.

Crystal form

A mineral grows or enlarges by addition of ions to its surfaces from a surrounding solution, melt, or gas. If the growing surfaces do not come into contact with other mineral grains, and if temperature, pressure, and the concentration of ions in the liquid or gas are suitable, then the mineral will acquire a characteristic crystal form. The faces of the crystals will be parallel to uniform sheets of atoms within the crystal.

In nature, the growing surfaces of crystals do come into contact with other mineral grains, and therefore perfect crystals are not abundant. More commonly, growing crystals interfere with one another during growth to form an interlocking system. Only occasional crystal faces are seen in such crystalline solids, but within individual mineral grains, the ions and atoms are nevertheless arranged in a uniform geometric pattern.

COMMON ROCK-FORMING MINERALS

The rocks that form the earth's crust are composed of aggregates of minerals. Although 2000 mineral species have been discovered and scientifically described, most of these comprise only a small percentage of the crust and are rarely encountered. For our purposes, it is important to consider only those minerals that compose the bulk of common rocks, or that are particularly useful in making interpretions about the earth's history.

Properties of minerals useful in identification

There are a large number of chemical and physical properties used by geologists in the identification of minerals. Some of these properties can be recognized only with the use of microscopes, x-ray equipment, or complex chemical tests. Fortunately, however, most of the common minerals can be identified with a knowledge of only a few more easily recognized properties. Color, cleavage, hardness, crystal form, luster, specific gravity, and magnetism are among the easily used clues for the identification of minerals.

Visual properties

Like most other properties of minerals, *color* depends largely on chemical composition and the arrangement of atoms. Color results from the selective absorption of certain wavelengths of light by the atoms. For example, the transmitted color (or the reflected color for opaque minerals) actually represents white light minus the wavelengths absorbed by particular atoms or ions. Some minerals in which the ions are relatively widely spaced and loosely bonded may tend to absorb nearly

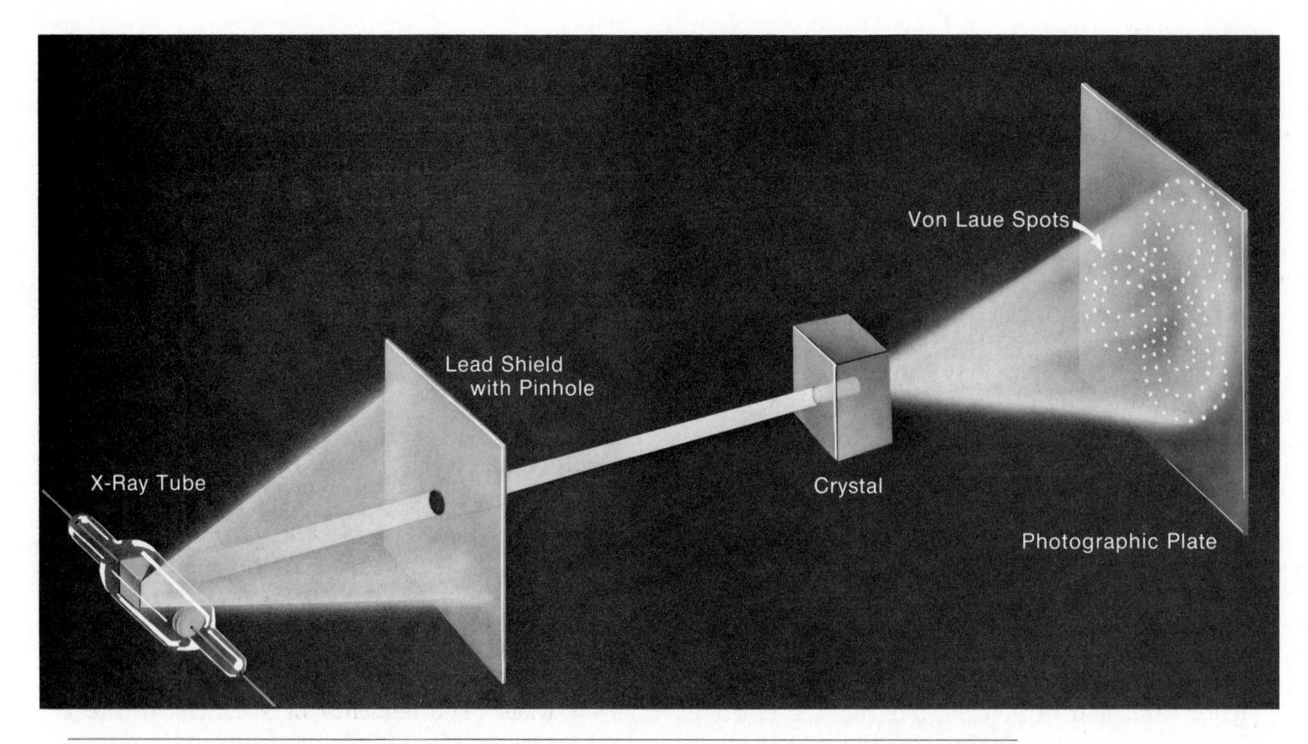

Figure 1–12 *In Von Laue's X-ray analysis, an X-ray beam is broken into a number of smaller beams by the internal arrangement of ions and atoms within a crystal. (The pattern of dots, however, does not depict the actual configuration of the atomic structure.)*

all of the light, and they are dark or black in color. This partially explains why graphite is black and opaque, whereas the more strongly bonded diamond is relatively clear. Iron and magnesium are ions that are strong absorbers of light. Therefore, minerals rich in iron and magnesium tend to be black or green.

Color is a mineral's most conspicuous characteristic. Some minerals possess colors that are constant, such as green in the mineral malachite, blue in lapis lazuli, or red in almandite garnet. Color, however, can be altered by even small amounts of impurities. Quartz and fluorite, for example, may be colorless, white, pink, purple, green, or blue. For this reason, color should always be used in combination with other properties in identifying a mineral (see Color plates 1 to 3).

To avoid errors in identification caused by superficial color variations, it is often useful to observe the mineral's **streak,** or color in powdered form. This can be quickly accomplished by rubbing the mineral across the surface of an unglazed porcelain "streak plate" and noting the color of the streak of powder that results.

The manner in which a mineral shines in reflected light is its **luster.** Those minerals that reflect light like metals are said to have *metallic luster,* whereas all others have nonmetallic luster. If the luster of a nonmetallic mineral is hard and brilliant like that of a diamond, it is said to have an *adamantine luster.* A glassy appearance is termed *vitreous luster,* whereas minerals with very low reflectance may simply be called *dull* or, if appropriate, *earthy.*

Hardness

The **hardness** of a mineral refers to its resistance to scratching by another substance of known hardness. To ensure uniformity in measuring hardness, geologists employ a scale formulated by the German mineralogist Frederich Mohs (1773–1839). Mohs arranged 10 relatively common minerals in order of increasing

TABLE 1–4 **SCALE OF MINERAL HARDNESS***

1	Talc
2	Gypsum
3	Calcite
4	Fluorite
5	Apatite
6	Orthoclase
7	Quartz
8	Topaz
9	Corundum
10	Diamond

* The scale of mineral hardness used by geologists was formulated in 1822 by Frederich Mohs. The scale begins with talc, a very soft mineral, and continues to diamond, the hardest of all minerals.

hardness (Table 1–4). If an unknown mineral can be scratched by a mineral in the scale, it has a lesser hardness than that mineral; if it cannot be scratched, it has a greater hardness. Mohs' scale is arithmetic from 1 to 9, but there is a 40-fold increase in hardness between 9 (corundum) and 10 (diamond). In the field it is often expedient to make hardness tests with a penny (hardness of 3), fingernail (hardness of 2 to 2.5), or a steel nail (hardness of 5).

A mineral's hardness depends largely upon the strength of bonds between the atoms or ions of its space-lattice. For example, in diamonds, every carbon atom forms sturdy single bonds with four other carbon atoms to produce a strong three-dimensional lattice. In contrast, carbon atoms in graphite crystals are arranged in layers, and the forces between layers are quite weak (Fig. 1–13). Graphite has a hardness of only 1 to 2 on Mohs' scale, whereas diamond has a value of 10.

Cleavage and fracture

We have discussed how ions and atoms in a crystal are arranged in definite layers or planes, all of which have a definite angular relationship to one another. If the bond strength in one set of planes is greater than that of other planes, then it is likely that the structure will separate more readily along the directions of weakest bonding, or where the spacing between planes is relatively wider. The tendency of a mineral to break

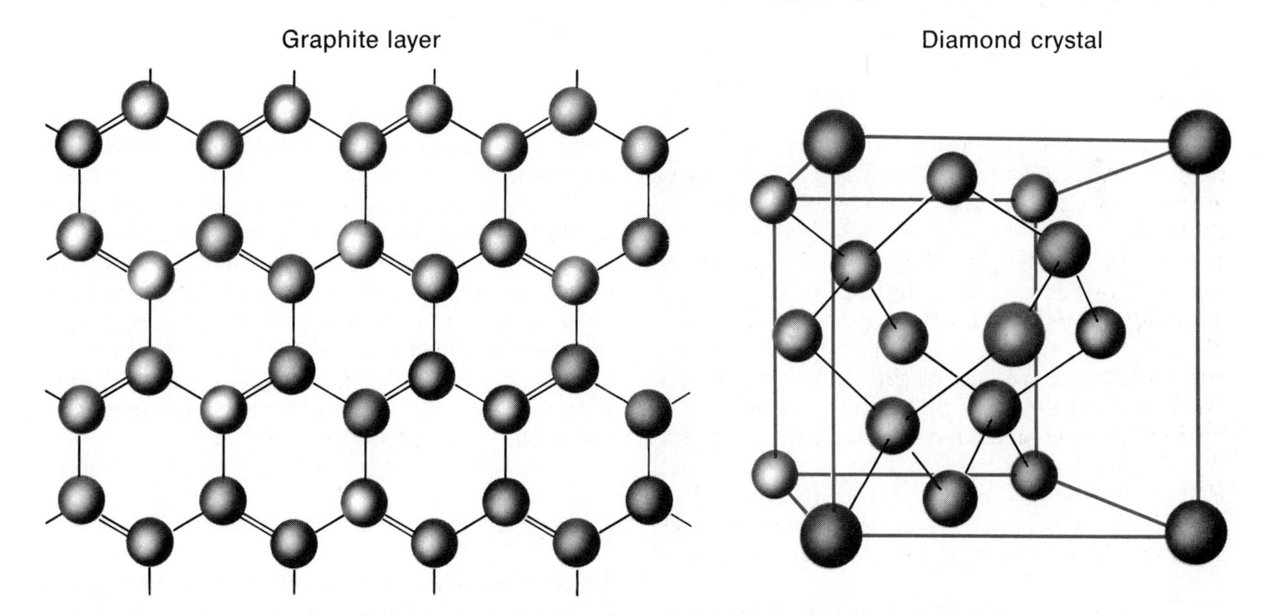

Figure 1–13 *Lattice structures of diamond and graphite. In diamonds, each carbon atom is at the center of a tetrahedron, on each corner of which is another carbon atom. In graphite, the carbon atoms are linked in planes of hexagons, the layers of which are loosely bonded. (From Masterton, W.L. and Slowinski, E.J.,* Chemical Principles, *5th ed. Philadelphia, W.B. Saunders Co., 1981.)*

Figure 1–14 Mica, showing characteristic cleavage along one-directional plane. (Courtesy of Wards Natural Science Establishment, Inc., Rochester, N.Y.)

smoothly along certain directions parallel to those planes of ions across which bonding is weakest is called **cleavage.** The surface along which the mineral breaks is the **cleavage surface.** Cleavage is usually described as poor, fair, or good.

The number and direction of cleavage planes are important aids to identification. Minerals like mica and graphite, for example, have good cleavage in one direction (Fig. 1-14). If a crystal of halite is broken, it will commonly produce three smooth lustrous surfaces at right angles to each other (Fig. 1-15). Thus, halite is a mineral having three good cleavage directions. Cleavage planes of minerals may superficially resemble crystal faces. Cleavage planes, however, are formed by breakage, not by the orderly addition of atoms or ions at the surface of a growing crystal.

Figure 1–15 Piece of halite broken from large crystal and exhibiting good cleavage in three mutually perpendicular directions.

Figure 1–16 The splintery fracture characteristic of the mineral actinolite is the result of the mineral's chain structure. (Courtesy of Wards Natural Science Establishment, Inc., Rochester, N.Y.)

A surface of breakage in a mineral other than a cleavage plane is termed a **fracture** (Fig. 1-16). Fractures tend to be irregular and randomly oriented. They are described by such terms as *splintery, fibrous, uneven,* or *conchoidal.* Opal and glass frequently exhibit the shell-like arcs of conchoidal fractures. A *hackly* fracture is a ragged surface with sharp points and edges resembling the broken surfaces of cast iron.

Specific gravity

Some minerals seem denser than others, and this can help us to identify them. To assure uniformity in determining whether one mineral is denser than another, geologists utilize a property called **specific gravity.** Specific gravity is the number of times a mineral is as heavy as an equal volume of water. The iron sul-

Figure 1–17 Crystals of pyrite (FeS_2). (Courtesy of the Institute of Geological Sciences, London.)

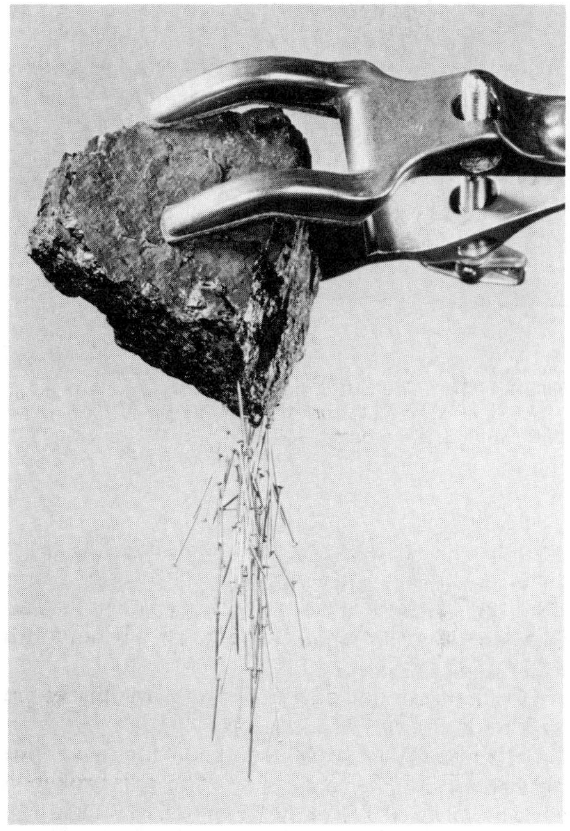

Figure 1–18 Magnetite, a mineral that serves as a natural magnet in attracting iron objects.

phide mineral known as pyrite (Fig. 1–17), for example, has a specific gravity of 5 and is therefore five times as heavy as an equal volume of water. Pyrite is sometimes called fool's gold. Real gold has a specific gravity of 19.3 and is thus readily distinguished from pyrite on the basis of specific gravity.

The **density** of a mineral or other body is its mass per unit volume.

$$\text{Density} = \frac{\text{Mass}}{\text{Volume}}$$

Unlike specific gravity, which (like other ratios) is not expressed in units, density must always be expressed in units, such as grams per cubic centimeter. Because one cubic centimeter of water has a mass of one gram, the density of a substance as expressed in the metric system has the same value as its specific gravity. For example, quartz with a specific gravity of 2.65 has a density of 2.65 g/cm^3.

As is true for all other physical properties of minerals, density and specific gravity depend upon the composition and the manner in which atoms and ions are arranged in the space-lattice. The more tightly packed the atoms, and the more elements of greater mass, the greater also will be a mineral's specific gravity and density.

Magnetism and other properties

In addition to the above characteristics by which a mineral may be recognized, certain minerals have other rather distinctive properties. The salty taste of halite, the malleability of gold, the flexibility of mica, the soapy feeling of talc, and the earthy odor of certain clays when moistened are all useful properties. A few minerals are attracted by a magnet, and if magnetized, the mineral magnetite acts as a magnet itself (Fig. 1–18).

Silicates

Atomic structure of silicates

About 75 percent by weight of the earth's crust is composed of the two elements oxygen and silicon (see Table 1–1). For the most part, oxygen and silicon occur in combination with other abundant elements, such as aluminum, iron, calcium, sodium, potassium, and magnesium, to form an important group of minerals called the **silicates.** A single family of silicate

Figure 1–19 *Quartz veins (light colored) intruded into darker colored mass of igneous rock.*

minerals, the **feldspars,** accounts for about one half of the material of the crust, and a single mineral species called **quartz** (Fig. 1 - 19) represents a sizable portion of the remainder. The fundamental unit in the crystal structure of silicates is called a silica tetrahedron (Fig. 1 - 20). It consists of a compact tetrahedral arrangement of four oxygen ions around a central silicon ion.

This tetrahedral shape is a consequence of the very small size and high charge of the silicon atom and of the relatively large size of the oxygen atoms. The silica tetrahedron, however, is not an electrically neutral unit. The combining of four oxygen ions (each with two negative charges) and one silicon ion (with four positive charges) leaves the resultant tetrahedron with four unpaired electrons. To correct the imbalance, the tetrahedral unit must either bond to one or more additional positive ions (such as magnesium or iron) or covalently share oxygen atoms with neighboring tetrahedra. Among the abundant silicates, olivine is a mineral that has all of its tetrahedra linked by metallic ions. In the other abundant silicates, the tetrahedra are strongly linked by sharing oxygens. The pattern developed by the connected tetrahedra not only forms the atomic structure of the mineral but also determines many of its properties. The various silicate structural types are as follows:

1 *Single Tetrahedra Structure.* These minerals are constructed of individual tetrahedra linked by pos-

Figure 1–20 *Model of the silicon-oxygen tetrahedron* $(SiO_4)^{4-}$ *viewed from above.*

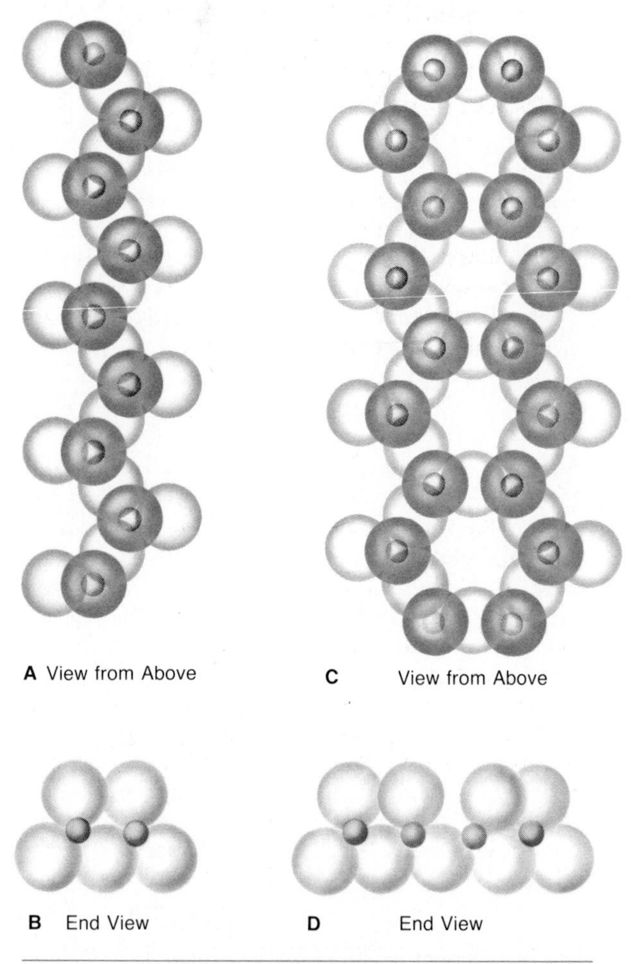

Figure 1–21 *Single-chain tetrahedra viewed from above (A) and from an end (B). C and D are similar views of double-chain structures. Individual chains are bonded to adjacent chains by cations. In the overhead views, the small silicon ions are located beneath the uppermost oxygen.*

itive (metallic) ions. Olivine, a mineral common to the lavas of the Hawaiian Islands, has this structure.

2 *Single- or Double-Chain Structure.* In single-chain structure, each tetrahedron shares two corner oxygen atoms of its base with two other tetrahedra; in the double chains, two single chains are combined by further sharing of oxygens (Fig. 1–21). Either kind of chain is bonded to adjacent chains by positive metallic ions. Because the bonds holding adjacent chains together are weaker than the silicon to oxygen bonds within each chain, cleavage develops parallel to the chains. The fibrous nature of amphibole varieties of asbestos results from chain structure. The fibers in serpentine varieties, however, are actually sheets that curl tightly around an axis.

3 *Sheet Structure.* In minerals having sheet structure, each tetrahedron shares three corners of its base with three other tetrahedra to form continuous sheets (Fig. 1-22).

4 *Framework Structure.* Each tetrahedron shares all four corners with other tetrahedra to form a continuous three-dimensional framework. *Quartz* is a mineral having a framework structure. Because the bonds are so strong in every direction in quartz, the mineral is exceptionally hard and has no tendency to break along preferred directions.

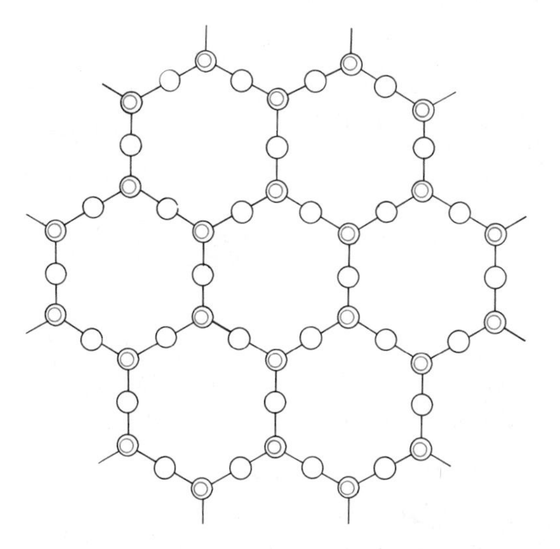

Figure 1–22 Lattice diagram of silicate sheet structure viewed from above. The small spheres represent silicon ions beneath oxygen ions.

Common silicate minerals

The most abundant rock-forming silicates are included in one or more of the groups described below and in Table 1–5.

Quartz

The mineral *quartz* (SiO_2) is one of the most familiar and important of all the silicate minerals (Fig. 1–23). It is common in many families of rocks. As mentioned earlier, quartz represents the ultimate in cross-linkage of silica tetrahedra; it therefore will not break along smooth planes. In quartz, the tetrahedra are joined only at the corners and in a relatively open arrangement. It is thus not a dense mineral, but it is quite hard because of the strong bonding in its framework structure. When quartz crystals are permitted to grow in an open cavity, they may develop the hexagonal prisms topped by pyramids that are prized by crystal collectors (see Fig. I–6). More frequently, the crystal faces cannot be discerned because the orderly addition of atoms had been interrupted by contact with other growing crystals.

Such minerals as chert, flint, jasper, and agate (Fig. 1–24) are varieties of a form of quartz called chalcedony. *Chalcedony* is composed of extremely small fibrous crystals of quartz. The crystals are so tiny that their study often requires the use of an electron microscope. Spaces between the small crystals are usually occupied by water molecules. Among the varieties of chalcedony, *chert* is exceptionally abundant in many sedimentary rock units. It is a dense, hard, usually

Figure 1–23 Cavity in rock which has become lined with quartz crystals. Such hollow, crystal-lined rocks are called geodes, and are prized for their beauty. (Institute of Geological Sciences, London.)

TABLE 1-5 **COMMON ROCK-FORMING SILICATE MINERALS**

Silicate mineral	*Composition*	*Physical properties*
Quartz	Silicon dioxide (silica, SiO_2)	Hardness of 7 (on scale of 1 to 10); will not cleave (fractures unevenly); specific gravity: 2.65
Potassium Feldspar Group	Aluminosilicates of potassium	Hardness of 6.0-6.5; cleaves well in two directions; pink or white; specific gravity: 2.5-2.6
Plagioclase Feldspar Group	Aluminosilicates of sodium and calcium	Hardness of 6.0-6.5; cleaves well in two directions; white or gray; may show striations on cleavage planes; specific gravity: 2.6-2.7
Muscovite Mica	Aluminosilicates of potassium with water	Hardness of 2-3; cleaves perfectly in one direction, yielding flexible thin plates; colorless; transparent in thin sheets; specific gravity: 2.8-3.0
Biotite Mica	Aluminosilicates of magnesium, iron, potassium, with water	Hardness of 2.5-3.0; cleaves perfectly in one direction, yielding flexible, thin plates; black to dark brown; specific gravity: 2.7-3.2
Pyroxene Group	Silicates of aluminum, calcium, magnesium, and iron	Hardness of 5-6; cleaves in two directions at 90°; black to dark green; specific gravity: 3.1-3.5
Amphibole Group	Silicates of aluminum, calcium, magnesium, and iron	Hardness of 5-6; cleaves in two directions at 56° and 124°; black to dark green; specific gravity: 3.0-33
Olivine	Silicate of magnesium and iron	Hardness of 6.5-7.0; light green; transparent to translucent; specific gravity: 3.2-3.6
Garnet Group	Aluminosilicates of iron, calcium, magnesium, and manganese	Hardness of 6.5-7.5; uneven fracture, red, brown, or yellow; specific gravity: 3.5-4.3

white mineral or rock. *Flint* is the popular name for the dark gray or black variety of chalcedony much used by Stone Age humans for making tools. *Jasper* is recognized by its opaque appearance and red or yellow color derived from iron-oxide impurities. The term *agate* is used for chalcedony that exhibits bands of differing color or texture.

The feldspars

Feldspars are the most abundant constituents of rocks, composing about 60 percent of the total weight of the earth's crust. There are two major families of feldspars: the *orthoclase* or *potassium feldspar* group (Fig. 1-25), which are the potassium aluminosilicates, and the *plagioclase group*, which are the aluminosilicates of sodium and calcium. Members of the plagioclase group exhibit a wide range in composition—from the calcium-rich *anorthite* ($CaAl_2Si_2O_8$) to the sodium-rich *albite* ($NaAlSi_3O_8$). Between these two extremes, plagioclase minerals containing both sodium and calcium occur. Feldspars are nearly as hard as quartz and range in color from white or pink to bluish-gray. Silica tetrahedra in the feldspars are joined in a strong three-dimensional lattice that is characterized by planes of weaker bonding in two directions at (or nearly at) right angles to each other. Because of this, the feldspars have good *cleavage* (break along smooth planes) in two directions. The resulting rectangular cleavage surfaces and a hardness of 6 are properties useful in the identification of feldspars.

The mica group

Mica, a silicate mineral having sheet structure, is easily recognized by its perfect and conspicuous cleavage in

Figure 1–24 Agate. (Courtesy of Wards Natural Science Establishment, Inc., Rochester, N.Y.)

Figure 1–26 Muscovite mica. (Courtesy of the Institute of Geological Sciences, London.)

one directional plane. The two chief varieties are the colorless or pale-colored **muscovite** mica (Fig. 1-26), which is a hydrous potassium aluminum silicate [$(KAl_2(Al_3Si_3O_0)(OH)_2)$] and the dark-colored **biotite** mica (see Fig. 1-14), which also contains iron and magnesium [$(K(Mg,Fe)_3AlSi_3O_{10}(OH)_2)$]. In muscovite mica, two sheets of tetrahedra are strongly held together along their inner surfaces by positively charged ions of aluminum. These sandwich-like paired sheets are in turn weakly joined to others by positively charged ions of potassium. When muscovite is cleaved into paper-thin layers, the separation occurs primarily along the weaker plane where the potassium ions are located. In biotite, magnesium and iron ions hold the inner surfaces of the sheets together, but once again potassium ions serve to weakly join each basic set of paired sheets to its neighbor.

Figure 1–25 A cluster of orthoclase feldspar crystals. The specimen measures 5 inches across.

Identification of large specimens of mica is rarely a problem because of its planar perfect cleavage and the way cleavage flakes snap back into place when they are bent and suddenly released. The micas are common constituents of igneous and metamorphic rocks, where they can be recognized by their shiny surfaces and the ease with which they can be plucked loose with a pin or pen knife. Before the manufacture of glass, one of the chief uses of muscovite mica was as window panes. This clear mica was quarried in Muscovy (an early name for Russia), and thus came to be known as Muscovy glass, and eventually muscovite. Today, mica is used in the manufacture of electrical insulators and as a filler in plaster, roofing products, and rubber.

Hornblende

Hornblende is a vitreous black or very dark green mineral. It is the most common member of a larger family of minerals called **amphiboles,** which

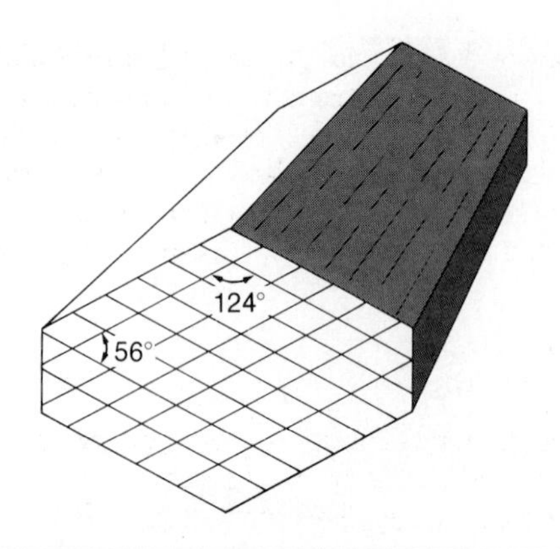

Figure 1-27 Typical shape of a hornblende crystal. Planes of cleavage intersect at 56° and 124°.

have generally similar properties. As can be seen from its chemical formula, $NaCa_2(Mg,Fe,Al)_5(Si,Al)_8O_{22}(OH)_2$, hornblende contains a relatively large number of elements. Because of the presence of iron and magnesium, hornblende (along with biotite, augite, and olivine) is designated a ferromagnesian mineral. Crystals of hornblende tend to be long and narrow. Two good cleavages are developed parallel to the long axis and intersect each other at angles of 56° and 124° (Fig. 1-27). The cleavage is a reflection of the location of planes of weaker bonds that exist between the double-chain units of silica tetrahedra in the atomic lattice.

Augite

Augite is an important member of the **pyroxene** family. Its chemical formula, $Ca(Mg,Fe,Al)(SiAl)_2O_6$, indicates that it too is a ferromagnesian mineral and thus dark colored. An augite crystal (Fig. 1-28) is typically rather stumpy in shape, with good cleavages developed along two planes that are nearly at right angles (87° and 93°). Thus, the cross section of a crystal appears nearly square (rather than rhombic as in hornblende). Unlike hornblende, which has a double-chain silicate structure, augite is constructed of single chains, and this accounts for its having differently shaped cleavage fragments.

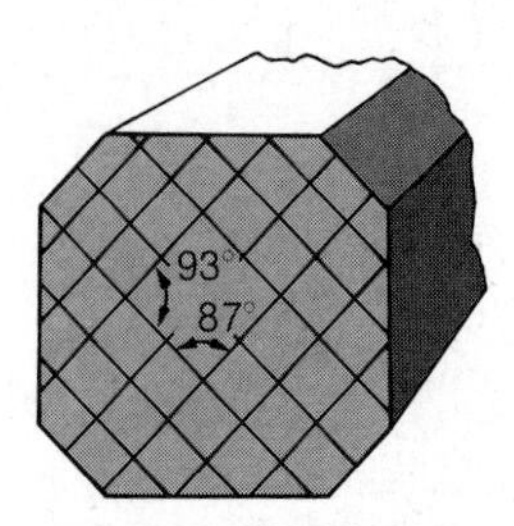

Figure 1-28 Typical shape of an augite crystal. Planes of cleavage intersect at 87° and 93°.

Olivine

Olivine, another ferromagnesian mineral, has been mentioned earlier as having isolated silicon-oxygen tetrahedra bonded by iron and/or magnesium ions. Its formula, $(Fe,Mg)_2SiO_4$, indicates that it is a solid solution mineral, containing variable proportions of iron and magnesium. The substitution of these ions for each other is facilitated by their having similar ionic radii and two electrons in their outer electron shell. The ions in olivine are so strongly held by ionic bonding that the mineral has a hardness of 6.5 to 7 on Mohs' scale. As you might guess from its name, this glassy-looking mineral often has a green color. Frequently, it occurs as masses of small sugary grains (Fig. 1-29) or as tiny vitreous crystals in black lavas. It is also an important constituent of stony meteorites. Large, unblemished crystals of magnesium-rich olivine may be cut and polished into attractive gemstones called *peridot.*

Figure 1-29 Granular mass of olivine. (Courtesy of Wards Natural Science Establishment, Inc., Rochester, N.Y.)

Garnet

Garnet is the name used for a group of mineral species closely related in composition, crystal form, and physical properties. In general composition, these minerals are silicates of aluminum with iron, calcium, or magnesium. The iron-rich variety, almandite $Fe_3Al_2(SiO_4)_3$, is widely used in jewelry because of its rich, dark-red color. However, depending on their composition, garnets are found in all colors except blue. For example, if calcium fills the place of iron in the lattice, the garnet is called grossularite and may have a pale green color, although other hues may also develop. Garnets can usually be identified by their well-formed equidimensional crystals (Fig. 1 – 30) and their resinous to vitreous luster. Crystals will not cleave, but rather break with a conchoidal fracture. Because of the sharp edges produced on fracturing and its hardness (7 to 7.5 on Mohs' scale), garnet is widely mined and sold as an abrasive material.

Chlorite

The chlorites are a group of greenish minerals having a platy form similar to that of micas. In composition, the minerals of the chlorite group are hydrous aluminum silicates of iron and magnesium. Chlorite can usually be identified by its perfect cleavage in one direction, greenish color and streak, and relative softness. It is a common constituent of metamorphic rocks and also occurs as an alteration product of ferromagnesian minerals.

Clay minerals

The clay minerals are silicates of hydrogen and aluminum with additions of magnesium, iron, and potassium. At one time, clays were assumed to be amorphous mixtures of metal oxides and hydroxides. Through the use of x-ray equipment, however, it was demonstrated that clays are composed of crystalline units that, like mica, have sheet structure. The individual flakes are extremely small and require the magnification provided by the electron microscope for adequate examination (Fig. 1 – 31). Clays are formed through the decomposition of other aluminum silicate minerals, especially the feldspars. Clay minerals are characteristically very soft, have low density, occur in earthy masses, and become rather plastic when wet.

The common nonsilicate minerals

Approximately 8 percent of the earth's crust is composed of nonsilicate minerals. These include a host of

Figure 1 – 30 Well-formed garnet crystal. The crystal form is termed a dodecohedron. (Courtesy of the Institute of Geological Sciences, London.)

Figure 1 – 31 Electron micrograph of the clay mineral kaolinite. The flaky, stack-of-cards character of the clay crystals is a manifestation of their silicate sheet structure. Magnified about ×2000. (Courtesy of the Kevex Corporation.)

carbonates, sulfides, sulfates, chlorides, and oxides. The nonsilicates that are most abundant at the earth's surface are the **carbonates,** the sulfates and chlorides known collectively as **evaporites,** and oxides of iron. of iron.

Carbonates

Calcite and **dolomite** are the two most important members of the carbonate mineral group. Calcite ($CaCO_3$), which is the main constituent of limestone and marble, forms in many ways. It is secreted as skeletal material by certain invertebrate animals, precipitated directly from seawater, or formed as dripstone in caverns. Calcite is easily recognized by its rhombohedron-shaped cleaved fragments and by the fact that an application of hydrochloric acid on its surface will cause effervescence. Clear cleavage rhombs of calcite, called Iceland spar, exhibit a property called **double refraction** (Fig. 1-32). That is, if a transparent piece of calcite is held over a dot on white paper, two dots will appear when the calcite is held in certain positions. As the mineral is slowly rotated, one dot will appear to rotate around the other. The explanation for this bit of "magic" is that the light entering the specimen is bent, or *refracted,* and divided into two polarized rays, one of which is bent more than the other (Fig. 1-33). The image from the more refracted ray appears to move around the image of the less refracted ray as the piece of calcite is rotated.

Dolomite, $CaMg(CO_3)_2$, is a carbonate of calcium and magnesium that occurs as extensive layers of a rock called *dolostone.* In the field, geologists often distinguish dolomite by its failure to effervesce in dilute hydrochloric acid unless it is powdered. Many ancient dolostones are thought to have been formed by recrystallization of still older limestones. Because calcium and magnesium ions are nearly the same size, it is possible for the magnesium to replace calcium in the atomic structure.

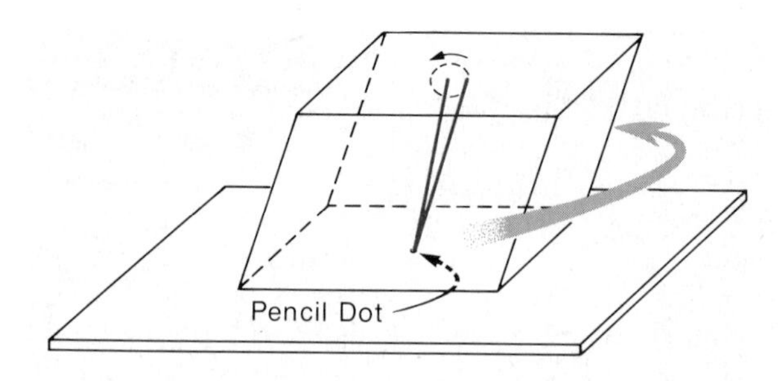

Figure 1-33 *If a piece of clear calcite is placed as shown over a pencil dot, two dots will be viewed from a point above the specimen. If one then rotates the calcite as indicated by the arrow, the image of the dot from the more refracted ray will appear to move in a circular path around the other dot.*

Figure 1-32 *Cleavage fragment of exceptionally transparent calcite. The mineral's ability to cause double refraction is clearly exhibited. (Courtesy of the Institute of Geological Sciences, London.)*

Evaporites (mostly gypsum and halite)

Evaporites are minerals that have precipitated from bodies of water subjected to intense evaporation. They include common rock salt, or halite ($NaCl$), and gypsum. *Halite* is easily identified from its salty taste and perfect cubic cleavage. *Gypsum* is a soft (2 on Mohs' scale) hydrous calcium sulfate ($CaSO_4 \cdot 2H_2O$). The variety of gypsum called satinspar has a fine fibrous structure, whereas *selenite* (Fig. 1-34) will split into thin plates. The finely crystalline massive variety known as *alabaster* is widely used in carvings and sculpture because of its uniform texture and softness.

Iron oxide minerals

The iron oxide minerals that occur abundantly at or near the earth's surface are **hematite** (Fe_2O_3) and a weathering product of iron-bearing minerals called **goethite** ($HFeO_2$). Goethite is the principal mineral in the amorphous hydrous iron oxide material known as **limonite** ($Fe_2O_3 \cdot 2H_2O$). **Magnetite** (Fe_3O_4) is of less importance because it is less abundant.

Bedded deposits of hematite supply the bulk of the world's iron ore. The mineral varies in color from steel gray to brownish red but always exhibits a distinctive

Figure 1–34 Selenite gypsum. This mineral crystallizes in the monoclinic crystal system. (Courtesy of the Institute of Geological Sciences, London.)

reddish streak. In contrast, limonite has a yellow-brown streak and is generally yellow or brown in color.

One is not likely to confuse magnetite with the other common iron oxides. Its magnetic properties, high specific gravity, and black streak are distinctive (see Fig. 1–18). Magnetite occurs in concentrations that can be mined as iron ore, but is also common as disseminated crystals in rocks rich in ferromagnesian minerals.

Mineralogic clues to geological events

Minerals are studied by geologists not as mere constituents of the crust but as clues to the history of the rocks in which they occur. Some minerals develop exclusively in ocean water and provide the geologist with evidence that a particular sediment was deposited in the sea rather than in a freshwater lake. A thick bed of halite indicates aridity and evaporation so extreme that the brine had become ten times saltier than ordinary seawater. The magnetic properties of magnetite can, in certain situations, provide clues to the position of continents relative to the earth's magnetic poles. Certain minerals form within a narrow range of conditions and can therefore be used to diagnose the pressures and temperatures involved in the formation of crustal rocks and mountains. Both diamonds and graphite, for example, are crystalline varieties of carbon, but diamonds form only at high temperatures and extremely high pressures, and graphite forms at lower temperatures and pressures. Other minerals contain radioactive isotopes that permit us to know the age of the parent rocks. By their size, crystals of feldspar give the geologist information about the rate at which ancient molten mass congealed. Even ordinary grains of quartz may hold clues as to what has happened to them since they were eroded from some ancient granitic terrain. The ancient buried mineral products of weathering may provide information about past climatic conditions in a region.

Minerals as gems

Some occurrences of silicate minerals introduced in this chapter are so attractive, durable, and rare that they are highly valued as gems. Their beauty depends upon color, luster, and brilliance—properties that are enhanced when the rough stones are cut, faceted, or shaped and polished. Gems are sold by weight, and the basic unit of weight is called a gem carat. One gem carat is equal to ⅕ of a gram. (The gem carat is not to be confused with the gold carat, which indicates the proportion of gold in an alloy and is not a unit of weight.)

Such familiar minerals as quartz, corundum, feldspar, olivine, and garnet occur sporadically as materials of gem quality. Gems in the quartz family are particularly well known and include the purple variety known as amethyst, yellow quartz or citrine, rose quartz, opal, the banded microscrystalline quartz called agate, and the mysterious tiger's-eye quartz, which results from the replacement of asbestos by microcrystalline quartz. Jasper is an opaque to semiopaque microcrystalline quartz that contains an abundance of iron oxide.

The two most famous gems in the corundum family are blue sapphire and red ruby. The color differences in these corundum minerals result from small traces of metal oxides. For example, ruby is colored by chromium oxide, and sapphire by oxides of iron and titanium. These and other impurities also produce exquisite hues of green, purple, and gold in sapphires. Feldspar gems include the attractive white opalescent moonstone and the green orthoclase called amazonite. Peridot is the gem variety of olivine. In the garnet family, almandite is prized for its clear wine-red color. There are, however, many other beautifully colored garnets, including pink, orange, and green varieties. Topaz, the third hardest mineral in Mohs' scale, may occur as a lovely yellow gem, although stones colored pale blue and light green are also prized.

Diamond, the most highly treasured of gems, is a nonsilicate composed only of carbon. As mentioned earlier, diamonds are known for their brilliance and hardness. They are derived from basic plutonic rocks called kimberlites. The principal factors involved in determining the value of a cut and faceted diamond are its weight, its color, and its degree of perfection. A so-called perfect diamond is one that shows no conspicuous inclusions or flaws when viewed at a magnification of 10× by an experienced person. The weight of a diamond is usually expressed in a unit called a *point*, which is 1/100 of a carat. Thus, a 25-point diamond would weigh 0.25 carat.

Summary Minerals and rocks are the documents that contain the evidence of all the forces and processes that have affected the earth in the past. They are the basis for nearly all geological studies, and are therefore of fundamental importance. In this chapter, we have examined the manner in which atoms of elements are arranged so as to form minerals, and we have considered some simple observations that permit one to recognize the more common minerals.

Centuries ago, a Greek philosopher named Democritus speculated that water and rocks were made of invisibly tiny pieces of matter. Democritus coined the term atom to describe these particles. The present understanding of atoms is that they have a central nucleus containing positively charged protons and electrically neutral neutrons. Circling the nucleus are negatively charged electrons. The arrangement of electrons into levels, the number of atomic particles, the size of the atoms, and their electrical characteristics determine how the atoms will combine into molecules and ultimately into minerals.

Minerals are naturally occurring compounds that exist as crystals or crystalline solids, have either a definite chemical composition or restricted range of chemical compositions, and possess a systematic three-dimensional atomic order. Knowledge of their crystallinity is now well established as a result of x-ray diffraction studies. X-ray analyses are also useful in the identification of minerals, but usually the more common minerals can be recognized without the aid of special instruments by their characteristic physical properties. The most common physical properties are color, luster, cleavage, crystal form, hardness, specific gravity, and magnetism.

Because the elements oxygen and silicon make up nearly 75 percent of the earth's surface rocks, the silicates are the most important of the rock-forming minerals. The fundamental unit in silicate mineral structures is a tetrahedral arrangement of four oxygen ions around a central silicon ion. The tetrahedra are joined by chemical bonds to form chains, sheets, or three-dimensional frameworks; these structures are the underlying causes for many of the specific properties of a mineral.

The principal rock-forming silicate mineral families are relatively few in number. They include the quartz family, potassium and plagioclase feldspars, muscovite and biotite micas, hornblende, augite, olivine, garnet, chlorite, and clay. Those silicate minerals, such as biotite, augite, hornblende, and olivine, that are rich in iron and magnesium (and usually black or green in color) are designated ferromagnesian minerals. Common nonsilicate rock-forming minerals are the carbonates (calcite and dolomite), evaporites (gypsum and halite), and iron oxide minerals (hematite, limonite, and magnetite).

Certain relatively rare specimens of otherwise common minerals are so beautiful that they have been used since ancient times as ornamental gems. The beauty of these gem varieties results from their color, luster, and the manner in which they transmit or reflect rays of light. Gems should be relatively hard, for otherwise wear will destroy their beauty. Their value is also influenced by their rarity.

QUESTIONS FOR REVIEW

1 What is a crystal? Why are complete and perfectly formed mineral crystals rare in nature?

2 In terms of atomic structure, how does a crystalline rock (like granite) differ from an amorphous rock (like volcanic glass)?

3 What is the difference between an atom and an ion?

4 In terms of atomic structure, explain why asbestos is fibrous and mica is platy, and why does the mineral quartz not break into fragments with smooth surfaces?

5 Explain how covalent bonding differs from ionic bonding.

6 Name four common rock-forming silicate minerals rich in iron and magnesium. Which of these have good cleavage in two directions? Which is the softest? Which is constructed of individual tetrahedra linked at their corners by metallic ions?

7 The minerals halite and calcite look somewhat alike. Without tasting the specimens, how would you identify each?

8 Both diamond and graphite are minerals composed of carbon. Why are they so different in specific gravity, color, and hardness?

9 If you were a Stone Age human and had to choose between calcite and chert for a stone axhead, which would you choose? Explain your choice.

10 Under what circumstances can one element substitute for another in a mineral? What mineral family is characterized by such substitution?

11 List, in order of abundance by weight, the eight most abundant elements in the earth's crust. What is the chemical symbol of each? Which has the largest ionic radius?

SUPPLEMENTAL READINGS AND REFERENCES

Dietrich, R.V., and Skinner, B., 1979. Rocks and Minerals. New York, John Wiley & Sons, Inc.

Ernst, W.G., 1969. Earth Materials. Englewood Cliffs, N.J., Prentice-Hall Inc.

Gait, R.L., 1972. Exploring Minerals and Crystals. Toronto, McGraw-Hill Ryerson Ltd.

Holden, A., and Singer, P., 1960. Crystals and Crystal Growing. Garden City, N.Y., Doubleday & Co.

Hurlbut, C.S., Jr., 1969. Minerals and Man. New York, Random House Inc.

Pough, F.H., 1960. A Field Guide to Rocks and Minerals. Cambridge, Mass., Houghton Mifflin Co.

Shaub, B.M., 1975. Treasures from the Earth. New York, Crown Publishers, Inc.

Tennisson, A.C., 1974. The Nature of Earth Materials. Englewood Cliffs, N.J., Prentice-Hall Inc.

Zoltai, T., and Stout, J.H., 1984. Mineralogy. Minneapolis, Burgess Publishing Co.

2 igneous geology

The Montagne Pelée presents no more danger to the inhabitants of Saint Pierre than does Vesuvius to those of Naples.

PROFESSOR LANDES, 1902
of St. Pierre College the day before he and 30,000 residents of the town perished in the eruption of Mont Pelée.

Prelude Geology recognizes three major divisions of rock according to their mode of origin. These divisions are igneous, metamorphic, and sedimentary. In this chapter, we will examine igneous rocks and the processes responsible for their origin. *Igneous rocks* are those that have been formed by the solidification of molten rock material. The molten matter from which they are formed is termed *magma.* Magma may exist for long periods of time beneath the surface of the earth, intruding here and there into overlying rock yet never reaching the surface. When it has finally congealed underground, such magmas form the so-called *plutonic* or *intrusive igneous rocks.* Magmas may also penetrate to the earth's surface, where the molten matter pours out of openings in the form of *lava.* Lavas and solid particles that have congealed from lavas constitute the *extrusive igneous rocks.*

The eruption of Mount St. Helens volcano, May 18, 1980. (U.S. Geological Survey photograph by Austin Post.)

VOLCANIC ACTIVITY

Volcanic activity (see Color plate 6) has been an important process on this planet throughout the long span of geologic time. In the earliest stages of the earth's history, volcanoes contributed their products to form parts of the earth's crust. The gases and vapors emitted by the blistering array of primordial volcanoes provided the substance of an earlier atmosphere and of the oceans. Fortunately, volcanic eruptions are probably less frequent today, yet eruptions still occur on nearly all continents, as well as on the floors of the oceans. We view some of these eruptions on television and are able to witness in safety the devastation they cause. For geologists, these eruptions provide valuable information about the earth's interior and the way in which rocks form. Volcanoes are evidence of the enormous forces that operate uneasily beneath the frail crust on which we live. Our planet is not a great lifeless blob of cold rock making an annual trip around the sun. It is constantly changing and vibrant with igneous activity. During 1983 alone, the Scientific Event Alert Network at the Smithsonian Institution in Washington, D.C., provided daily observations of eruptions in Japan, the Philippines, Guatemala, the Aleutians, Hawaii, and Indonesia, and from submarine volcanoes in the south-central Pacific Ocean.

Four notable eruptions

Civilized humans have lived on this planet for only a few thousand years. During this time, many noteworthy volcanic eruptions have occurred. Nearly all of the recent eruptions and a few of the older ones have been documented and are a part of recorded history. Many of the most ancient accounts of volcanic activity are associated with superstition and mythology. Some, like Plato's story of the eruption of Santorini, contain just enough information to be puzzling.

Santorini

The islands of Santorini in the Aegean Sea mark the location of an eruption of particular interest to historians. This eruption, which occurred about 1470 B.C., abruptly destroyed the highly developed Minoan culture that had spread from Crete. Because the eruption resulted in explosive destruction of most of the island of Santorini, it may very well have given rise to the ancient legends of Atlantis, the "lost continent." The only known account of this event was written by Plato in 355 B.C. He described a city called Atlantis that was built on a circular plan about 24 km in diameter. The city was on an island that sank beneath the sea during a war with the Athenians. During this war, wrote Plato, "there occurred portentious earthquakes and floods, and one grievous day and night befell them when the whole body . . . of warriors was swallowed up by the earth, and the island of Atlantis in like manner was swallowed up by the sea and vanished."

Today, the name Santorini applies to a small group of islands that enclose a nearly circular bay about 10 km in diameter (Fig. 2 - 1). The islands are remnants of the large volcanic cone that was the probable source for Plato's story, and the bay is a submerged area that formed by collapse of the central part of the volcano. Careful investigations by archaeologists working in this area have revealed a story of volcanic and earthquake activity to rival that of Pompeii. The major eruptions were apparently preceded by earthquakes. These tremors so terrified the local inhabitants that they gathered their possessions and fled to safer terrain. For this reason, human skeletons have not been found in the excavations of the ruins of the principal town, Akrotira, which is thought to have had a population of about 30,000. The eruption itself was of enormous magnitude. Towns and villages were completely buried in ash, and ocean waves generated by associated earthquakes may have ravaged towns all around Crete. Many believe these events contributed to the eventual demise of the Minoan civilization.

Vesuvius

Prior to A.D. 79, Vesuvius was a cone-shaped mountain with a rounded summit reaching 1200 meters above the shoreline of the Bay of Naples (Fig. 2 - 2). Fertile vineyards clothed its lower slopes, and the people who lived nearby considered the volcano — then known as Monte Somma — to be dead. It was, however, only sleeping. Its awakening was signaled by an increase in earthquakes for several years before the now-famous eruption. On August 24, A.D. 79, the volcano blasted away an entire side of Monte Somma, covered the surrounding countryside with a thick gray blanket of ashes and cinders, and formed a huge caldera. Beneath the blanket of ash lay the city of Pompeii. Many Romans in Pompeii, Herculaneum, and lesser towns were killed by the suffocating fall of ash and blasts of poisonous gases. Entombed in ash, the bodies of many of these hapless victims left molds, which when filled with plaster, provide casts that show even the tortured expressions of the victims.

In the centuries following the A.D. 79 eruption, the

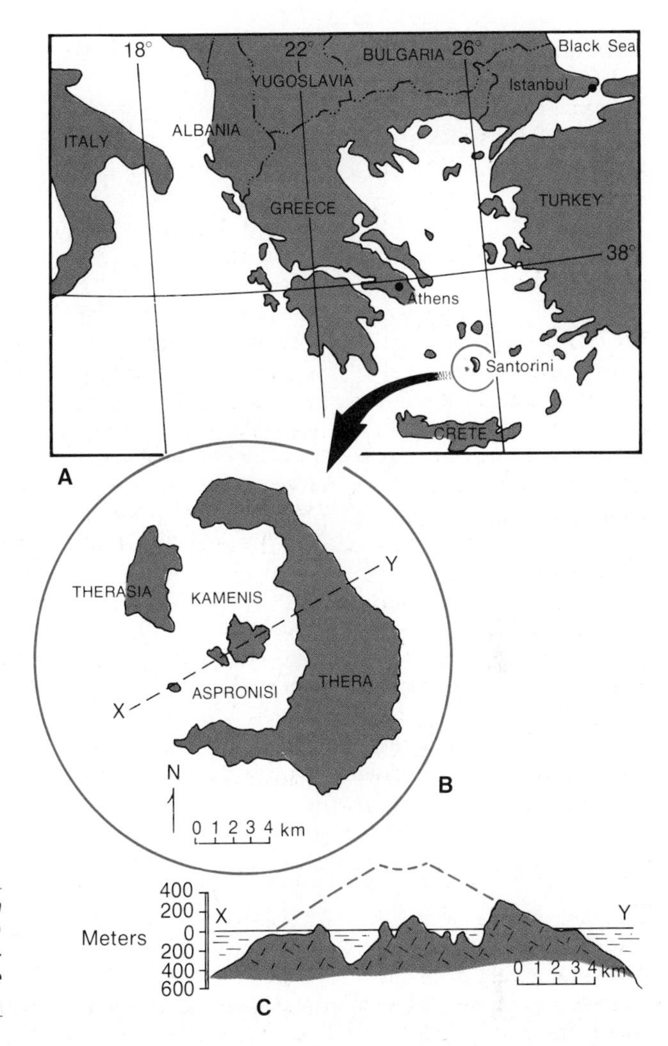

Figure 2–1 (A) Map of part of the Mediterranean area showing the location of the Santorini volcanic complex. (B) Map of the islands of the Santorini complex. (C) Cross-section along line x–y. Dashed line suggests probable profile of the volcanic cone before its explosive destruction.

present cone of Vesuvius was constructed inside the Somma caldera by periodic bursts of activity that continued until 1139. The volcano remained dormant until 1631, but since that time it has been almost continuously active (Fig. 2–3).

Krakatoa

Krakatoa is an island in the Strait of Sunda between Java and Sumatra (Fig. 2–4). It was originally constructed of the materials ejected from several volcanoes that had grown upward from the sea floor. On Sunday, May 20, 1883, Krakatoa came suddenly to life, announcing its awakening with a series of explosive blasts that could be heard by the inhabitants of islands nearly 150 km away. On the following day, ships passing through the Sunda Strait could easily see the tall column of steam and ash that rose more than 11 km into the atmosphere and showered debris over a radius of 480 km. Activity continued with gradually diminishing intensity until the final week in August. At that time, a series of tremendous explosions shook the islands, and ash began to fall in quantities never before recorded in human experience. Indeed, the sound of the accompanying barrages was loud enough to wake sleeping people in South Australia, more than 3000 km away. In islands far across the Pacific, people ran to vantage points along the beach, thinking they had heard a ship in distress signalling for help by firing its guns. A series of destructive **tsunami** (great oce-

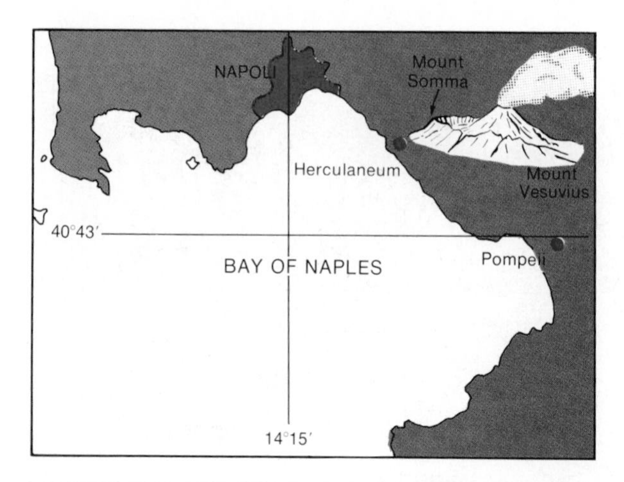

Figure 2–2 The Bay of Naples and surrounding area showing location of Herculaneum and Pompei, destroyed during the A.D. 79 eruption of Vesuvius.

anic waves triggered by earthquakes) tore through the Sunda Straits, crushing towns and killing more than 36,000 people. The cloud of ash over Krakatoa climbed 80 km, and terrified sailors on ships over 40 km away were pelted with volcanic bombs the size of pumpkins. For two days, the sun in that area was completely blotted out by ash and dust. When a vessel of the Royal Dutch Navy was eventually able to reach Krakatoa, its crew discovered that about 65 percent of the island had disappeared. The immense crater that had formed now lay submerged beneath 300 meters of water. For several years following the eruption, dust thrown into the upper atmosphere caused brilliant red sunsets around the earth. By blocking part of the sun's radiation, the dust also caused a slight lowering of mean annual temperatures.

Mount St. Helens

Until recently, many mainland Americans preferred to believe that volcanic eruptions were hazardous geologic events that occurred mostly elsewhere. This belief was shattered at 8:31 A.M. on May 18, 1980, when Mount St. Helens in Washington State (Fig. 2–5) exploded with the energy equivalent of 50 million tons of TNT and a roar that was heard 300 km away from the once-scenic mountain (Fig. 2–6 A–E). A great turbulent cloud of hot gases, steam, pulverized rock, and ash burst laterally from the north side of the mountain, rose 20,000 meters into the atmosphere, and began to drift slowly toward the east. An estimated one cubic kilometer of air-borne ash and other rock debris from the explosion blocked out the sun's light and caused automatic street lights to switch on in towns hundreds of kilometers downwind from the rumbling volcano. The ash fell like a dismal gray snow, blanketing streets, dangerously burdening the roofs of buildings, choking the engines of vehicles, and covering the leaves of trees and crops. Hot gas and ash from the volcano melted

Figure 2–3 An eruption of Mount Vesuvius as seen from the city of Naples. (Photograph by Frederick Lewis Inc.)

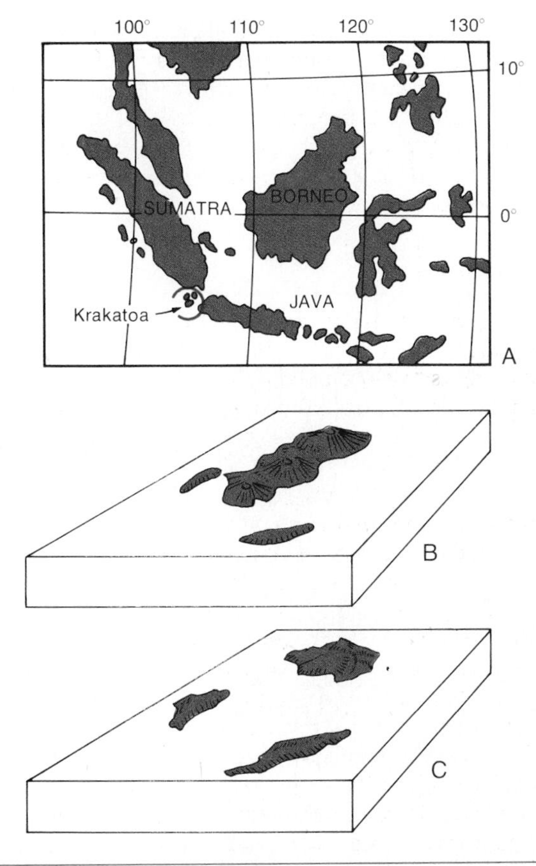

Figure 2–4 *(A) Map showing location of Krakatoa. (B) Appearance of the Krakatoa complex just before the 1883 eruption. (C) Island remnants of Krakatoa after the 1883 explosive eruption.*

part of Mount St. Helens' snow and ice cap, and the resulting meltwater mixed with ash and formed mudflows that surged down the mountain slopes at speeds as great as 80 km per hour. In the nearby town of Toutle, the mudflows destroyed 123 homes. In addition, 22 people camping or working nearby were killed by heat, gases, or burial under the downpour of ash. The May 18th eruption was clearly a major volcanic event. The fact that it was actually rather puny when compared with the famous eruptions of Krakatoa or Santorini (described earlier) was no consolation to the citizens of Washington as they viewed the devastated gray landscape, the hundreds of square kilometers of valuable trees toppled like matchsticks, buried roads, harbors choked with ash-sludge, and smothered fields of wheat.

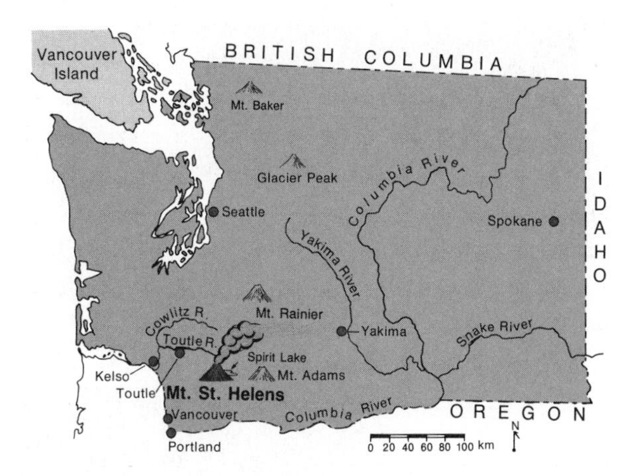

Figure 2–5 *Location map, Mount St. Helens volcano in southeastern Washington State.*

The recent activity at Mount St. Helens, and the older eruptions that gave us the volcanic peaks of the Cascades, are surface manifestations of an ongoing collision between two of the earth's crustal plates. One of these is the American Plate and the other is the small Juan de Fuca plate of the eastern Pacific, which moves eastward on a collision course toward the coasts of Oregon and Washington. The Juan de Fuca plate plunges beneath the coastline, and molten rock generated as the plate moves downward rises to supply the lava of the volcanoes.

Volcanoes and plate tectonics

The vast majority of the volcanoes that have been active during recorded history are located either along particular border regions of continents or within ocean basins. Relatively few are found in the central regions of continents. It is also evident that most volcanoes are distributed along somewhat discontinuous linear zones or chains. One of the largest of these chains encircles the basin of the Pacific Ocean in what has been described as a "ring of fire." More than 75 percent of the world's active volcanoes are located on this ring of fire. On the east side of the Pacific, the list includes such familiar volcanoes as Popacatepetl and Paricutín in Mexico; Fuego in Guatemala; Lassen, Shasta, and Rainier (Fig. 2–7) in the continental United States; and Katmai in Alaska.

Although continental margin volcanoes are very abundant, there are also impressive numbers of midocean volcanoes. Some of these erupt deep beneath

A

B

C

D

E

Figure 2–7 Mount Rainier, Washington. This volcano is composed of interbedded lava flows and layers of ash. The top of this famous peak is about 2200 meters above the nearby peaks of the Cascade Range. (Courtesy of U.S. Park Service.)

the ocean surface and may or may not build up above the waves. Surtsey, an island south of Iceland, is composed of the materials built up by an oceanic volcano that first appeared at the ocean surface in 1963. Iceland has a total of 22 similar volcanoes. Farther to the south in the Atlantic, the Azores continue to experience periodic eruptions. Due west of the southern tip of Africa lies Tristan da Cunha, whose human inhabitants were evacuated to Britain to provide safety from a 1962 eruption. Finally, subsidiary volcanic zones extend eastward through the Mediterranean region and along the rift zones of Africa.

What is the explanation for the linear distribution patterns of volcanoes? The answer to that question is provided by a relatively simple theory called **plate tectonics.** The theory will be described fully in Chapter 5, but it requires our brief attention here because of its relationship to the global pattern of volcanoes.

The central idea of plate tectonics is that the litho-

Figure 2–6 Mount St. Helens erupts. (A) An aerial view of ash-laden Mount St. Helens on March 27, 1980. The craters visible at this time were destroyed during the subsequent eruption of May 18, 1980. (B) Photograph taken on April 10, 1980, showing the volcano from the east at a time when several steam and ash eruptions were in progress. This activity occurred about 2 weeks after the initial eruption of March 27 and 5 weeks before the great blast of May 18. (C) The massive May 18, 1980, eruption, during which clouds of steam and ash ascended to an altitude of over 20,000 meters. (D) A view from the northwest photographed 9 hours after the initial explosion shows the destruction of part of the rim of the crater. (E) The new crater of Mount St. Helens as it appeared on May 23, 5 days after the great May 18th eruption. By projecting the slopes diagonally upward, one can obtain a rough idea of the amount of the mountain's northern side that was obliterated. (Photographs A–D, Austin Post, U.S. Geological Survey; E, Keith and Dorothy Stoffel, Washington State Division of Geology and Earth Resources.)

sphere of the earth is composed of six or seven large "plates" and several smaller ones (Fig. 2-8). The plates, which are 75 to 150 km thick, consist of both crust and part of the uppermost mantle (see Fig. 1-11). Lithospheric plates rest upon a weak plastic layer of the mantle that we defined in Chapter 1 as the asthenosphere. Two characteristics of plates are that they move relative to one another, and that volcanic eruptions and earthquakes frequently occur along their borders. Geologists are uncertain about what causes them to move. Most believe the plates are conveyed by the drag of slowly moving convection currents in the underlying mantle.

If one visualizes a globe sheathed in moving lithospheric plates, it becomes apparent that the boundaries between plates will differ according to the directions of plate movement. For example, zones along which plates move away from one another are called **divergent boundaries.** Along these boundaries, black lavas pour from volcanoes and fill the area between the diverging plates. The result is the great linear chains of mostly subsea volcanoes that constitute the midoceanic ridges (see Fig. I-4). Iceland and the Azores represent small segments of the mid-Atlantic Ridge that have become emergent.

On a sphere such as the earth, it is not possible for lithospheric plates to move apart from one another in some places unless they converge or slide past one another along other boundaries. **Convergent boundaries** are those along which plates move toward one another. They are also places where part of the crust is consumed in order to make room for new crust produced at divergent boundaries. This consumption may occur when one of the converging plates (usually the denser oceanic plate) sinks beneath the other plate to be melted and recycled in the un-

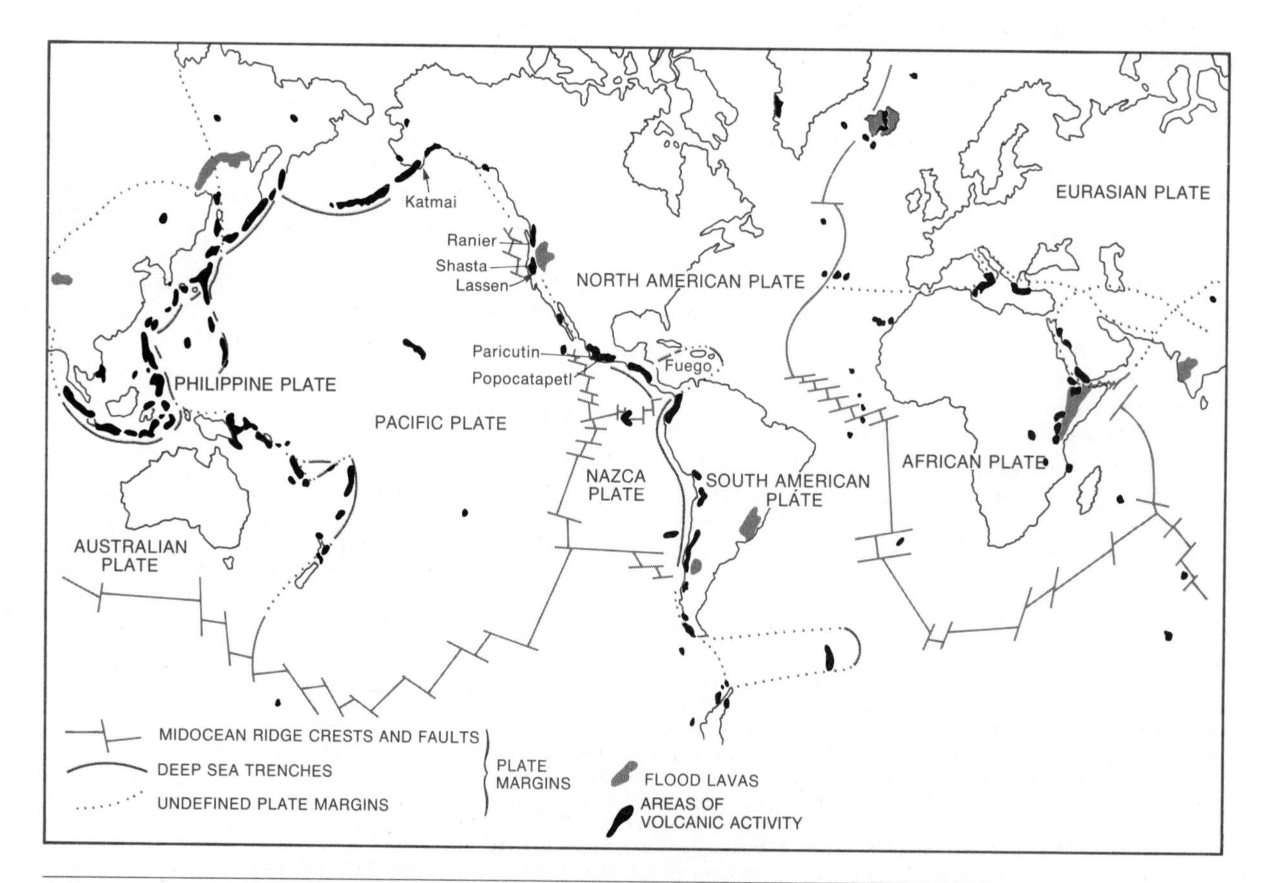

Figure 2-8 *Tectonic plate boundaries and the distribution of active volcanoes on the earth. Note that volcanoes are not randomly distributed, but rather tend to be associated with plate boundaries.*

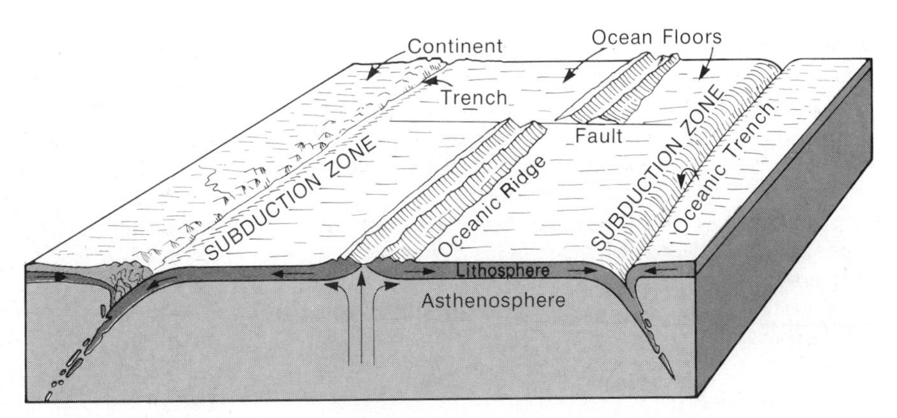

Figure 2–9 *Conceptualized diagram showing lithospheric plate motions at divergent and convergent boundaries. The plates move somewhat like conveyor belts from a divergent boundary (oceanic ridge) to a convergent boundary where subduction occurs.*

derlying mantle (Fig. 2–9). The process is called subduction, and the tops of the subduction zones are recognized by the presence of deep-sea trenches. Convergent boundaries are also zones of intense volcanic and earthquake activity. The lavas produced by convergent zone volcanoes, however, are not exclusively the dark lavas of midocean ridge volcanoes. Rather, they tend to be more diverse in composition, are somewhat richer in silica, and poorer in iron and magnesium. This more varied composition probably results from partial melting of the descending plate with its blanket of sediment, and from partial melting of the upper mantle as well.

The third possibility for movement along a plate border is neither convergent nor divergent but rather lateral motion as two plates slide past each other. Such zones of horizontal movement are called **shear boundaries.** They are locations where crustal material is neither added nor consumed. Shear boundaries are marked by great faults. Movements along these faults generate frequent and often intense earthquakes. The San Andreas fault zone in California marks a shear boundary.

The products of volcanism

The gases, liquids, and solids emitted in the course of volcanic eruptions are ultimately derived from magma. Magmas are hot complex melts that contain not only the elements that combine to form minerals but also water and such gases as carbon dioxide, nitrogen, and sulphur dioxide. Water, which can make up about 1 to 9 percent of a magma, is particularly important. When water changes from a liquid to a gas, its volume increases 1700 times. Pressures resulting from this expansion are ample to cause spectacular eruptions. At the great depths where magmas are generated, they tend to hold water and gases in solution. However, when circumstances permit magmas to work their way upward near the earth's surface, water and gases may come out of solution somewhat in the manner of the bubbles in a carbonated soft drink. If the released gases and vapors become trapped as by a congealed plug of lava, pressures may build up underground until a volcanic explosion occurs. Vast quantities of steam, gases, and fragmented rock may be thrown skyward, as was the case during eruptions of Krakatoa and Pelée. The liquid that flows from vents or fissures is lava, and it soon cools into solid rock. Before discussing the nature of volcanic eruptions, it will be useful to describe some of the rocks formed from lava and the fragmented solids ejected during eruptions.

Pyroclastics

Even relatively quiet volcanic eruptions produce large quantities of ash, cinders, congealed blobs of lava, and solid fragments torn from the walls of vents. Such ejected materials are collectively termed **pyroclastics** (also *tephra*), and they are classified according to their size and shape. **Bombs** (Fig. 2–10) for example, are rounded or ellipsoidal masses ranging in diameter from only 4 mm to several meters. They frequently display a twisted appearance resulting from the flow of still molten material during ejection. **Blocks** are solid, angular pieces of rock greater than 32 mm in diameter. They consist of pieces of the crustal layers beneath the volcano or older lavas broken from the walls of the vent or edge of the crater. Blocks weighing hundreds of tons have been thrown several kilometers during particularly explosive eruptions. In contrast to bombs and blocks, smaller pyroclastics include *lapilli*

Figure 2–10 Volcanic bombs. The specimen in the upper left part of the photograph is about 30 cm long. These objects are ejected during eruption as incandescent clots of liquid lava and largely consolidated during flight. (Institute of Geological Sciences, London.)

Figure 2–11 Eruption of Paricutín Volcano, Michoacán, Mexico. This eruption was accompanied by great outpourings of cinders and ash. Small pond in foreground was formed by lava dam. Photograph taken July 25, 1943. (Courtesy of U.S. Geological Survey, photograph by W.F. Foshag.)

(Italian, meaning little stones) which are mostly the size of gravel, *cinders* (particles of pulverized lava the size of sand grains), and *ash* (Fig. 2–11). All of these fragments of volcanic ejecta can be consolidated or lithified to form the so-called pyroclastic rocks. If round in shape, the larger fragments form **agglomerates,** and if angular, **volcanic breccias.** The rocks from cinders and ash are called **volcanic tuffs. Welded tuffs,** or **ignimbrites,** are developed from ash and cinders that have congealed while still hot.

Accumulations of volcanic ash have been known to bury entire towns and villages. When mixed with water, cinders can also contribute to the formation of mudflows, called **lahars.** In tropical areas with heavy seasonal rainfall, such as Indonesia and the Philippines, lahars develop on the sides of active volcanoes and may flow downward at velocities of more than 80 km per hour. They kill and destroy everything in their paths. Lahars sometimes form when an eruption blasts up through the rim of a crater lake, expelling a lethal avalanche of boiling water, mud, and boulders.

Glasses

Most volcanic rocks, even though they cool rapidly from lavas, are crystalline solids. However, in such rocks, the individual mineral grains are usually too small to be seen without a microscope. But what about those lavas that have been chilled so quickly that there has not been time for the atoms in the melt to order themselves into the structures necessary to form crystals? The atoms and unlinked silica tetrahedra are frozen into the random positions they held in the liquid. The result is a noncrystalline mixture of silicates and silica called glass. The familiar rock **obsidian** (Fig. 2–12) is a natural glass that, if it had cooled more slowly, would have developed crystals of feldspar, quartz, and ferromagnesians. Most obsidian is black as a result of its content of impurities and its light-absorbing properties. Less commonly, such impurities as iron oxide may color obsidian red or brown. It is easily recognized by its glassy appearance and distinctive conchoidal fracture.

Obsidian most commonly forms from lavas that

Figure 2–12 Obsidian, a common type of volcanic glass. The conchoidal fracture typical of obsidian is shown well in this specimen. (Courtesy of the Institute of Geological Sciences, London.)

Figure 2–13 Pumice, a type of glass produced by silica-rich lavas in which the gas content is so great as to cause the lava to "froth" as it rises in the chimney of the volcano and experiences rapid decrease in pressure. Some pieces of pumice will float on water because of the many air spaces formed by expanding gases. (Institute of Geological Sciences, London.)

have lost most of their dissolved gases. Like a bottle of beer that has been left open too long, such lavas have gone "flat." However, if the lava still contains abundant gas, bubbles may continue to escape while the melt cools. Such bubbles are preserved in the rock as cavities called **vesicles. Pumice** (Fig. 2–13) is an extremely vesicular glass. In a very real sense, it is frozen silicate froth. Usually whitish or gray, pumice is light in weight and, because of its many sealed cavities, can sometimes actually float on water.

Crystalline volcanic rocks

Basalt

By far the most abundant volcanic rock is **basalt.** It underlies the oceans and is the rock from which Iceland and the Hawaiian Islands are constructed. The lavas from which basalts solidify have relatively low viscosity, or resistance to flow. As a result, they may spread over great distances before they congeal. The basalt flows of the Columbia Plateau (Fig. 2–14), the Deccan Plateau of India, and the maria of the moon cover tens of thousands of square kilometers.

Basalt is a dense dark-gray or black rock composed of pyroxene, calcium-rich plagioclase feldspar, and minor amounts of magnetite and olivine. Basaltic lava pours from vents and fissures at temperatures of about 1200°C and becomes solid by the time the lava has cooled to about 750°C. Crystallization is relatively rapid, resulting in a very finely crystalline texture in which individual minerals cannot be discerned with the unaided eye. This type of rock texture is termed **aphanitic.** However, one can view the uniformly sized crystals under the microscope by preparing a thin section of the rock. With the aid of a diamond saw, a rectangular piece of the basalt is cut and cemented onto a microscope slide. The mounted piece is then ground down until it is so thin that light can be transmitted through it. A cover glass may be added to complete the slide, and the myriads of tiny feldspar and pyroxene grains can then be examined microscopically (Fig. 2–15).

The origin of basalt is a subject that has puzzled geologists for many years. Many believed that basaltic lavas originate from melts in the lower part of the earth's basaltic crust. However, several lines of evidence now suggest that these lavas may have come from molten pockets of upper mantle material. Evidence for this idea is derived from present-day vol-

Figure 2–14 Basalt lava flows exposed on either side of Summer Falls, Columbia Plateau. (Courtesy of U.S. Department of the Interior, Bureau of Reclamation, Columbia Basin Project.)

canic activity associated with deep earthquakes occurring within the mantle, far beneath the crust. It is likely that fractures produced by these earthquakes serve as passages for the escape of molten material to the surface. A detailed study of earthquake shocks from particular volcanic eruptions in Hawaii indicates that the erupting lavas were derived from pockets of molten material within the upper mantle at depths of 65 to 100 km. However, mantle materials are denser, richer in iron and magnesium, and more deficient in silica than are oceanic basalts. How then can the melting of mantle material give rise to basaltic rocks that are less dense and contain more silica as a constituent of their abundant feldspars? **Partial melting** provides a tentative answer. Partial melting is a process by which a rock mass subjected to high temperature and pressure is partly melted and the liquid component is moved to another location. At the new location, the separated liquid may solidify into rocks having different composition from the parent mass. The word "partial" refers to the fact that some minerals melt at lower temperatures than others, and so for a time the material being melted resembles a hot slush composed partially of liquid and partially of solid crystals. The molten fraction is usually less dense than the solid from which it was derived, and thus it tends to separate from the parent mass and work its way toward the surface. In this way melts of basaltic composition may separate from denser rocks of the upper mantle and make their way to the surface.

The lavas from which basalts are formed consist chemically of about 50 percent silica (SiO_2), with lesser percentages of oxides of aluminum, iron, calcium, magnesium, sodium, and potassium. Nonbasaltic lavas differ in their silica content, and this difference provides the basis for recognition of such other volcanic rocks as **rhyolite** and **andesite.**

Rhyolite

Rhyolite is a gray or pink aphanitic rock derived from a highly siliceous lava. Rhyolitic lavas may consist

Figure 2–15 Photograph taken through a microscope of part of a thin section of basalt. (Width of field is 2 mm.) Larger crystals, if exhibiting shaded zones, are plagioclase. Smaller crystals in groundmass are plagioclase and pyroxene (mostly augite). Glass is also present in the groundmass. Basalt shown here is from Guam. (Courtesy of U.S. Geological Survey, photograph by D. Carroll.)

of as much as 75 percent silica. Quartz, potassium feldspar, sodium plagioclase, and lesser amounts of ferromagnesian minerals occur in rhyolite (Fig. 2–16). Rhyolitic lavas erupt at temperatures of 800° to 1000°C. Usually they are far more viscous than basaltic lavas and flow very sluggishly.

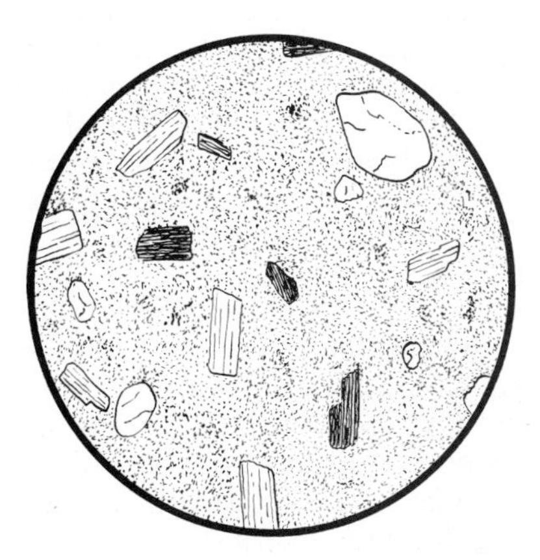

Figure 2–16 Thin section of rhyolite viewed through a microscope. Clear grains are quartz, elongate grains with parallel lines are orthoclase and sodium-rich plagioclase, and dark grains are biotite. Smaller grains are indistinguishable in the fine groundmass. (Diameter of field is 1.2 mm.)

Rhyolites are the products of continental volcanism rather than volcanoes that form on the ocean floor. This suggests that at least some rhyolites are formed by remelting of parts of the silica-rich continental crust. Another theory proposes that rhyolites are formed by the melting of sediments that had been eroded from the continents and deposited along continental margins. Highly silicic sediment, if caught in the vise of converging tectonic plates and melted, could produce magmas of rhyolitic composition.

It may also be possible to produce rhyolitic lavas by means of a process known as **fractional crystallization.** Imagine a deep reservoir of magma in which olivine and other ferromagnesian minerals have formed early in the cooling period and settled to the bottom of the magma chamber. Early crystallization of these minerals leaves the original liquid changed in composition. The proportion of silica in the melt relative to oxides of iron and magnesium, for example, may have been increased significantly. If so, the melt might be compositionally suitable for the formation of rhyolite.

Andesite

Andesite (Fig. 2–17) is a fine-grained, light-gray rock intermediate in density and silica content (about 60 percent) between basalt and rhyolite. Andesitic lavas are also intermediate in viscosity and thus form thicker flows than do basaltic lavas. Volcanoes of the andesitic type are common around the edges of continents bordering the Pacific Ocean and are generally more explosive than midoceanic volcanoes, like those of Hawaii.

Andesitic lavas may originate in more than one way. Some appear to have been formed from originally basaltic magmas by fractional crystallization. Evidence for this mode of origin is provided by Iceland's volcanoes. Iceland is a volcanic island formed on oceanic basaltic crust. It has been observed that the longer the quiet period between eruptions of Icelandic volcanoes, the more siliceous is the lava that is extruded. Apparently, longer periods of quiescence provide time for fractional crystallization and settling of ferromagnesian minerals out of the residual liquid.

Because andesites are roughly intermediate in composition between rhyolites and basalts, it would appear that at least some andesitic lavas are the products of mixing of silica-rich and silica-poor melts. For example, a basaltic magma might work its way into pockets of rhyolitic magma and then blend with the more siliceous liquid. As another possibility, a basaltic magma might engulf, melt, and assimilate surrounding

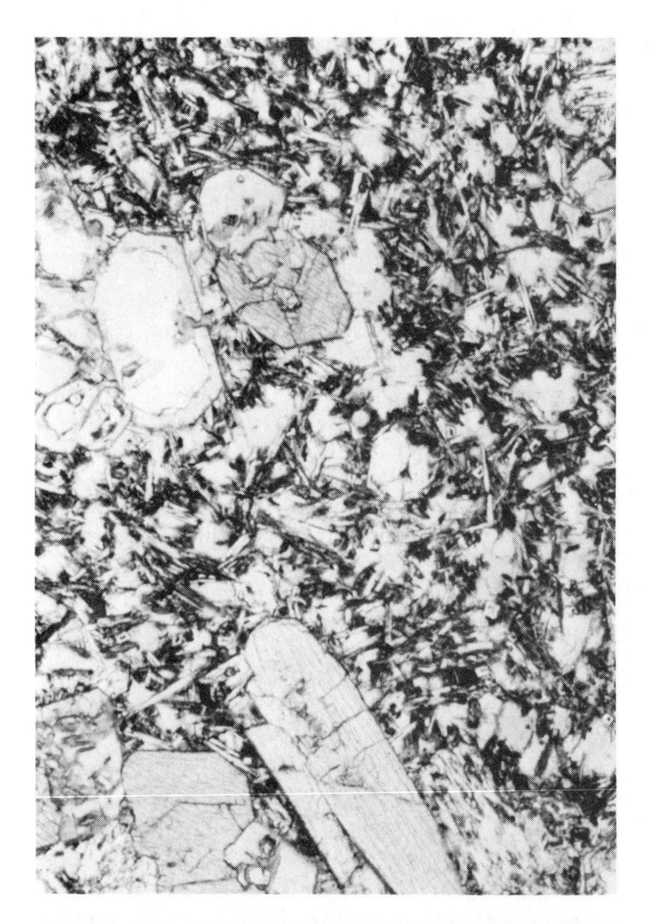

Figure 2–17 Thin section of andesite, viewed through a microscope. Larger crystals are augite and feldspar in a matrix of calcium-rich plagioclase, augite, magnetite, glass, and alteration products. From a lava flow on Guam, Mariana Islands. (Width of field is 3 mm.) (Courtesy of U.S. Geological Survey, photograph by D. Carroll.)

silica-rich rhyolitic rocks. Yet another method for infusing silica into basaltic materials to produce andesite is provided by plate tectonics. In this method, the required amount of silica is derived from the thin layer of oceanic sediment that rests upon the basalt of an oceanic plate as it descends along a subduction zone (Fig. 2–18). As this plate plunges diagonally downward, heat rising from the asthenosphere and generated by friction melts masses of basalt and sediment to produce fluids of andesitic composition. These melts then rise buoyantly and erupt to form the large volumes of andesite frequently found adjacent to subduction zones.

Features of volcanic rocks

As lavas cool and begin to congeal, they may develop a variety of textural features. We have already noted how gas bubbles produce vesicles. Vesicles are the most characteristic feature of the volcanic rock called **scoria** (Fig. 2–19), a dark clinker-like rock that is partly crystalline and partly glassy. In some scorias, vesicles become filled with minerals deposited from watery solutions that percolate through the rock. The fillings are termed **amygdules** because of their resemblance in shape to almonds (Latin *amygdala*, almond).

The surfaces of solidified lavas may also develop distinctive appearances. **Pahoehoe** is ropy lava (Fig. 2–20). The ropy texture is produced in the plastic surface of lava by the drag of more rapidly flowing liquid lava below. **Aa** is a blocky, fragmented lava. It is derived from a more viscous and slower-moving liquid. As this thicker lava flows down a slope, the upper layer hardens and is carried along in conveyor-belt fashion. On reaching the front of the lava flow, the

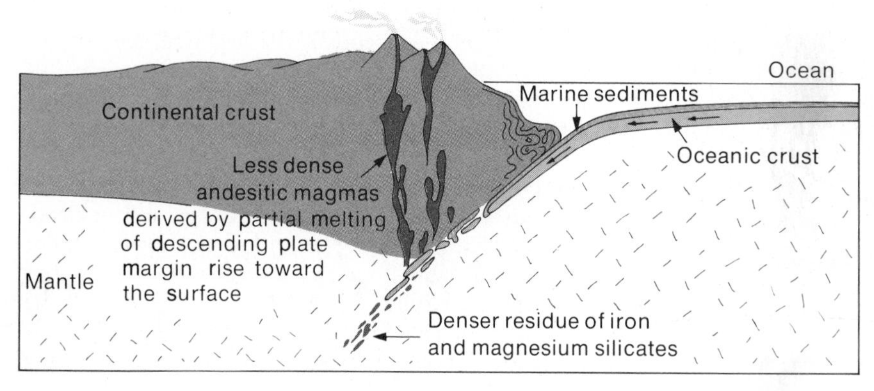

Figure 2–18 Subduction of sediment-bearing leading edge of an oceanic plate with partial melting at depth to produce andesitic magmas.

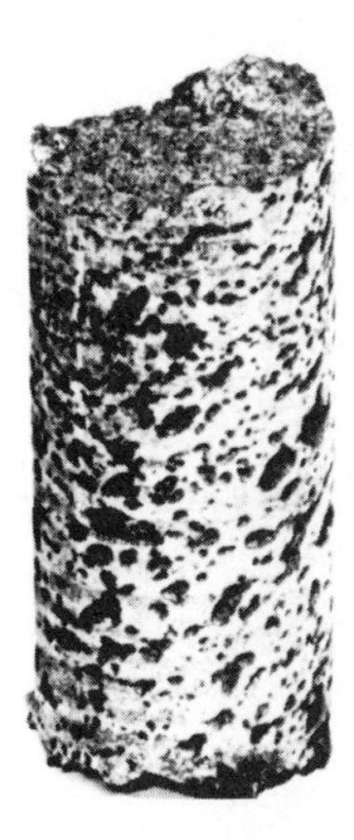

Figure 2–19 A core cut in scoria from the Kana-a flow in northern Arizona. Note the abundant vesicles. The core is 6 cm in diameter. (Courtesy of U.S. Geological Survey.)

brittle top layer is broken into a chaotic jumble of jagged fragments. If you were to walk upon this material in your bare feet, you might easily guess why native Hawaiians named it aa.

Lava that is extruded under water bulges outward into bulbous pillow-shaped masses that average about a meter in diameter (Figs. 2–21, 2–22). Scuba-diving geologists have filmed these pillows as they form from tongues of lava that are intermittently extruded from submarine fissures. Pillow lavas are useful indicators of underwater volcanic eruptions and also provide a means of locating the original top of deformed sequences of layered rocks that contain lava flows. The convex surfaces of the pillows indicate which way was "up" in the sequence.

Figure 2–20 Pahoehoe lava. (Courtesy of D.A. Swanson, U.S. Geological Survey.)

Frequently, igneous masses contain fragments that differ in mineral composition, texture, and color from the enclosing rock. These inclusions are either fragments of preexisting rocks that magma has torn loose from the walls of the vent, or they are pieces of the roof or sides of the magma chamber that spalled off into the melt. Such fragments of preexisting foreign rock included within an igneous mass are called **xenoliths,** meaning "stranger rocks" (Fig. 2–23).

Lava tunnels, tubes, and caverns are among the most frequently visited tourist attractions on Hawaii. They are formed as the outer zone of large lava flows solidify, while an inner zone is still molten and flowing. After a period of time, the molten lava will drain out of the flow so that empty space, in the form of a long tube, remains. Such lava tunnels are not uncommon, and some extend for many kilometers. Frequently, during eruptions, parts of the ceiling of a lava tube collapse. The rills observed on the surface of the moon, for example, are thought to be collapsed lava tunnels. **Lava fountains** (Fig. 2–24) may erupt at the surface above the tube and form **spatter cones** up to a few meters high. In addition, however, lava fountains are often developed along fissures rather than from ruptures in tunnels or tubes.

Perhaps the most distinctive feature formed in cooling lava flows are **columnar joints.** Joints are shrinkage cracks that form in rather regular patterns when lava cools and contracts. Columnar joints are perpendicular to the cooling surface and divide the rock into polygonal prisms or columns (Fig. 2–25). From a distance, the columns resemble bunches of giant fence posts. When viewed from above, they exhibit a pattern not unlike polygonal bathroom tiles. The columnar jointing of the Giants' Causeway in Ireland and the Devil's Post Pile in California are well-known examples.

Volatiles released during eruptions

Volatiles are elements or compounds that become gaseous at relatively low temperatures. By far the most abundant volatile emitted during volcanic eruptions is

Figure 2–21 *Pillow lavas on ocean floor, southwestern Pacific Ocean. Photograph taken from camera on oceanographic research submersible craft. (Courtesy of R. Batiza, Department of Earth and Planetary Sciences, Washington University.)*

Figure 2–22 *Pillow lava of Middle Cambrian age, Trinity Bay, Newfoundland. (Geological Survey of Canada.)*

Figure 2–23 *Xenolith of schist in a granitic matrix, British Columbia. (Geological Survey of Canada.)*

Figure 2–24 *Night view of lava "fountaining" to a height of 400 meters above a vent at Mauna Ulu on Kilauea's east rift zone. Photograph was taken August 22, 1969. (Courtesy of D.A. Swanson, U.S. Geological Survey.)*

water vapor. Many people are surprised to learn that silicate melts actually contain water. Provided there is sufficient pressure, water in contact with molten rock passes into solution in much the same way that CO_2 is dissolved in a soft drink. Experimental studies of high-silica melts have indicated that at depths of 1220 to 3650 meters and temperatures not over 870°C, these melts can hold 6 to 9 percent water. When eruptions occur, the water is given off as steam and constitutes over 70 percent of the gaseous cloud that rises from the vent. Carbon dioxide is next in abundance, and there are lesser amounts of carbon monoxide, sulfur dioxide, hydrogen sulfide, and other gases. Relatively small amounts of other elements also occur in elemental form or as compounds within the cloud of volatiles rising above a volcano. These include nitrogen, argon, boron, chlorine, fluorine, and such metallic elements as iron, copper, zinc, and mercury.

In the course of geologic time, volcanoes have brought huge quantities of gases to the surface. The water vapor in these emissions condensed and was precipitated to form the water bodies of the earth, and other gases formed the planet's early atmosphere. The process can be observed today not only in major eruptions but also in the **fumaroles,** or "smoke holes," that are common near volcanoes and in areas underlain by cooling magmas, such as Yellowstone National Park.

Landforms produced by volcanoes

The viscosity of a magma is a critical factor in determining the eruptive style and ultimate shape of a volcano. The most important factors that influence viscosity are composition (including volatiles), pressure, and temperature. In regard to composition, the generalization can be made that the more silica in a magma, the more resistant it will be to flow. This is because with an abundance of silica in the melt, tetrahedra readily form and begin to group into network structures that resist flow. An increase in pressure within the magma chamber also tends to increase viscosity. Heating has the reverse effect. Other things being

Figure 2–25 Waterfall formed at an exposure of basalt which shows well-developed columnar jointing. Svartifoss Waterfall, Iceland. (Photograph by Terry Plank, with permission.)

equal, a rise in temperature results in a decrease in the viscosity of the magma.

In addition to the silica content of a melt, dissolved water can also effect viscosity. In general, water tends to reduce viscosity by inhibiting linkage of tetrahedra in the melt.

Landforms associated with fissure eruptions

In the canyon walls of the Snake River of southern Washington one can see scores of layers of flat-lying basaltic lava flows (see Fig. 2–14). They are termed **flood basalts,** or **plateau basalts.** The lavas extruded were of low viscosity and therefore were able to spread widely before solidifying. Characteristically, flood basalts occur in areas where lavas emerge from one or more long, narrow fissures. The basalts that built up the Columbia Plateau completely buried preexisting topography over an area of 130,000 km. Other famous examples of such flood basalts occur in the Deccan Plateau of India and the Parana Plateau of Brazil and Paraguay.

Fissure eruptions tend to be more prevalent in areas where the earth's crust is subjected to tensional forces; that is, forces that tend to pull the crust apart. Crustal tension may result in vertical cracks through which magma may force its way to the surface. Such fissures are particularly characteristic of the mid-oceanic ridge (and fissure) systems. Truly prodigious amounts of lavas have poured from these features.

Central eruptions and volcanic cones

Shield volcanoes

We are accustomed to thinking of volcanic eruptions as the discharge of lavas and gases from some sort of central vent or group of vents. The results of such eruptions are the familiar volcanic cones. Depending on the nature of the lava, the landforms produced from central eruptions differ in shape. Viscous lavas form cones with steep slopes. However, if lavas are relatively thin and free flowing, they spread widely, producing a broadly convex or shield-shaped volcano, such as Mauna Loa (Fig. 2–26). Hawaii (the "Big Island") is the upper part of five coalesced **shield volcanoes** that include Mauna Loa, Mauna Kea, Kilauea, Kohala, and Hualalai. Hawaii, which rises 9.6 km above its subsea base, has been built from the accumulation of more than 50,000 cubic km of lava flows and pyroclastics. It is the largest pile of geologically young volcanic material on our planet.

Cinder cones

Cinder cones (see Fig. 2–11) are volcanic peaks built entirely of cinders and other pyroclastics that have been explosively ejected from a vent. Cinder cones rarely attain heights in excess of 500 meters. Their slopes vary between 30° and 35°, depending upon the maximum angle at which the debris remains stable and does not slide downhill—that is, the *angle*

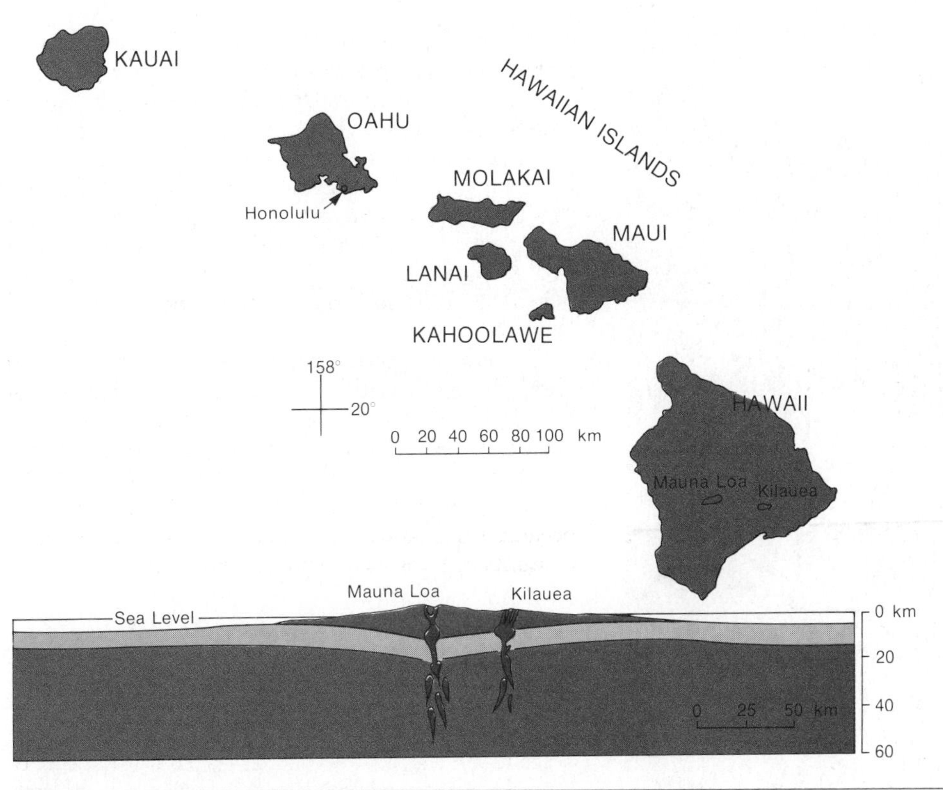

Figure 2–26 Map of the Hawaiian Islands and cross-section through Hawaii showing the broad, gently convex profile of Mauna Loa.

of repose. Paricutín in Mexico is a particularly well known cinder cone whose growth was witnessed by dozens of geologists called quickly to the scene soon after the eruption started. Paricutín began as a crack that opened up in the corn field of a farmer named Dionisio Pulidio on February 22, 1943. By the next day, the eruption had built a cone 8 meters high. Within a month, the glassy ash, cinders, and lapilli had built a mountain of debris that towered 300 meters above the surrounding countryside.

When both lava and pyroclastics are alternately emitted through a vent, a **composite cone,** or **stratovolcano,** is constructed of alternating pyroclastic layers and lava flows. These are the most common of the large continental volcanoes. They are represented by such familiar peaks as Mount Rainier and Mount St. Helens in Washington, California's Mount Shasta (Fig. 2–27), Mayon (on Luzon in the Philippines), Fujiyama (Japan), Vesuvius (Italy), Etna (Sicily), and Stromboli (on the Aeolian Islands north of Sicily).

Craters and calderas

Most people would have no difficulty defining a **crater** as a circular depression at the summit of a volcanic vent. By definition, craters also have steep inner sides and a total diameter less than three times their depth. They are seldom more than 1.6 km in diameter. Some craters form when eruption ceases and the lava that has not flowed down the slopes sinks slowly back into the vent. Craters may also form when the material at the top of the vent is blasted out by the force of one or more volcanic explosions.

Crater Lake in Oregon is a familiar example of a **caldera** (Fig. 2–28). Calderas are roughly circular in outline, have relatively flat floors, and steep circumventing walls, and they are generally more than 1.6 km in diameter. In contrast to craters, the diameter of calderas is three or four times their depth. Calderas take their name from a Spanish word meaning kettle. They may be formed by collapse, by explosion, or

Figure 2–27 *Mt. Shasta, California, a prominent stratovolcano in the Cascade range. The smaller peak is called Shastina.*

Figure 2–28 *A view of the west wall of Crater Lake, a water-filled caldera. Projecting above the lake is Wizard Island, a small cinder cone built during the last eruption. (Courtesy of U.S. National Park Service, photograph by G.A. Grant.)*

possibly by a combination of these mechanisms. Many geologists believe that calderas like Crater Lake form when eruptions drain a large portion of the magma from the magma chamber beneath the volcano. Unsupported by underlying molten rock, the magma chamber roof might collapse to form a caldera.

The caldera that formed on Bandai-san volcano in Japan is an example of a caldera formed by explosive activity. In 1888, a stupendous explosion blew off the summit and part of the northern slope of Bandai-san, leaving a caldera more than 2.5 km wide. The inner walls of Bandai-san rise an impressive 350 meters above its floor.

Extraterrestrial volcanism

Volcanic activity is by no means a phenomenon restricted to the earth. Along with meteorite impact, volcanism has been an important process in the development of other planetary surfaces and atmospheres. About 60 percent of Mars, for example, appears to be covered with volcanic rocks. At least 20 separate centers of volcanic activity have been recognized on its surface, and there are many isolated volcanoes as well.

Figure 2–30 *High-resolution Viking Orbiter I spacecraft photograph of a large lava flow on Mars. Analysis of the height of the flows suggests the lavas were probably basaltic. Distance across the photograph is approximately 10 km. (Courtesy of R. Arvidson, NASA Space Imagery Center, Department of Earth and Planetary Sciences, Washington University.)*

Figure 2–29 *The giant Martian volcano, Olympus Mons. This enormous volcano has a diameter of 600 km (375 miles) and the rim rises 24 km (15 miles) above the surrounding terrain. (Courtesy of R. Arvidson, Department of Earth and Planetary Sciences, Washington University.)*

All the volcanoes on Mars are usually interpreted to be shield volcanoes whose gentle slopes are composed of thin flows of what was once low-viscosity lava. Mars also has the distinction of having the largest known volcano in the solar system. Olympus Mons (Fig. 2–29), as it is named, is 600 km in diameter at its base and rises 23 km above the surrounding plain. It is easily large enough to contain all of Manhattan Island, and its diameter measures five times that of Hawaii. Although often larger, Martian volcanoes have generally the same shape as those on earth. Tentatively, they can be interpreted as having been formed by similar processes. Mars also has huge lava flows (Fig. 2–30), some of which extend over 1000 km from their apparent source vents.

Venus is a planet rather similar to the earth in size and mean density, and therefore it may have an interior somewhat like that of the earth. Volcanoes also exist on Venus. One huge peak topped by a crater

Figure 2–31 Volcanic cones, craters, and lava flows provide abundant evidence of past volcanic activity on the moon. (Courtesy of R. Arvidson, Department of Earth and Planetary Sciences, Washington University.)

40 km across has been located by radar, and clusters of smaller peaks, as well as what appear to be areas of extensive lava flows, have also been detected.

Hundreds of observations derived from the Apollo missions have furthered immensely our understanding of the moon's history. Volcanic activity is very much a part of that history (Fig. 2–31). The moon formed, as did the earth, about 4.6 billion years ago. From about 4.2 to 3.9 billion years ago meteorite bombardment produced the majority of the lunar craters. By about 3.8 billion years ago, the interior of the moon had become sufficiently hot as a result of radioactivity to initiate a 700-million-year episode of extensive volcanic activity. Lava flowed onto the lunar surface, filling the large basins developed earlier by meteorite bombardment and forming the basalt-floored maria. The era of volcanism appears to have ended about 3.1 billion years ago. Our only satellite has been relatively quiet since then.

INTRUSIVE IGNEOUS ACTIVITY

We have described some of the events that occur when magmas are able to penetrate to the surface of the earth and are extruded as lavas. But what is the origin and nature of magmas that congeal as intrusive igneous rocks far beneath the earth's surface? Although the solidification of these melts has never been observed in nature, geologists have nevertheless been able to ascertain the conditions under which they originated by carefully studying their composition and texture and by conducting laboratory experiments that approximate rock-forming processes. It was not always necessary to obtain the samples needed for study from well-borings or mines, for intrusive rocks that once were beneath the ground are now widely exposed at the surface as a result of uplift and erosion.

Occurrences of intrusive igneous rocks

Bodies of intrusive igneous rocks that have solidified from magmas located deep within the crust are called **plutons.** The shape, size, and relation of plutons to surrounding rock are quite varied. Some plutons represent injections of once-molten rock parallel to preexisting strata. Such plutons are said to be **concordant.** In contrast, a pluton that cuts across layering is termed **discordant** (Fig. 2–32).

Tabular plutons

A tabular body of rock is one shaped like a book or board; that is, having small thickness relative to its length and width. There are several kinds of tabular plutons, some of which are more uniform in shape than others. The concordant ones, which have been injected along bedding planes of sediment or between older lava flows, include **sills** and **lopoliths** (see Fig. 2–32). Sills may be either horizontal or tilted, depending upon the attitude of the enclosing beds. They range in thickness from only a centimeter to hundreds of meters. The well-known Palisades of the Hudson River owe their distinctive appearance to columnar jointing in the exposed edge of the 300-meter-thick Palisades Sill (Fig. 2–33).

Because both sills and lava flows are sheetlike and parallel to adjacent beds, the distinction between a sill and a flow that has been buried by later strata can be difficult. There are, however, a few characteristics that are useful in distinguishing between the two features (Fig. 2–34). Flows bake only the stratum beneath them, may fill the cracks only of the underlying bed, and are often vesicular in their upper portions. In contrast, sills cause heat metamorphism in the stratum above as well as in the one below, penetrate cracks in underlying and overlying beds, and rarely have vesicular borders.

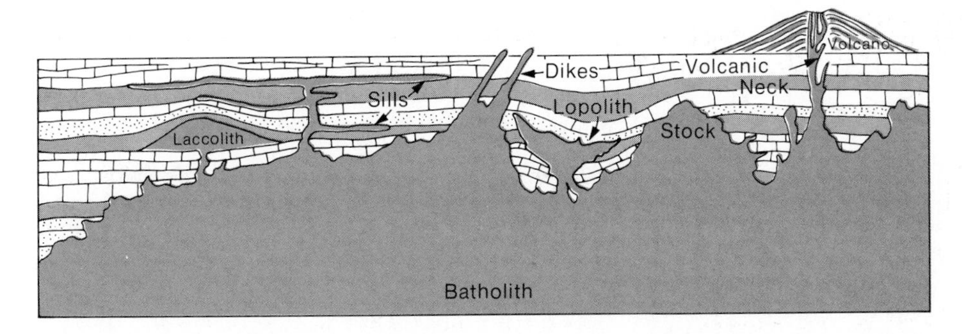

Figure 2–32 *Nomenclature for bodies of intrusive igneous rocks. The discordant bodies include dikes, necks, stocks, and batholiths. Concordant bodies include sills, laccoliths, and lopoliths.*

Figure 2–33 *The Palisades Sill exposed along the banks of the Hudson River in New York. Note the well-developed vertical columnar jointing. Overlying beds have been removed by erosion. (Courtesy of Palisades Interstate Park Commission.)*

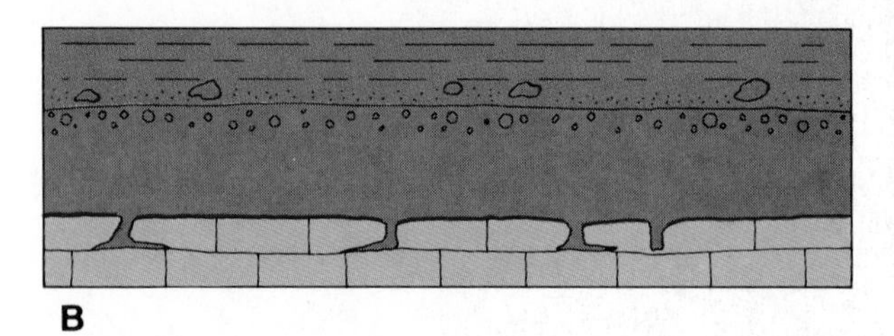

Figure 2–34 Differences between a sill (A) and lava flow (B): Sills contain few if any vesicles, thermally alter overlying and underlying rock, and send extensions of once-molten rock into beds above and below. Flows are vesicular, especially in their upper portion, and often bake the underlying layer.

Figure 2–35 Basaltic dike (dark colored) intruded into light-colored rhyolitic rocks in road cut, southeastern Missouri.

Lopoliths are tabular plutons in which the roof and floor sag downward so as to give the overall shape of a bowl. Examples of these large-scale structures occur in the Bushveld Igneous Complex of South Africa, as well as in the region around Duluth, Minnesota. The Duluth lopolith is estimated to have a diameter of 250 km and thickness of 15 km. Lopoliths are composed of coarsely crystalline igneous rocks, as is to be expected in deeply buried bodies that have cooled slowly.

Dikes (Fig. 2–35) are tabular bodies that are *discordant* in that they cut across preexisting rock layers. Dikes may be many kilometers long and range in thickness from paper thin to tens of meters. Many dikes branch and cross one another so as to form a complex system called a **dike swarm.** They also occur in circular patterns called **ring dikes.** Ring dikes are thought to develop when lava rises into concentric fractures created around an area where overlying rocks sink into a magma chamber below. Although dikes may be composed of nearly any kind of igneous rock, most are basaltic in composition and were derived from magmas of low viscosity.

Batholiths

In many areas of former igneous activity, great discordant nontabular masses of intrusive rock occur. The largest of such bodies form the cores of great mountain ranges. They are known as batholiths. Among the better-known examples are the Idaho Batholith, the Coast Range Batholith of British Columbia, and the Sierra Nevada Batholith of California (Fig. 2–36). Each of these immense bodies of igneous rock actually represents the coalescence of several smaller intrusions. Granitic rocks (i.e., granites and granodiorites) form the bulk of most of the world's batholiths. They are rather coarsely crystalline masses composed mostly of quartz, orthoclase, plagioclase, biotite, and hornblende. Granitic rocks are derived from magmas richer in silica than the melts that form basalts.

To be termed a **batholith,** the intrusive body must have a horizontal cross-sectional area greater than 100 square km. It may be irregular or roughly cylindrical. It is difficult to determine the depth to the base of existing batholiths, for at no place on earth is the bottom of a batholith exposed for study. However, gravity measurements and earthquake data suggest batholiths begin to form at depths of about 30 km. All batholiths were once deeply covered. We see them at the surface of the earth today only because great thicknesses of covering rocks have been removed by erosion. Batholiths are usually located in present or former mountain belts; such regions are subject to frequent episodes of uplift, which in turn greatly increase the rates of erosion needed to expose the batholith.

Batholiths intrude into **country rock** (preexisting rock) in a variety of ways. In some places there is evidence that the melt has been forcefully injected. Elsewhere, the heat of the magma appears to have been

Figure 2–36 Impressive exposures of granitic rocks of the Sierra Nevada Batholith. View is of "Half Dome," Yosemite National Park, California. (Courtesy of U.S. National Park Service, photograph by R.N. Anderson.)

sufficient to melt surrounding country rock so that the magma melted its way upward. Such a process would, of course, change the composition of the magma. **Stoping,** a process by which magma moves upward as blocks of country rock are wedged loose and fall into the magma chamber, provides another mechanism by which batholiths work their way upward. The upward migration of magma may also be aided if it is less dense than the enclosing rock and rises bouyantly.

Laccoliths

Laccoliths (see Fig. 2–32) are massive concordant plutons that have a mushroom shape. Like sills, they are formed by magma that has been injected into bedding planes. However, unlike sills, they dome up the overlying rock layers. Some laccoliths represent bulges on the surface of a sill, whereas others are self-contained structures fed by a vent from an underlying magma.

The nature of magmas

Composition

The kinds and amounts of different minerals that will crystallize to form an igneous rock are determined by the chemical composition of the parent magma. There are not a large number of chemical elements that are abundant in igneous rocks. Only the elements oxygen, silicon, aluminum, iron, magnesium, calcium, sodium, potassium, and titanium are abundant. Of these elements, oxygen and silicon are essential. These two elements may either combine to form quartz or bond with other elements in specific ways to form the feldspars, ferromagnesian minerals, and micas that are the major constituents of igneous rocks.

Although the average abundance of silicon in igneous rocks of the crust is 27.72 percent, that proportion has a *range* of 24 to 34 percent. Furthermore, petrologists have noted that the amount of other cations in igneous rocks has a definite relationship to silicon content. For example, as the weight percentage of silicon increases, the weight percentages of iron, magnesium, and calcium necessarily decrease. Thus, rocks rich in ferromagnesian minerals are less likely to contain the mineral quartz, since there would be insufficient silica "left over" to form quartz after the ferromagnesian minerals had crystallized. The amount of potassium, on the other hand, is usually proportional to the amount of silica. Rocks rich in potassium (contained in such minerals as orthoclase and muscovite) are likely also to contain quartz.

Crystallization of magmas

During the crystallization of magmas, not all minerals form simultaneously. If, for example, the magma originally has a basaltic composition, olivine, pyroxene, and calcium plagioclase will be among the earliest minerals to crystallize. Often these first-formed crystals are larger and more perfect than those formed later because when they formed, there was ample space and an abundance of elements required for growth. Those minerals that crystallize later must fit into the remaining spaces and thus tend to be smaller and less perfect. Also, minerals enclosed in other minerals must have formed before the enclosing mineral. By using these observations of size and spatial relationships of minerals as viewed in thin sections of the rock, it is sometimes possible to make a reliable estimate of the order of crystallization of minerals in a magma.

During the early part of the present century, an eminent petrologist named N.L. Bowen studied the

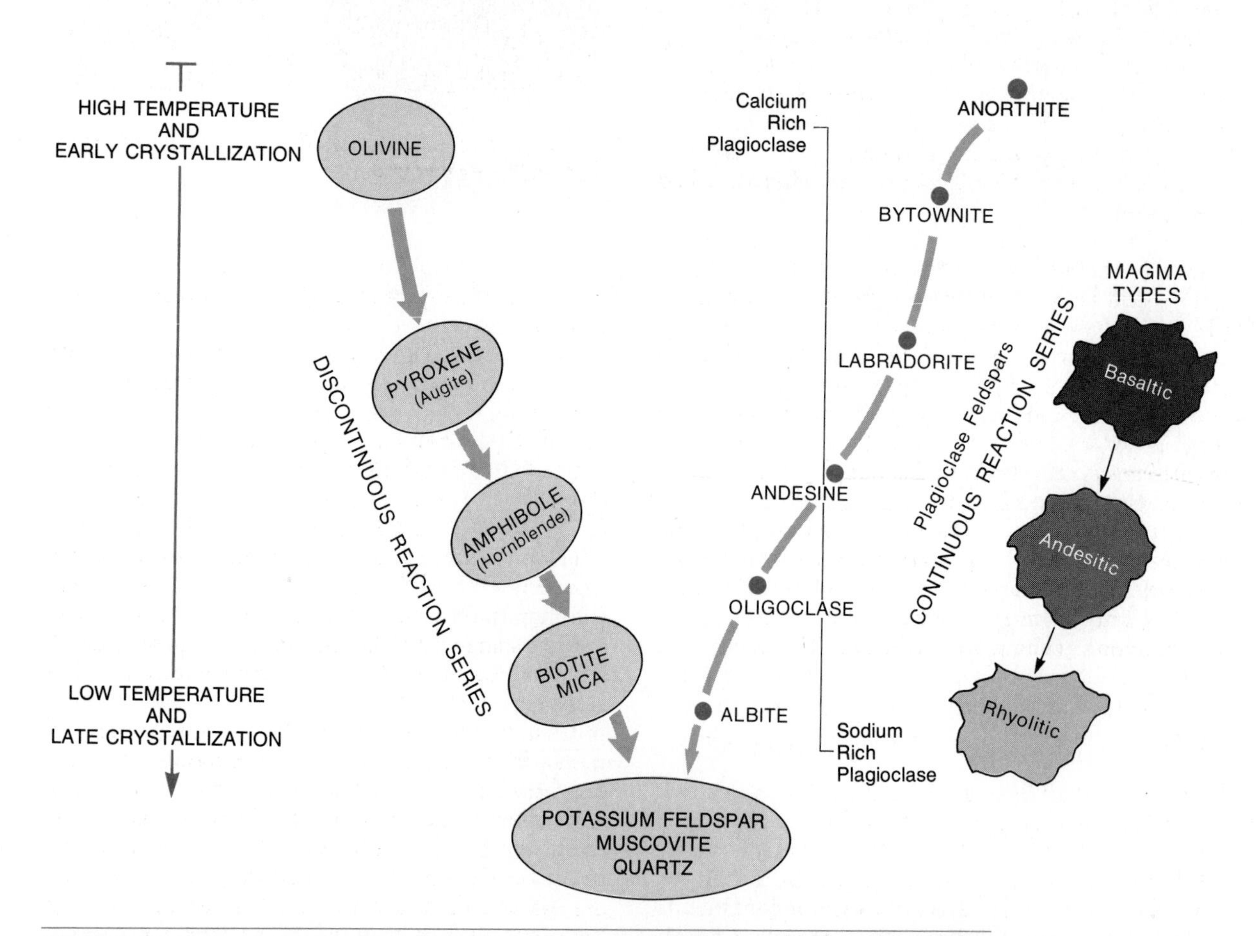

Figure 2–37 *Bowen's Reaction Series. Note that the earliest minerals to crystallize from a cooling magma are olivine and calcium-rich plagioclase minerals like bytownite. As crystallization proceeds, olivine reacts with the melt to form pyroxene, pyroxene reacts to form amphibole, etc. The plagioclase minerals also react with the remaining liquid, but rather than forming new minerals, change continuously in composition.*

crystallization of artificial silicate melts in the laboratory and provided further evidence of the approximate order of crystallization in gradually cooling magmas. Bowen used melts of basaltic composition for his experiments and was able to ascertain the specific temperature range over which particular silicate minerals would begin to form. Olivine, for example, would crystallize first and at the highest temperature. Pyroxenes and the more calcium-rich feldspars would form next, followed by hornblende, biotite, and the more sodic feldspars. The order of crystallization is shown in Figure 2-37. This now-famous diagram has been named the *Bowen Reaction Series*, because after the early-formed minerals have crystallized, the magma will *react* with some of them and will change some of them to the next lower mineral in the series. For example, the magma would react with the olivine crystals and change some of them to pyroxene (augite), and at a somewhat lower temperature the melt would react with pyroxene to form hornblende. Reaction would be prevented, of course, if minerals were removed from contact with the magma shortly after they had crystallized.

On the right branch of Bowen's chart one can trace the changes that occur as calcium-rich plagioclase reacts with the magma to produce varieties of plagioclase that are successively richer in sodium. Because the plagioclase minerals maintain the same basic structure (triclinic) but change continuously in their content of calcium and sodium, the right side of the diagram was named the *continuous series*. The ferromagnesian or left side of the diagram depicts reactions that result in minerals with distinctly different structure. It was therefore termed a *discontinuous series*. The three minerals at the base of the chart do not react with the melt. By the time they crystallize, little liquid remains, and residues of silicon, aluminum, oxygen, and potassium join to form orthoclase, muscovite, and quartz.

Causes of diversity among igneous rocks

In formulating his reaction series, Bowen experimented with a melt of basaltic composition and then theorized that the different kinds of igneous rocks might be formed from a single basaltic magma. As we have noted, this process requires the sequential removal of earlier formed minerals as the magma gradually cooled. Bowen termed the process **fractional crystallization.** If the early formed crystals were not removed, a basaltic magma would yield only basaltic rocks. By removing the earlier iron and magne-

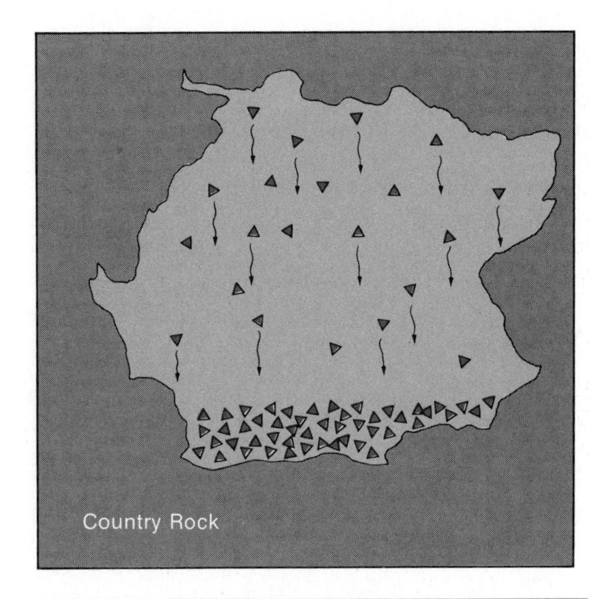

Figure 2-38 Fractional crystallization and accumulation of early-formed crystals by gravity settling.

sium-rich minerals, the magma is gradually changed toward a more rhyolitic composition.

The removal of early-formed minerals may be accomplished in various ways. In one process, early-formed crystals having a specific gravity greater than that of the remaining magma may sink toward the bottom of the chamber. The mechanism has been termed **gravity settling** (Fig. 2-38). Gravity settling of minerals at the top of the reaction series would leave the remaining magma poorer in iron, magnesium, and calcium. This altered melt would be likely to form a rock composed of hornblende and plagioclases that are somewhat richer in sodium. If these minerals were then removed, yet another kind of rock would be formed.

Another mechanism for differentiating minerals in a magma has been called **filter pressing.** At some point in its crystallization, magma exists as a mixture of solid crystals and hot liquid. If crustal forces were to compress this magmatic "slush," the crystals would be compressed to form one kind of igneous rock, and the liquid would be squeezed into pockets and fissures to solidify as rocks of different composition (Fig. 2-39).

Although fractional crystallization of a basaltic melt can indeed produce different kinds of igneous rocks, the process does not adequately explain all ig-

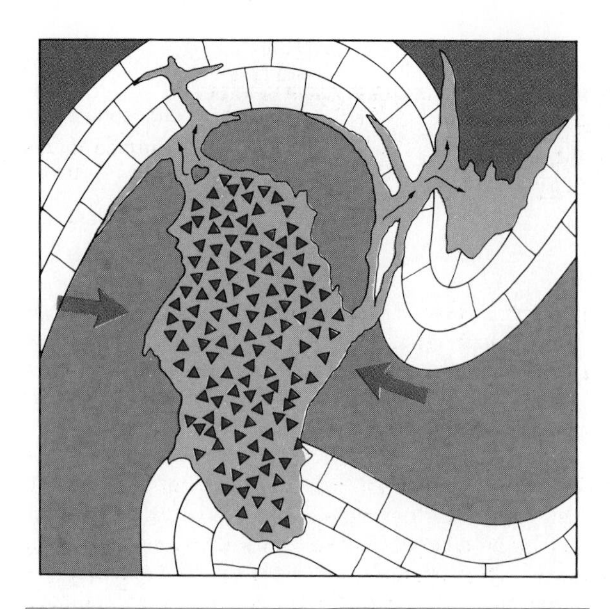

Figure 2–39 Filter pressing. Compression accompanying crystal deformation may squeeze residual liquid from a magma chamber. The residual liquid may then crystallize into one kind of igneous rock, while the early-formed minerals are left behind to form another.

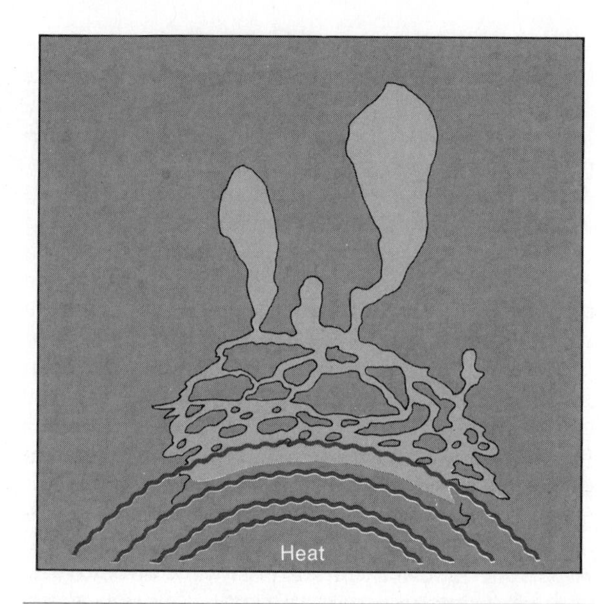

Figure 2–40 Conceptualized diagram of partial melting. First portion of parent rock to be melted separates because of density difference, leaving residue of parent rock depleted of the separated chemical components.

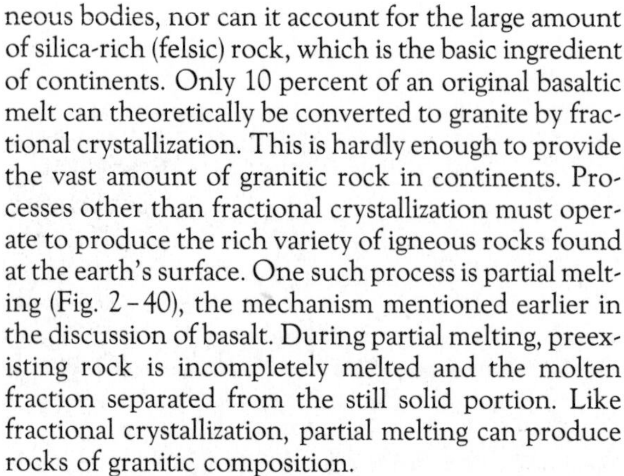

neous bodies, nor can it account for the large amount of silica-rich (felsic) rock, which is the basic ingredient of continents. Only 10 percent of an original basaltic melt can theoretically be converted to granite by fractional crystallization. This is hardly enough to provide the vast amount of granitic rock in continents. Processes other than fractional crystallization must operate to produce the rich variety of igneous rocks found at the earth's surface. One such process is partial melting (Fig. 2–40), the mechanism mentioned earlier in the discussion of basalt. During partial melting, preexisting rock is incompletely melted and the molten fraction separated from the still solid portion. Like fractional crystallization, partial melting can produce rocks of granitic composition.

Yet another major factor affecting the composition of magma is assimilation. **Assimilation** is a general term used for the many reactions that occur between a rising body of magma and the country rocks with which it comes into contact. Unlike either fractional crystallization or partial melting, which may begin with a single parent rock, assimilation involves two source materials, namely, the magma and the surrounding roof or wall rock. These two sources combine to produce magmas of intermediate composition.

The origin of granitic rocks

As noted in Chapter 1, the crust of the earth really has two major components. The crust beneath the oceans is basaltic in composition, whereas the continental crust is composed of granitic rocks rich in silicon, aluminum, and potassium (Fig. 2–41). Field and experimental studies suggest that the granitic continental crust developed from an original, thin, world-encircling basaltic crust, and that the volume of granitic rock has gradually increased through the operation of such processes as partial melting, assimilation, and the melting of silica and aluminum-rich sedimentary materials. Relatively smaller amounts of granitic rock may have resulted from fractional crystallization, although this mode of origin alone is considered inadequate to account for the enormous amounts of granitic rocks in the continents. Another process that may produce granitic rocks is called granitization.

Granitization is a process whereby chemically potent solutions of magmatic origin move through deeply buried crustal rocks, exchange ions and atoms with them, and gradually convert them to rocks of granitic character without their ever having become molten. This is a very different idea than the older

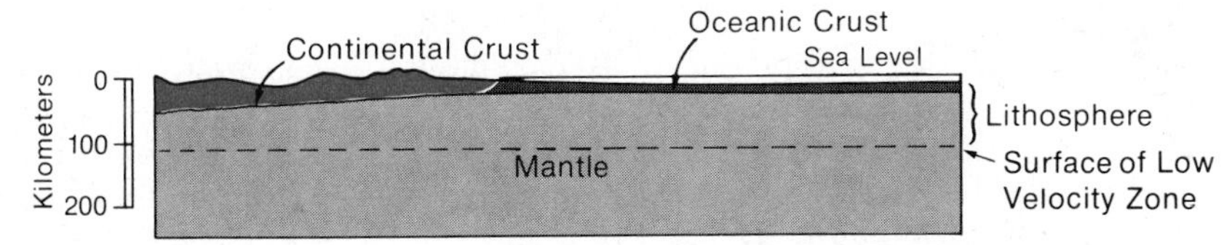

Figure 2–41 *Cross-section of the crust and lithosphere.*

concepts, which stipulated that granites are produced only from magma.

The significance of texture and color

Igneous rocks (see Color plate 4) are an excellent example of a group of natural objects that can be classified and named on the basis of descriptive properties that have genetic significance. In the preceding section on the crystallization of magmas, we noted that the mineral composition of an intrusive rock can provide a clue to the temperature of its formation as well as the chemistry of the melt from which it solidified. Rocks that are light-colored (white, light gray, pink) have a high proportion of such light-colored minerals as quartz, orthoclase, sodium-rich plagioclase, and muscovite. They must have been derived from a melt having a high silica content, perhaps as high as 65 to 75 percent. Dark-colored rocks are derived from magmas having a lower silica content (40 to 55 percent) and are typically composed of a predominance of the dark (gray, black, green) ferromagnesian minerals, as well as calcium plagioclases.

The texture of an igneous rock can provide clues to that rock's cooling history. Texture refers to the size, shape, and arrangement of the component mineral grains in a rock. We have already noted that a rock having *aphanitic* texture is so finely crystalline that the individual minerals cannot be seen without the aid of a magnifying lens or microscope. In general, an aphanitic rock is considered to have experienced cooling so rapid that there was not sufficient time for the growth of large crystals. Rapid cooling, of course, characterizes volcanic rocks, but some intrusive rocks may also be cooled quickly if they are near the surface or have a broad thin shape and hence a large surface area from

Figure 2–42 *Phaneritic texture. The rock is the Pikes Peak Granite, Lake County, Colorado. The majority of white minerals are orthoclase and sodic-plagioclase. Gray grains are quartz, and black minerals are mostly biotite.*

which heat may be quickly lost to the surrounding country rock. The extreme product of rapid chilling is represented by a natural glass in which few, if any, crystals have had time to form.

Most large bodies of magma are well insulated by the rocks that surround them and therefore retain their heat for long periods of time. This slow cooling favors the development of relatively fewer seed crystals. Only a few of these crystal nuclei can withstand the disruptive forces associated with the high temperatures of melts. The fewer number of crystal nuclei that survive disruption have the wealth of ions in the magma to ultimately gather into large crystals. In lava flows that cool quickly, many crystal nuclei quickly form, leaving fewer ions for each of the nuclei. The result is an aphanitic rock composed of multitudes of tiny crystals.

The more coarsely crystalline igneous rocks are said to have a **phaneritic** texture. Crystals in phaneritic rocks are clearly visible without magnification (Fig. 2-42). Magmas capable of retaining heat long enough to develop phaneritic textures are characteristically those that are deeply buried and rather equidimensional in shape. During the late stages of crystallization of such magmas, they develop a very high content of volatiles and water. The water inhibits crystal nucleation by weakening the bonds between silica tetrahedra. At the same time water lowers viscosity, which permits a rapid movement of ions to the relatively few nuclei. The exceptionally coarsely crystalline rocks that form as a result of these conditions are known as **pegmatites** (Fig. 2-43).

As can be readily imagined, the cooling rates of particular magmas vary. A deep-seated magma may begin its history by cooling slowly and developing few but large crystals. Then, while only partly crystallized, the magma may move into another environment in which cooling is more rapid and the remaining melt congeals quickly. The result is a rock having large, earlier-formed crystals that are surrounded by a later-formed groundmass of smaller crystals. This type of texture is called **porphyritic** (Fig. 2-44), and the large crystals are called **phenocrysts**.

Figure 2-43 Pegmatite. Large crystals of hornblende (black), scaly masses of muscovite mica (upper left), and large intergrown crystals of gray, glassy-looking quartz. (Courtesy of the Institute of Geological Sciences, London.)

Classification of igneous rocks

Texture and mineral composition are the most useful properties for inferring the origin of igneous rocks and are the basis for classification and identification. Although there are hundreds of names used for igneous rocks, most of these represent variations of the relatively few major groups shown in Figure 2-45. In a

Figure 2-44 Porphyritic texture. Large white crystals of calcium feldspar are immersed in a dark, finely crystalline groundmass.

widely used simple scheme, there are nine groups of crystalline igneous rocks: granite, granodiorite, diorite, gabbro, peridotite, rhyolite, dacite, andesite, and basalt.

To identify rocks in each of these groups, it is necessary to recognize texture and a few of the mineral constituents as they appear in the rock itself. To better identify the more common minerals, hold the rock in strong light and turn it back and forth so that you can see reflections from cleavage planes and crystal faces. Quartz can be recognized by its glassy appearance and mostly irregular reflecting surfaces. Feldspars show cleavage faces, may be white, gray, or pink, and often have rectangular outlines. Ferromagnesians are black or green.

Rocks of the granitic clan

The light-colored igneous rocks, as we have already noted, are derived from high-silica magmas. When such magmas crystallize, they yield a high proportion of feldspar and quartz. Granite (see Fig. 2-42) is a familiar representative of this group and is recognized by its light color, phaneritic texture, and the presence of about 25 percent quartz grains along with larger quantities of potassium feldspar (orthoclase) and sodium plagioclase. Muscovite and biotite are present, as are small quantities of hornblende. Granite, of course, is well known because it has been used for centuries in the construction of buildings, statues, and monuments. It is not, however, the most abundant granitic rock. Granodiorite (Fig. 2-46), a somewhat less siliceous granitic rock, is more abundant. The Sierra Nevada Batholith is composed largely of granodiorite.

Rocks of the dioritic clan

Igneous rocks of the diorite clan include diorite itself and its aphanitic equivalent, known as andesite. **Diorite** is a phaneritic rock containing less silica than granite or granodiorite but more than gabbro. Thus, in the gradations of composition across the rock chart, it is an intermediate rock type. The most abundant minerals are plagioclases, and these are compositionally in the mid range and contain both calcium and sodium. Quartz and orthoclase are very rare or absent in dio-

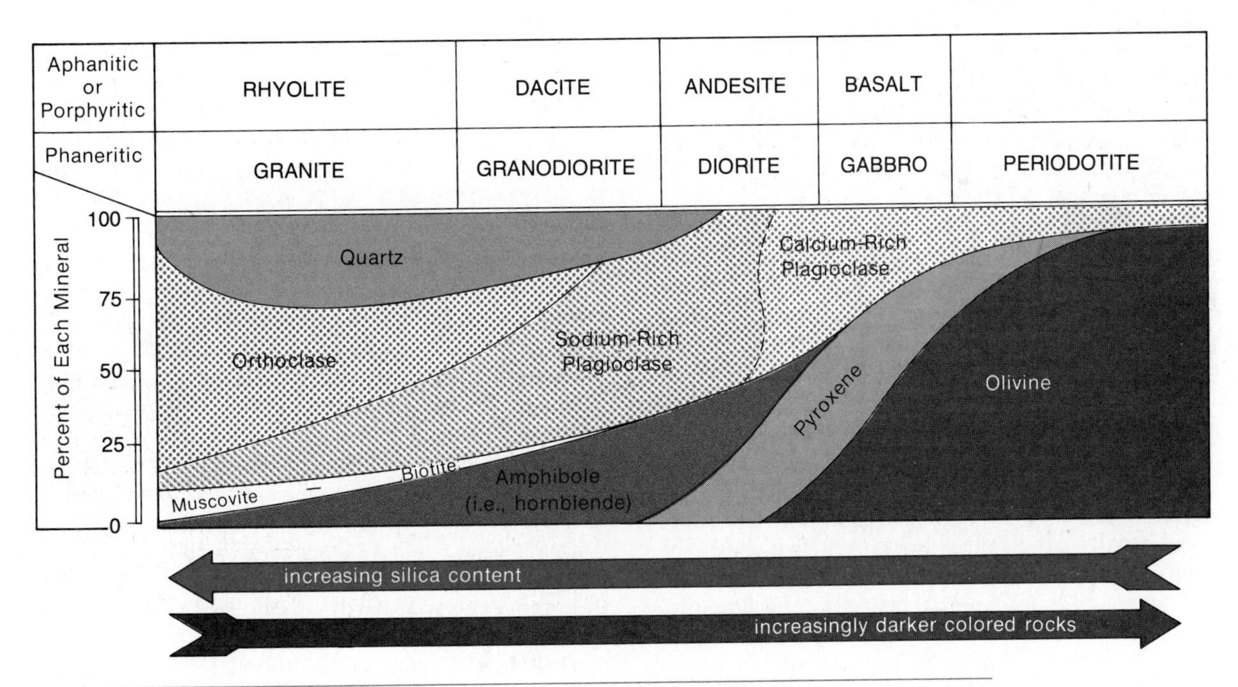

Figure 2-45 Diagram illustrating the mineralogic composition and textures of eight common igneous rocks. In order to ascertain the composition of a rock, estimate the percentages of each mineral beneath its name by reference to the percentage scale. (Adapted from Dietrich, R.V., Virginia Minerals and Rocks, *Virginia Polytechnical Institute.)*

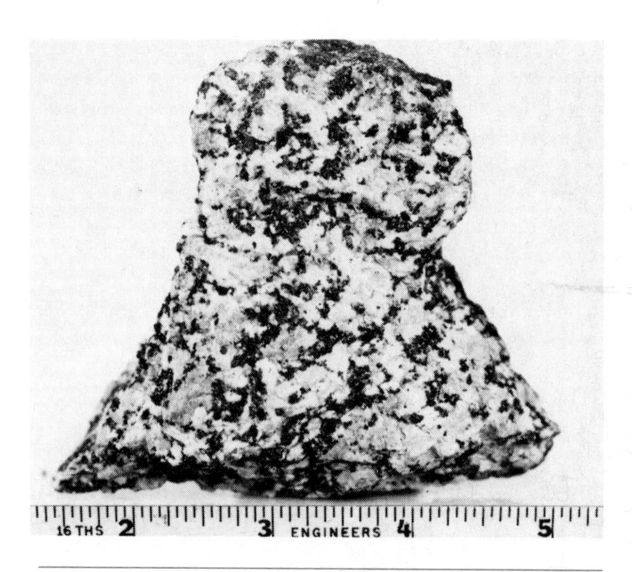

Figure 2–46 Granodiorite from Ferry County, Washington, showing phaneritic texture. The principal minerals are white plagioclase, pink orthoclase, quartz, biotite, and hornblende. (Courtesy of R.L. Parker, U.S. Geological Survey.)

rites. Among the ferromagnesian constituents, hornblende occurs in amounts about equal to plagioclase, and biotite mica is a common accessory mineral. The color of diorite is determined by its composition of nearly equal amounts of black ferromagnesians and grayish plagioclase. Thus, it tends to be rather drab gray or greenish gray in color.

The somber color of diorite prevails also in andesite—its aphanitic equivalent. Andesites, which were first identified in the summit volcanoes of the Andes Mountains of South America, are mostly found along the more mountainous continental borders and island arcs that surround the Pacific Ocean.

Rocks of the gabbro clan

In **gabbro** the chief mineral is the labradorite form of plagioclase (Fig. 2-47), which is darker and more calcium-rich than the plagioclase found in diorite. The typical ferromagnesian constituents are pyroxenes and olivine. Gabbros are typically coarsely crystalline plutonic rocks. A variety called anorthosite is characteristic of the lighter-colored crustal areas of the moon. Because of their rich dark hue and the iridescent play of colors reflected by the labradorite crystals, some gabbros are used as ornamental stone.

Basalt is now familiar to us as the rock name for the aphanitic equivalent of gabbro. We have previously noted its importance as the most abundant of extrusive igneous rocks.

The ultramafic clan

Ultramafic rocks are characterized by having a high density (3.3 g/cm^3), low percentage of silica (45 percent), and a major mineral composition of pyroxenes and olivine. **Peridotite,** in fact, is an ultramafic rock composed of 70 to 95 percent olivine; **pyroxenite** is almost entirely an intergrowth of pyroxene crystals. The ultramafic rocks are not common near or on the surface of the earth. They are occasionally encountered in the lower parts of thick sills, laccoliths, and lopoliths where dense ferromagnesian minerals accumulated by fractional crystallization and gravity settling. At several locations around the world, peridotites have pushed their way upward from great depths and penetrated near surface rocks. Geologists are keenly interested in these peridotites, for they may represent samples of the earth's mantle.

The global distribution of granitic and basaltic rocks

One of the most interesting observations about the global occurrence of igneous rocks is that granitic plutons occur within or on the edges (or former edges) of continents, and that the crust beneath the ocean basins is practically devoid of granites and kindred intrusive rocks. This indicates that the crust of the continents and ocean basins originates by fundamentally different processes. The igneous rocks of the ocean basins are predominantly basalts that have apparently been generated by partial melting within the lower crust or upper mantle. This is indicated by the high temperatures at which these basaltic melts are extruded and their relatively rich ferromagnesian composition. The process of building the oceanic crust is thought to be going on continuously as prodigious amounts of basalt pour out of fissures along the network of midoceanic ridges. These ridges have developed where tectonic plates separate and permit magma to well up and form new oceanic crust, which then moves away from the ridges at the rate of a few centimeters a year. Ultimately, rocks that were extruded along the ridges reach subduction zones where they, along with the cover of sediment they have accumulated, are plunged downward into the mantle and consumed. By this process, the basaltic ocean crust is being continuously renewed.

Figure 2–47 *Thin section of gabbro viewed through the petrographic microscope. The elongate crystals with gray "stripes" are plagioclase. Specimen collected in Gem Park, Fremont County, Colorado. (Width of field is 4 mm.) (Courtesy of R.L. Parker, U.S. Geological Survey.)*

The plate tectonic model may also provide an explanation for the origin of large granitic batholiths located along some plate margins. Because of their abundant silica and aluminum and relative deficiency of iron and magnesium, it is unlikely these granitic bodies could have been derived directly from the upper mantle. They might, however, be formed from the melting of mixtures of ocean basalts and sediments. Weathering processes tend to remove iron and magnesium and concentrate silica and aluminum in sediments, and if the sedimentary materials can be melted, they may generate silica- and aluminum-rich granitic rocks. A likely place for this to occur is along subduction zones where a lithospheric plate moves downward carrying with it not only basalts of the ocean floor but water-saturated marine sediments as well. When these descending slabs reach the asthenosphere, partial melting occurs, and magmas of the kind that form granodiorites and andesites are produced (see Fig. 2–18).

Summary

Igneous rocks are those that have cooled from a molten state. The melts from which igneous rocks are derived are called magmas while still beneath the earth's surface, and lavas when they find their way to the surface through fissures and cylindrical openings. Such openings, and the resulting build-up of extruded materials, form volcanoes. Rocks that solidify from lavas are frequently called *extrusive,* whereas those that crystallize at depth constitute the *intrusive* or plutonic family of igneous rocks.

Volcanoes vary considerably in their style of eruption. In some, the eruptions are characterized by great explosions that blast huge billowing clouds of gases, steam, and ash into the atmosphere. In such volcanoes, the flow of lava may be minimal. Other volcanoes, such as those of Hawaii, have little explosive activity, and are noted for their relatively quiet extrusions of lava. There are, of course, many volcanoes that behave explosively at some times and are relatively quiet at other times. The principal factors that determine whether an eruption will be quiet or explosive are temperature, the viscosity of the magma, and the amount of dissolved gases it contains. In general, magmas with a high percentage of silica are more viscous than low-silica magmas. If there is also a high gas content in the magma, explosive eruptions are likely. Silica-poor magmas rich in iron and magnesium (i.e., *mafic* magmas) are more characteristic of quiet eruptions.

The intrusive igneous rocks differ from volcanic rocks primarily by their coarser texture. They solidified deep beneath the earth's surface in variously shaped masses called plutons. We are able to examine plutons today only because uplift and subsequent deep erosion have uncovered some of them.

The magma from which igneous rocks are formed consists mainly of the elements oxygen, silicon, aluminum, iron, calcium, sodium, potassium, magnesium, and titanium, as well as dissolved gases and water. From these basic ingredients are derived the common minerals of igneous rocks. These include feldspars, quartz, micas, hornblende, augite, and olivine. Quartz, feldspars, and ferromagnesian minerals are especially useful in the classification of intrusive rocks. For example, a coarsely crystalline (phaneritic) rock composed of quartz orthoclase, sodium plagioclase, and a ferromagnesian mineral like hornblende or biotite would be termed granitic. A phaneritic rock lacking quartz in which ferromagnesian minerals and calcium plagioclase predominated would be called gabbroic. Rocks having a composition intermediate between gabbro and granite might be classified as dioritic. Granite, diorite, and gabbro are compositionally the plutonic equivalents of rhyolite, andesite, and basalt.

Among the mechanisms that provide for a variety of rock types are fractional crystallization, partial melting, and assimilation. Fractional crystallization is a process whereby minerals forming in a cooling magma are successively removed at progressively lower temperatures. The process by which a preexisting rock is incompletely melted and molten fractions removed is called partial melting. Assimilation involves modification of the chemistry of a magma as a result of melting and assimilation of surrounding rocks.

QUESTIONS FOR REVIEW

1 What is the sequence of minerals in the Bowen Reaction Series? If an igneous rock contains quartz, what other minerals would you expect it to contain by reference to the Bowen Reaction Series? If it contained pyroxene, what companion minerals might be expected?

2 Why is a high-silica lava more viscous than a low-silica lava? Why is explosive volcanic activity frequently associated with high-silica lava?

3 How may the mineral composition of an igneous rock reveal whether it was derived from a high-silica or a low-silica magma?

4 For each of the following rocks, what inferences can be made as to cooling history and type of magma from which it is derived?

- **a** Rock consisting of crystals of pyroxene in a dark, fine-grained groundmass.
- **b** Rock consisting of crystals of feldspar in a reddish, fine-grained groundmass; no ferromagnesian minerals visible.
- **c** Rock having a coarsely crystalline texture, consisting of feldspar, quartz, and mica.

5 What is the difference between basalt and gabbro? What chemical elements account for the dark color in these rocks?

6 What kind of extrusive rock is most abundant in ocean basins? What kind of intrusive rock is most abundant on the continents?

7 What is the probable original source of the water in the oceans and elsewhere at the earth's surface?

8 In terms of their origin and structure, why do shield volcanoes, cinder cones, and strato-volcanoes differ in appearance?

9 Distinguish between a dike and a sill, and between a sill and a lava flow.

10 Describe a theory that would account for the existence of large granitic batholiths, such as the Sierra Nevada Batholith.

11 By what two mechanisms may volcanic caldera originate?

12 Upon what two factors does the classification of igneous rocks depend? Which of these factors is more important in distinguishing andesite from basalt? Granite from rhyolite?

13 Summarize several possible mechanisms that would permit different kinds of igneous rocks to form from an initially basaltic melt.

14 How do fissure flows differ from volcanoes? Have fissure flows produced an appreciable amount of igneous rocks during the geologic past? Cite an example.

15 How would you explain the olivine-rich layer that commonly occurs near the base of thick sills?

SUPPLEMENTAL READINGS AND REFERENCES

Barker, D.S., 1983. *Igneous Rocks.* Englewood Cliffs, N.J., Prentice-Hall Inc.

Bullard, F.M., 1976. *Volcanoes of the Earth.* Austin, Tex., University of Texas Press.

Ernst, W.G., 1969. *Earth Materials.* Englewood Cliffs, N.J., Prentice-Hall Inc.

Lipman, P.W., and Mullineaux, D.R., 1980. *Eruption of Mount St. Helens.* U.S. Geological Survey Professional Paper No. 1250

Simpson, B., 1966. *Rocks and Minerals.* London, Pergamon Press Ltd.

Tindall, J.R., and Thornhill, R., 1975. *The Collector's Guide to Rocks and Minerals.* New York, Van Nostrand Reinhold Co.

3

sedimentary and metamorphic rocks

In the high mountains, I have seen shells. They are embedded in rocks. The rocks must have been earthy materials in days of old, and the shells must have lived in water. The low places are now lifted high, and the soft material turned to hard stone.

CHU-HSI, A.D. 1200

Prelude Ever since the earth has had an atmosphere and hydrosphere, the altered and decomposed products of weathering called sediments have been accumulating. When consolidated and cemented, sediments have become sedimentary rocks (see Color plates 7 and 8). Sediments are usually deposited in layers; hence the majority of sedimentary rocks have a layered or stratified appearance (Fig. 3–1). Stratifica-

Tunnel in cliff of sandstone of the Triassic Windgate Formation, Colorado National Monument, Mesa County, Colorado. (Courtesy of U.S. Geological Survey.)

Figure 3–1 *Stratified sedimentary rocks. In this road cut south of St. Louis, Missouri, stratification results from compositional and textural differences between limestones (more resistant to weathering) and shale beds. (Photograph by Stephen D. Levin.)*

tion results from differences in rock composition and texture and is the most obvious characteristic of sedimentary rocks.

Although sedimentary rocks constitute only about 8 percent by volume of the earth's crust, they cover an impressive 75 percent of total land area. When properly interpreted, these rocks tell us about environmental conditions in the geologic past. By careful study of successively higher strata, one can decipher a part of the geologic history of the earth. Sedimentary rocks are also important as sources of building materials, coal, ceramic products, and certain metals. Within them, one sometimes finds the petroleum and natural gas that are so important to the economic health of nations.

Rocks changed or metamorphosed from pre-existing sedimentary, metamorphic, or igneous rocks by the action of heat, pressure, and chemically active solutions are termed metamorphic rocks. In some instances, the changes may only involve deformation of mineral grains, whereas in other cases there may be recrystallization and reorganization of the parent elements into new minerals and an altered texture. Either way, metamorphism involves little or no loss or gain in bulk chemical composition. The type of metamorphic rock produced during metamorphism depends upon the kinds and intensities of metamorphic agencies acting upon the parent rock, as well as the nature of the parent rock.

SEDIMENTARY ROCKS

The derivation of sediments

What was the origin of the many kinds of materials found in sedimentary rocks? A partial answer to that question is apparent when we observe the decayed surface of rocks in the field or on old buildings and monuments. These rock surfaces are slowly being destroyed by the action of rain, ice, sun, wind, and pollutants. Weathering, the breakdown and alteration of rocks at the earth's surface, includes processes that cause the fragmentation of rock without chemical change (as when water freezes and expands in crevices so as to cause fracturing), as well as the chemical decomposition of rocks when their constituent minerals react with water and air. Whatever the specific process, weathering is an initial step in a series of events that leads ultimately to sedimentary rocks. As an illustration of some of these events, consider the history of granite as it undergoes weathering in a relatively moist climate (Fig. 3-2). The most abundant minerals in granite are quartz, orthoclase, calcium and sodium feldspars, and ferromagnesian minerals. As weathering proceeds, sediment derived from the granite will gradually accumulate as loose, sandy, and clayey material. On close examination, this material is found to consist of quartz grains, partly decayed and clay-coated feldspars, rust-colored grains of partly decayed ferromagnesian minerals, and clay. The quartz grains in the weathered material are present because of quartz's resistance to chemical attack. As other minerals in the granite decay around them, the quartz grains are freed from their rock matrix. Later, these quartz grains may be transported and deposited as sand.

As we noted in Chapter 1, the feldspars are silicates of potassium, calcium, and sodium. During chemical weathering, feldspars and other silicates decompose, producing insoluble products (mostly clay), and liberating ions of potassium, calcium, and sodium, which may be carried away in solution. Subsequently, they may be combined with other elements and precipitated to form such rocks as limestone or rock salt. The remaining elements of the original feldspar are aluminum and silicon. These two elements become the chief ingredients of clay.

The decay of the ferromagnesian minerals in an igneous rock is similar to that described above. Again, potassium, sodium, and calcium are dissolved and carried away, perhaps to form sedimentary rocks. Aluminum and silicon go into the making of clay minerals. The ions remaining are iron and magnesium. Iron ions, freed from the parent ferromagnesian mineral, combine readily with oxygen to form iron oxides, such as hematite (Fe_2O_3) and limonite, $FeO(OH)$. These iron minerals color sedimentary rocks in tints of yellow, orange, and brown. They enhance the beauty of natural rock formations in many parts of the world. The magnesium that is derived from parent ferromagnesian minerals commonly finds its way into limey sediments or becomes incorporated into clay minerals.

Thus, we see that it is possible to obtain most of the common constituents of sedimentary rocks by weathering older rocks. Of course, not all parent rocks will provide the same kinds or proportions of products. Basalt, which has no quartz, cannot yield quartz on weathering. Similarly, the only clay that can be derived from limestone (composed of $CaCO_3$) would be clay already present as an impurity. Weathering is important for many other reasons than the production of sediment and will be discussed further in Chapter 7.

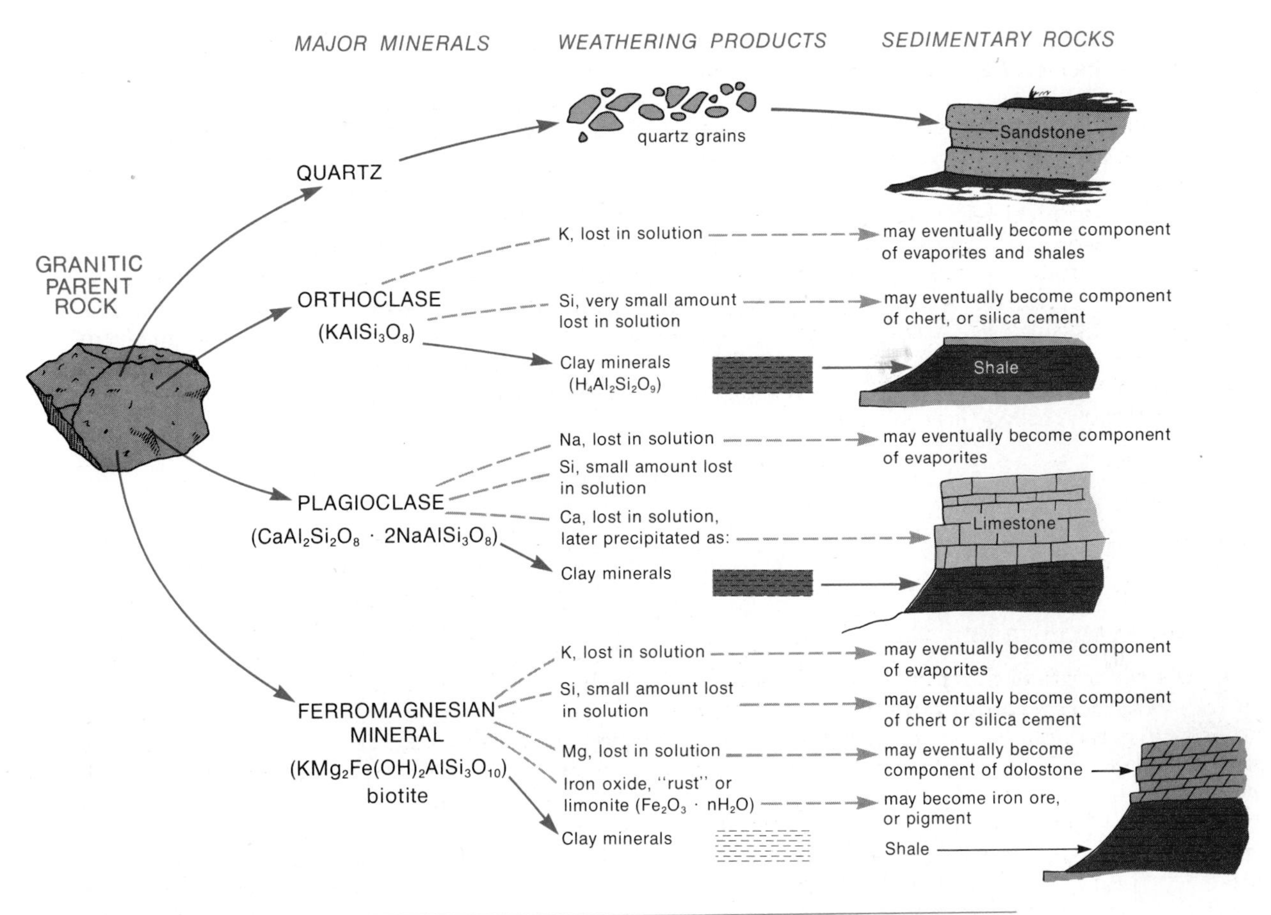

Figure 3–2 Conceptual diagram showing how the weathering of granitic rock yields quartz grains for quartz sandstone, clay for shale, and calcium for limestone. If weathering is not too severe, detrital grains of feldspar will also be included in sands and sandstones. For simplicity, minor mineral components are not included.

From sediment to rock

Sediment transport and deposition

Once weathering products have been formed from preexisting rocks, the next stage in the sequence of events leading to sedimentary rock is the removal and transport of those products. Running water, moving ice, and wind assist in this removal. The wind is an effective agent in picking up and blowing away the smaller and lighter particles. Glacial ice (Fig. 3–3) can move very large pieces of rock and carry an immense load of coarse sediment. Streams (Fig. 3–4) are also exceptionally effective in carrying not only solid particles of sediment but invisible dissolved salts as well. Ultimately, sediment-laden streams flow into lakes or the sea and deposit their load of sediment. It may form sandy beaches, silty floodplains, and sometimes the muddy bogs of estuaries and deltas.

The solid particles carried by wind or water will be deposited whenever there is insufficient energy to carry them further. If the velocity of dust-laden wind abates, there will be insufficient energy to carry particles of a given size, and those particles will be dropped. Similarly, if a stream's velocity is checked, as when entering a standing body of water, the stream also loses energy and is unable to continue to carry the material formerly carried at the higher velocity.

A reduction in a stream's velocity does not, of course, affect the dissolved materials. Material carried in solution is deposited by a process called **precipitation,** in which dissolved material is changed to a solid and separated from the liquid in which it was formerly

Figure 3–3 Tikke Glacier. This glacier, located north of Glacier Bay in British Columbia, is characterized by periodic rapid advances accompanied by increased rates of erosion. (United States Geological Survey, photograph by A. Post.)

Figure 3–4 Sediment-laden Tanana River, Alaska. River has so great a load of sediment that it has deposited a portion in its own bed, resulting in a braided stream pattern. (Courtesy of James C. Brice.)

dissolved. For example, calcium carbonate, the principal substance in limestone, may be precipitated from water that contains calcium in solution as indicated in the simple relationship below:

$$Ca^{+2} + 2HCO_3^- \underset{\text{yields}}{\overset{\text{yields}}{\rightleftharpoons}} CaCO_3\downarrow + \overbrace{H_2O + CO^2}^{H_2CO_3}$$

Ca^{+2}	$2HCO_3^-$	$CaCO_3$	H_2O	CO^2
(dissolved calcium ions)	(dissolved bicarbonate ions)	(calcium carbonate)	(water)	(carbon dioxide)

The bicarbonate ions that participate in the above reaction can be derived from the ionization of carbonic acid. As indicated by the arrows, the reaction is reversible. If carbon dioxide is removed from sea water, the reaction will proceed toward the right and calcium carbonate will be precipitated. If, however, carbon dioxide is added to sea water, the amount of carbonic acid in the water would increase and the reaction would proceed to the left. This would result in a chemical environment not conducive to calcium carbonate precipitation, in which existing calcium carbonate might begin to dissolve. The precipitation of calcium carbonate is a complex and delicate process in nature. It is influenced by organisms that utilize or liberate carbon dioxide, by processes that alter the acidity or alkalinity of the water, by the presence of organic compounds, by ions of sulfur, phosphorus, and magnesium that may be present, by temperature, and by other factors.

Lithification

Many changes take place in sediment after it has been deposited. Mineral grains may be dissolved away, some may grow by additions of new mineral matter, and the shapes of particles may be distorted by compaction. The result of some of these changes is to convert sediment into sedimentary rock. The conversion process is called **lithification.** *Cementation, compaction,* and *crystallization* are the principal means by which unconsolidated sediment is lithified.

Cementation involves the precipitation of minerals in the pore spaces between larger particles of sediment. The precipitated mineral, which most frequently is either calcium carbonate ($CaCO_3$) or silica (SiO_2), is called the **cement** (Fig. 3-5). Cement is added to a sediment *after* deposition. It differs from a rock's **matrix** (Fig. 3-6), which consists of clastic particles (often clay) that are deposited at the same time as the larger grains and help to hold the grains together.

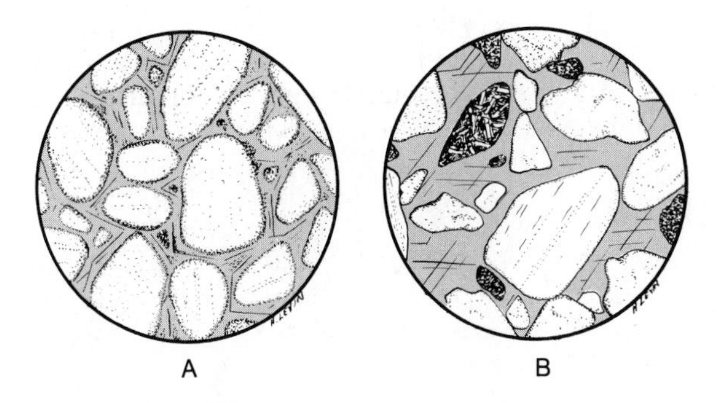

Figure 3-5 *Two common types of cement in sandstones. (A) Quartz sandstone composed of well-sorted, rounded quartz grains tightly cemented by quartz (SiO_2) outgrowths. (B) Sandstone composed of quartz, feldspar, and rock fragments cemented by coarse sparry calcite. Both are drawings of thin sections as viewed under the microscope. Diameter of areas each 1.0 mm.*

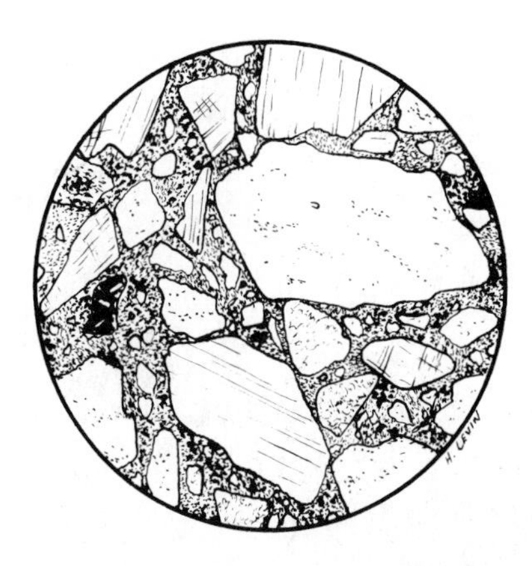

Figure 3-6 *Thin section of sandstone composed of poorly sorted angular grains of quartz (clear), feldspars (thin stripes), and rock fragments. This sandstone lacks cement. Spaces between grains are filled with a* matrix *of clay and silt.*

The reduction in pore spaces in a rock as a result of the pressure of overlying rocks or pressures of earth movements is termed **compaction.** During compaction, individual grains are pressed tightly against one another, causing the reduction in pore space and rearrangement of particles. Compaction may expedite cementation and help to convert loose sediment into hard, indurated rock.

Lithification by **crystallization** may begin with an initial chemical precipitate in which the developing crystals grow together to form a crystalline solid. The process of crystallization, however, may also result in a change in the form of grains that have already been deposited. For example, quartz may be precipitated onto rounded quartz grains to form a strong interlocking mosaic of crystals, clay may be converted to a matted aggregate of tiny mica crystals, or calcium carbonate skeletal debris may be reorganized into crystalline calcite. The migration of solutions through sediment favors crystallization, as does deep burial and consequent increases in temperature and pressure.

Kinds of sedimentary rocks

The basis for identification: composition and texture

Sedimentary rocks are identified and named according to their composition and texture (Table 3-1). In re-

TABLE 3–1 **CLASSIFICATION OF SEDIMENTARY ROCKS**

Origin	*Rock name*	*Texture*	*Composition*
Clastic	Conglomerate	Larger particles greater than 2 mm, particles rounded	Mostly gravel-size pieces of quartz, chert, quartzite, or any rock type
	Breccia	Larger particles greater than 2 mm, particles angular	Particles of quartz or any rock type
	Sandstone	Particles range in size from 0.0625 to 2 mm in diameter	Varies according to kind of sandstone. Quartz usually dominant. Arkose must contain 25% feldspar. Composition of graywacke includes considerable clay
	Siltstone	Particles range in size from 0.0039 to 0.0625 mm in diameter	Generally, composition similar to that of associated sandstones
	Shale	Particles less than 0.0039 mm in diameter	Quartz and clay minerals
Biogenic and/or Chemical	Limestone	Varied textures, including crystalline, micritic, oolitic, bioclastic, etc.	Calcite ($CaCO_3$)
	Dolostone	Varied textures, often porous, fine-grained	Dolomite ($CaMg(CO_3)_2$)
	Evaporites	Usually crystalline	Gypsum ($CaSO_4 \cdot 2H_2O$) Halite ($NaCl$) Sylvite (KCl)
	Coal	Compressed particles of partly decomposed plant matter	A complex of organic compounds

gard to composition, the three most abundant mineral components of sedimentary rocks are clay minerals, quartz, and calcite. Evaporite minerals and dolomite, although less abundant, nevertheless form an appreciable portion of many sedimentary sequences. Sedimentary rocks nearly always also contain variable amounts of limonite and hematite. A generalization one can make about the mineral composition of sedimentary rocks is that most are mixtures of two or more components in which one mineral may predominate. Thus, sandstones composed mostly of quartz grains nearly always contain some clay or calcite, and limestones made mostly of calcite nearly always contain clay and quartz grains (Fig. 3–7).

Texture refers to the size and shape of individual mineral grains and to their arrangement in the rock. A rock that has a **clastic** texture (from the Greek *klastos,* broken) is composed of particles of clay, silt, sand, and gravel, or fragments of parent rock or fossils that have been moved individually from their place of origin. In contrast, *nonclastic* rocks form by chemical or biochemical precipitation within a sedimentary basin. Most nonclastic rocks are crystalline and include certain limestones and evaporites to be discussed later in this chapter.

Sedimentary rocks made of the remains of plants and animals are categorized as **biogenic.** Coal, for example, can be considered a sedimentary rock derived from the accumulation of plant remains. Limestones composed predominantly of the skeletal remains of invertebrate animals would also be considered biogenic.

Clastic sedimentary rocks

Categories of clastic rocks. The component fragments of clastic sedimentary rocks range in size from microscopic particles to huge boulders (Table 3–2).

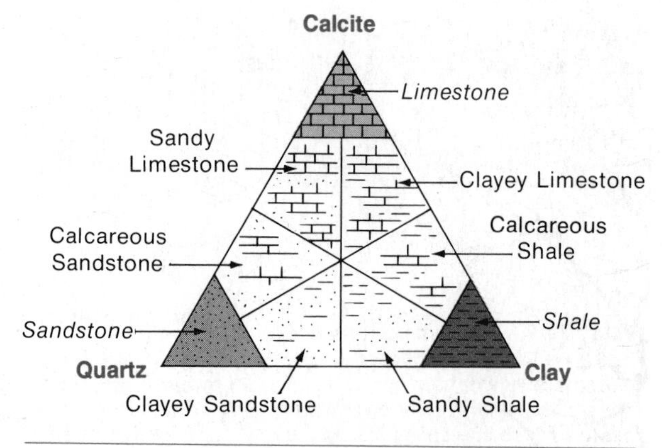

Figure 3–7 Triangular diagram suggesting the gradations that can exist between pure shales, sandstones, and limestones.

TABLE 3-2 **SIZE RANGE OF SEDIMENTARY PARTICLES**

Wentworth scale (in millimeters)	*Particle name*
	Boulders
256	
128	Cobbles
64	
32	
16	Pebbles
8	
4	
	Granules
2	
	Very coarse sand
1.0	
	Coarse sand
0.5	
	Medium sand
0.25	
	Fine sand
0.125	Very fine sand
0.0625	
0.0313	
0.0156	
	Silt
0.0078	
0.0039	
	Clay

The clastic group includes such familiar rocks as conglomerates, sandstones, siltstones, and shales. Texture, and particularly particle size, is the key to naming the clastic rocks. Thus, **conglomerate** is a rock composed of rounded particles larger than 2 mm in diameter (Figs. 3-8, 3-9). **Breccias** (Fig. 3-10) are composed of particles that are angular but similar in size to conglomerates. In sandstones, grains range between 0.0625 and 2.0 mm in diameter. Siltstones, which tend to resemble sandstones in composition, are made of particles in the 0.0039 to 0.0625 mm range. Shales are the finest of clastic rocks. They are composed of particles of silt and clay too small to be seen without magnification.

Grain size and sorting in clastic rocks. Geologists are interested in the sizes of component grains in clastic rocks, not only as a means of identifying them but also because the size of particles can provide clues to the environment of deposition. For example, it is obvious that a stronger current of water (or wind) is required to move a big particle than a small one. Therefore, the size distribution of grains tells the geologist something about the turbulence and velocity of currents. It can also be an indicator of the mode and extent of transportation. If sand, silt, and clay are supplied by streams to a coastline, the turbulent inshore waters often winnow out the finer particles, so that gradation from sandy nearshore deposits to offshore silty and clayey deposits may result (Fig. 3-11). Sandstones formed from such inshore sands may retain considerable porosity and provide void space for pe-

Figure 3-8 *Conglomerate exposed near the southeast end of the Malone Mountains, Hudspeth County, Texas. (Courtesy of C.C. Albritton, Jr., U.S. Geological Survey.)*

Figure 3–9 Cobbles and pebbles such as these exposed along a California coastline are the materials from which conglomerates may be produced. (Courtesy of E.D. McKee, U.S. Geological Survey.)

troleum accumulations. For this reason, one often finds petroleum geologists assiduously making maps that show average grain size of deeply buried ancient formations.

Another aspect of a clastic rock's texture that involves grain size is sorting. **Sorting** is an expression of the range of particle sizes deviating from the average size. Rocks composed of particles that are all about the same size are said to be "well sorted" (see Fig. 3–5A), and those that include grains with a wide range of sizes are termed "poorly sorted" (see Fig. 3–6). Sorting often provides clues to conditions of transportation and deposition. Wind, for example, winnows the dust particles from sand, leaving grains that are all about the same size. Windblown deposits are usually better sorted than deposits formed in an area of wave action, and wave-washed sediments are better sorted than stream deposits. It must be kept in mind, however, that if a source sediment is already well sorted, the resulting deposit will be similarly well sorted. Poor

Figure 3–10 Because it consists mostly of angular fragments greater than 2 mm in diameter, this rock is termed a breccia. It is part of the Old Red Sandstone Formation and is exposed in Scotland. (Courtesy of the Institute of Geological Sciences, London.)

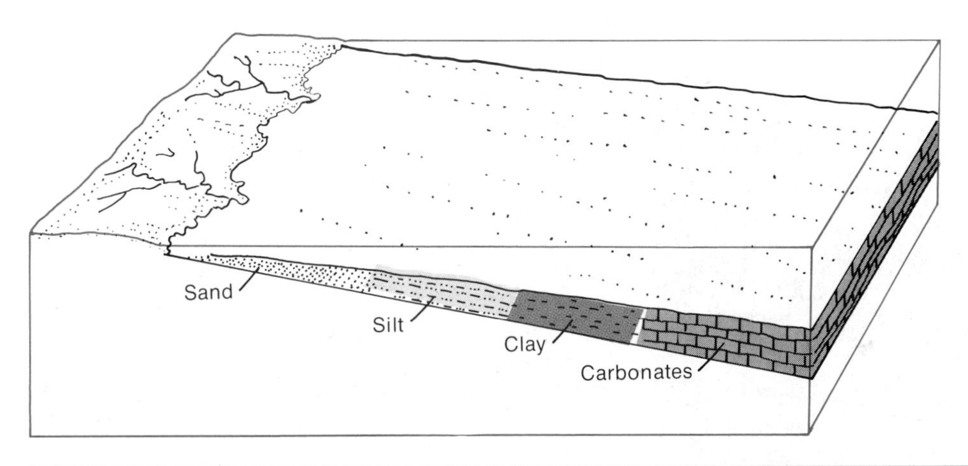

Figure 3–11 *Gradation from coarser sediments near shore to finer sediments farther offshore.*

sorting occurs when sediment is rapidly deposited without being selectively separated by currents into sizes. Poorly sorted conglomerates and sandstones are deposited at the foot of mountains where stream velocity is suddenly checked, or where streams deliver large volumes of sediment to a deep offshore basin where there is little opportunity for sorting by waves.

The shape of particles in a clastic sedimentary rock can also be useful in determining its history. Shape can be described in terms of rounding of particle edges and sphericity, or how closely the grain approaches the shape of a sphere (Figs. 3–12, 3–13). A particle becomes rounded by having sharp corners and edges removed by impact with other particles. The roundness of grains is a clue to how much abrasion a sediment has experienced during transport and repeated cycles of erosion and deposition.

Sandstones. Of all the clastic sedimentary rocks, sandstones have been studied the most completely and provide the greatest amount of information about ancient environmental conditions. The mineral composition of the grains in sandstones provides important information about the source of the sediment as well as the history of the sediment prior to deposition. Often

Figure 3–12 *Well-rounded grains of quartz viewed under the microscope. From the St. Peter Formation near Pacific, Missouri. The median grain diameter is 0.22 mm.*

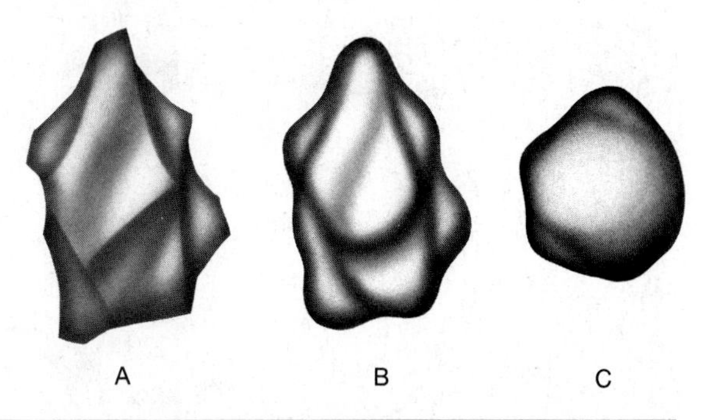

Figure 3–13 *Shape of sediment particles. (A) An angular particle (all edges sharp). (B) A rounded grain that has little sphericity. (C) A well-rounded highly spherical grain.* Roundness *refers to the smoothing of edges and corners, whereas* sphericity *measures the degree of approach of a particle to a sphere.*

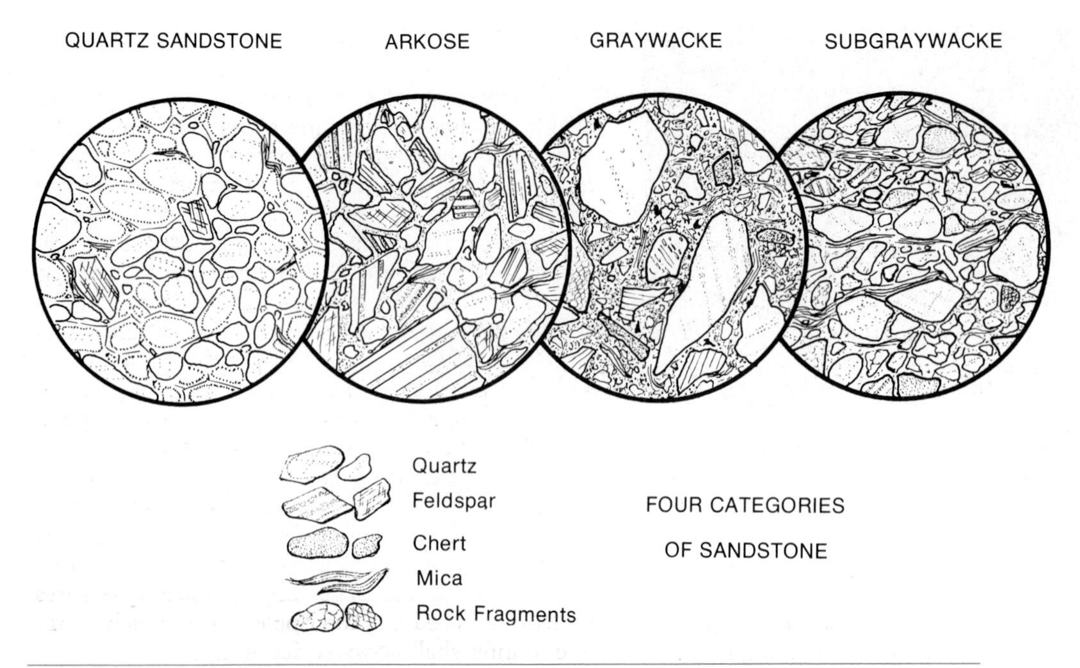

Figure 3–14 *Four categories of sandstone as seen in thin section under the microscope. (Diameter of field about 4 mm.)*

by studying the grains closely, one can ascertain whether the source material was metamorphic, igneous, or sedimentary. The mineral content also provides a rough estimate of the amount of transport and erosion experienced by the sand grains. Rigorous weathering and long transport tend to destroy less stable feldspars and ferromagnesian minerals and cause rounding and sorting of grains. Hence, one can assume that a sandstone rich in these less durable and angular components underwent relatively little transport and other forms of geologic duress. Such sediments are termed *immature* and are most frequently deposited close to their source areas. On the other hand, quartz can be used as an indicator of a sandstone's maturity; the higher the percentage of quartz, the greater the maturity. In addition to providing an indication of a rock's maturity, composition is an important factor in the classification of sandstones into graywackes, quartz sandstones, arkoses, and subgraywackes (Fig. 3–14).

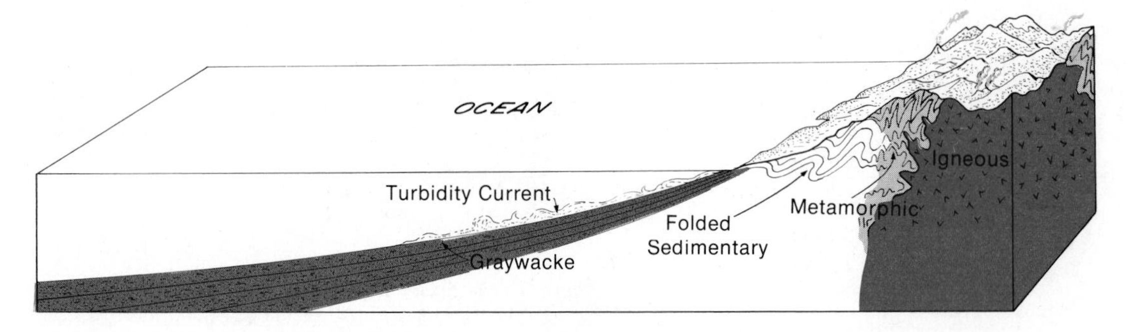

Figure 3–15 *Tectonic setting in which graywacke is deposited. Frequently, graywackes are transported and deposited by masses of water highly charged with suspended sediment. Because of the suspended matter, the mass is denser than surrounding water and moves along the sea floor as a "turbidity current."*

Figure 3–16 Quartz sandstone (Galesville Formation) exposed in quarry near Portage, Wisconsin. (Courtesy of J.C. Brice.)

Graywackes are immature, poorly sorted sandstones having significant quantities of very-fine-grained matrix and angular grains of quartz, feldspar, and rock fragments. The poor sorting, angularity of grains, great thicknesses of these sandstones, and their association with volcanic rocks indicate dynamic source and depositional areas in which debris from rapid erosion of highlands was quickly carried to a subsiding depositional basin (Fig. 3–15). Many graywackes appear to have been deposited by **turbidity currents.** These are masses of dense, sediment-laden water that, because of their greater density, flow down submarine slopes beneath less dense clear water. As the turbidity current loses its velocity, it deposits its load of sand and silt.

Quartz sandstones (Fig. 3–16) are clastic rocks characterized by dominance of quartz with little or no feldspar, mica, or fine clastic matrix. The quartz grains are well sorted and well rounded. They are most commonly held together by such cements as calcite and silica.

Calcite cement may develop between the grains as a uniform, finely crystalline filling, or large crystals may form, and each may incorporate hundreds of quartz grains. Silica cement in quartz sandstones commonly develops as overgrowths on the original quartz grains (see Fig. 3–5A). During this process, ions are added in crystallographic continuity with the host grain even when the host grain is well rounded. The crystal faces of the overgrowths are clearly visible when the rocks are examined with the microscope, and the boundary that separates the original grain and the overgrowth is often marked by a thin zone of impurities that once coated the host grain.

Quartz sandstones reflect deposition in stable, quiet, shallow water, such as the shallow seas that once inundated large parts of low-lying continental regions, or some parts of our modern continental shelves (Fig. 3–17). These sandstones exhibit sedimentary features, such as cross-bedding and ripple marks, which permit one to infer shallow-water deposition.

Cross-bedding is an arrangement of beds or laminations in which one set of layers is inclined relative to the others (Figs. 3–18, 3–19). The cross-bedding units are often formed when sediment slides down the lee slope of dunes or small deltas, either of which has inclined beds at the front. A depositional environment dominated by currents is inferred from cross-bedding. The currents may be wind or water. In either medium, however, the direction of inclination of the sloping beds is a useful criterion for determining the ancient current directions.

Ripple marks are commonly seen sedimentary features that developed along the surfaces of bedding planes (Figs. 3–20, 3–21). Symmetrical ripple marks (Fig. 3–22) are formed by the oscillatory motion of

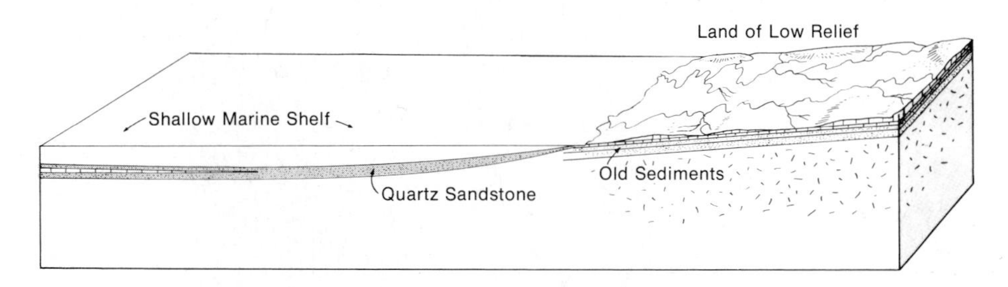

Figure 3–17 Idealized geologic conditions under which quartz sandstone may be deposited. There is little tectonic movement in this environment. Water depth is shallow and the basin subsides only very slowly.

Figure 3–18 Cross-bedded waterlaid sandstone. Tar Springs formation of Mississippian Age, southern Illinois. This is an example of a planar cross-bedding in that units are neither wedge-shaped, nor very thick.

water beneath waves. Asymmetrical ripple marks are formed by air or water currents and are useful in indicating the direction of movement of currents. For example, ripple marks form at right angles to current directions; the steeper side of the asymmetric variety

Figure 3–19 Cross-bedding accentuated by weathering in dune sandstones of Navajo Formation. Checkerboard Mesa, Zion National Park. (Courtesy of U.S. National Park Service.)

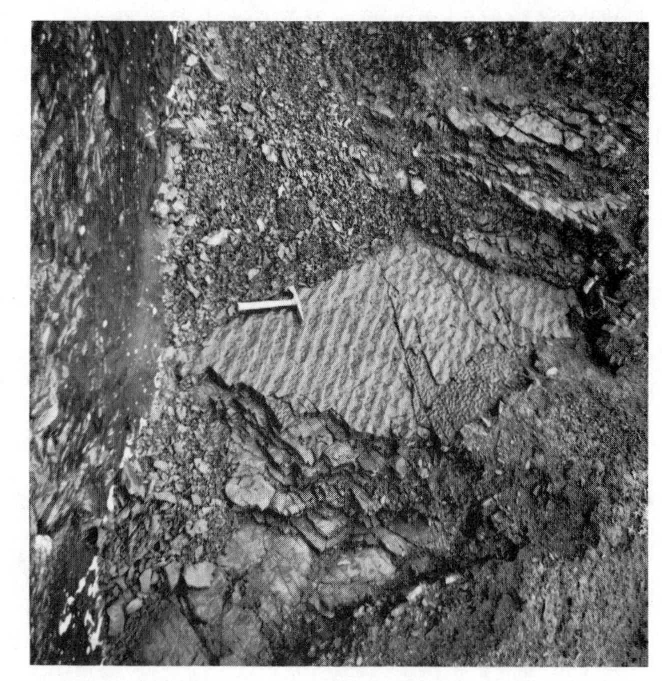

Figure 3–20 Ripple marks on bedding planes of nearly vertical beds of sandstone. Lisburne District, Northern Alaska. (Courtesy of U.S. Geological Survey.)

Figure 3–21 Ripple marks formed in sand on intertidal zone at Puerto Penasco, Sonora, Mexico. (Courtesy of Guillermo A. Salas, Universidad De Sonora.)

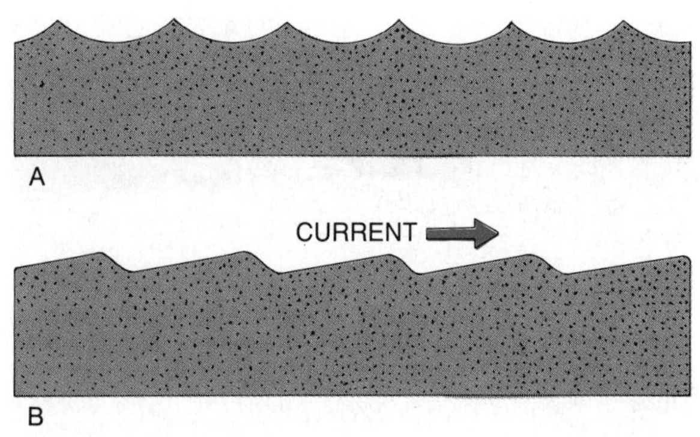

Figure 3–22 Profiles of ripple marks. (A) Oscillatory ripples. (B) Aqueous current ripples.

faces the direction in which the wind blows or water is flowing.

Sandstones containing 25 percent or more of feldspar (derived from erosion of a granitic source area) are called **arkoses** (see Fig. 3–14). Quartz is the most abundant mineral, and the angular-to-subangular grains are bonded together by calcareous cement, clay minerals, or iron oxide. The presence of abundant feldspars and iron imparts a pinkish-gray coloration to many arkoses. In general, arkoses are coarse, moderately-well-sorted sandstones. They may develop as basal sandstones derived from the erosion of a granitic coastal area experiencing an advance of the sea. Such blanketlike deposits may exhibit cross-bedding and ripple marks. Arkosic sandstones are more frequently nonmarine in origin, accumulating as feldspathic sands in fault troughs or other areas adjacent to granitic highlands (Fig. 3–23).

Sandstones that are transitional in composition and texture between quartz sandstones and graywackes are called **subgraywackes.** Feldspars in these sandstones are relatively scarce, quartz and chert are abundant, and the fine-grained detrital matrix rarely exceeds 15 percent. The voids are filled with mineral cement and clay. In subgraywackes, the quartz grains are better rounded and more abundant, the sorting is better, the quantity of matrix material is lower, and the porosity is greater than in graywackes. These differences are the result of more reworking and transporting of the subgraywacke sediment. Subgraywackes often exhibit cross-bedding and ripple marks. Deltaic coastal plains are the most frequent environment for

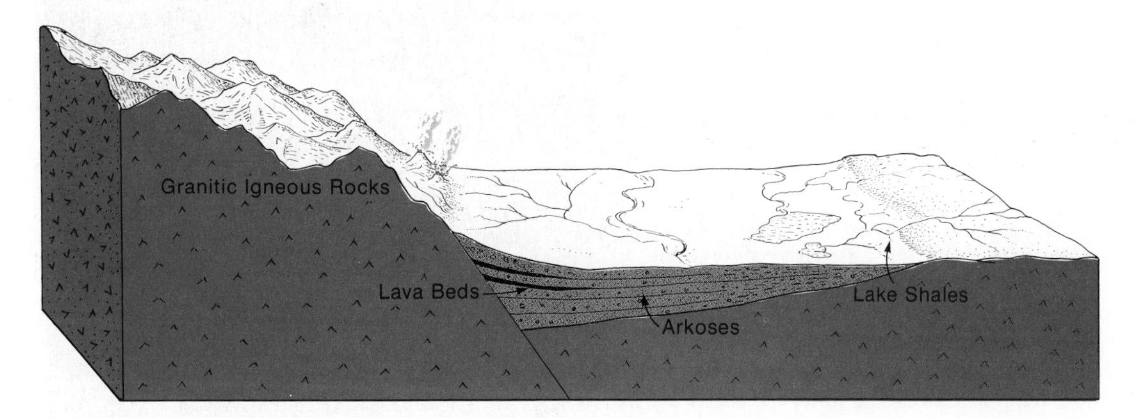

Figure 3–23 Geologic environment in which arkose may be deposited.

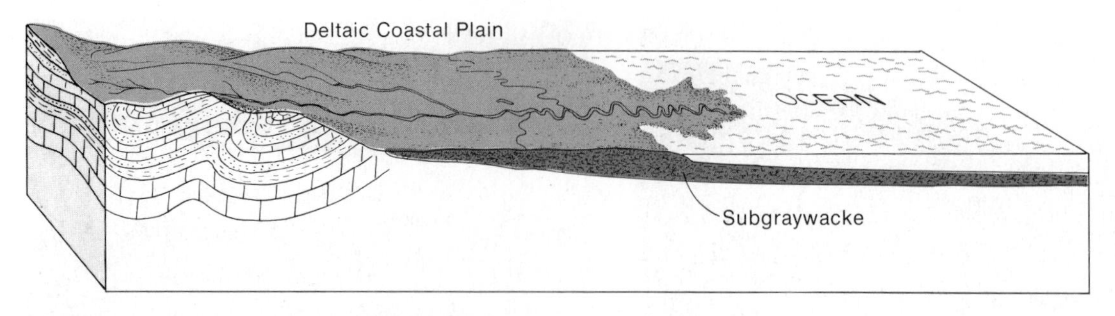

Figure 3–24 Deltaic environment in which subgraywacke may be deposited.

such rocks (Fig. 3–24). Coal beds and micaceous shales are frequently associated with subgraywackes. These sandstones really warrant a more distinctive name, for they are the most abundant of all the sandstone groups.

Shale, mudstone, and siltstone. Shale (Figs. 3–25, 3–26) is a fine-grained, laminated sedimentary rock in which an appreciable quantity of the constituent particles are in the clay and silt size range. Shales take their name from the French word *escale,* meaning scale. The name refers to the tendency of shales to split readily into thin layers parallel to bedding planes. This property, termed **fissility** (Latin, *fissilis,* split), probably results from the parallel alignment of tiny flat flakes of mica and clay minerals. Such particles are naturally deposited with their flat surfaces parallel to the floor of the depositional basin. Under compaction, even particles that were not originally in horizontal alignment may be brought into this position by the vertical pressure from overlying beds.

Some clay-rich rocks do not exhibit fissility and are often thick-bedded and massive. The general name for such nonfissile clayey rocks is **mudstone.** Another fine-grained sedimentary rock that generally lacks fissility is **siltstone.** Siltstones are composed largely of silt-size particles of quartz and feldspar and contain less clay than shales or mudstones.

Bedding surfaces of clay-rich rocks like shales and mudstones sometimes exhibit a distinctive polygonal pattern of cracks. These so-called **mud cracks** (Fig. 3–27) develop when fine-grained sediments shrink on drying. They are a feature commonly preserved not only in clay-rich rocks but also in rocks formed from carbonate mud. Mud cracks can be observed today on floodplains (Fig. 3–28), the subaerial parts of deltas, and on the floors of dried-up lakes and ponds. Because they cannot develop under water, mud cracks are a feature mostly associated with nonmarine deposition.

Carbonate sedimentary rocks

Limestone. Limestones are the most abundant of carbonate sedimentary rocks. Although limestone

Figure 3–25 Sedimentary sequence of shale, coal, and limestone. (Courtesy of Illinois Geological Survey.)

Figure 3–26 Exposures of shale form slopes beneath the prominent sandstone cliffs in the canyon of the Dolores River, south of Grand Junction, Colorado. (Courtesy of J.C. Brice.)

Figure 3–27 Ancient mudcracks in silty shale, Isle Royale National Park, Keweenaw County, Michigan. Polygonal crack system was caused by shrinkage following loss of water. Cracks were subsequently filled with the darker sediment. (Courtesy of U.S. Geological Survey.)

lake deposits do occur, most limestones originated in the seas. Nearly always, the formation of these marine limestones appears to have been either directly or indirectly associated with biological processes. In some limestones, the importance of biology is obvious, for the bulk of the rock (Fig. 3–29) is composed of shells of molluscs and skeletal remains of corals and other marine organisms. In other limestones, skeletal remains are not present, but nevertheless the calcium carbonate that forms the bulk of the deposit was precipitated from seawater because of the life processes of organisms living in that water. For example, the relatively warm, clear ocean waters of tropical regions are usually saturated with calcium carbonate. In this condition, only a slight increase in temperature, loss of dissolved carbon dioxide, or influx of supersaturated water containing $CaCO_3$ "seeds" can bring about the precipitation of tiny crystals of calcium carbonate. Also through photosynthesis, myriads of microscopic marine plants remove carbon dioxide from the water and thus may trigger the precipitation of calcium carbonate. The precipitate is thus an inorganic product of organic processes.

Limestones show considerable variety in texture and distinctive components (Fig. 3–30). Fine carbonate muds, for example, form a very fine-grained and uniform limestone called *micrite*. Lithographic limestone is a variety of micrite once widely used in printing. Some limestones are called clastic limestones because they are composed of particles of shells and older limestones that have been reworked by currents and waves. A cement of clear crystalline calcite called sparry calcite may cement these so-called clasts. Small spherical grains called oolites (Fig. 3–31) occur in some limestones. Oolites are formed by the precipitation of carbonate around a nucleus. The carbonate in oolites is added in concentric layers as the tiny spheres are rolled about on the sea floor by currents. Tidal areas off the coasts of Florida and the Bahamas have oolites forming today.

Chalk (Fig. 3–32) is a soft, porous variety of limestone composed largely of the tiny calcareous skeletal parts of unicellular marine organisms. Still other varieties of limestone include travertine (Fig. 3–33), which forms stalactites in caverns, and calcareous tufa, which is a limy crust frequently found around springs.

Figure 3–28 Modern mud cracks along the side of the Missouri River in Montana. (Courtesy of W.G. Pierce, U.S. Geological Survey.)

Figure 3–29 This prominent cliff called El Capitan stands at the south end of the Guadalupe Mountains in Texas. The cliff is an ancient reef composed of corals and associated marine invertebrates. Sandstones compose the slope below the cliff. (Courtesy of U.S. National Park Service.)

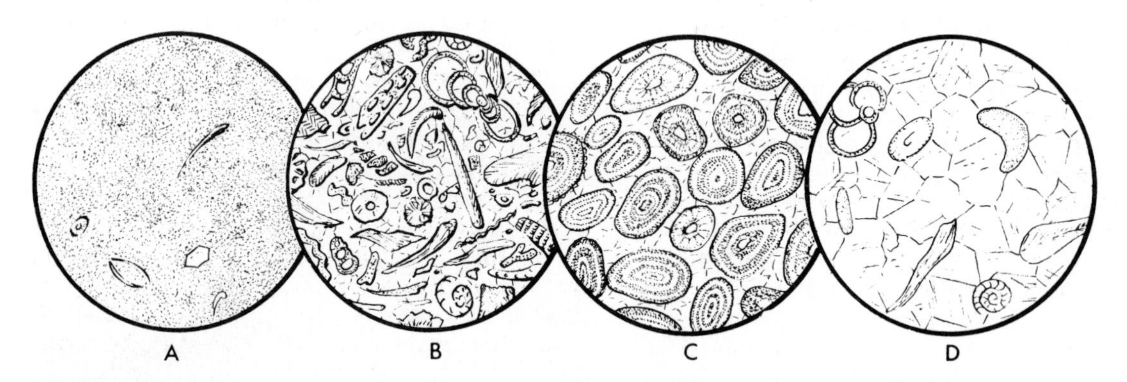

Figure 3–30 Limestones as seen in thin section under the microscope. (A) Micrite. (B) Bioclastic limestone with sparry calcite cement. (C) Oolitic limestone with sparry calcite cement. (D) Sparry or crystalline limestone. (Diameter of each field is about 2.0 mm.)

Figure 3–31 Modern oolites from the Bahama Banks. (Diameter of field is 5 mm.) (Courtesy of J.C. Brice.)

Figure 3–32 (A) Cliffs composed of chalk near Dorset, England. (B) Close-up of chalk containing gray nodules of flint.

Dolostone. *Dolostone* is nonclastic rock composed largely of the mineral dolomite, which is a calcium and magnesium carbonate. As found in exposures, dolostone is not easily distinguished from limestone. The usual field test for distinguishing dolostone from limestone is to apply cold dilute hydrochloric acid. Unlike limestone, which bubbles readily, dolostone will effervesce only slightly, if at all. Some dolostones have a pale yellow color and appear to be more porous than crystalline limestones.

The origin of dolostones is somewhat problematic. The mineral dolomite is not secreted by organisms in shell-building. Direct precipitation from seawater does not normally occur today, except in a few environments where the sediment is steeped in abnormally saline water. Such an origin is not considered adequate to explain the thick sequences of dolomitic rock commonly found in the geologic record. The most widely believed theory for the origin of dolostones is that they result from partial replacement of calcium by magnesium in an original calcareous sediment. Indeed, the porosity of dolostone may be the result of the slightly smaller ionic radius of magnesium ions that replaced calcium ions in the original sediment.

Chert

Chert is a very finely crystalline (microcrystalline) form of quartz that occurs in layers (Fig. 3–34) or as nodules (Fig. 3–35) in limestones. Most cherts are at least 90 percent silica, with the remaining 10 percent consisting of clay, calcite, and/or hematite impurities. Water is usually present in amounts less than 1 percent.

The origin of nodular chert is a subject of debate among geologists. A favored theory is that the nodules form as replacements of carbonate sediment by silica-rich seawater trapped in the sediment. Other cherts are not nodular but rather occur as stratified rocks. These so-called bedded cherts are thought to have formed from accumulations of siliceous remains of diatoms and radiolaria and from subsequent reorganization of the silica into a microcrystalline quartz. Additional silica for the formation of chert is also sometimes derived from the dissolution of volcanic ash. In fact, many sequences of bedded cherts are found in association with layers of ash and submarine lava flows.

Figure 3–33 *Travertine (dripstone) in the Dome Room of Carlsbad Caverns National Park, New Mexico. (Courtesy of T. Swan, Jr., and U.S. National Park Service.)*

Evaporites

Evaporites are chemically precipitated rocks that are formed as a result of evaporation of saline water bodies. Only about 1 percent of all sedimentary rocks consist of evaporites. Evaporite sequences of strata are composed chiefly of such minerals as gypsum ($CaSO_4 \cdot 2H_2O$), anhydrite ($CaSO_4$), halite (NaCl), and sylvite (KCl), and associated calcite and dolomite. Ancient deposits of evaporites are currently being commercially worked in Michigan, Kansas, Texas, New Mexico, Germany, and Israel.

Evaporites are important raw materials for the chemical and construction industries. Gypsum, for

Figure 3–34 Folded beds of chert along an Arkansas highway cut. (Arkansas Geological Commission.)

example, is used in the manufacture of plaster and plaster products (Fig. 3–36) and as an additive in Portland Cement to control the setting time. Halite is mined in large quantities for use in melting snow and ice on streets, in seasoning foods, and in manufacturing hydrochloric acid, which is essential to many industrial processes. Sylvite, called potash in the industry, is used along with phosphates and nitrates to manufacture fertilizers.

Figure 3–35 The chert nodules in this bioclastic limestone are the darker bodies which, because of their greater resistance to weathering, stand out in relief. Mississippian limestone, Hardin County, Illinois. (Courtesy of C. Butts and the U.S. Geological Survey.)

The ideal conditions for precipitation of evaporites are a warm, relatively arid climate and periodic additions of seawater to the evaporating marine basin. These conditions are needed to account for the great thickness of many ancient evaporite deposits. Evaporites are of special interest in the Gulf Coast of the United States, where as a result of the pressure of overlying rocks, deeply buried deposits of salt have flowed upward to form underground salt domes (Fig. 3–37). In the process, the salt arched the overlying strata, thereby producing folds and faults for trapping petroleum.

Coal

Coal (see Fig. 3–25) is a carbonaceous rock resulting from the accumulation of plant matter in a swampy environment. That plant tissue is altered by both biochemical and physical processes until it is converted to

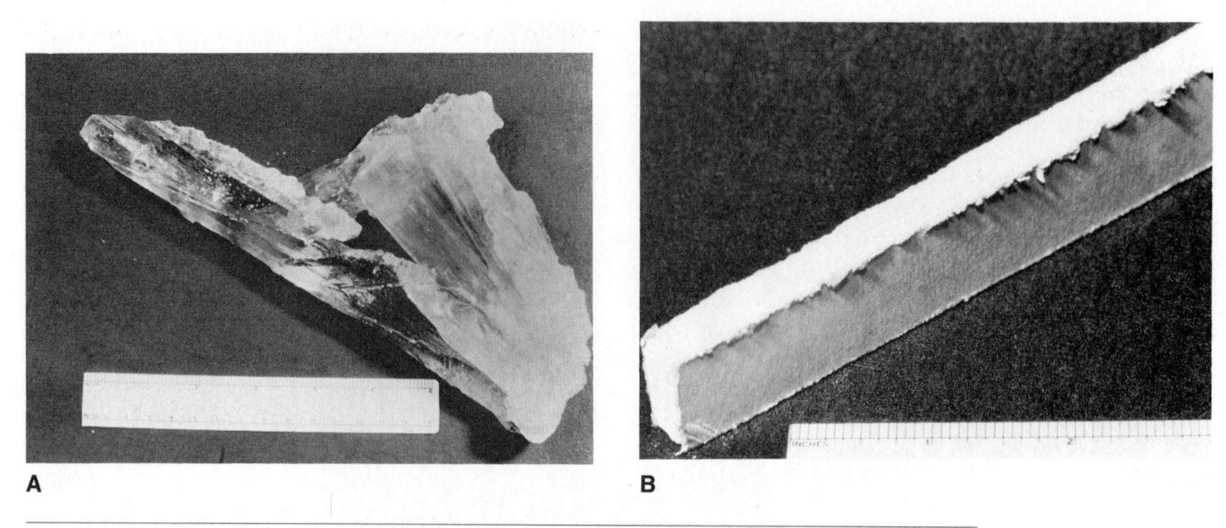

Figure 3–36 (A) Selenite gypsum ($CaSO_4 \cdot 2H_2O$). (B) "Drywall" wallboard used in construction consists of plaster of Paris prepared from gypsum, sand, and water, and sandwiched between two pieces of cardboard. (Courtesy of J.C. Brice.)

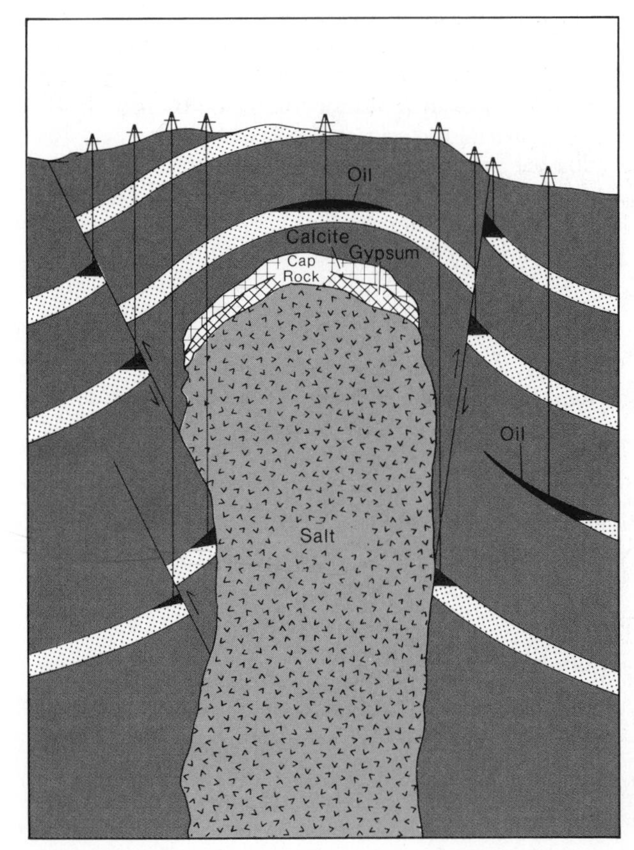

Figure 3–37 Salt dome, illustrating possibilities for oil entrapment in dome-like structures (top center) and by faults, and the pinchout of oil-bearing strata.

a consolidated carbon-rich material. The biochemical and physical changes may produce a series of products ranging from peat and lignite to bituminous and anthracite coal. For coal to form, plant tissue must accumulate under water or be quickly buried, because vegetable matter, if left exposed to air, is readily oxidized to water and carbon dioxide. With underwater accumulation or quick burial of plant material, a major part of the carbon can be retained.

Color in sedimentary rocks

The color of a sedimentary rock is one of the first things noticed by an observer. Yet for sedimentary rocks, color is not very useful in identification. Color may, however, provide clues to a rock's composition and environment of deposition.

Hues of brown, red, and green are frequently formed in sedimentary rocks as a result of their content of iron oxides. Few, if any, sedimentary rocks are free of iron, and less than 0.1 percent of this metal can color a sediment a deep red. The iron pigments are not only ubiquitous in sediments but also difficult to remove by natural processes.

Strata colored in shades of red, brown, or purple by ferric iron are designated **red beds** by geologists. Oxidizing conditions required for the development of ferric compounds are more typical of nonmarine than marine environments. Thus, most red beds are floodplain, alluvial fan, or deltaic deposits. Some, however, are originally reddish-colored sediment carried to and deposited in the sea.

Black and dark-gray coloration in sedimentary rocks—especially shales—usually results from the presence of organic carbon compounds and/or iron sulfides. The occurrence of an amount of organic carbon sufficient to result in black coloration implies an abundance of organisms in the depositional areas, as well as environmental circumstances that kept the remains of those organisms from being completely destroyed by oxidation or bacterial action. As the carbon is not oxidized and allowed to escape as CO_2, it accumulates to form the dark pigment in the sediment.

Fossils in sedimentary rocks

Just as modern environments of deposition include animals and plants, so also did ancient depositional sites. The remains or traces of those organisms are frequently preserved in sedimentary rocks as **fossils.** They may constitute only a minor feature of the rock, or, as is the case with many limestones, the fossils may be the dominant constituents.

Former living organisms may become fossils as the result of several natural processes. The largest number of fossils are remains of marine creatures (Fig. 3-38)

Figure 3-38 *A slab of rock containing specimens of* Ellipsocephalus, *a Cambrian trilobite. The slab is about 16 cm long. (Courtesy of Wards Natural Science Establishment, Inc., Rochester, N.Y.)*

that died and were covered by the rain of sediment that falls continuously to the sea floors. Less commonly, plant and animal remains are covered with windblown silt or volcanic ash, immersed in tar or quicksand, or engulfed by lava flows. For preservation, quick burial and the possession of a mineralized skeleton are usually required. Once enclosed in sediment, the original hard parts may remain largely unaltered, or they may be replaced by mineral matter of a different composition (Fig. 3-39). Pores may be filled with lime or silica brought in by underground water. Parts or organs may be dissolved completely, leaving behind empty molds. Molds may later be filled so as to form casts that resemble the original organic part. All of these processes for changing shells, bones, wood, and other organic matter into durable traces of former life are termed **petrifaction.**

For the student of sedimentary rocks, fossils are exceptionally useful as indicators of ancient environments of deposition. A limestone containing the remains of corals, for example, must have been deposited in an ocean; preserved impressions of oak leaves in a sandstone would suggest the enclosing rock was of continental origin. Fossils may shed light on temperature, salinity, depth of water, and presence or absence of bottom currents in the depositional environment. In studies of broader scope, fossils provide information about the distribution of seas, shores, and mountains as they existed hundreds of millions of years ago. They inform geologists about past changes in climate, the location of former land connections between continents, and changes in the positions of continents. Fossils document the long history of life and are indices to the age of the rock in which they are found.

Figure 3-39 *Petrified wood. The original wood has been replaced by silica. (Length of specimen is 20 cm.) (Courtesy of Wards Natural Science Establishment, Inc., Rochester, N.Y.)*

Formations and geologic maps

An understanding of the geology of an area begins with the study of the actual rocks exposed at the earth's surface. These rocks are grouped by geologists into units that differ from rocks above and below because of differences in color, texture, or composition. The units of distinctive lithology and appearance are called *formations.* Formations are the fundamental units for describing and mapping the geology of any particular area around the world (Fig. 3-40).

Formations are given two names: (1) a geographic name that refers to a locality where the formation is well exposed or where it was first described, and (2) a rock name if the formation is primarily of one lithologic type. For example, the "St. Louis Limestone" is a carbonate formation named after exposures at St. Louis. Where formations have several lithologic types within them, the locality name may be followed by the word *formation.*

In mapping formations and other rock units, geologists recognize that the unit may or may not be the same age everywhere it is encountered. The nearshore sands deposited by a sea slowly transgressing across a coastal plain may deposit a single blanket of quartz sandstone. However, it will be older where the transgression began and younger where it ended (Fig. 3-41).

Maps that depict the distribution and nature of formations, and the occurrence of geologic features like folds and faults are called **geologic maps** (Fig. 3-42). Geologic maps show the characteristics of the bedrock beneath the covering of loose soil and rock debris; that is, beneath the *regolith.* Assume for a moment that the regolith were miraculously removed from a certain region, so that bedrock would be ex-

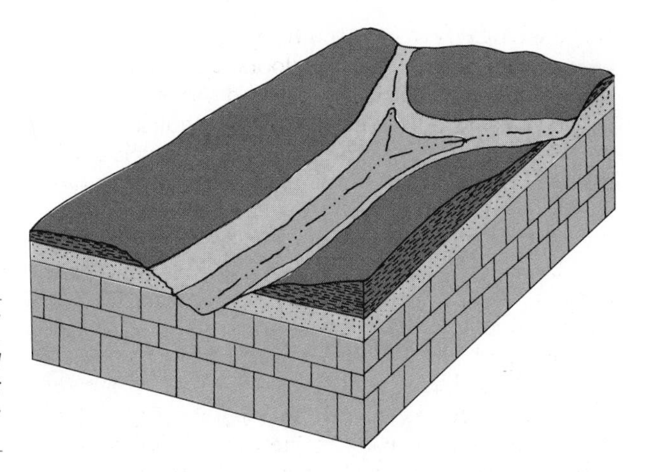

Figure 3–40 *Formations. The diagram shows three formations. In practice, these formations would be formally named, often after a geographic location near which they are well exposed. For example, the three formations shown here might be designated the "Cedar City Limestone," "Big Springs Sandstone," and "Plattsburgh Shale."*

posed everywhere. Next imagine that the surfaces of each rock unit were painted a different color and the entire region photographed vertically from an airplane. The resulting photograph would constitute a simple geologic map.

The actual preparation of a geologic map is basically similar to the construction of most other kinds of maps. The essential ingredients are a series of observations made at particular geographic locations. For geologic maps, the observations are available in natural exposures, boreholes, quarries, and roadcuts. The geologist notes not only the locations of individual exposures but also the locations of formation boundaries ("contacts"). Places where rocks are disrupted and displaced are recorded, along with measurements of the angles of beds that are tilted and bent. Symbols are added for further clarification. In most geologic mapping today, the field observations are plotted directly on aerial photographs or on the topographic maps that were prepared from such photographs. A recent technique called radar imagery is capable of penetrating dense vegetation to accurately reveal the patterns of rock units at the ground surface.

Once the geologic map is completed, the geologist can tell a good deal about the geologic history of an area. The rock formations can be considered "chapters" in the sequence of geologic events. From the geologic map shown in Figure 3–42, one can deduce that the rock formations had been compressed into a folded pattern and then broken (faulted), and that the folded layers were of greater age than the level limestone layer that covers them in the southeast area of the map.

METAMORPHIC ROCKS

The agents of metamorphism

Rocks that have been altered from previously existing rocks by the action of heat, pressure, and chemically

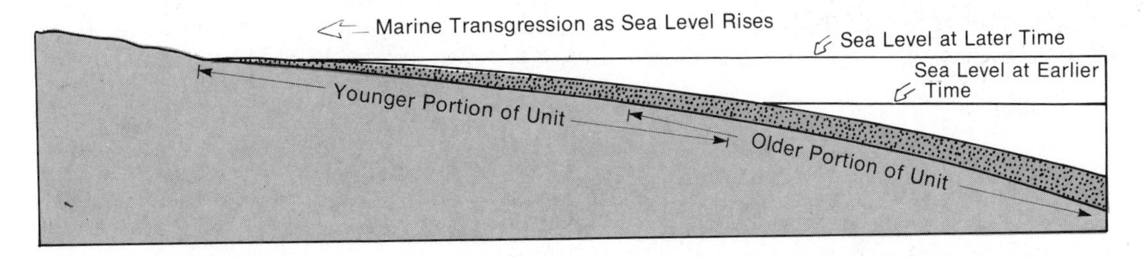

Figure 3–41 *Diagram showing how the original deposits of a formation may vary in age from place to place.*

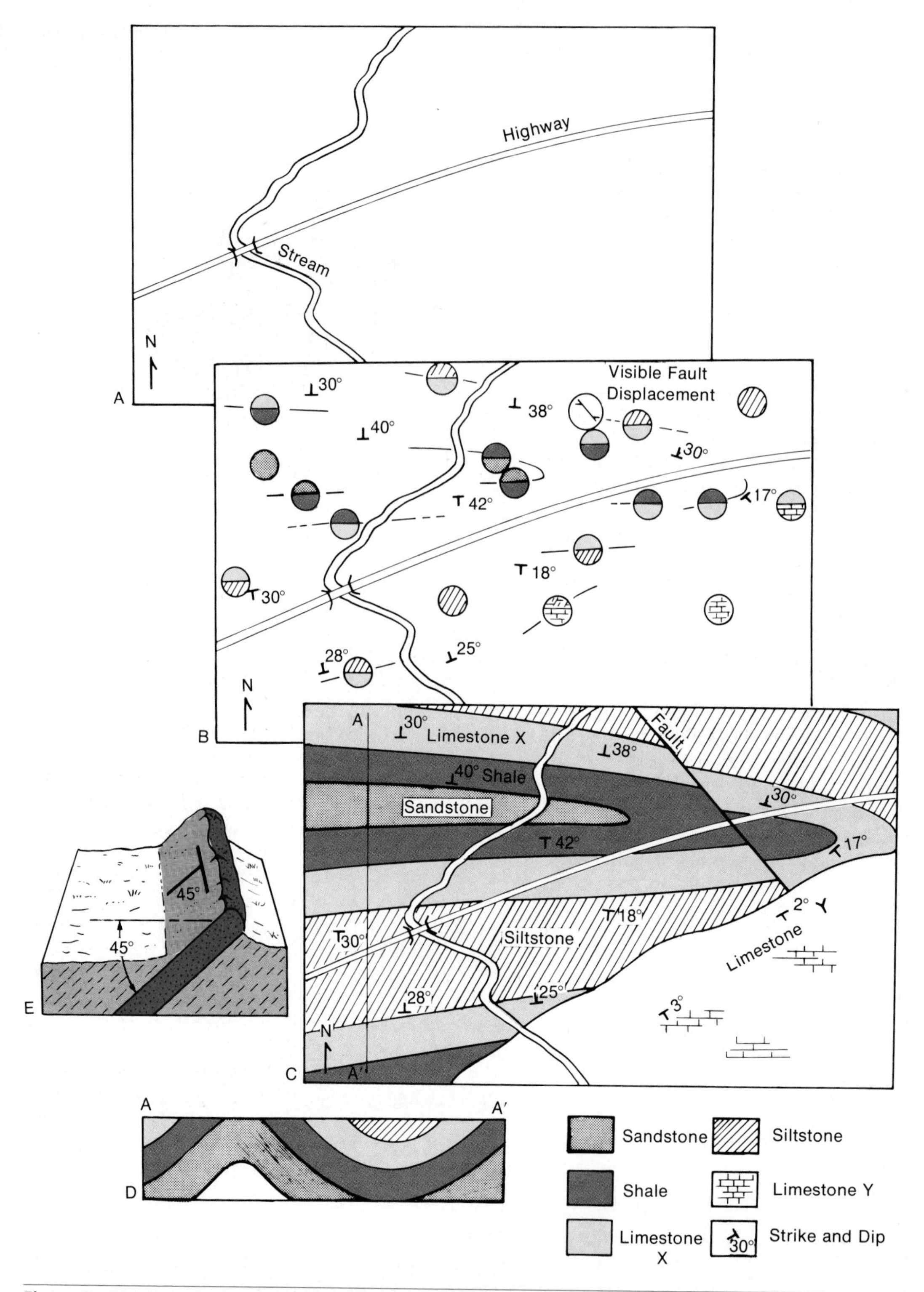

Figure 3–42 *Steps in the preparation of a geologic map. A suitable base map (A) is selected. The locations of rock exposures of the various formations are then plotted on the base map as indicated in (B). Special attention is given to exposures that include contacts between formations; where they can be followed horizontally, they are traced onto the base map also.* Strike (the compass direction of a line formed by the intersection of the surface of a bed and a horizontal plane) and dip *are measured wherever possible and added to the data on the base map. After careful field study and synthesis of all the available information, formation boundaries are drawn so as to best fit the data. On the completed map (C), color patterns are used to show the appearance of the area if there were no soil cover. A cross-section along line A–A′ is shown in (D); (E) is a block diagram illustrating strike and dip.*

active solutions are called metamorphic rocks. The heat, pressure, and natural chemical solutions constitute the agents of metamorphism. Alone or in combination, these agents operate at various intensities to produce metamorphic rocks having distinctive textures and compositions.

Heat

As an important agent of metamorphism, heat reduces the ability of a rock to resist deformation and causes an increase in the rate of most chemical reactions. The heat for metamorphism may be provided by nearby intrusions of magma. It may be associated with the compression of the crust in regions experiencing mountain building, or it may be induced by increases in pressure resulting from deep burial. The rate of increase in temperature at increasing depths in the earth is called the **geothermal gradient.** Measurements made in deep mines and wells indicate that the geothermal gradient for the crust averages about 30°C per kilometer of depth. This rate varies from place to place and is (not unexpectedly) greater near centers of active volcanism. The geothermal gradient is such that at depths of about 35 km, temperatures are high enough to melt rock (Fig. 3–43).

Because particular mineral-forming chemical reactions occur only within a specific range of temperature, heat influences the ultimate mineral composition of metamorphic rocks. Also, as temperature rises, ions within the atomic lattice of minerals become increasingly agitated. Eventually, the original mineral structure may become disassembled and the ions subsequently reassembled into structural forms stable under the newer thermal conditions. As an example, consider what might happen to clay minerals in shale strata invaded by a granitic magma. Clay minerals are hydrous aluminum silicates. Heat applied to these minerals from the hot magma would cause a loss of the water contained in the clay and a conversion of clay into a common metamorphic mineral named andalusite (Fig. 3–44). Andalusite is a nonhydrous aluminum silicate having an atomic structure that is stable under the new conditions. The released water in the above example, together with such gases as carbon dioxide, may also participate in the many chemical reactions associated with metamorphism.

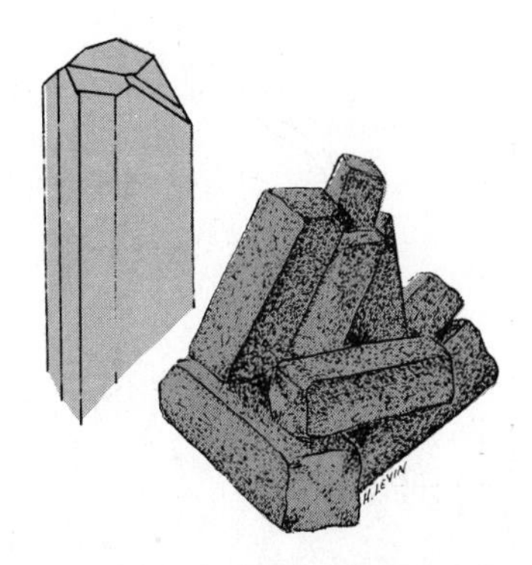

Figure 3–44 Andalusite. Orthorhombic crystal form and aggregate of natural crystals showing usual development of imperfect blunt prisms.

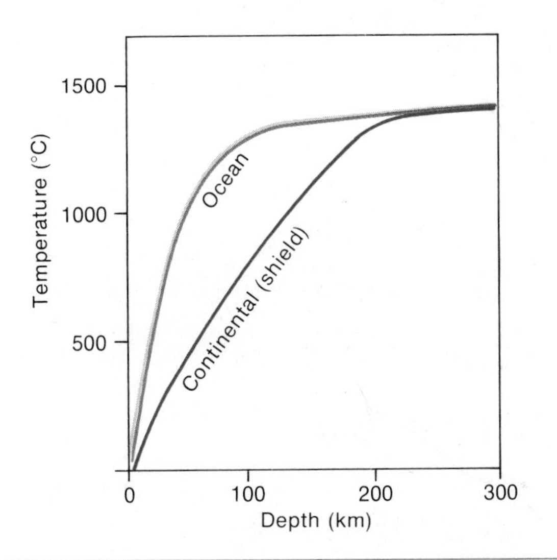

Figure 3–43 The increase in temperature with depth beneath a continental shield area (orange) and ocean basin (brown). (Courtesy of Geoffrey F. Davies.)

As noted in Chapter 2, if temperatures in a rock mass rise high enough, the melting points of constituent minerals will be exceeded and magma will begin to form. If the melting becomes pervasive, igneous rather than metamorphic rocks will result.

Pressure

The tremendous pressures that exist several kilometers below the earth's surface (about 1000 kg/cm^2 at a depth of 4 km), or like those associated with collisions of tectonic plates, can cause profound changes in deeply buried rocks. Mineral grains, for example, may recrystallize into new minerals with more tightly packed atomic structure and therefore greater density.

Figure 3–45 Metamorphic rock showing banding. The rock is a gneiss exposed in the valley of Humbug Creek, Summit County, Colorado. Darker bands are composed of hornblende, plagioclase, and small amounts of quartz. Lighter bands are predominantly quartz, orthoclase, and light-colored plagioclase. (Courtesy of M.H. Bergendahl and the U.S. Geological Survey.)

Where mineral grains are in contact, the squeezing action may cause melting or solution at the points of contact and precipitation of material along the sides of grains that experience less pressure.

Pressure may also result in reorientation of grains, the development of tiny shear planes in the rock, and recrystallization. These processes are largely responsible for the development of the lineated and banded textures characteristic of many metamorphic rocks (Fig. 3–45).

Chemically active solutions and gases

In varying amounts, liquids and gases are always present in regions where metamorphism takes place. They play an important role in metamorphism by increasing the efficiency of recrystallization, serving as solvents, and accelerating the rate of chemical reactions. Laboratory experiments have shown that most minerals react so slowly to increases in temperature that metamorphic reactions would require lengthy spans of time to run their course. By adding only a minute amount of water to a laboratory capsule containing the experimental minerals, however, reaction rates are dramatically increased. One reason for this is that ions in a fluid medium can quickly be brought into close proximity with each other and reach appropriate sites in the atomic structure of minerals.

In addition to water, the gas carbon dioxide also promotes metamorphism. Carbon dioxide is readily liberated during the heating of limestone. This leaves the remaining oxide of calcium free to combine with silica or other impurities in the limestone.

Kinds of metamorphism

In general, one can recognize two basic kinds of metamorphism. In one kind the most important agent is heat induced by a nearby magma. This is called *contact*

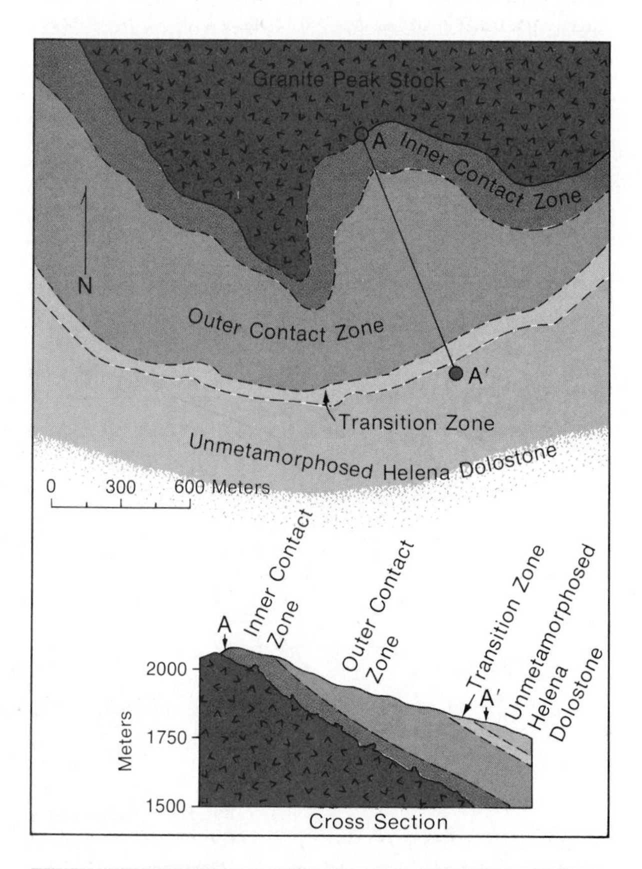

Figure 3–46 The Granite Peak aureole. This metamorphic aureole is developed around a granitic stock that has intruded into dolostone country rock. Particular assemblages of metamorphic minerals occur within each contact metamorphic zone. (Simplified from Melson, W.G., American Mineralogist, *51:404, 1966, Fig. 1.)*

metamorphism. In contrast, the metamorphic rocks of some regions have suffered the effects not only of high temperature but also of strong directional pressures. These rocks have experienced *regional metamorphism*.

Contact metamorphism

Contact metamorphism takes its name from its occurrence at or near the contacts between hot igneous material and country rock. Heat is the most important agent in contact metamorphism, although pressure and chemical activity assist in producing alterations. Temperatures would, of course, be highest immediately adjacent to the igneous mass. Thus, as the distance increases between the contact of the igneous body and the country rock, the effect of the metamorphism decreases. Around many intrusions, roughly concentric zones of decreasing metamorphism, called **metamorphic aureoles,** can be recognized and mapped (Fig. 3-46).

The width of the metamorphic aureole and the amount of change that occurs during contact metamorphism depend upon the composition, temperature, size, and shape of the hot igneous material, its viscosity and gas content, and the composition and permeability of the country rock (Fig. 3-47). In general, the larger and more acidic the igneous mass, the greater the effects of contact metamorphism. Not all rocks are uniformly affected by intruding magmas.

Regional metamorphism

Regional metamorphism may be defined as metamorphism that has developed over areas of many thousands of square kilometers as a result of intense compression associated with the convergence of tectonic plates. Most tracts of regionally metamorphosed rocks contain evidence of a complex history that began with thick accumulations of sedimentary and volcanic rocks within trenches or troughs at the margins of continents. As these materials accumulated, they were exposed to increasing temperatures and pressures that promoted extensive deformation and recrystallization. Ultimately, batholithic masses of magma were generated, adjacent to which contact metamorphism oc-

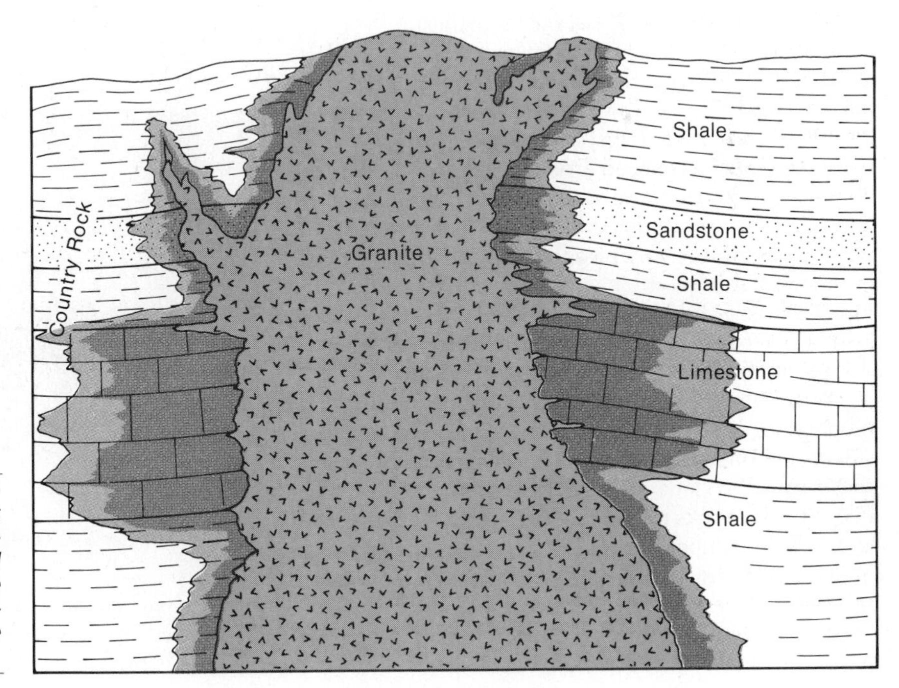

Figure 3-47 Metamorphic aureole developed around a granitic intrusion. The intensity of metamorphism diminished outward from the granite margin, and the width of aureole increased adjacent to more chemically reactive and permeable country rock.

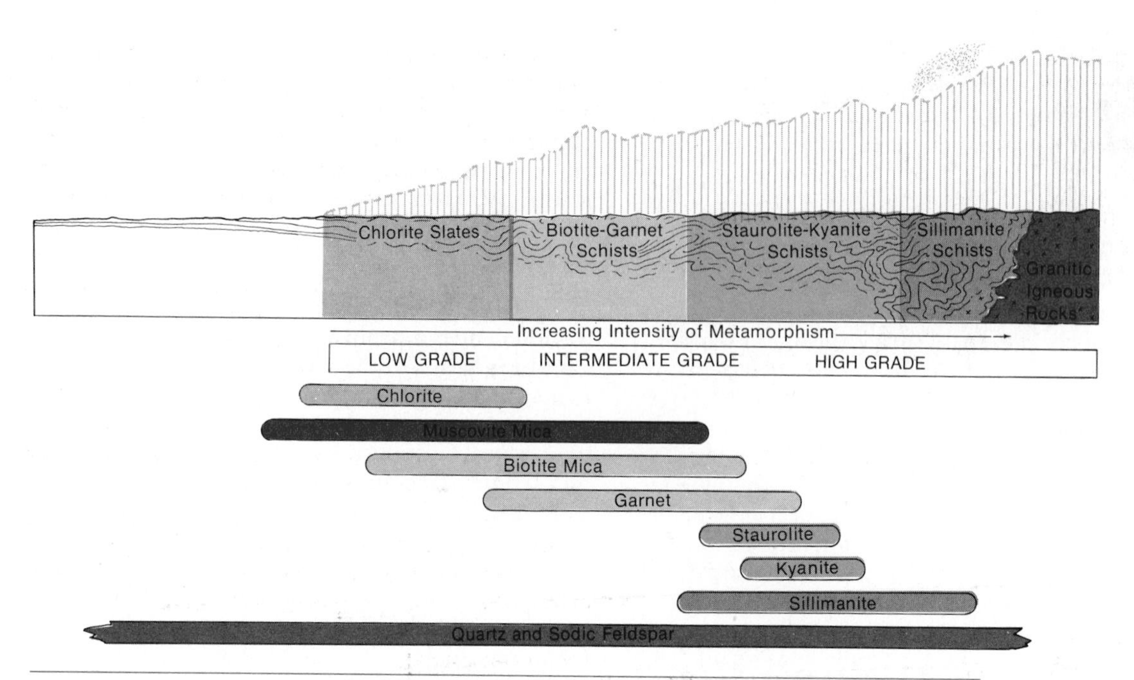

Figure 3–48 *Changes in the mineralogic composition of a shield area exhibiting progressive regional metamorphism.*

curred. Clearly, all three of the agents of metamorphism participate in regional metamorphism.

Today, tracts of regionally metamorphosed rocks attest to the occurrence of ancient periods of mountain building. To account for the exposure of such rocks at the earth's surface, several kilometers of overlying rock must have been eroded. Nearly every continent has such metamorphic regions. Where they are exceptionally broad and composed of rocks metamorphosed during several episodes of deformation, they are termed **shields.** Shields are ancient terrains that may include former belts of deformation that have been reduced to low-lying, regionally convex plains during repeated cycles of uplift and erosion.

With the help of metamorphic index minerals, geologists attempting to unravel the history of ancient shields are often able to discern associations of rocks that formed under specific metamorphic conditions. For example, rocks rich in chlorite formed under relatively low temperature and pressure conditions. Those containing biotite and garnet usually represent an intermediate grade of metamorphism, and the presence of certain other metamorphic minerals indicates severe regional metamorphism (Fig. 3–48).

Figure 3–49 *Chlorite gives this metamorphic rock (a chlorite schist) a scaly texture and greenish color. (Courtesy of Wards Natural Science Establishment, Inc., Rochester, N.Y.)*

Figure 3–50 Crystals of almandite garnet in mica schist. (Courtesy of the Institute of Geological Sciences, London.)

Metamorphic minerals

Quartz, hornblende, feldspar, and mica are all minerals that compose ordinary igneous rocks. These same minerals are among the most abundant in metamorphic rocks as well. There are also minerals like chlorite (Fig. 3–49) and garnet that occur in both igneous and metamorphic rocks, but they are far more common in the latter. Some minerals ordinarily occur only in metamorphic rocks. Examples are **andalusite, sillimanite,** and **kyanite,** all of which have the composition Al_2SiO_5. Although these three minerals have the same composition, they vary in the shape of the crystals they form, and each develops within its own characteristic pressure and temperature range. The hydrous iron aluminum silicate known as **staurolite** ($FeAl_4Si_2O_{10}(OH)_2$) as well as **almandite garnet** ($Fe_3Al_2(SiO_4)_3$) (Fig. 3–50) and the pink garnet **grossularite** ($Ca_3Al_2(SiO_4)_3$) are also common in metamorphic rocks. **Spinel** ($MgAl_2O_4$) and **wollastonite** ($CaSiO_3$) are found in calcareous rocks that have experienced metamorphism. Parent rocks rich in magnesium and silica may yield the metamorphic mineral **talc** ($Mg_3SiO_{10}(OH)_2$) as indicated by the reaction below.

$$\underset{\text{(dolomite)}}{3\,CaMg(CO_3)_2} + \underset{\text{(quartz)}}{4\,SiO_2} + \underset{\text{(water)}}{H_2O} \xrightarrow[\text{yields}]{} \underset{\text{(talc)}}{Mg_3Si_4O_{10}(OH)_2} + \underset{\text{(calcite)}}{3\,CaCO_3} + \underset{\text{(gas)}}{3\,CO_2}$$

Whether chlorite, sillimanite, or some other mineral forms in a rock undergoing metamorphism depends not only on the composition of the parent rock but also on the intensity of heat and pressure. Particular minerals develop under specific conditions. The amount of heat and pressure required for the formation of some of the simpler metamorphic minerals can be determined in the laboratory. In such experiments, silicate powders are mixed with water vapors and gases and raised to high temperatures and pressures within specially designed containers called *hydrothermal bombs.* Different metamorphic minerals form in the bomb at specific levels of temperature and pressure. Experiments of this kind have helped geologists to designate metamorphic index minerals as indicative of a particular degree of metamorphism or metamorphic grade. The presence of the index mineral chlorite in a metamorphic rock, for example, would suggest that the rock formed at a lower temperature than a rock containing the mineral sillimanite.

Metamorphic rocks

Metamorphic rocks are at least as varied as are igneous and sedimentary rocks. Their variety is related to the diversity of parent materials, as well as the kind and intensity of metamorphic agencies. As is true with igneous and sedimentary rocks, the basis for the classification of metamorphic rocks is composition and texture (Table 3–3). As far as possible, compositional and textural characters are selected that permit placement of metamorphic rocks into categories that reflect similar parentage, history, and kind of metamorphism.

Texture of metamorphic rocks

Most metamorphic rocks can be distinguished from igneous and clastic sedimentary rocks on the basis of texture. Because mineral grains in metamorphic rocks have not developed in sequence from a melt, they do not exhibit the same kind of disoriented or chaotic interlocking fabric characteristic of igneous rocks. Also, because they are not composed of detrital grains, they do not resemble clastic sedimentary rocks. Metamorphic minerals grow in the solid state and often are aligned parallel to one another so as to fit together like a wall constructed of well-fitted, irregular elongated stones. Sometimes certain mineral grains will grow considerably larger than the average grain size. Such

TABLE 3–3 **ORIGIN AND CHARACTERISTICS OF METAMORPHIC ROCKS**

Category	Metamorphic rock	Texture and appearance	Typical mineral composition	Type of metamorphism	Parent rock
Foliated	Slate	Aphanitic, smooth, dull foliation planes; gray, black, green, or purple	Clay minerals, mica, chlorite	Regional	Shale
	Phyllite	Aphanitic, although some grains may be visible. Foliation planes commonly wrinkled and more lustrous than slate	Mica, chlorite, quartz	Regional	Shale
	Schist	Phaneritic, distinctly foliated, platy (mica, chlorite) or needle-like (hornblende) minerals commonly segregated into layers	Mica, chlorite, talc, graphite, hornblende, garnet, staurolite	Regional	Shale, volcanic rocks
	Gneiss	Phaneritic, irregular foliation composed of relatively robust minerals. Foliation less distinct than in schists	Quartz, feldspars, garnet, mica, hornblende, staurolite	Regional	High-silica igneous rocks and sandstones
Nonfoliated or Weakly Foliated	Marble	Phaneritic, fine to coarsely crystalline; often variegated (marbled)	Calcite, dolomite	Regional or contact	Limestones or dolostones
	Quartzite	Phaneritic, sugary textured	Quartz	Regional or contact	Quartz sandstones
	Greenstone	Aphanitic, scattered dark visible crystals, dark green	Chlorite, epidole, amphiboles	Regional	Low-silica volcanic rocks
	Amphibolite	Phaneritic, similar to amphibole schist but foliation less apparent	Hornblende and plagioclase	Regional	Low-silica volcanic rocks

larger grains are called **porphyroblasts** (Fig. 3–50) and the texture is called **porphyroblastic** (Fig. 3–51).

An important textural attribute of most but not all metamorphic rocks is **foliation** (Fig. 3–52). Foliation (Latin, *foliatus*, leaved) can be described as a parallel arrangement or distribution of minerals in a metamorphic rock. This alignment of mineral grains may cause the rock to split into more or less flat pieces. Whether foliation is fine or coarse depends upon the size and shape of the component grains. Conspicuously foliated rocks usually contain abundant flat minerals, such as mica or chlorite. Coarse or imperfect foliation is usually the result of the growth and segregation of rather blocky minerals, such as feldspar and quartz, into roughly parallel bands and lenses (Fig. 3–53).

Kinds of metamorphic rock

There are hundreds of kinds of metamorphic rocks, but those that occur extensively at the earth's surface constitute a relatively short list. It is convenient to

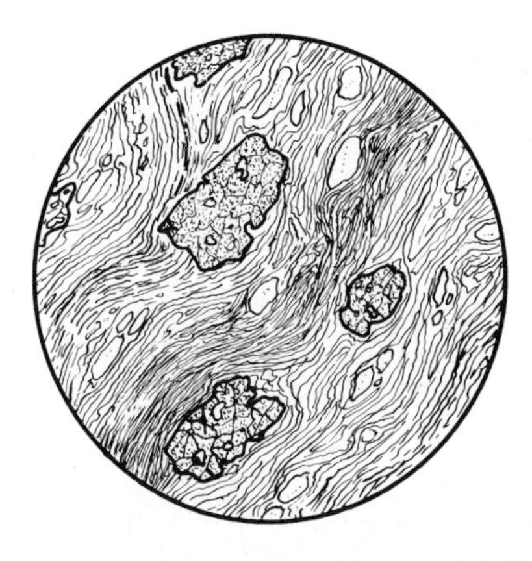

Figure 3–51 Porphyroblastic texture as viewed in thin section with the aid of a microscope. The large, dark, ragged porphyroblasts are staurolite grains. Clear grains are quartz and the flaky minerals consist of biotite and muscovite. The rock is a staurolite-mica schist. Area of field 5 mm.

Figure 3–53 Coarse foliation developed in a gneiss. The lenselike lighter colored areas are feldspar and quartz. The darker bands are composed mostly of ferromagnesian minerals. (Courtesy of Wards Natural Science Establishment, Inc., Rochester, N.Y.)

divide these common varieties into two groups, namely, those that exhibit foliation and those that do not.

Foliated metamorphic rocks

Slate. Slate (Fig. 3–54) is a product of regional metamorphism of clayey rocks, particularly shale. Its origin involves the transformation of clay minerals into microscopic crystals of sheet-structure silicates, such as muscovite. The small size of these flakey minerals and the persistence of relic planes of bedding attest to the relatively low grade of metamorphism under which slate forms.

As a result of the parallel growth of the minute flakes of mica in slate, the rock characteristically will split into rather smooth, thin slabs. This trait is called **slaty cleavage.** Slaty cleavage is caused by the growth of the sheet-structure silicates at right angles to the

Figure 3–52 Well-developed foliation resulting from parallel alignment of biotite grains. The rock type is biotite schist and occurs as a xenolith in granite (the surrounding lighter colored rock). (From an exposure at Wādī al 'Arabah, Jordan, courtesy of F. Bender and the U.S. Geological Survey.)

Figure 3–54 Slate. (Courtesy of the Institute of Geological Sciences, London.)

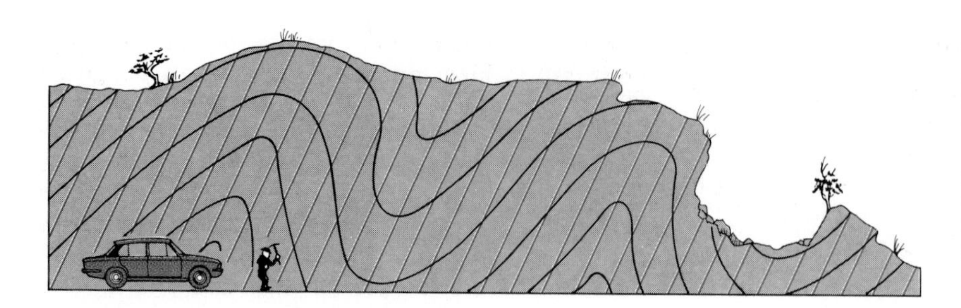

Figure 3–55 The planes of slaty cleavage (shown by diagonal orange lines) do not necessarily have a relation to the original bedding planes. In most folded slates, the cleavage planes cut across the bedding at various angles. In the arches of upfolds (anticlines) and troughs of downfolds, slaty cleavage often cuts across the bedding at right angles.

directions from which pressures were applied (Fig. 3–55). Planes of slaty cleavage need have no relationship to bedding planes in the shales from which slates are usually derived.

Because of its uniform cleavage and impermeable nature, slate is an ideal material for classroom blackboards, pool table tops, and roofing tiles. Carbon and iron sulfide are the main coloring agents in black slates, whereas red and green varieties derive their color from iron oxide impurities.

Phyllite. Phyllite is a metamorphic rock that is similar to slate in composition. It forms as the result of metamorphism of slates or clayey rocks. Tiny glistening flakes of chlorite and mica that have formed from the recrystallization of clay minerals under pressure can be discerned on the parting surfaces of phyllites if one looks closely. These parting planes characteristically develop a wrinkled appearance and are lustrous compared with the dull surfaces of slate. Phyllite represents an intermediate degree of metamorphism between slate and schist.

Schist. The flakey or needle-like minerals in schist are sufficiently large to be readily visible to the unaided eye. They impart a kind of foliation that is described as schistose. Schistose foliation is normally associated with medium-grade metamorphism. The foliation in schists is distinctly less uniform than in either slate or phyllite, and traces of relic bedding are rarely present. Thin bands of flakey minerals, such as mica, chlorite, or talc, are visible to the unaided eye and cause the schistosity (see Fig. 3–52). In some schists, lath-like, needle-like, or fibrous minerals may also cause foliation. Schists may develop during the regional metamorphism of igneous, sedimentary, or lower-grade metamorphic rocks. The majority of schists, however, appear to have been derived from rocks rich in clay. Schists are named according to the most conspicuous mineral in the rock. Thus, there are mica schists, chlorite schists, amphibole schists, and many others.

Gneiss. Gneisses are coarse-grained, coarsely banded metamorphic rocks (see Fig. 3–53). Their foliation results from the segregation of minerals into bands rich in quartz, feldspar, biotite, and amphibolites. Usually the ferromagnesian minerals are concentrated into darker bands, leaving feldspars and quartz as the principal components of the lighter-colored layers. Mica is usually present but is commonly not as conspicuous as in schists. There is also far less tendency for the rock to split along foliation planes as in schists. Gneiss may be formed from the severe or high-grade metamorphism of a variety of preexisting igneous, metamorphic, or sedimentary rocks, although high-silica igneous rocks and sandstones are the common parent materials.

Figure 3–56 Polished slab of white marble from Massa, Italy. (Courtesy of the Institute of Geological Sciences, London.)

Nonfoliated or weakly foliated metamorphic rocks

As is implied by their name, the nonfoliated metamorphic rocks show little, if any, preferred orientation of grains. Their occurrence is the result of the absence of directed pressures, as in the aureoles of igneous intrusions, and/or the prevalence of such minerals as quartz and calcite, which are rather equidimensional in form and thus unlikely to contribute to foliation.

Marble. Marble (Fig. 3 – 56) is a fine to coarsely crystalline rock composed of calcite or dolomite. It is a relatively soft rock and can be scratched easily with a steel nail. Limestones and dolostones are the parent rocks for marbles. The fossils often present in such carbonate sedimentary rocks are nearly always destroyed during the metamorphism that produces marble. In its purest form, marble may be snowy white, but many varieties are beautifully variegated (marbled) with colors resulting from impurities derived from the parent rock. Small amounts of iron oxide or hydroxide produce red, brown, and yellow hues, whereas greens are imparted by amphiboles, talc, and serpentine. Organic matter may tint marbles gray or black. It is the lack of foliation, the uniformly crystalline character, the beauty, and the relative softness of marbles that have caused them to be used by sculptors and architects down through the centuries.

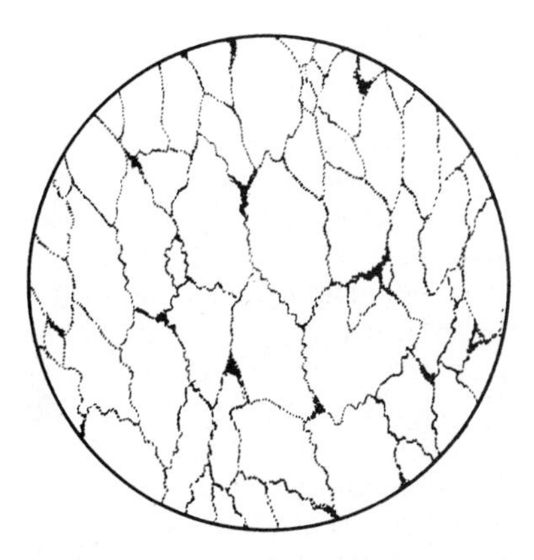

Figure 3–57 *Quartzite as seen in thin section under the microscope. The shapes of the original quartz grains have been obscured by the recrystallization which has resulted in the interlocking texture. Pressure has forced grains together so that contact between grains has become finely crenulate or sutured. Hematite (black color) is sparingly distributed along grain boundaries. (Width of field is 1.2 mm.)*

Quartzite. Quartzite is a fine-grained, often sugary-textured rock composed of intergrown quartz grains (Fig. 3 – 57). Quartzites generally lack pore space. They are very hard rocks, and a hefty blow of a rock hammer is needed to break them. If the broken fragments are closely scrutinized, it will be apparent that the rock was broken through, rather than around, the quartz grains. Quartzites are derived from quartz sands and sandstones. They can develop under either regional or contact metamorphic conditions.

Greenstone. Greenstones (Fig. 3 – 58) are dark-green rocks having a generally aphanitic texture interrupted by scattered dark porphyroblasts. The minerals largely responsible for the green in these rocks are chlorite, epidote, and actinolite. Greenstones are derived from

Figure 3–58 *Outcrop of greenstone showing well-developed pillow structure, near Jackfish, Ontario, along north shore of Lake Superior. (From Pye, E.G., Geology and scenery North Shore of Lake Superior, Ontario Department of Mines Guidebook No. 2, 1969, p. 4; reprinted with permission.)*

low-silica igneous rocks, such as basalts, that have been subjected to relatively low-grade metamorphism. Great elongate outcrops of these rocks occur in ancient igneous terrains and have come to be known as *greenstone belts*. Geologists speculate that these greenstone belts may be the metamorphosed relics of a basaltic crust that was later partially engulfed by intrusions of granitic magma.

Amphibolite. Amphibolites are dark-colored, medium- to coarse-grained metamorphic rocks. The two essential and most abundant minerals in amphibolites are hornblende and plagioclase. In some amphibolites, the longer axes of these minerals are in parallel arrangement (Fig. 3–59). More commonly, little or no parallel orientation is visible in hand specimens. Rocks rich in ferromagnesian minerals, such as basalt, are the usual parent rocks for amphibolites.

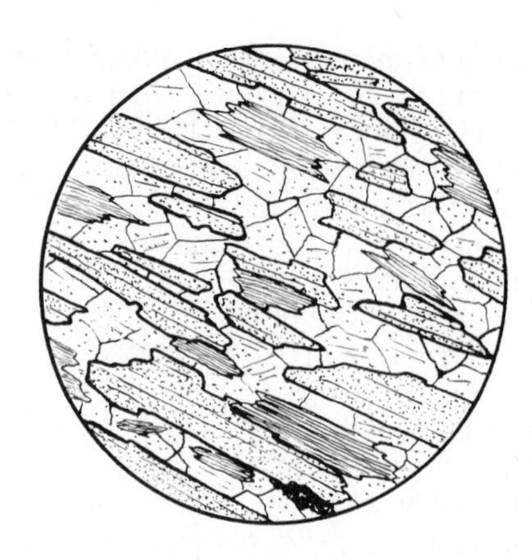

Figure 3–59 *Amphibolite from rocks of Precambrian age, Grand Canyon of the Colorado. The section has been cut parallel to lineation. The heavily outlined darkly stippled grains are hornblende. Grains having line pattern are biotite. Lightly stippled grains are plagioclase, and clear grains are quartz.*

Summary

The weathering of igneous and other rocks through chemical and mechanical action produces gravel, sand, silt, clay, and solutions of calcium, sodium, potassium, magnesium, and iron. These products, after they have been deposited in the sea or in continental environments, are bound together by rock lithification processes, such as cementation, compaction, and crystallization, to form sandstones, shales, limestones, and evaporites.

Sedimentary rocks represent the material record of environments that once existed on the earth. Their composition and texture offer clues to the nature of the source areas that provided the sediment, the kinds of transporting media, and the general environment of deposition. Fossils in sedimentary rocks are particularly instructive as environmental indicators. They tell us the age of strata and whether they are marine or nonmarine, whether the water at the depositional site was deep or shallow, and whether the climate was warm or cool. From the size and shape of grains in clastic rocks, geologists can sometimes infer whether the sediment was deposited in quiet water or turbulent currents, whether it was sorted by wind or wave action, and whether it experienced many or few cycles of erosion and deposition. Similar kinds of information can sometimes be elucidated by the study of such sedimentary features as graded bedding, cross-bedding, and current ripple marks. Sandstones are particularly useful in investigations of ancient depositional environments. Quartz sandstones, graywackes, arkoses, and subgraywackes accumulate in particular paleogeographic and tectonic situations.

Geologists usually divide successions of sedimentary rocks into rock units that are sufficiently distinctive in color, texture, or composition to be recognized easily and mapped. Such rock units are called formations.

Metamorphism is a process by which previously existing rocks of any kind undergo mineralogical and textural changes in the solid state in response to changing chemical and physical conditions.

The term metamorphism is not intended to include changes associated with weathering or the ordinary lithification of sedimentary rocks. Rather, the metamorphism of rocks involves recrystallization, chemical change, and mechanical deformation. In high-pressure metamorphic environments, atoms in older minerals tend to regroup into denser, more compact minerals that are stable under the new conditions. Particular minerals will form only at very high pressures and temperatures; others form at moderate or low pressures or temperatures. Thus, metamorphic minerals can be used to infer the conditions responsible for producing specific metamorphic rocks. Those minerals that are particularly useful in this way are called metamorphic index minerals.

The agents of metamorphism include heat, pressure, and chemically active solutions. Certain chemical reactions involved in the recrystallization of minerals occur only at specific temperatures. For this reason, and also because it accelerates chemical reactions, heat plays an essential role in metamorphism.

The enormous pressures associated with mountain building or deep burial can result in plastic deformation of rocks, pulverize mineral components, and promote the growth of denser minerals. The texture developed by a metamorphic rock may result from the manner in which pressure has been applied to the parent rock. If it is applied uniformly to all sides, it is not likely to cause a preferred alignment of growing crystals. Directed pressure promotes the growth of denser minerals in the direction of least stress and results in a layering of minerals that is called foliation.

Chemically potent solutions are the great promoters of metamorphism. Even a small amount of water facilitates the movement of ions to appropriate sites in the atomic structures of metamorphic minerals.

The two major kinds of metamorphism are contact and regional. As implied by its name, contact metamorphism occurs around the margins of magmatic bodies. Beginning at the contact, progressively less intense zones of metamorphism may develop, resulting in a metamorphic aureole.

In contrast to contact metamorphism, which is relatively local, regional metamorphism may occur across thousands of square kilometers. Metamorphic rocks in these regional belts are well foliated and divisible into distinct metamorphic zones, or facies, characterized by minerals that developed in response to particular conditions of pressure and temperature.

QUESTIONS FOR REVIEW

1 What characteristics of sedimentary rocks are most useful in their classification and identification?

2 What are some of the changes that take place in a sediment after deposition that lead ultimately to lithification?

3 Explain the distinction between the following:
- **a** Clastic and nonclastic (chemical) sedimentary rocks.
- **b** Conglomerate and breccia.
- **c** Sandstone and clastic limestone.
- **d** Mature and immature sandstones.

4 What interpretations about the history of a sedimentary rock can sometimes be obtained from an examination of the following:
- **a** Degree of grain rounding.
- **b** Cross-bedding.
- **c** Grain size.

5 How does matrix in a rock differ from cement? What kinds of cement bind sedimentary rock together?

6 Compare the probable tectonic environments in which a graywacke and a quartz sandstone might be deposited.

7 Why is fine-grained sediment often deposited farther from a shoreline than coarse-grained sediment?

8 A geologist studying a sequence of strata discovers that limestone is overlain by shale, which in turn is overlain by sandstone. What might this signify with regard to the *advance* or *retreat* of a shoreline?

9 What features or properties of sedimentary rocks indicate relatively shallow-water deposition?

10 What are turbidity currents? How might turbidity current deposits be recognized?

11 Are carbonate sediments more likely to accumulate in warmer or colder parts of the ocean? Why?

12 What are the major causes or agents of metamorphism? Do these agents operate independently or are they interrelated?

13 Why does the presence of even small amounts of water generally increase the rate and amount of metamorphism in rocks already subjected to high temperature and pressure?

14 What is slaty cleavage? How does the origin of slaty cleavage differ from the origin of fissility in shale?

15 A metamorphic aureole is composed of three successive shells or zones. One zone contains schists rich in sillimanite and kyanite, another contains chlorite schists, and the third contains well-developed garnet crystals and hornblende. Which zone is located adjacent to the intrusion? Which is on the periphery of the aureole?

16 What factors influence the amount of alteration resulting from contact metamorphism?

SUPPLEMENTAL READINGS AND REFERENCES

Blatt, H., 1982. Sedimentary Petrology. San Francisco, W.H. Freeman & Co.

Ehlers, E.G., and Blatt, H., 1982. Petrology: Igneous, Sedimentary, and Metamorphic. San Francisco, W.H. Freeman & Co.

Friedman, G.M., and Sanders, J.E., 1978. Principles of Sedimentology. New York, John Wiley & Sons.

Krumbein, W.C., and Sloss, L.L., 1963. Stratigraphy and Sedimentation. (2nd ed.). San Francisco, W.H. Freeman & Co.

Leeder, M.R., 1982. Sedimentology. London, George Allen & Unwin.

Matthews, R.K., 1984. Dynamic Stratigraphy. Englewood Cliffs, N.J., Prentice-Hall, Inc. (2nd ed.).

Pettijohn, F.J., 1975. Sedimentary Rocks. New York, Harper & Row (3rd ed.).

Selley, R.C., 1978. Ancient Sedimentary Environments. Ithaca, N.Y., Cornell University Press (2nd ed.).

Winkler, H.G.F., 1979. Petrogenesis of Metamorphic Rocks. New York, Springer-Verlag (5th ed.).

4

earthquakes and earth structure

. . . The strong-based promontory
Have I made shake, and by the spurs pluck'd up
The pine and cedar.

SHAKESPEARE
The Tempest

Prelude In this age of artificial satellites and spacecraft, our attention is often directed toward the remarkable discoveries resulting from the exploration of outer space. We tend to forget that there is still much to learn about the "inner space" of our own planet. This is particularly true for that part of the earth that lies beneath the crust. Our deepest wells penetrate only about 6 of the 6300 kilometers that separate us from the center of the planet. Basaltic lavas sometimes provide a glimpse of materials that originated 50 or 60 kilometers below the surface, and some diamond-bearing kimberlites may have risen from depths of 200 km. For the most part, however, our knowledge of the earth's insides has been indirectly inferred from the study of earthquake waves (seismic waves). In this chapter, we shall examine the cause and effects of earthquakes and review what they tell us about the hidden interior of our planet.

Collapsed overpass connecting Foothill Boulevard and Golden State Freeway. Collapse was a consequence of the San Fernando, California, earthquake of February 9, 1971. (Courtesy of U.S. Geological Survey.)

WHEN THE EARTH SHAKES

We often think of the earth as a firm and stable planet. Actually, both its surface and its interior are in dynamic, changing states. The evidence of our planet's energetic condition is amply provided by the more than 1 million earthquakes that jostle its surface each year. At least 50 of these cause loss of life and significant damage to property (Figs. 4–1, 4–2). A fewer number of giant earthquakes each year are terrifying catastrophes. The lives lost from earthquakes over the past 500 years number in the millions and far exceed the loss of lives from volcanic eruptions. What happens during an earthquake? What does it feel like to be in one? Perhaps the best way to answer these questions is to describe the nightmarish events of two noteworthy earthquakes.

The Good Friday earthquake, Alaska, 1964

It was 5:36 in the afternoon on March 27th in Anchorage, Alaska. Except for a few snowflakes that floated down from partly cloudy skies, the weather had been pleasant, and there was a general mood of geniality as Alaskans looked forward to their Easter holiday and the coming of spring. A housewife was preparing dinner while her two children played in the neighborhood park across the street. Robert B. Atwood, editor of the Anchorage *Daily Times*, was relaxing at home by improving his skill at playing the trumpet. In a real estate office nearby, a new resident of the city was preparing to purchase a home in a tract called Turnagain Heights. Moments later the newcomer and the other unsuspecting citizens of Anchorage were jarred by the strongest earthquake ever recorded in North America. In the home of the young housewife, dishes rattled and walls began to sway at awkward angles. Terrified, she rushed out of the house to retrieve her two children, but fell headlong into a crack that had unexpectedly opened in the loose sliding soil. She was certain she would lose her life in the next moment, but the floor of the trench heaved upward so that her children and a neighbor were able to bring her to safety.

Across town, Robert Atwood ran outside his shaking house and watched in panic as his driveway and yard broke into large, dizzily sliding blocks. He tried to leap from block to block, but a fissure opened beneath him, and he tumbled downward. Sand and clay rained down upon him. He tried to free himself but found his right arm anchored in the sand. Sud-

Figure 4–1 Earthquake damage to the Government Hill School near Anchorage, Alaska. During 1964 this area experienced chaotic surficial faulting and sliding. The south wing of the building, shown here, collapsed into a down faulted block. (Courtesy of U.S. Geological Survey.)

Figure 4–2 *Rails twisted away from their bed as underlying soil shifted laterally during 1965 Alaskan earthquake. (Courtesy of U.S. Geological Survey.)*

denly he realized that his hand still clutched the trumpet he had been playing moments before. He released his hold on the instrument and scrambled out of the ditch. Stunned, he watched his neighbor's house collapse and slide into a fissure. Not far away, the customer for a new home in Turnagain Heights watched with disbelief as the new home he had hoped to purchase slid down the hillside and splintered into a mass of ruin.

The Good Friday earthquake was centered not in Anchorage but rather about 130 km southeast of the city beneath the frigid waters of Prince William Sound (Fig. 4–3). Other cities, such as Valdez and Seward, were also damaged, although not as severely. The greater damage at Anchorage can be attributed to the fact that the city is built upon a thick blanket of loose glacial sediment — gravel, sand, and clay. These unstable materials are weakly held together by frozen water. Beneath the glacial deposits, one finds a thick bed of weak plastic clay known to geologists as the Bootlegger Cove Clay. When the tremendous shaking began, the Bootlegger Cove Clay began to flow downward toward Cook Inlet, causing great cracks and landslides in the overlying glacial material. Indeed, most of the damage to Anchorage was either directly or indirectly due to slumps, slides, and settling of the ground.

The 1906 San Francisco earthquake

The most infamous of American earthquakes struck San Francisco at 5:12 on the morning of April 18, 1906.

Citizens awoke in stark terror amid the awful roar of collapsing buildings and clanging of jostled church bells. For many, crushed by collapsing roofs and masonry that crashed through ceilings, their conscious moments were brief. Those who managed to make their way out into the open and away from falling objects recalled that the earthquake, which had begun with relatively small movement, increased to a jarring crescendo at the end of about 40 seconds. The shocks abruptly stopped for a few moments and then struck again even more violently for about half a minute. This main shock was then followed by smaller earthquakes, termed aftershocks. Although the destruction to buildings by shaking was extensive, the main damage came from fire, as gas from broken gas mains was ignited. Water lines were snapped by the earth movements, and firemen could not effectively battle the conflagration that raged through the city. For three days, however, firemen and volunteers worked with wet mops, dynamite, and shovels until they had finally put out the fires. Slowly, the thousands that had fled

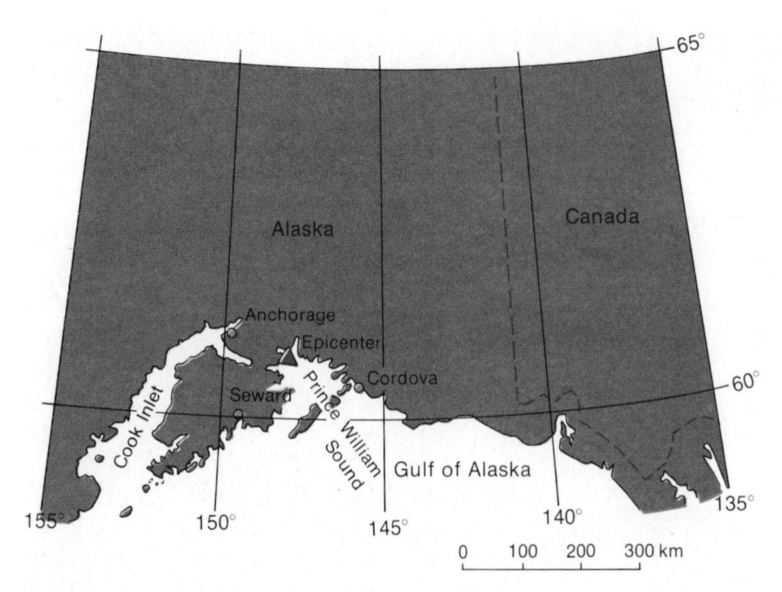

Figure 4–3 Approximate location of epicenter of the Alaskan earthquake of 1964.

the city returned. It was then apparent that the parts of the city that had experienced the greatest destruction from the earthquake itself had been built upon loose soils used to fill ravines and swampy areas.

The cause of the San Francisco earthquake was sudden movement along the notorious San Andreas fault (Figs. 4–4, 4–5). This great scar in the earth's crust can be traced nearly 1000 km from Cape Mendocino to the Gulf of California. A fault is simply a rock fracture along which movement had occurred, and the San Andreas fault, along with adjoining networks of faults, is responsible for most of the earthquakes felt in California. The damage to San Francisco (as well as San Jose, Santa Rosa, and Stanford University) during the 1906 quake was primarily the result of horizontal displacement along the San Andreas fault. The amount of horizontal movement varied from place to place up to a maximum of 6.5 meters, most of which occurred as the oceanward side of the fault was jolted northwestward.

In the Introduction, we alluded to the theory of plate tectonics, in which the earth's outer shell is believed to be composed of moving rigid plates. Where such plates slip past one another, faulting can be expected. The San Andreas fault is one such plate boundary (Fig. 4–6). It marks the boundary between the North American and the Pacific plates. The Pacific Plate carries a splinter of North America with it as it creeps northwestward at a rate of about 7.5 centimeters a year (relative to the main body of the continent). Thus, if movement continues for the next 20 million years, Los Angeles will have traveled past San Francisco. Californians are well aware that movement along the San Andreas is neither continuous nor smooth. The masses of rock on either side of the fault tend to lock and then slip suddenly, thereby releasing the stress that has accumulated and sending severe jolts throughout the land.

THE CAUSE OF EARTHQUAKES

Stress, strain, and rupture

Children of the Wanyamwasi tribe of Africa learn from their elders that the earth is a great disc supported on one side by a lofty mountain and on the other side by an amorous giant. The giant's lovely wife stands beside him holding up the sky. Occasionally, the giant gets the notion to hug his wife, thereby causing the momentarily neglected earth to totter. Thus, a romantic impulse, say the Wanyamwasi, can cause an earthquake.

The Wanyamwasi legend is only one expression of the centuries-old search by peoples of every culture for the cause of earthquakes. Today, of course, most people would say earthquakes result when rocks break or there are sudden movements along "locked seg-

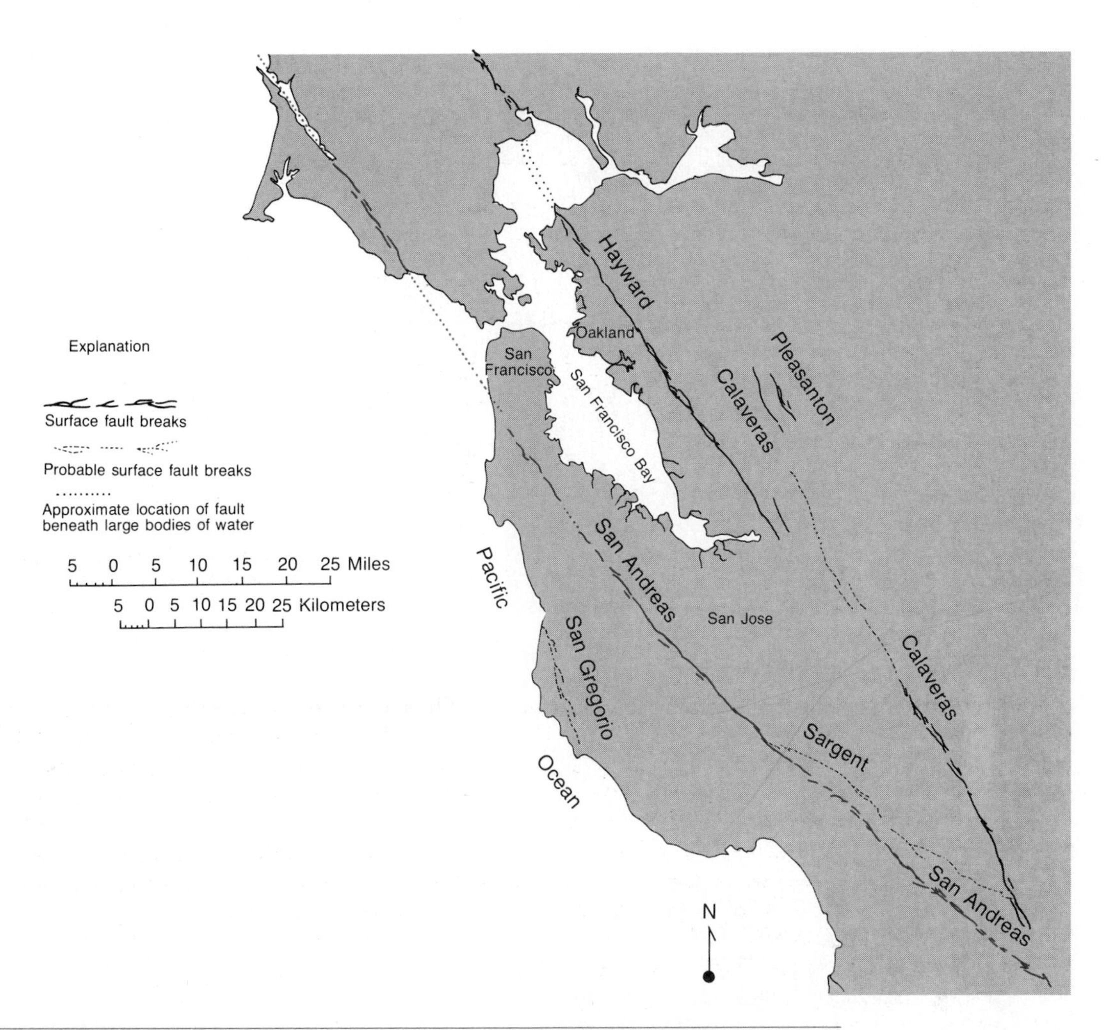

Figure 4–4 *The San Andreas and other historically active faults in the San Francisco Bay region.*

ments of faults." But what determines when a rock will break or rupture? Are the forces that break rocks applied instantaneously or gradually over a long period of time? How does a rock resist the forces that tend to break or distort it?

In regions where earthquakes occur, rocks are being acted upon by enormous forces. Geologists refer to these as **stress,** defined as the force per unit area applied to any given body of rock to either compress it (compressional stress) or pull it apart (tensional stress). The effect of stress on a rock mass is to cause **strain,** which is the actual change in shape or volume of the rock that has experienced stress. Rocks can withstand certain amounts of stress without breaking or distorting, and this ability is a measure of their **strength.** Thus, strength can be considered the stress level at which the rock either ruptures or deforms.

The way rocks behave when subjected to stress is illustrated in Figure 4–7. Initially, while the stress is just beginning to build up, the rock may adapt by deforming elastically. Elastic deformation means that if the stress were removed, the rock mass would recover its original shape. In the region of the graph where elastic deformation occurs, the amount of strain produced is directly proportional to the amount of stress applied. As the stress is increased, a point called

Figure 4–5 The surface trace of the San Andreas fault. The mountains on the left are the Temblor range which form the western border of the southern part of the San Joaquin Valley. (Courtesy of U.S. Geological Survey and J.R. Balsey.)

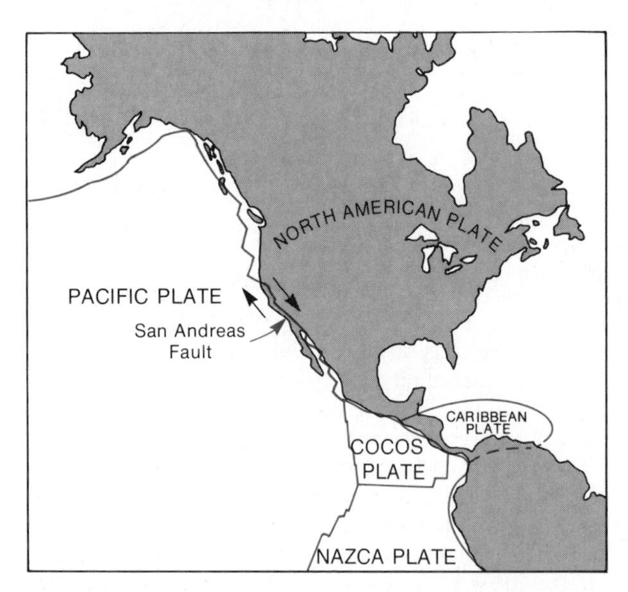

Figure 4–6 The San Andreas fault in California is a transform fault along which the Pacific Plate slides northwestward relative to the North American Plate at a rate of about 7½ cm per year.

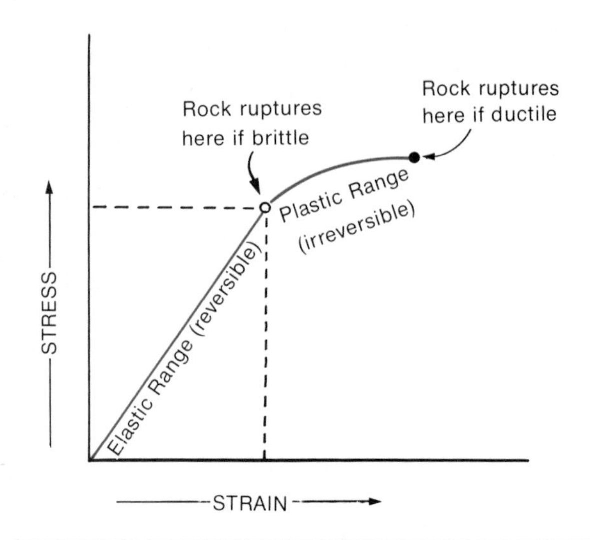

Figure 4–7 Graph of stress vs. strain on hypothetical earth materials.

the **elastic limit** is reached. Beyond the elastic limit, the rock does not recover its original shape after removal of the stress, but rather experiences permanent distortion. The leveling off of the curve indicates that small increases in stress beyond the elastic limit will now cause relatively greater deformation. Finally, the rock will no longer be able to adapt to the stress directed against it, and it will break. In brittle rocks, breakage occurs almost at the elastic limit, whereas in ductile rocks, considerable plastic deformation may occur prior to breaking or **rupture.** At the moment the rock breaks, it rebounds into positions that relieve the stresses. This movement generates the shocks we recognize as earthquakes.

Elastic rebound

The mechanism involved in **elastic rebound** was originally described by the California geologist H.F. Reid, who based his model on surveys obtained for land adjacent to the San Andreas fault both before and after the 1906 earthquake. In some ways elastic rebound resembles what happens when one slowly bends a wooden yardstick. The energy applied to bending the stick is stored within it as the wood responds to the stress by bending elastically. When the elastic limit of the wood is finally exceeded, it breaks and the splintered ends whip violently back and forth.

Figure 4–8 depicts a more pertinent example of the elastic rebound mechanism. Block A shows a tract

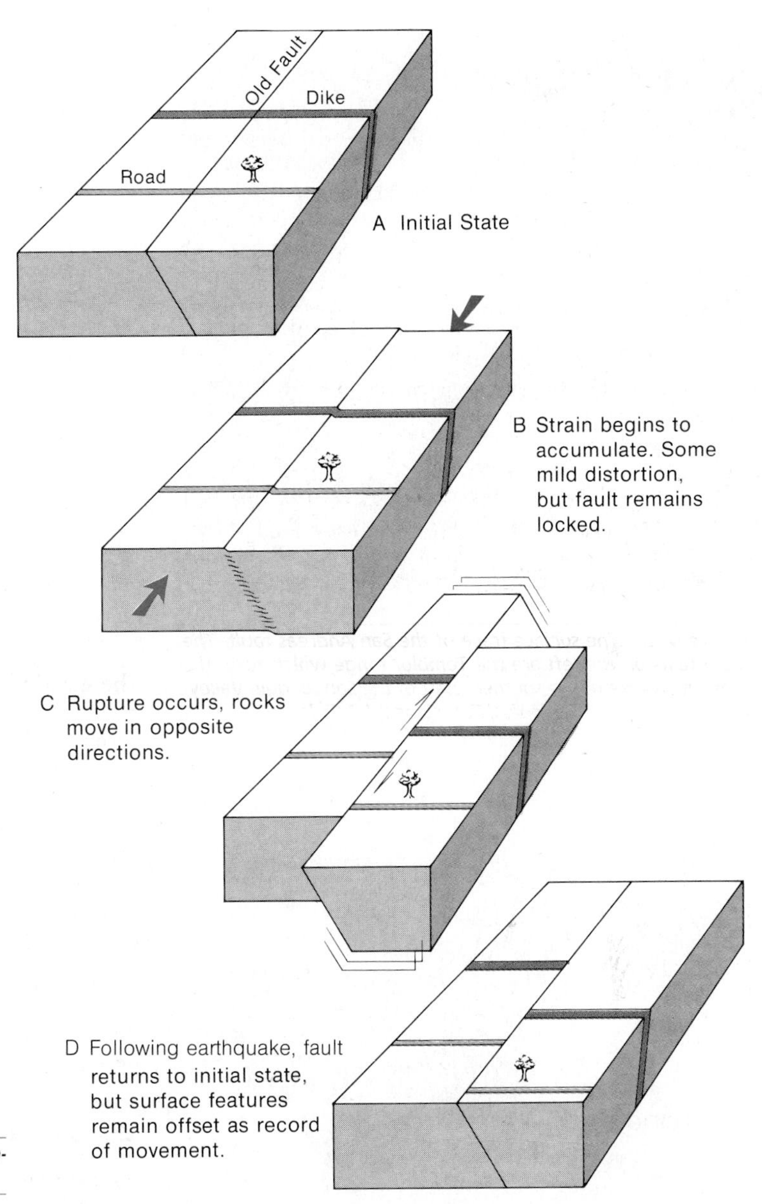

Figure 4–8 *Model of the elastic rebound mechanism for earthquakes.*

of land that has a fault, but there has been no movement along that fault for some time. A road and fence have been constructed across the fault and provide convenient landmarks for measuring displacement. Gradually, stress begins to build up along the fault, and the rocks adapt by bending slightly (Block B). For a while, because of friction or cementation, the opposite sides of the fault remain locked together. They may begin to bend by amounts measurable with instruments called strain gauges. Eventually, so much stress has accumulated that rupture occurs (Block C), and the rocks are moved violently in opposite directions. The sudden jolt is like a giant hammer blow to the earth and starts a series of vibrations, or earthquake waves, that travel outward in all directions. These vibrations are the earthquakes we feel underfoot or detect with instruments. The release of stress at one place may cause rupture at other places along a large fault, so earthquake activity may continue for days or weeks following the main shock. An existing fault is not always a requirement for an earthquake. Stresses may build up in unfaulted rocks, and earthquakes may then be generated with initial faulting. Most earthquakes, however, do occur along existing faults. Deep-focus earthquakes may result from sporadic slippage in rocks that are simultaneously undergoing plastic deformation.

SEISMOGRAPHS AND EARTHQUAKE WAVES

Anyone who was gone through the harrowing experience of a major earthquake will recall sensing the confusion of side-to-side and up-and-down vibrations. However, it is difficult to differentiate particular kinds of movements within the jumble of jerks and jolts. Actually, the various movements are manifestations of different kinds of elastic waves that travel through and around the earth from the point where rock breakage or slippage occurred. With the help of seismographs (Fig. 4-9), it is possible to isolate the major kinds of elastic wave motion involved in earthquakes.

Recording earthquakes

Seismographs are constructed with weights suspended from springs. The springs dampen earth movements and thereby allow the weights to remain fixed in space. For recording vertical movements, the seismograph weight is suspended by springs from an overhead support (Fig. 4-10A). Horizontal movements are recorded by seismographs containing a horizontal pendulum having a weight on a rigid arm supported by a wire. The pendulum is freely attached to a socket joint from the supporting column (Fig. 4-10B). In either type of seismograph, the weight tends to remain at rest while the support vibrates with the earth. The difference in motion can then be recorded and measured in many different ways. One simple device employs a delicate pen attached to the weight. The pen traces a continuous line on a revolving drum that is itself fixed to the supporting framework of the instrument. The drum turns as a clock so that the arrival times of a vibration can be determined. Of course, the actual seismographs placed in earthquake-monitoring stations are far more complex than the simple instruments illustrated in Figure 4-10A and B. Rather than a pen for recording, tiny rays of light are beamed onto photographic paper. The ray is reflected off a series of mirrors to amplify even very small mo-

Figure 4-9 Interior of a modern seismograph station. The pier on the left holds three short-period seismometers (rear corner) driving galvanometers which reflect onto the drum in the foreground. Long-period seismometers and galvanometers are on the pier to the right. Both piers are isolated from the floor of the vault. (Photograph courtesy of the Earth Physics Branch, Department of Energy, Mines, and Resources, Canada.)

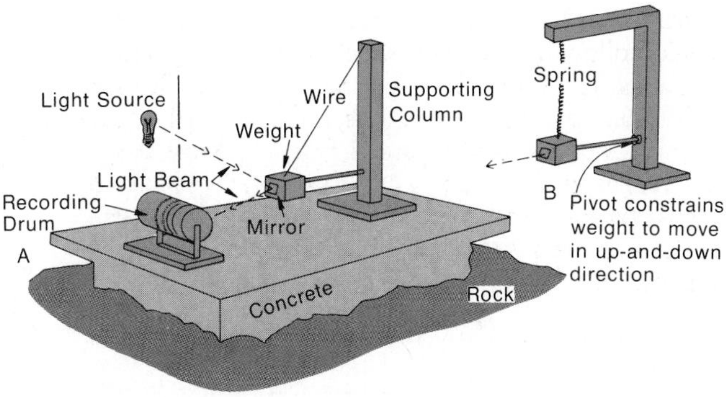

Figure 4–10 (A) Sketch of a seismograph that is recording horizontal earth motion. A light spot on the boom moves across photographic paper on the recording drum as the boom oscillates. (Springs are omitted for clarity.) By reorienting the drum and modifying apparatus as shown in B, vertical ground motion may be recorded.

tions of the ground. Other devices serve to dampen the natural swinging motion of the pendulum and filter out unrelated background motions. To analyze earthquakes fully, at least three seismographs must be in simultaneous operation.

Earthquake intensity and magnitude

Soon after an earthquake has been reported, people want to know something about its intensity. Therefore, attempts have been made to formulate standard scales of intensity and magnitude. **Magnitude** describes the total amount of energy released during an earthquake, whereas **intensity** is a measure of the effect of the earthquake on structures and people. Intensity scales are qualitative. They record the observable effects of earthquakes as ranging from slight tremors detected by only a few people (ranked I on a scale of I to XII) to total destruction (XII on the intensity scale). A widely used standard for judging earthquake intensity is the modified Mercalli Scale provided in Table 4-1.

Intensity scales have rather obvious limitations. They cannot be used in uninhabited areas because there would be no structures subjected to earthquake damage. Also, the intensity assigned to an area is likely to be influenced by the psychology of the inhabitants, the quality of construction, and geological conditions. People living far from the earthquake center would perceive the earthquake as of low intensity, whereas those nearby would consider it of higher intensity. The intensity scale does not adequately measure the force involved at the source of the earthquake.

In an attempt to remedy the problems associated with intensity scales, C.F. Richter devised a system that measures the magnitude of an earthquake in terms of the motion recorded by a seismograph of certain specifications at a standard distance (100 km) from the earthquake source. Calculations permit one to determine the equivalent responses of other seismographs at any distance. In this quantitative scale, one measures the seismic wave amplitude released by the shock rather than the intensity or degree of destructiveness. The Richter Scale has a logarithmic basis, so an increase in one whole number corresponds to an earthquake 10 times stronger than one indicated by the next lower number. Thus, magnitude 7 represents an energy release about 10 times that of magnitude 6 and about 100 times that of magnitude 5. The 8.5 magnitude of the 1964 Alaskan earthquake was the greatest ever recorded for an earthquake in North America. An earthquake rated at 2.5 on the Richter Scale would hardly be noticed, but one of magnitude 4.5 may cause local damage. An earthquake with magnitude larger than 8.9 has never been recorded, and this suggests that there is a limit to which rocks can accumulate strain energy before they break or slip.

Seismic waves

From the study of seismograms (the records of seismographs), geologists have recognized that earthquakes move through the earth as waves. Although there are several different kinds of waves, the three that are of most importance are *primary*, *secondary*, and *surface* waves. They are defined by describing the motion of a "particle" of rock that lies in the path of the wave.

Primary waves

Primary waves take their name from the fact that they are the speediest of the three kinds of earthquake

TABLE 4-1 **MODIFIED MERCALLI SCALE OF EARTHQUAKE INTENSITY***

I. Not felt except by a very few persons under especially favorable circumstances.

II. Felt only by a few persons at rest, especially on upper floors of buildings. Delicately suspended objects may swing.

III. Felt quite noticeably indoors, especially on upper floors, but many people do not recognize it as an earthquake. Standing motor cars may rock slightly. Vibration like passing truck.

IV. During the day, felt indoors by many, outdoors by few. At night, some awakened. Dishes, windows, doors disturbed; walls make creaking sound. Sensation like heavy truck striking building. Standing motor cars rocked noticeably.

V. Felt by nearly everyone; many awakened. Some dishes, windows, etc., broken; a few instances of cracked plaster; unstable objects overturned. Disturbances of trees, poles, and other tall objects sometimes noticed. Pendulum clocks may stop.

VI. Felt by all, many frightened and run outdoors. Some heavy furniture moved; a few instances of fallen plaster or damaged chimneys. Damage slight.

VII. Everybody runs outdoors. Damage negligible in buildings of good design and construction; slight to moderate in well-built ordinary structures; considerable in poorly built or badly designed structures; some chimneys broken. Noticed by persons driving cars.

VIII. Damage slight in specially designed structures; considerable in ordinary substantial buildings, with partial collapse; great in poorly built structures. Panel walls thrown out of frame structures. Fall of chimneys, factory stacks, columns, monuments, walls. Heavy furniture overturned. Sand and mud ejected in small amounts. Changes in well-water levels. Disturbs persons driving motor cars.

IX. Damage considerable in specially designed structures; well-designed frame structures thrown out of plumb; damage great in substantial buildings, with partial collapse. Buildings shifted off foundations. Ground cracked conspicuously. Underground pipes broken.

X. Some well-built wooden structures destroyed; most masonry and frame structures destroyed with foundations; ground badly cracked. Rails bent. Landslides considerable from river banks and steep slopes. Shifted sand and mud. Water splashed over banks.

XI. Few, if any, masonry structures remain standing. Bridges destroyed. Broad fissures in ground. Underground pipelines completely out of service. Earth slumps and land slips in soft ground. Rails bent greatly.

XII. Damage total. Waves seen on ground surfaces. Lines of sight and level distorted. Objects thrown upward into the air.

* Modified from Richter, C.F., 1958. *Elementary Seismology*. San Francisco and London, W.H. Freeman Co.

waves and therefore the first to arrive at a seismograph station after there has been an earthquake (Fig. 4-11). They travel through the upper crust of the earth at speeds of 4 to 5 km per second. Near the base of the crust they speed along at 6 or 7 km per second. In these primary waves (also called P-waves for brevity), pulses of energy are transmitted in such a way that the movement of rock particles is parallel to the direction of propagation of the wave itself. Thus, a given particle of rock set in motion during an earthquake is driven into its neighbor and bounces back. The neighbor strikes the next particle and rebounds, and subsequent particles continue the motion. The result is a series of alternate compressions and expansions that speed away from the source of shock. Thus, P-waves are similar to sound waves in that they are *longitudinal* and travel by compression and rarefaction (Fig. 4-12). It is an accordion-like "push-pull" movement that can be transmitted through solids, liquids, and gases. Of course, the speed of P-wave transmission will differ in materials of different density and elastic properties. They tend to die out with increasing distance from the earthquake source and will echo or reflect off rock masses of differing physical properties.

Secondary waves

Another part of the energy released at an earthquake source is carried away by slower-moving waves called **secondary waves.** These waves are also called shear, transverse, or S-waves. They travel 1 or 2 km per second slower than do P-waves. The movement of rock particles in secondary waves is at right angles to the direction of propagation of the energy (Fig. 4-13). A demonstration of this type of wave is easily managed by tying a length of rope to a hook and then shaking the free end. A series of undulations will develop in the rope and move toward the hook — that is, in the direction of propagation. Any given particle or point along the rope, however, will move up and down in a

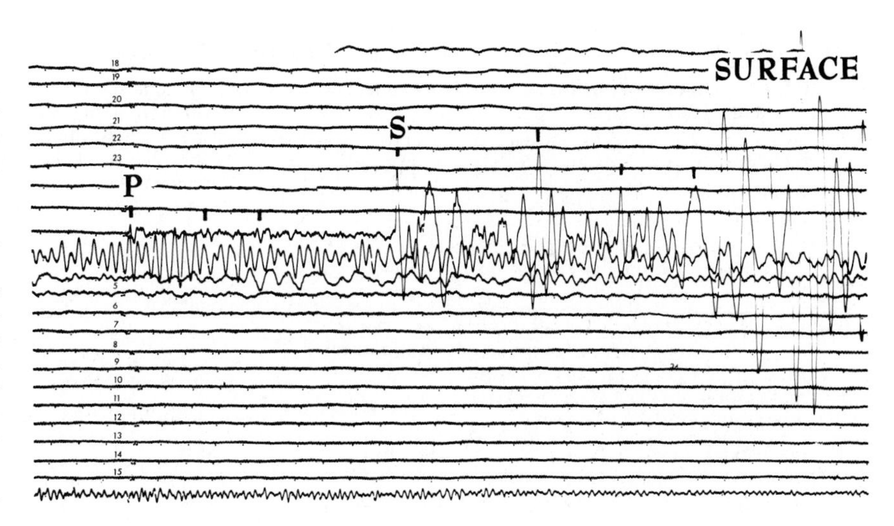

Figure 4–11 *Record of a magnitude 6 earthquake that occurred in Turkey, March 28, 1964 and was recorded by the vertical component seismograph located in northwestern Canada. Time increases from left to right. Small tick marks represent one-minute intervals. P means primary waves, S indicates secondary waves. Notice that the first S-waves arrive about ten minutes after the arrival of the first P-waves. (Courtesy of the Earth Physics Branch, Department of Energy, Mines, and Resources, Canada.)*

direction perpendicular to the direction of propagation. It is because of their more complex motion that S-waves travel slower than P-waves. They are the second group of oscillations to appear on the seismogram (see Figure 4–11). Unlike P-waves, secondary waves will not pass through liquids or gases.

Both P- and S-waves are sometimes also termed *body waves* because they are able to penetrate deep into the interior or body of our planet. Body waves travel faster in rocks of greater elasticity, and their speeds therefore increase steadily as they move downward into more elastic zones of the earth's interior and then decrease as they begin to make their ascent toward the earth's surface. The change in velocity that occurs as body waves invade rocks of different elasticity results in a bending or refraction of the wave. The many small refractions cause the body waves to assume a curved travel path through the earth (Fig. 4–14).

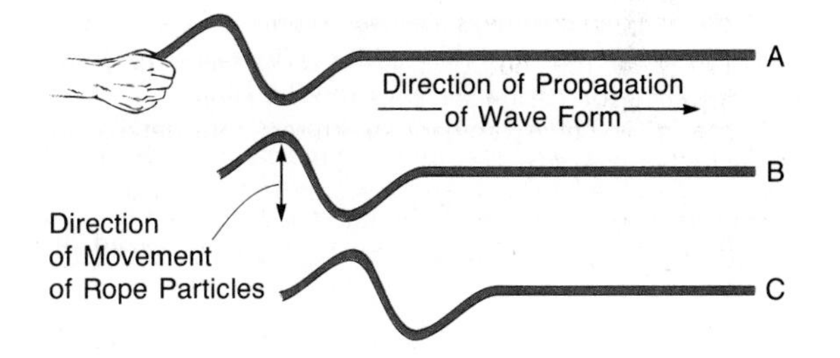

Figure 4–13 *Analogy of propagation of S-waves by displacement of a rope. In rocks, as in this rope, particle movement is at right angles to the direction of propagation of the wave. A, B, and C show the displacement of the crest from left to right at successive increments of time.*

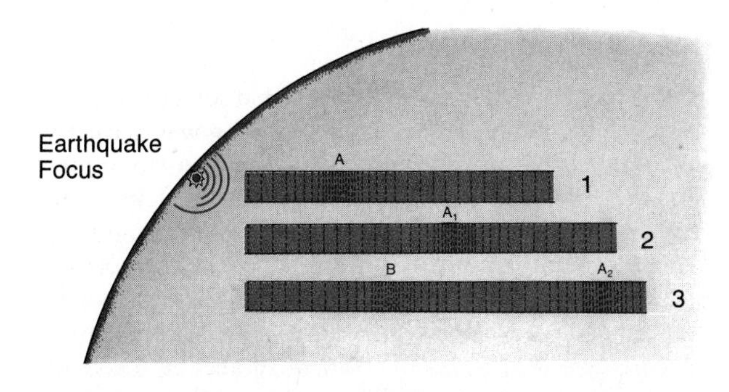

Figure 4–12 *Movement of primary wave. In 1, the compression has moved the particles closer together at A. In 2, the zone of compression has moved to A_1. In 3, the zone of compression has moved to A_2 and a second compressional zone (B) has moved in from the left.*

Not only are body waves subjected to refraction, but they may also be partially reflected off the surface of a dense rock layer in much the same way as light is reflected off a polished surface. Many factors influence the behavior of body waves. An increase in the temperature of rocks through which body waves are traveling will cause a decrease in velocity, whereas an increase in confining pressure will cause a

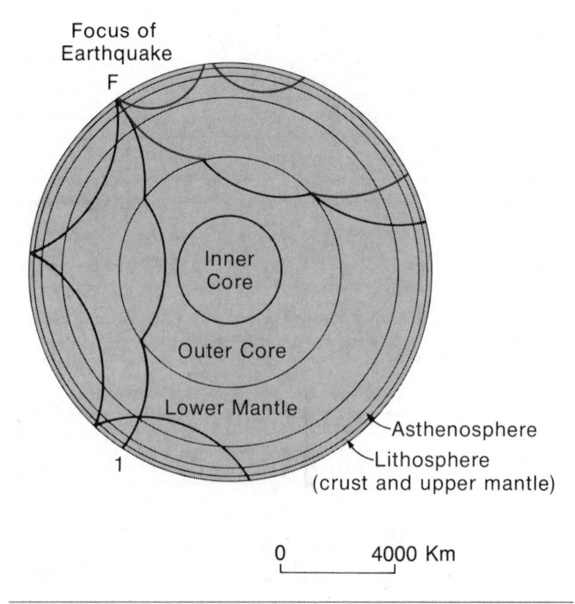

Figure 4–14 Cross-section of the earth showing paths of some earthquake waves. P-waves (black) that penetrate to the core are sharply refracted as shown in the path from F to 1. S-waves (orange) end at the core, although they may be converted to P-waves, traverse the core, and emerge in the mantle again as P- and S-waves. Both P- and S-waves may also be reflected back into the earth again at the surface. (From U.S. Geological Survey publication, The Interior of the Earth.*)*

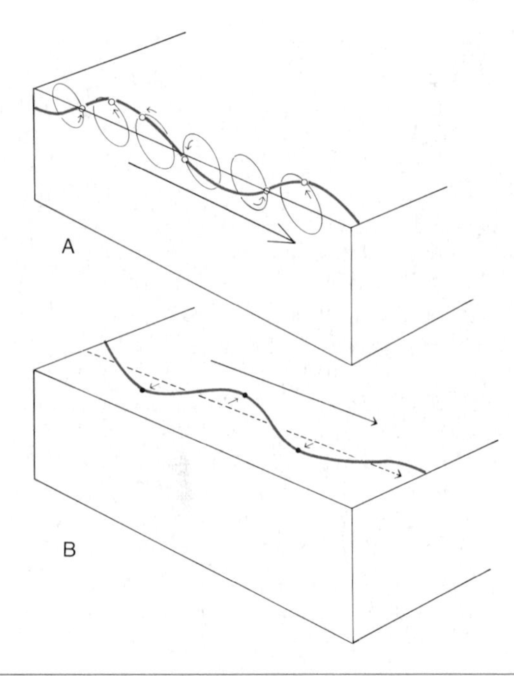

Figure 4–15 (A) Elliptical particle motion in Rayleigh surface waves (ground surface shown without displacement). (B) Particle motion in Love wave is in horizontal plane and perpendicular to direction of wave propagation.

corresponding increase in wave velocity. As mentioned earlier, in a fluid where no rigidity exists, S-waves cannot propagate, and P-waves are markedly slowed.

Surface waves

Surface waves are large-motion waves that travel through the outer crust of the earth. Their pattern of movement resembles that of waves caused when a pebble is tossed into the center of a pond. They develop whenever P- or S-waves disturb the surface of the earth as they emerge from the interior. There are actually several different types of motion in surface waves. *Rayleigh surface waves,* for example, have an elliptical motion that is opposite in direction to that of propagation, whereas *Love surface waves* vibrate horizontally and perpendicular to wave propagation (Fig. 4–15). Surface waves are the last to arrive at a seismograph station. They are usually the primary cause of the destruction that can result from earthquakes affecting densely populated areas. This destruction results because surface waves are channeled through the thin outer region of the earth, and their energy is less rapidly dissipated into the large volumes of rock traversed by body waves. Indeed, surface waves may circle the earth several times before friction causes them to fade.

Locating earthquakes

Epicenter and focus

The point within the earth at which an earthquake disturbance is initiated is called the **focus.** Directly over the focus on the surface of the earth is a point, specified by latitude and longitude, that is termed the **epicenter** (Fig. 4–16). The depth to the focus may vary from nearly 0 to more than 600 km. If the depth of focus is less than about 8 km, the earthquake is usually not felt at any appreciable distance from the epicenter. Deep-focus earthquakes, however, are detected at great distances from their epicenters. Seismologists have developed a classification of earthquakes according to their depth of focus. A *shallow-focus* earthquake has its focus at a depth between 0 and 70 km. Foci that lie between 71 and 300 km define *intermediate-focus*

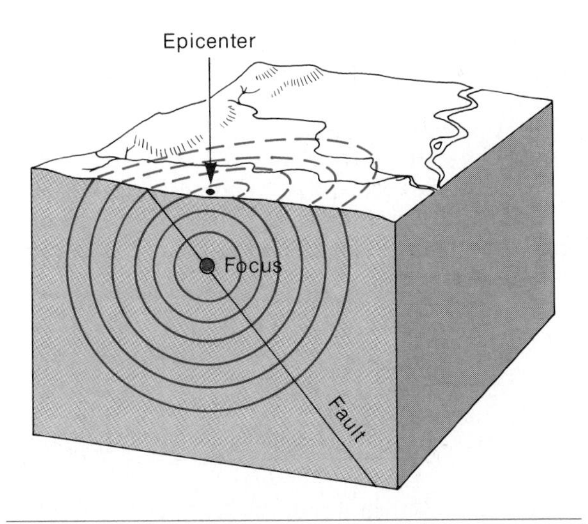

Figure 4–16 *The relation between the focus and epicenter of an earthquake. The focus is the point of initial movement and the epicenter is the point on the surface directly above the focus. The location of an earthquake is usually described by the geographic position of its epicenter, and by its focal depth.*

earthquakes, and *deep-focus* earthquakes have a focal depth greater than 300 km. Earthquakes originating at depths greater than 700 km rarely occur because the rocks at such depths are relatively plastic and do not accumulate strain.

Distance to the epicenter

As a first step in determining the location of an earthquake, one must find the distance between the recording station and the epicenter. The study of seismograms provides the key to determining this distance. Typically, the record of an earthquake can be divided into three major parts. The record begins with relatively simple oscillations produced by the arrival of primary waves. These smaller-scale tracings are followed abruptly by waves of somewhat greater amplitude representing incoming secondary waves. The surface waves come along last and produce the largest and most complex patterns of all (see Figure 4–11). How might the pattern of seismic traces be used to find the distance to the epicenter? The answer to that question was provided in the early 1900s by John Milne, the founder of modern seismology. Milne discovered that the time separation on a seismogram between the starting points of the P- and S-waves is greater for a distant earthquake than for one located nearby. For example, the time that elapses between the arrival of the P-wave and the arrival of an S-wave (sometimes termed the P minus S interval) for a focus 500 km away is about 30 seconds. For a focus 1000 km away, 100 seconds would separate the arrival times of P- and S-waves. The situation is comparable to a phenomenon we have all experienced when we see lightning in the sky but hear the clap of thunder a moment or two later. The time that elapses between seeing the lightning and hearing the thunder would be greater at greater distances between ourselves and the location of the lightning bolt.

Geophysicists have determined how rapidly seismic waves travel by observation of earthquakes whose source locations and times of occurrence are well known. The information obtained from these observations is used to construct **time-distance graphs** that can then be employed in locating earthquake epicenters. Figure 4–17 is such a graph. Assume that a seismograph records the arrival of a P-wave at 8:00 A.M.

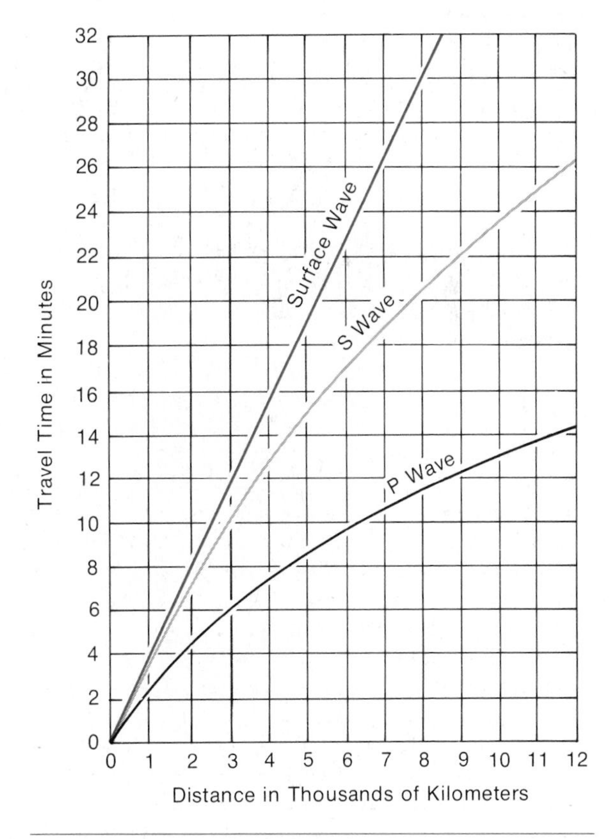

Figure 4–17 *Time-distance graph.*

and the arrival of the S-wave two minutes later. The P minus S interval of two minutes can be marked off on the time-travel scale on the left side of the time-distance chart. This interval can be vertically fitted into just the right size space between the P- and S-wave curve. Then, by reading down onto the horizontal scale, one can ascertain that the earthquake was 1400 km away from the seismograph station. A line extended horizontally from the point where the two-minute mark touches the P-wave curve to the left edge of the graph indicates that the P-wave required 3 minutes and 15 seconds to reach the seismograph. Thus, the time of the actual shock at the focus occurred 3 minutes and 15 seconds before 8:00 A.M. Newspapers might now announce the time of occurrence of the earthquake and its distance, but precisely *where* did it occur?

Direction to epicenter

To find the precise location of the epicenter, distance determinations from three or more seismograph stations are needed. On a map, the location of each seismograph station is plotted. A circle is drawn around the location that has the radius of the seismograph to epicenter distance (Fig. 4-18). The earthquake must have occurred somewhere on that circle. Communications from two or more other stations are used as the basis for plotting similar circles, and the point where three circles intersect locates the epicenter. Large earthquakes are recorded at hundreds of earthquake stations. Therefore, there are an abundance of measurements to use in checking the validity of the epicenter's location.

EARTHQUAKE-RELATED HAZARDS

The vibrations that accompany earthquakes trigger landslides, cause the collapse of buildings, roads, and bridges, generate tsunami (seismic sea waves), and contribute to outbreaks of fires. There are also hazards of disease, panic, and the breakdown of order caused by the interruption of vital services. The severity of any of the hazards depends not only on the magnitude of the earthquake but also on the general geology of the areas affected and the time and place of occurrence.

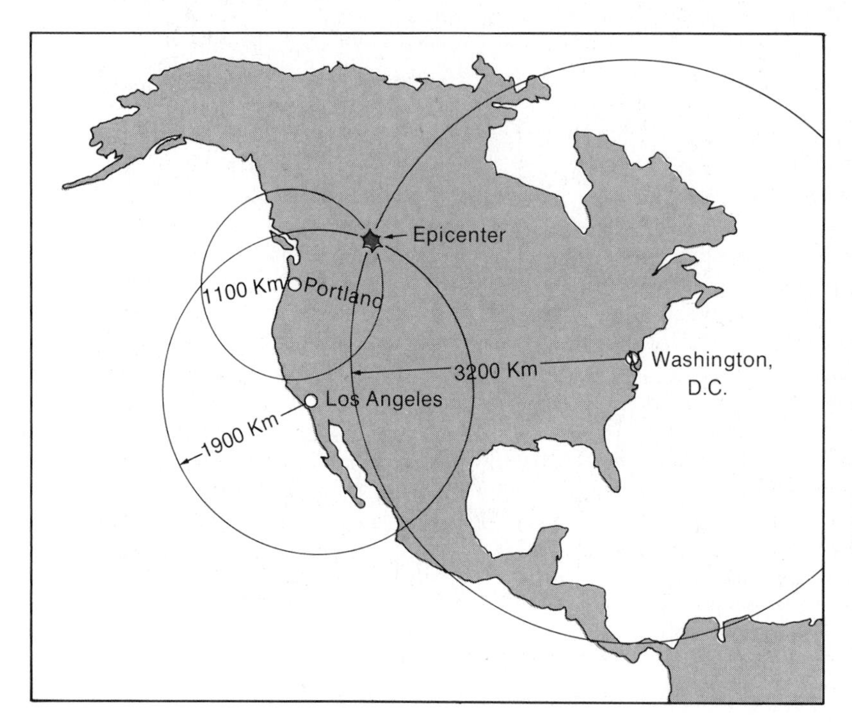

Figure 4–18 Method used for locating the epicenter of an earthquake. The distance to the epicenter from seismograph stations in Washington, D.C., Los Angeles, and Portland was first determined from the P minus S time interval. This distance is used as the radius to draw circles around each station in order to find the intersection of all three. The epicenter was determined to be near Saskatoon, Saskatchewan, Canada.

Tsunami

November 1, 1755, was a day of unprecedented terror for the people of Lisbon. With a thunderous terrifying roar, the city was convulsed by an earthquake that was to claim 60,000 lives and demolish five of every six buildings. Panic-stricken citizens rushed down to the open wharfs to avoid falling masonry and fire, but within minutes they were hurled into oblivion as a 15-meter wall of water smashed into the harbor, destroying everything in its path. That wall of water was the advancing front of a seismic sea wave, or **tsunami.**

Tsunami frequently accompany earthquakes that occur near coastlines (Fig. 4–19). The tsunami associated with the 1964 Alaskan earthquake was triggered by the sudden movement of the sea floor in Prince William Sound. Seventy miles away from the quake center at the Port of Valdez, onrushing tsunami lifted the steamship *Chena* nine meters skyward and then plunged it to the floor of the bay. Miraculously, the steamer was able to right itself. Only three crewmen were killed (one suffered a heart attack and two were crushed by lurching cargo). People on shore were less fortunate, as the initial and then second waves swept them away and reduced the waterfront facilities to a mass of splintered wood and twisted metal.

Although tsunami may be caused by landslides along a coastline and by volcanic eruptions, the majority are generated by earthquakes. Often they develop when segments of the sea floor subside abruptly or are suddenly thrust upward along faults. Waves are generated in the water mass in much the same way as they

Figure 4–19 Tsunami resulting from the 1964 Alaskan earthquake washed these vessels into the center of Kodiak, Alaska. (Courtesy of U.S. Navy.)

would be when a water-filled metal tub is struck by a sharp blow. In the open ocean, tsunami are known to travel faster than 400 km per hour. They are scarcely noticed on the open ocean, however, because the wave length of the tsunami is increased to as much as 100 km and the wave height over so broad a distance is not visibly evident. On approaching the shallow zone near a coastline, however, the bottom of the tsunami is retarded by friction. As the wave is slowed in the shallower zone, the enormous volume of water tends to pile up. The result is an increase in wave height to 30 or 40 meters.

Tsunami travel rapidly, but because of the great distances involved, they may require several hours to cross an ocean. Therefore, it is feasible to set up a warning system to alert coastal areas of the possibility of property damage and loss of life from an approaching tsunami. Such a warning system for the Pacific, called the SSWWS (Seismic Sea Wave Warning System), has been developed by the U.S. Coast and Geodetic Survey. A network of seismograph stations alert scientists in Honolulu of the location and magnitude of earthquakes. If the interpretation of these data indicates the possibility of a tsunami, its estimated arrival times at different locations are calculated with the use of tsunami travel-time charts, and warnings are issued. Of course, for locations close to the earthquake, there may be insufficient time to provide a warning. It is also difficult to predict the force and destructiveness of a tsunami when it arrives at a particular locality, as these factors are related to the geographic and bathymetric configuration of the coastline, the direction of the approach of the tsunami, and the state of the tides.

Inhabitants of coastal areas that lie in the path of a tsunami would be well advised to evacuate to high ground. They should also be informed that the arrival of a tsunami may sometimes be signaled by a temporary withdrawal of the sea along a coastline. Foolishly venturing out on the vacated sea floor to gather stranded fish might be a fatal mistake.

Ground displacements

One of the most pervasive misconceptions about the effects of earthquakes is the terrifying notion that bottomless canyons suddenly open in solid rock, swallowing up screaming victims. Actually, there have been very few authenticated deaths caused by people falling into fissures. Fissures do sometimes open up in the unconsolidated materials overlying bedrock, but not in solid rock. Usually these cracks appear as a result of settling, slumping, or sliding triggered by the much deeper ruptures in solid rock that caused the earthquake. As noted previously, the extreme fracturing at Anchorage in 1964 and San Francisco in 1906 was largely in the surface layer of unconsolidated materials.

Landslides

Earthquake-induced landslides are more dangerous to humans than cracks in the ground. As an example of the magnitude of this hazard, one need only recall the 1970 catastrophe in Peru that was responsible for burying tens of thousands of villagers in Yungay and Ranrahirca. Another well-known landslide caused by an earthquake occurred near midnight in the Madison River Canyon, Montana, on August 17, 1959. An earthquake dislodged a great chunk of the mountain above the Madison River, and a chaotic jumble of fragmented rock roared across the valley, blocking the river and burying vacationers at the Rock Creek Campsite under more than 60 meters of debris. The landslide effectively dammed the Madison River, and a lake, appropriately dubbed Earthquake Lake, formed behind the slide.

Liquefaction

Often during earthquakes, fine-grained, water-saturated sediments may lose their former strength and form into a thick, mobile mudlike material. The process is called **liquefaction.** The liquefied sediment not only moves about beneath the surface but may also rise through fissures and "erupt" as mud boils and mud "volcanoes." As described earlier, most of the destruction to buildings at Turnagain Heights during the 1964 Alaskan earthquake could be related to the flow of liquefied soils.

Fire

Injury from falling objects, burial in landslides or collapsed buildings, and fire pose the greatest dangers to life during earthquakes. We have already noted the ravages of fire during the San Francisco earthquake. For many years that disaster was referred to as the Great San Francisco Fire, as if the earthquake itself was of lesser importance. Yet another example of this hazard is provided by the earthquake that devastated the cities of Yokahama and Tokyo in 1923. The quake struck during the noon hour when midday meals were being prepared. Hundreds of fires broke out almost instantaneously. The panic-stricken citizens of Tokyo crowded into small open areas, only to die of the heat and suffocation as the flames from surrounding buildings consumed most of the available oxygen. When the fires were finally extinguished, there were more than 100,000 dead, 40,000 seriously injured, and nearly a half million houses demolished. The Japanese had learned a costly lesson. The rebuilt cities have broad streets to accommodate fire trucks, fire-fighting systems designed to function even during earthquakes, auxiliary water systems, and buildings constructed to resist damage.

DEFENSE AGAINST EARTHQUAKES

If there is an earthquake near your home this year, will you know what to do to lessen the danger to your family and yourself (Table 4–2)? Is it possible to reduce the hazards of earthquakes? Can anything be done in earthquake-prone areas to lessen the intensity of future earthquakes? Geophysicists and engineers are aggressively examining these questions, and their research holds great promise.

Planning for earthquakes

Every major earthquake stimulates interest in revising building codes, initiating more rigorous zoning rules, and providing for emergency water and power. The newer buildings in earthquake-prone areas of the United States, Russia, China, and Japan are designed to resist damage from shaking. The effectiveness of these efforts can be judged by comparing the damage from earthquakes that occur in underdeveloped countries with the lesser damage caused by recent earthquakes of the same magnitude in Japan and the United States. Even in the world's most modern cities, however, there is no reason for complacency. Many of

TABLE 4–2 **EARTHQUAKE SAFETY TIPS**

The following checklist of action to take in the event of an earthquake may be clipped and posted for handy reference.

Before

1. Store emergency supplies: food, water, first-aid kit, flashlight, and battery-powered radio.
2. Take a practical first-aid course.
3. Locate main switches and valves that control the flow of water, gas, and electricity into your house. Know how to operate them.
4. Support community programs that inform the public and emergency personnel about earthquake preparedness.
5. Take action to help strengthen or eliminate structures that are not earthquake-resistant.
6. Support "parapet ordinances" that would remove dangerous, unreinforced overhangs and cornices from buildings.
7. Support building codes that require earthquake-resistant construction and careful foundation preparation and grading.
8. Support land-use policies that recognize and allow for the potential dangers of active fault zones.
9. Bolt to the floor heavy furniture above the fifth floor in tall buildings.
10. Require guard rails across the inside of plate glass windows that extend to the floor.
11. Support basic research into the cause and mechanism of earthquakes and fault movement.

During

1. Don't panic even if you are frightened.
2. If you are indoors, stay there. Get under a desk, table, or doorway.
3. Do not rush outside. Falling debris has caused many deaths.
4. Watch for falling plaster, bricks, and other objects.
5. If you are outside, move away from buildings and power lines; stay in the open.
6. If you are in a moving car, stop as soon as it is safe. Remain in the car.

After

1. Check your family and the people near you for injuries.
2. Inspect your utilities for damage to water, gas, or electrical conduits. If they are damaged, turn them off.
3. Extinguish open flames.
4. Do not use the telephone except to report an emergency.
5. Turn on your battery-powered radio for emergency information.
6. Don't go sightseeing.
7. Stay away from damaged structures; aftershocks can cause the collapse of weakened structures.
8. Stay away from beaches and waterfront areas subject to seismic sea waves (commonly called "tidal waves").

From *California Geology, 24*(11):216, 1971, published by the California Division of Mines and Geology.

Figure 4–20 Housing developments obliterate the surface expression of the San Andreas fault near San Francisco. The solid line traces the location of the fault. (Courtesy of U.S. Geological Survey.)

our most cherished structures were built long before the advent of seismology, and far too many more recently constructed buildings lack the necessary safeguards, because of costs or public apathy. Zoning plans frequently fail to include considerations of the nature, water content, and strength of the materials upon which construction is planned. In general, solid rock is a much safer foundation material than granular soil or sediment because such loose material tends to magnify seismic wave amplitudes. Construction should not be planned over active faults (Fig. 4–20).

The prediction of earthquakes

If it were possible to predict early enough the precise time, size, and place of earthquakes, then thousands of lives could be saved by evacuation. Although the prediction of earthquakes within months or days has been reported in China and Japan, such predictions are still rarely possible for most areas. Seismic risk maps (Fig. 4–21) developed from compilations of the distribution and intensity of earthquakes cannot be used for prediction, but they are at least helpful in evaluating seismic risks for purposes of building design and insurance.

Long-term earthquake prediction of a rather vague nature can be made, and such forecasts help to increase the public's awareness of the need for precautions. For example, studies conducted along the San Andreas fault in California suggest a break will occur within the next two decades. Forces have been building up along the fault since the 1906 earthquake. Calculations suggest there is now enough of this stored energy to propel one side of the fault at least 5 meters—an amount approaching the 6.5 meters of slippage that occurred in 1906.

To predict a major earthquake more precisely, scientists must learn to detect and evaluate the often subtle changes that occur in the physical characteristics of the rocks that are experiencing earthquake-inducing stress. Sometimes, the rocks respond to the buildup of stress by a series of staccato-like slippages that generate a cluster of preliminary small earthquakes called foreshocks. It has been shown that foreshocks have indeed preceded some intense earthquakes. There are, however, other ways that the crust may signal the buildup of stress. Many of these clues are measurable. For example, if parts of the crust are either compressed or stretched, rocks will experience changes in density, water content, and magnetism. They may be tilted, raised, or lowered by amounts that can be measured with the help of sensitive instruments. In the investigations of areas having a high seismic risk, instrument stations are arranged in a network so as to record automatically these crustal changes.

SEISMIC WAVES AND THE STRUCTURE OF THE EARTH'S INTERIOR

The divisions of inner space

Our knowledge of the earth's interior has been derived from the study of countless seismograms documenting earthquakes and nuclear explosions all around the globe. The seismograms show not only that the velocities of body waves generally increase with depth but also that there are two levels within the earth where body waves experience an abrupt jump in velocity. Such rapid changes in velocity are called *seismic discontinuities*. They indicate that seismic waves have passed into materials that are significantly different in physical properties. One of these major jumps in wave velocity is called the *Mohorovičić Discontinuity*

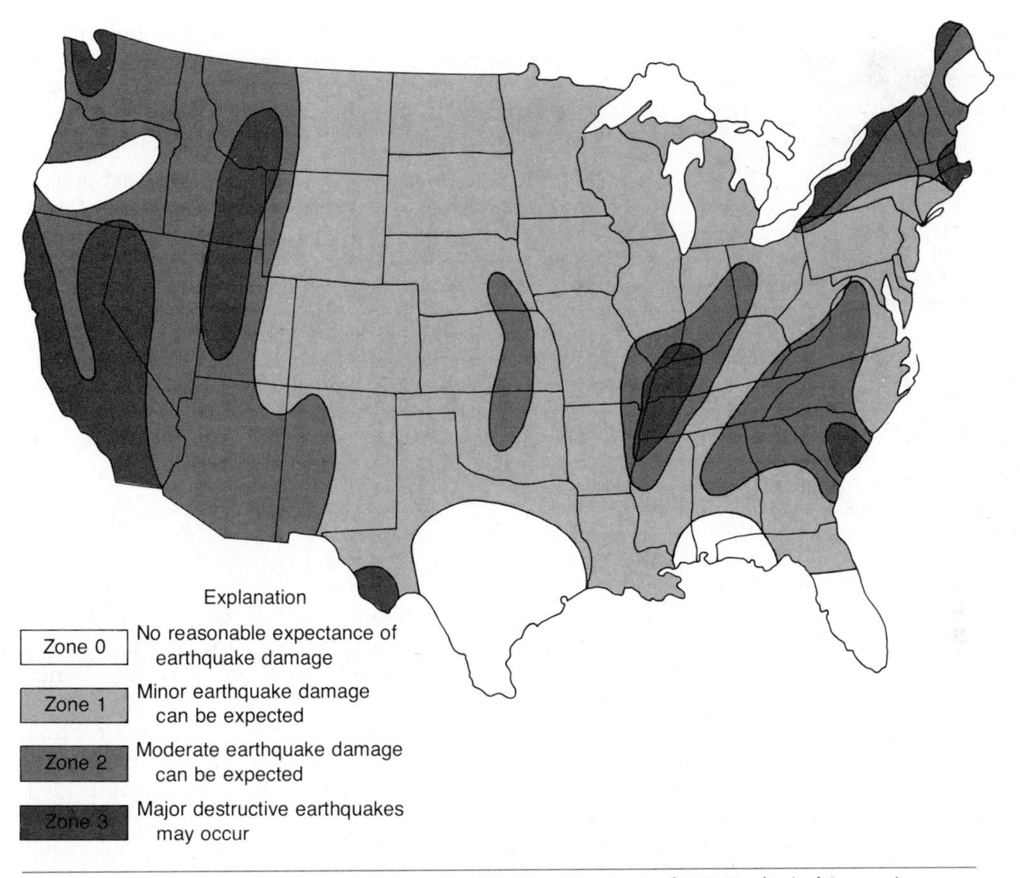

Figure 4–21 *Earthquake risk map of the United States. (Courtesy of U.S. Geological Survey.)*

(or more simply "Moho") after its discoverer, Andrija Mohorovičić. The Moho (Fig. 4–22) lies about 30 or 40 km below the surface of the continents and at lesser depths beneath the ocean floors. The second major discontinuity occurs nearly halfway to the center of the earth at a depth of 2900 km. It is called the *Gutenberg Discontinuity* in honor of the geophysicist Beno Gutenberg, whose work defined its location. These two discontinuities make it possible to think of the earth as being divided into the *crust,* which is the thin surface layer extending downward to the Mohorovičić Discontinuity, the *mantle,* which extends from the base of the crust to the Gutenberg Discontinuity, and the *core,* which extends from 2900 km to the center of the earth (Fig. 4–23). Let us next examine what seismic waves tell us about the nature of the earth materials that lie within these major divisions of inner space.

The core

Inferences from body waves

For many years before the development of modern seismology, geologists correctly inferred that the earth had a very dense central core. Such an interpretation was indicated by the overall high density of the earth compared with the relatively low density of surface rocks. As noted above, the precise boundary of the core was determined by the study of earthquake waves. Seismology has also provided a means for discerning subdivisions of the core and deciphering some of its physical properties. For example, at a depth of 2900 km (the core boundary) the S-waves meet an impenetrable barrier; at the same time P-wave velocity is drastically reduced from about 13.6 to 8.1 km per second. Earlier, we noted that S-waves are unable to travel through fluids. Thus, if they were to enter a

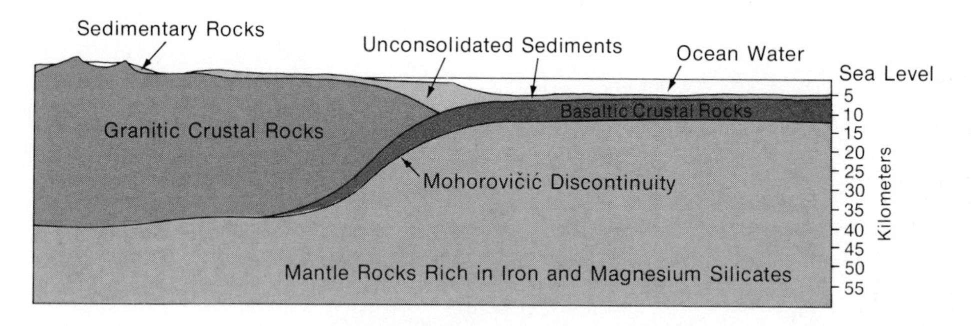

***Figure 4–22** Generalized cross-section of a segment of the earth's crust showing location of the Mohorovičić Discontinuity.*

fluid region of the earth's interior, they would be absorbed there and would not be able to continue. Geophysicists believe this is what happens to S-waves as they enter the outer part of the core. As a result, the secondary waves generated on one side of the earth fail to appear at seismograph stations on the opposite side, and this observation is the principal evidence for an outer core that behaves as a fluid. The outer core barrier to S-waves results in an *S-wave shadow zone* on the side of the earth opposite an earthquake. Within the **shadow zone,** which begins 105 degrees from the earthquake focus, S-waves do not appear.

Unlike S-waves, primary waves are able to pass through liquids. They are, however, abruptly slowed and sharply refracted as they enter a fluid medium. Therefore, as primary seismic waves encounter the molten outer core of the earth, their velocity is checked and they are refracted downward. The result is a *P-wave shadow zone* that extends about 105 to 143 degrees from the focus. Beyond 143 degrees, P-waves are so tardy in their arrival as to further validate the inference that they have passed through a liquid medium. At the upper boundary of the core P-waves are also reflected back toward the earth's surface (Fig. 4–24). Such P-wave echos are clearly observed on seismograms.

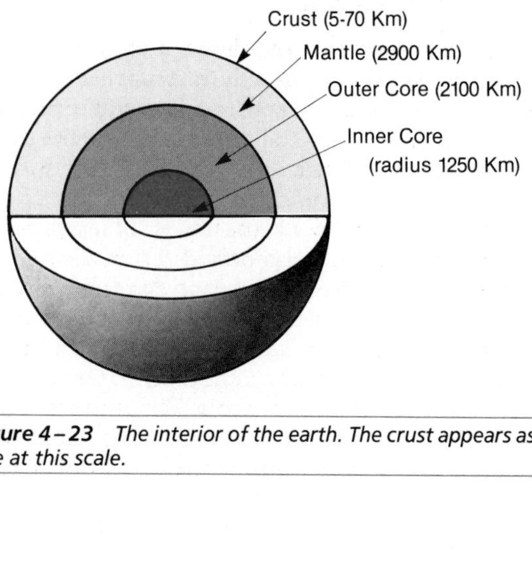

***Figure 4–23** The interior of the earth. The crust appears as a line at this scale.*

The radius of the core is about 3500 km. The inner core is solid and has a radius of about 1220 km, which makes this inner core slightly larger than the moon. A transition zone approximately 500 km thick surrounds the inner core. Most geologists believe that the inner core has the same composition as the outer core and that it can only exist as a solid because of the enormous pressure at the center of the earth.

Evidence for the existence of a solid inner core is derived from the study of hundreds of seismograms produced over several years. These studies showed that weak, late-arriving primary waves were somehow penetrating to stations that were within the P-wave shadow zone. Geophysicists recognized that this could be explained by assuming the inner core behaved seismically as if it were solid.

Core composition

The earth has an overall density of 5.5 g/cm^3, yet the average density of rocks at the surface is less than 3.0 g/cm^3. This indicates that materials of high density must exist in the deep interior of the planet to achieve the 5.5 g/cm^3 overall density. Calculations indicate that the rocks of the mantle have a density of about 4.5 g/cm^3 and that the average density of the core is about

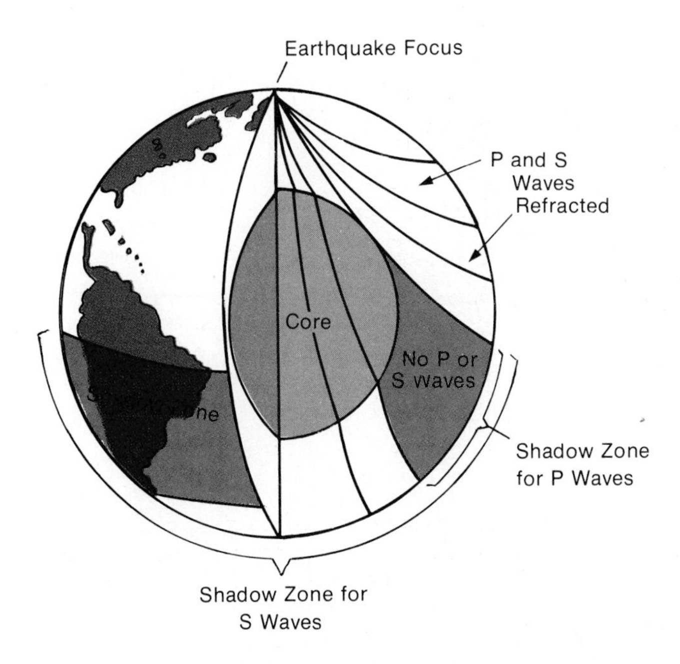

Figure 4–24 *Refraction of seismic body waves. The S-wave shadow zone is believed to be the result of absorption of S-waves in the liquid outer core. When the P-wave enters the liquid outer core, it slows and is bent downward, giving rise to the P-wave shadow zone within which neither P- nor S-waves are received.*

10.7 g/cm^3. Under the extreme pressures that exist in the region of the core, iron mixed with nickel would very likely have the required high density. In fact, laboratory experiments suggest that a highly pressurized iron-nickel alloy might be too dense, and that minor amounts of such elements as silicon, sulfur, or oxygen may also be present to "lighten" the core material.

Support for the theory that the core is composed of iron (85 percent) with lesser amounts of nickel has come from the study of meteorites. A large number of these samples of solar system materials are iron meteorites that consist of metallic iron alloyed with a small percentage of nickel. They appear to have formed under conditions of extremely high temperature and pressure, such as exist in planetary interiors. Iron meteorites may very well be fragments from the core of a shattered planet. Their abundance in our solar system suggests that the existence of an iron-nickel core for the earth is a quite plausible concept.

There is another kind of evidence for the earth's having a metallic core. Anyone who understands the functioning of an ordinary compass is aware that the earth has a magnetic field. The planet itself behaves as though there were a great bar magnet embedded within it at a small angle to its rotational axis. Geophysicists believe that the earth's magnetic field may, in some way, be associated with electric currents. This interpretation is favored by the discovery more than 60 years ago that a magnetic field is produced by an electric current flowing through a wire. The silicate rocks of the lithosphere and mantle are not good conductors, however, and are therefore unlikely materials for the development of electromagnetism. In contrast, iron is an excellent conductor. If scientists are correct in the inference made from density considerations and study of meteorites that the core is mostly iron, then the magnetic field lends credence to that inference.

The origin of the core

How does the existence of a dense metallic core relate to theories on the origin of the earth? A currently favored theory would have the earth and other planets created from the atoms, molecules, and particles circulating within a turbulent cosmic dust cloud. According to this concept, the earth would have acquired most of its materials while relatively cool, and its various elements would have been mixed and dispersed from its surface to its center. How, then, did layering develop? The answer that comes immediately to mind is that after the earth had accumulated most of its matter, it became partially or entirely molten. Somewhat as in a great blast furnace, much of the iron and nickel percolated downward to form the core. Remaining iron and other metals combined with silicon and oxygen and separated into the overlying less dense mantle. Still lighter components may have separated out of the mantle to form an uppermost crustal layer. Geologists do not insist on complete melting for this differentiation of materials into layers. Under conditions of high pressure and temperature, elements might have migrated to their appropriate levels by solid diffusion. Indeed, it would be difficult to account for the present abundance of volatile elements on earth if it had melted completely, for under such circumstances these materials would be driven off. Even for partial melting and solid diffusion to have occurred, however, heat would have been required. That heat may have been supplied in various ways. Certainly, the decay of radioactive elements must have been an important thermal source. In addition, heat was probably supplied by the kinetic energy of debris still showering the protoplanet, from gravitational compression, from solar radiation, and even from tidal friction generated by the nearby moon.

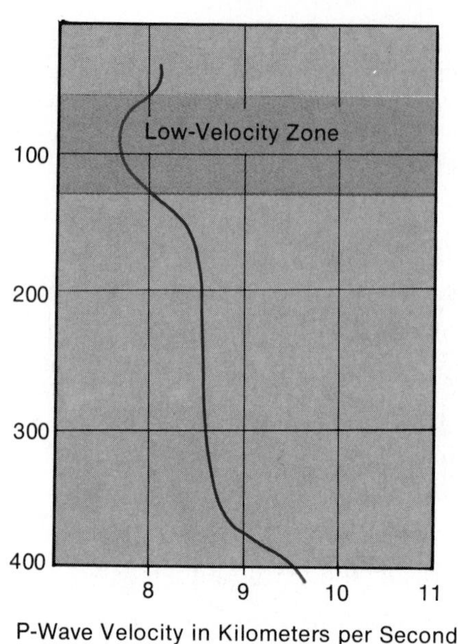

Figure 4–25 *Profile of P-wave velocity for part of the upper mantle showing low-velocity zone.*

The mantle

As was the case with the earth's core, our understanding of the composition and structure of the mantle is based on indirect evidence. As inferred from seismic data, the mantle's average density is about 4.5 g/cm^3, and it is believed to have a stony, rather than metallic, composition. Oxygen and silicon probably predominate and are accompanied by iron and magnesium as the most abundant metallic ions. The iron-rich rock **peridotite** approximates fairly well the kind of material inferred for the mantle. A peridotitic rock not only would be appropriate for the mantle's density but also is similar in composition to stony meteorites as well as rocks that are thought to have reached the earth's surface from the upper part of the mantle itself. Such suspected mantle rocks are indeed rare. They are rich in olivine and pyroxenes and contain small amounts of certain minerals, including diamonds, that can form only under pressures greater than those characteristic of the crust.

The mantle is not a homogeneous layer, but is composed of a lower mantle, a transitional zone, and an upper mantle. The upper mantle is of particular importance because its characteristics and movements affect the crust. The most remarkable feature of the upper mantle is the **low-velocity zone** (Fig. 4–25). As suggested by its name, this is a region in which there is a decrease in the speed of S- and P-waves. The low-velocity zone occupies an upper region of the much larger **asthenosphere.** Directly beneath the asthenosphere lies the **mesosphere.** This zone of the upper mantle is composed of rocks that are again sufficiently strong and rigid to cause an increase in P-wave velocity.

Geophysicists believe the seismic waves are slowed in the asthenosphere not because of a decrease in density but rather because they enter a region that is in the state of a crystalline-liquid mixture. In such a "hot slush," perhaps 1 to 10 percent of the material would consist of pockets and droplets of molten silicates. This interpretation is strengthened by the observation that in certain regions of the asthenosphere, S-waves are absorbed as if by large bodies of magma. If this interpretation of the physical state of the asthenosphere is correct, then it is capable of considerable motion and flow. Such a slippery mobile layer would enhance movement of tectonic plates in the overlying lithosphere.

THE CRUST OF THE EARTH

The crust of the earth is seismically defined as all of the generally solid earth above the Mohorovičić discontinuity. It is the thin, rocky veneer that constitutes the continents and the floors of the oceans. The crust is not a homogeneous shell in which low places were filled with water to make oceans and higher places are continents. Rather, there are two distinct kinds of crust, which, because of their distinctive compositions and physical properties, determine the very existence of separate continents and ocean basins.

The oceanic crust

Beneath the varied topography of the ocean floors lies an oceanic crust that is approximately 5 to 12 km thick and has an average density of about 3.0 g/cm^3. The oceanic crust consists of basalts that were extruded under water. They are covered by a thin layer of geologically young sediment.

The continental crust

Properties of the continental crust

At the boundaries of the ocean basins, the Mohorovičić discontinuity plunges sharply beneath the thicker continental crust. Depth to the Moho beneath the continents averages about 35 km, although it may be considerably deeper or shallower in particular regions. The continental crust is not only thicker than its oceanic counterpart but also less dense, averaging about 2.7 g/cm^3. As a result, continents "float" higher on the denser mantle than the adjacent oceanic crustal segments. Somewhat like great stony icebergs, the roots of continents extend downward into the mantle.

The concept of light crustal rocks floating on denser mantle rocks has been given the name **isostacy.** Were it not for isostacy, mountain ranges would gradually subside, for there are no rocks having sufficient strength to bear the heavy load of mountain ranges. Thus, mountains are not supported by the strength of the crust, but rather are in a state of flotational equilibrium with denser underlying rocks.

Although the continental crust is referred to as being granitic, it is really composed of a variety of rocks that approximate granite in composition. Igneous continental rocks are richer in silicon and potassium and poorer in iron, magnesium, and calcium than igneous oceanic rocks. Also, extensive regions of the continents are blanketed by sedimentary rocks.

Origin of the continental crust

One of the major questions relating to the history of the earth is how the lighter granitic masses that constitute the continents developed. Many geologists believe the development of the crust involved a series of events that began about 4 billion years ago with upwellings of lava derived from the partially molten upper mantle. This initial crust of lava was then subjected to repeated episodes of remelting, during which time lighter components were separated out and distributed near the earth's surface. Wherever uplands existed, as along volcanic island arcs, the solidified lavas were subjected to erosion and chemical weathering. The products of this weathering and erosion were the earth's earliest sediments, which were then altered by rising hot gases and silica-rich solutions from below. Recycling and melting of this well-cooked,

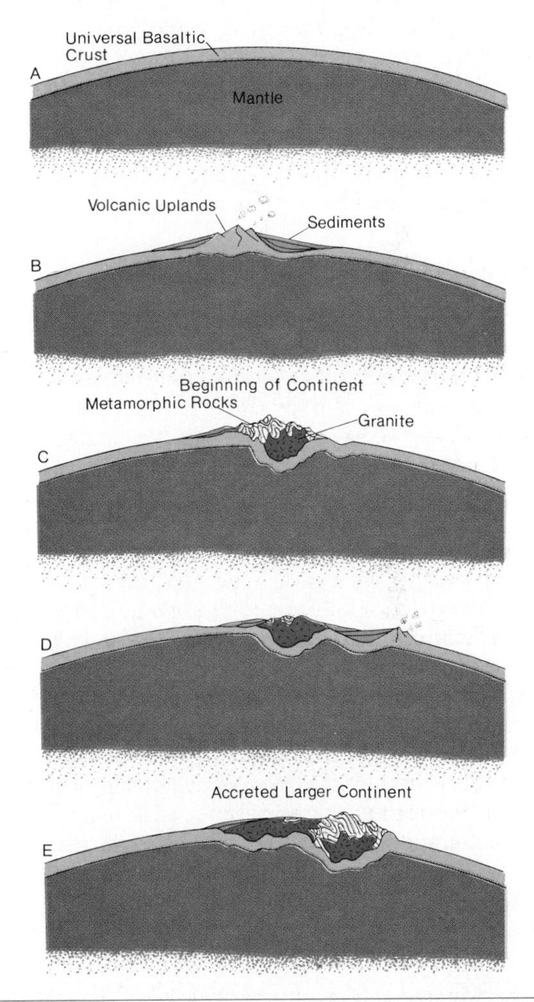

Figure 4–26 Model for the gradual evolution of continents by orogeny and conversion of sedimentary to granitic igneous rocks. (A and B) Original basaltic crust with volcanic uplands as source for sediments. (C) Orogeny of sedimentary accumulation to produce metamorphic and granite rocks. (D) New sediments accumulate along margins of continental nucleus. (E) The process is repeated.

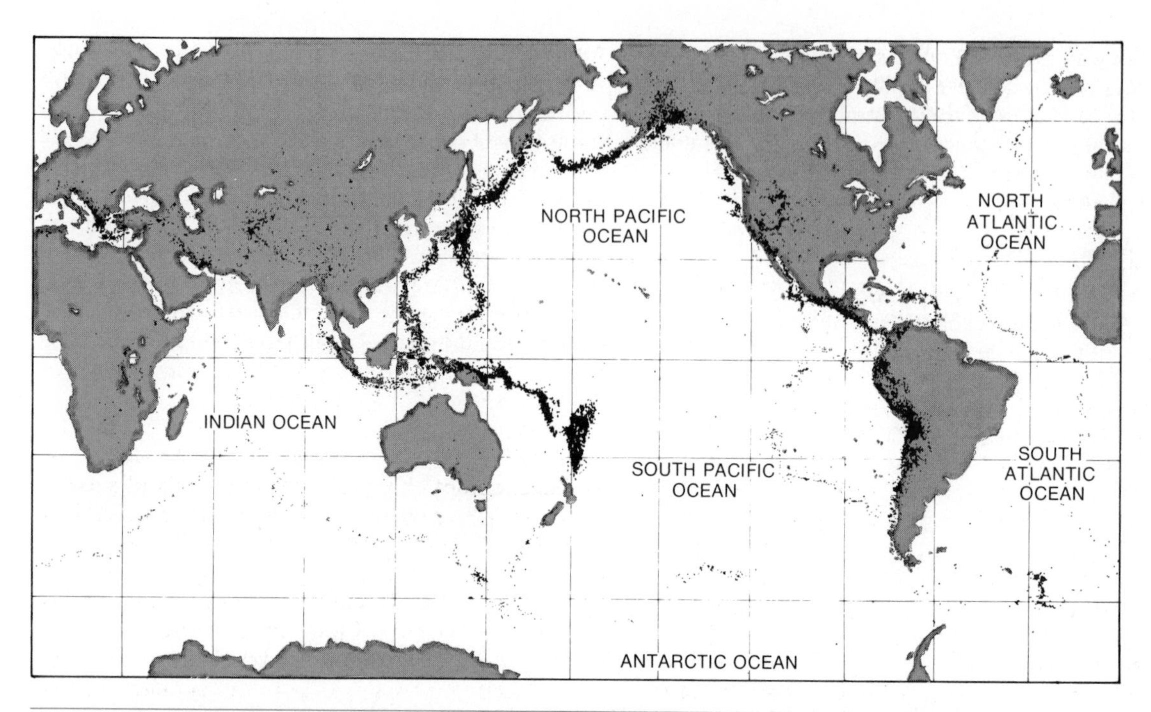

Figure 4–27 World distribution of earthquakes occurring at depths of 0 to 700 km. Notice the major earthquake belt that encircles the Pacific and another that extends eastward from the Mediterranean toward the Himalayas and East Indies. The midocean ridges are also the site of many earthquakes. (From Turk, J., and Turk, A., Physical Science with Environmental and Other Practical Applications, *Philadelphia, W.B. Saunders Co., 1977.)*

lighter mix of earth materials led ultimately to the rocks of granitic character that formed the nuclei of continents. The new continents might then have provided a source for additional sediment that would have collected along continental margins. Subsequently, these sediments also might have been metamorphosed and melted during orogenic events, so that successive bands of granitic crust would have become welded or "accreted" onto the initial continental nuclei (Fig. 4–26).

EARTHQUAKE DISTRIBUTION AND TECTONICS

If one examines a map showing the location of earthquake epicenters (Fig. 4–27), it is readily apparent that earthquakes do not occur randomly around the globe. Rather, earthquake epicenters tend to be concentrated along relatively narrow zones. In fact, more than 80 percent of the total earthquake energy for the earth is released in the zones that border the Pacific Ocean. A second set of zones extends from southern Europe through Turkey, Iran, northern India, and on toward Burma, and it accounts for nearly another 15 percent of earthquake energy. Clearly, there is some sort of disruptive activity along these great earthquake trends. According to the concepts of plate tectonics, which will be examined in detail in the next chapter, the zones of high seismicity are believed to delineate the margins of great moving plates of lithosphere. The plates collide, pull apart, or slide past neighboring plates, and these interactions along plate margins generate the stress in crustal rocks that causes earthquakes.

Summary Earthquakes are vibrations of the earth's surface caused by sudden rupture of rock masses or displacements along faults within and below the crust. They are frequent phenomena. It has been estimated that at least a million earthquakes occur each year, and at least 150,000 of these are felt by persons nearby. Their effects range from insignificant to serious structural damage or complete destruction of a large percentage of the buildings in cities and towns. In large cities where gas and water lines are broken by earthquakes, fire may result in a far greater loss of life and property than the actual ground movements. Damage from earthquakes is intensified in areas where buildings have been erected on marshy or loose fill material. In hilly regions, heavy loss of life and property may result from landslides triggered by earthquakes. Coastal areas hundreds of kilometers from earthquake centers may also experience danger, as tsunami (seismic sea waves) strike shoreline and wharf facilities.

Seismology is the study of earthquakes, and the principal tool of the seismologist is the seismograph. These instruments are constructed with a heavy weight suspended so that it will exist as independently of earth movement as possible. Earth movements are recorded on a time-recording rotating drum so that the arrival of earthquake waves can be determined to a fraction of a second.

The three most important kinds of seismic waves propagated during an earthquake are primary waves, secondary waves, and surface waves. A seismic wave is defined by describing the motion of a particle in its path. In primary waves, for example, the direction of vibration is in the same direction as the path of propagation of the wave. They have a "push-pull" or compressional progression through the rock. In secondary waves, the direction of vibration is perpendicular to the direction of propagation. Thus, secondary waves have a sinuous shearing or transverse motion. Secondary waves can travel in solids only, as there is no shear resistance in liquids or gases. Surface waves, the third type, spread out along the surface of the earth only. They are slower moving than either primary or secondary waves and have very complex motion.

Primary waves travel faster than secondary waves and thus arrive first at a seismograph station. On the seismogram, the time that has elapsed between the arrival of the primary and secondary waves is readily determined. This interval is different for each distance between the seismograph and earthquake center, and it therefore can be used to calculate the distance from the seismograph station to the source of the waves. Using the calculated distance to the source of waves from three or more different seismograph stations, the location of the earthquake can be determined. Circles with radii equal to the calculated distances are drawn around each of the three stations, and the intersection of the three circles locates the earthquake center. The point within the earth at which the waves originated is called the focus, whereas the location directly above the focus is the earthquake's epicenter.

Much of what we infer about the earth's deep interior is derived from the study of earthquake waves. For example, abrupt changes in the velocities of seismic waves at different depths provide the basis for a threefold division of the earth into a central core, a thick overlying mantle, and a thin enveloping crust. Sudden changes in earthquake wave velocities are termed discontinuities. The core of the earth, as indicated by the Gutenberg discontinuity, begins at a depth of 2900 km. It is likely that the outer core is molten and that the entire core is composed of iron with small amounts of nickel and possibly either silicon or sulfur. The core probably originated during an episode of heating, when heavier constituents of protoplanet were drawn by gravity to the center and lighter components rose to the surface. Constituting about 80 percent of the earth's volume, the mantle is composed of iron and magnesium silicates, such as olivine and pyroxene. A warm, rather plastic zone in the mantle is recognized by lower seismic wave velocities. The presence of this low-velocity layer, or asthenosphere, is an important element in the modern theory of plate tectonics. The asthenosphere may function as a weak plastic layer upon which horizontal motions in more rigid surface layers can occur.

The seismic boundary that separates the mantle from the overlying crust is the Mohorovičić discontinuity. It lies far deeper under the continents than under the ocean basins. Thus, the continental crust is thicker than the oceanic crust. There are compositional and density differences as well. The continental crust has an overall granitic composition and is less dense than the oceanic crust, which is composed of basaltic rocks.

QUESTIONS FOR REVIEW

1. What is the cause of earthquakes? Include in your answer a description of the elastic rebound theory.
2. What is the difference between the terms *stress* and *strain* as used in the studies of the strength of rocks? What is meant by the terms *elasticity* and *elastic limit?*
3. What are the three major kinds of earthquake waves? Which are also termed "body waves"? Describe the motion of the two kinds of body waves.
4. A seismogram records an interval of three minutes between the arrival of P-waves and S-waves. With the use of the time-travel curves (Fig. 4–17), determine the distance between the seismograph station and the epicenter.
5. What major internal zones of the earth would be penetrated if one were able to drill a well from the North Pole to the center of the earth? What major seismic discontinuities would be crossed?
6. What is a tsunami? Why are they rarely detected by ships at sea?
7. What is the evidence for the fluidity of the earth's outer core and for the solidity of the inner core?
8. What is the likely composition of the core and of the mantle? What is the basis for inferring these compositional characteristics?
9. How does the continental crust differ from the oceanic crust in composition, thickness, and density?
10. What is the asthenosphere, and how may it be related to the cause of earthquakes?

SUPPLEMENTAL READINGS AND REFERENCES

Bolt, B.A., 1982. *Inside the Earth.* San Francisco, W.H. Freeman Co.

Bolt, B.A., 1978. *Earthquakes—A Primer.* San Francisco, W.H. Freeman Co.

Bolt, B.A., Horn, W.L., Macdonald, G.A., 1977. *Geological Hazards.* 2d ed. New York, Springer-Verlag.

Hodgson, John H., 1964. *Earthquakes and Earth Structure.* Englewood Cliffs, N.J., Prentice-Hall Inc.

Oakeschott, G.B., 1975. *Volcanoes and Earthquakes: Geologic Violence.* New York, McGraw-Hill Book Co.

Press, F., 1975. Earthquake prediction. *Scientific American,* 232 (5):14–23.

Raleigh, C.B., Healy, J.H., Bredehoeft, J.D., 1976. An experiment in earthquake control at Rangely, Colorado. *Science,* 191:1230–37.

Verney, P., 1979. *The Earthquake Handbook.* London, Paddington Press.

5

crustal deformation and plate tectonics

What was solid earth has become the sea,
and solid ground has issued from the
bosom of the waters.
OVID (43 B.C.–A.D. 18)
Metamorphoses

Prelude As implied in the above epigram from the work of the poet Ovid, nearly 2000 years ago, Roman intellectuals recognized that parts of the sea floor had risen from the sea, and that dry land had sunk beneath the waves. Much later, Leonardo da Vinci (1452–1519), Robert Hooke (1635–1703), and Nicholaus Steno (1638–1687) also wrote of such geologic changes. Steno, a Danish scientist working in Italy, taught that because marine sediments are laid down in horizontal layers, wherever they are found as folded strata, crustal movements must have taken place.

In the three centuries since Steno's death, geologists have mapped the world's mountain ranges and arrived at an understanding of the nature of the deformational forces that produced mountains. They have shown that each mountain range did not originate during a unique and isolated event, but rather developed as part of a system of interrelated changes in the earth's lithosphere. These changes, their causes, and the features they produce at the earth's surface are the important topics to be examined in this chapter.

Mount Owen and Symmetry Spire of Wyoming's Grand Teton Mountains. The Grand Tetons are the result of movement along high-angle normal faults. (Courtesy of U.S. National Park Service.)

A CHANGEABLE CRUST

Mountains and plains, ocean basins and the continents, all seem everlasting. The human life span is far too short to permit a view of lands heaving upward or the slow migration of continents and ocean floors. Nevertheless, we are reminded of the planet's restlessness by earthquakes, volcanoes, and many more subtle indications of instability.

Evidence of change

Vertical movements of the earth's crust are more readily recognized along coastlines, where sea level serves as a reference horizon. Consider the interpretation of a feature formed offshore by wave erosion and found

Figure 5–1 Elevated marine terrace provides evidence for change in sea level at Rhossili Bay, Wales. (Courtesy of J.C. Brice.)

Figure 5–2 The Grand Canyon of the Colorado River. View across O'Neill Butte to Bright Angel Creek. (Photograph courtesy of National Park Service.)

today several meters above sea level. Three explanations for elevated position are possible. The solid crust along the coastline may have been lifted above the waves, the crust itself may have remained motionless during a fall in sea level, or there may have been some combination of fluctuating sea and land levels. Movements of land level independent of the sea are termed **tectonic.** Whenever a change in sea level occurs more or less independently of continental tracts, the change is termed **eustatic.** Eustatic changes occurred during ice ages when the accumulation or melting of great volumes of glacial ice alternately reduced or increased the amount of water in the ocean basins.

There are many geologic clues to tectonic uplift around the world. Along certain coasts, rocks and coral reefs that were once completely submerged now stand well above sea level. Sections of the Scandinavian Penninsula have been rising by nearly 75 meters a year for the past several decades. Along the shorelines of regions that have experienced such uplift, one often finds flat tracts cut by wave action now located high above the level of the waves that formed them (Fig. 5 - 1). In addition to uplift, subsidence may also occur, so that ocean water enters river valleys to form elongate bays. Drowned valleys of this kind have formed the harbors of San Francisco and New York.

Evidence for crustal unrest is also found in the interiors of continents. Sedimentary rocks originally deposited on the ocean floor (as is evident from the marine fossils they contain) are now found in mountains and plateaus. For example, the flat-lying strata along the rim of the Grand Canyon of the Colorado River (Fig. 5 - 2) were laid down at or near sea level, yet today they stand 2000 meters above sea level. **Incised meanders** (Fig. 5 - 3) may also suggest the possibility of uplift in some regions of the continents. Meanders may become incised as a result of regional uplift, which increases the slope or gradient of the river channel. The river flows faster down the steeper gradient and more vigorously erodes its channel. Incised meanders are frequently developed in uplifted plateaus. They are not, however, *prima-facie* evidence for uplift, as they may also form where a stream flows over rocks that differ in their resistance to erosion, or in streams near shorelines where there has been a eustatic lowering of sea level.

Even three centuries ago, scientists recognized that the mere existence of folded strata was in itself evidence of crustal change. In the earth's great mountain ranges, uplifted, folded, and faulted rocks document the passage of quiet seas into lofty peaks (Fig. 5 - 4). Igneous rocks that formed deep within the cores of these mountains are now exposed at the surface be-

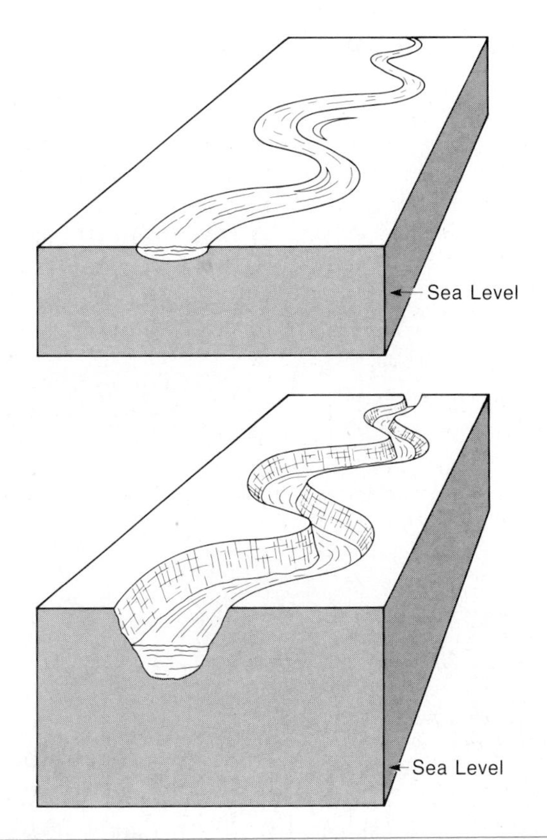

Figure 5 - 3 *Incised bends in a stream may be the result of either uplift of land or lowering of sea level.*

cause of repeated episodes of uplift and erosion. In some parts of the globe, the mountains themselves are gone, having succumbed to the relentless attack of wind, water, and ice. Only the deep, twisted roots of the ranges remain as testimonials to geologic change (Fig. 5 - 5).

GEOLOGIC STRUCTURES

Almost from the moment they solidify, rocks begin to respond to a variety of deformational forces. They are compressed by a heavy burden of overlying rocks, squeezed or stretched along the margins of moving tectonic plates, and subside or rise as they strive to reach equilibrium with denser rocks that lie far beneath the earth's surface. Therefore, we are not surprised to find rocks bent into folds and broken along

Figure 5–4 Folded strata of marine rocks on the side of Scapegoat Mountain, Montana. The rocks are Cambrian in age. (Courtesy of U.S. Geological Survey.)

Figure 5–5 Aerial view of old mountains nearly buried in their own debris. Dark areas are mostly igneous and metamorphic rocks. Linear features are joints, faults, and dikes. Southwestern Saudi Arabia (N 22°, E 44°). (Courtesy of U.S. Geological Survey.)

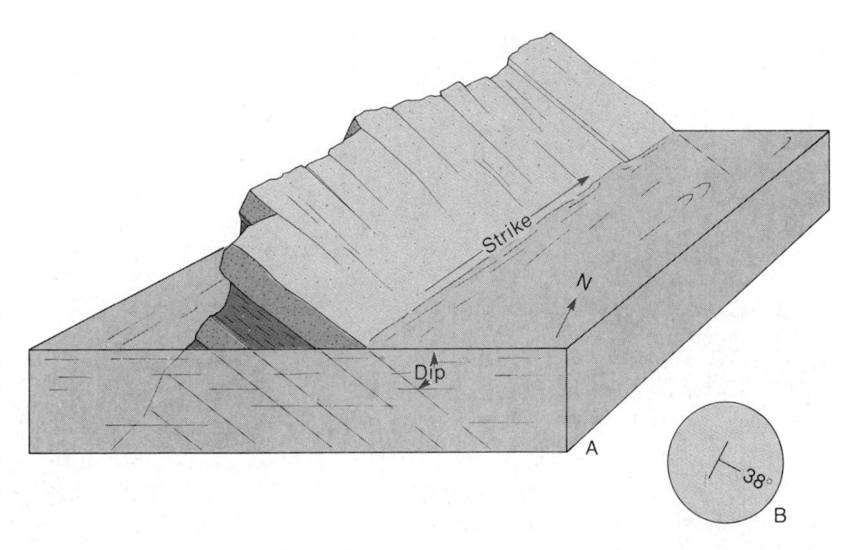

Figure 5-6 Strike and dip of tilted strata. The strike is represented by the line of intersection of a horizontal plane and the inclined (dipping) bed. In this illustration the horizontal plane is represented by the water surface of a lake. The dip is perpendicular to the strike. The symbol for strike and dip is shown in B.

faults. Such deformational features constitute **geologic structures.**

Structural terms

Geologists employ the word **attitude** whenever they wish to refer to the way in which a stratum, fault surface, or other geologic feature is oriented in space. Are the rocks horizontal or tilted? If they are tilted, how much and in what compass direction? The two terms *strike* and *dip* are essential in clearly conveying answers to these questions. **Strike** is the compass direction of a line formed by the intersection of the surface of an inclined stratum with an imaginary horizontal plane (Fig. 5-6). Thus, the strike notation N 35°W indicates a stratum having a strike of 35° west of true north.

Dip refers to the maximum angle of slope of a tilted stratum measured directly downward from the horizontal plane. In addition to providing a measure of the steepness at which a bed is tilted, dip also refers to the direction toward which the bed is tilted. The *direction of dip* is perpendicular to the strike. A ball placed on the exposed surface of a dipping bed would roll directly down the bed's direction of dip. In recording the attitude of an inclined stratum on a map, a symbol such as that shown in Figure 5-6 is used to express the strike as well as the angle and direction of dip. In a geologic report, the geologist might abbreviate the attitude of a stratum with the notation N 15°E, 30°SE. Readers would then know that the stratum in question had a strike of 15° east of north and was sloping (dipping) at an angle of 30° toward the southeast.

Folds

A **fold** is a bend or flexure in layered rocks (Fig. 5-7). Although folds may occur in igneous and metamorphic rocks, they are more commonly noticed in layers of sedimentary rock that were originally laid down in relatively horizontal layers. Long ago geologists recognized this simple fact and correctly inferred that if strata were found in other than a horizontal attitude, then crustal movements had affected them. Folds are found in all sizes, from great sweeping flexures mea-

Figure 5-7 Folded layers of sandstone and siltstone. Old Red Sandstone Formation, St. Ann's Head, Pembroke, England. Both a syncline and an anticline are clearly visible in this exposure. (Courtesy of the Institute of Geological Sciences, London.)

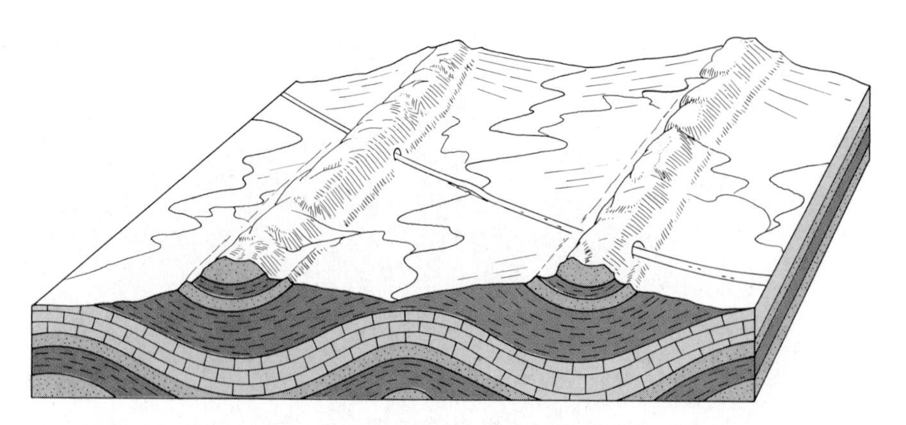

Figure 5–8 Anticlines (upfolded strata) do not necessarily correspond to ridges on the earth's surface. Here the ridges are formed by resistant layers of sandstone that are part of synclinal (downfolded) strata.

suring many tens of kilometers across to tiny accordion-like crenulations. Inclined strata of folds, found here and there in isolated exposures, are carefully located on geologic maps. Data collected in this way are used to reconstruct the original appearance and nature of deformation of various parts of the crust.

It should be remembered that folds are geologic structures existing below the surface. They do not necessarily correspond to topographic forms. A level plain may exist above highly folded strata, a valley above an uparching of beds, or a hill above a downfold (Fig. 5–8).

The principal categories of folds are anticlines, synclines, domes, basins, and monoclines (Fig. 5–9). **Anticlines** are uparched rocks in which successively older rocks are encountered toward the center of curvature (Fig. 5–10). **Synclines** are folds that are concave upward and have younger rocks toward the center of curvature. Folds tend to occur together, rather like a series of petrified wave crests and troughs. The erosion of anticlines and synclines characteristically produces a topographic pattern of ridges and valleys (Fig. 5–11). The ridges develop where resistant rocks project at the surface, whereas valleys develop along

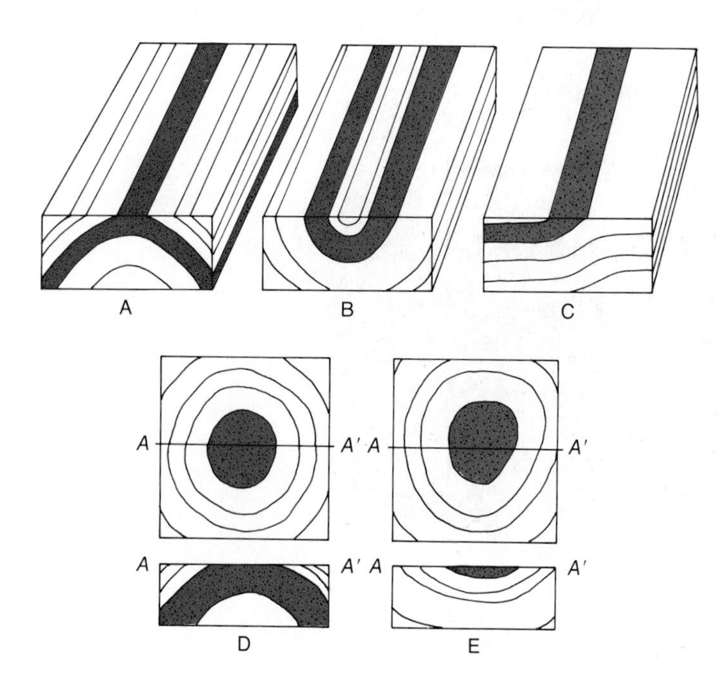

Figure 5–9 Kinds of folds. (A) Anticline. (B) Syncline. (C) Monocline. (D) Map view and cross-section of a dome. Notice older strata are in center of outcrop pattern. (E) Map view and cross-section of a basin. Younger rocks occur in central area of outcrop pattern.

Figure 5–10 Anticline in Early Paleozoic strata in the Neptune Range, Pensacola Mountains, Antarctica. (Photograph courtesy of J.R. Ege, U.S. Geological Survey.)

zones underlain by more easily eroded rocks (Fig. 5–12).

The axis of a fold is a line drawn along its maximum curvature (Fig. 5–13). The hinge may be horizontal or it may be inclined, in which case it is said to **plunge.** If the axis is horizontal, ridges and valleys tend to be parallel. In plunging folds, zigzag patterns of ridges and valleys are more characteristic. The **axial plane** of a

Figure 5–11 Ouachita Mountains of Arkansas. The ridges are underlain by dipping layers of rock of greater relative resistance to erosion than the rocks beneath valleys. (Courtesy of U.S. Geological Survey.)

Figure 5–12 Aerial view of the tilted, resistant layers of rock, located between Rawlins and Laramie, Wyoming. (Courtesy of U.S. Geological Survey and J.R. Balsey.)

fold is an imaginary surface that divides the fold as symmetrically as possible.

In deciphering the configuration and location of a fold, geologists carefully note strike and dip on a base map at as many localities as feasible (see Fig. 3–42). It is then possible to reconstruct the appearance of the folds prior to erosion. Where rock exposures are sparse, an understanding of where to expect relatively older or younger layers of an erosionally beveled anticline or syncline can be very useful. As illustrated in Figure 5–9, sedimentary beds are necessarily deposited so that the oldest beds are at the bottom of a sequence, and sequentially younger beds occur above the initial layer. If the layers were then arched into an anticline, and the top "sliced off" by erosion, the geologist mapping the anticline would encounter successively younger beds when walking from the axis toward the limbs. In the case of a syncline, younger beds would be found along the axis and older beds toward the limbs.

Folds may vary in symmetry as well as size. The curve of the arch may be broad and rounded or sharp and angular. **Symmetrical folds** are ones in which the angle of dip on both limbs is the same, so that the sides are mirror images. As implied by their name, **asymmetrical folds** are those in which the angle of inclination of beds on either side of the hinge is different. Compressional forces may push a fold over and onto its side so that the axial plane is inclined and both limbs dip in the same direction but at different angles. Such structures are called **overturned folds** (Fig. 5–14). If the overturning is so complete that the axial plane is horizontal, then the fold is called **recumbent.** During the formation of the Alps, immense recumbent folds were developed. Large and complex recumbent folds of the Alpine type are also called **nappes.** A special kind of fold in which there is a single steplike bend in

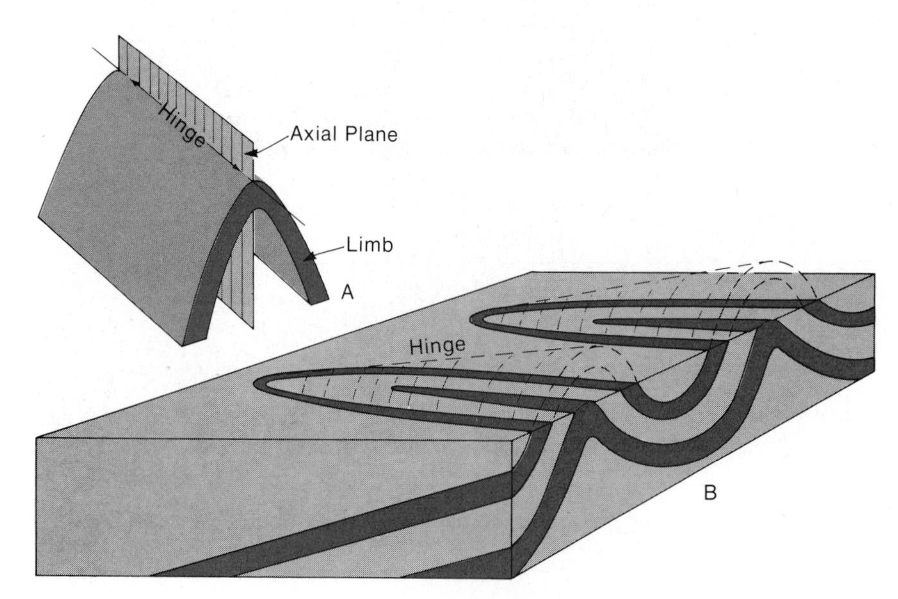

Figure 5–13 (A) Diagram of an anticline illustrating nomenclature of hinge, axial plane, *and* limb. *(B) Map pattern of plunging folds idealized on a level surface. These folds have an inclined hinge.*

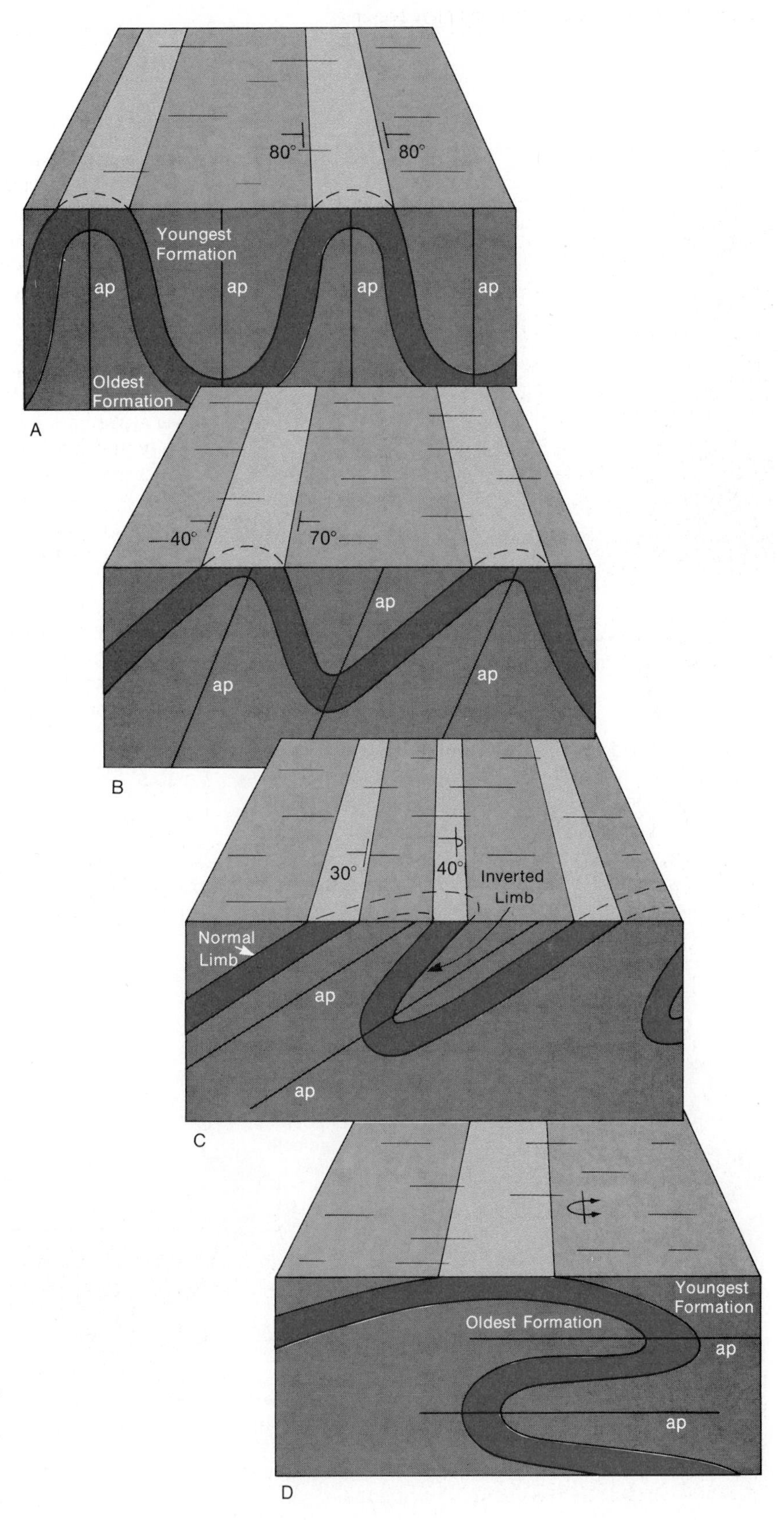

Figure 5–14 *Four varieties of folds. (A) Symmetrical folds. (B) Asymmetrical folds. (C) Overturned folds. (D) Recumbent folds. (ap = axial plane)*

Figure 5–15 Structural dome at Sinclair, Carbon County, Wyoming. (Courtesy of U.S. Geological Survey and J.R. Balsey.)

otherwise horizontal beds is termed a **monocline** (see Fig. 5–9).

Domes and **basins** are similar to anticlines and synclines except that they have a roughly circular outcrop pattern (Fig. 5–15). A dome consists of uparched rocks in which the beds dip in all directions away from the center point. In contrast, a basin is a downwarp in which beds dip from all sides toward the center of the structure. The truncated beds of simple domes and basins may form a circular pattern of ridges and valleys at the earth's surface. Older beds are found at the surface in the center of domes, and younger strata at the center of basins.

Anticlines and synclines are mapped by geologists to better understand the past geologic events of a region and to decipher the principal strain directions acting on an area of the crust. There are, however, more practical reasons for determining the precise locations and geometry of folds. The attitude of strata has an influence on the stability of the bedrock on which buildings and roads are constructed. Further, folded rocks may serve to concentrate supplies of petroleum and provide underground reservoirs of artesian water. Geologists tracing the locations of coal seams and other mineral deposits must understand the pattern of folding in order to predict the presence of resources at locations where erosion, younger beds, or soil cover has obscured their presence (Fig. 5–16).

Faults

Recognition of faults

In the previous chapter, we briefly defined **faults** as fractures along which there has been movement. That is, adjacent rock masses have slipped past one another

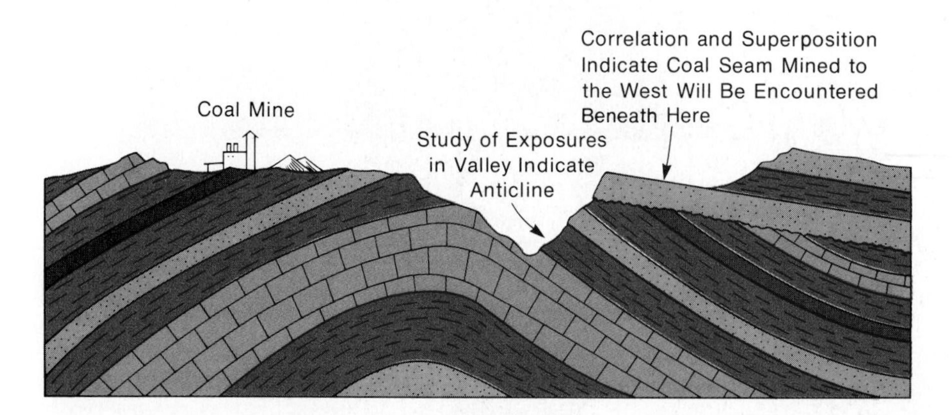

Figure 5–16 Knowledge of the pattern of folding in an area is useful in the exploration for coal beds and layered ore deposits. The geologist, noting angle and direction of dip and superpositional sequence of strata, can predict the location of coal seams or mineral deposits where they are not exposed at the surface.

Figure 5–17 *Fault scarp exposed north of Bitterwater, California. (Courtesy of J.C. Brice.)*

in response to tension, compression, or shearing stress. The various categories of faults are named according to the relative directions in which the slippage occurs. It is not always an easy matter to identify the direction of movement, especially in faults that have been inactive for perhaps tens of millions of years. Geologists depend primarily on offsets in strata, veins, or other geologic features to determine the character of ancient faults. For faults that have experienced recent movements, escarpments (fault scarps) may show the displacement (Fig. 5–17). There may also be displacements of physiographic features, such as ridges and stream channels (Fig. 5–18).

Figure 5–18 *Surface trace of San Andreas fault adjacent to Temblor Mountains, California. Note offset of stream at arrow. (Courtesy of U.S. Geological Survey and J.R. Balsey.)*

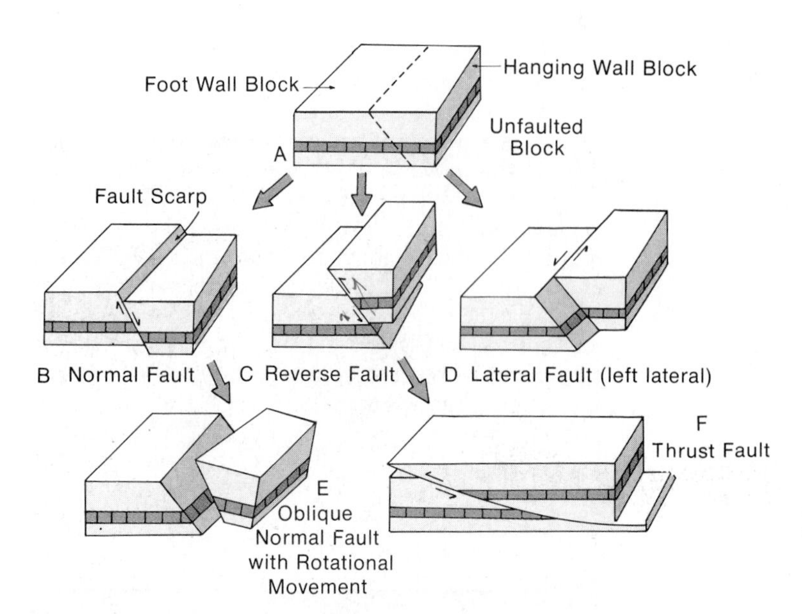

Figure 5–19 Kinds of faults. A shows the unfaulted block with the position of the potential fault shown by dashed line. In nature, movements along faults may vary in direction, as shown in E. A thrust fault is a type of reverse fault that is inclined at a low angle from the horizontal.

Categories of faults

Fault planes are permeable zones along which hot mineralized water solutions can penetrate and may, in relatively rare instances, deposit valuable ores. Long ago, miners tunneling along sloping fault planes helped to provide a nomenclature that still serves to identify certain kinds of faults. The side of the fault on which the miners stood was appropriately dubbed the *footwall*. Above their heads was the *hanging wall*. These terms are put to work in the definition of two major categories of faults, *normal* and *reverse faults*. In a normal fault the hanging wall appears to have moved downward relative to the footwall (Fig. 5–19). Normal faulting occurs most frequently in rocks that have been subjected to tensional forces—that is, forces that tend to stretch the earth's crust. A system of normal faults with parallel fault planes may have topographic expression as linear uplands separated by troughlike valleys, somewhat like a piano keyboard in which some of the keys are depressed. The topographically depressed down-faulted segments that are bounded by normal faults are termed *grabens* (Fig. 5–20), after the German word for grave, and the high-standing blocks that generally stand above the grabens are called *horsts*.

In reverse faults, the hanging wall appears to have moved upward relative to the footwall. If one holds two blocks of wood cut like those in Figure 5–19, then it is easy to visualize that they may be pushed together or compressed in order to cause reverse faulting. Similarly, regions of the earth's crust containing numerous reverse faults (and folded strata as well) are likely to have experienced compression.

Reverse faults in which the fault plane is inclined 45° or less from the horizontal plane are termed **thrust faults** (Fig. 5–21). In other words, thrust faults are simply low-angle reverse faults. Many thrust faults, such as the Lewis Overthrust in Montana and the Pine Mountain thrust fault in Tennessee, have displacements of many kilometers.

Another type of faulting involves primarily horizontal movements along the strike of the fault plane. Such features are called **strike-slip faults,** although the informal term lateral fault is also sometimes applied. The fault plane in a strike-slip fault is usually nearly vertical, and the displacement along the fault planes may be only a few meters to tens or even hundreds of kilometers. The San Andreas fault is a strike-slip fault. It is more than 1000 km long, and the total lateral movement along the fault has been estimated by geologists to be more than 500 km.

Strike-slip faults are designated as either "right lateral" or "left lateral." To decide which kind is being observed, one simply must look across the fault zone to see whether the opposite block moved to the right or the left. The San Andreas is a right lateral fault (see

Figure 5–20 *Photograph taken from Earth Resources Technology Satellite at an altitude of 900 kilometers. Near the center is the Dead Sea. The smaller body of water to the north is Lake Tiberias. To the northwest is the Mediterranean. The biblical cities of Sodom and Gomorrah were probably located at a site now flooded at the south end of the Dead Sea. Both Lake Tiberias and the Dead Sea are located along a north-south trending system of faults and occur in down-faulted blocks or grabens. The steep sides of the faults form the rugged terrain on either side of the Dead Sea rift valley. The fault system was apparently formed by stresses associated with the separation of Arabia from Africa.*

Figure 5–21 *A thrust fault. The rocks lying above the fault plane have been moved to the left. Notice the rupture and displacement of bed A. Thrust faulting has produced the drag folds in bed A and adjacent beds. Exposure is in Pembrokeshire, England. (Courtesy of the Institute of Geological Sciences, London.)*

Figure 5–22 *Vertical and horizontal joints in granite. Such joints weaken the rock and, along with wave action, facilitate the development of sea cliffs. Acadia National Park, along the coast of Maine. (Courtesy of U.S. National Park Service and C.W. Stoughton.)*

Fig. 5-18). Of course, there is nothing in nature that restricts fault movements to either a purely vertical or a horizontal direction, and so there are many faults that are formed as a result of oblique and rotational movements.

Joints

Rocks at or near the earth's surface are nearly always broken by cracks or fractures. Such fracturing tends to break a rock mass into a network of blocks. If there has been little or no slipping of one block relative to its neighbor, then the crack between the blocks is termed a **joint** (Fig. 5-22). Joints occur most frequently as a consequence of either crustal tension or shearing forces. Tensional joints may develop in rocks whenever they experience a shrinkage in volume as a result of the loss of moisture or heat. Usually, the contraction takes place around many centers within the rock mass, and fractures develop at right angles to the directions of tension. Everyone has seen such tensional joints developed where layers of mud have dried out. The joint pattern is polygonal (many-sided). Earlier we noted that polygonal jointing is also characteristic of basaltic lava flows, sills, and dikes. More coarsely crystalline rocks, such as granite, also develop jointing. Although in many granite bodies this jointing appears to be related to cooling, it is evident from the increase in the number of joints near the surface of granite outcrops that jointing may result from the relief of pressure that occurs when a weighty burden of overlying rock has been removed by erosion.

PLATE TECTONICS

Perhaps more than other scientists, geologists are accustomed to viewing the earth in its entirety. It is their task to assemble the multitude of observations about minerals, earthquakes, and landforms and then to weave them into a coherent view of the whole earth. Recently, a new theory has been developed that provides a comprehensive base for associating a great variety of geologic information. The new concept has been named **plate tectonics,** and it has taken on the dimensions of a scientific revolution. The new ideas were late in coming. Perhaps the most important reason for their tardy arrival was simply that, until recently, there were large regions of the globe that could not be adequately studied. This was particularly true of the vast realms beneath the oceans. For two centuries the continents, which constitute only about 30 percent of the earth's surface, received most of the attention.

The lack of adequate information about the ocean floors began to be corrected in the years following World War II. Research related to naval operations produced submarine detection devices that also proved useful in measuring magnetic properties of rocks. From the mid-1940s to the late 1950s, the need to monitor atomic explosions resulted in the establishment of a worldwide network of seismometers. This network provided the precise information about the global pattern of earthquakes. The magnetic field over large portions of the sea floor was soon to be charted by the use of newly developed, sensitive *magnetometers.* Other technologic advances ultimately permitted scientists to examine rock that had been carefully dated by radiometric methods and then to determine the nature of the earth's magnetic field at various times in the geologic past. Geologically recent reversals of the magnetic field were detected and correlated. A massive, federally funded program to map the bottom of the oceans was launched, and depth information from improved echo depth sounding devices was translated into maps and charts.

A new picture of the ocean floor began to emerge (see Figure I-4). It was at once awesome, alien, and majestic. Great chasms, flat-topped submerged mountains, boundless abyssal plains, and interminable volcanic ranges appeared on the new maps and begged an explanation. How did the volcanic midocean ridges and deep-sea trenches originate? Why were both so prone to earthquake activity? Why was the mid-Atlantic Ridge so nicely centered and parallel to the coastlines of the continents on either side? As the topographic, magnetic, and geochronologic data accumulated, the relationship of these questions became apparent. An old theory called *continental drift* was reexamined, and the new, more encompassing theory of *plate tectonics* was formulated. It was an idea whose time had come.

Fragmented continents

It requires only a brief examination of the world map to notice the remarkable parallelism of the continental shorelines on either side of the Atlantic Ocean. If the continents were pieces of a jigsaw puzzle, it would seem easy to fit the great "nose" of Brazil into the reentrant of the African coastline. Similarly, Greenland might be inserted between North America and northwestern Europe. It is not surprising, therefore, that earlier generations of map gazers also noticed the fit and formulated theories involving the breakup of an ancient supercontinent.

In 1858 there appeared a work titled *La Création et ses Mystères Devoilés*. Its author, A. Snider, postulated that before the time of Noah and the biblical flood, there existed a great region of dry land. This antique land developed great cracks encrusted with volcanoes, and during the Great Deluge, a portion separated at a north-south trending crack and drifted westward. Thus, North America came into existence.

Near the close of the nineteenth century, the Austrian scientist Eduard Suess became particularly fascinated by the many geologic similarities shared by India, Africa, and South America. He formulated a more complete theory of a supercontinent that drifted apart following fragmentation. Suess called that great land mass **Gondwanaland** after Gondwana, a geologic province in east-central India.

The next effort to convince scientists of the validity of these ideas was made in the early decades of the twentieth century by the energetic German scientist Alfred Wegener. His book, *Die Entstehung der Kontinente und Ozeane* ("The Origin of the Continents and Oceans"), is considered a milestone in the development of the concept of continental drift.

Wegener's hypothesis was straightforward. Building on the earlier notions of Eduard Suess, he argued again for the existence in the past of a supercontinent that he dubbed **Pangaea.** That portion of Pangaea that was to separate and form North America and Eurasia came to be known as **Laurasia,** whereas the southern portion retained the earlier designation of Gondwanaland. According to Wegener's perception, Pangaea was surrounded by a universal ocean named **Panthalassa,** which opened to receive the shifting continents when they began to split apart some 200 million years ago (Fig. 5-23). The fragments of Pangaea drifted along like great stony rafts on the denser material below. In Wegener's view, the bulldozing forward edge of the slab might be expected to crumple and produce mountain ranges like the Andes.

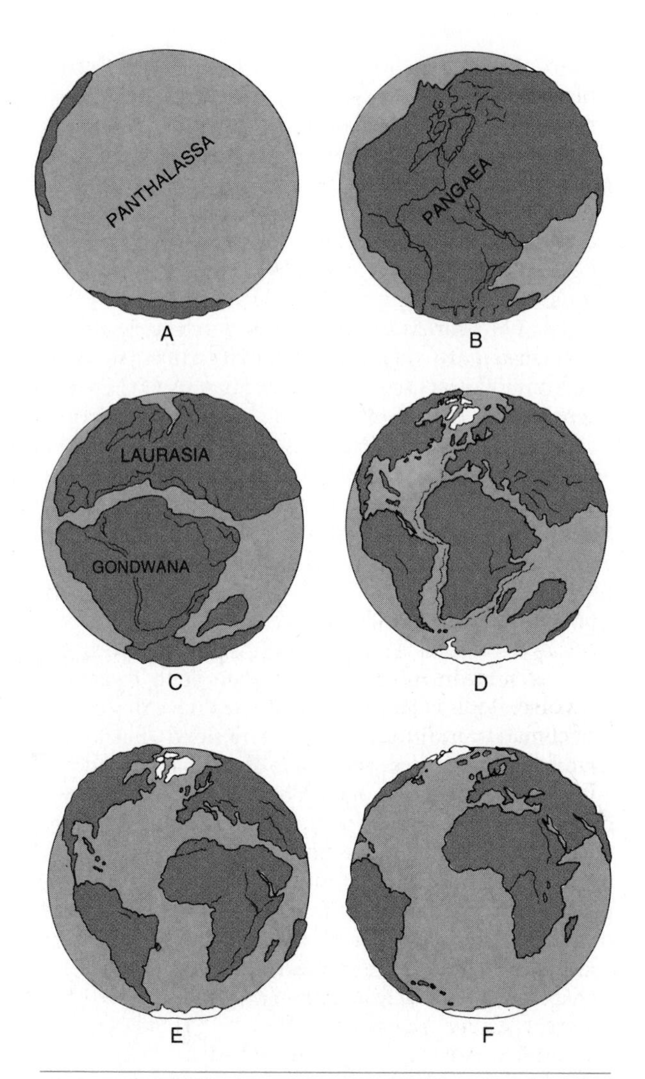

Figure 5-23 Alfred Wegner's view of the earth as it existed 200 million years ago is shown in A and B above. At this time there was one ocean (Panthalassa) and one continent (Pangaea). By about 20 million years later, the supercontinent had begun to split into a northern Laurasia and southern Gondwana (C). Dismemberment continues (D), and by about 65 million years ago the earth resembled drawing E. Further widening of the Atlantic and northward migration of India brings the earth to its present state, shown in F.

The ideas presented by Wegener were not to go unchallenged. Criticism was leveled chiefly against his notion that the continents slid along through denser oceanic crust in the manner of giant granitic icebergs. The scientifically formidable physicist Sir Harold Jeffries calculated that the ocean floor was far too rigid to allow for the passage of continents, no matter what the imagined driving mechanism. It is now known that Jeffries was correct in asserting that continents cannot—and do not—plow through oceanic crust. Shortly, it will be shown that continents do move, but they move only as passive passengers on large rafts of lithosphere that glide over a comparatively soft and plastic upper layer of the earth's mantle. Nevertheless, much of the evidence Wegener and others had assembled can be used to substantiate the newer concepts. We should have a look at this evidence.

The most convincing evidence for continental drift remains the geographic fit of the continents. Indeed, the correspondence is far too good to be fortuitous, even when one considers the expected modifications of shorelines resulting from erosion and deformation following the breakup of Pangaea some 200 million years ago. A still closer match results when one fits the continents together at the outer edge of the continental shelves, which are really the true edges of the continents. Such a computerized and error-tested fitting of continents was carried out by Sir Edward Bullard, J.E. Everett, and A.G. Smith of the University of Cambridge (Fig. 5-24). Remarkably, this work showed that over most of the boundary the average mismatch was no more than a degree — a snug fit indeed.

Another line of evidence favoring the drift theory involves sedimentologic criteria indicating similarity of climatic conditions for widely separated parts of the world that were once closely adjacent to one another. For example, in such widely separated places as South America, southern Africa, India, Antarctica, Australia, and Tasmania, one finds glacially grooved rock surfaces and glacial deposits that are relics of late Paleozoic continental glaciers. When these glacial deposits become lithified, they are called **tillites.** They, along with the grooves and scratches on rock surfaces that were apparently beneath the moving ice, attest to a great ice age that affected Gondwanaland at a time when it was as yet unfragmented and lying at or near the south polar region (Fig. 5-25). Furthermore, if the directions of the grooves on the bedrock are plotted on a map, they indicate the center of ice accumulation and the directions in which it moved. Unless the southern continents are reassembled into Gondwanaland, this center would be located in the ocean, and great ice sheets do not develop centers of accumulation in the ocean. Hence, the existence of Gondwanaland seems plausible. In a few instances, oddly foreign boulders in the tillites of one continent are found in the deposits of another continent now located thousands of miles across the oceans.

There are additional climatic clues as well. Trees of tropical regions have been found as fossils in coal seams now located far away from their place of origin in temperate or even polar locations. Similarly, thick

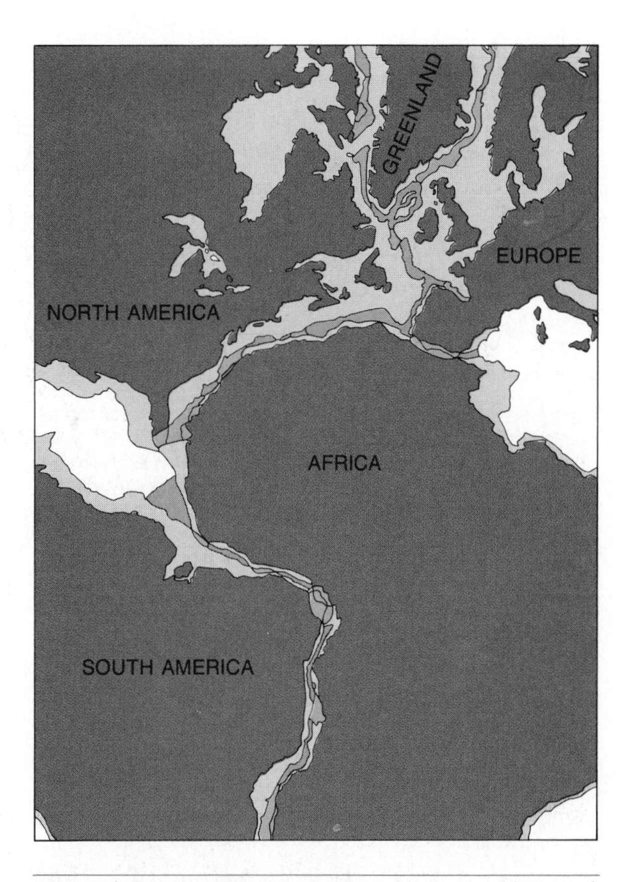

Figure 5-24 Fit of the continents as determined by Sir Edward Bullard, J.E. Everett, and A.G. Smith. The fit was made along the continental slope (light gray) at the 500-fathom contour line. Overlaps (light orange) and gaps (dark gray) are probably the result of deformation and sedimentation after rifting. (Adapted from Bullard, E.C., et al., Philos. Trans. R. Soc. Lond., *VA 258:41, 1965.)*

deposits of salt and gypsum that characteristically form in warm equatorial regions are now found far to the north of the equator (Fig. 5-26). One can only infer that they developed at lower latitudes and were subsequently transported to their present locations as cargo on moving continents.

A somewhat similar kind of evidence can be obtained by examining the locations of Permian reef deposits. Modern reef corals are restricted to a band around the earth that is within 30° of the equator, whereas ancient reef deposits are now found far to the north of the latitudes at which they had originated.

At least some of the paleontologic support for continental drift was well known to Suess and Wegener.

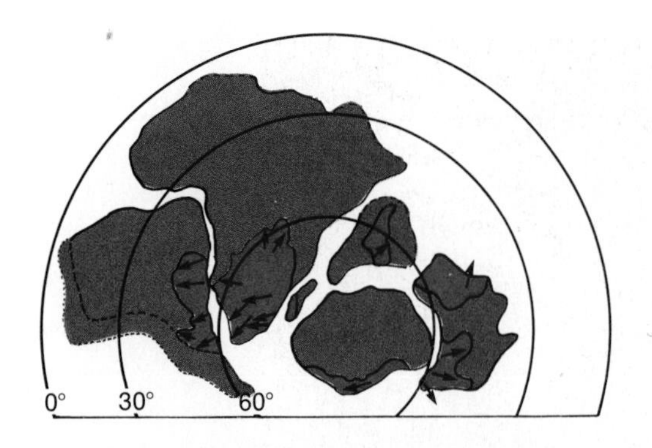

Figure 5–25 Reconstruction of Gondwanaland near the beginning of Permian time, showing the distribution of glacial deposits (shown in orange tint). Arrows show direction of ice movement as determined from glacial scratches on bedrock. (Modified from Hamilton, W., and Krinsley, D., Geol. Soc. Am. Bull., *78:783–799, 1967.)*

Usually, in the Gondwana strata overlying the tillites or glaciated surfaces, there can be found nonmarine sedimentary rocks and coal beds containing a distinctive assemblage of fossil plants. Named after a prominent member of the assemblage, the plants are referred to as the **Glossopteris flora** (Fig. 5–27). Paleobotanists who have supported the idea of shifting continents have argued that it would be virtually impossible for this complex temperate flora to have developed in identical ways on the southern continents as they are separated today. The seeds of *Glossopteris* were far too heavy to have been blown over such great distances of ocean by the wind.

Another element of paleontologic corroboration for the concept of moving land masses is provided by the distribution in the Southern Hemisphere of fossils of a small aquatic reptile named *Mesosaurus* (Fig. 5–28). This reptile lived in lakes and estuaries and was not adapted to life in the open ocean. The discovery of fossil remains of *Mesosaurus* both in Africa and in South America lends credence to the notion that these continents were once attached. Perhaps for a time, as they were just beginning to separate, the location of the present coastlines became dotted with lakes and protected bodies of water that harbored *Mesosaurus*. This seems a more plausible explanation than one requiring the creature to navigate the South Atlantic or

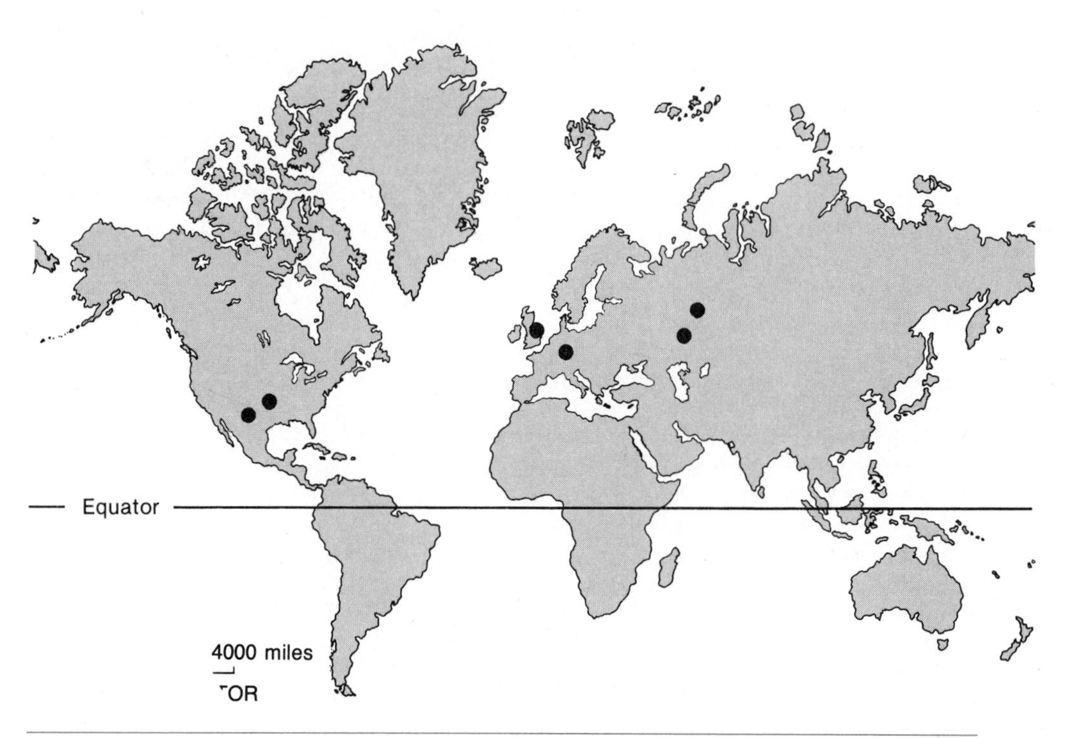

Figure 5–26 Location of prominent Permian evaporite deposits.

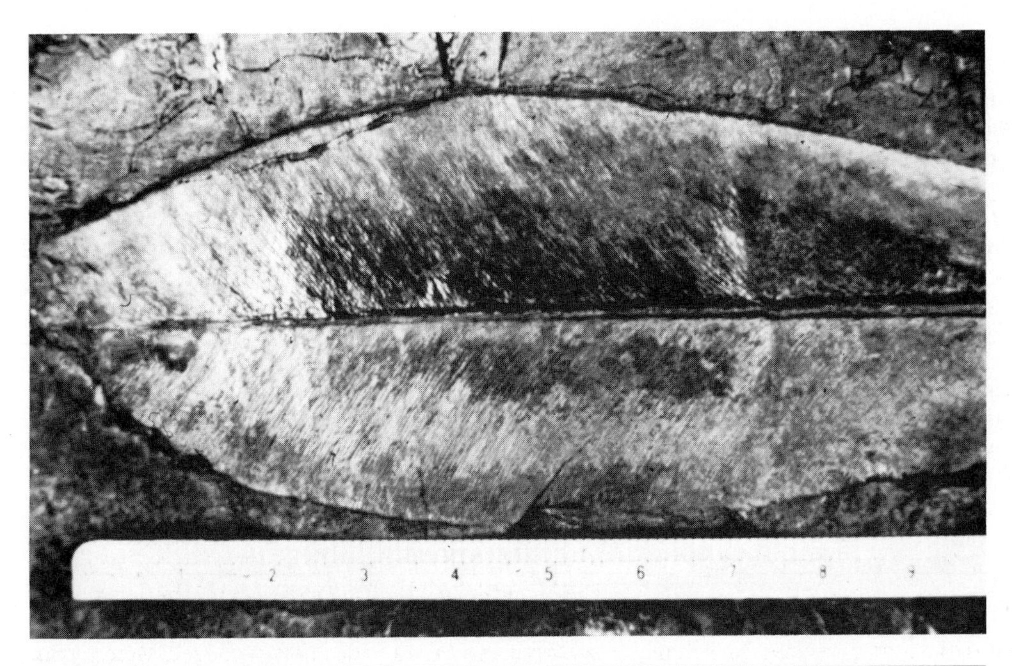

Figure 5–27 *Fossil* Glossopteris *leaf associated with coal deposits and derived from glossopterid forests of Permian age. This fossil was found on Polarstar Peak, Ellsworth Land, Antarctica. (Courtesy of U.S. Geological Survey, J.M. Schopf and C.J. Craddock.)*

to wend its way across formidable climatic barriers by following the shorelines northward and around the perimeter of the Atlantic.

Before fragmentation of a supercontinent such as Laurasia, one would expect to find numerous similar plants and animals living at corresponding latitudes on either side of the line of future separation. Such similarities do occur in the fossil record of Northern Hemisphere continents now separated by wide expanses of ocean. This, of course, indicates these lands were once

Figure 5–28 *Mesosaurus.*

adjoined and had uniform climate along similar latitudinal zones.

During the Cenozoic Era, separation of continents became pronounced, and therefore fossil faunas are distinctly different on different continents. As lands became separated from each other by ocean barriers, genetic isolation resulted in morphologic divergence. The modern world's enormous biologic diversity is at least partially a result of evolutionary processes operating on more or less isolated continents.

The character, sequence, age, and distribution of rock units have also been examined for evidence of drift. One might presume that locations close to one another on the hypothetic supercontinent would have environmental similarities that would result in similarities in the kinds of rocks deposited. As indicated by the correlation chart of southern continents (Table 5-1), there is such a similarity in the geologic sections of now widely separated land masses. There are also similarities in rock outcrops. Precambrian strata in central Gabon, Africa, for example, can be traced into the Bahia Province of Brazil. Also, isotopically dated Precambrian rocks of West Africa can be correlated with rocks of similar age in northeastern Brazil.

The testimony of paleomagnetism

One can sympathize with Alfred Wegener in his losing battle with the imposing Sir Harold Jeffries. As pre-

TABLE 5-1 **GONDWANA CORRELATIONS**

System	Southern Brazil	South Africa	Peninsular India
Cretaceous	Basalt	Marine Sediments	Volcanics Marine Sediments
Jurassic	(Jurassic Rocks Not Present)	Basalt	Sandstone and Shale Volcanics Sandstone and Shale
Triassic	Sandstone and Shale with Reptiles	Sandstone and Shale with Reptiles	Sandstone and Shale with Reptiles
Permian	Shale and Sandstones with *Glossopteris* Flora Shale with *Mesosaurus*	Shale and Sandstones with Coal and *Glossopteris* Shale with *Mesosaurus*	Shale and Sandstones with Coal and *Glossopteris* Shale
Carboniferous	Sandstone Shale and Coal with *Glossopteris* Tillite	Sandstone Shale and Coal with *Glossopteris* Tillite	Tillite

viously noted, Wegener's ideas came along a half century too early. The scientific discoveries of the past two decades would have provided him with evidence that even Jeffries would have had difficulty in refuting. The new information came from the study of magnetism that had been imparted to ancient minerals and rocks and preserved to the present time. To understand this **paleomagnetism,** as it is called, it is necessary to digress for a moment and consider the general nature of the earth's present magnetic field.

The earth's present magnetism

It is common knowledge that the earth has a magnetic field. It is this field that causes the alignment of a compass needle. The origin of the magnetic field is still a question that has not been fully resolved, but many geophysicists believe it is generated as the rotation of the earth causes slow movements in the liquid outer core. The magnetic lines of force resemble those that would be formed if there were an imaginary bar magnet extending through the earth's interior. The long axis of the magnet would be the conceptual equivalent of the earth's magnetic axis, and the ends would correspond to the north and south geomagnetic poles (Fig. 5–29). Although today the geomagnetic poles are located about 11° of latitude from the rotational axis, they slowly shift position. When averaged over several thousand years, the geomagnetic poles and the geographic poles do coincide. If we assume that this relationship has always held true, then by calculating ancient magnetic pole positions from paleomagnetism in rocks, we have coincidentally located the earth's geographic poles. It should be kept in mind, however, that such interpretations are based on the supposition that the rotational and magnetic poles have always been

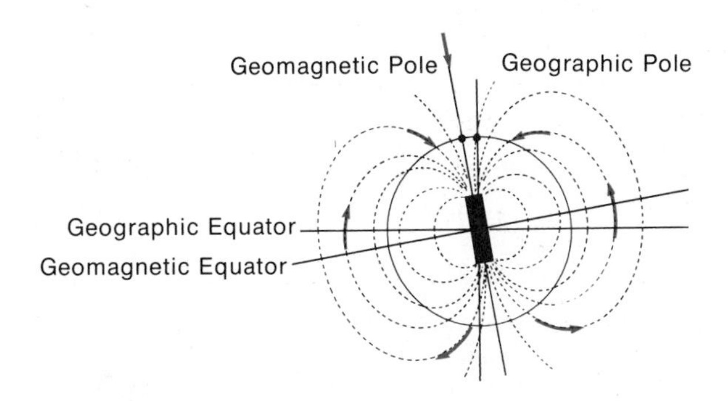

Figure 5–29 *Magnetic lines of force for a simple dipole model of the earth's magnetic field.*

relatively close together. This seems a reasonable assumption based on the modern condition, as well as on paleontologic studies that have shown inferred ancient climatic zones in plausible locations relative to ancient pole positions. Another assumption is that the earth has always been dipolar. Paleomagnetic studies from around the world support this supposition as well.

Remanent magnetism

The magnetic information frozen into rocks, called **remanent magnetism,** or **RM,** may originate in several ways. Imagine for a moment the outpouring of lava from a volcano. As the lava begins to cool, magnetic iron oxide minerals form and align their polarity with the earth's magnetic field. The alignment is then retained in the rock as it solidifies. In a simple analogy, the magnetic orientations of the minerals respond as if they were tiny compass needles immersed in a viscous liquid. Because they are aligned parallel to the magnetic lines of force surrounding the earth, they not only point the way toward the poles (magnetic declination) but also become increasingly more inclined from the horizontal as the poles are approached (Fig. 5–30). This inclination, when detected in paleomagnetic analysis, can be used to determine the latitude at which an igneous body containing magnetic minerals cooled and solidified.

Igneous rocks are not the only kinds of earth materials that can acquire remanent magnetism. In lakes and seas that receive sediments from the erosion of nearby land areas, detrital grains of magnetite settle slowly through the water and rotate so that their directions of magnetization parallel the earth's magnetic field. They may continue to move into alignment while the sediment is still wet and uncompacted, but once the sediment is cemented or compacted, the depositional remanent magnetism is locked in.

Over the past two decades, geophysicists have been measuring and accumulating paleomagnetic data for all the major divisions of geologic time. Their results are partly responsible for the recent revival of interest in drift hypotheses. For example, when ancient pole positions were located on maps, it appeared that they were in different positions relative to a particular continent at different periods of time in the geologic past. Either the poles had moved relative to stationary continents, or the poles had remained in fixed positions while the continents shifted about. If the poles were wandering and the continents "stayed put," then a geophysicist working on the paleomagne-

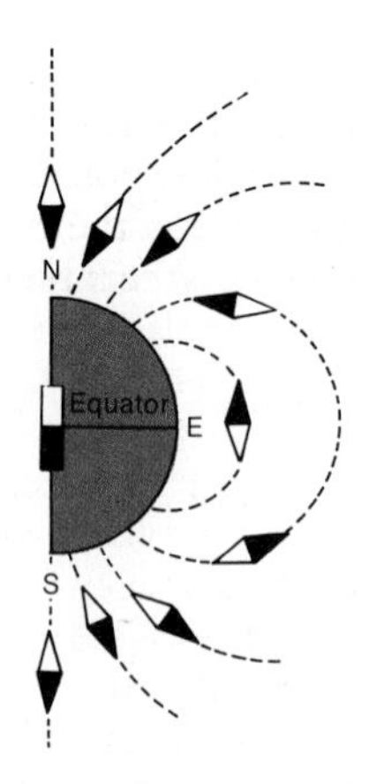

Figure 5–30 A freely suspended compass needle aligns itself in the direction of the earth's magnetic field. The inclination of the needle will vary from horizontal at the equator to vertical at the poles.

tism of Ordovician rocks in France should arrive at the same location for the Ordovician poles as a geophysicist doing similar work on Ordovician rocks from the United States. In short, the paleomagnetically determined pole positions for a particular age should be the same for all continents. On the other hand, if the continents had moved and the poles were fixed, then we should find that pole positions for a particular geologic time would be different for different continents. The data suggest that this latter situation is the more valid.

Another way to view the data of paleomagnetism is to examine what are called **polar wandering curves** (the name implies that the poles are moving, but as we have just noted, this is not likely). These curves are merely lines on a map connecting pole positions relative to a specific continent for various times during the geologic past. As shown in Figure 5–31, the curves for North America and Europe meet in recent time at the present North Pole. This means that the paleomagnetic data from recently formed rocks from both continents indicate the same pole position. A plot of the more ancient poles results in two similarly shaped but increasingly divergent curves. If this divergence resulted from a drifting apart of Europe and North America, then one should be able to reverse the movements mentally and see if the curves do not come together. Indeed, the Paleozoic portions of the polar wandering curves can be brought into close accord if North America and its curve were to be slid eastward

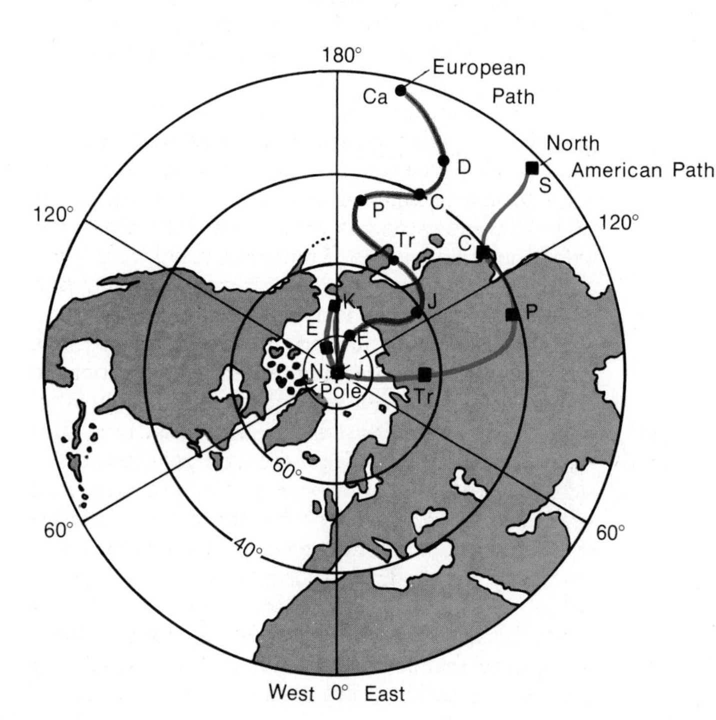

Figure 5–31 Apparent paths of polar wandering for Europe and North America. A scatter of points have been averaged to a single point for each geologic period: Ca, *Cambrian;* S, *Silurian;* D, *Devonian;* C, *Carboniferous;* P, *Permian;* Tr, *Triassic;* J, *Jurassic;* K, *Cretaceous;* E, *Eocene. (After Bott, M.H.P.,* The Interior of the Earth: Structure and Processes. *New York, St. Martin's Press, 1971.)*

about 30° toward Europe. These paleomagnetic data gave enormous credibility to the new models of plate tectonics.

The basic plate tectonics concept

Tectonics is a remarkably simple concept. The lithosphere, or outer shell of the earth (Fig. 5-32), is constructed of 6 or 7 huge slabs and about 20 smaller plates that are squeezed in between them. The larger plates (Fig. 5-33) are approximately 75 to 125 km thick. Movement of the plates causes them to converge, diverge, or slide past one another, and this results in frequent earthquakes along plate margins. When the locations of earthquakes are plotted on the world map, they clearly define the plate boundaries (Fig. 5-34).

A tectonic plate containing a continent would have the configuration shown in Figure 5-32. Plates "float" upon a weak, partially molten region of the upper mantle called the **asthenosphere** (from the Greek *asthenos,* meaning "weak"). Geophysicists view the asthenosphere as a region of rock plasticity and

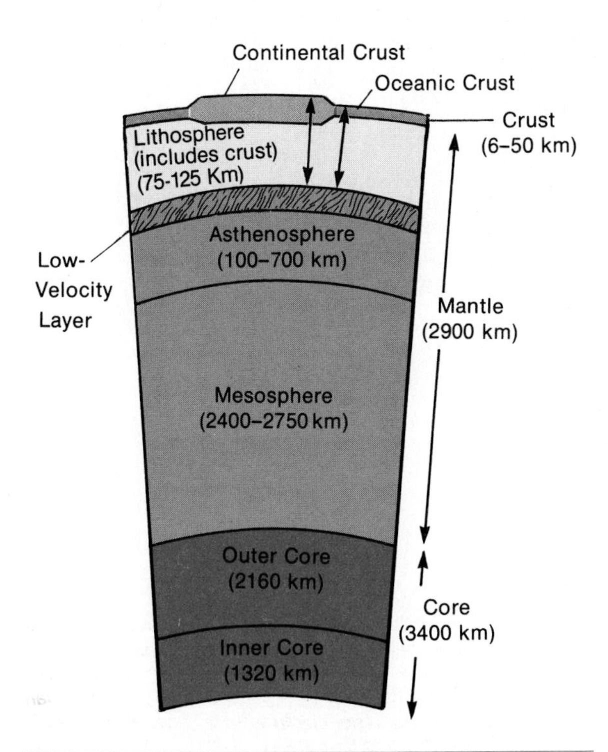

Figure 5-32 *Divisions of the earth's interior. (For clarity, the divisions have not been drawn to scale.)*

flowage. Its presence was first detected by Beno Gutenberg in the 1950s on the basis of changes in seismic wave velocities. It should be noted that the boundary between the lithosphere and the asthenosphere does not coincide with the crust-mantle boundary.

Plate boundaries and sea floor spreading

Central to the idea of plate tectonics is the differential movement of lithospheric plates. For example, plates tend to move apart at **divergent plate boundaries,** which may manifest themselves as midocean ridges complete with tensional ("pull apart") geologic structures. Indeed, the mid-Atlantic Ridge approximates the line of separation between the Eurasian and African plate on the one hand and the American plate on the other.

As is to be expected, such a rending of the crust is accompanied by earthquakes and enormous outpourings of volcanic materials that are piled high to produce the ridge itself. The void between the separating plates is also filled with this molten rock, which rises from below the lithosphere and solidifies in the fissure. Thus, new crust (i.e., new sea floor) is added to the **trailing edge** of each separating plate as it moves slowly away from the midocean ridge (Fig. 5-35). The process has been appropriately named **sea floor spreading.** Zones of divergence may originate beneath continents, rupturing the overlying land mass and producing rift features like the Red Sea and Gulf of Aden.

The axis of spreading is not a smoothly curving line. Rather, variations in spreading rates require that the plates be offset by numerous faults. These features are **transform faults** and are an expected consequence of horizontal spreading of the sea floor along the earth's curved surface. Transform faults take their name from the fact that the fault is transformed into something different at its two ends. The relative motions of transform faults are shown in Figure 5-36. The ridge acts as a spreading center that exists both to the north and to the south of the fault. The rate of relative movement on the opposite sides along fault segment X to X′ depends upon the rate of extrusion of new crust at the ridge. Because of the spreading of the sea floor outward from the ridge, the apparent relative movement is opposite to that expected by ordinary fault movement. Thus, at first glance, the ridge-to-ridge transform fault (Fig. 5-36A) appears to be a left lateral fault, but the actual movement along segment X-X′ is really right lateral. Notice, further, that only the X-X′ is really right lateral and that only the X-X′

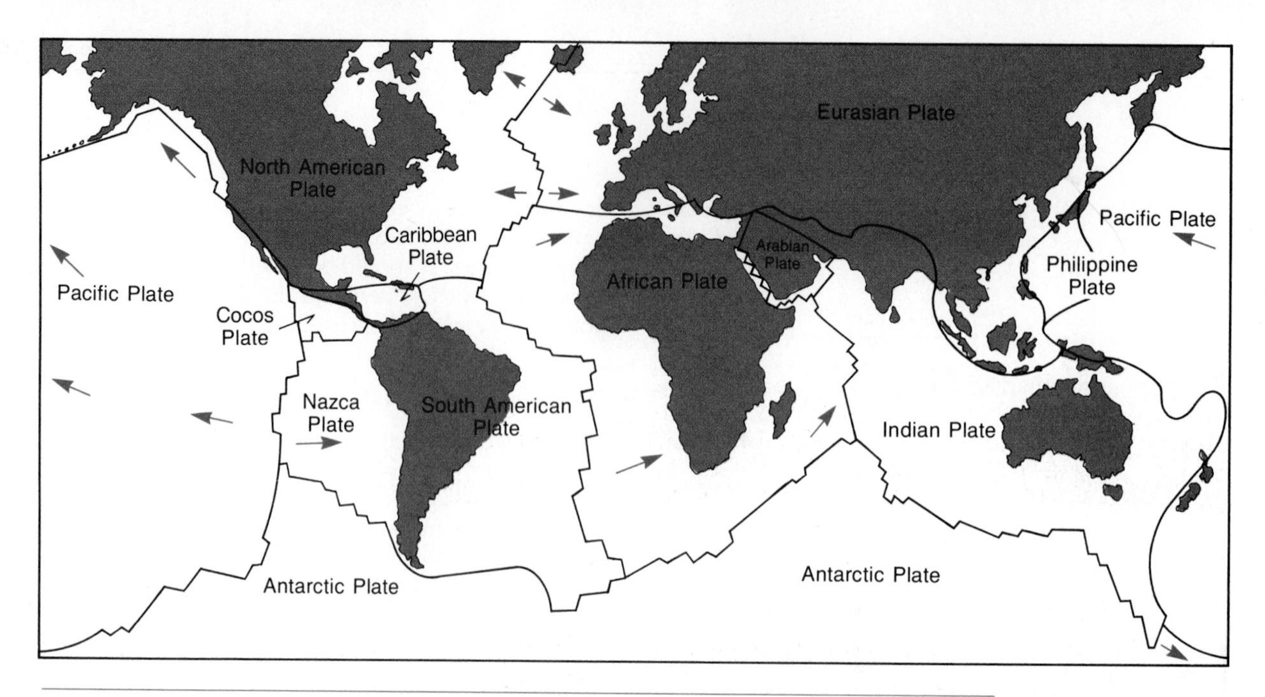

Figure 5–33 The earth's major tectonic plates. Arrows indicate general direction of movement of plates.

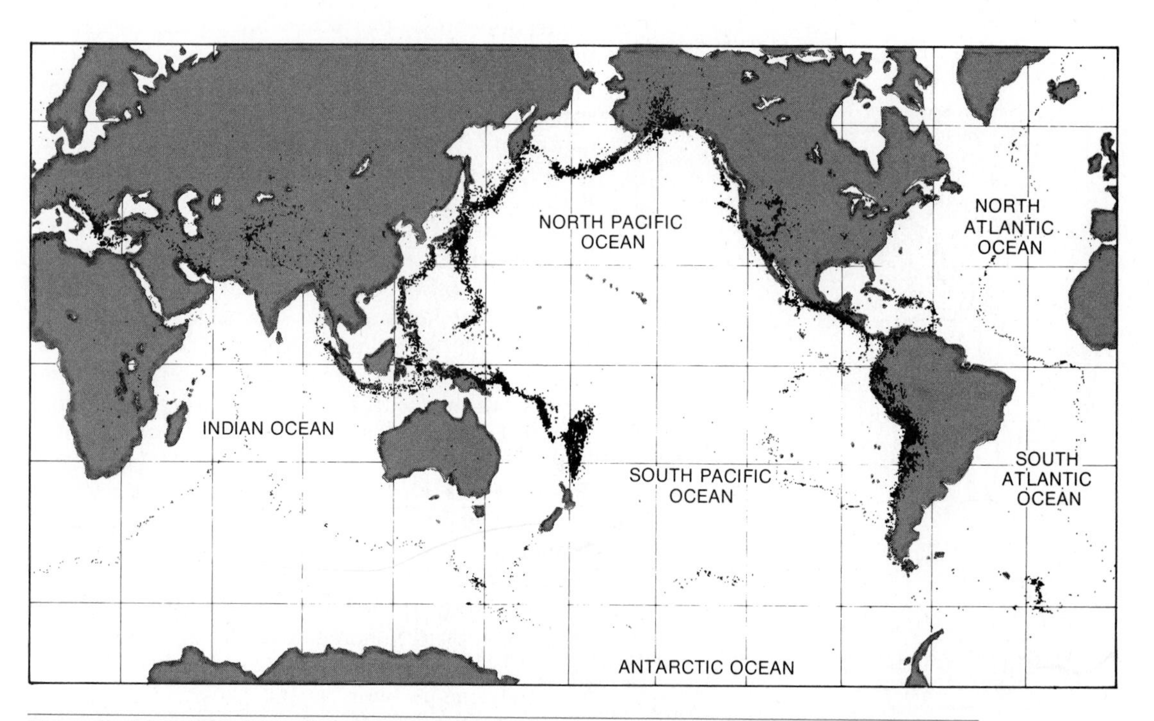

Figure 5–34 World distribution of earthquakes. Notice the major earthquake belt that encircles the Pacific and another that extends eastward from the Mediterranean toward the Himalayas and East Indies. The midocean ridges are also the site of many earthquakes. (From Turk, J., and Turk, A., Physical Science, *2nd ed. Philadelphia, Saunders College Publishing, 1981.)*

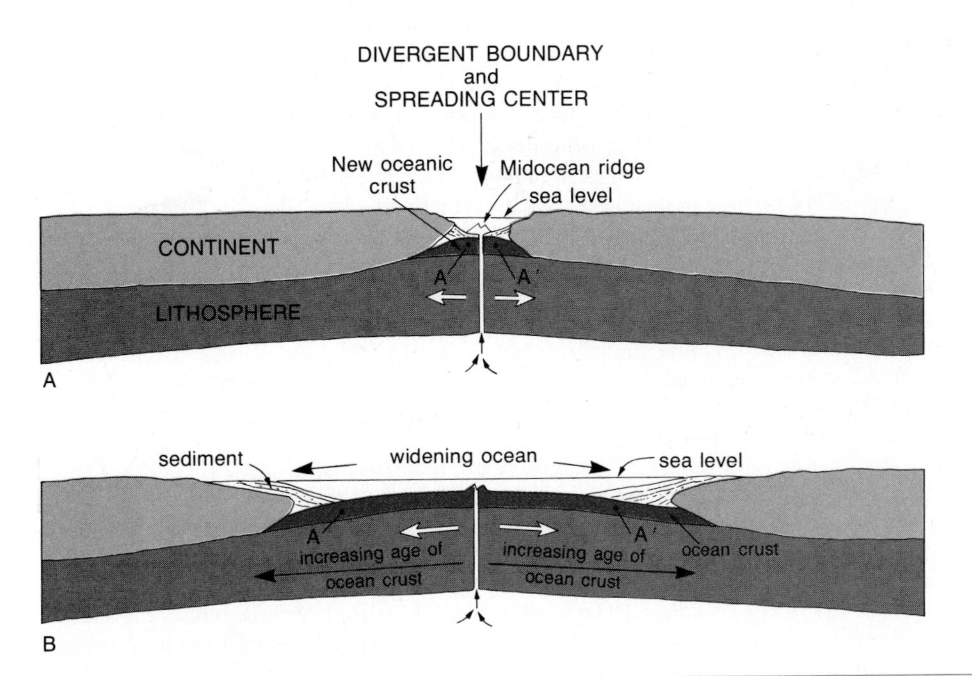

Figure 5–35 *Sea floor spreading. As the rift widens, basaltic lavas rise to fill the space and solidify to become a new part of the trailing edge of a tectonic plate. A and A′ are reference points.*

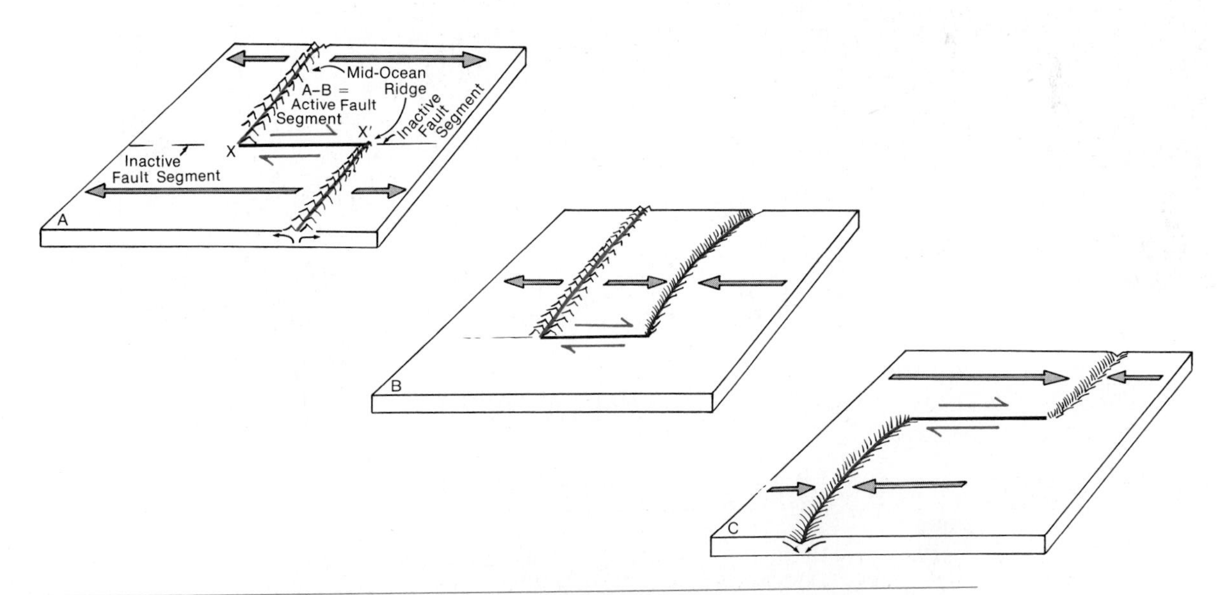

Figure 5–36 *Three types of transform faults. (A) Ridge-ridge transform. (B) Ridge-trench. (C) Trench-trench.*

segment is active. Along this segment, seismic activity is particularly frequent. To the west of X and the east of X′ there is little or no relative movement except for sinking as crustal materials cool.

If plates are receding from one another at one boundary, they may be expected to collide or slide past other plates at other boundaries. Thus, in addition to the divergent plate boundaries that occur along mid-ocean ridges, there are **convergent** and **shear boundaries.** Convergent plate boundaries develop when two plates move toward one another and collide. As one might guess, these convergent junctions are characterized by a high frequency of earthquakes. In addition, they are thought to be the zones along which folded mountain ranges or deep-sea trenches may develop (Fig. 5-37). The structural configuration of the convergent boundary is likely to vary according to the rate of spreading and whether the leading edges of the plates are composed of oceanic or continental crust. Geophysicists speculate that when the plates collide, one slab may slip and plunge below the other, producing what is called a **subduction zone.** The sediments and other rocks of this plunging plate are pulled downward (subducted), melted at depth, and, much later, rise to become lavas and plutons incorporated into the materials of the upper mantle and crust.

An example of a shear plate boundary is the well-known San Andreas fault in California. Along this great fault, part of the Pacific plate moves laterally against the American plate (Fig. 5-38). Shear plate boundaries are decidedly earthquake-prone but are less likely to develop intense igneous activity. They are the active segments of transform faults along which no new surface is formed or old surface consumed.

Crustal behavior at plate boundaries

We have noted that there are three basic kinds of plate boundaries: divergent, convergent, and shear. There are also three kinds of crustal behavior that may occur at colliding plate edges; however, the behavior may change from one region of a boundary to another. If the leading edge of a plate happens to be composed of continental crust, and it collides with a similarly continental opposing plate margin, the result is a folded mountain range in which are developed igneous rocks of granitic composition (see Figure 5-37A). In this

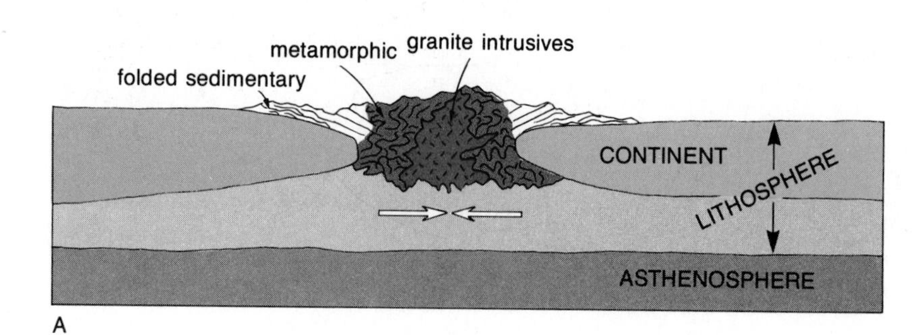

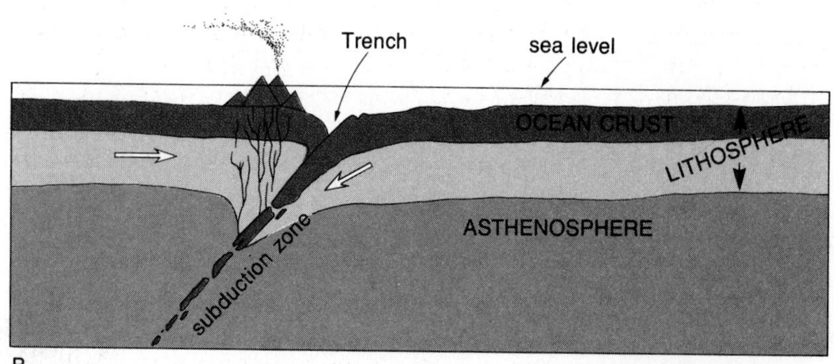

Figure 5-37 Two types of convergent plate boundaries. (A) Convergence of two plates, both bearing continents. (B) Convergence of two plates, both bearing ocean crust.

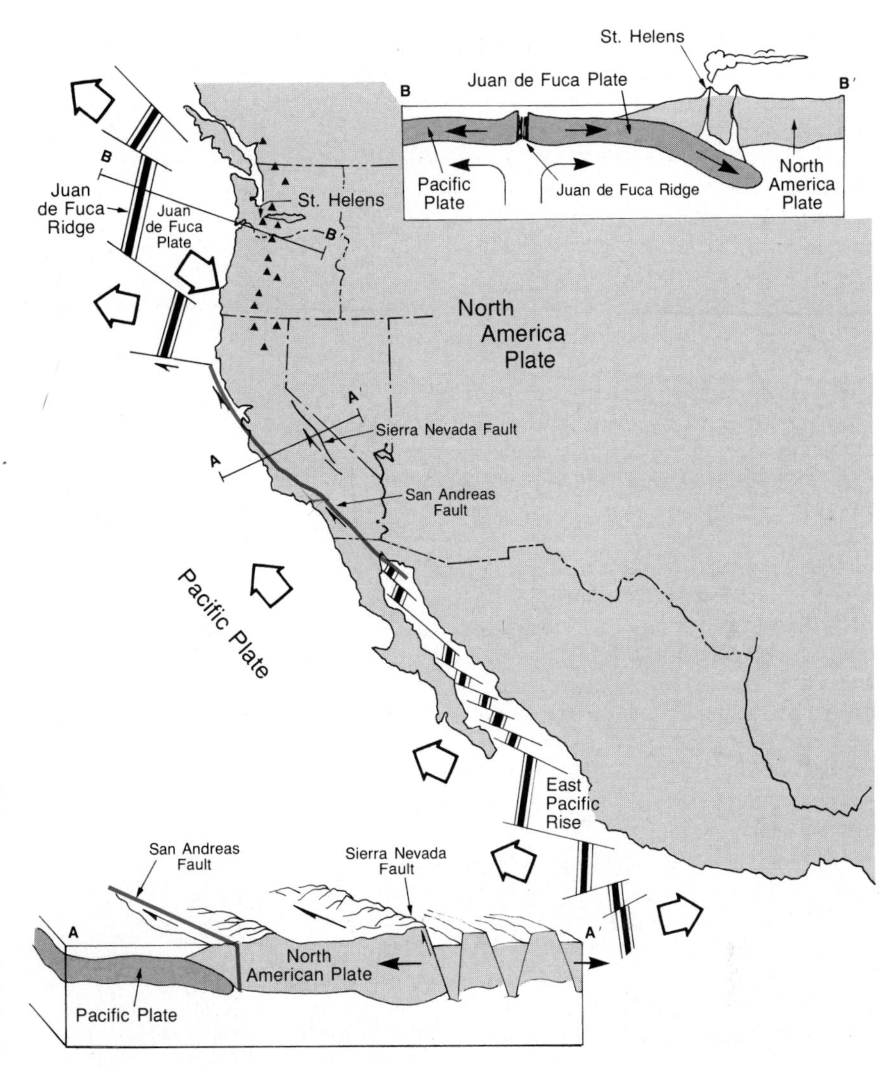

Figure 5–38 The juncture of the North American and Pacific Tectonic Plates. The heavy lines are spreading centers. Note the trace of the San Andreas fault. To the north in Oregon and Washington, the small Juan de Fuca Plate plunges beneath the North American Continent to form the Cascades. (Courtesy of U.S. Geological Survey.)

kind of collision, subduction does not occur because the continental plates are too buoyant to be carried down into the asthenosphere. The Himalayan and related ranges were formed when the plates carrying Africa, Arabia, and India converged on the Eurasian plate. The zone of convergence between the once-separate continental masses is called a **suture zone.** The Appalachian and Ural mountains may also have formed in such a smash-up of continental plate margins. It is likely, for example, that there was an earlier supercontinent than the one envisioned by Wegener. This earlier supercontinent was probably present at least by the end of the Precambrian. It evidently broke apart early in the Paleozoic, and as the segments began to move away from one another, they created an Appalachian Ocean and Uralian Ocean along the widening rifts. Much later the segments moved back toward one another and converged along a suture line to form the Appalachians and Urals. The recombined continent was the Pangaea of Alfred Wegener.

The second kind of convergence situation at plate boundaries involves the meeting of two plates that both have oceanic crust at their converging margins (Fig. 5–37B). Although the rate of plate movement in an ocean-ocean convergence will affect the kinds of structures produced, it is likely that such locations will develop deep-sea trenches with bordering volcanic arcs, such as those of the southwestern Pacific.

Finally, there is the third possibility. It involves the collision of a continental (granitic) plate boundary

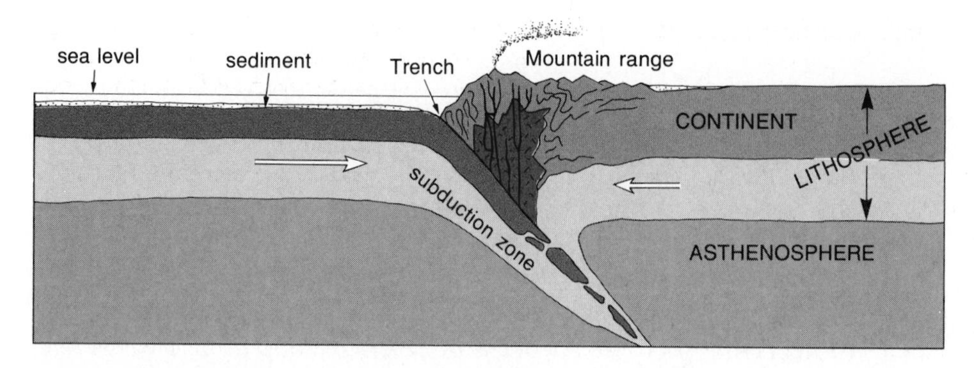

Figure 5–39 Collision of a plate bearing a continent with an ocean plate. The leading edge of the overriding continental plate is crumpled, whereas the oceanic plate buckles downward, creating a trench offshore. The situation generalized here is similar to that off the west coast of South America, where the Nazca Plate plunges beneath the South American Plate.

with an oceanic (basaltic) one (Fig. 5–39). The result of such a collision might be a deep-sea trench located offshore from an associated range of mountains. Numerous volcanoes would be expected to develop, and the lavas pouring from their eruptions would probably be a compositional blend of granite and basalt. The Andes are such a mountain range, and the igneous rock *andesite,* named for its prevalence in the Andes, may represent such a silicate blend.

Regions of ocean-continent convergence are characterized by rather distinctive rock assemblages and geologic structure. As we have seen, the convergence of two lithospheric slabs results in subduction of the oceanic plate, whereas the more buoyant continental plate maintains its position at the surface but experiences intense deformation, metamorphism, and melting. A great mountain range begins to take form as a result of all of this dynamic activity. At the same time, sediments and submarine volcanic rocks along the subduction zone are squeezed, sheared, and shoved into a gigantic, chaotic medley of complexly disturbed rocks called a **mélange.**

Associated with the tectonic mélange of active or once-active continental margins one frequently finds a distinctive rock assemblage consisting of serpentinized peridotite at the base, overlain successively by gabbros, basaltic dikes, pillow lavas, fossiliferous deep sediment (e.g., radiolarian cherts), and turbidites (Fig. 5–40). These rocks constitute an **ophiolite suite.** Ophiolites are actually splinters of the oceanic crust that were scraped off during subduction and plastered against the overriding continental plate. They mark the location of plate convergence and former subduction

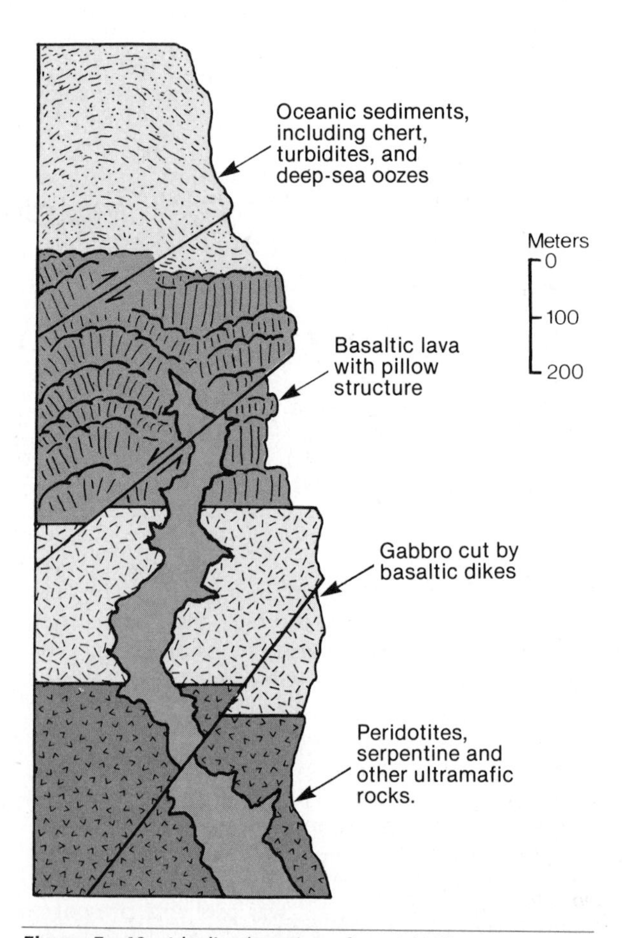

Figure 5–40 Idealized section of an ophiolite suite. Ophiolites are thought to be splinters of the ocean floor squeezed into the continental margin during plate convergence.

zones in ancient mountain ranges like the Alps, Appalachians, and Himalayas.

Another clue to the presence of ancient subduction zones is a distinctive kind of metamorphic rock containing blue amphiboles. These rocks are called blueschists. They form at high pressures but relatively low temperatures. This rather unusual combination of conditions is characteristic of subduction zones where the relatively cool oceanic plate plunges rapidly into deep zones of high pressure. Blueschists are found today in such geologically young mountains as the Alps, Himalayas, and the mountains around the margins of the Pacific Ocean.

What drives it all?

If we accept the premise that plates of lithosphere do move across the surface of the globe, then the next question is how they are moved. The propelling mechanism is thought by some scientists to consist of large thermal convection cells that flow like currents of thick liquid and are provided with heat from the decay of radioactive minerals (Fig. 5-41A). The convection cells are believed to be located in the asthenosphere or, less likely, in the deeper parts of the mantle. Currents rise in response to heating from below, diverge, and spread to either side. As they move laterally, they carry along with them the overlying slab of lithosphere and its surficial layers of sediment. Mantle material upwelling along the line of separation would join the trailing edges of the plates on either side. Ultimately, the convecting current would encounter a similar current coming from the opposite direction, and both viscous streams would descend into the deeper parts of the mantle to be reheated and moved toward the direction of an upwelling. Above the descending flow, one might expect to find subduction zones and deep-sea trenches, whereas midocean ridges would mark the locations of ascending flows.

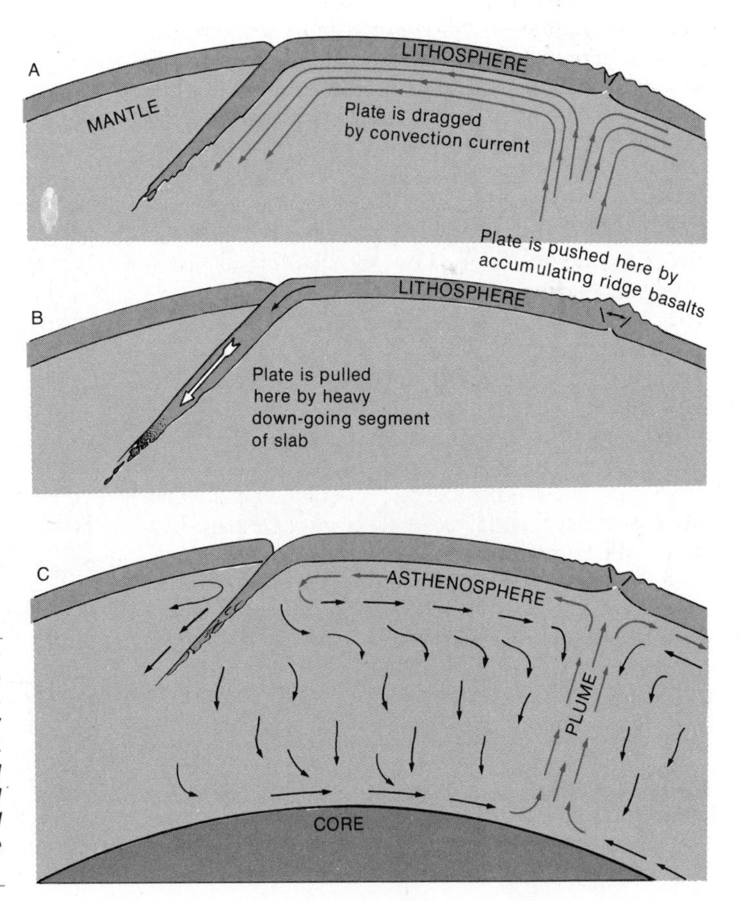

Figure 5-41 Three models that have been suggested as possible driving mechanisms for plate movements. (A) Plate dragged by roll-like convection movements in the mantle. (B) The "push-pull" model in which plates are pushed laterally at spreading centers, and pulled by the plunging cool and dense leading segment of the plate. (C) Thermal plume model in which upward movement is confined to thermal plumes that spread laterally and drag the lithosphere along.

The thermal convection mechanism for moving lithospheric plates is only a theory and should not be regarded as established fact. As yet, there has not been an entirely satisfactory way to test the concept, although such currents may be physically plausible and there is some tenuous evidence that they exist. For example, heat flows from the earth's interior at a greater rate along midocean ridges than from adjacent abyssal plains. Yet analyses of hundreds of measurements by geophysicists at Columbia University suggest that although heat flow along the mid-Atlantic Ridge is 20 percent greater than on the adjacent floors of the ocean basins, it is still not great enough to move the sea floor at the rates suggested by paleomagnetic studies. Another problem relates to the layering in the upper mantle that has been detected in seismologic studies. The mixing that accompanies convectional overturn would seem to preclude such layering.

The search for a mechanism to drive lithospheric plates has produced another convection-related model. In this so-called *thermal plume model* (Fig. 5-41C), mantle material does not circulate in great rolls but rather rises from near the core-mantle boundary in a manner suggestive of the shape and motion of a thundercloud. Proponents of this model suggest there may be as many as 20 of these thermal plumes, each a few hundred kilometers in diameter. According to the model, when a plume nears the lithosphere, it spreads laterally, doming surficial zones of the earth and moving them along in the directions of radial flow. The center of the Afar Triangle in Ethiopia (Fig. 5-42) has been suggested as the site of a plume that flowed upward and outward, carrying the Arabian, African, and Somali plates away from the center of the triangle.

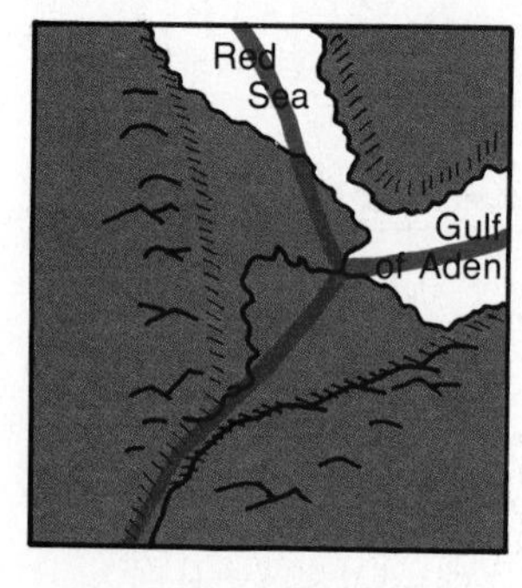

Figure 5-42 Rising plumes of hot mantle material may cause severe rifts, often forming 120-degree angles with one another. An example is the Afar Triangle, shown at the south end of the Red Sea on the small map of Africa.

Faced with the uncertainties in all existing theories for plate-moving forces, one should keep an open mind about other, perhaps less popular, hypotheses. Among other mechanisms proposed are "slab-pull" and "ridge-push." Slab-pull is thought to operate at the subduction zone, where the subducting oceanic plate, being colder and denser than the surrounding mantle, sinks actively through the less dense mantle, *pulling* the rest of the slab along as it does so (see Fig. 5-41B). The ridge-push mechanism, which may operate independently of or along with slab-pull, results from the fact that the spreading centers stand high on the ocean floor and have low-density roots. This causes them to spread out on either side of the midocean ridge and transmit this "push" to the tectonic plate. Other hypotheses suggest rafting of plates in response to stresses induced by differential rotation between the core and mantle, which might then be transmitted to the crust. In the years ahead, studies of rock behavior under high temperature and pressure, geophysical probings of the mantle, and careful field studies will reveal which of the proposed mechanisms can be most confidently accepted.

Tests of the theory of plate tectonics

Paleomagnetic evidence

We have examined the various lines of evidence supporting Wegener's notion of continental drift. Nearly all of these clues also support the newer concepts embodied in the theory of plate tectonics. The earlier clues were based mostly on evidence found on land. The new theory, with its keystone concept of sea floor spreading, was developed from evidence gleaned from the sea floor. That evidence was derived from sensitive magnetometers that were being carried back and forth across the oceans by research vessels. These instruments were able to detect not only the earth's main geomagnetic field but also local magnetic differences or magnetic anomalies frozen into the rocks of the sea floor. Maps of the anomalies exhibited linear bands of high and low field-magnetic intensities parallel to midocean ridges (Fig. 5–43). Directly over the ridge, the earth's magnetic field was 1 percent stronger than expected. Adjacent to this zone the field was somewhat weaker than would have been predicted, and the changes from weaker to stronger and back again often occurred in distances of only a few score miles.

In 1963, F.J. Vine, a 23-year-old research student at Cambridge University, and a senior colleague, Drummond Mathews, suggested that these variations in intensity were caused by reversals in the polarity of the earth's magnetic field. The magnetometers towed behind the research vessels provided measurements that were the sum of the earth's present magnetic field strength and the paleomagnetism frozen in the crustal rocks of the ocean floor. If the paleomagnetic polarity was opposite in sign to that of the earth's present magnetic field, the sum would be less than the present magnetic field strength. This would indicate that the crust over which the ship was passing had reversed paleomagnetic polarity. Conversely, where the paleomagnetic polarity of the sea floor basalts was the same as that of the present magnetic field, the sum would be greater, and a normal polarity would be indicated.

Since 1963, geophysicists have learned that these irregularly occurring reversals have not been at all infrequent. During the past 70 million years, the earth's magnetic field has seen many episodes when the polarity was opposite to that of today. These changes in polarity were incorporated into the remanent magnetism of the lavas extruded at the midocean ridges. The lava acquired the magnetic polarity present at the time of extrusion and then moved out laterally, as has been previously described. If during that time the earth's polarity was as it is today, then the stripe is said to represent "normal" polarity. If the earth's normal polarity reversed, then the band of extruded lavas that followed behind the previous band would acquire "reverse" polarity as it cooled. As the process repeated itself through time, the result would be symmetrical, mirror-image patterns of normal and reverse bands on either side of spreading centers like the midocean ridges (Fig. 5–44).

The stripes show that geologists were correct in depicting the sea floor as spreading to both sides from the ridges, and they even provide a way of measuring the rates of spreading. For example, a specific anomaly on the sea floor can be correlated with anomalies obtained from rocks of known absolute age on land. Having this correlation with which to date the oceanic rocks of a particular anomaly, one may use the distance of those sea floor rocks from the crest of the mid-

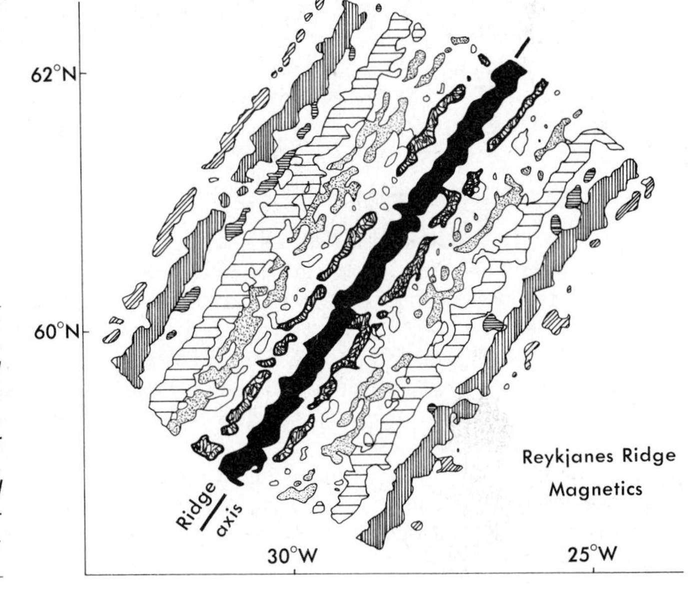

Figure 5–43 The magnetic field produced by the ocean-floor rocks over the Reykjanes Ridge near Iceland. Note the similarity to Figure 5–44. The outlined areas represent predominantly normal magnetization. The age of the rocks increases away from the ridge. (After Heirtzler, J.R., et al., Magnetic anomalies over the Keykjanes ridge. Deep Sea Res., *13:427–443, 1966; and Vine, F.J., Magnetic anomalies associated with mid-ocean ridges.* In *Phinney, R.A. (ed.),* The History of the Earth's Crust, A Symposium. *Princeton, Princeton University Press, 1968.)*

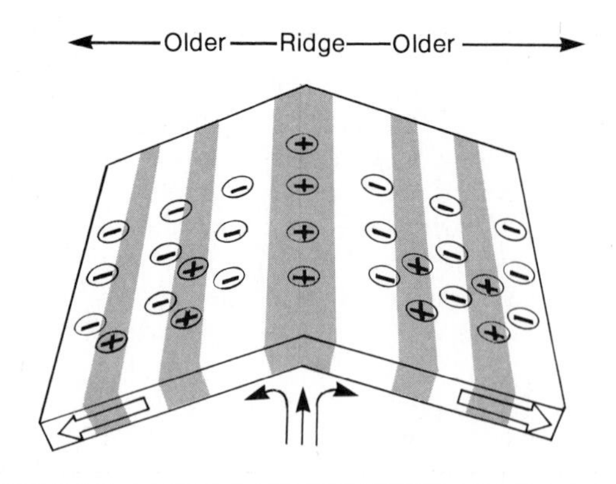

Figure 5–44 The normal (+) and reversed (−) magnetizations of the sea floor. Note the symmetry of the magnetizations with respect to the ridge. (From McCormick, J.M., and Thiruvathukal, J.V., Elements of Oceanography, *2nd ed. Philadelphia, Saunders College Publishing, 1981.)*

oceanic ridge to calculate the rate of spreading and sometimes even to reconstruct the former positions of continents. Figure 5–45 provides an illustration of how this can be done. For example, if one wishes to know the distance between the eastern coast of the United States and the northeastern coast of Africa about 81 million years ago, one simply brings together the traces of the two 81-million-year-old magnetic stripes, being careful to move the "sea floor" parallel to the transform faults, which indicate the direction of movement.

The velocity of plate movement is not uniform around the world. Plates that include large continents tend to move slowly. Their velocity relative to the underlying mantle rarely exceeds 2 cm per year. Plates that are largely devoid of large continents have average velocities of 6 to 9 cm per year.

Evidence from ocean sediment

If newly formed crust joined the ocean floor at spreading centers and then moved outward to either side, then the sediments, dated by the fossil planktonic organisms they contain, could be no older than the surface on which they came to rest. Near a midocean ridge, the sediments directly over basalt should be relatively young. Samples of the first sedimentary layer above the basalt but progressively farther away from the spreading center should be older (Fig. 5–46).

In a succession of cruises that began in 1969, the American drilling ship *Glomar Challenger* has collected ample evidence to prove that sediments above the basalt increase in age as distance from the ridge increases. The studies carried out by scientists associated with the drilling project confirmed earlier assumptions that there were no sediments older than about 200 million years on the sea floor, that the sediments on the sea floor were relatively thin, and that they became thinnest closer to the midocean ridges. Spreading provided a logical reason for these observations. Oceanic plate surfaces were simply not in existence long enough to accumulate a thick section of sediments. The sea floor clays and oozes were conveyed to subduction zones, dragged back down into the mantle, and resorbed. Thus, the entire world ocean is virtually swept free of deep-sea sediments every 200 or 300 million years.

Measurements from outer space

The space program has provided yet another method for measuring the movement of tectonic plates. This new method utilizes laser devices, which emit an intense beam of light of definite wavelength. The beam is capable of traveling across an enormous distance without becoming dispersed. During the Apollo missions, astronauts placed three clusters of special reflectors on the moon. Laser beams from earth are "fired" toward the moon, and by recording the time required for the signal to travel to the moon and return to earth, one can determine the distance between the laser device and the lunar reflector. The method is accurate to within only 3 cm. Distances from the moon to stations on two moving tectonic plates will, of course, change over a period of time, and such measurements can be used to calculate how fast continents are moving toward or away from one another. Currently, measurements indicate that the Atlantic Ocean is widening by about 1.4 cm per year.

Seismologic evidence

The evidence for the actual movement of lithospheric plates appears to be fairly direct and susceptible to several methods of testing. It is more difficult to prove that plates of the lithosphere are dragged back down into the mantle and resorbed. The best indications that subduction actually occurs come from the study of earthquakes generated along the seismically active colliding edges of plates.

Well known among these studies are the investigations during the 1940s by Hugo Benioff, a seismologist and designer of improved seismographs. Benioff and subsequent investigators compiled records of

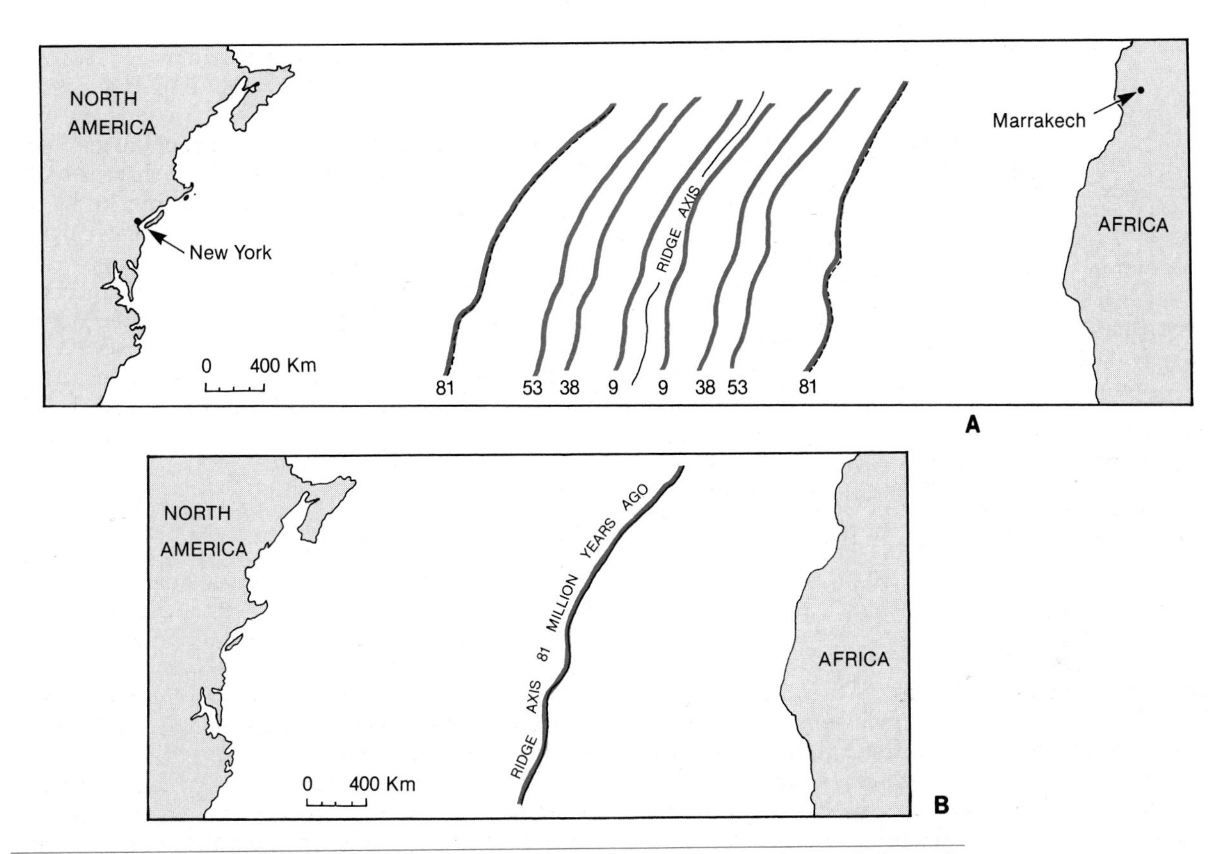

Figure 5–45 *A method for determining the paleogeographic relations of continents. A shows the present North Atlantic with ages plotted for some of the magnetic stripes. To see the location of North America relative to Africa, say, 81 million years ago, the 81-million-year-old bands are brought together. The result is the view of the much narrower North Atlantic of 81 million years ago shown in B. (Modified from Pitman, W.C., III, and Talwani, M.,* Geol. Soc. Am. Bull., *83:619–644, 1972.)*

earthquake foci that occurred along presumed plate boundaries. The charting of the data showed that the deeper earthquake foci occur along a narrow zone that was tilted at an angle of 45° under the adjacent island arc or continent (Fig. 5–47). This inclined seismic shear zone was traced to depths as great as 700 km. It was named the **Benioff Seismic Zone** after its discoverer. These inclined earthquake zones are believed to define the positions of the subducted plates where they plunge into the mantle beneath the overriding plate. The earthquakes are the result of fracturing and subsequent rupture as the cool, brittle lithospheric plate descends into the hot mantle.

In addition to the deep and intermediate focus earthquakes along the Benioff Zone, there are intense and often destructive earthquakes at shallow depths

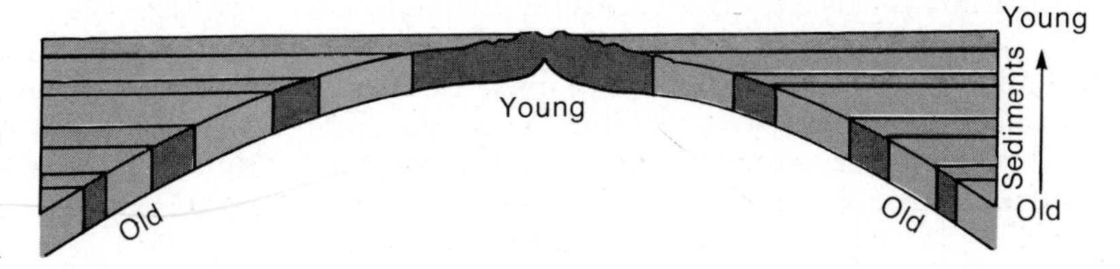

Figure 5–46 *As a result of sea floor spreading, sediments near the midocean ridge are youngest, and sediments directly above the basaltic crust but progressively farther away from the ridge are sequentially older. (Normal polarity = gray; reverse polarity = orange.)*

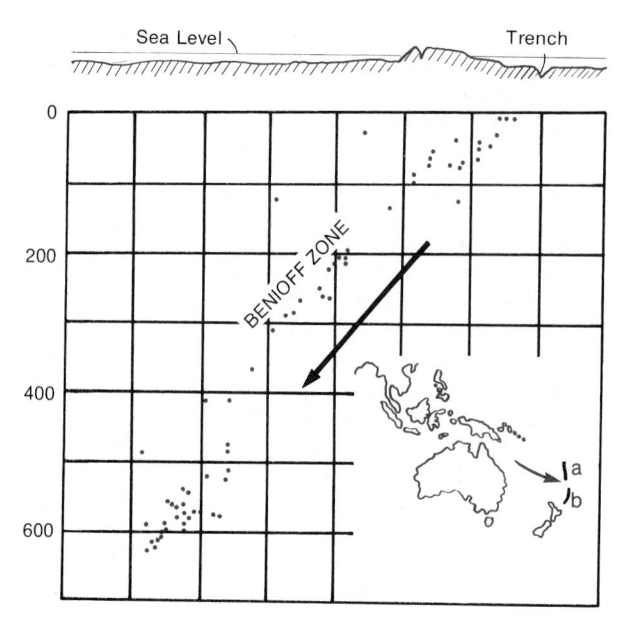

Figure 5–47 Vertical cross-section showing spatial distribution of earthquakes along a line perpendicular to the Tonga Trench and volcanic island arc. The Tonga Trench (a on inset map) lies just to the north of the Kermodec Trench b). (Simplified from Isacks, B., Oliver, J., and Sykes, R., J. Geophys. Res., 73:5855, 1968.)

where the edges of two rigid plates press against one another. The 1964 Alaskan earthquake originated at this shallow level. Such quakes are thought to occur along the shear plane between the subducting oceanic lithosphere and the continental lithosphere.

Gravity plays a role

Gravity measured over deep-sea trenches is characteristically lower than that measured in areas adjacent to the trenches. Geophysicists refer to such phenomena as negative gravity anomalies. A **gravity anomaly** is the difference between the observed value of gravity at any point on the earth and the computed theoretic value. **Negative gravity anomalies** occur where there is an excess of low-density rock beneath the surface. The very strong negative gravity anomalies extending along the margins of the deep-sea trenches can mean only that such belts are underlain by rocks much lighter than those at depth on either side. Because the zone of lower gravity values is narrow in most places, it is believed to mark trends where lighter rocks dip steeply into the denser mantle. Geophysicists assume that the less dense rocks of the negative zone must be held down by some force to prevent them from floating upward to a level appropriate to their density. That force might be provided by a descending convection current, or the plate may be pushed by rising magma along midocean ridges until it encounters an opposing plate and plunges into the asthenosphere.

Hot spots

Anyone examining the superb topographic maps (see Figure I–4) of the sea floor that have been produced over the past decade cannot help but notice the chains of volcanic islands and seamounts (submerged volcanoes) on the sea floor. Because these volcanoes occur at great distances from plate margins, they must have originated differently from the volcanoes associated with midocean ridges or deep-sea trenches. Their striking alignment has been explained as a consequence of sea floor spreading. According to this notion, intraocean volcanoes develop over a "hot spot" in the asthenosphere. The hot spot is fixed in location and may be a manifestation of one of the deep plumes of upwelling mantle rock described earlier. Lava from the plume may work its way to the surface to erupt as a volcano on the sea floor. As the sea floor moves (at rates as high as 10 cm per year in the Pacific), volcanoes form over the hot spot and expire as they are conveyed away, and new volcanoes form at the original location. The process may be repeated indefinitely, resulting in a linear succession of volcanoes that may extend for thousands of kilometers in the direction that the sea floor has moved. The Hawaiian volcanic chain is believed to have formed in this manner from a single source of lava over which the Pacific plate has passed on a northwesterly course. In support of this concept are radioactive dates obtained from rocks of volcanoes that clearly indicate that volcanoes farthest from the source are the oldest.

At the western end of the Hawaiian Islands is a string of submerged peaks called the Emperor Seamounts (Fig. 5-48). These submerged volcanoes lie in a more northerly direction than the Hawaiian Islands. However, both the seamounts and the islands are part of a single chain that has been bent as a result of a change in the direction of plate movement. Such an interpretation is supported by age determinations. The oldest of the Hawaiian Islands (near the bend) was formed about 40 million years ago. The seamounts continue the age sequence toward the end of the chain, where the peaks are about 80 million years old. Thus, 40 million years ago the Pacific plate changed course and put a kink in the Hawaiian chain.

The ocean around Hawaii is not the only part of the globe having hot spots. As indicated in Figure 5-49, hot spots are widely dispersed and occur beneath both continental and oceanic crust.

Lost continents and alien terrains

We are accustomed to thinking of continental crust in terms of large land masses like North America or Eurasia. There are, however, many relatively small patches of continental crust scattered about on the lithosphere. As long ago as 1915, Alfred Wegener described the Seychelles Bank (Fig. 5-50) in the Indian Ocean as a small continental fragment that had broken away from Africa. The higher parts of the Seychelles Bank project above sea level as islands, but many other such small patches of continental crust are totally submerged. Geologists call these bits of continental crust **microcontinents.** They are recognized by their granitic composition, by the velocity with which com-

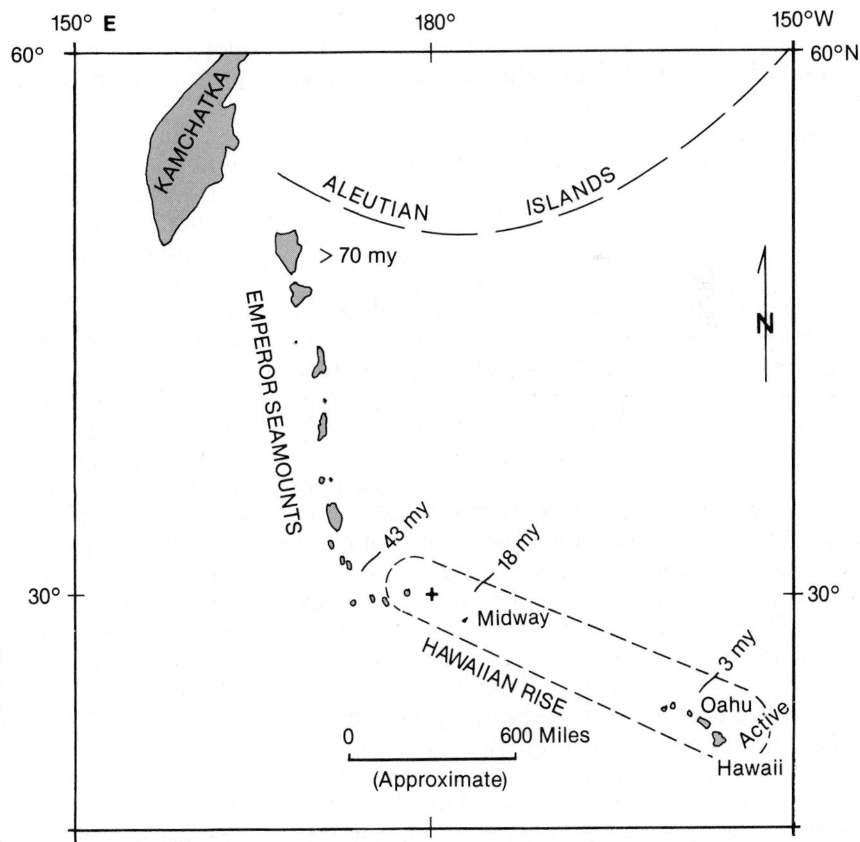

Figure 5-48 Bend in trend of Hawaiian Island-Emperor Seamount Chain probably caused by change in direction of movement of the Pacific Tectonic Plate. (From Watkins, J.S., Bottino, M.L., and Morisawa, M., Our Geological Environment. Philadelphia, Saunders College Publishing, 1975.)

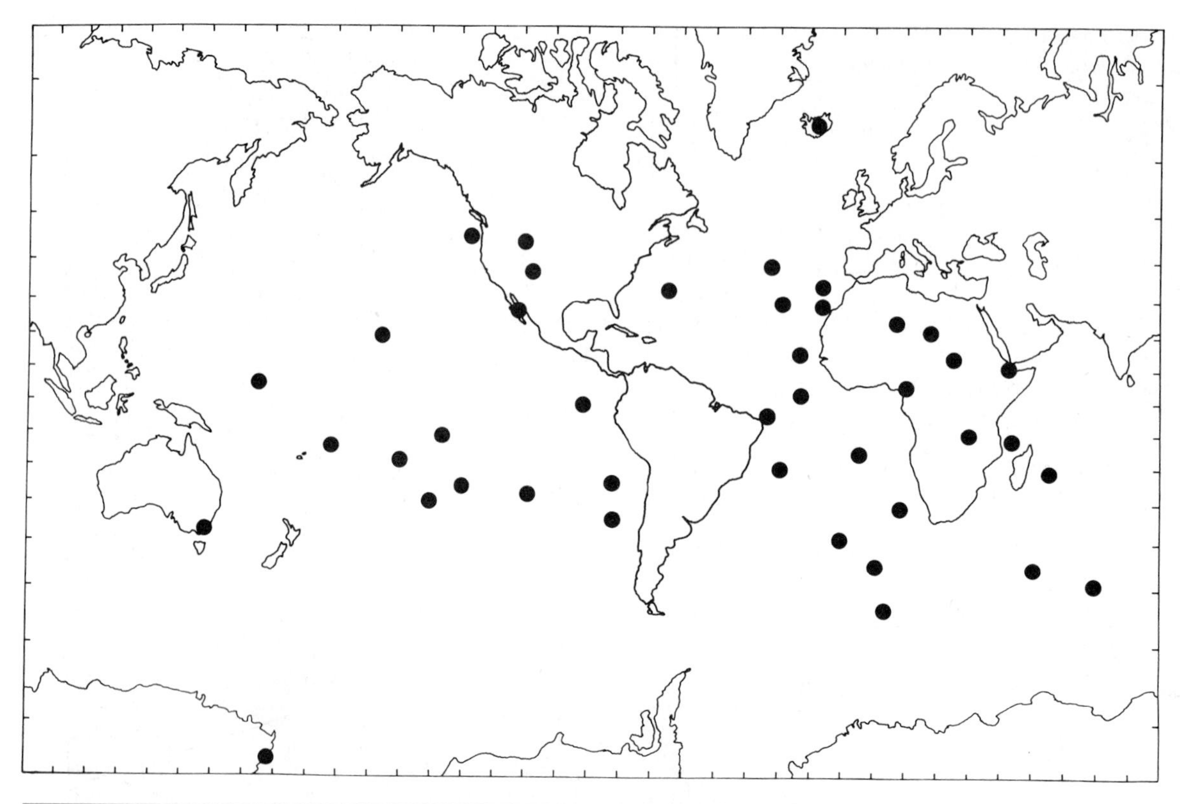

Figure 5–49 Locations of some of the major hot spots around the earth. (Courtesy of Tom Crough, Department of Geologic and Geophysical Sciences, Princeton University.)

pressional seismic waves traverse them (6.0 to 6.4 km per second), by their general elevation above the surrounding oceanic crust, and by their comparatively quiet seismic nature. Submarine plateaus differ from microcontinents in seismic properties and composition.

It is apparent that microcontinents are merely small pieces of larger continents that have experienced fragmentation. As these smaller pieces of continental crust are moved along by sea floor spreading, they may ultimately converge upon the subduction zone at the margin of a large continent. Because they are composed of relatively low-density rock, they are difficult for the subduction zone to swallow. Their buoyancy prevents their being carried down into the mantle and assimilated. Indeed, the small patch of crust may become incorporated into the crumpled margin of the larger continent as an exotic or alien block. It is interesting to note that geologists found evidence of this process even before much was known about microcontinents. While mapping in Precambrian shields and old mountain belts, they discovered outcrop areas that were incongruous in structure, age, fossil content, and lithology when compared with the surrounding geology. These areas were given the name **allochthonous terrains,** indicating that they had not originated in the places where they now rested. Allochthonous terrains have been recognized on every major land mass, with well-studied examples in the northeastern U.S.S.R., in the Appalachians, and in many parts of western North America (Fig. 5–51).

Plate tectonics and mountain building

Generations of geologists have speculated about the origin of mountain ranges, many of which contain thousands of meters of marine sedimentary rocks. How were they raised from the sea floor, deformed, and intruded by igneous rocks? Plate tectonics provides an answer. A tectonic plate has a forward, leading

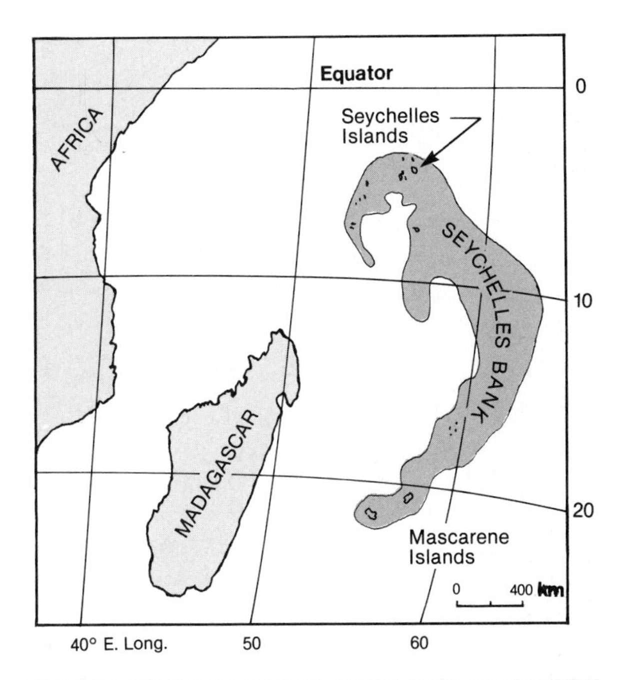

Figure 5–50 Location map of the Seychelles Bank.

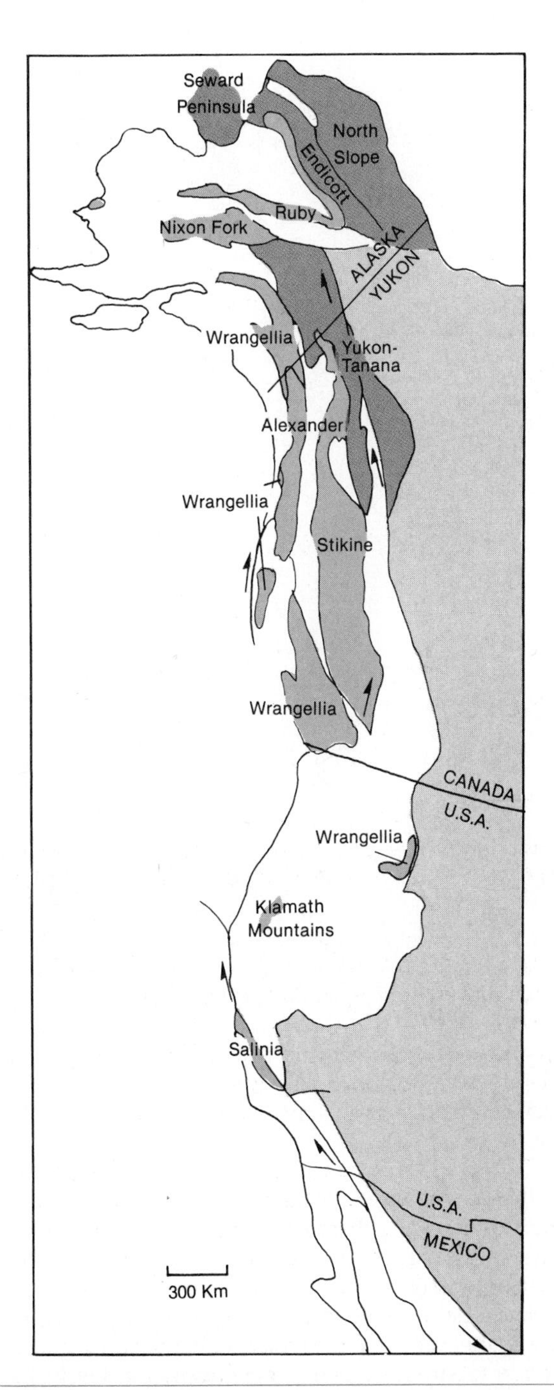

Figure 5–51 *Allochthonous continental terrains in western North America that contain Paleozoic or older rocks. At some time prior to reaching their present locations, these terrains were continental fragments embedded in oceanic crust, or microcontinents. Those colored light orange probably originated as parts of continents other than North America, whereas the darker colored brown blocks are possibly displaced parts of the North American continent. (From Ben-Avraham, Z.,* Am. Sci., *69:298, 1981.)*

edge and a more passive, trailing edge. The history of a mountain range begins with the accumulation of great thicknesses of sediment on the trailing edge of a plate that bears a continent. This marginal depositional tract, which includes the continental shelf and slope, is called a geosyncline, or more simply, geocline. Next, the trailing edge is overtaken by the leading edge of an adjacent tectonic plate. The encounter would crumple the marginal tract of sediments, cause melting at depth, and gradually convert the former elongate depositional basin into mountains, as shown in Figure 5-52.

The American geologist Robert Dietz believes that a plate tectonics model can adequately explain most of the structural and sedimentologic features of ranges like the Appalachians. Figure 5-53 illustrates the model proposed by Dietz. The sequence of events begins with thick (geosynclinal) bodies of sediment accumulating on the trailing edges of diverging plates. The ocean basin is presumably widened as the continents drift apart. At a later period, the movement of the crustal plates is reversed, the oceans narrow, and they are ultimately squeezed out of existence as the plates collide and crumple the accumulated sediment.

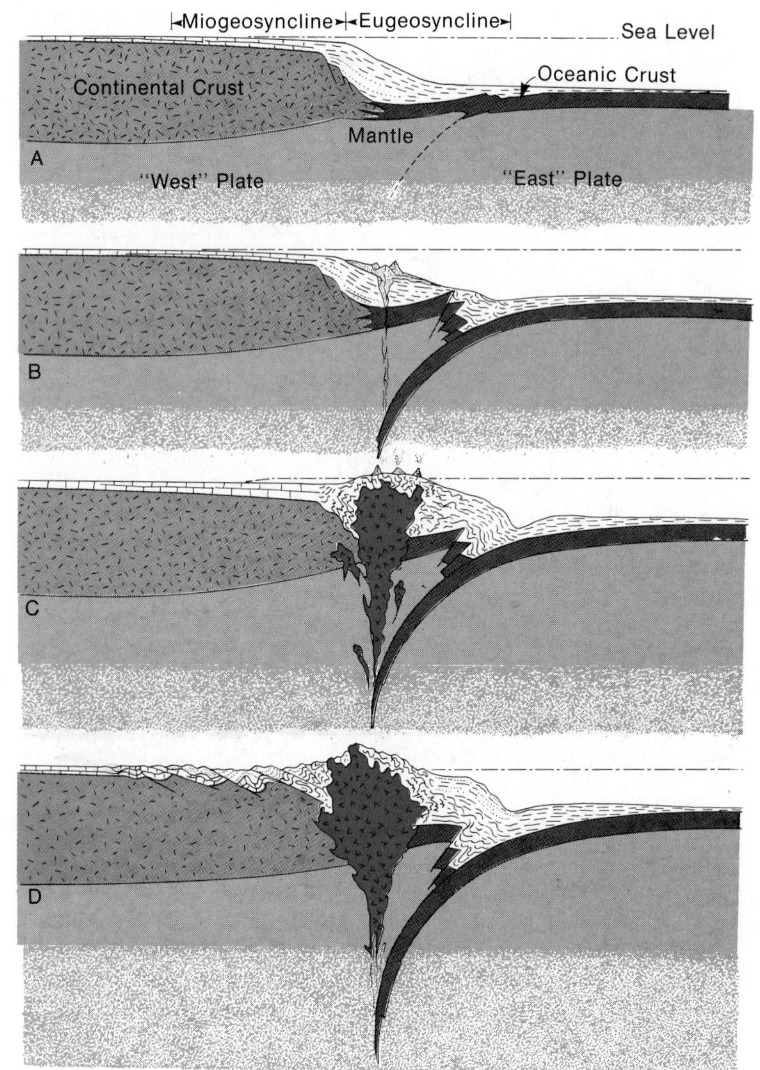

Figure 5–52 *Plate-tectonic model for the origin of a mountain belt. (A) Trailing margin of a continent sags and receives thick accumulation of sediments as it drifts to the left. (B) Lithospheric plate bearing oceanic crust collides and is subducted beneath plate bearing continent. When subducted plate has descended to about 100 km, basaltic-type lavas, generated by melting of the lithosphere, rise and are extruded as submarine volcanics. A trench begins to form in the zone of subduction. (C) As subduction continues, granitic type magmas, formed from molten subducted materials, rise. Their buoyancy and heat cause doming at the surface, and this is accompanied by metamorphism and deformation. Graywackes accumulate adjacent to the rising mountain chain. (D) Subduction and mountain building continue. (Modified from Dewey, J.F., and Bird, J.M., Mountain belts and the new global tectonics.* J. Geophys. Res., *75:2625–2647, 1970. Copyright, The American Geophysical Union.)*

Summary The earth has a constantly changing surface. We are reminded of this by earthquakes, by evidence of elevated and submerged coastlines, by high-standing plateaus of marine sedimentary rocks, and by the many faults and folds that are found in once-horizontal strata. Studies of the magnetic properties of crustal rocks have revealed that even the floors of the oceans, once thought to be the most changeless of all earth features, are in slow and constant movement.

The structural features formed by movements in the earth's crust include faults and folds. Faults are breaks in crustal rocks along which there has been a displacement of one side relative to the other. According to the directions of that movement, faults are classified as normal, reverse (thrust), or lateral. Intermediate types include rotational and oblique faults. No less important than faults as evidence of the earth's instability are folds, the principal categories of which are anticlines, synclines, domes, basins, and monoclines.

According to the views embodied in the theory of plate tectonics, the crust of the earth and part of the upper mantle compose a brittle shell called the litho-

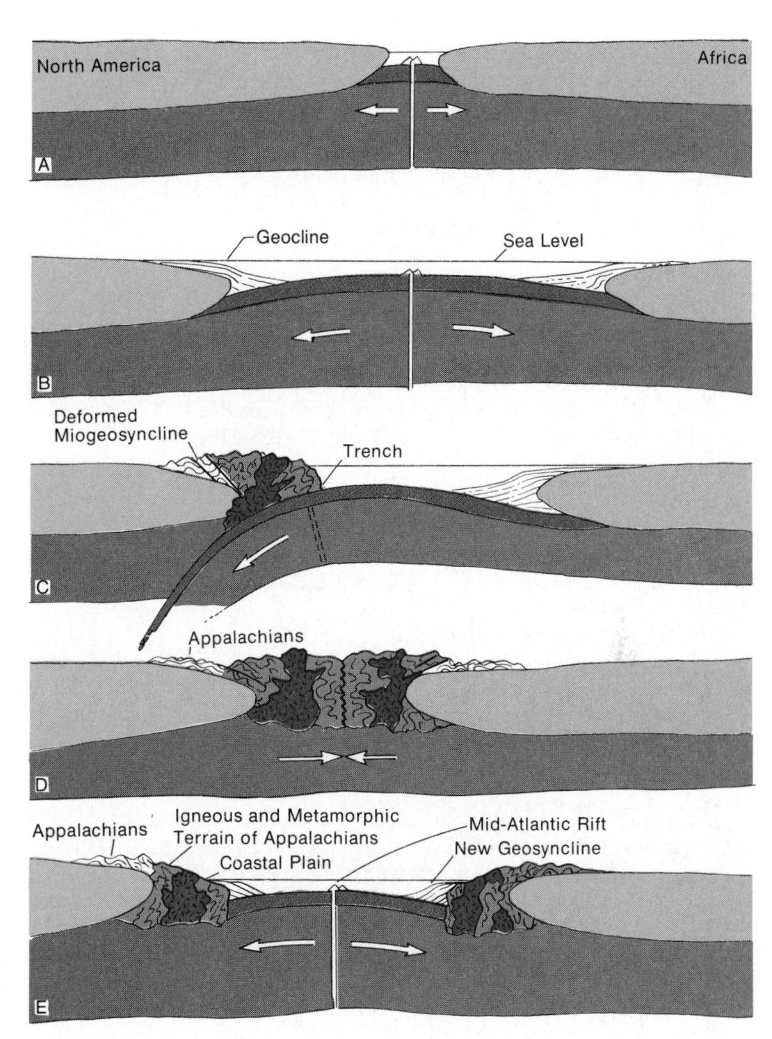

Figure 5–53 *Plate tectonic model for events leading to the formation of the Appalachian Mountains and present Atlantic Ocean, as proposed by Robert S. Dietz. (A, B) Continents separate, and the ancestral Atlantic ocean is inserted; geosynclinal sediments form on the margins of continents. (C, D) Plates converge, the subduction zone is formed, and the geocline is collapsed. Continents are sutured together. (E) Rifting occurs again, and new geoclines form along continental margins.*

sphere. The lithosphere is broken into a number of plates that override a soft plastic layer in the mantle called the asthenosphere. Heat from within the earth may create convection currents in the mantle, and such currents may provide a propelling force to move the plates. However, sliding of plates in response to gravity may also prove to be a mechanism for their movement.

Materials for the lithospheric plates originate along fracture zones typified by midocean ridges. Along these tensional features, basaltic lavas rise and become incorporated into the trailing edges of the plates, simultaneously taking on the magnetism of the earth's field as they crystallize. In recent geologic time, this process has provided a record of geomagnetic reversals that in turn permit geophysicists to determine the rates of sea floor spreading.

When plates that do not have continents on opposing edges collide, one of them may slide beneath the other and enter the mantle at an angle, to be remelted at depth. The zone of collision between plates may be marked by systems of volcanoes, deep-sea trenches, and great mountain ranges. The continental crustal segments of the lithosphere ride passively on the plates. They have a lower density than oceanic crust and do not sink into subduction zones. This explains why continents are older than the ocean floors, and why, when continents collide, they produce mountain ranges rather than trenches. Where continents straddle zones of divergent movements in the asthenosphere, the land mass may break and produce rift features like the Red Sea and Gulf of Aden. These fracture zones may then widen further, resulting in the formation of new tracts of ocean floor.

QUESTIONS FOR REVIEW

1. What are the principal categories of faults? Which kinds of faults result primarily from tension in the earth's crust and which come from compression?
2. What are folds? What are the principal kinds of folds? Are tracts of closely folded rocks the result of compressional or tensional forces? How are the topographic ridges of the Appalachian Mountains related to folds?
3. What large-scale features of the sea floor seem to have developed in response to crustal tension? What sea floor features seem related to crustal compression?
4. What is remanent magnetism? What is its origin, and how is it used in finding ancient pole positions?
5. How do the data of paleomagnetism support the concept of moving continents?
6. How has the detection of reversals in the earth's geomagnetic field been used as support for the concept of sea floor spreading?
7. How does the age distribution of chains of intraoceanic volcanoes and seamounts enhance the concept of sea floor spreading?
8. What relationship might thermal convection cells in the mantle have to regions of crustal tension and compression?
9. Why are the most ancient rocks on earth found on the continents, and only relatively young rocks found on the ocean floors?
10. Compile a list of items that Alfred Wegener might have used to convince a skeptic of the validity of the theory of continental drift.
11. Do midocean volcanoes have lavas of granitic or basaltic composition? What is the composition of lavas in continental mountain ranges adjacent to deep-sea trenches? How might one account for these compositions?
12. What is a gravity anomaly? Where in the ocean basins have negative gravity anomalies been detected? What is the meaning of these anomalies with regard to deep-sea trenches?
13. According to plate tectonics theory, how might the Himalayan Mountains have originated? the San Andreas fault? the Dead and Red Seas?
14. According to the theory of plate tectonics, where is new material added to the sea floor, and where is older material consumed?
15. A moving tectonic plate must logically have sides, a leading edge, and a trailing edge. Where are the leading and trailing edges of the American plate located?

SUPPLEMENTAL READINGS AND REFERENCES

Ben-Avraham, Z., 1981. The movement of continents. *Am. Sci.* 69:291-299.

Continents adrift and continents aground, 1976. *Readings from Scientific American.* San Francisco, W.H. Freeman Co.

Dietz, R.S., and Holden, J., 1970. Reconstruction of Pangaea; break-up and dispersion of continents, Permian to Recent. *J. Geophys. Res.* 75:4939-4956.

Hallam, A., 1973. *A Revolution in the Earth Sciences.* New York, Oxford Press.

Jones, D.L., Cox, A., Coney, P., and Beck, M., 1982. The growth of western North America. *Sci. Am.* 247:70-127.

Miller, R., 1983. *Continents in Collision.* Alexandria, Virginia, Time-Life Books.

Pitman, W.C., III, and Talwani, M., 1972. Sea-floor spreading in the North Atlantic. *Geol. Soc. Amer. Bull.* 83:619-646.

Sullivan, W., 1974. *Continents in Motion.* New York, McGraw-Hill Book Co.

Wegener, A., 1929 (1966 translation). *The Origin of the Continents and Oceans.* New York, Dover Publications.

Weyman, D., 1981. *Tectonic Processes.* London, Allen and Unwin.

6
geologic time

Each grain of sand, each minute crystal in the rocks about us is a tiny clock, ticking off the years since it was formed. It is not always easy to read them, and we need complex instruments to do it, but they are true clocks or chronometers. The story they tell numbers the pages of earth history.

PATRICK M. HURLEY
How Old Is the Earth? 1959

Prelude We humans are fascinated by time. Geology instructors are particularly aware of this fascination, for they are regularly asked the age of various rock and mineral specimens brought to the university by students and returning vacationers. If informed that the samples are tens or even hundreds of millions of years old, the collectors are often pleased, but they are also perplexed. "How can this fellow know the age of this specimen by just looking at it?" they think. If they insist on knowing the answer to that question, they may next receive a short discourse on the subject of geologic time. It is explained that the rock exposures from which the specimens were obtained have long ago been organized into a standard chronologic sequence based largely on superposition, evolution as indicated by fossils, and actual rock ages in years obtained from the study of radioactive elements in the rock. The geologist's initial estimate of age is based on *experience.* He may have spent a few hours on his knees at those same collecting localities and had a background of information to draw upon. Thus, at least sometimes, he can recognize particular rocks as being of a certain age. The science that permits him to accomplish this feat is called *geochronology.* With an understanding of geochronology, many formerly isolated bits of information about the geologic past can be placed in proper sequence, and a history of the earth can be written.

The continuum of time and life on earth as represented by a living tree standing beside a log bridge formed by a 200-million-year-old petrified conifer.

DEVELOPMENT OF CONCEPTS ABOUT GEOLOGIC TIME

Relative and quantitative geologic time

There are two ways of thinking about geologic time. On the one hand, we may wish to know the actual number of years ago that a particular geologic event (such as an intrusion of granite or deposition of limestone) occurred. This kind of specific time information is called *quantitative geologic time. Relative geologic time,* on the other hand, does not require that one know a precise age, but rather whether or not a given rock body is older or younger than another. Before the discovery of radioactivity, relative dating provided the only method available to early geologists as they began to piece together episodes in the history of the earth. Even with the advent of quantitative methods, the principles of relative dating continue to be widely employed in the day-to-day geologic study of the earth and other planets in our solar system.

Superposition

The concept of relative geologic dating of rocks was introduced about three centuries ago by a Danish physician named Nils Stensen. Following the custom of his day, Stensen latinized his name to Nicolaus Steno. He settled in Florence, Italy, where he became physician to the Grand Duke of Tuscany. Because the duke did not require Steno's constant attention, the physician had ample time to explore the countryside, visit quarries, and examine strata. Around the year 1669, his geologic studies led him to formulate a basic principle of geochronology. It is called the **principle of superposition.**

The principle of superposition states that in any sequence of undisturbed strata, such as those in Figure 6-8, the oldest layer is at the bottom, and successively higher layers are successively younger. It is a rather obvious axiom, which probably had been understood by many naturalists even before Steno. Yet Steno, based on his observations of strata in northern Italy, was first to explain the concept formally. The fact that his principle was self-evident does not diminish its geological importance.

James Hutton on time and process

The factor of time enters into every geological process. It requires time to weather rocks, to form crystals, to raise mountains, and to erode canyons. The length of time available for the action of such geologic processes as erosion influences the configuration of landforms resulting from those processes (Fig. 6-1). When the science of geology was in its infancy, however, all rocks were thought to be of about the same age, and all geologic features were believed to have been formed at the same time. Furthermore, the earth was generally believed to be only several thousand years old. It was James Hutton, known as the founder of modern geology, who first recognized the immensity of geologic time and its importance as a partner in geologic processes. In his *Theory of the Earth,* which he presented to the Royal Edinburgh Society in 1785, Hutton showed irrefutably that hills and valleys were not everlasting, but that the surface of the earth undergoes constant change as a result of mountain-building forces, erosion, and sedimentation. The earth, said Hutton, "is thus destroyed in one part, but it is renewed in another." Hutton watched as sediment was transported to sites of deposition and noted the development of bodies of clay and sand. It became apparent to him that ancient sedimentary rocks had formed in the same way and must have required an enormous length of time to have accumulated to such great thicknesses. His understanding of the vastness of geologic time stands as a milestone in the history of geology.

Figure 6-1 A view of Monument Valley, Colorado Plateau, southern Utah. At one time the surface of the plateau in this area was approximately at the level of the flat-topped summit of the prominent butte. Erosion over a long interval of time has steadily removed the surrounding rock. Here and there, a remnant of that former surface remains, still protected by a once more extensive upper layer of rock.

Hutton was a logical naturalist who built his geological theories around a belief that "the past history of our globe must be explained by what can be seen to be happening now." This simple but powerful idea was eventually named **uniformitarianism.**

Uniformitarianism

Uniformitarianism is often paraphrased as "the present is the key to the past." Thus, geologic features formed long ago were produced by processes that are still at work today. These processes — whether deposition, erosion, or volcanism — are governed by unchanging natural laws. Today, geologists are careful to point out that although processes long ago were the same as those today, they may have differed in rate and intensity. At times, continents may have stood relatively higher above the oceans. This higher elevation would have resulted in higher rates of erosion and relatively harsher climatic conditions compared with intervening episodes when the lands were low and partially covered with inland seas.

Unconformities: Gaps in the time record

On the Isle of Arran (an island of Scotland) and also at Jedburgh, Scotland, Hutton came across exposures of rock where steeply inclined older strata had been beveled by erosion and covered by gently inclined younger layers (Figs. 6–2, 6–3). It was clear to Hutton that the older sequence was not only tilted but also partly removed by erosion before the younger rocks were deposited. The erosional surface meant that there was an interval of time not represented by strata. It represented a chronological gap or hiatus. Hutton did not suggest a name for this feature, which was later termed an **unconformity.** More specifically, the exposure studied by Hutton was an **angular unconformity** because the lower beds were tilted at an angle to the upper. This and other unconformities (Fig. 6–4) provided Hutton with ample evidence for the episodes of geologic change he described in *Theory of the Earth.* The lower strata had been deposited, uplifted, and tilted, and then subjected to erosion. Much later, sedimentation was renewed on the old erosional surface. There were, in Hutton's words, "a succession of

Figure 6–2 *Angular unconformity at Siccar Point, southeastern Scotland. It was here that the historical significance of an unconformity was first realized by James Hutton in 1788. (Courtesy of the Institute of Geological Sciences, London.)*

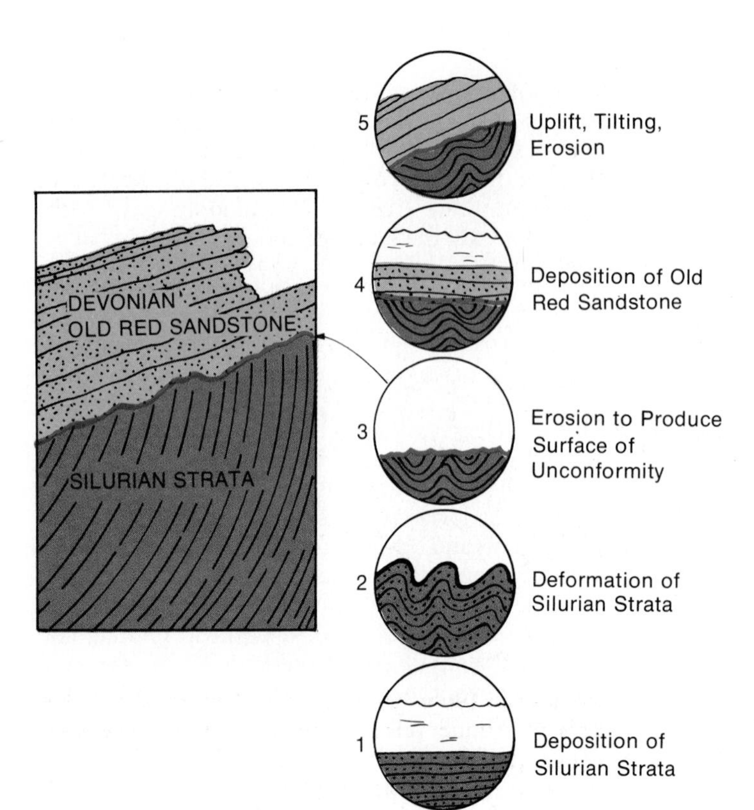

Figure 6–3 *Simplified sketch of part of "Hutton's unconformity" (compare with photograph of Siccar Point) and a diagrammatic representation of the historical sequence of events that produced the geologic structure at Siccar Point, Scotland.*

worlds" that could be revealed in the careful interpretation of unconformities.

Faunal succession

Until about 1800, geologists did not fully grasp how to use their knowledge of local geology to develop a wider view of the relative sequence of events for an entire region, or even a continent. They were able to understand the sequence of older to younger rocks at local exposures of strata, but could not relate their findings to lithologically different strata that were exposed at distant locations. This problem was to be resolved by a geologist named William Smith (1769–1839).

William Smith was an English surveyor and engineer who devoted 24 years to the task of tracing the strata of England and representing them on a map. Small wonder that he acquired the nickname "Strata." He was employed to locate the routes of canals, to design drainage for marshes, and to restore springs. In the course of this work, he independently came to understand the principles of stratigraphy, for they were of immediate use to him. By knowing that different types of stratified rocks occur in a definite sequence and that they can be identified by their composition and texture (lithology), the soils they form, and the fossils they contain, he was able to predict the kinds and thicknesses of rock that would have to be excavated in future engineering projects. His use of fossils was particularly significant. Prior to Smith's time, collectors rarely noted the precise beds from which fossils were taken. Smith, on the other hand, carefully recorded the occurrence of fossils and quickly became aware that certain rock units could be identified by the particular assemblages of fossils they contained. He used this knowledge first to trace strata over relatively short distances and then to extend his "correlations" over greater distances to strata of different lithology, but inferred to be the same age because they contained the same fossils. Ultimately, this knowledge led to the **principle of faunal and floral succession,** which stipulates that the life of each age in the earth's long history was unique for particular periods, that the fossil remains of life permit geologists to

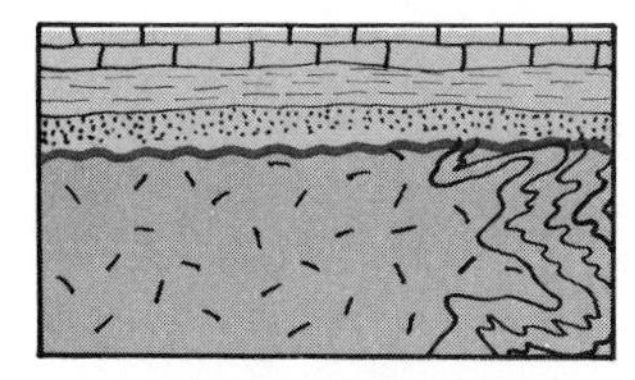

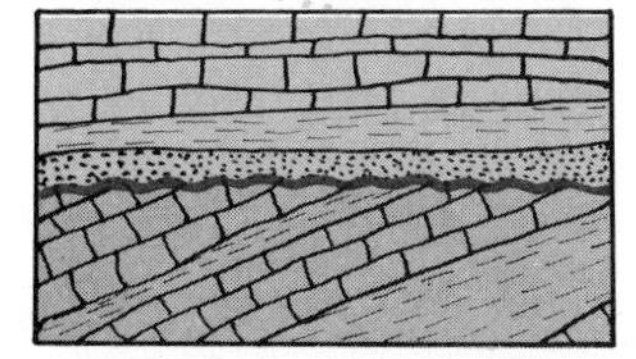

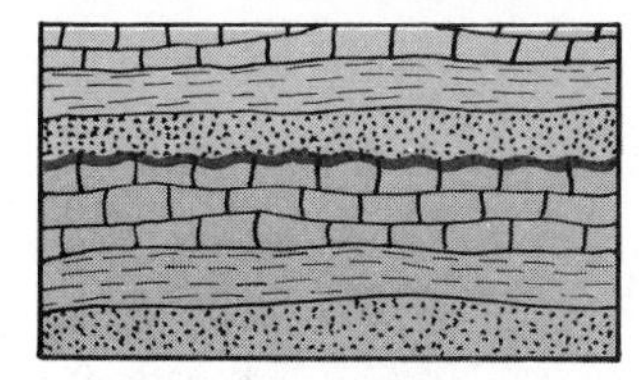

Figure 6–4 Cross-sections illustrating three types of unconformities. A nonconformity *is characterized by stratified rocks that rest upon unstratified rocks such as igneous or metamorphic. In an* angular unconformity *the older strata are inclined at a distinctly different angle than overlying strata. In a* disconformity, *the strata of the older and younger strata are parallel.*

recognize contemporaneous deposits around the world, and that fossils of ancient animals and plants could be used to assemble the scattered fragments of the record into a chronological sequence.

William Smith did not know why each unit of rock had a particular fauna. Today, we recognize that different kinds of animals and plants succeed one another in time because life has evolved continuously. Because of this continuous change, or evolution, only rocks formed during the same age could contain similar assemblages of fossils. Beginning with the work of Smith, geologists determined the succession of fossil animals and plants in many areas around the world where the strata are undeformed and their superposition clearly apparent. The faunal and floral succession determined for these areas was then verified at many additional localities. Ultimately, this knowledge of the correct succession was used to ascertain the relative ages of rocks even in areas where the original superposition of strata was uncertain because of folding and faulting.

Inclusions and cross-cutting relations

In the early 19th century, a geologist wrote a book that presented under one title the most important geologic concepts of the day. His name was Charles Lyell, and his book was the classic *Principles of Geology*. In this work can be found additional criteria useful in establishing relative ages of rock units. For example, Lyell discussed the general principle that a geologic feature that cuts across or penetrates another body of rock must be younger than the rock mass penetrated. In other words, the feature that is cut is older than the feature that crosses it. This observation is now termed the **principle of cross-cutting relationships.** It applies not only to rock units but also to geologic structures like faults and unconformities. Thus, in Figure 6–5, the break in rocks represented by fault *b* was formed after the deposition of strata *d*. Because the

Figure 6–5 Time sequence of events from spatial relations of strata, faults, and intrusions. See text for details.

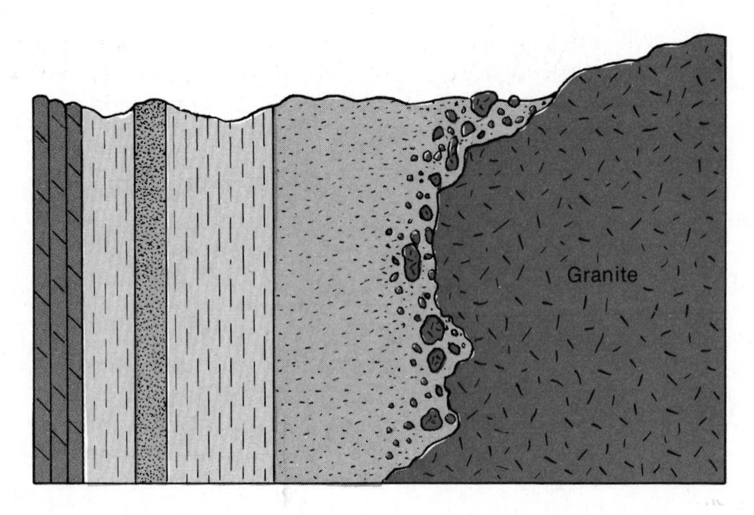

Figure 6–6 Granite inclusions in sandstone indicate granite is older unit. (Strata were originally horizontal.)

intrusion of magma *c* cuts across *b* and *d*, it is younger (later) than both of these features. By superposition, deposition of beds *e* was the last event.

Another generalization to be found in Lyell's *Principles* relates to *inclusions*. Lyell, like Hutton before him, logically discerned that fragments within larger rock masses are older than the rock masses in which they are enclosed. Thus, whenever two rock masses are in contact, the one containing pieces of the other will be the younger of the two.

In Figure 6–6, the fragments of granite within the sandstone tell us that the granite is older and that the eroded granite fragments were incorporated into the sandstone. In Figure 6–7, the granite was intruded as a melt into the sandstone. Because there are inclusions of country rock in the granite, the granite must be the younger of the two units.

CORRELATION OF SEDIMENTARY ROCKS

We have already alluded to William Smith's skills in matching the strata at one end of the country to those of the other. Without such skills we would know only the history of individual localities, not one of which contains a record of all of geologic time. A history of the entire earth could not be written, nor could a

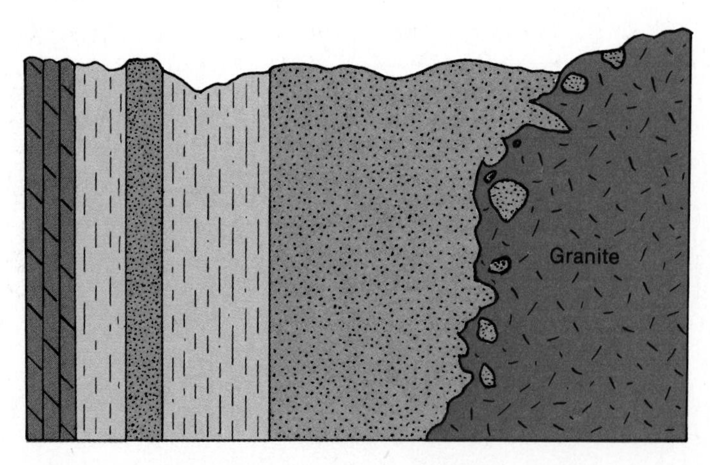

Figure 6–7 Inclusions of sandstone in granite indicate sandstone is older unit. (Strata were originally horizontal.)

Figure 6–8 *Correlation of lithologically distinctive strata by physically tracing them across an area is easier in regions where vegetative cover is sparse and rocks are boldly exposed, as in this view of northeastern Utah's Book Cliffs. (Courtesy of J.C. Brice.)*

standard geologic time scale be assembled without an ability to match strata from locality to locality.

The process of matching strata at one place to those of another is called **stratigraphic correlation.** The term correlation is derived from the Latin words meaning "relate" and "together." Formations of rock may be correlated based on their physical features, the fossils they contain, or their position in a stratigraphic sequence.

Correlation based on physical features

In the most obvious kind of correlation, formations from widely separated outcrops are matched on the basis of composition, texture, color, and weathered appearance. Of course, such lithologic features must be used with caution as they may change from place to place within the same formation. In many arid regions the problem of lateral change of a lithologic feature causes less concern, because little soil and vegetation cover the strata, and one can physically trace the formation by walking continuously along the outcrop (Fig. 6–8). Sometimes the lithology of the rock is not sufficiently distinctive to permit correlation from place to place. In such situations, the position of the rock unit relative to strata immediately above or below may suffice to make the correlation (Fig. 6–9).

A simple illustration of how correlations are used to build a composite picture of the rock record is provided in Figure 6–10. A geologist working along

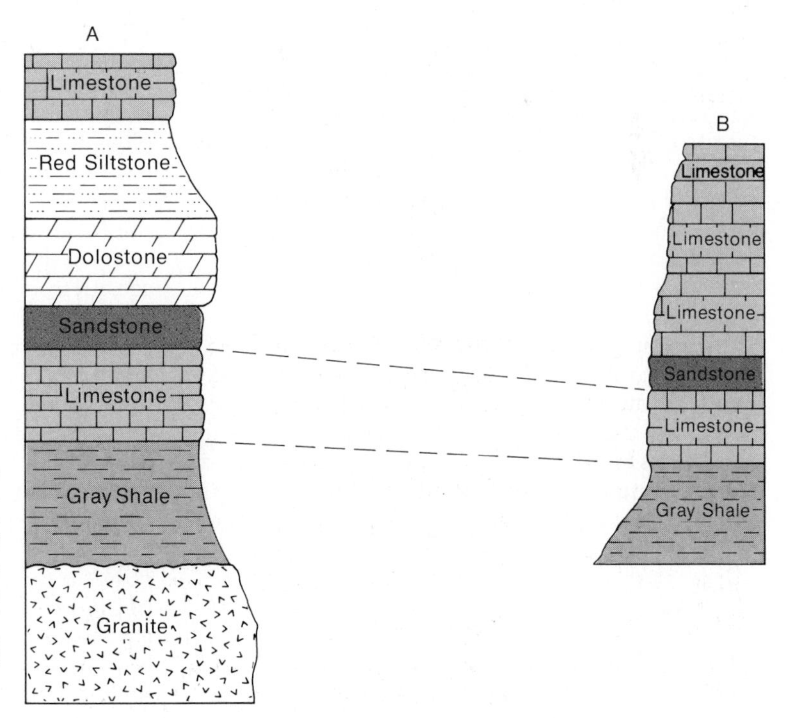

Figure 6–9 *If the lithology of a rock is not sufficiently distinctive to permit its correlation from one locality to another, its position in relation to distinctive rock units above and below may aid in correlation. In the example shown here, the limestone unit at locality A can be correlated to the lowest of the four limestone units at locality B because of its stratal position between the gray shale and sandstone units.*

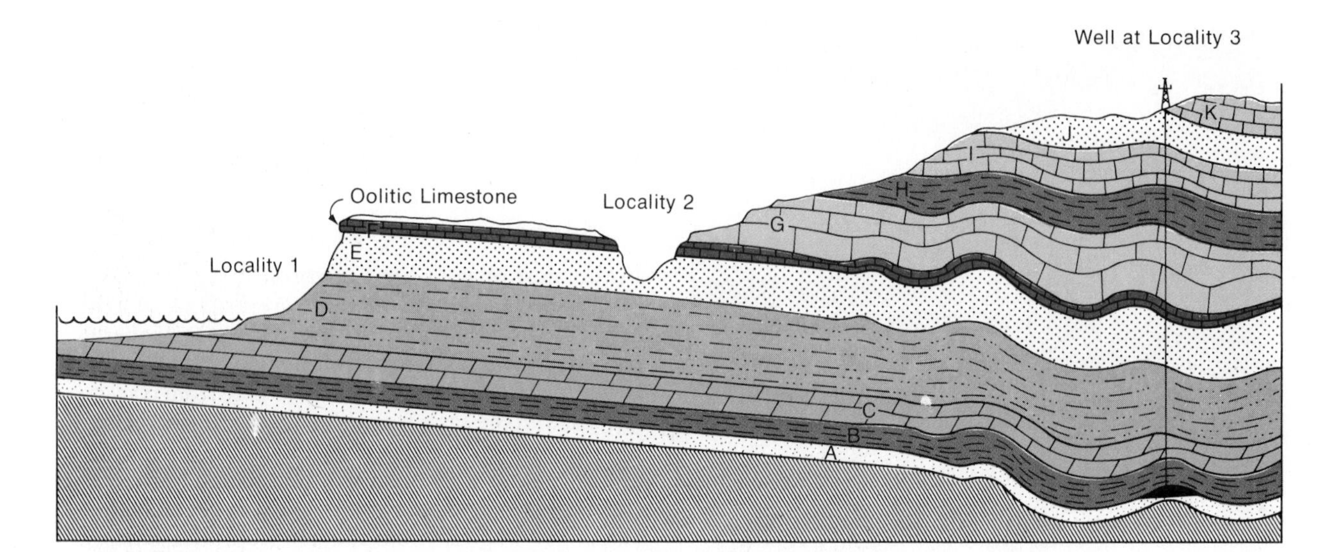

Figure 6–10 An understanding of the sequence of formations in an area usually begins with examination of surface rocks and correlation between isolated exposures. Study of samples from deep wells permit the geologist to expand the known sequence of formations and to verify the areal extent and thickness of both surface and subsurface formations.

the sea cliffs at location 1 recognizes a dense oölitic limestone (formation F) at the lip of the cliff. The limestone is underlain by formations E and D. Months later, the geologist continues his survey in the canyon at locality 2. Because of its distinctive character, he recognizes the oölitic limestone in the canyon as the same formation seen earlier along the coast and makes this correlation. The formation below F in the canyon is somewhat more clayey than at locality 1, but is inferred to be the same because it occurs right under the oölitic limestone. Working his way upward toward locality 3, the geologist maps the sequence of formations from G to K. Questions still remain, however. What lies below the lowest formation thus far found? Perhaps years later an oil well, such as that at locality 3, might provide the answer. Drilling reveals that formations C, B, and A lie beneath D. Petroleum geologists monitoring the drilling of the well would add to the correlations by matching all the formations penetrated by the drill to those found earlier in outcrop. In this way, piece by piece, a network of correlations across an entire region is built up.

Correlation based on fossils

One of the limitations inherent in correlating rocks by their lithologic attributes is that rocks of similar lithology have been deposited repeatedly over the long span of geologic time. Thus, there is the danger of correlating two apparently similar units that were deposited at different times. Fortunately, there is another method of correlation that helps to prevent mismatching, which is particularly useful in correlating strata that crop out in widely separated areas. The method involves the use of fossils (Fig. 6–11) and employs the same principle of faunal succession introduced earlier. Because of the continuous change in living things through long intervals of geologic time, the fossil remains of life are recognizably different in rocks of different ages. Conversely, rocks of the same age but from widely separated regions can be expected to contain similar assemblages of fossils.

Of course, there are complications to these generalizations. In order for two strata to have similar fossils, they would have to have been deposited in similar environments. A sandstone formed on a river floodplain would have different fossils from one formed at the same time in a nearshore marine environment. How might one go about establishing that the floodplain deposit could be correlated to the marine deposit? In some cases, this could be done by physically tracing the beds along a cliff or valley side. Occasionally, one is able to find fossils that actually occur in both deposits. Pollen grains, for example, could have been wafted by the wind into both environments. Possibly, both deposits occur directly above a distinctive,

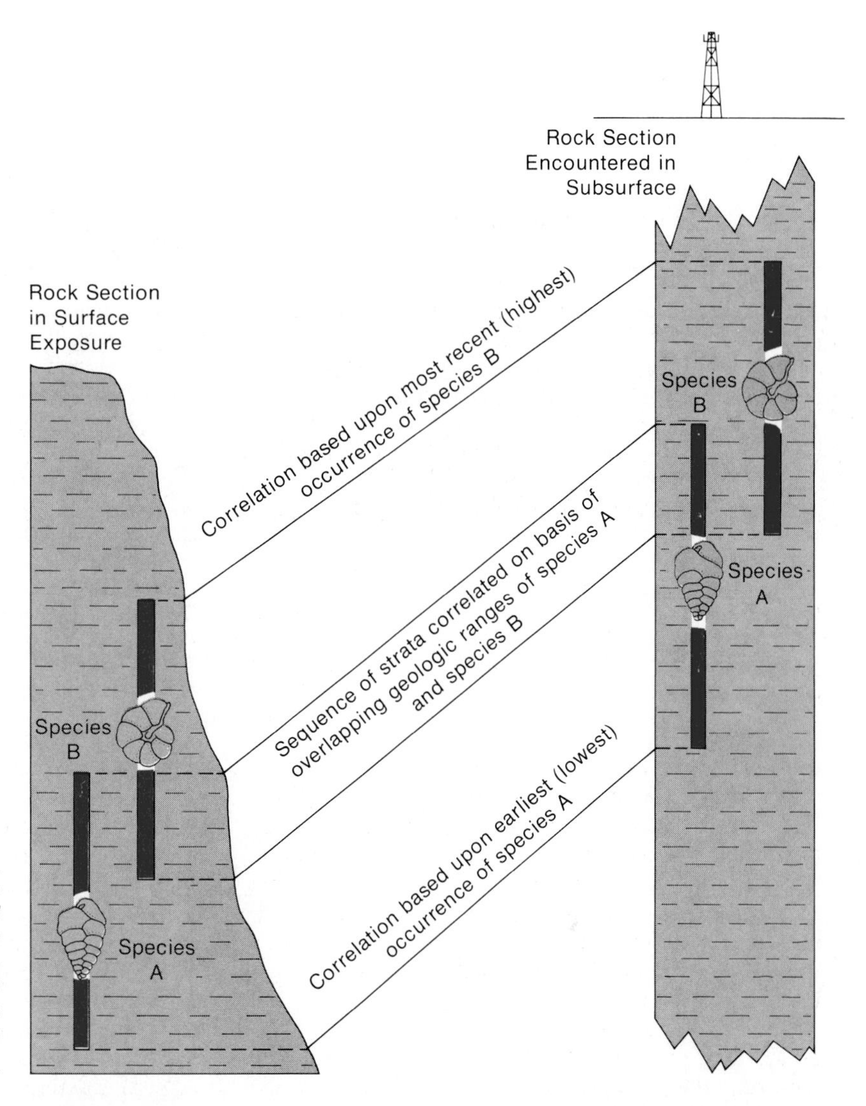

Figure 6–11 Correlation by fossils. In this simplified illustration, it is assumed that the geologic ranges of fossils have been verified at numerous other localities, and do not represent fortuitous faunal migrations into or out of the two localities. (The fossil species shown are protozoans called foraminifers.)

firmly correlated stratum, such as a layer of volcanic ash. Ash beds are particularly good time markers because they are usually deposited over a wide area during a relatively brief interval of time. Such key beds are exceptionally useful in establishing the correlation of overlying strata. Finally, the geologist may be able to obtain absolute dates from radioactive components of the rocks in question, and this, too, can be used to establish the correlation.

Some fossils are rare, restricted to a few localities, or are derived from organisms that existed over a great length of geologic time. Such fossils have only limited value in correlation. There are, however, other fossils that are abundant, easy to identify, widely dispersed around the globe, and are the remains of animals or plants that lived during a relatively short span of geologic time. Such short-lived but widespread fossils, because they are especially useful in correlating strata, are called **guide fossils.**

THE STANDARD GEOLOGIC TIME SCALE

The early geologists had no way of knowing how many time units would be represented in the completed geologic time scale, nor could they know which fossils would be useful in correlation, or which new strata might be discovered at a future time in some distant corner of the globe. Consequently, the time scale grew piecemeal, in an unsystematic manner. Units were named as they were discovered and studied. Sometimes the name for a unit was borrowed from local geography, from a mountain range in which rocks of a particular age were well exposed, or even from an ancient tribe of Welshmen. Sometimes the name was suggested by the kind of rocks that predominated. When a relatively well-exposed and complete sequence of previously unnamed strata was discovered, it was carefully defined and designated the **type section.** The location of the type section was then termed the **type locality.** This terminology continues to be in use today. The type section is a standard to which strata from other localities can be correlated (Fig. 6-12).

Divisions in the geologic time scale

The two major divisions in the geologic time scale are termed *eons.* Approximately seven eighths of all earth history is expended in the first eon—the Cryptozoic Eon (informally termed "Precambrian"). To a geologist, the biblical phrase "In the beginning" alludes to this long interval of time that began about 4600 million years ago and ended about 570 million years ago.

It was during the **Cryptozoic Eon** (literally meaning "hidden life") that the earth had completed the process of gathering together most of its rocky substance, possibly from what was part of a much older cloud of turbulent cosmic dust. In addition, it was the interval during which life on earth first appeared. Cryptozoic sequences around the world include great tracts of igneous and metamorphic rocks, although sedimentary rocks also occur.

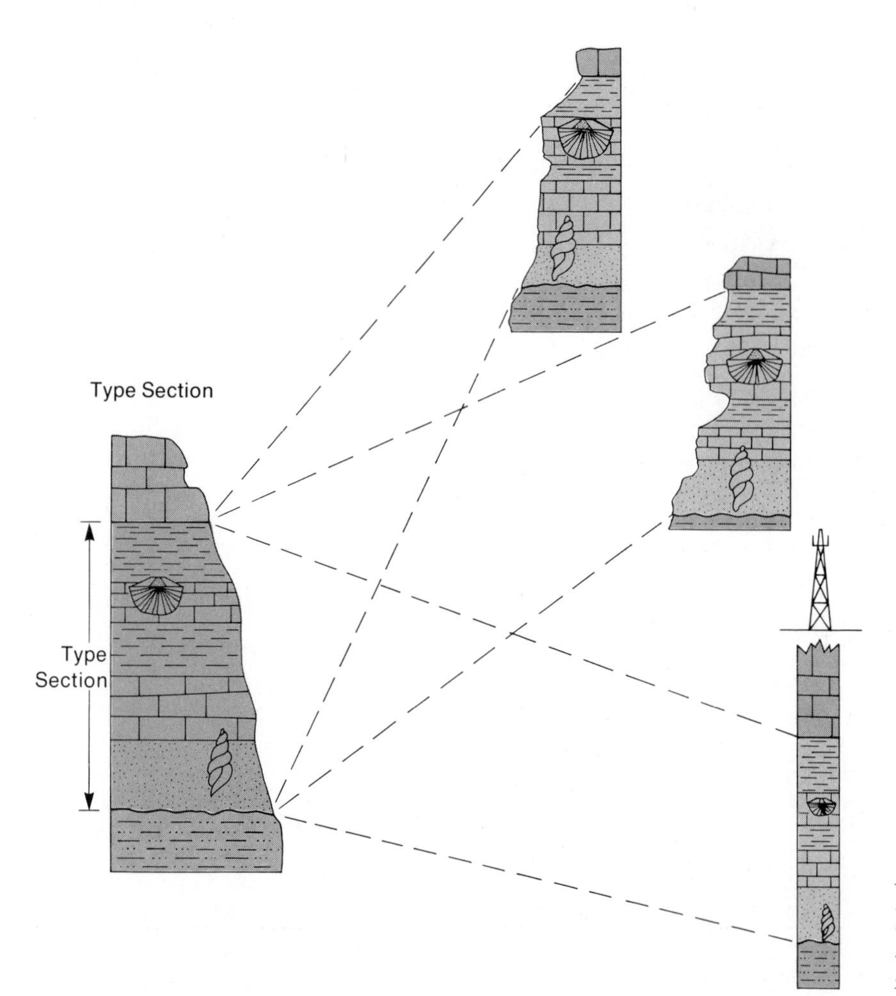

Figure 6-12 *The* type section *is a carefully defined and described standard to which strata at other locations can be compared.*

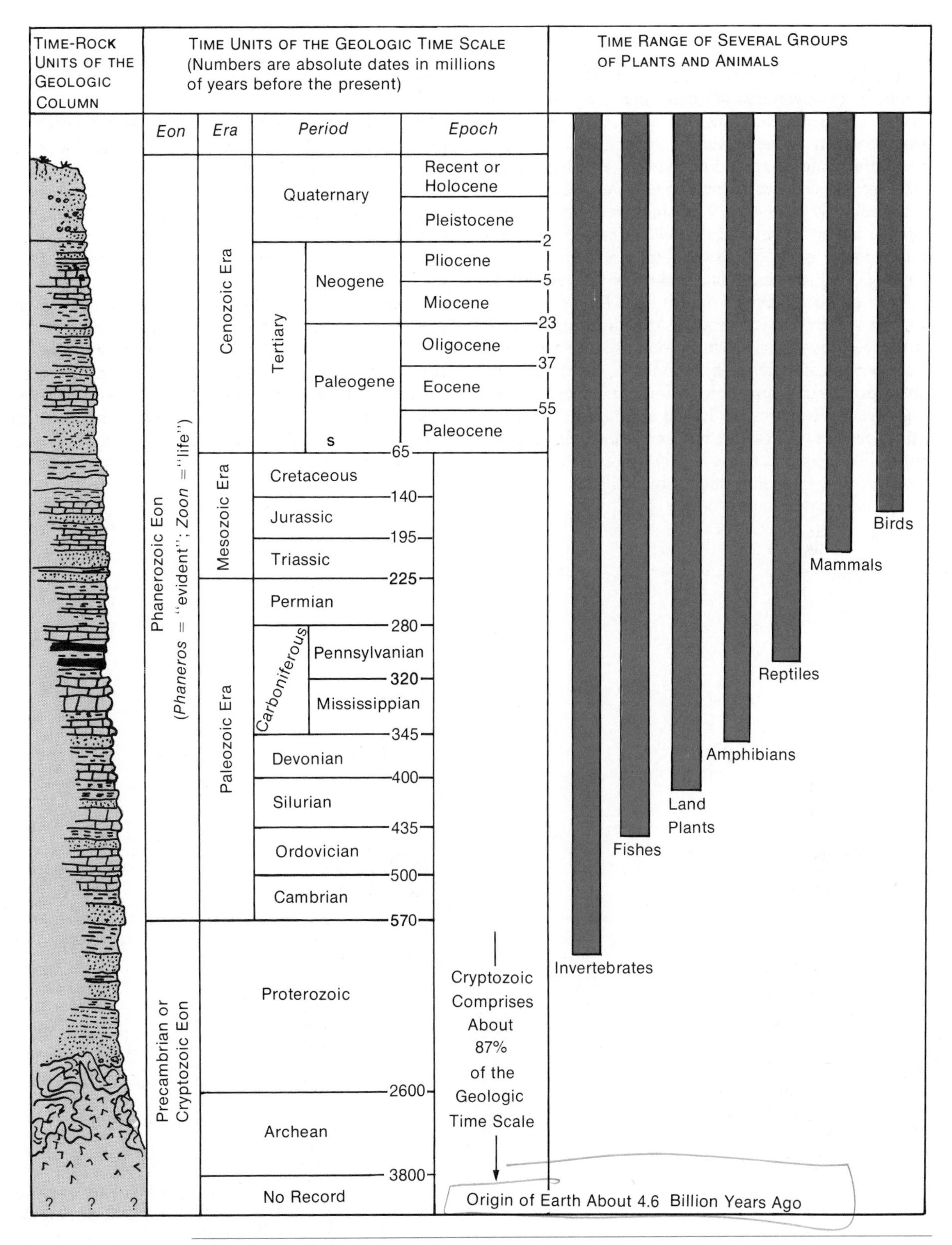

Figure 6–13 *Geologic time scale.*

All of the remainder of geologic time (Fig. 6–13) is included in the second, or **Phanerozoic Eon** (literally meaning "evident life"). As a result of careful study of superposition, accompanied by correlations based on the abundant fossil record of the Phanerozoic, geologists have divided this interval of time into three major subdivisions, termed *eras*. The oldest era is the **Paleozoic,** which lasted about 345 million years. Following the Paleozoic is the **Mesozoic Era,** which continued about 160 million years. The **Cenozoic Era,** in which we are now living, began about 65 million years ago.

Eras are divided into shorter time units called **periods.** Periods may in turn be divided into **epochs.** Eras, periods, epochs, and divisions of epochs called **ages** all represent intangible increments of pure time. They are geologic **time units.** The rocks formed during a specified interval of time are called **time-rock units.** For example, strata laid down during a given *period* compose a time-rock unit called a **system.** Thus, one may properly speak of climatic changes during the Cambrian *Period* as indicated by fossils found in rocks of the Cambrian *System*. Each of the geologic systems is recognized largely by its distinctive assemblage of fossil organisms. The fossils are different in stage of evolution from other fossils in both older and younger systems. *Series* is the time-rock term used for rocks deposited during an epoch; *stage* represents the tangible rock record of an age.

Recognition of time units

Units of geologic time bear the same names as the time-rock units to which they correspond. Thus, we may speak of the "Jurassic System" or the "Jurassic Period" according to whether we are referring to the rocks themselves or to the time during which they accumulated.

Time terms have come into use as a matter of convenience. Their definition necessarily depends upon the existence of tangible time-rock units. The steps

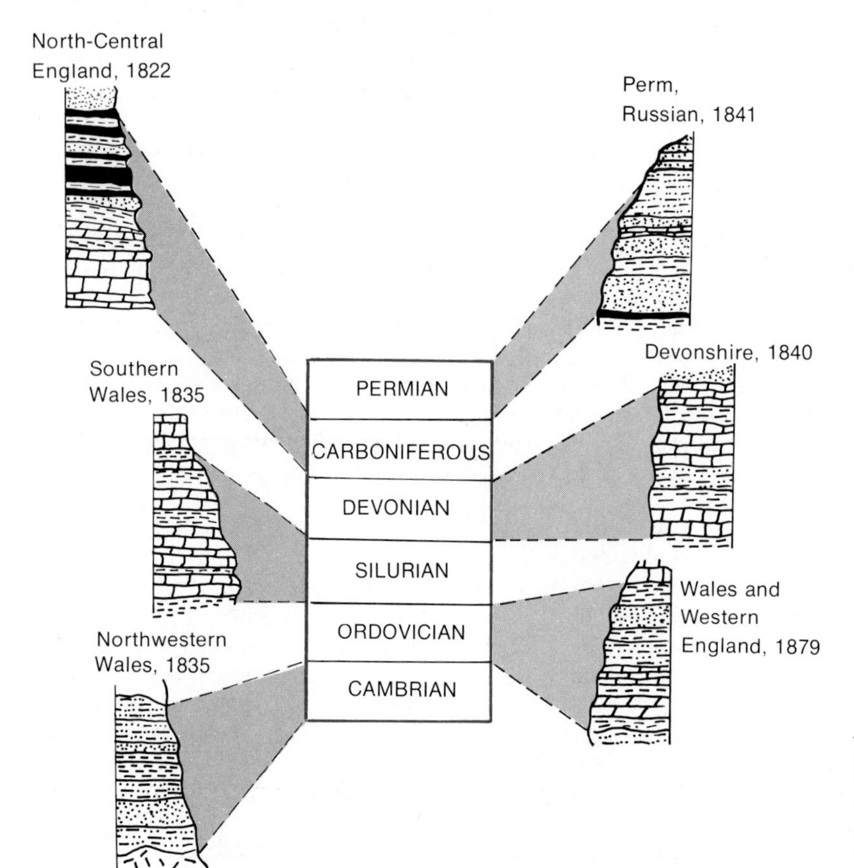

Figure 6–14 The standard geologic time scale for the Paleozoic and other eras developed without benefit of a grand plan, but rather by the compilation of "type sections" for each of the systems.

leading to the recognition of time-rock units began with the use of superposition in establishing age relationships. Local sections of strata were used by early geologists to recognize beds of successively different age and, thereby, to record successive evolutionary changes in fauna and flora. (The order and nature of these evolutionary changes could be determined because higher layers are successively younger.) Once the faunal and floral succession was deciphered, fossils provided an additional tool for establishing the order of events. They could also be used for correlation, so that strata at one locality could be related to the strata of various other localities. No single place on earth contains a complete sequence of strata from all geologic ages. Hence, correlation to the type sections of many widely distributed local sections was necessary in constructing the geologic time scale (Fig. 6-14). Clearly, the time scale was not conceived as a coherent whole, but rather evolved part by part as a result of the individual investigations of many earth scientists. That process continues even today.

QUANTITATIVE GEOLOGIC TIME

Radiometric methods for dating the earth

After having constructed a geologic time scale on the basis of relative age, it is understandable that geologists would seek some way to assign absolute ages in millions of years to the various periods and epochs. From the time of Hutton, leaders in the scientific community were convinced that the earth was indeed very old. But how old? And how might one quantify the geologic time scale? Various methods were tested, but it was not until the discovery of radioactivity that accurate determination of a rock's age could be made.

Radioactivity

Discovered by Antoine Henri Becquerel in 1896, **radioactivity** is the emission of radiation by atomic nuclei. It occurs when some elements, such as uranium and thorium, decay to form other elements or other isotopes of the same element. To understand what is meant by "decay," let us consider what happens to a radioactive element like uranium 238. Uranium 238 has an atomic weight of 238. The "238" represents the sum of the atom's protons and neutrons (each proton and neutron having a "weight" of 1). Uranium has an atomic number (number of protons) of 92. Such atoms with specific atomic number and weight are sometimes termed **nuclides.** Sooner or later (and entirely spontaneously) the uranium 238 atom will fire off a particle from the nucleus called an **alpha particle.** Alpha particles are positively charged ions of helium. They have an atomic weight of 4 and an atomic number of 2. Thus, when an alpha particle is emitted, the new atom will have an atomic weight of 234 and an atomic number of 90. From the decay of the parent nuclide, uranium 238, the daughter thorium 234 is obtained (Fig. 6-15). A shorthand equation for this change is written:

$$^{238}_{92}U \rightarrow ^{234}_{90}Th + ^{4}_{2}He$$

This change is not, however, the end of the process, for the nucleus of thorium 234 is not stable. It eventually emits a **beta particle** (an electron discharged from the nucleus when a neutron splits into a proton and an electron). There is now an extra proton in the nucleus but no loss of atomic weight because electrons are essentially weightless. Thus, from $^{234}_{90}Th$ the daughter element $^{234}_{91}Pa$ (protactinium) is formed. In this case, the atomic number has been increased by one. In other instances, the beta particle may be captured by the nucleus, where it combines with a proton to form a neutron. The loss of the proton would decrease the atomic number by one.

A third kind of emission in the radioactive decay process is called **gamma radiation.** It consists of a form of invisible electromagnetic waves having even shorter wavelengths than do x-rays.

As alpha and beta particles, as well as gamma radiation, move through the surrounding materials, their energy is transformed into increased activity of the electrons in the atoms of the surrounding medium. The result is *heat.* This radiogenic source of heat may be released during igneous activity.

The clocks in the rocks (radiometric dating)

There are many steps in the nuclear decay process before a final, stable daughter element, such as lead, is formed. The rate at which the steps in the process take place is unaffected by changes in temperature, pressure, or the chemical environment, since these do not involve the nucleus. Indeed, one can confidently assume that the rate of decay of long-lived isotopes has not varied since the earth came into existence. Therefore, once a quantity of radioactive nuclides has been incorporated into a growing mineral crystal, that quantity will begin to decay at a steady rate with a

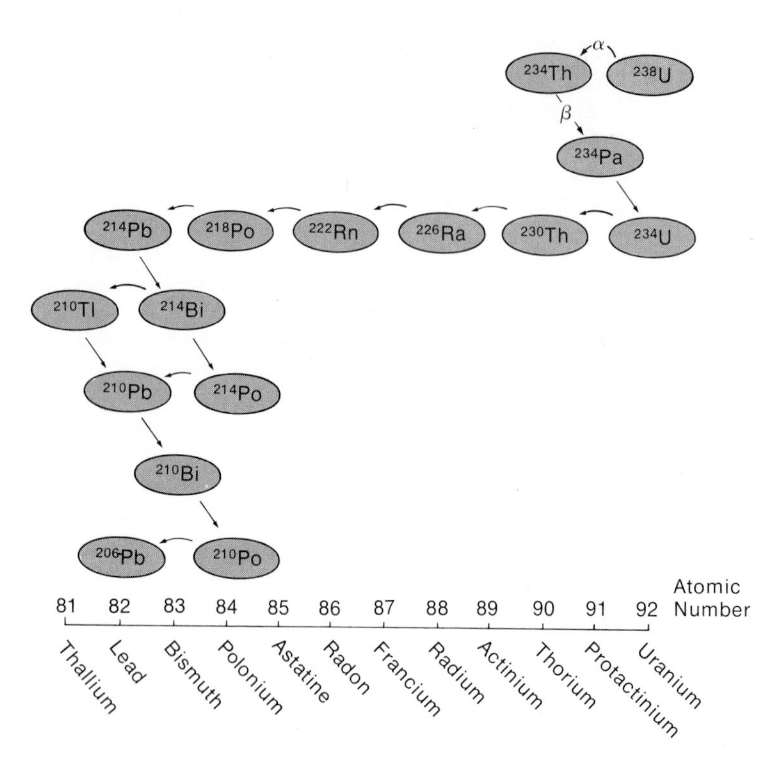

Figure 6–15 *Radioactive decay series of uranium 238 (^{238}U) to lead 206 (^{206}Pb).*

definite percentage of the radiogenic atoms undergoing decay in each increment of time. Each radioactive element has a particular mode of decay and a unique decay rate. As time passes, the quantity of the parent nuclide diminishes and the amount of daughter atoms increases, thereby indicating how much time has elapsed since the clock began its time-keeping. The beginning, or "time zero," for any mineral containing radioactive nuclides would be the moment when the radioactive parent atoms became part of a mineral from which daughter elements could not escape. The retention of daughter elements is essential, for they must be counted to determine the original quantity of the parent nuclide (Fig. 6–16).

The device used to determine the ratio of parent to daughter nuclides is a **mass spectrometer,** an analytical instrument capable of separating and measuring the proportions of minute particles according to their mass differences. In the mass spectrometer, samples of elements are vaporized in an evacuated chamber, where they are bombarded by a stream of electrons. This bombardment knocks electrons off the atoms, leaving them positively charged. A stream of these positively charged ions is deflected as it passes between plates that bear opposite charges of electricity. The degree of deflection (Fig. 6–17) depends upon the mass of the ions (the heavier the ion, the less their deflection), as well as their charge.

Not all radioactive decays are measured by means of a mass spectrometer. In the case of carbon 14, which decays by beta particle emission, the measurement of nuclides is accomplished indirectly by the use of a very sensitive **geiger counter.**

Of the three major families of rocks, the igneous clan lends itself best to radiometric dating. The dates obtained from such rocks indicate the time that a silicate melt containing radioactive elements solidified. In contrast to igneous rocks, sedimentary rocks can only rarely be dated radiometrically. Some dates for sedimentary strata have been obtained from a mineral called *glauconite,* which is believed to form "in place" at the time of deposition. This greenish mineral contains radioactive potassium 40, which decays to argon 40 and can be used in geochronology. Because of possible losses in the daughter element argon, care must be taken in interpreting dates, however; in most instances, potassium-argon dates derived from glauconite are considered minimal ages for the enclosing strata. As for classic sedimentary rocks that contain radioactive elements in their detrital mineral grains,

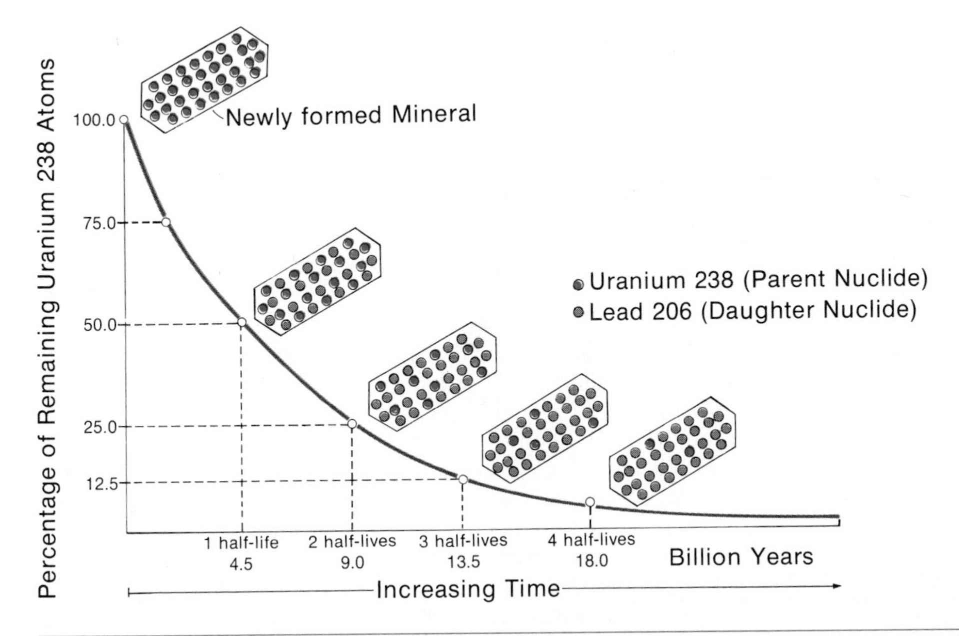

Figure 6–16 Rate of radioactive decay of uranium 238 to lead 206. During each half-life, one half of the remaining amount of the radioactive element decays to its daughter element. In this simplified diagram only the parent and daughter nuclides are shown, and the assumption is made that there was no contamination by daughter nuclides at the time the mineral formed.

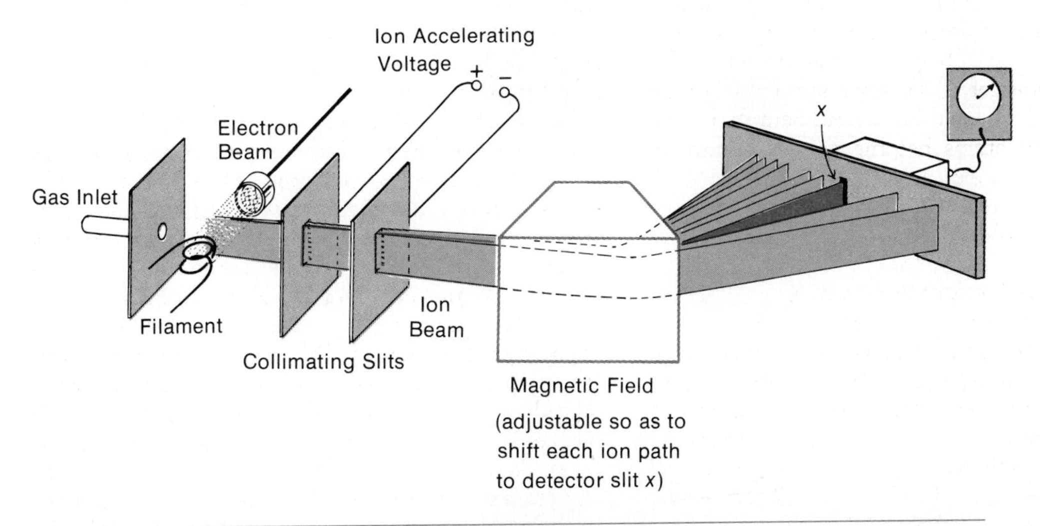

Figure 6–17 Schematic drawing of a mass spectrometer. In this type of spectrograph the intensity of each ion beam is measured electrically (rather than recorded photographically) to permit determination of the isotopic abundances required for radiometric dating.

the ages obtained refer to the parent rock that was eroded and is older than the sedimentary layer.

Dates obtained from metamorphic rocks may also require special care in interpretation. The age of a particular mineral may record the time the rock first formed or any one of a number of subsequent metamorphic recrystallizations.

There are many other problems that can affect the validity of a radiometric age. If some of the daughter products are removed from the sample by weathering or leaching, its age would be underestimated. If the element being analyzed was a gas, some of that gas (as with argon in glauconite) might have diffused out of the rock. The heat accompanying burial or mountain building might enhance such losses. There is also the possibility that at a later time older rocks were partially remelted so that the age would be that of the second, rather than the initial, melting event. Clearly, great care must be taken in understanding the field relationships of the rock masses under investigation and in selecting samples.

Once an age has been determined for a particular rock unit, it is sometimes possible to use that date to approximate the age of adjacent rock masses. A shale lying below a lava flow that is 110 million years old and above another flow dated at 180 million years old must be between 110 and 180 million years of age (Fig. 6-18). Fossils within the shale might permit one to assign it to a particular geologic system or series and, by correlation, to extend the age data around the world.

Half-life

There is no way that one can predict with certainty the moment of disintegration for any individual radioactive atom in a mineral. We do know that it would take an infinitely long time for all of the atoms in a quantity of radioactive elements to be entirely transformed to stable daughter products. Experimenters have also shown that the decline in the number of atoms is rapid in the early stages but becomes progressively slower in the later stages (see Fig. 6-19). One can statistically forecast what percentage of a large population of atoms will decay in a certain amount of time.

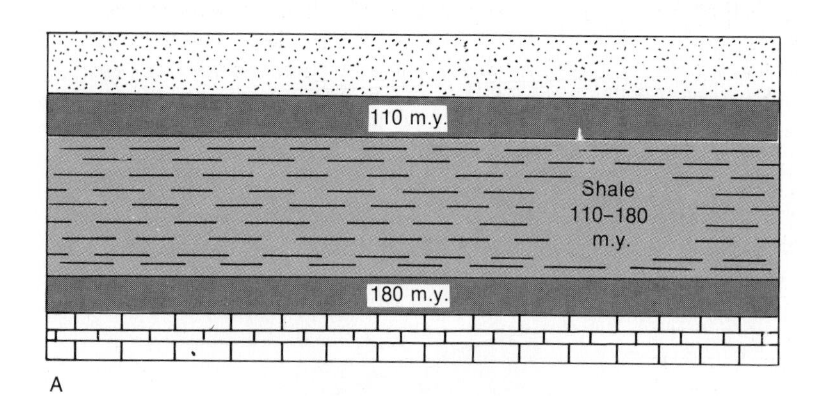

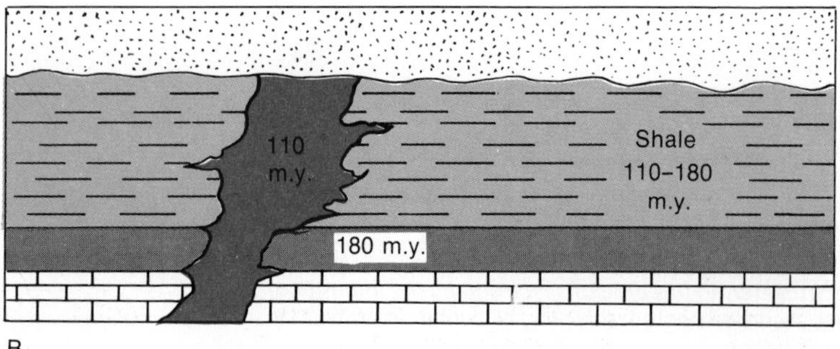

Figure 6-18 Igneous rocks that have provided absolute radiogenic ages can often be used to date sedimentary layers. In (A) the shale is bracketed by two lava flows. In (B) the shale lies above the older flow and is intruded by a younger igneous body. (Note: m.y. = million years.)

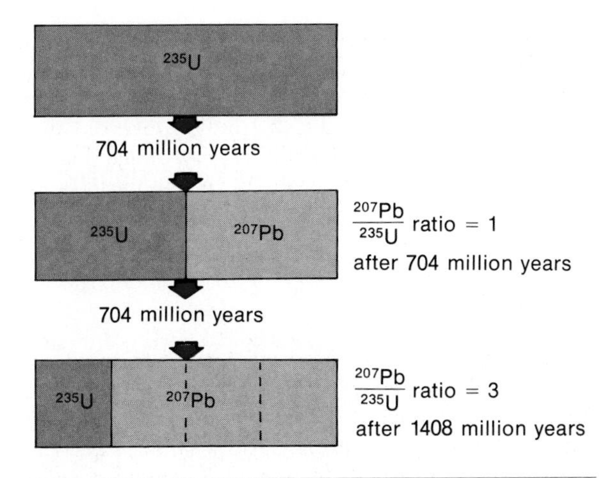

Figure 6–19 Radioactive decay of uranium 235 to lead 207.

Because of these features of radioactivity, it is convenient to consider the number of years needed for half of the original quantity of atoms to decay. This span of years is termed the **half-life.** Thus, at the end of the years constituting one half-life, one half of the original quantity of radioactive element has undergone decay. After another half-life, one half of what was left remains, leaving one fourth of the original quantity. After a third half-life, only one eighth would remain, and so on.

Every radioactive nuclide has its own unique half-life. Uranium 235, for example, has a half-life of 704 million years. Thus, if a sample contains 50 percent of the original amount of uranium 235 and 50 percent of its daughter product, lead 207, then that sample is 704 million years old. If the analyses indicate 25 percent uranium 235 and 75 percent lead 207, two half-lives would have elapsed, and the sample would be 1408 million years old (Fig. 6–19).

The principal geologic timekeepers

At one time, there were many more radioactive nuclides present on earth than there are now. Many of these had short half-lives and have long since decayed to undetectable quantities. Fortunately for those interested in dating the earth's most ancient rocks, there remain a few long-lived radioactive nuclides. The most useful of these are uranium 238, uranium 235, rubidium 87, and potassium 40 (Table 6–1). A few short-lived radioactive elements are used for dating more recent events. Carbon 14 is an example of such a short-lived isotope. There are also short-lived nuclides that represent segments of a uranium or thorium decay series.

Timekeepers that produce lead

Dating methods involving lead require radioactive nuclides of uranium or thorium that were incorporated into the earth's crust when it congealed. To determine the age of a sample of mineral or rock, one must know the original quantity of parent nuclides as well as the quantity remaining at the present time. The original number of parent atoms should be equal to the sum of the present quantity of parent atoms and daughter atoms. This raises the question of whether some of the lead may not have already been in the mineral and, if not detected, cause its radiometric age to exceed its true age. Lead 204, which is never produced by decay, provides a means of detecting original lead. All common lead contains a mixture of four lead isotopes. In most minerals used for dating, the proportions of the lead isotopes are nearly constant, so lead 204 can be used to calculate the quantities of *original* lead 206 and lead 207. These quantities can then be subtracted from the total to give the amount due to radioactivity.

As we have seen, isotopes decay at different rates. Geochronologists take advantage of this fact by simultaneously analyzing two or three isotope pairs to cross-

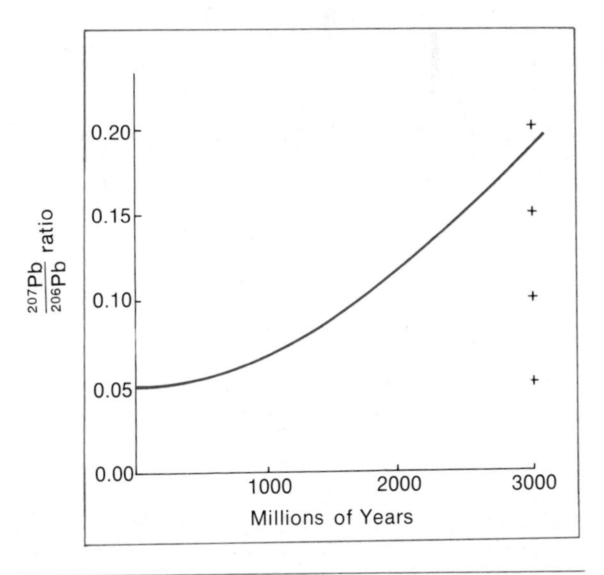

Figure 6–20 Graph showing how the ratio of lead 207 to lead 206 can be used as a measure of age.

TABLE 6–1 **SOME OF THE MORE USEFUL NUCLIDES FOR RADIOACTIVE DATING**

*Parent nuclide**	*Half-life (in years)*†	*Daughter nuclide*	*Materials dated*
Carbon 14	5730	Nitrogen 14	Organic materials
Uranium 235	704 million (7.04×10^8)	Lead 207 (and helium)	Zircon, uraninite, pitchblende
Potassium 40	1251 million (1.25×10^9)	Argon 40 (and calcium 40‡)	Muscovite, biotite, hornblende, volcanic rock, glauconite, K-feldspar
Uranium 238	4468 million (4.47×10^9)	Lead 206 (and helium)	Zircon, uraninite, pitchblende
Rubidium 87	48,800 million (4.88×10^{10})	Strontium 87	K-micas, K-feldspars, biotite, metamorphic rock, glauconite

* A *nuclide* is a convenient term for any particular atom (recognized by its particular combination of neutrons and protons).
† Half-life data from Steiger, R. H., and Jäger, E., 1977. Subcommission on geochronology: Convention on the use of decay constants in geo- and cosmochronology. *Earth and Planetary Science Letters* 36:359–362.
‡ Although potassium 40 decays to argon 40 and calcium 40, only argon is used in the dating method because most minerals contain considerable calcium 40 even before decay has begun.

check ages and detect errors. For example, if the $^{235}U/^{207}Pb$ radiometric ages and the $^{238}U/^{206}Pb$ ages agree, then they are said to be **concordant;** there then exists a high probability that the radiometric age is valid.

Radiometric ages that depend upon uranium/lead ratios may also be checked against ages derived from the ratio of lead 207 to lead 206. Because the half-life of uranium 235 is much less than the half-life of uranium 238, the ratio of lead 207 (produced by the decay of uranium 235) to lead 206 will change regularly with age and can be used as a radioactive timekeeper (Fig. 6–20).

The potassium-argon method

Potassium and argon are another radioactive pair widely used for dating rocks. By means of **electron capture** (causing a proton to be transformed into a neutron), about 11 percent of the potassium 40 in a mineral decays to argon 40, which may then be retained within the parent mineral. The remaining potassium 40 decays to calcium 40 (by emission of a beta particle from a neutron, thereby transforming it into a proton). The decay of potassium 40 to calcium 40 is not useful for obtaining radiometric ages because radiogenic calcium cannot be distinguished from original calcium in a rock. Thus, geochronologists concentrate their efforts on the 11 percent of potassium 40 atoms that decay to argon. One advantage of using argon is that it is inert—that is, it does not combine chemically with other elements. Argon 40 found in a mineral is very likely to have originated there following the decay of adjacent potassium 40 atoms in the mineral. Also, potassium 40 is an abundant constituent of many common minerals, including micas, feldspars, and hornblendes.

Like all radiometric methods, potassium-argon dating has limitations. A sample will yield a valid age only if none of the argon has leaked out of the mineral being analyzed. Leakage may indeed occur if the rock has experienced temperatures above about 125°C. In specific localities, the ages of rocks dated by this method reflect the last episode of heating rather than the time of origin of the rock itself. A less serious problem is mechanical entrapment of atmospheric argon in flowing lavas.

The half-life of potassium 40 is 1251 million years. As illustrated in Figure 6–21, if the ratio of potassium 40 to argon 40 is found to be 1 to 1, then the age of the sample is 1251 million years. If the ratio is 3 to 1, then yet another half-life has elapsed, and the rock would have a radiogenic age of two half-lives, or 2502 million years.

Among the many uses for the potassium-argon method of dating is the determination of ages of basalts erupting over oceanic hot spots. This was the method used in finding the age of the volcanoes and seamounts of the Hawaiian chain cited in the previous chapter. Without potassium-argon dating of sea floor rocks, it would be far more difficult to ascertain rates of sea floor spreading.

The rubidium-strontium method

The dating method based on the disintegration by beta decay of rubidium 87 to strontium 87 can sometimes be used as a check on potassium-argon dates because

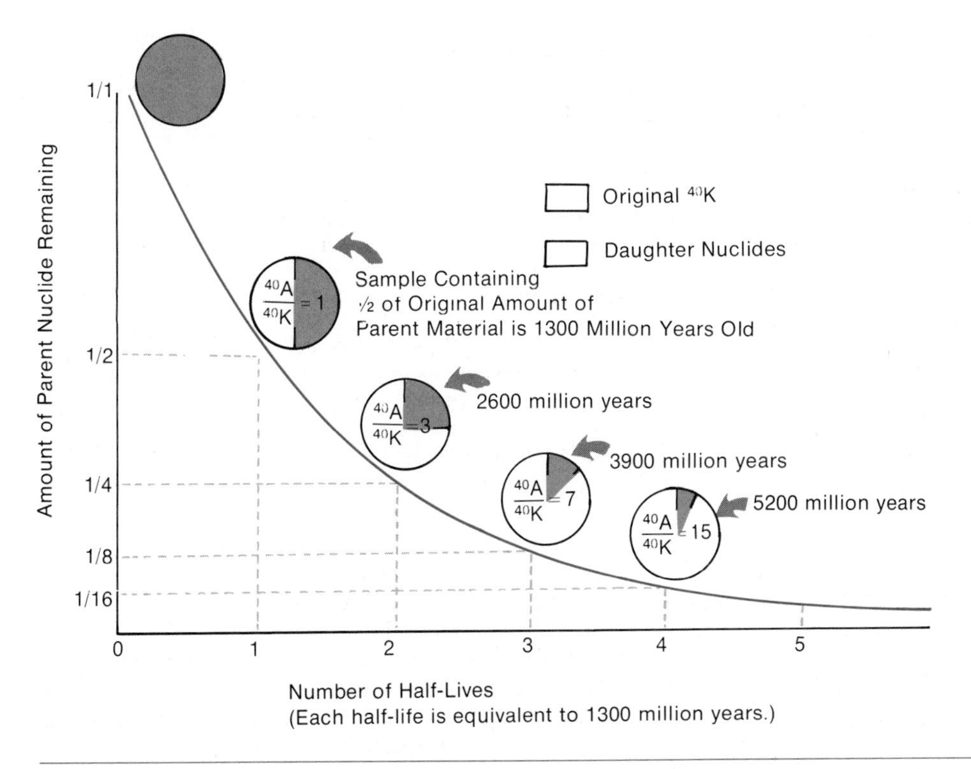

Figure 6–21 *Decay curve for potassium 40.*

rubidium and potassium are often found in the same minerals. The rubidium-strontium scheme has a further advantage in that the strontium daughter nuclide is not diffused by relatively mild heating events, as is the case with argon.

In the rubidium-strontium method, a number of samples are collected from the rock body to be dated. With the aid of the mass spectrometer, the amounts of radioactive rubidium 87, its daughter product strontium 87, and strontium 86 are calculated for each sample. Strontium 86 is an isotope not derived from radioactive decay. A graph is then prepared in which the $^{87}Rb/^{86}Sr$ ratio in each sample is plotted against the $^{87}Sr/^{86}Sr$ ratio (Figure 6–22). From the points on the graph, a straight line is constructed that is termed an **isochron.** The slope of the isochron results from the fact that, with the passage of time, there is continuous decay of rubidium 87, which causes the rubidium 87/strontium 86 ratio to decrease. Conversely, the strontium 87/strontium 86 ratio increases as strontium 87 is produced by the decay of rubidium 87. The older the rocks being investigated, the more the original isotope ratios will have been changed, and the greater will be the inclination of the isochron. The slope of the isochron permits a computation of the age of the rock.

The rubidium-strontium and potassium-argon methods need not always depend upon the collection of discrete mineral grains containing the required isotopes. Sometimes the rock under investigation is so finely crystalline and the critical minerals so tiny and dispersed that it is difficult or impossible to obtain a suitable collection of minerals. In such instances, large samples of the entire rock may be used for age determination. This method is called **whole-rock analysis.** It is useful not only for fine-grained rocks but also for rocks in which the yield of useful isotopes from mineral separates is too low for analysis. Whole-rock analysis has also been useful in determining the age of rocks that have been so severely metamorphosed that the potassium-argon or rubidium-strontium radiometric clocks of individual minerals have been reset. In such cases, the age obtained from the minerals would be that of the episode of metamorphism, not the total age of the rock itself. The required isotopes and their decay products, however, may have merely moved to nearby locations within the same rock body,

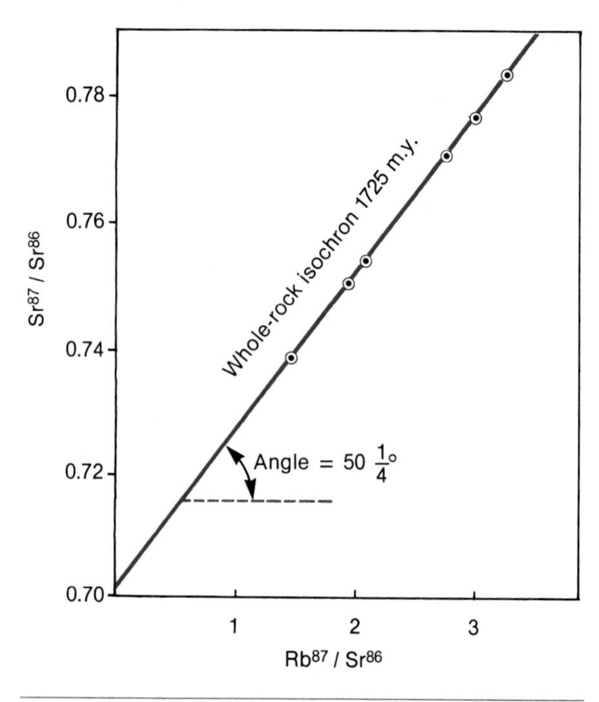

Figure 6–22 Whole-rock rubidium-strontium isochron for a set of samples of a Precambrian granite body exposed near Sudbury, Ontario. (Modified from Krogh, T.E., et al., Carnegie Institute Washington Year Book, *Vol. 66, 1968, p. 530.)*

and therefore analyses of large chunks of the whole rock may provide valid radiometric age determinations.

The carbon 14 method

Techniques for age determination based on content of radiocarbon were first devised by W.F. Libby and his associates at the University of Chicago in 1947. It has become an indispensable aid to archaeologic research and frequently is useful in deciphering the very recent events in geologic history. Because of the short half-life of carbon 14—a mere 5730 years—organic substances older than about 60,000 years no longer contain carbon 14 in measurable amounts.

Unlike uranium 238 and rubidium 87, carbon 14 is created continuously in the earth's upper atmosphere. The story of its origin begins with **cosmic rays,** which are extremely-high-energy particles (mostly protons) that bombard the earth continuously. Such particles strike atoms in the upper atmosphere and split their nuclei into smaller particles, among which are neutrons. Carbon 14 is formed when a neutron strikes an atom of nitrogen 14. As a result of the collision, the nitrogen atom emits a proton and becomes carbon 14 (Fig. 6–23). Radioactive carbon is being created by this process at the rate of about two atoms per second for every square centimeter of the earth's surface. The newly created carbon 14 combines quickly with oxygen to form CO_2, which is then distributed by wind and water currents around the globe. It soon finds its way into photosynthetic plants, since they utilize carbon dioxide from the atmosphere to build tissues. Plants containing carbon 14 are ingested by animals, and the isotope becomes a part of their tissues as well.

Eventually, carbon 14 decays back to nitrogen 14 by emission of a beta particle. A plant removing CO_2 from the atmosphere should receive a share of carbon 14 proportional to that in the atmosphere. A state of equilibrium is reached in which the gain in newly produced carbon 14 is balanced by the decay loss. The rate of production of carbon 14 has varied somewhat over the past several thousand years (Fig. 6–24). As a result, corrections in age calculations must be made. Such corrections are derived from analyses of such standards as wood samples, whose exact age is known.

The age of some ancient bit of organic material is not determined from the ratio of parent to daughter nuclides, as is done with previously discussed dating schemes. Rather, the age is estimated from the ratio of carbon 14 to all other carbon in the sample. After an animal or plant dies, there can be no further replacement of carbon from atmospheric CO_2, and the amount of carbon 14 already present in the organism begins to diminish in accordance with the rate of carbon 14 decay. Thus, if the carbon 14 fraction of the total carbon in a piece of pine tree buried in volcanic ash were found to be about 25 percent of the quantity in living pines, then the age of the wood (and the volcanic activity) would be two half-lives, or 11,460 years. To allow for unavoidable error, the age of the wood might be expressed as 11,460 $\pm$ 250 years.

The age of the earth

Anyone interested in the total age of the earth must decide what event constitutes its "birth." Most geologists assume "year 1" commenced as soon as the earth had collected most of its present mass and had developed a solid crust. Unfortunately, rocks that date from those earliest years have not been found on earth. They have long since been altered and converted to other rocks by various geologic processes. The oldest earth materials known are grains of the mineral zircon

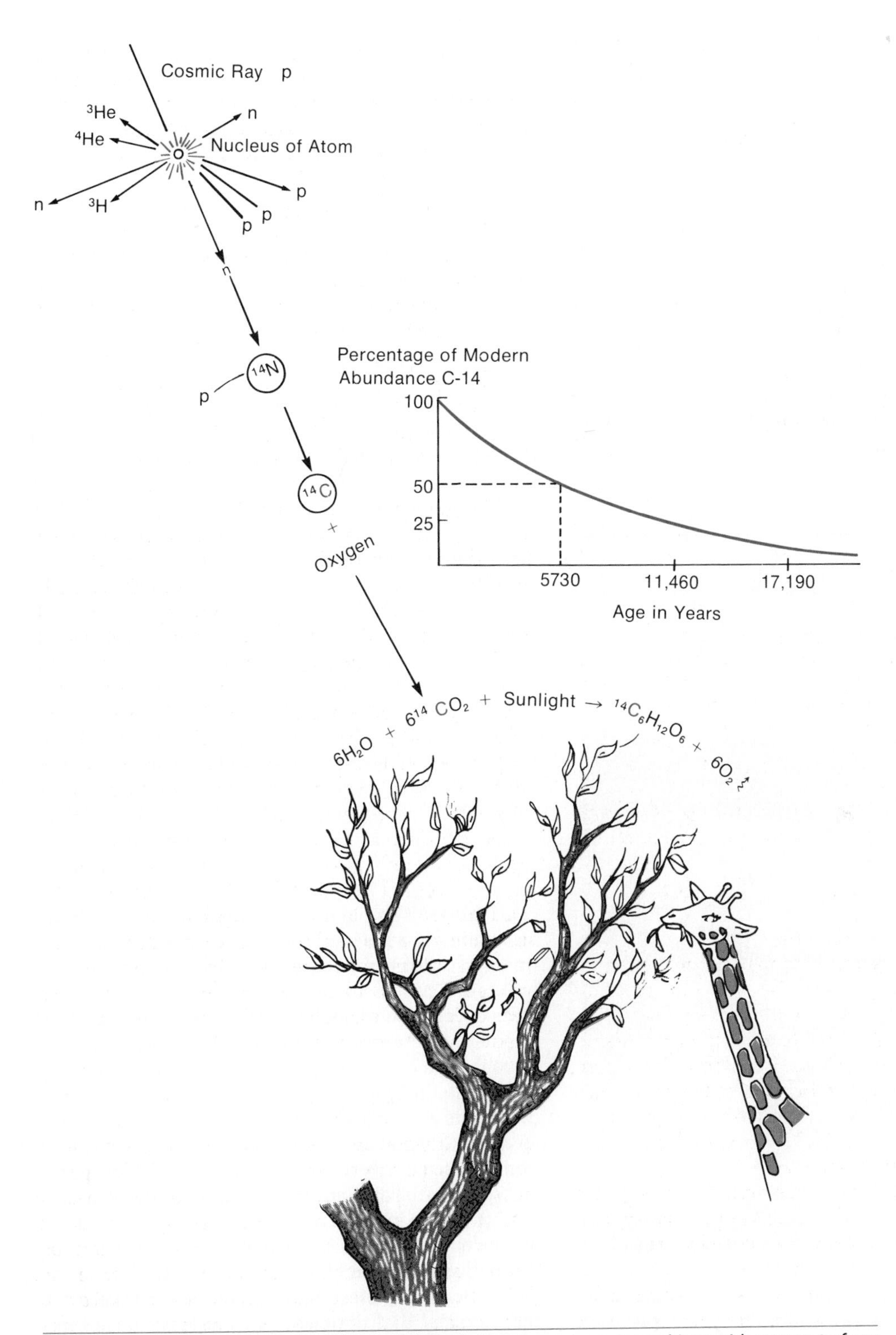

Figure 6–23 *Carbon 14 is formed from nitrogen in the atmosphere. It combines with oxygen to form radioactive carbon dioxide and is then incorporated into all living things.*

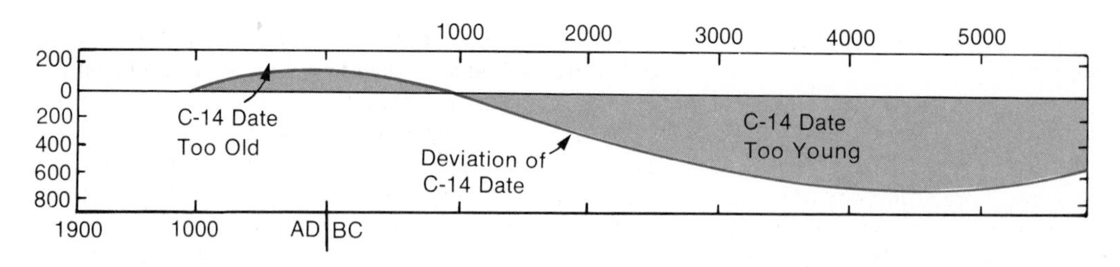

Figure 6–24 Deviation of carbon 14 ages to true ages from the present back to about 5000 B.C. Data are obtained from analysis of bristle cone pines from the western United States. Calculations of ^{14}C deviations are based on half-life of 5730 years. (Adapted from Ralph, E.K., Michael, H.N., and Han, M.C., Radiocarbon dates and reality, MASCA Newsletter, *9:1, 1973.)*

taken from a sandstone formation in western Australia. The zircon grains are 4.1 to 4.2 billion years old. They are thought to have been derived from granitic crust and deposited, along with other detrital grains, as a river sand. Other very old rocks on earth include 3.4-billion-year-old granites from the Barberton Mountains of South Africa, 3.7-billion-year-old granites of southwestern Greenland, and metamorphic rocks of about the same age from Minnesota (Fig. 6–25).

Figure 6–25 An exposure of gneiss from southwestern Minnesota that is 3.6 to 3.7 billion years old. (Minnesota Geological Survey.)

Meteorites, which many consider to be remnants of a disrupted planet that originally formed at about the same time as the earth, have provided uranium-lead and rubidium-strontium ages of about 4.6 billion years. From such data, and from estimates of how long it would take to produce the quantities of various lead isotopes now found on the earth, geochronologists feel that the 4.6-billion-year age for the earth can be accepted with confidence. Substantiating evidence for this conclusion comes from moon rocks. The ages of these rocks range from 3.3 to about 4.6 billion years. The older age determinations are derived from rocks collected on the lunar highland, which may represent the original lunar crust. Certainly, the moons and planets of our solar system originated as a result of the same cosmic processes and at about the same time.

Summary

In no other science does time play as significant a part as in geology. Time provides the frame of reference necessary to the interpretation of events and processes on the earth. In the earlier stages of its development, geology was dependent upon relative dating of events. James Hutton helped scientists visualize the enormous periods needed to accomplish the events indicated in sequences of strata, and the geologists who followed him pieced together the many local stratigraphic sections, using fossils and superposition. A scale of relative geologic time gradually emerged. Initial attempts to decide what the rock succession meant in terms of years were made by estimating the amount of salt in the ocean, the average rate of deposition of sediment, and the rate of cooling of the earth. However, these early schemes did little more than suggest that the planet was at least tens of millions of years old and that the traditional concept of a

6000-year-old earth did not agree with what could be observed geologically.

An adequate means of measuring geologic time was achieved only after the discovery of radioactivity at about the turn of the twentieth century. Scientists found that the rate of decay by radioactivity of certain elements is constant and can be measured, and that the proportion of parent and daughter elements can be used to reveal how long they had been present in a rock. Over the years, continuing efforts by investigators as well as improvements in instrumentation (particularly of the mass spectrometer) have provided many thousands of age determinations. These radiometric dates have shed light on difficult geologic problems, provided a way to determine rates of movement of crustal rocks, and permitted geologists to date mountain building or to determine the time of volcanic eruptions. In a few highly important regions, radiometric dates have been related to particular fossiliferous strata and have thereby helped to quantify the geologic time scale and to permit estimation of rates of organic evolution. Radiometric dating has also changed the way humans view their place in the totality of time.

The radiometric transformations most widely used in determining absolute ages are uranium 238 to lead 206, uranium 235 to lead 207, thorium 232 to lead 208, potassium 40 to argon 40, rubidium 87 to strontium 87, and carbon 14 to nitrogen 14. Methods involving uranium/lead ratios are of importance in dating the earth's oldest rocks. The short-lived carbon 14 isotope that is created by cosmic ray bombardment of the atmosphere provides a means to date the most recent events in earth history. For rocks of intermediate age, schemes involving potassium/argon ratios, those utilizing intermediate elements in decay series, or those employing fission fragment tracks are most useful. A figure of 4.6 billion years for the earth's total age is now supported by ages based on meteorites and on lead ratios from terrestrial samples.

Improvements in radiometric geochronology are being made daily and will provide further calibration of the standard geologic time scale in the future. Some of the time boundaries in the scale, such as that between the Cretaceous and Tertiary systems, are already well validated. Others, like the boundary between the Paleozoic and Mesozoic, require additional refinement. Additional efforts to further incorporate radiometric ages into sections of sedimentary rocks are among the continuing tasks of historical geologists. The usual methods for determining the age of strata involve the dating of intrusions that penetrate these sediments, or the dating of interbedded volcanic layers. Less frequently, strata can be dated by means of radiogenic isotopes incorporated within sedimentary minerals that formed in place at the time of sedimentation.

QUESTIONS FOR REVIEW

1 Define "half-life." Why is this term used in radiometric dating instead of an expression like "whole-life"? What are the half-lives of uranium 238, potassium 40, and carbon 14?

2 What types of radiation accompany the decay of radioactive isotopes?

3 In making an age determination based upon the uranium-lead method, why should an investigator select an unweathered sample for analysis?

4 How do the isotopes carbon 12 and carbon 14 differ from one another in regard to the following:
- **a** number of protons
- **b** number of electrons
- **c** number of neutrons

What is the origin of carbon 14?

5 Pebbles of basalt within a conglomerate yield a radiometric age of 300 million years. What can be said about the age of the conglomerate? Several miles away, the same conglomerate strata are bisected by a 200-million-year-old dike. What now can be said about the age of the conglomerate in this location?

6 Has the amount of uranium in the earth increased, decreased, or remained about the same over the past 4.6 billion years? What can be stated about changes in the amount of lead?

7 State the estimated age of a sample of mummified skin from a prehistoric human that contained 12.5 percent of an original quantity of carbon 14.

8 State the effect on the radiometric age of a zircon crystal being dated by the potassium-argon method if a small amount of argon 40 escaped from the crystal.

9 What is the advantage of having both uranium and thorium present in a mineral being used for a radiometric age determination?

10 If an intrusion of granite cuts into or across several strata, which is older—the granite or the strata?

11 Minerals suitable for radiometric age determinations are usually components of igneous rocks. How, then, can absolute ages be obtained for sedimentary formations?

12 If an erosional surface (unconformity) cuts across or truncates folded strata, did the erosion occur before or after the strata were folded?

SUPPLEMENTAL READINGS AND REFERENCES

Berry, W.B.N., 1968. *Growth of a Prehistoric Time Scale.* San Francisco, W.H. Freeman Co.

Eicher, D.L., 1976. *Geologic Time,* 2nd ed. Englewood Cliffs, N.J., Prentice-Hall Inc.

Faul, H., 1966. *Ages of Rocks, Planets, and Stars.* New York, McGraw-Hill Book Co.

Harbaugh, J.W., 1968. *Stratigraphy and Geologic Time.* Dubuque, Iowa, William C. Brown Co.

Ojakangas, R.W., and Darby, D.G., 1976. *The Earth, Past and Present.* New York, McGraw-Hill Book Co.

Toulmin, S., and Goodfield, J., 1965. *The Discovery of Time.* New York, McGraw-Hill Book Co.

7 weathering and mass wasting

The earth, like the body of an animal, is wasted at the same time that it is replaced. It has a state of growth and augmentation; it has another state which is that of diminution and decay.

JAMES HUTTON
Theory of the Earth, 1795

Prelude Over 3 billion years ago, rain and snow fell upon the earth much as today. Solid rocks were reduced to grains of sand by the action of water, ice, and chemically active solutions. We know this from clues left in ancient rocks whose great antiquity has been established by radioactive dating. The processes by which rocks are broken into smaller particles and chemically decomposed have continued unabated through the eons of geologic time. We term these most enduring of geologic processes *weathering.* Weathering is of enormous importance to all of us, for without it we would have no soils with which to grow our foods.

By reducing solid rock masses to loose clastic debris, weathering prepares the earth's surface for another important geologic process, mass wasting. *Mass wasting* is the downslope movement of earth materials under the influence of gravity. Such movements include landslides, mudflows, avalanches, and a variety of less obvious events. Along with running water and moving ice, mass wasting accounts for more than 90 percent of the erosion that has given us our landscapes.

The Wall of Windows as viewed from Peek-a-boo Trail, Bryce Canyon National Park, Utah. The spectacular natural sculpturing is the result of variation in resistance to weathering and erosion of the different strata, as well as preferential erosion along the joint system. (Courtesy of U.S. National Park Service.)

MECHANICAL WEATHERING

The chief agents of **mechanical weathering,** or **disintegration,** are frost action, temperature changes, unloading, crystal growth, and the wedging action of plant roots. Many different factors influence the efficiency of these agents in causing disintegration. The composition and texture of the rocks being weathered and the presence of joints, fractures, and voids clearly affect the rate at which solid rock can be reduced to rubble. Mechanical weathering is also affected by climate, topography, and the length of time over which weathering agents have been operating. In general, mechanical weathering predominates over chemical weathering in regions characterized by extreme cold or in warm, arid regions. By definition, mechanical weathering involves the physical disintegration of solid masses of rock into loose fragments. There is little chemical change in the rock itself. Thus, a chemical analysis of the disintegrated rock would be similar to that of the parent rock. Frost action, unloading, crystal growth, and root wedging are the principal mechanisms of mechanical weathering.

Frost action

Water expands by about 9 percent when it freezes. If the water freezes in a confined space, pressure caused by the expansion may cause rock masses to be pushed apart and ruptured. Such frost action is an important effect in mechanical weathering in that the freezing water is capable of exerting thousands of pounds of pressure per square inch. Many of us have become aware of the effects of frost action during severe winters when concrete roads and sidewalks are cracked by alternate freezing and thawing. The way freezing water disrupts solid rock is similar to the manner in which it breaks up streets and sidewalks. After water has filled a crack in the rock, the water at the lip of the crack may freeze (Fig. 7-1). Once this has occurred, the water deeper in the crack is sealed off, and as it begins to freeze and expand, it pushes against the walls of the fracture, and thereby widens the space between those walls. In a subsequent thaw, fragments of rock may slip down into the crack and act as wedges to hold it open. These processes are most prevalent in areas where water is abundant and where temperatures drop below freezing at night and then rise during the day. Mountainous regions in temperate zones have this kind of daily temperature change, and in such rugged regions, frost action is an important weathering process.

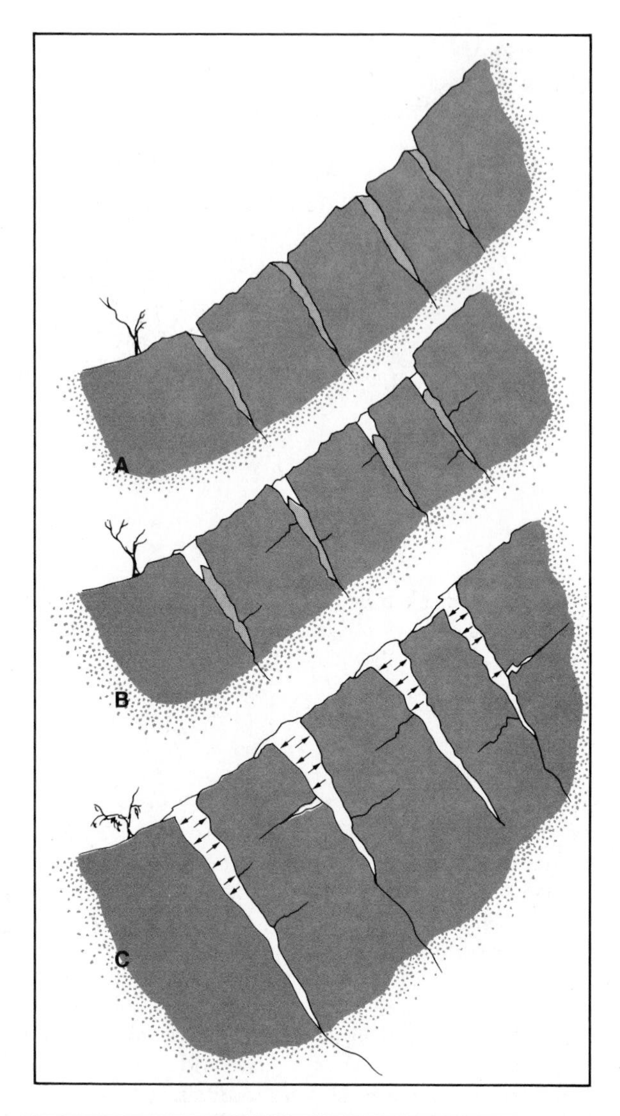

Figure 7-1 Three steps in the process of widening fractures and joints by the action of freezing water. (A) Water enters and fills rock fractures and joints. (B) As temperatures decline, ice forms at the surface and thereby forms a seal over underlying water. (C) Continued cooling eventually causes water at deeper levels to form ice. The transformation involves an increase in volume of 9 percent, resulting in pressure on the walls of fractures. Joints and fractures are pried open and widened.

Unloading

Quarrymen and miners are well aware that when the heavy weight of rock is removed from a mine with consequent loss of support to the surrounding rock, the outside pressure may sometimes cause a veritable

Figure 7–2 Sheet structure or sheeting in granite. Anza Desert east of San Diego, California. (Courtesy of J.C. Brice.)

explosion of rock fragments into the excavations. In nature, erosion may similarly remove large volumes of surface rock, thus lightening the load on deeper rock and allowing it to expand. Because the mass of rock is still confined on all sides, it can only respond to the release of pressure by expanding upward. As it expands it will rupture and form joints that are roughly parallel to the surface topography. The process is called **unloading,** and the joint systems are referred to as **sheeting** (Fig. 7–2). Sheeting is often seen in granite and other massive crystalline rocks that have been laid bare by glaciation or running water. It can also be readily observed along the upper walls of quarries, where it may even facilitate the removal of the stone. Sheeting provides planar passageways along which solution and frost action may proceed vigorously. Half Dome at Yosemite National Park and Stone Mountain, Georgia, are examples of mountains whose shape has been controlled, at least partially, by sheeting.

Crystal growth

The growth of crystals, especially crystals of sodium chloride, calcium sulfate, or magnesium sulfate, has been found to cause scaling in exposed rock surfaces and to dislodge grains and crystals from their parent rock. The process is particularly effective in porous rocks, where crystals exert large expansive stresses as they grow. Disintegration of building stone because of the growth of crystals in pore spaces has been a vexing problem for architects involved in the preservation of historically important buildings. The temples at Luxor, Egypt, for example, have been seriously damaged by alternate solution and crystallization of salt. During dry spells, magnesium (and calcium) sulfate has been known to crystallize along zones of weakness in the dolomitic building stones of London's Houses of Parliament, thereby causing spalling and crumbling at an alarming rate. The magnesium sulfate is formed as a result of a chemical reaction between coal smoke and the magnesium in the dolomite.

Root wedging

When one observes the growth of a plant from a seed, it is apparent that even a frail seedling is capable of exerting sufficient force to push aside relatively firm soil. As they grow into rock fractures and expand, the roots of large plants can exert correspondingly greater forces and are capable of widening cracks and accelerating the rate of disintegration by significant amounts (Fig. 7–3). When the plant dies and the roots decay, they leave openings in which freezing water may accumulate and further widen the void. Plants also react with rocks chemically, as will be described in the following section.

Figure 7–3 The roots of this tree have cracked this large boulder apart. Root action is a powerful weathering agent. (Photograph by Grant Lashbrook.)

CHEMICAL WEATHERING

Chemical weathering refers to the decomposition of rocks at or near the earth's surface and under relatively low temperatures (Fig. 7-4). During chemical weathering, water and chemically active water solutions as well as oxygen and carbon dioxide from the atmosphere attack the minerals in rocks. The susceptible parent minerals are frequently altered to softer secondary minerals, as when feldspars are converted to clay minerals and lose certain ions to solution. Chemical weathering is an exceedingly important process, for it leads to the formation of soils, and in some parts of the world it has resulted in the concentration of iron, aluminum, uranium, gold, and tin.

Relation of disintegration to decomposition

The processes of chemical weathering and mechanical weathering are not truly independent of one another. Because of climatic factors, especially the amount and frequency of rainfall, one or the other process may be dominant in a particular location. For the most part, however, decomposition and disintegration are interacting processes. For example, disintegration greatly promotes decomposition by reducing large intact rock masses to accumulations of smaller fragments. This fragmentation provides more total surface area for chemical attack than was originally present in the larger mass of rock. One can see what happens readily by imagining a block of rock (Fig. 7-5) measuring 2 meters on each edge. Such a block would have a total of 24 square meters of surface area. If the block is split once along each of three perpendicular planes, each resulting block would have 6 square meters of surface area, or a total of 48 square meters for all eight pieces. Chemically reactive solutions are only able to decompose surfaces they can reach, and so the rate of chemical attack is enhanced by increasing surface area. One can demonstrate this fact by applying a few drops of dilute hydrochloric acid first to an unfragmented piece of limestone, and then to a similar piece of limestone that has been pulverized. Because of the greater amount of total surface area, the pulverized rock will display the more vigorous effervescence.

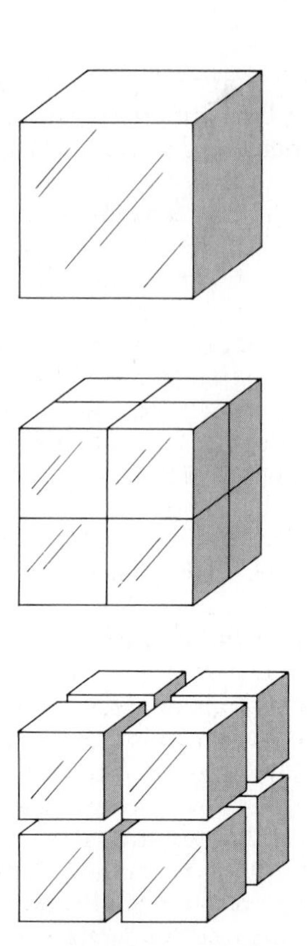

Figure 7-5 By splitting a cube (of rock) the surface area exposed to attack by chemical agents is doubled. It is for this reason that finer-grained materials decompose more rapidly than coarser-grained material of identical composition.

Figure 7-4 Chemical weathering has produced these solution pits on solid granite, Graniteville, Missouri.

Processes of chemical weathering

People sometimes are surprised to learn that many rocks exposed at the earth's surface are really not in chemical equilibrium with their environment, but rather are slowly but continuously reacting with atmospheric components to dissolve or change to new substances that are more nearly stable at the earth's surface. Weathering proceeds in accordance with a generalization called the "rule of stability," which states that a mineral approaches stability most closely in an environment similar to that in which it formed. Intrusive igneous rocks, of course, are formed under conditions of high temperature, high pressure, and a deficiency of free oxygen and fresh water. When such rocks are laid bare on the continents, they find themselves in an environment of low temperature, low pressure, and abundant oxygen and water. Chemical change is inevitable, and the minerals most susceptible to change are those that formed under physical and chemical conditions most removed from conditions at the earth's surface. For example, among the common rock-forming silicate minerals, olivine forms at high temperatures and pressures early in the crystallization of a magma. Consequently, it rapidly weathers in the environments that exist at the earth's surface. Quartz forms much later under less extreme conditions of temperature and pressure and is less susceptible to weathering. In reference to the Bowen Reaction Series (discussed in Chapter 2), it is apparent that minerals nearer the top of the series generally weather more rapidly than those near the base (Fig. 7-6).

In Chapter 1, we noted that silicon-oxygen tetrahedra are joined together by sharing some oxygen ions, and that the number of metallic ions needed to neutralize the crystal is reduced by this sharing. This increased oxygen sharing by silicon results in a greater number of strong covalent-like bonds between silicon and oxygen, and this in turn greatly increases the ability of the mineral to resist weathering. For example, the ratio of oxygen to silicon in olivine is 4, in pyroxene 3, in hornblende 2.7, in biotite 2.5, and in quartz 2. The diminishing ratios correlate nicely with greater resistance to weathering. All of the oxygen atoms are shared by silicon atoms in quartz, and this partially accounts for its great resistance to weathering.

The chemical weathering of rocks involves reactions that are interrelated, that may occur simultaneously, and that utilize water, oxygen, carbon dioxide, and organic acids. These processes include hydrolysis, oxidation, carbonation, solution, hydration, and chemical changes induced by the growth of plants.

Hydrolysis

Hydrolysis is a chemical reaction between a mineral and water. It involves a reaction between the H or OH ions in the water and relatively active metallic ions, such as sodium, calcium, potassium, and magnesium. Hydrolysis is particularly important in causing the decomposition of silicate minerals. Althrough it may occur in the presence of pure water, in nature hydrolysis nearly always involves carbon dioxide. To illustrate, small quantities of carbon dioxide from the atmosphere or soil are dissolved in water to form carbonic acid.

$$\underset{\text{water}}{H_2O} + \underset{\text{carbon dioxide}}{CO_2} \rightarrow \underset{\text{carbonic acid}}{H_2CO_3}$$

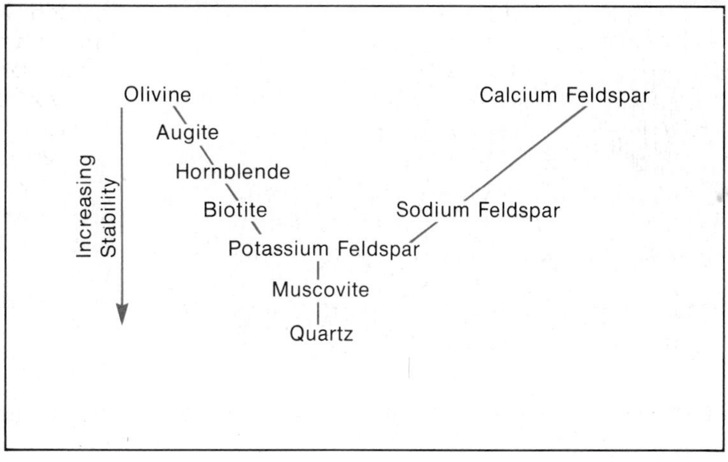

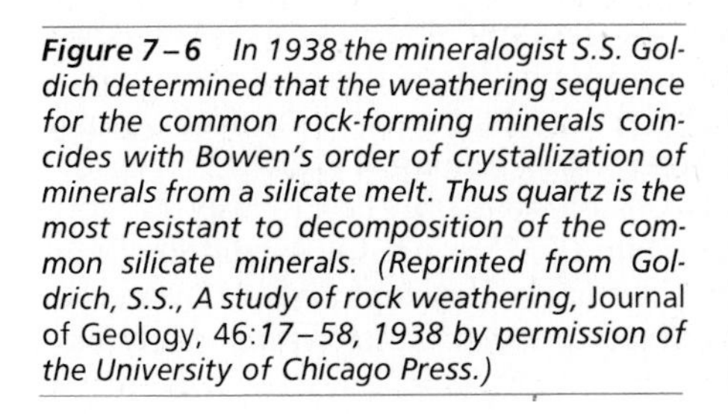
Figure 7-6 *In 1938 the mineralogist S.S. Goldich determined that the weathering sequence for the common rock-forming minerals coincides with Bowen's order of crystallization of minerals from a silicate melt. Thus quartz is the most resistant to decomposition of the common silicate minerals. (Reprinted from Goldrich, S.S., A study of rock weathering,* Journal of Geology, *46:17-58, 1938 by permission of the University of Chicago Press.)*

The carbonic acid may then ionize to form hydrogen ions and bicarbonate ions.

$$\underset{\text{carbonic acid}}{H_2CO_3} \rightarrow \underset{\text{hydrogen ion}}{H^{+1}} + \underset{\text{bicarbonate ion}}{(HCO_3)^-}$$

Next, the hydrogen ions penetrate into the crystal lattices of silicate minerals, such as potassium feldspar. They have little difficulty in gaining admission because of their small size, and once within the lattice, they disrupt the charge balance as they displace cations, such as potassium.

$$\underset{\text{water}}{H_2O} + \underset{\text{hydrogen ions}}{2H^{+1}} + \underset{\text{potassium feldspar}}{2(K\,Al\,Si_3O_8)} \rightarrow \underset{\text{potassium ions}}{2K^{+1}} + \underset{\text{kaolinite clay}}{Al_2Si_2O_5(OH)_4} + \underset{\text{silica}}{4(SiO_2)}$$

The potassium ions released from the feldspar may be carried away in solution, utilized by plants, or become incorporated into clay minerals. A small part of the silica is removed in solution, although the greater part remains in the clay-rich weathered residue.

Oxidation

Oxidation, the addition of oxygen to a compound, is one of the main kinds of changes produced in rocks by chemical weathering. Oxygen has a strong affinity for iron, which may be present in such silicate minerals as hornblende, and olivine, as well as such sulfides as pyrite (FeS_2). The oxidation of the iron (essentially what we call rusting) takes place chiefly in the presence of atmospheric moisture and results in the range of red and brown colorations we see in soils and weathered rocks. In the oxidation process, oxygen gas dissolved in water reacts with iron to form hematite (Fe_2O_3) or limonite ($Fe_2O_3 \cdot H_2O$). The process is illustrated by the following formula:

$$\underset{\text{iron}}{4Fe} + \underset{\text{oxygen}}{3O_2} + \underset{\text{water}}{nH_2O} \rightarrow \underset{\text{limonite (iron hydroxide or "rust")}}{2(Fe_2O_3) \cdot nH_2O}$$

(n means a variable amount)

The oxidation of a common nonsilicate, such as pyrite ("fool's gold"), involves combining oxygen with both iron and sulfur as follows:

$$\underset{\text{pyrite}}{4FeS_2} + \underset{\text{water}}{nH_2O} + \underset{\text{oxygen}}{15O_2} \rightarrow \underset{\text{limonite}}{2Fe_2O_3 \cdot nH_2O} + \underset{\text{sulfuric acid}}{8H_2SO_4}$$

Here, the relatively insoluble iron compounds may remain as a coating on the rock, whereas the sulfuric acid is leached away and becomes available for chemical reactions with other minerals.

Carbonation

As implied by the term, carbonation involves the chemical addition of carbon dioxide to earth materials. Carbon dioxide in the atmosphere (and in the air trapped within soils) is readily absorbed by water to form carbonic acid. Although relatively weak, carbonic acid nevertheless has a pervasive cumulative effect in the chemical weathering of a variety of different kinds of rocks. It is involved in the chemical attack on common silicate minerals and is particularly effective in dissolving limestones (Fig. 7-7) and dolostones. The reaction for limestone is indicated below.

$$\underset{\text{water}}{H_2O} + \underset{\text{carbon dioxide}}{CO_2} \rightarrow \underset{\text{carbonic acid}}{H_2CO_3}$$

$$\underset{\text{carbonic acid}}{H_2CO_3} + \underset{\text{limestone}}{CaCO_3} \rightarrow \underset{\text{soluble calcium bicarbonate}}{Ca(HCO_3)_2}$$

For carbonation to occur, water must be readily available. For this reason carbonation is most vigorous in moist climates.

Hydration

Hydration is a process whereby water is absorbed by a mineral and incorporated into the weathering product. For example, the mineral anhydrite ($CaSO_4$) may take in water to become alabaster gypsum ($CaSO_4 \cdot nH_2O$), or hematite (Fe_2O_3) may be converted to limonite ($Fe_2O_3 \cdot nH_2O$). Hydration is an important process in the development of clay and accounts for the presence of water within many clay

Figure 7–7 *A Dinosaur* (Triceratops) *carved in limestone peers from the ivy covering part of the geology building at Washington University in St. Louis. The original smooth surface of the limestone is now rough and pitted because of differential chemical weathering enhanced by carbon dioxide in rainwater. The particles of calcite which stand out in relief are relatively dense, nonporous shell fragments of marine invertebrates. The carving and the building it decorates are about 70 years old. (Photograph by Herb Weitman.)*

minerals. Another aspect of hydration is that the hydrated mineral, because of the water it has taken up, is larger than the parent mineral. The increase in volume causes growing hydrated crystals to exert pressure on the walls of the spaces they occupy, and such pressure may contribute to rock disintegration.

Solution

We have seen how the dissolving power of water for certain rocks and minerals is increased when carbon dioxide is present. Even without the addition of carbon dioxide, however, water has the ability to dissolve rocks and minerals. The dissolution of thick beds of salt and gypsum is an example of simple solution weathering. A quartz sandstone also may weather by solution, although the process is exceedingly slow because of the low solubility of quartz.

The ability of water to dissolve substances is related to the configuration and electrical properties of the water molecule itself. A water molecule consists of a large negative oxygen atom and two hydrogen atoms. The hydrogen atoms have contributed their electrons to the molecule and thus exist as two small, positively charged protons. Both of the protons are located on the same side of the oxygen atom, so the molecule has an electrically positive side and an electrically negative side (see Figs. 1–7 and 1–8). Stated differently, the water molecule is strongly dipolar and is highly reactive. Minerals are dissolved as the water molecules react with constituent ions in the mineral's crystal structure. Depending upon the basic atomic structure and composition of the mineral, solution will proceed at either a slow or a relatively fast rate.

Biochemical weathering

Rooted plants also play a role in the weathering of materials at the surface of the earth. During photosynthesis, plants utilize the energy of the sun's radiation to combine water and carbon dioxide and produce tissue necessary for growth and maintenance. Part of the hydrogen from moisture taken in by the plant is released at the surface of rootlets (Fig. 7-8), where it is exchanged for ions of calcium, magnesium, and sodium from adjacent particles of clay and other minerals. These exchanged ions are required by the plant for nutrition. The hydrogen ions transferred to the clay particles render the clay slightly acidic, and this triggers weathering reactions with neighboring feldspars and other silicates. Eventually, some of these minerals also decompose to clay, thus perpetuating the weathering process.

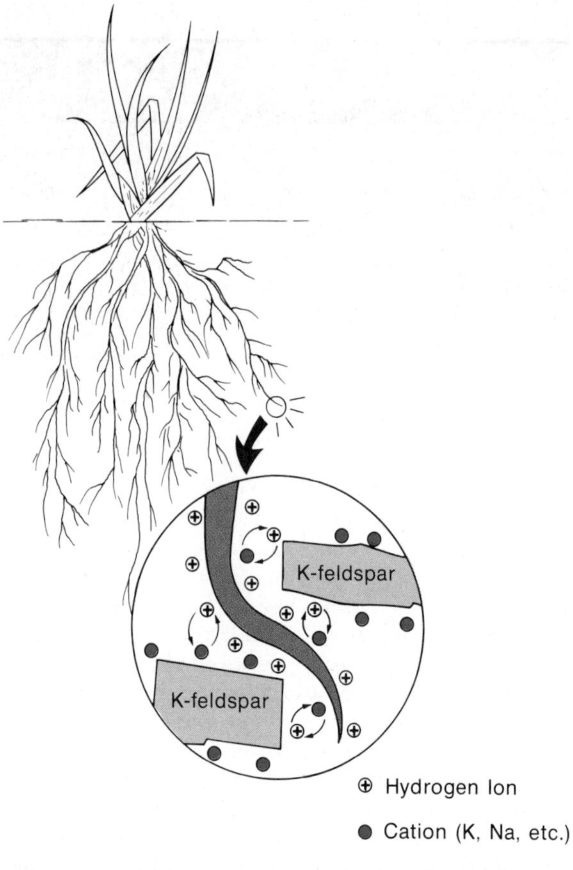

K-feldspar
K-feldspar
⊕ Hydrogen Ion
● Cation (K, Na, etc.)

Figure 7-8 Plant rootlets provide a continuing supply of hydrogen ions which are exchanged for cations from adjacent mineral particles. These exchanged ions are required for plant nutrition. The processes accelerate weathering of minerals in soils. (Modified from Keller, W.D., Principles of Chemical Weathering. *Columbia, Missouri, Lucas Bros. Publishers, 1957.)*

Spheroidal weathering

Spheroidal weathering is a term used to describe the spalling away of concentric surficial shells of the rounded surface of a boulder or rock mass. Such "onion-skin" weathering is believed to result primarily from the mechanical effects of chemical weathering. When feldspars decompose, the clay product has a greater volume than the parent feldspar. The increase in volume disrupts the interlocking texture of mineral grains in the rock and causes breakage and separation of the layer of partially weathered rock near the surface. Corners of roughly rectangular blocks broken loose along joints are decomposed along three surfaces simultaneously and are progressively rounded by spalling until the rock mass assumes the spheroidal form (Fig. 7-9).

Spheroidal weathering operates most effectively on relatively small rock masses, such as boulders. Larger-scale breaking or peeling off of concentric plates from large bare rock surfaces is referred to as **exfoliation** (Fig. 7-10). Although decomposition may assist in the process, exfoliation is primarily mechanical in nature and results from unloading, as well as pressures developed by the growth of crystals in confined intergranular spaces and rock crevices.

THE DECAY OF GRANITE

As described in Chapter 2, granite is a common igneous rock consisting of about two parts orthoclase, one part quartz, one part plagioclase and small amounts of ferromagnesian minerals. The weathering of granitic rocks has produced a large portion of the materials constituting sedimentary rocks now exposed on the continents. What happens to granite when it has experienced a long episode of chemical weathering in a moist, relatively warm climate? One can readily see at least some of the transformation if one visits an area where granite has been weathering and the erosional products have not been swept away by flowing water (Fig. 7-11). In such areas, the exposed surfaces of the rock may appear soft and clayey. If examined with a magnifying lens, one can recognize quartz grains and

Figure 7–9 *Spheroidal weathering in granite boulder (A) near Virginia Dale, Colorado; in massive sandstones (B) along highway 150 east of Bessemer, Alabama; in the Balolatchie Conglomerate (C) near Girven; Ayrshire, Great Britain; and in the basalt (D) of Ireland's Giant's Causeway. (Courtesy of J.C. Brice.)*

corroded particles of feldspar and mica. The finest material seems to be mostly clay, and a yellowish or brownish coloration indicates the presence of iron oxide. During the weathering of granite, ferromagnesian minerals are the first to show signs of decay. Brown stains or pockets of brownish powder appear at sites once occupied by grains of hornblende or biotite. Thus, the weathering of ferromagnesian minerals produces iron oxides and hydroxides as well as clay minerals.

Potassium and sodium feldspars weather more slowly than most ferromagnesian minerals, but the effects of weathering can sometimes be seen as a thin cloudy film around the outside of feldspar grains. As described earlier, the aluminum and silica in the feldspar are utilized in the formation of clay minerals, whereas the soluble sodium and calcium ions (as well as small amounts of dissolved silica) are carried away by surface and subsurface waters. As for potassium, some is carried away in solution and some is retained in the newly formed clay minerals.

As noted earlier, quartz is a mineral that is highly resistant to solution and chemical alteration. As a result, quartz grains will persist without apparent change in loose material weathered from granite. The quartz grains originally derived from granitic rocks may experience many cycles of erosion, transportation, and deposition before they are reduced to particles no smaller than silt size by abrasion and nearly imperceptible amounts of dissolution.

Figure 7–10 *A view of Moro Rock. Moro Rock is an* exfoliated dome, *which means that layers of rock weather and peel off like layers of an onion.*

Figure 7–11 *Exposure of weathered granodiorite exhibiting an accumulation of grains of feldspar and quartz resulting from disintegration and decomposition of the parent rock. This sandy material is called* grus. *The exposure is in the Sierra Nevada of California.*

RATES OF WEATHERING

The role of climate

How rapidly a particular exposure of rock will weather depends upon several interacting factors in the environment. Climate is important because it controls temperature and rainfall. High temperature, of course, promotes chemical reaction. Even an increase as little as 20°C can double reaction rates. Similarly, high rainfall accelerates rates of weathering by increasing the possibilities for solution and reaction, and by helping to wash away surface debris and expose fresh surfaces for chemical attack. Both higher temperatures and an abundance of moisture promote the growth of plants, and luxuriant plant growth assists in weathering by extracting ions from minerals and adding chemically reactive compounds.

Parent rock

Depending upon their composition, texture, and such features as fractures and joints, rocks exposed at the earth's surface may exhibit a wide range of resistance to weathering. Limestones and dolostones, for example, are highly susceptible to solution. In regions having high rainfall, these rocks weather more rapidly than in arid regions. This observation suggests that it is difficult to list rocks in the order of their resistance to weathering because a rock might be more durable than another under one set of conditions and less durable under different conditions. As a general rule, igneous and metamorphic rocks are more resistant to weathering than sedimentary rocks. A silica-cemented quartz sandstone, however, is among the most enduring of rock types.

Statues and other sculptured edifices that have actual dates carved into the stone permit one to assess rates of weathering. In cities, because of the higher levels of carbon and sulfur dioxides, weathering rates are generally higher than in nonindustrialized areas. Figure 7–12 provides an example of damage to statues in cities. The bust of Beethoven was erected in a St. Louis park in 1884. Despite many attempts to treat the Italian Carrara marble from which the bust was carved, it has continued to decompose. The com-

Figure 7–12 *The ravages of chemical weathering are clearly visible on this statue of the composer Ludwig van Beethoven erected in 1884 in Tower Grove Park, St. Louis, Missouri. The statue was carved in Italian Carrara marble. (Courtesy of the Washington University Center for Archaeometry.)*

poser's gruff expression has been so altered that St. Louisans refer to the statue as "poor Beethoven."

SOIL

Components of soil

In terms of importance to human populations, the most significant aspect of weathering is in the development of soil. **Soil** can be defined as weathered material that will support the growth of rooted plants. A heap of quartz sand is not soil according to this definition, for it will not support vegetation. Nor is the so-called lunar soil of the moon a true soil. True soil is not a mere accumulation of particles derived from mechanical and chemical weathering. It is a vital natural material that would not develop were it not for the activities of a myriad of bacteria, fungi, worms, and insects. Actually, soil can be considered an intricate mixture of mineral solids, water, gases, dissolved substances, remains of dead organisms, and multitudes of thriving organisms.

Why is a certain amount of organic material necessary in order for weathered material to support the growth of rooted plants? One reason is that the organic material supplies nutrients required for vigorous plant growth. Plants need nitrogen to synthesize protein, but they are unable to take free nitrogen (N_2) from the air. Certain soil bacteria, however, are able to use this free nitrogen and convert it into usable fixed nitrogen compounds (nitrates). Also, the decaying vegetation and animal waste contain ammonia (NH_3), the nitrogen of which can be utilized by most plants.

The material that colors the upper parts of many soils gray or black is called **humus.** Humus is organic material that is so thoroughly decayed that one cannot discern the nature of the parent organisms. This essential organic mixture increases the porosity and water-holding capacity of the soil, provides a buffer against rapid changes in acidity, and assists in the retention of chemicals needed as plant nutrients. Carbon dioxide liberated by humus will, as we have noted, combine with water to form a weak acid capable of dissolving calcium carbonate. Once the calcium is separated from the carbonate part of a compound, chemicals in humus known as chelating agents bond to the calcium and prevent it from again forming $CaCO_3$. Most plants are unable to extract the calcium they need directly from $CaCO_3$, but they have evolved mechanisms for releasing the chelated calcium that exists in humus.

Factors governing soil development

Five factors are involved in soil formation. These are climate, parent material, topography, time, and biological activity. Of these factors, climate is of greatest importance. Indeed, soil types and climate are so closely related that maps showing global distribution of climates resemble maps showing the occurrence of different soil types (Fig. 7–13). Temperature and rainfall, of course, influence not only rates of weathering but also the nature of vegetation. Given enough time,

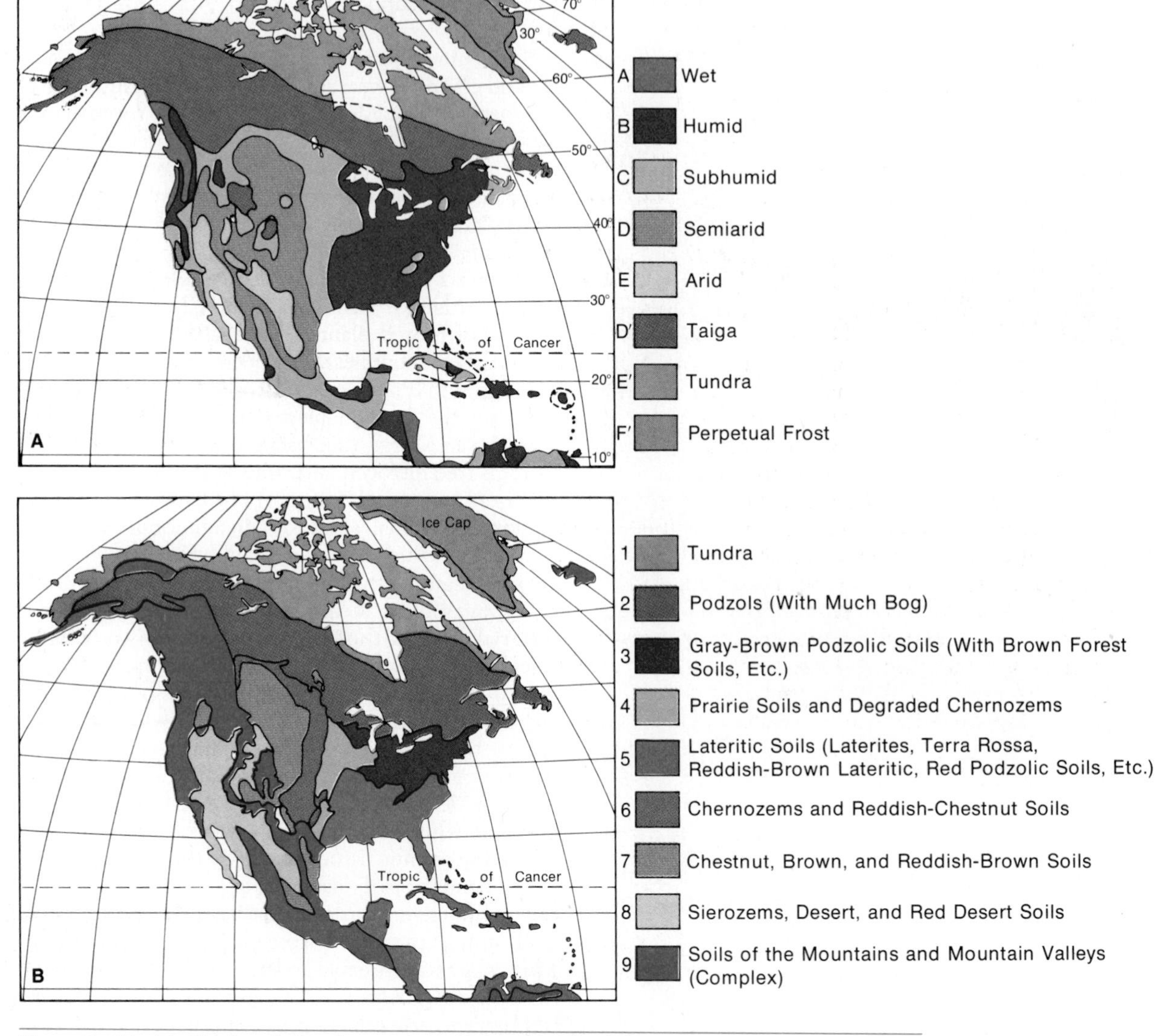

Figure 7–13 *Distribution of climatic zones in North America (A) exhibits a rough correspondence to the distribution of soil types (B). (From U.S. Department of Agriculture Yearbook, 1941.)*

soil-forming processes operating within a given climate will prevail over differences in parent material and provide a basically climate-controlled soil type. The reverse is also true. Identical rocks in temperate, high-rainfall regions and in semitropical arid regions will produce quite different soils.

The influence of parent material on the formation of soils is of secondary importance but can be observed in soils that have recently developed from unaltered bedrock. Such soils still retain textural and mineral components that are derived from the parent rock, but these may become obscured after the soil has evolved to a further stage. In some cases, various influences of parent material will persist in the developing soil. Soils developed above quartz sandstones may tend to be more acidic than soils formed on limestones, and these differences will be reflected by the natural vegetation living on each soil type. Not all soils are developed

Figure 7–14 Fertile topsoil being washed away in gullies in an untended field.

from underlying consolidated bedrock. Many fine agricultural soils are formed on loosely consolidated sediment laid down by streams, winds, glaciers, or the waters of lakes.

Topography also influences the ultimate nature of soils. On steep slopes, for example, erosion is usually more vigorous and the soil layer generally thinner. Also, water falling on slopes is partially lost to runoff and therefore there is less water available for percolation into deeper layers of soil. Even the direction of slope may have a bearing on soil development. In the Northern Hemisphere, slopes facing south tend to be warmer and dryer and therefore support different kinds of plants than do the northern slopes.

The importance of time in soil formation is often not fully appreciated by those who clear away soil for construction or mining and expect nature to quickly restore this important resource. The development of soil is an exceedingly slow process. Depending on climatic conditions, several hundred to several thousands of years are required to produce a fertile soil. Agricultural soils formed during this lengthy natural process may be lost to erosion in only a few years, particularly in areas where soil conservation measures are not followed. The problem is not trivial, for an estimated 3 billion metric tons of good agricultural soil is lost each year from agricultural fields in the United States alone (Fig. 7–14).

Soil horizons

Soils are not uniform in texture and composition from bedrock to the surface, but rather consist of a number of fairly distinct layers, or **soil horizons,** which differ in composition and physical characteristics. The stack of soil horizons is called a **soil profile** (Fig. 7–15). A simple profile, such as might develop in a humid re-

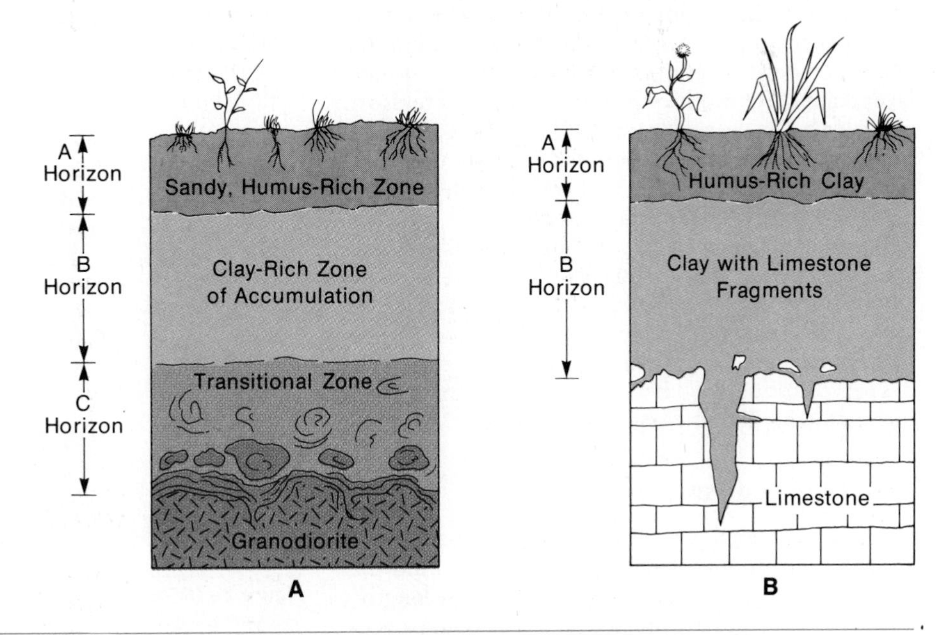

Figure 7–15 Soil profiles developed on granodiorite (A) and limestone (B).

gion on granitic rock, has three distinct divisions. Soil scientists have named these layers the A, B, and C horizons. The A horizon, also known as topsoil, which lies immediately beneath the surface, is characterized by a high content of organic matter. It is also a zone in which soluble compounds are dissolved and, along with fine clay particles, carried downward to be deposited in the underlying B horizon. Because of these losses, the A horizon is referred to as leached or eluviated, whereas the B layer is sometimes called the washed-in or illuviated zone. Iron and clay minerals tend to accumulate in the B layer. In humid regions one can often recognize the B horizon by its more brownish color and clayey texture. The C horizon of the soil profile consists of partially altered parent material. In it one finds evidence of chemical weathering, but soil development has not progressed to the level at which the original characteristics of the bedrock are unrecognizable. Beneath the C horizon lies unaltered parent material.

Soils developed on limestones in humid regions often do not exhibit a clear division between the B and C horizons. In such soils, the A horizon consists of dark, humus-rich material that changes downward into relatively light-colored clay, and then stained and partly dissolved limestone. The clay in these soils is the residue of clay impurities left when the limestone was dissolved.

How does the sequence of soil horizons develop in a humid region underlain by granitic rock? Imagine that glaciation has recently stripped away all loose sediment over a granite outcrop. Initially, disintegration and decomposition of the granite will produce a sandy layer composed mostly of grains of quartz and feldspar along with clay derived from the weathering of silicate minerals. As the layer of granular material thickens, plants establish themselves, and organic material from a variety of sources becomes incorporated into the surface layer. Meanwhile, with each rainfall, soluble ions and clay particles are flushed downward from the surface layer into a lower level. This lower level soon takes on the clayey character of the B horizon. While these events are taking place, solutions reaching still deeper continue the chemical attack on the parent material, thus perpetuating the C horizon.

Major kinds of soils

Soil scientists employ a rather complex nomenclature for the great variety of soil types present around the globe. Fortunately, most of the names in these lengthy classifications can be avoided if one seeks only a general understanding of a few major kinds of soils. For example, the soils that predominate in the eastern United States and southeastern Canada are called **pedalfers** (Greek *pedon*, ground). The name was fabricated to emphasize the fact that in such soils aluminum (*al*) and iron (*fe*) have been leached from the A horizon and deposited in the underlying B horizon. Pedalfers are clayey soils that develop in regions having an abundance of rainfall. As this water percolates through the humus it becomes acid, and this accounts for its ability to leach aluminum and iron from the topsoil. It is also effective in dissolving carbonates, which are then carried away by groundwater. The acidity of some pedalfers may diminish their fertility, and so farmers often spread finely ground limestone (agricultural lime) on their fields to combat the acidity.

Since the topsoil or A horizon in pedalfers is leached, it is usually a lighter color than the underlying subsoil, where iron has accumulated. Pedalfers include several lesser categories of soils that differ in their profiles. Most important among these are *podzols*, which typically have an ashy gray A horizon.

In the more arid regions of the world, less humus develops in soils and there is less opportunity for solution and leaching. In the absence of water, chemical weathering is slower and less clay is produced. Usually, there is insufficient groundwater to flush such soluble materials as calcite out of the B horizon, so it accumulates there as soil moisture is lost by evaporation. Because of the persistence of calcite in these soils, they are called **pedocals** (*pedon*, soil, and *cal* for calcium carbonate) (Fig. 7–16).

Frequently in dry areas, there is not only insuffi-

Figure 7–16 *Pedocal soil exposed by gully erosion, western Missouri. (Courtesy of J.C. Brice.)*

cient water to cause downward leaching but even an upward movement of water because of the high rate of evaporation at the surface. Mineral matter dissolved in the diminishing soil water is precipitated at the surface as a hardpan or caliche layer.

In the hot and humid regions of the tropics, the characteristic soils are **laterites.** The term laterite is derived from the Latin word *later*, brick, and originally referred to the use of this material to make bricks in India and Cambodia. Laterites have a relatively thin organic layer covering a reddish leached layer, which is often underlain by a still darker red layer. In laterites, oxidation and hydrolysis have been so intense that feldspars and ferromagnesian minerals are completely decayed. Not only is calcium carbonate removed but also silica. Only the most insoluble compounds, mainly aluminum and iron oxides, accumulate in these soils.

Although laterites may support a lush growth of natural tropical vegetation, they are not good agricultural soils. When forested areas with lateritic soils are cleared and plowed, the thin organic cover tends to be rapidly oxidized in the prevailing warm climates. A thick accumulation of organic matter like that found in rich black soils of more temperate regions cannot develop. After a few years of tilling and planting, the organic component of the soil is so depleted that fields must be abandoned.

In some laterites, the concentration of either iron or aluminum may reach levels that permit the deposits to be profitably mined. The iron-rich laterites originate from the weathering of parent materials that also contained iron, although not concentrated into ore bodies. Extensive lateritic ores of iron occur in Cuba, Colombia, Venezuela, and the Philippines. If there is little iron in the parent material and an abundance of aluminum, then laterites rich in hydrated aluminum oxides may form. Such materials are called **bauxites.** Ancient bauxite deposits are mined in Guyana, Ghana, northern Queensland, and Arkansas. Concentration of bauxite takes place in tropical areas of low relief where temperatures exceed 25°C most of the time, and where there is an abundance of water for leaching and chemical reactions. At the present time, bauxite is the only ore from which it is economically practical to extract aluminum.

Soil erosion

It takes nature thousands of years to produce a fertile soil. That same soil can be lost to erosion in only a few decades. Clearly, as we attempt to feed our expanding world population, care must be taken to prevent unnecessary erosional losses of this vital resource.

Soil erosion results primarily from uncontrolled runoff of surface water. In areas that have never been tilled, soil is protected by a cover of plants, and there is an approximate balance between the slow loss of soil by erosion and the development of new soil from underlying parent material. Whenever the protective shield of vegetation is broken, however, erosion will follow (Fig. 7-17). There are, of course, natural events that can destroy the vegetative cover. These include droughts, plagues of insects, epidemics of plant disease, and fires caused by lightning.

Agriculture, however, has been a far more potent factor in causing erosional loss of soils. Some of the erosion resulting from farming is unavoidable. We must have food, and so we must break into the natural vegetative cover. There are, however, farming methods that reduce the loss of valuable topsoil to erosion, and these methods are practiced widely. For example, farmers now plow and plant their crops in rows that follow contours so that rainwater is retained rather than allowed to run off (Fig. 7-18). Efforts are made to reduce the amount of time between preparation of the soil for planting and planting itself so that the bare soil will not remain exposed to erosion for long periods. *Strip-cropping* is also practiced (Fig. 7-19). This technique involves planting alternate bands of erosion-resistant crops like clover and alfalfa with open-spaced crops like corn. Soil losses in the erosion-susceptible corn strip are trapped in the adjacent

Figure 7-17 *Gullying and consequent loss of soil on land which has been cleared of its protective plant cover as part of an urban development project.*

Figure 7–18 *A field in which crops have been planted in rows which parallel topographic contours and in which terraces have been built to help retain rainwater. (From USDA, SCS Ia-2707.)*

strip of denser crops. Where fields are temporarily not utilized, soil-holding "cover" crops are planted, not only to retard erosion but also to improve the soil by adding nitrogen. Because ordinary grass is an effective retardant for erosion, overgrazing by cattle and sheep must be prevented.

MASS WASTING

At the beginning of this chapter, we defined mass wasting (Fig. 7–20) as the process by which earth materials are moved downslope by gravity. Although these

Figure 7–19 *Strip cropping. (From USDA.)*

Figure 7–20 The January 9, 1965, landslide near Hope, British Columbia, provides a dramatic example of mass wasting. (Courtesy of British Columbia Ministry of Environment.)

movements are indeed ultimately induced by gravity, other factors influence the susceptibility of an area to mass wasting. These include steepness of slopes, kind of materials being moved, water content, bedrock attitude, susceptibility to earthquakes, and vegetation. Usually, two or more of these factors operate in unison to cause mass wasting.

Factors governing mass wasting

Gravity and slope

As noted above, gravity is the force behind all mass wasting. As masses of earth materials are detached and undermined because of weathering and erosion, gravity moves these rocks and soils to places where equilibrium is restored. The loss of the cover of unconsolidated material on slopes exposes fresh rock that can then be weathered in continuing cycles of degradation.

Rock fragments are attracted directly to the center of the earth by gravity. Components of gravity also act on a given rock mass on a hill in the downslope direction as well as perpendicular to the slope.

Whether a given block of rock will be able to overcome friction and move downslope will be influenced most directly by the angle of the slope itself. Compare, for example, slopes X and Y in Figure 7–21. On both the X and the Y, the gravitational force drawing the boulder directly to the center of the earth is the same. On the steeper slope, however, the component of gravity directed along the slope has increased, whereas the component of gravity that assists in holding the rock against the slope has decreased. This shifting in the magnitude of force is the reason for an observation all of us have made; namely, that objects slide down steeper slopes more readily than gentle ones. On slopes steeper than 45°, most unconsolidated materials will not remain in position but will begin to slide or flow downward.

Earth materials

Some kinds of rocks, and even certain unconsolidated earth materials, are inherently better able to stay in

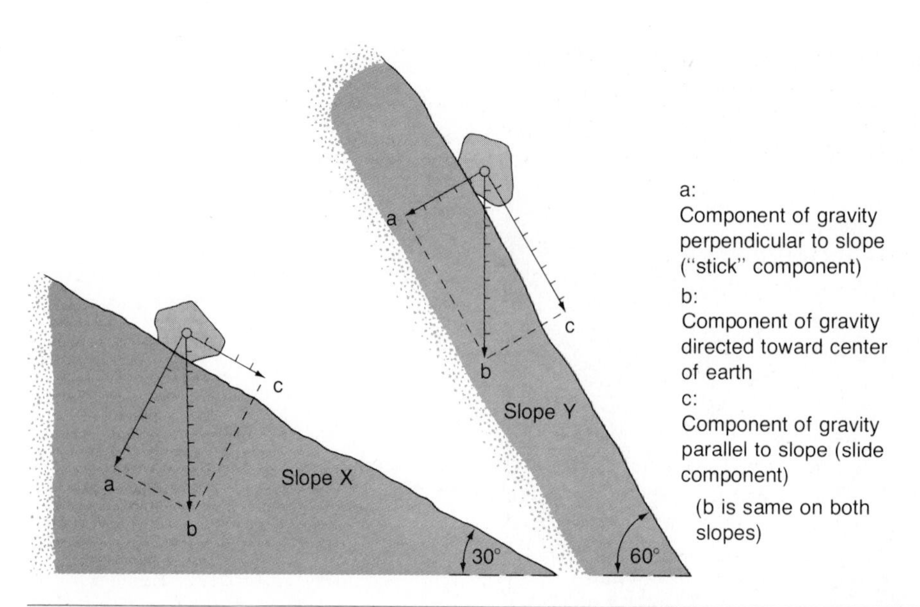

Figure 7–21 Illustration indicating how gravity acts upon a boulder on a slope. The boulders are identical, and therefore the b *component of gravitational force is the same, as indicated by similar length of the arrows directed toward the earth's center. Note that in slope Y, the* a *component (which helps keep the boulder in place) is shorter than on slope X, and the* c *component (which promotes sliding downslope) is longer than on slope X.*

position on a slope than others. Massive, uniform-textured rocks like granite, basalt, and quartzite have interlocking grains that give them nearly equal strength in all directions. Such rocks will hold their position against gravity until separated from the unaltered rock mass by weathering and the development of joints. Once loosened, of course, the rock will tumble or slide downslope.

Not all kinds of crystalline rocks have equal strength in all directions. Planes of schistosity and slaty cleavage in metamorphic rocks, as well as clayey seams separating the strata of massive limestones, serve as slip zones along which the overlying rock may move gravitationally upon the lower material. If these planes of potential slippage lie parallel to the surface of a slope, great tabular masses of rock may lose their hold and slide down the valley side. This indicates that, in addition to the kind of rock involved in a mass movement, the composition and structural attitude of underlying material must also be considered in assessing the potential for mass wasting at any given location.

Unconsolidated earth materials such as soil, clay, sand, and gravel are also susceptible to mass wasting. The fact that clay becomes plastic and weak when it absorbs water is well known. Wet clays not only act as lubricants for mass movements, but tend to flow out laterally from beneath burdens, thus causing subsidence and slumping. Some clays may resist mass movements for centuries, either because of weak cements or a disorderly arrangement of mineral particles. Such clays, however, may suddenly become unstable when disturbed by vibrations or leaching. For particles of sediment coarser than clay, stability on slopes is influenced by grain shape (whether angular or round) and grain roughness. The degree of uniformity of grain size (sorting) and the presence of root networks and mineral cements also affect the ability of coarse materials to keep their position on a slope.

Water

Water induces mass wasting in a variety of ways. With clay, it is transformed into a slippery mixture that can lubricate surfaces over which earth materials move and form a heavy liquid for flotation and transport of coarser sediment and debris. A slope may fail simply because of the added mass of water itself. Water also contributes to mass wasting by helping to widen joints and bedding planes through solution, and by freeing resistant materials like chert from their more readily weathered parent rocks. Even the unseen hydrostatic pressures exerted by underground water on subsurface materials may disrupt equilibrium on a slope and cause earth materials to move to lower locations.

Attitude of bedrock

As described in Chapter 5, the term attitude in a geologic sense refers to the way in which geologic features like strata or fault planes are positioned in the earth's crust. The attitude of a sandstone stratum, for example, may be horizontal, vertical, or inclined at any angle. As is apparent from Figure 7-22, the inclination of layered rocks relative to a slope can have great influence on the susceptibility of the slope to mass wasting. If strata dip parallel to the slope surface, the chances of failure are considerably greater than if beds dip in a direction opposite to the direction of slope. Fault surfaces, joints, and large fractures trending parallel to valley sides may also contribute to mass movements of overlying material.

Vegetation

Plants protect soil and regolith against erosion and contribute to the stability of slopes by binding rocks together within root systems. In areas where plant cover is sparse, mass wasting may be enhanced, particularly if slopes are steep and water readily available.

Initiating causes of mass movements

Conditions favoring mass wasting, such as those described in the preceding sections, may exist for a long time without any movement actually occurring. Like the proverbial straw that broke the camel's back, some additional ingredient is needed to initiate movement. Among the many situations or events that trigger mass movements are removal of support for material on a slope, overloading, reduction of friction, and vibrations from earthquakes.

Removal of support

Slopes that have remained stable for centuries can become susceptible to mass wasting simply by the removal of supporting materials near the base of the slope. Engineers are aware of this and attempt to avoid oversteepening slopes or excavating into the base of a hillside composed of weak materials. The base of a slope, after all, acts as a buttress supporting material farther uphill. The removal of that base during road construction is the cause of the failures seen along highways (Fig. 7-23). Although slope failures resulting from poor construction practices are common, geologists are especially interested in slope failures resulting from such natural causes as the continuous erosion and undermining of hillsides by streams (Fig. 7-24), waves, and glaciers.

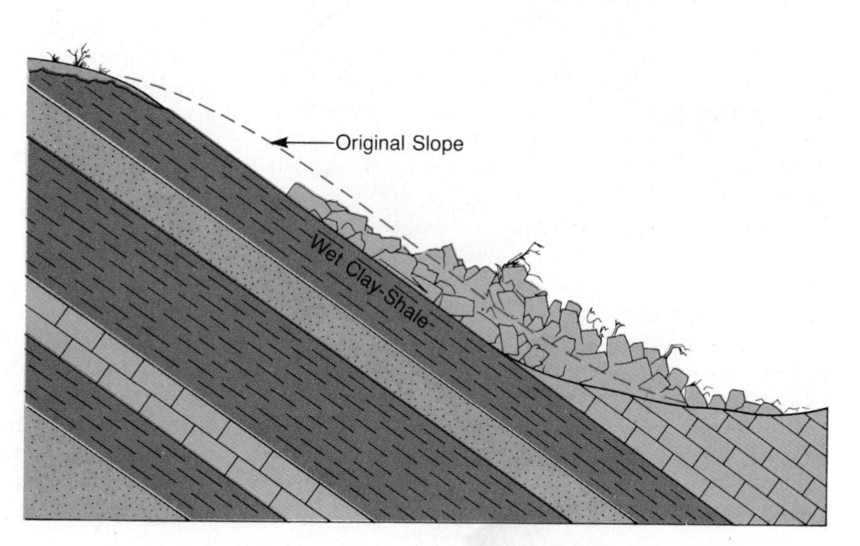

Figure 7-22 *The possibility of downslope movement of stratified rocks is increased if the strata dip parallel to slopes and are underlain by weak materials such as saturated shales.*

Figure 7-23 Mass wasting (rotational slump) resulting from removal of lower part of slope during road construction, Marin County, California. (Courtesy of U.S. Geological Survey and J. Schlocker.)

Figure 7-24 Mass wasting (slump) in shale caused by undercutting by the Bad River, Stanley County, South Dakota. (Courtesy of U.S. Geological Survey and D.R. Crandell.)

Overloading

The costly and frustrating problems of earth movements during the construction of the Panama Canal are almost legendary. Slumping of material along the sides of the excavation was partly the result of cutting the sides of the canal too steeply, as well as having to excavate through the highly plastic clays of the Cucaracha Formation. The unstable conditions were further aggravated, however, when materials excavated from the canal were foolishly dumped at the upper edge of the cut. The added weight was more than the already weak underlying materials could support, and slumping occurred with nerve-racking frequency. Overloading may also occur naturally, as when material is transferred onto a slope by landslides, avalanches, and snowfall. Rainwater seeping into soils adds its weight to that of the soil and thus may contribute to overloading of underlying materials.

Reduction of friction

Mass wasting can result from a reduction in the frictional forces that hold particles together. Water is the usual culprit in reducing the cohesiveness of materials and causing them to fail (Fig. 7-25). In addition to water from rain and melting snow, leakage from reservoirs, canals, and septic tanks may trigger mass movements.

Earthquakes

Landslides and other forms of mass wasting are frequently initiated by the jolt or vibrations of earthquakes. As described in Chapter 4, earthquakes preceded the Madison Canyon landslide of 1959, as well as the destructive landslides accompanying the 1964 Alaskan earthquake. In addition to the landslide-triggering effects of earthquakes, mass movements may be initiated by vibrations from blasting in mines, quarries, and road cuts, from gunfire, and even from the passage of heavy vehicles.

Types of mass wasting

Mass movements of earth materials are classified according to their speed of movement, whether the mass moves as a coherent body or is disturbed during movement, and the amount of water in the mass. Table 7-1 is a simple classification of the kinds of mass wasting. In this classification, movement of earth material along a clearly defined plane of slippage is referred to as a

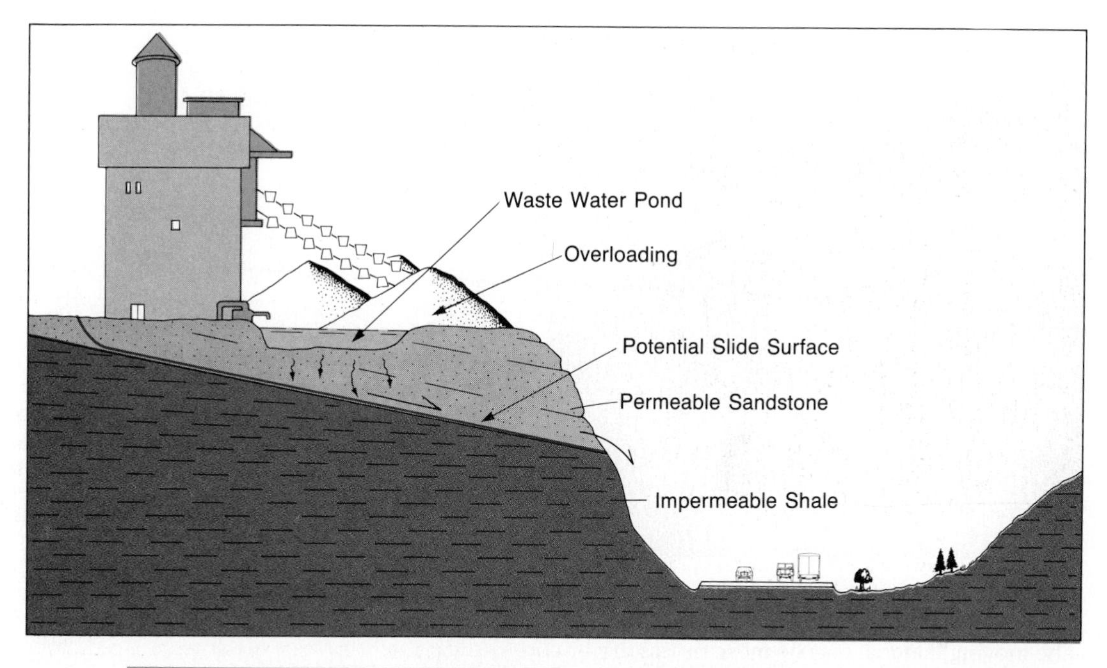

Figure 7–25 *Overloading near the edge of a slope or road cut, as well as seepage of water, provides the potential for slope failure.*

slide. In a **flow,** the material near the top of the moving mass moves more rapidly than successively lower layers, and as a result there is considerable internal deformation. The popular term **landslide** is used in a general sense for all forms of rapid downslope movement.

Creep

The slowest form of flow is **creep.** Creep is measured in mere centimeters per year. It can be defined as the slow downhill movement of regolith that results from continuous rearrangement of its constituent particles.

TABLE 7–1 **KINDS OF MASS WASTING**

Kind and rate of movement		*Materials include ice*	*Earth materials with minor amounts of ice or water*	*Materials include water*
Flow	Usually imperceptible	Rock glacier Solifluction	Creep	Solifluction
Flow	Slow to rapid	Avalanche		Earthflow Mudflow Debris avalanche
Slide	Slow to rapid		Slump Rock slide Rock fall	

Simplified from Sharpe, C.F.S., *Landslides and Related Phenomena,* New York, Columbia University Press, 1938, p. 96.

Figure 7–26 Indications of creep include downslope bending and dragging of rock layers as well as tilted and displaced posts, poles, and graveyard markers.

The moving material may be more or less dry or contain only minor amounts of water. As is usual in flow, the rate of movement is greatest near the surface and decreases with depth. Depending on the material, one can differentiate soil creep and rock creep.

One cannot stand in a field and observe creep in action. It is much too slow a process. One can, however, observe the accumulative effects of creep. These effects include pavements and stone walls that have been displaced, fences forced out of alignment, and tilted fence posts and telephone poles (Fig. 7–26). If a slope is underlain by steeply dipping strata, the upper edges of the beds may be broken and bent as a result of creep.

Creep may be regarded as the result of three mechanisms. The first of these is called **heaving** and involves repeated expansion of surface layers in a direction normal to the slope. The expansions may result from the freezing of interstitial water or simply periodic wetting, which may cause certain clays to swell. When the creep material thaws or dries, the material drawn out perpendicular to the slope during the expansion is lowered by a component of gravity into a position slightly downslope from the initial position (Fig. 7–27). The second mechanism involved in creep is the plastic flow of moist clays as they respond to the pull of gravity and weight of overburden. Finally, biological agencies contribute to creep in several ways. The burrowing activities of soil animals and trampling of the surface by cattle and sheep may disturb the regolith sufficiently to offset its stability.

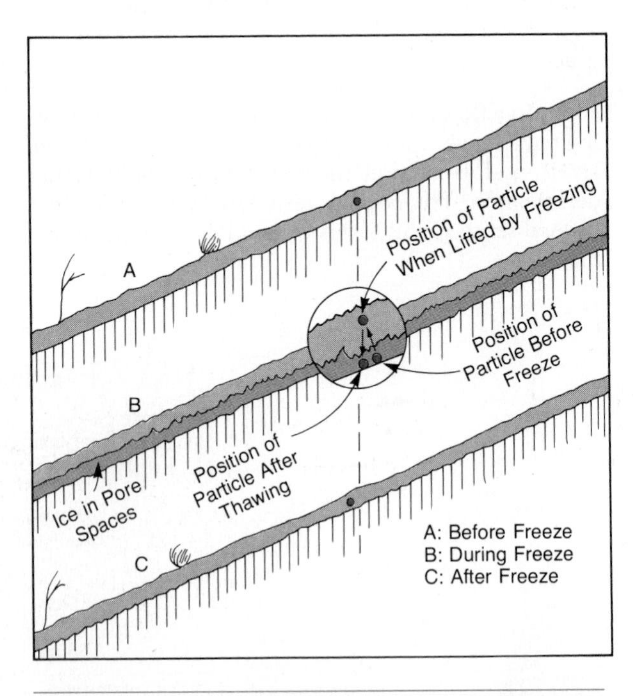

Figure 7–27 The effect of frost heaving on creep. Expansion of the water in porous material during freezing raises the material perpendicular to the slope. On thawing, the material is lowered to a position slightly downslope of the original position.

Solifluction

The term **solifluction** (Latin *solum,* soil; *fluere,* to flow) literally means "soil flow." It applies to the slow flow of soil and rock debris that is saturated with water and may be subjected to alternate freezing and thawing. Solifluction may occur along slopes as gentle as only 5°. It is a process particularly prevalent in the saturated soils of colder regions of the earth. During the relatively short summer thaw of subpolar regions, meltwater may thoroughly saturate the upper mobile layer of regolith. This mucky surface layer may then slowly flow down even very gentle slopes on top of the still-frozen underlying zone.

The permanently frozen earth material on which solifluction may occur is called **permafrost.** Permafrost is not at all an unusual condition, for it is widespread on lands within the Arctic Circle. The frozen layer of permafrost may reach a thickness of more than 300 meters (Fig. 7–28).

Permafrost is an important reason why much of the Arctic tundra is so water soaked. The subsurface layer of solid frozen ground is essentially impermeable, and so water in the mobile surface zone cannot escape underground. It remains at the surface, where it saturates the regolith and facilitates movements like solifluction.

Rock glaciers

Near or above the timberline in high mountains one can find tongues of coarse rock debris having an overall form suggestive of a valley glacier. These glacier-like masses of angular, unsorted rock waste are called **rock glaciers.** Unlike true glaciers which are composed of ice containing rock fragments, rock glaciers are primarily composed of rock debris with ice and meltwater filling the spaces between rock fragments. The rock glacier moves by means of internal deformation of the rock and ice mixture, as well as the mobility resulting from seasonal freezing and thawing. Some rock glaciers in Alaska move at a rate of about one meter per year.

Earthflows

Earthflows are rather erratic movements of clayey or silty regolith down relatively gentle slopes. Water is present but usually not to the point of saturation. Earthflows can sometimes be recognized by a curved scarp that develops at the breakaway line on the slope and by the crescent-shaped bulges at the convex "toe" of the flow.

Earthflows are frequent occurrences along the valley of the St. Lawrence River and in parts of Scandinavia. In these areas, one finds relatively thick deposits of fine-grained sediments deposited in former marine embayments that existed shortly after the last advance of continental glaciers. Initially, the fine-grained sediments were relatively firm and stable. The clay particles had been rapidly deposited in salty ocean water and quickly collected into aggregates having a chaotic lack of alignment. Cohesiveness, resulting from elec-

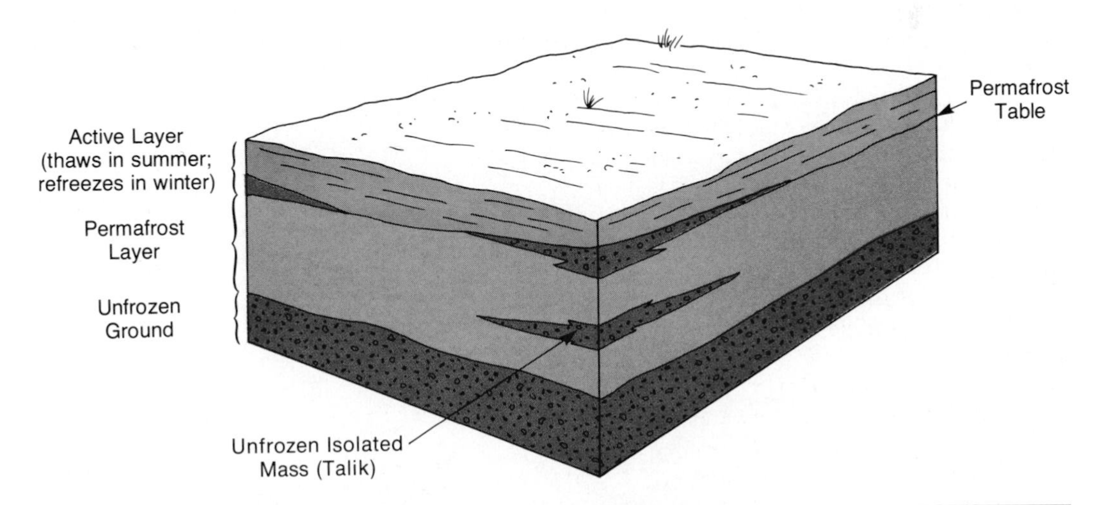

Figure 7–28 *Typical section of permafrost terrain. (Modified from U.S. Geological Survey circular* Permafrost, *Stock No. 2401-02433.)*

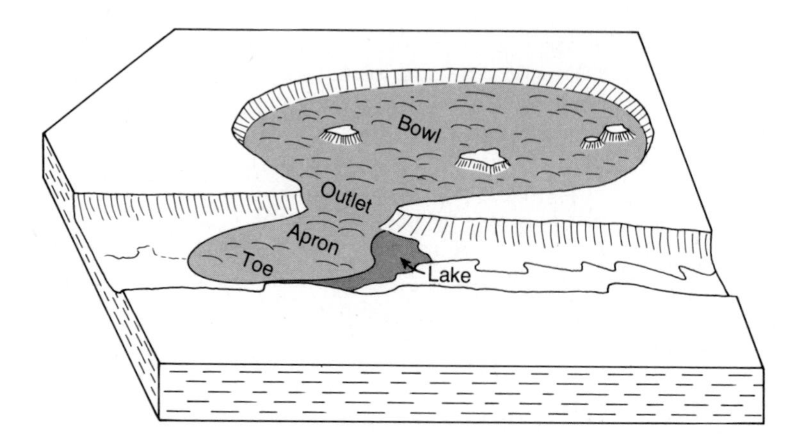

Figure 7–29 *Typical features of an earthflow resulting from the mobilization of thixotropic clays.*

trostatic forces between clay particles and saline pore water, helped to stabilize the clay. After thousands of years of subaerial leaching by fresh water, however, the soluble salts were largely removed and the clays became susceptible to earthflow. Clay that becomes fluid when disturbed by shock or vibration from earthquakes, explosions, overloading, or the passage of heavy vehicles is termed **thixotropic.** When such disturbances occur, water in the pores of the sediment mobilizes, and the former disorderly framework of mineral grains collapses. The once relatively firm clay becomes a free-flowing liquid, sometimes called quick clay. Road cuts and escarpments several meters high founder and flow downslope. As they do so, they create pear-shaped areas of disturbed regolith (Fig. 7–29). Earthflows resulting from the mobilization of thixotropic clays can cause considerable damage and loss of life. Canadians still recall the 1971 earthflow that engulfed houses in the village of Saint-Jean-Vianney, Quebec, and killed 30 townspeople.

Mudflows

In general, mudflows are even more mobile than earthflows. They can be defined as the relatively fast flow of regolith that has the consistency of mud and behaves rather like a viscous fluid. Mudflows are so supersaturated with water that they have a "soupy" consistency. They tend to travel in streamlike masses within the confines of valleys and canyons. On reaching the mouth of the canyon, they characteristically spread out in a broad apron (Fig. 7–30) covering fields and houses. In addition to a sudden abundance of water (as after a heavy thunderstorm), mudflow development is favored by a lack of vegetation and an abundance of loose regolith. Often these conditions are

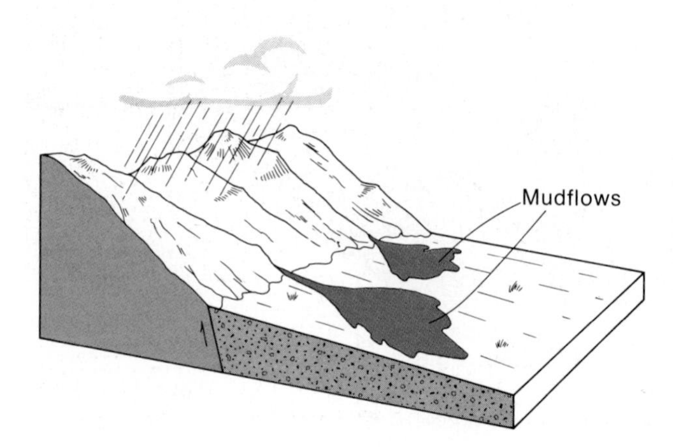

Figure 7–30 *Setting for typical mudflows at the front of a mountain range or hilly terrain in an arid region.*

found in arid regions that are subjected to periodic torrential rains.

Avalanches

We usually think of an **avalanche** as a large mass of snow and ice in swift motion down a mountainside. An avalanche, however, need not consist only of snow and ice, but can incorporate unspecified amounts of regolith. Ice and snow may be entirely lacking in so called *debris avalanches.*

Avalanches (Fig. 7–31) are the most frightening, unpredictable, and fast moving of all forms of mass wasting. Certainly the fear that mountain dwellers have for avalanches is well justified. A dreadful example of their destructive power occurred in 1970, when a mass of ice weighing about three million tons broke away from a glacier on the sides of Mount Huascarian in Peru. The great mass of ice plummeted down the mountainside, ricocheted off valley walls, and gathered within its surging mass additional millions of tons of rock, soil, and trees. Within only seven minutes, the avalanche had traveled 12 miles. It roared into the densely populated farming area at the base of the mountain, killing at least 3500 people instantly.

Slump

Slump is a form of slide motion in which a mass of rock or regolith slips along a concave upward, curved surface while rotating backward upon a horizontal axis (Fig. 7–32). As a result of this movement, the upper surface of the slump block may be tilted toward the slope. Earthflows may develop at the toe of the slump. Slumping is a common type of slide movement, and slumped regolith can frequently be seen along the steep banks of streams, highway cuts (Fig. 7–33), and cliffs.

Rockslides and rockfalls

The free sliding of recently detached segments of bedrock on a sloping joint or bedding plane constitutes a rockslide. Rockslides are likely to take place along the steep sides of mountains and may cause great damage. A particularly interesting example of a rockslide occurred in 1925 in Wyoming's Gros Ventre River Valley. The stratified rocks on one side of the valley consisted of wet clay interbedded with limestones and sandstones. These strata dip toward the valley at an angle of approximately 20°. Thus, conditions were ideal for downdip and downslope slippage of overlying beds on the weak clay layers. The 1925 slide occurred

A

B

Figure 7–31 *(A) Snow and rock avalanche at the head of Emerald Bay, Lake Tahoe, California, 1956. (B) Long narrow scars of a debris avalanche at Franconia Notch, New Hampshire. (A, Courtesy of California Division of Highways; B, Courtesy of Department of Public Works and Highways, New Hampshire.)*

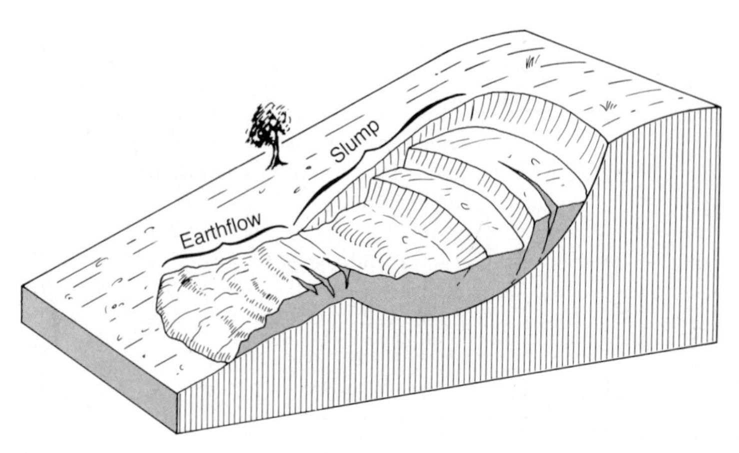

Figure 7–32 *Idealized diagram showing characteristics of slump feature with earthflow at the toe.*

in the spring as meltwater from snow and ice saturated the clayey strata. Thus weakened, the side of the mountain broke loose, releasing nearly 48,000,000 cubic meters of rock waste, which swept across the valley floor and up the opposite slope. When the dust had settled, the rock debris had created a 70-meter-high natural dam across the Gros Ventre River. That dam failed two years later, causing a flood downstream that destroyed most of the town of Kelly, Wyoming.

When earth materials experience nearly vertical free fall, as from cliffs, the descriptive term used is **rockfall** (Fig. 7–34). The more or less continuous rain of loosened rock fragments from cliff faces results in an accumulation of shattered rock at the base of cliffs. These piles of debris or **talus** form the talus slopes frequently seen at the base of rocky bluffs (Fig. 7–35). The angle of repose for talus slopes is remarkably constant at 34° to 37°.

Relation of mass wasting to valley development

Streams are the most important agents in the development of valleys. The erosive powers of running water, aided by the multitudinous impacts and abrasion of pebbles and other transported rock debris within the flowing stream, effectively excavate the channel. Yet, if a stream were the only mechanism involved in forming valleys, they would have vertical rather than sloping valley sides (Fig. 7–36). A stream, after all, can only

Figure 7–33 *Slump along highway, St. Louis, Missouri.*

Figure 7–34 *Rockfalls near mouth of Bay d'Espoir, Newfoundland. (Courtesy of Geological Survey of Canada.)*

Figure 7–35 *(A) Talus slopes on the inner rim of Crater Lake, Oregon. (B) Talus slopes along the canyon of the Salmon River in Idaho. (Courtesy of J.C. Brice.)*

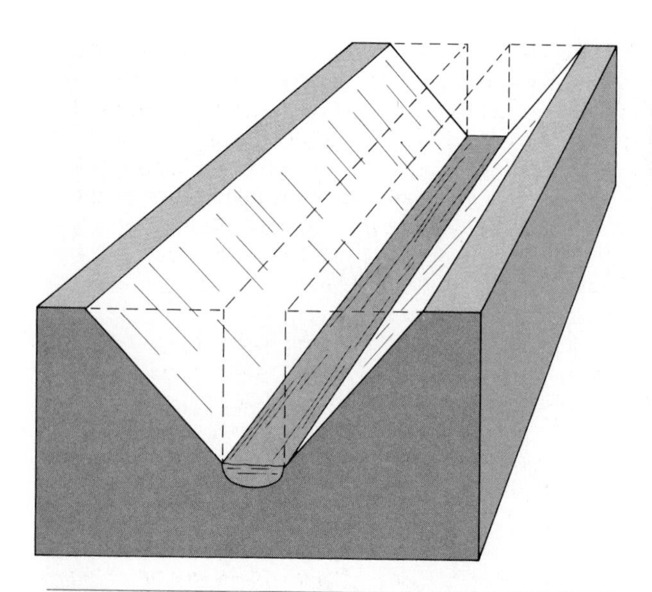

Figure 7–36 *Streams do not act alone in the formation of valleys. They erode their own channel, but also serve like conveyor belts in carrying away material derived from the valley sides by mass wasting and slope wash. The volume of material contributed by mass wasting is represented by the space enclosed within the dashed lines.*

Figure 7–37 *The splash of falling raindrops on unconsolidated sediment effectively loosens it and detaches particles. (Courtesy of U.S. Dept. of Agriculture.)*

erode the surfaces with which it is in direct contact. Therefore, other agencies, especially mass wasting, are constantly at work widening and shaping valleys. The stream acts as a conveyor belt, carrying away its own load as well as the debris supplied to it by mass wasting. Streams even contribute to the effectiveness of mass wasting by removing material at the base of the valley so as to undermine the slopes and cause slumping (see Fig. 7–24) and debris slides. Weathering also assists mass wasting by reducing coherent rock to loose regolith, which is then susceptible to such processes as creep, solifluction, and earth flow.

Erosion by sheetwash

Another process that contributes significantly to valley widening is sheetwash. Sheetwash (also called sheet flow) is a thin sheet of flowing water that forms on slopes when rain falls faster than the soil can absorb it, or when the ground is already saturated. Such a sheet of water flowing down a sloping surface will pick up and transport small particles in its path, much as rainfall will remove dust from a sloping roof. The effectiveness of sheetwash erosion depends upon several factors, among which are the kind and thickness of plant cover; the porosity, permeability, and cohesiveness of the soil; the duration and intensity of the rainfall; and dimensions and steepness of the slope over which the sheet of water is moving.

Sheetwash will not flow indefinitely as a planar unit of uniform thickness. Friction and irregularities on the ground surface will cause water to be concentrated into threads of current that in turn erode narrow channels called **rills.** Rills in turn widen and deepen as water moves into them. Eventually, they develop into gullies and are integrated into a nearby stream system.

Erosion by rainsplash

The thin film of water moving down a slope as sheetwash has limited velocity and mass and therefore is not of itself always able to appreciably erode and transport soil particles. Help is provided by irregularities in the ground surface and by clots of vegetation that temporarily dam small areas and subsequently result in rapid surges of water as the obstacle is surmounted or circumvented. Another potent erosional aid to sheetwash is the often-overlooked but significant power of raindrops. When a raindrop strikes the ground, it splashes not only water into the air but suspended particles of sediment as well (Fig. 7–37). On a slope, the sediment derived from one point on the ground and thrown into the air will return to the ground at a point downslope from its point of origin. Thus, there is a net downhill transport of material.

Summary Rocks exposed at the earth's surface are continuously subjected to the destructive forces of weathering. Weathering is a term used to describe the changes in surface materials that occur when they interact with water, air, and organisms. In general, two categories of processes operate during weathering. The first of these is called disintegration, or mechanical weathering, and acts primarily to break rock into smaller sizes without directly causing a change in composition. Disintegration is accomplished in many ways. When water in rock crevices expands upon freezing, for example, it widens existing cracks and opens new ones. Erosion is constantly at work removing rock at the surface, and this unloading causes rocks to rupture and form joints as they expand under the lessened pressure. The growth of mineral crystals and moisture-seeking plant roots may also help to wedge rocks apart. Acting together, all these forms of disintegration reduce great blocks of rock to small particles and thereby greatly increase the total amount of surface area available for attack by chemically potent solutions. This chemical activity comprises the second kind of weathering, known as decomposition. The essential substance for decomposition is water. Water is an effective solvent for some minerals and can chemically react with many more. For example, when water becomes acidic by dissolving carbon dioxide and sulfur dioxide from the air or soil, it is capable of decomposing a variety of silicate and nonsilicate minerals. Limestones are particularly susceptible and may experience extensive solution in the formation of caverns. Feldspars attacked by rainwater will yield clay, soluble carbonate, and residual silica. Iron oxides and clay minerals as well as soluble carbonates are the products of decomposition of ferromagnesian minerals. The main chemical processes operating during decomposition are oxidation, carbonation, hydration, hydrolysis, and ordinary solution.

The mechanically fragmented and chemically decomposed rock and soil formed by weathering are ultimately moved to lower elevations under the impetus of gravity. This gravitational transfer of earth materials constitutes mass wasting. Although often imperceptible, mass wasting is of enormous importance in the degradation and leveling of the lands. Loose material is delivered by mass-wasting processes to streams, glaciers, and ocean waves, and then transported and redistributed by those agents.

The most beneficial effect of weathering for humanity is that it produces the nonorganic components of soil. Soil is weathered, unconsolidated material capable of supporting the growth of plants. It develops as a result of a complex interaction of weathering processes and biologic activity. Climate, the parent rock materials, degree of slope, amount of time, and soil organisms are all factors in determining the characteristics of the soils that develop in any particular region of the globe.

Mass movements may be either rapid or slow, depending upon the nature of the conditioning factors noted above. Slow movements include creep, earthflow, mudflow, and solifluction. Creep is the slowest but most important of these movements. It can sometimes be detected by tilted or displaced fence posts and bent or broken upper edges of steeply dipping strata. Frost heaving is an important contributor to creep. If the amount of water in the overburden increases, creep may give way to an earthflow, and if water content in the moving material reaches the point of saturation, mudflows are a likely result. Solifluction is a kind of mass wasting that occurs primarily in regions having deeply frozen ground that does not thaw completely in the summer (permafrost). Solifluction may take place when such frozen ground melts from the top downward. The water-saturated surficial regolith then moves along on top of the still-frozen underlying layer.

QUESTIONS FOR REVIEW

1 In what way does an increase in the mechanical weathering of a rock mass increase its susceptibility to chemical weathering?

2 Describe what happens during the chemical weathering of plagioclase feldspar.

3 What are the differences between pedalfer and pedocal soils? Account for the origin of these differences.

4 In what way is the susceptibility of a mineral to weathering related to its position in the Bowen Reaction Series?

5 It has been observed that cemetery headstones made of the same marble and erected at the same time weather more rapidly in urban than in country cemeteries. What the explanation for this?

6 A molecule of ordinary water is described as a dipolar molecule. What is the meaning of this expression? Which "ends" of the water molecule would attract a sodium ion in solution?

7 What is creep? Why is creep more rapid on a steep slope than on a gentle slope? What is the difference between creep and earthflow?

8 How does falling rain accomplish erosion? Under what circumstances is falling rain most effective in causing erosion?

9 Under what geologic conditions would there be a high probability of landslides?

10 In what ways may massive downslope movements be triggered?

11 Suppose that building sites on a hillside with a fine view are restricted to two locations. Both are underlain by shale strata, but at one site the strata dip steeply into the hill, and at the other the strata dip roughly parallel to the slope of the hill. Which homesite would you select and why?

12 How does the mass movement called mudflow differ from slump?

13 Explain how streams and mass wasting work together to erode the land surface.

14 Mass movements and erosion have been prevalent on earth for billions of years. Why then do we still have regions of high mountains and irregular topography?

15 In what way are solifluction and permafrost related? What special problems associated with permafrost affect people and communities located in polar or near-polar regions?

SUPPLEMENTAL READINGS AND REFERENCES

Birkeland, P.W., 1974. *Pedology, Weathering, and Geomorphology.* New York, Oxford University Press.

Bloom, A.L., 1978. *Geomorphology.* Englewood Cliffs, New Jersey, Prentice-Hall, Inc.

Carroll, D., 1970. *Rock Weathering.* New York, Plenum Press.

Hunt, C.B., 1972. *Geology of Soils.* San Francisco, W.H. Freeman Co.

Keller, W.D., 1969. *Chemistry in Introductory Geology.* 4th ed. Columbia, Mo., Lucas Bros.

Paton, T.R., 1978. *Formation of Soil Material.* Boston, Geo. Allen & Unwin.

Sharpe, C.F.S., 1938. *Landslides and Related Phenomena.* New York, Columbia University Press.

Winkler, E.M., 1973. *Stone: Properties and Durability in Man's Environment.* New York, Springer-Verlag.

8
streams and valleys

Rivers are among the most powerful of the agents of topographical development, and it is important to understand something of their modes of change and adjustment.
WILLIAM B. SCOTT, 1904

Salmon River, Lemhi County, Idaho. The Salmon River at this locality has cut a V-shaped valley eroded through pegmatite rock. (Courtesy of U.S. Geological Survey and C.P. Ross.)

Prelude In southeastern Australia, there is a river that flows for a distance of about 2000 km from the highlands of Victoria to Encounter Bay south of Adelaide. The aborigines who live nearby tell an interesting tale about the origin of this stream and its valley. They describe how a narrow crevice was opened by an earthquake and how subsequently, whenever it rained, the crevice would develop a thin trickle of water. The tale does not end at that point, however, for it is necessary to account for the considerable width and graceful bends of the stream and its valley. A second earthquake, along with the best "fish story" of all time, is invoked to completely account for the characteristics of the stream: During the second quake, an immense fish squeezed upward into the cleft from somewhere deep within the earth. The trickle of water in the rift was quite inadequate for the great fish, and so with mighty strokes of its powerful tail it wriggled its way to the ocean, leaving behind the wide, sinuous valley of the Murray River.

Actually, of course, the valley described above was eroded over a long period of time by the river that flows within it. The concept that, given sufficient time, even puny streams have such erosive powers would have been difficult for the aborigines to grasp. It was also a concept that was not understood by intellectuals in 16th-century Europe. Most naturalists of that time believed valleys formed as a consequence of unknown catastrophic events. An exception was a Saxon professor of mineralogy named Agricola (1494–1555). Agricola taught his students that streams cut the valleys in which they flow and even were responsible for the development of fertile floodplains.

Today, there is no disagreement with the concept that, except for relatively few fault troughs, most valleys are formed by streams. The process of valley forma-

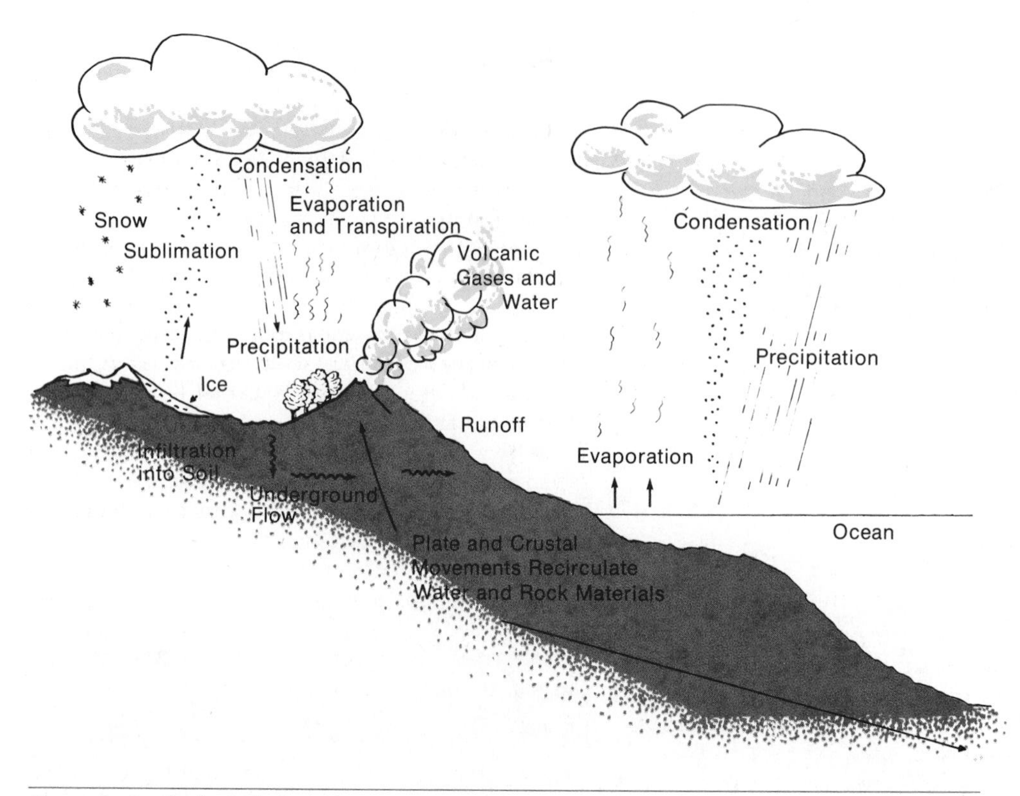

Figure 8–1 The hydrologic cycle.

tion begins when water flowing down a slope as sheetwash is separated by surface irregularities into small channels that then develop into gullies. With further additions of water from rain or snow, gullies are deepened by the running water and widened by mass wasting until they begin to take on the appearance of small valleys. Headward erosion results in the lengthening of the channels, and new channels are eroded on the slopes of the developing valleys. In their early stages of development, the streams are marked by many irregularities, such as waterfalls and rapids. In time, however, the slopes of channels become more gentle, and the streams begin to widen their valleys by cutting laterally. Little by little, a pattern of broader primary streams and secondary tributaries begins to emerge. If we were to inform the tribe of aborigines mentioned earlier that the real factor behind the development of stream valleys was the sun, they would probably be as skeptical of our sun story as we are of their fish story. Nevertheless, the sun is essential for the development of streams. The sun evaporates water that then falls from the atmosphere as rain or snow. Some of this water is returned to the atmosphere by evaporation or transpiration from plants. Another portion flows down slopes and in channels as runoff (Fig. 8–1). A third portion percolates into the ground and emerges later as springs at lower altitudes. This water also ultimately enters the stream system. Thus, streams are fed by water from runoff and springs (Fig. 8–2). The rate at which water is supplied to the stream determines its discharge or volume of flow. Discharge in turn affects a stream's velocity, width, depth, and, at least partially, the amount of sediment it can transport. In this chapter we will define and examine the relations among these interacting variables and attempt to explain the many facets of stream behavior. Streams, after all, are not unimportant to us humans. They provide us with drinking water, electric power, water for crops, fertile floodplains, and convenient routes for highways. Properly managed, they help us dispose of the waste products of our cities. Most important, they have sculptured esthetically pleasing landscapes to which we can escape whenever our view of the world becomes disagreeable because of too much concrete and too few trees.

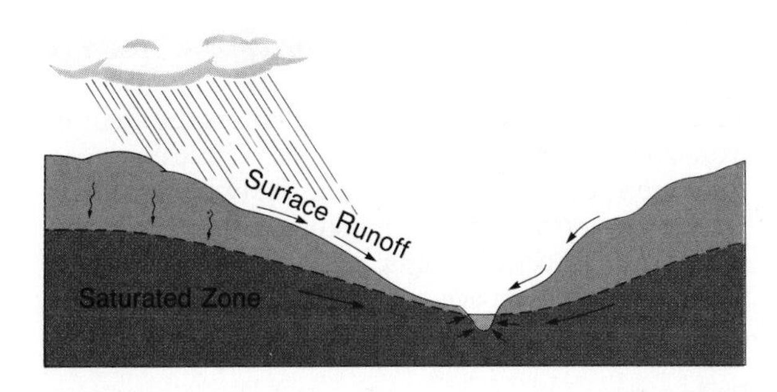

Figure 8–2 Streams are fed by water from runoff as well as from underground water which emerges above or within the channel.

GRAVITY: THE FORCE BEHIND THE FLOW

Everyone knows that water flows from high places to lower ones. Gravity, of course, is the cause of the movement. Every molecule of water in a stream is attracted directly downward toward the earth's center by gravity, but the molecules are unable to move rapidly downward because they are supported by surrounding water molecules and the stream bed. Water molecules in a stream therefore move along parallel to the slope of the stream bed under the influence of a component of gravity. As is the case in mass wasting, the proportion of the force of gravity acting on a water molecule depends upon the steepness of the slope. If the slope is vertical (as in a waterfall), the water will be pulled directly downward by the full force of gravity. As the angle of inclination of the slope from horizontal is reduced, the value of the component of gravity will also be lessened.

If the component of gravity acting on water molecules were unopposed in a stream, the water would quickly attain high rates of movement. Friction of the flowing water against the floor and walls of the channel, as well as internal friction among water molecules and between water molecules and suspended sediment, all serve to oppose the force of gravity. Because of these opposing forces, there is an upward limit to the velocity a stream can attain no matter how steep the slope over which it is flowing.

The slope of the stream channel between any two specified points (and along which the component of gravity operates) is called the **gradient.** Stream gradients are usually expressed as a ratio as follows:

$$\text{Gradient} = \frac{\text{Vertical Distance (in meters)}}{\text{Horizontal Distance (in meters)}}$$

One can determine the gradient of part of a stream by using data on a topographic map. To do this, one first locates two adjacent contour lines that cross the stream. Contour lines are simply lines connecting points of equal elevation. For example, Figure 8–3 depicts 180-meter and 190-meter contour lines crossing a stream at points that are 2000 meters apart. If we assume the channel floor is parallel to the water surface, then the stream has fallen 10 meters vertically over the 2000-meter horizontal distance. The gradient of this segment of the stream would be 10/2000, or 0.005 (this means the stream falls 0.005 meter per 1 meter of horizontal distance). The gradient of the stream in meters per kilometer can be obtained by

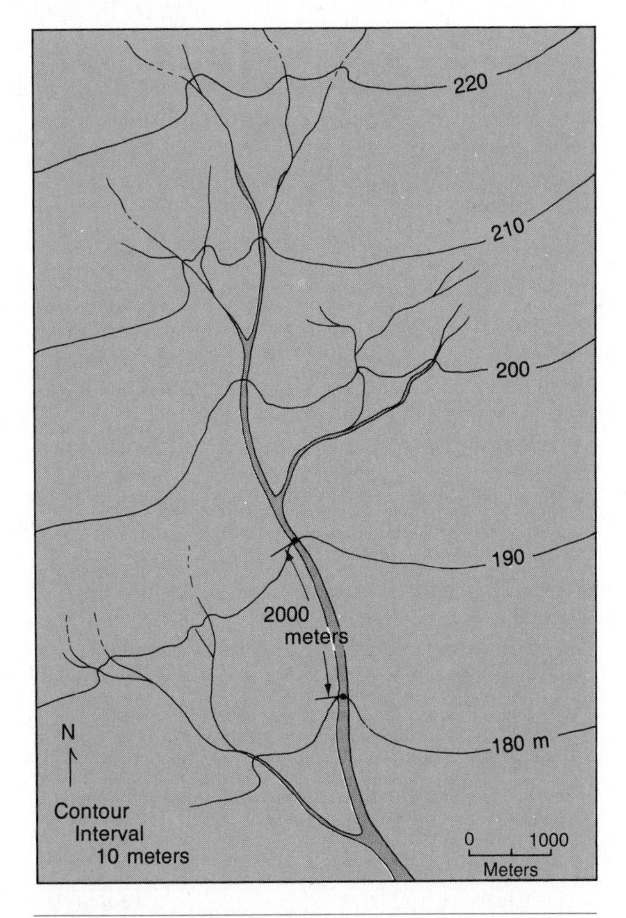

Figure 8–3 The gradient for a stream that descends 10 meters over a distance of 2000 meters would be 0.005.

multiplying the gradient by the number of meters in a kilometer. Thus, 1000×0.005 equals 5 meters per kilometer.

THE DYNAMICS OF STREAMS

Streams are highly complex systems affected by many interacting variables. As is often the case with natural systems, if any one of the variables is altered, changes are also produced in other dependent variables. The most important variables are *discharge, velocity, gradient, sediment load,* and *cross-sectional area* of the stream. As an example of the interdependence of variables, if the discharge of a stream suddenly increases after a rainfall, then the stream's velocity will increase, water may rise in the channel and provide a larger cross-sectional area, and the load of solid particles carried by the stream may increase. A stream's characteristics will also change along the length of its course. In most perennial streams, the quantity of water, quantity of transported sediment, the velocity, and cross-sectional area increase downstream, whereas the gradient decreases.

Discharge

The amount of water passing through a given cross-section of a stream in a given time is termed **discharge.** Discharge is usually expressed in terms of cubic meters of water per second. Thus, the quantity of discharge, by convention designated Q, is obtained by multiplying a stream's cross-sectional area in square meters (A) by the average velocity (V) in meters per second. The area of a stream's cross-section is derived from measurements of average width and depth. A stream's average velocity is taken to be the velocity at 0.4 meter above the channel bottom at its deepest point. It is measured with a special device called a *current meter,* and the average velocity is calculated from these readings. After the cross-sectional area and average velocity of a stream have been calculated, it is a simple matter to determine discharge by use of the equation $Q = AV$. A stream with an average velocity of 0.5 meters per second and a cross-sectional area of 20 square meters would have a discharge of 10 cubic meters per second. The discharge equation indicates that a river with constant water volume would have to increase its velocity if its channel were restricted by either geologic features or artificial constructions.

A stream has no inherent ability to increase or decrease its discharge. Discharge is determined for the stream by the amount of water that enters the channel from runoff, springs, and tributaries. These contributions are in turn influenced by climate (which controls rates of evaporation and precipitation), topography, infiltration capacity of soils, and the susceptibility of the riverbed to seepage.

Velocity

The **velocity** of a stream is the speed at which water moves down the channel in a direction perpendicular to the channel cross-section. Velocity is controlled by several factors, including discharge, slope of the riverbed, amount of the channel surface in contact with the water (the so-called wetted perimeter), and the roughness of the channel. The wetted perimeter and channel roughness influence frictional resistance to flow and therefore affect velocity as well. One can calculate the average velocity of a stream if its discharge and cross-sectional area are known from the relation $V_{av} = \frac{Q}{A}$. The average velocity may differ considerably from the actual velocity measured at a particular point in the channel. A stream is likely to be flowing fastest about a third of the distance down from its surface. Water would be moving a bit slower at the air-water interface because of friction with the overlying air, and slowest of all along the bottom and sides of the channel because of friction with the solid material in which the channel is eroded. A longitudinal segment of a stream between any two points is called a **reach.** If the section across which velocity is measured is in a straight reach, the greatest velocity will be toward the middle of the stream. Around curves, however, maximum velocity is shifted by the centrifugal component of force toward the outside of the bend. Figure 8-4 indicates the distribution of velocity along straight and curved reaches of the stream.

Flow pattern

When water is moving slowly and smoothly, the paths taken by individual water molecules may be either straight or gently curved and are parallel to the paths of neighboring molecules. This type of flow is called **laminar.** It was first demonstrated in 1883 by Osborne Reynolds, who introduced a few drops of dye to clear water flowing gently through a glass tube (Fig. 8-5). The dye moved with the water in smooth parallel

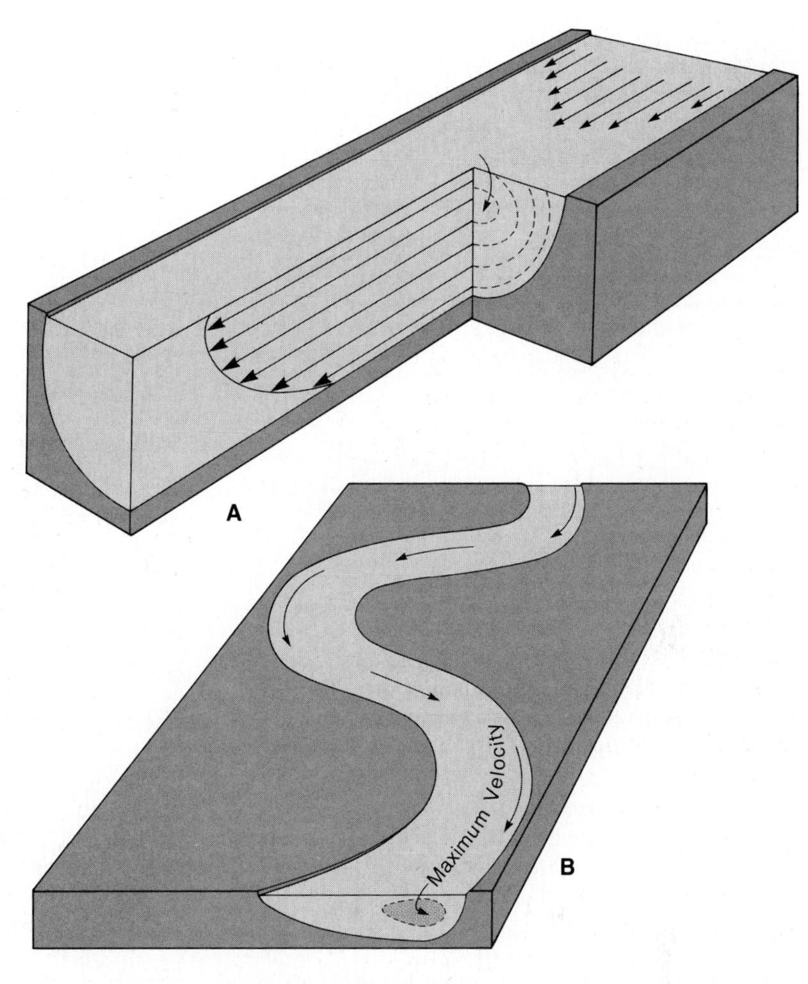

Figure 8–4 Velocity distribution along a straight section of an idealized river. Upper right part of block (A) indicates relative surface velocities, and sides of block indicate minimum velocity near the bottom of the stream where velocity is decreased because of friction. Velocity increases upward except for a decrease near the surface where velocity is again reduced because of frictional drag. (B) Maximum surface velocity is generally at the center in straight sections of a stream, and shifts toward the outer bank along curves in the river.

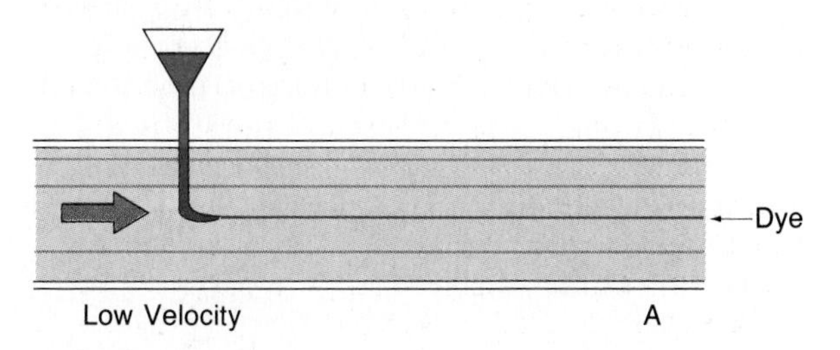

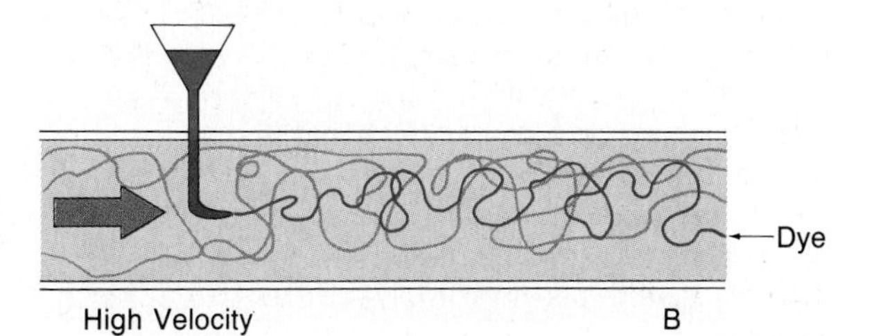

Figure 8–5 In laminar flow (A), molecules move in parallel paths and flow lines do not cross. Turbulence (B) is characterized by many irregular eddies superimposed on the general direction of flow.

lines. One can also produce this effect in a metal trough, which more closely resembles a straight reach of a river. If the velocity of flow in the trough is increased (by increasing the gradient or discharge), the flow lines of dye will become wavy and chaotic. Reynolds designated this kind of flow **turbulent.** Such swirling, eddying turbulence is the universal kind of flow in all natural streams. Turbulence is the result of high velocity and irregularities in the bed and banks of a stream. It is an important attribute of stream flow, for it enhances the ability of the stream to erode its bed and to keep particles suspended in the moving water mass.

TRANSPORTATION AND EROSION BY STREAMS

The load

The dissolved and solid rock waste carried by a stream constitutes its **load.** A stream's load is derived from several possible sources. A part of the load is carried to the channel by sheetwash and mass wasting. Additional load is provided by the stream itself as it undercuts its banks and erodes its bed. At locations where erosion by wind or ice is prevalent, these agencies may also contribute to the load carried by the stream.

Dissolved load

It is easy to overlook the dissolved load carried by a stream. It is invisible and does not significantly affect a stream's behavior. Nevertheless, transportation of dissolved mineral matter is an important function of rivers. It has been estimated that nearly 4000 metric tons of dissolved materials are carried to the ocean each year by streams. The most abundant components of this load are positive ions of calcium and sodium and negatively charged sulfate, chloride, and bicarbonate ions. These dissolved substances may precipitate as the thin mineral crust that forms on the inside of kettles. Soap does not lather well in mineral-rich "hard water" and will react to form calcium stearate. The latter compound is the bane of fastidious housekeepers, for it is the substance of which bathtub rings are composed.

Solid matter

The solid grains in a stream are transported by suspension, saltation, and traction. Particles in **suspension** are carried along within the body of flowing water. They are kept from being deposited by upward-moving turbulent eddies. Only relatively small grains, such as those in the silt and clay sizes, are able to resist the pull of gravity and remain in the suspended load. Except in extremely fast-flowing streams, sand particles do not remain suspended but collect on the surface of the stream bed, where they move along in a series of leaps (Fig. 8-6). The motion is termed **saltation.** Still larger sand grains, as well as granules, pebbles, and cobbles, are moved along by sliding and rolling motions. The process of transporting coarse sediment along the bottom of a stream, either by rolling or sliding, is called **traction.**

Precisely which range of particle sizes is transported by suspension, saltation, or traction is, within limits, dependent upon velocity. Should the velocity of a rapidly flowing stream be decreased, the larger particles in the suspended load might become part of

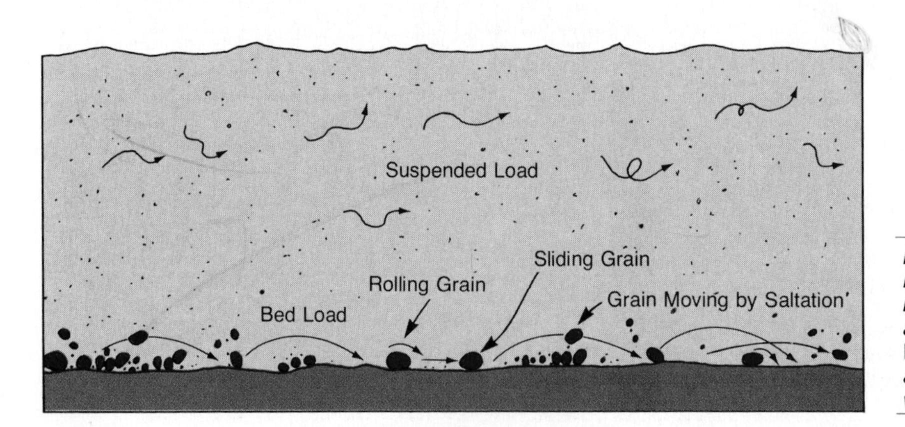

Figure 8-6 Transportation of sediment by a stream. The particles being moved by saltation, rolling, and/or sliding constitute the bed load. *Particles in the* suspended load *are carried above the stream bed within the body of flowing water.*

the saltation load, and particles formerly moving by saltation would continue their downstream progress by sliding and rolling.

Competence and capacity

The term **competence** is used to express the largest particle that a stream can transport under a given set of conditions. The competence of a river may change considerably from its headwaters to its mouth and even at any single point along its course. Streams can transport much larger particles during times of flood, when velocity of flow is at a maximum. In general, the diameter of a rock fragment that can be moved by the stream is proportional to the square of the velocity of flow. Thus, if the velocity is doubled, the stream can move particles four times as large as before. The explanation for this relationship is that each molecule of water not only strikes a grain of sediment twice as hard, but also twice as many molecules strike the grain during a given interval of time.

A second term used to describe sediment transport of a stream is capacity. **Capacity** refers to the maximum amount of material the stream can carry under a given set of conditions. Because there are limits to the amount of load that can be included within a given volume of water, it is evident that a stream's capacity will be markedly influenced by discharge.

Erosion by streams

Relation of velocity to erosion

Velocity is the most important factor in determining the rate of channel erosion by a stream. Erosion occurs whenever particles of sediment are removed from the stream bed. Particles may be pushed forward by the force of the moving water against their upstream sides. They may also be lifted because of the reduced pressure on their upper surfaces as water flows over them. The forces that attempt to lift grains at rest and bring them into the transported load are opposed by gravity as well as the natural cohesive force that tends to hold together small particles at rest. Of these two forces, gravity is clearly the more important, although cohesion does play an important role, at least for the finer clay-size particles. Once deposited on the stream bed, the flat clay particles tend to cling to the channel floor. Water, moving against the thin edges of the clay particles, has little surface to push against, and besides, it is flowing in the zone of lowest velocity. In general, if the channel surface is composed of clay, stream velocities must often exceed 500 cm/sec before erosion can begin, whereas sand on the stream bed is readily transported by stream velocities of only 10 cm/sec. This relationship is clearly shown in Figure 8-7. Notice that the velocity needed to initiate movement of particles is greater for clay than for coarse gravel because of

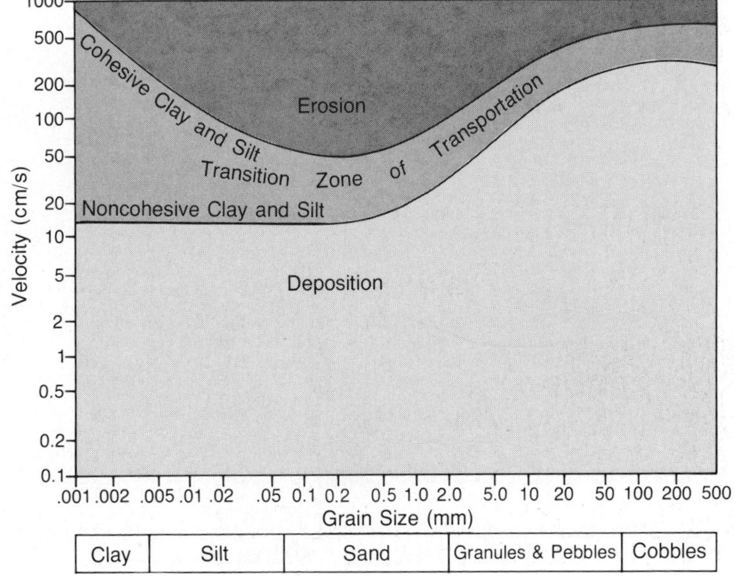

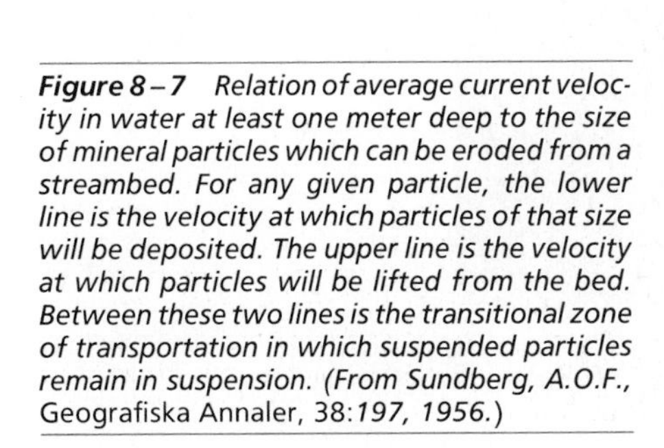

Figure 8-7 *Relation of average current velocity in water at least one meter deep to the size of mineral particles which can be eroded from a streambed. For any given particle, the lower line is the velocity at which particles of that size will be deposited. The upper line is the velocity at which particles will be lifted from the bed. Between these two lines is the transitional zone of transportation in which suspended particles remain in suspension. (From Sundberg, A.O.F.,* Geografiska Annaler, 38:*197, 1956.)*

the cohesiveness of clay. Intermediate particles are eroded at the lowest range of velocities.

Once clay particles from a channel are lifted into the mass of flowing water, the particles remain in suspension even at very low velocities. Figure 8–7 also indicates the velocity at which particles of various sizes will be deposited by a stream. For velocities in the region above the lower curve, grains of sediment already in motion will continue to be transported indefinitely.

Erosional processes

Stream erosion is the removal of rock and mineral matter from the bed of a river. The material in which the bed is cut may consist of bedrock, or, as is more common, unconsolidated river-laid sediment called **alluvium** (Fig. 8–8). Streams flowing on alluvium may erode their beds and channel sidewalls by lifting particles of sediment into the mass of flowing water. In addition, erosion of bedrock as well as alluvium in a stream's bed is accomplished by such processes as solution, corrasion, and impact.

Solution is a process that is most prevalent in regions where streams flow across particularly soluble rocks like dolostone or limestone. The dissolution of rock forming the bed of the stream contributes to the dissolved load. That contribution from the bed, however, is relatively minor when compared with the much larger amount of dissolved material derived from chemical weathering across the entire drainage area and carried to the streams by underground water or sheetwash.

In contrast to the largely chemical effects of solution, **corrasion** refers to the mechanical wearing away of the channel as a result of the collision and grinding action of the sand grains and pebbles carried by the stream. Acting together, corrasion and solution may produce basin-like depressions in the bedrock of streams. These smooth, rounded cavities are called

Figure 8–8 Sandy alluvium from a floodplain showing thin layers and lenses of coarser particles. (Courtesy of U.S. Geological Survey and G.P. Williams.)

potholes. Most potholes are formed by the relentless circular motion of eddies and whirlpools that contain loose rock and mineral fragments as tools.

Impact, an erosional process associated with wind action, is also important in stream erosion. It occurs whenever large particles carried by the stream are thrown forcibly against the sides and bottom of the bed. If the stream bed is cut in bedrock, the many continuous blows knock off fragments of rock and thereby contribute to the solid particles constituting the stream's load. In addition, fragments of rock in transport are themselves eroded by impact and abrasion. The process contributes to the rounding of pebbles and cobbles.

Base level

The degradational work of streams is truly immense. Each year, rivers carry an estimated 10 million tons of continental material to the oceans. This work, however, cannot continue indefinitely in a drainage basin that remains tectonically stable. Eventually, a stream would cut its valley to sea level and would be unable to accomplish further erosion. This lower limit of erosion is called **base level.**

The significance of base level was recognized in 1869 by the one-armed Civil War veteran and geologist J.W. Powell. While leading a perilous boat expedition down the unexplored gorges of the Colorado River, Powell was understandably impressed with the formidable erosive abilities of that great stream. "How long," he wondered, "could the river continue to actively cut downward?" The answer seemed readily apparent. Eventually, the stream would have eroded its bed nearly to sea level and, with little or no slope, would cease its geologic work. "We may consider the level of the sea," wrote Powell, "to be a grand base level, below which the dry lands cannot be eroded."

Powell also made studies of the mouths of tributaries entering the Colorado River and perceived that these confluences served as local base levels for each tributary. The tributary could cut its valley no deeper than the major stream into which it empties. Lakes, dams, and resistant rock formations may also form local base levels that temporarily provide a vertical limit to the erosive powers of streams.

Streams and tectonics

If we recall the statement made earlier that rivers carry approximately 10 million tons of solid and dissolved material to the oceans each year, and further remember that this transfer has been in progress for more than 3.6 billion years, then one might wonder why entire continents have not been reduced to what Powell described as the grand base level. The reason, of course, is that the lands have not been static. They have opposed degradational agencies by rising to various heights above sea level during episodes of isostatic adjustment and/or tectonic activity. As a result of these uplifts, old streams flowing sluggishly across lowlands have been occasionally rejuvenated and additional streams formed to begin anew the erosional assault on the lands.

THE CONCEPT OF A GRADED STREAM

If a stream of water begins to flow down a sloping surface, the water will cut a channel large enough to contain the flow. In the early stages of development, the gradient of the stream is likely to be relatively steep, and the valley will develop a V-shaped cross profile. The stream will work actively to erode its channel and eliminate the irregularities. As time passes, the gradient will be lessened while the load is steadily increased because of contributions from an expanding network of tributaries. Eventually, the stream attains a condition in which the supply of sediment fed into the stream approximately equals the stream's capacity to transport that load. Indeed, most of the river's available energy will be consumed in this transport, and rapid cutting of the channel ceases. The stream is now said to have reached the graded condition. A graded stream is ideally a stream in equilibrium. Any change in the supply of sediment, velocity, or discharge will result in a displacement of equilibrium in a manner that will tend to absorb the effects of the change. For example, a graded stream that becomes underloaded (as when sediment-trapping dams are constructed in its tributaries or headwaters) will have energy to spare and will compensate by deepening its bed through erosion so as to flatten the gradient and decrease velocity. As an alternative, the stream might increase the number of bends (i.e., increase its sinuosity) and thereby lessen its gradient. Should a stream become overloaded, it may build up its bed initially by depositing the surplus load and thereby steepen its gradient (Fig. 8–9), or it might increase overall gradient by reducing its sinuosity. In either case, the steepened gradient would result in increased velocity and capacity.

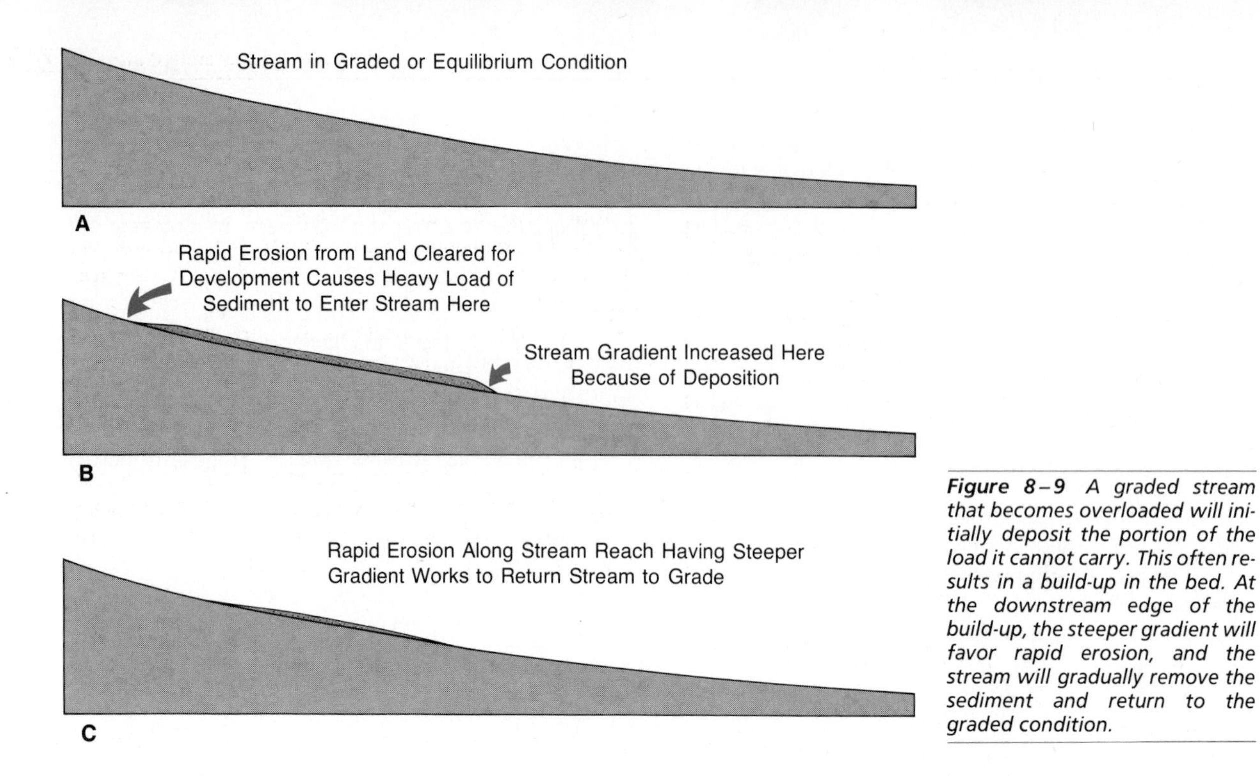

Figure 8–9 *A graded stream that becomes overloaded will initially deposit the portion of the load it cannot carry. This often results in a build-up in the bed. At the downstream edge of the build-up, the steeper gradient will favor rapid erosion, and the stream will gradually remove the sediment and return to the graded condition.*

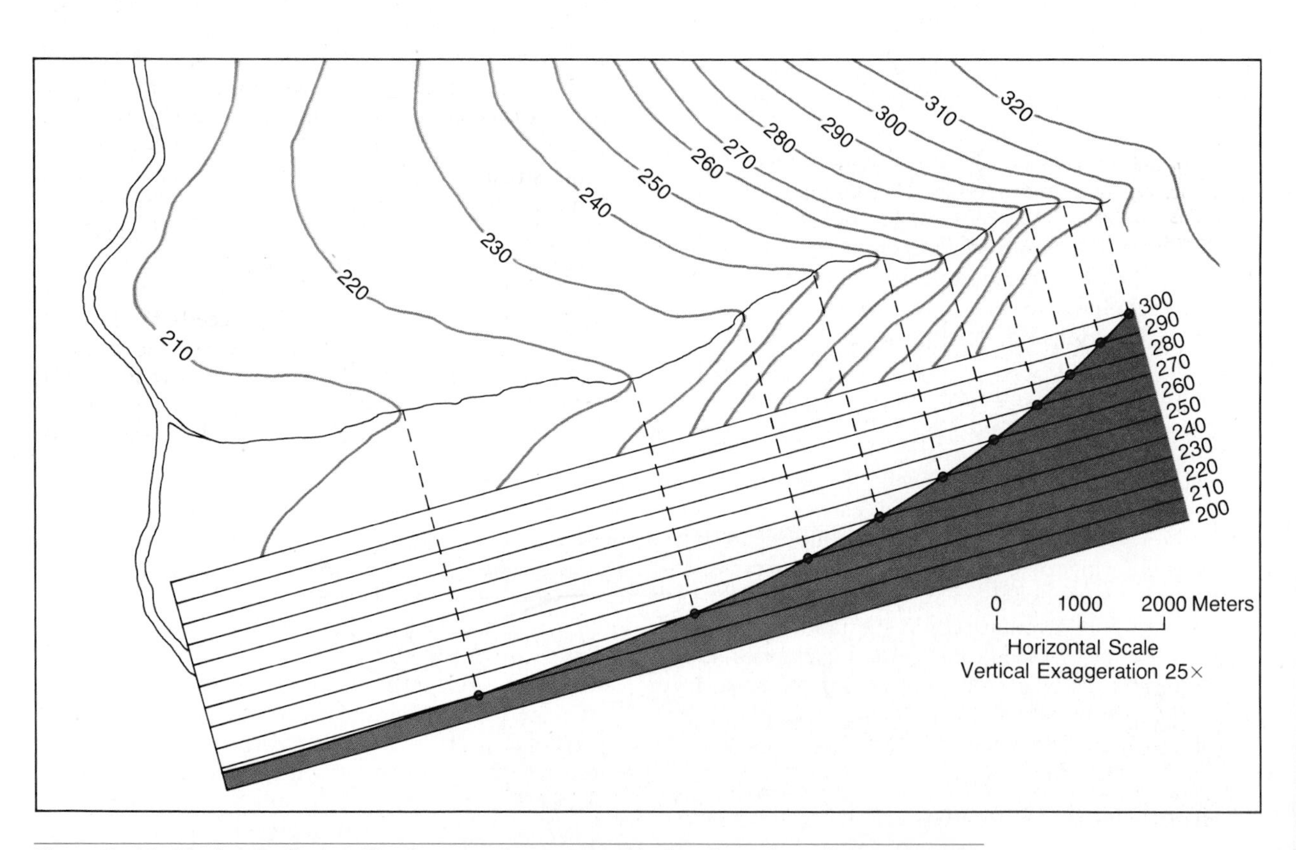

Figure 8–10 *The longitudinal profile of a stream is constructed by locating points where topographic contour lines cross a stream channel and projecting those points onto a piece of ruled paper. Graded streams typically develop the concave-up longitudinal profile shown here.*

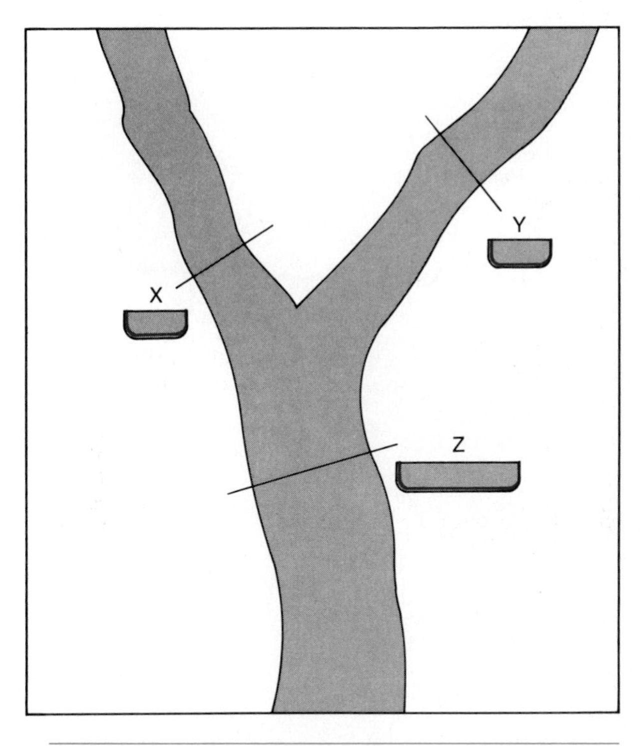

Figure 8–11 The cross-sectional area of tributary X plus Y equals that of the major stream Z. Yet the part of the channel surface in contact with flowing water (the wetted perimeter) of Z is less than the sum of X plus Y. Thus water flowing in Z experiences less frictional drag.

The overall result of these "cut and fill" events is that the stream develops a concave-upward profile of equilibrium. The profile, technically termed the **longitudinal profile,** can be constructed by connecting points of elevation along the length of the stream (Fig. 8–10). The concave slope is a fundamental characteristic of graded streams. Near the headwaters, the gradient of the stream is steep and the river can do its work with the relatively small discharge available. Nearer the mouth, the gradient is lower, yet the stream continues to accomplish its work because it has greater discharge, smaller particles to transport, and less friction-inducing channel surface. The reduced total friction from water contact with the channel in lower reaches means the stream can maintain a given velocity on a lesser gradient than can an upstream segment having a steeper gradient (Fig. 8–11).

CONSEQUENT AND SUBSEQUENT STREAMS

The first streams to develop on an undissected land surface will flow down the preexisting slope of the land according to the most favorable and unobstructed route. Such streams are termed **consequent streams** because their courses are a consequence of the original slope of the land surface (Fig. 8–12). The course of other streams may be determined not primarily by an original surface but rather by lithologic variations or structural differences. Such streams are called **subsequent.** Subsequent streams lengthen themselves by headward erosion along belts of easily eroded rock. As the process of headward erosion proceeds, the head of the subsequent stream may ultimately extend into the drainage territory of rival streams that flow at slightly higher levels. When such connections are completed, the flow of the higher-level stream will be diverted into

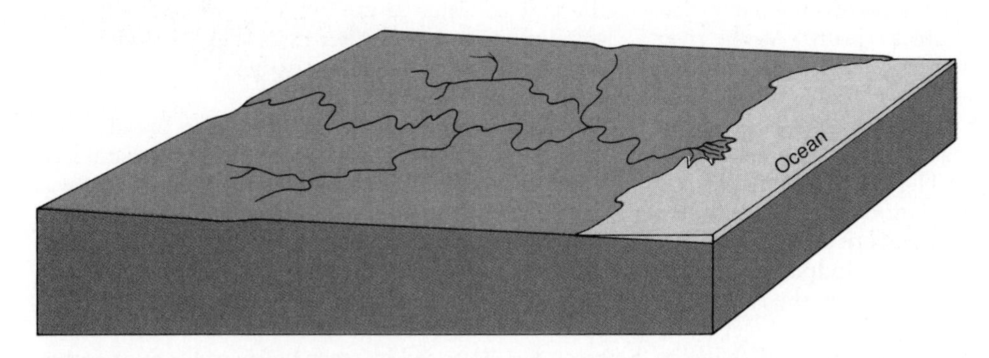

Figure 8–12 The uplifted portion of the continental shelf provides a sloping surface for the development of consequent *streams. The courses of such streams are a* consequence *of the original land surface.*

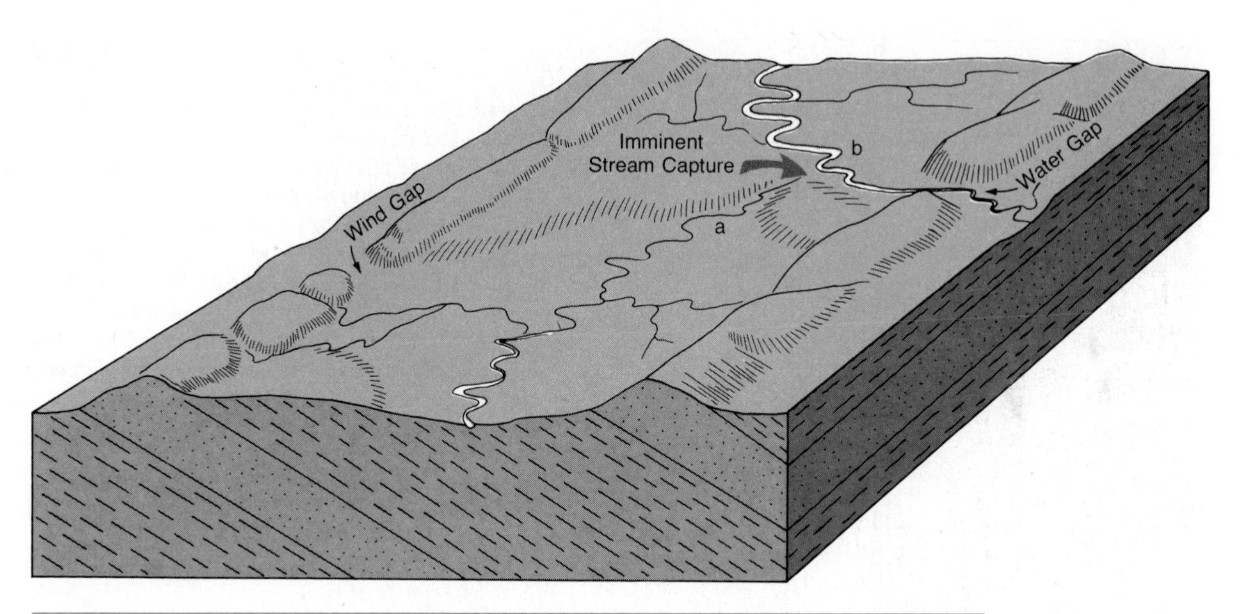

Figure 8–13 *Both streams* a *and* b *are subsequent streams. Stream* a *lies at a relatively lower level than stream* b*, and is extending its valley by headward erosion. Eventually, it will intersect the channel of stream* b *and capture its headwaters. A wind gap has been developed in the ridge on the left side of the block, whereas a water gap is present on the adjacent ridge.*

the channel of the subsequent stream, thereby increasing its discharge and erosive powers. This "capture" of one stream by another is termed **stream piracy,** and is a commonplace occurrence, especially in the early evolution of drainage systems (Fig. 8–13).

FEATURES FORMED BY STREAMS

Newly formed streams do not possess the valley flats and smooth gradients of older streams. As the river continues its work, however, gradients are reduced. The river will cease to cut downward as actively and begin to cut horizontally into the outside of bends. In the lower-velocity area that characterizes the inside of bends, deposition occurs. The process of cutting and filling is called **lateral planation** (Fig. 8–14) and is responsible for the relatively smooth areas bordering a stream. These areas are called **floodplains** because they are inundated during times when the river overflows its banks.

As the stream continues to widen its valley by lateral planation, the pattern of alternate left and right bends begins to assume some regularity. The bends are now termed **meanders,** from a Greek word meaning "bends." In general, the meandering pattern (Fig. 8–15) is developed by streams that flow on unconsolidated but relatively cohesive silt and clay and on flat valley floors. The cohesiveness of the fine alluvium helps to restrict the stream to a single channel, rather than permitting it to break up into numerous distributaries. In flume experiments, meanders are readily formed from originally straight channels. Although not essential, the sinuous pattern may be initiated when a part of the bank collapses and thereby diverts the flow of water to the opposite bank. The natural spiral motion of the water mass (Fig. 8–16) tends to perpetuate the development of meanders downstream from the initiating obstacle.

Most meandering streams traverse broad valley flats. Some, however, have steep valley walls and lack floodplains. The meander pattern in such streams was developed by the stream long ago when it was flowing on a valley flat. Subsequently, through regional uplift or some other geologic change, the river was caused to cut down into its former floodplain, but at the same time it maintained its meander pattern. The result is meanders bounded by steep valley walls. Such meanders are said to be incised or entrenched (see Figure 5–3). When accompanied by other substantiating

Figure 8–14 *The Ohio River near Cincinnati has developed a broad floodplain by processes of lateral planation involving erosion of banks on outside of meanders and deposition on the inner bends where velocity is reduced. (Courtesy of J.C. Brice.)*

Figure 8–15 *Symmetrical meander pattern developed along a reach of the Wood River near Fairbanks, Alaska. The three center meanders are complicated by chutes where water has broken through a swale or low area in the deposits of the inside of the bend. The sediment dropped on the inner curves of meanders form crescent-shaped buildouts called point-bar deposits. (Courtesy of J.C. Brice.)*

lines of evidence, incised meanders may be considered evidence for uplift of the earth's crust.

Not all streams or stream segments (reaches) develop a regular meandering pattern. Some divert their flow into numerous shallow dividing and reuniting subchannels separated by transient islands and bars. Such streams are said to have a **braided pattern** (Fig. 8–17). The braided pattern appears to develop in streams having highly variable discharge and readily erodible banks composed of noncohesive materials, such as sand and gravel. Because of the susceptibility of such bank material to erosion, the stream can easily divide into many small channels, called distributaries. Sporadic decrease in discharge among the distributaries may cause local deposition in channels, resulting in blockages and further separation into subchannels.

Features of floodplains

One may find a large number of secondary features formed along the valleys of meandering streams. These include point-bar deposits, cutoffs, oxbow lakes, and natural levees. **Point-bar** deposits are accumulations of sand and gravel along the insides of meander bends (Fig. 8–18). These deposits develop because

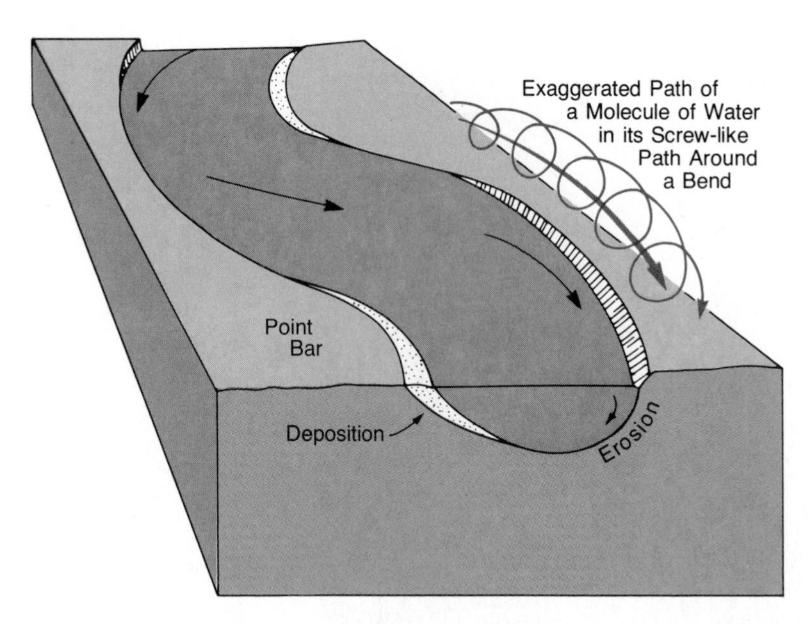

Figure 8–16 Water molecules in a sinuous channel, not only move downstream, but also follow a spiral-like path. (Modified from Leopold, L. B., and Langbein, W.B., A Primer on Water, *U.S. Geological Survey, U.S. Gov't. Printing Office 0–398–800, 1960.)*

of the lower velocity around the inner margin of a bend, so the stream deposits part of its load. Point-bars provide the frequent sandy havens for weary canoeists and vacationing bathers. On the outside bends of the meanders, the force of inertia, tending to cause the water to flow in a straight line, causes the water to impinge upon the outside banks, to "pile up," and to actively erode the banks.

The cutting along the outside bends of meanders with accompanying filling along inside bends results in slow migration of the meanders in a downstream direction (Fig. 8–19). The migration, if speeded up, might resemble the pattern produced where a snake had crossed an area of sand. Frequently, during the migration of meanders, two bends will come together to form a meander **cutoff** (Fig. 8–20). The abandoned loop of the channel is left as a curved remnant on the floodplain. Because of its resemblance to the device placed over the shoulders of oxen to assist them in pulling wagons, these abandoned meanders have been dubbed **oxbow lakes.** With time, the oxbow lake may become filled with sediment and disappear.

River floodplains are not completely level. Most floodplains have a raised crest or ridge adjacent to the

Figure 8–17 Braided stream pattern in Tanana River downstream from Fairbanks, Alaska. (Courtesy of J.C. Brice.)

Figure 8–18 Point bar developed on inside of a bend in the Salt River, northeastern Missouri. (Courtesy of J.C. Brice.)

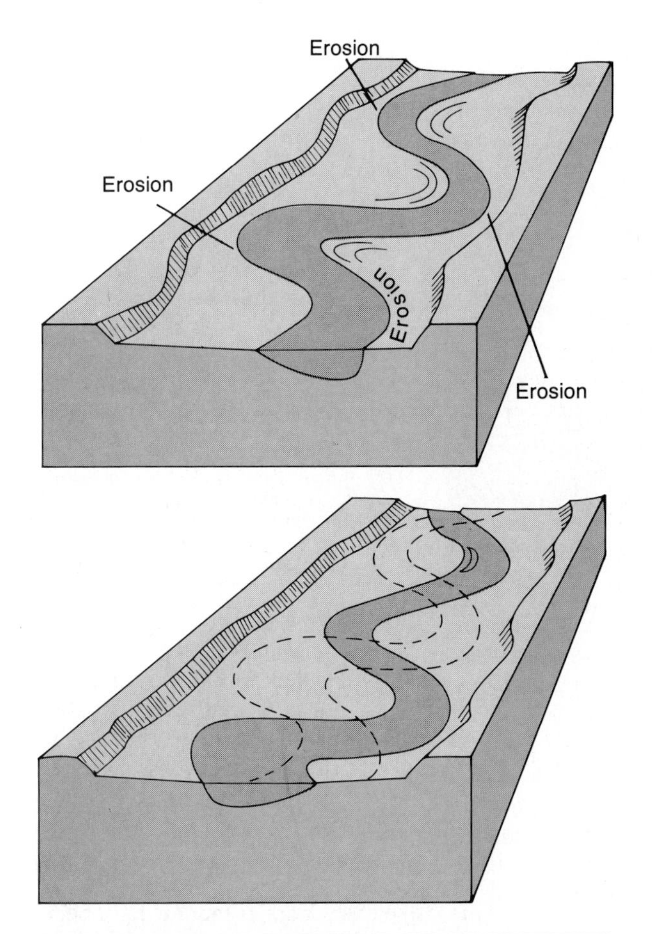

Figure 8–19 Migration of meanders in a downstream direction. Dashed line in lower block marks the earlier channel location.

channel. These ridges along the sides of streams are called **natural levees** (Fig. 8–21). They are formed during flood stages when the water overflows its banks, is suddenly slowed, and drops part of its load. Along parts of the lower Mississippi, the natural levee is the highest tract of land for miles around.

Alluvial fans

Alluvial fans are cone-shaped deposits composed of river-transported silt, sand, and gravel (Fig. 8–22). They are formed where streams emerge from a mountainous area and flow onto relatively level terrain. Stream velocity is reduced rather abruptly at such locations, largely because of loss of water to the porous

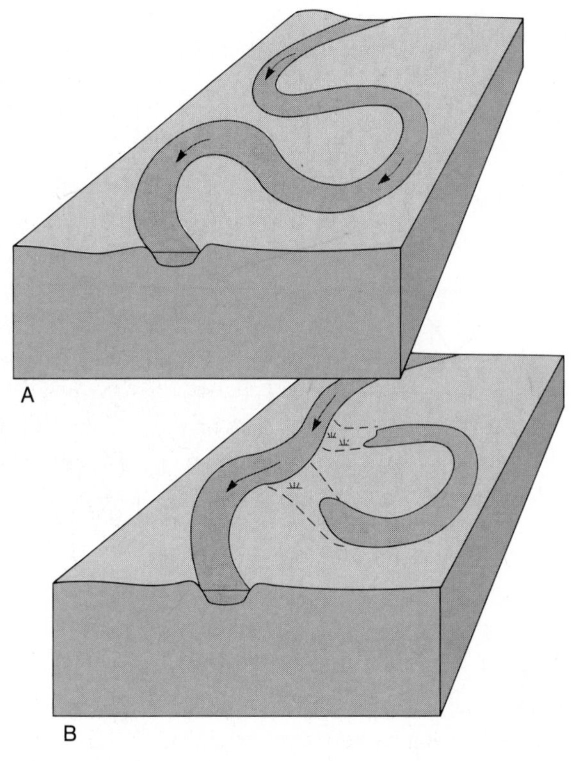

Figure 8–20 Development of a meander cutoff and oxbow lake.

sediment beneath the stream bed. The stream is unable to carry its former load under these new conditions. Sediment is deposited within channels, causing water to spill over banks and form new channels, many of which disappear before reaching the foot of the fan because of water loss by rapid seepage. These events are repeated endlessly back and forth across the depositional area, eventually building the alluvial fans.

Stream terraces

Whenever a stream is able to more actively erode its bed (because of uplift, lowering of sea level, or reduction of load), areas of the floodplain may be removed, leaving undissected flat remnants behind. These level remnants of former floodplain are called **stream terraces** (Fig. 8–23).

Deltas

Accumulations of sediment deposited by streams where they enter a large body of water like a lake or sea

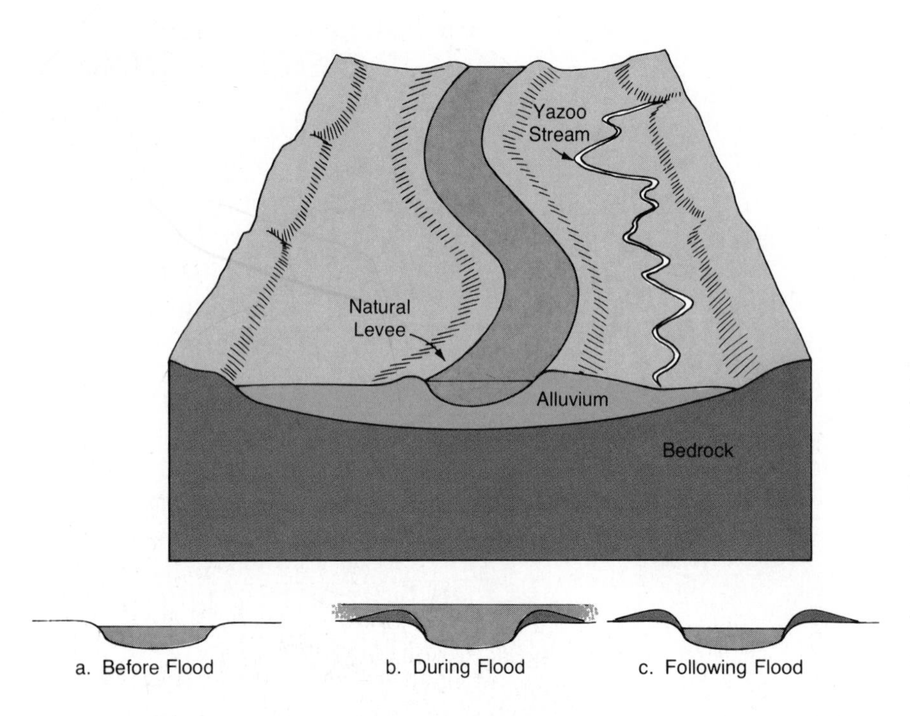

Figure 8–21 Floodplain of a stream having well-developed natural levees. The formation of the levees is indicated in the sequence of figures beneath the block diagram. Levees are developed when the river overflows its banks. Once outside the channel, velocity is reduced and deposition builds a submerged ridge. After the flood, this ridge is left standing higher than the surrounding floodplain.

Figure 8–22 Alluvial fans developed along the east face of the Panamint Range, Inyo County, California. The long dimension of the playa lake (dark area in northeast quadrant) is 1000 meters.

are called **deltas.** On encountering the body of standing water, the stream's velocity and transportive powers are quickly reduced and deposition occurs. Sediment accumulating in the channel reduces the gradient, blocks the channel, and causes the flow to be diverted and divided into new channels. Several environmental conditions favor the development of deltas. An abundant supply of sediment is the most important condition. In addition, delta formation is enhanced by shallow depth of water, tectonic stability, and lack of strong waves or currents in the area of delta growth.

One of the most distinctive sedimentological features of deltas is cross-stratification (Fig. 8–24). This feature begins to form as soon as the river enters the sea or lake and drops the coarser components of its load nearest the landward part of the delta. Usually, sediment is swept over the initial pile-up, coming to rest as seaward-sloping layers. These layers constitute the **foreset beds.** They make up the major portion of simple deltas. Finer sediments carried to the delta by the stream remain in suspension longer and are swept out farther into the basin. These horizontal layers constitute the **bottomset beds.** They are covered by foreset beds as the delta progrades seaward. Above the foreset beds, **topset beds** may develop as deposition occurs on top of the advancing foreset beds.

Because of the fertility of their soils and conve-

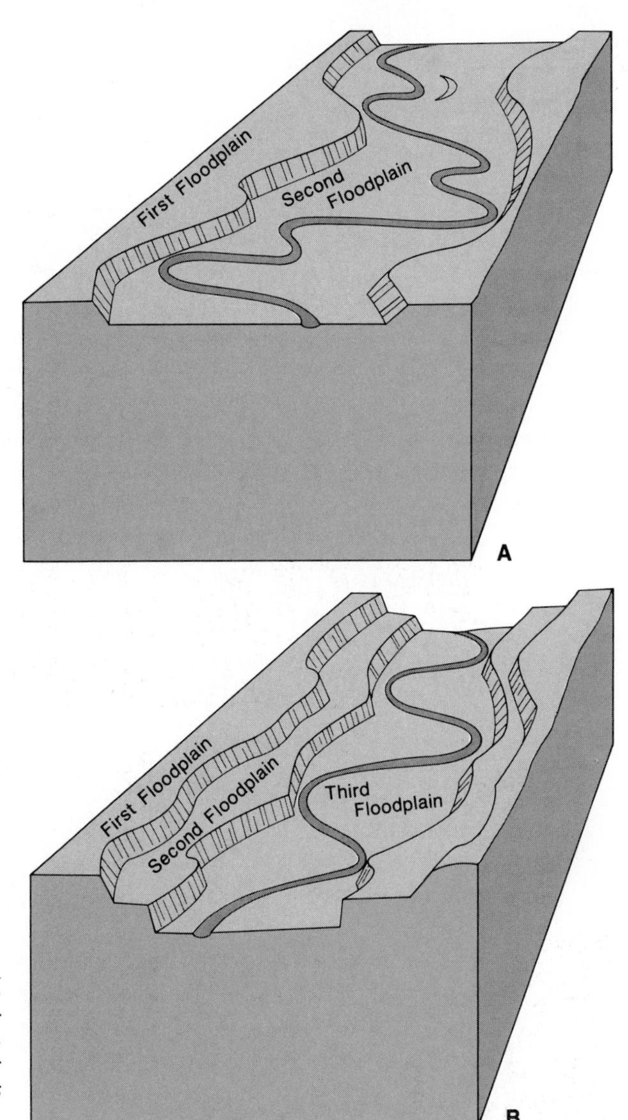

Figure 8–23 *Development of terraces by periodic uplift. In block A uplift has caused the stream to cut into its former floodplain and gradually develop a new floodplain. With another pulse of uplift, the process is repeated (B) and a yet lower floodplain is produced. Remnants of the former floodplains form the terraces that border the valley.*

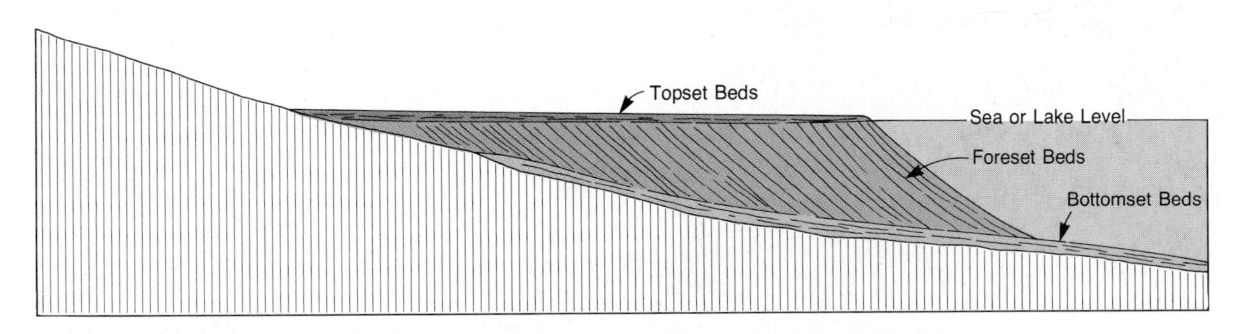

Figure 8–24 *The ideal arrangement of beds in a delta. Relatively coarse material is deposited initially to form the inclined layers or* foreset beds. *Somewhat finer sediment is swept farther toward the basin and deposited as* bottomset beds. *Foreset beds may prograde over these bottomset beds. The stream deposits the topset beds on top of the foreset beds.*

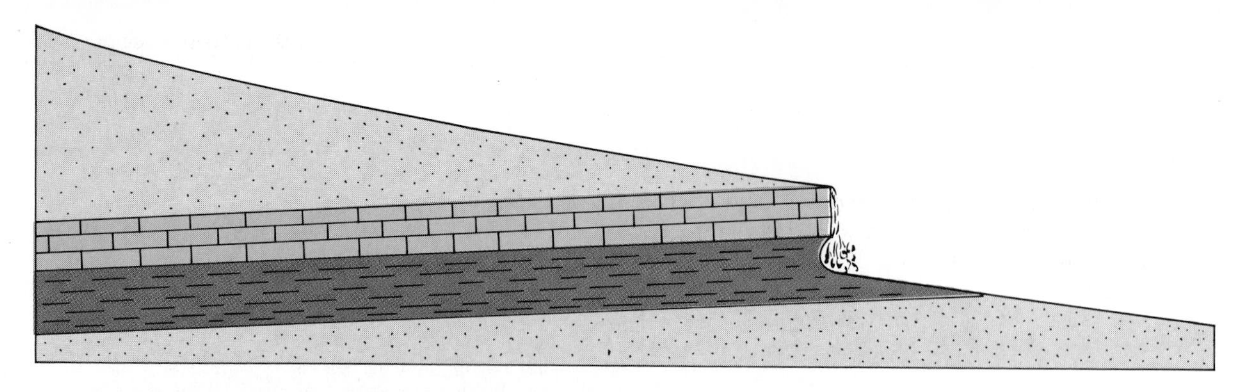

Figure 8–25 *Where an outcrop of relatively resistant rock is underlain by less resistant rock, the weaker rock may be rapidly eroded by the stream and the more resistant layer undercut. If the resistant rock collapses and leaves a vertical face, water will plunge over the crest as a waterfall.*

Figure 8–26 *Niagara Falls, formed where the Niagara River flows from Lake Erie into Lake Ontario. (Courtesy of Ontario Ministry of Industry and Tourism.)*

nience of their coastal locations, modern deltas have always been centers of commerce and agriculture. Most of the agricultural lands of Egypt, for example, are located on the delta and floodplains of the Nile. The fertile fields of the Mekong Delta are a vital food source for the peoples of Cambodia and Vietnam.

Waterfalls

Waterfalls are among the most esthetically pleasing and spectacular features of streams. A **waterfall** is a place in the stream where water descends vertically to a lower level. Waterfalls are usually a consequence of either bedrock characteristics or glaciation. Most commonly, waterfalls develop in places where the stream flows across an outcrop of a particularly resistant stratum that is underlain by one or more layers of more easily erodable rock. Downstream from the outcrop of hard rock, the less resistant layer is more rapidly eroded, resulting in an abrupt drop in the elevation of the stream bed (Fig. 8-25).

Erosion at the upper edge of the waterfall will proceed rapidly as water rushes over the rock surface and excavates the fragmented debris at the base of the fall. As a result, the waterfall itself gradually recedes upstream. Eventually, most waterfalls may degenerate into rapids and disappear altogether.

Niagara Falls (Fig. 8-26), located between Lake Erie and Lake Ontario, developed when the retreating ice of the last glacial age uncovered an escarpment formed by southwardly tilted resistant dolostone. Weak shales beneath the dolomite strata are continuously undermined in the pool beneath the falls, and this erosion contributes to the southward retreat of the falls.

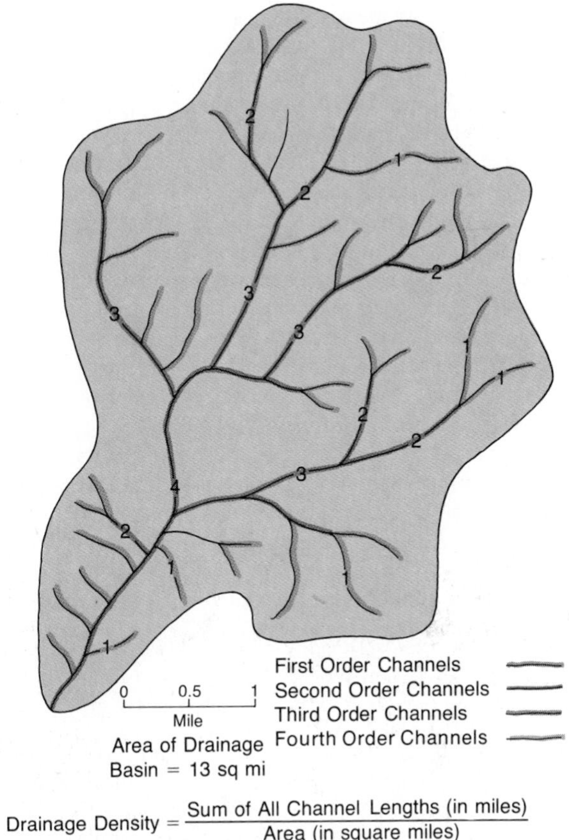

$$\text{Drainage Density} = \frac{\text{Sum of All Channel Lengths (in miles)}}{\text{Area (in square miles)}} = \frac{30.5}{13} = 2.34$$

Figure 8-27 Method of determining drainage density and ordering stream segments.

DRAINAGE BASINS AND PATTERNS

A river and its tributaries constitute a **drainage system.** Every such system drains an area called a **drainage basin,** each of which is separated from other drainage basins by upland or ridge tracts termed **drainage divides.** A tracing of a major river and its tributaries would reveal a definite hierarchy of channels which can be numbered and studied quantitatively. The first step in such an analysis is to outline the drainage basin, and to number the tributaries as indicated on Figure 8-27. The smallest unbranched headwater channels are designated as *first order*; the streams receiving water from first order streams are *second order*, and so on until the main trunk stream receives the highest designation. Using this scheme, it is possible to ascertain average stream length and number of streams in a drainage basin and drainage system area, and to make precise quantitative comparisons among different drainage systems.

Carried a step further, the cumulative length of all channels in a drainage system can be divided by the surface area of the basin to obtain the **drainage density.** Thus:

$$\text{Drainage Density} = \frac{\text{Length of Channels}}{\text{Area of Drainage Basin}}$$

Drainage density measures the abundance of channels in a given basin. It is influenced by climate, topographic relief, the nature of material exposed at the surface, and time. The highest values of drainage density are found in badlands that develop on weak clays in semiarid regions. In such areas, low permeability results in high runoff. Vegetative cover is minimal, and hence erosion proceeds toward the development of extreme gullying and numerous channels.

The arrangement of streams within a drainage basin is determined by such factors as initial slopes, inequalities in rock resistance to erosion, whether strata are level or tilted, and even tectonic movements. The particular plan or design of the river system constitutes the **drainage pattern.** By the study of drainage patterns, it is often possible to infer some of the characteristics of the structural geology of an area, as well as events in geological history. For example, if the rocks underlying an area are essentially homogeneous or consist of flat-lying strata (Fig. 8-28), the stream network will be characterized by an irregular branching of tributary streams. The overall drainage pattern is called **dendritic** (from the Greek *dendron*, tree) and rather resembles the branching of a tree. By contrast, **trellis patterns** consist of a system of nearly parallel streams, usually aligned between the outcrops of folded rocks that are relatively resistant to erosion (Fig. 8-29). Patterns of this type are well displayed in the drainage systems of parts of the Appalachian Mountains where the drainage divides are formed by elongate ridges of resistant sandstones. Two interesting features that are occasionally found in regions of trellis drainage are wind gaps and water gaps. **Water gaps** are gorgelike stream valleys cut directly through a well-defined mountain ridge. Most water gaps can be attributed to the erosional activities of a stream that had established its course long ago in rocks having one type of structure; the stream gradually eroded its way down to an underlying sequence of rocks having quite different structure (Fig. 8-30). The stream responsible for cutting the water gap is said to have been **superposed** (contraction of superimposed) upon the underlying structure. Because the stream had already established its course in the overlying rock sequence, it is able to continue in this course cutting across resistant and nonresistant strata indiscriminately.

Occasionally, a stream flowing through a water gap may be diverted into the more rapidly deepening channel of one of the longitudinal streams that flow across softer rocks. This event constitutes stream piracy and leaves the water gap without water. The notch cut into the ridge is now called a **wind gap** (see Fig. 8-13), even though the action of wind had nothing to do with its origin.

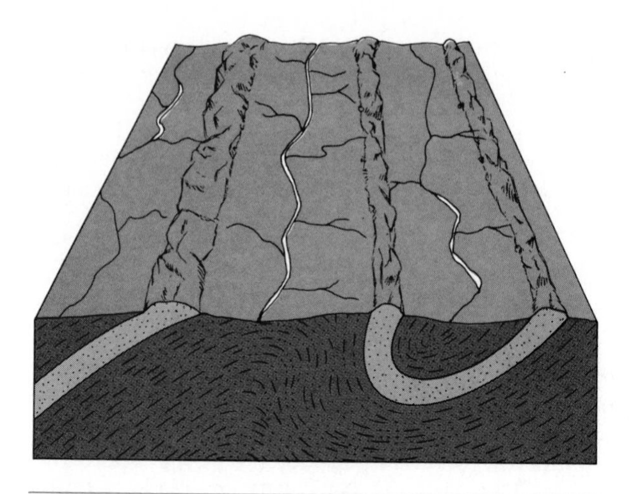

Figure 8-29 Trellis drainage pattern.

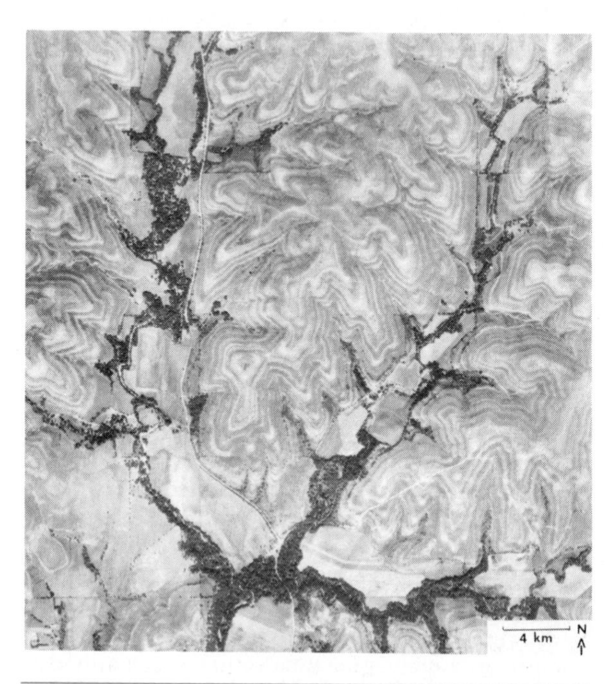

Figure 8-28 Dendritic drainage pattern developed on flat-lying limestone and shales in the Flint Hills of Pottawatomie County, Kansas. Area is a prairie, and trees grow only in valleys of Cedar Creek and its tributaries. (Courtesy of U.S. Geological Survey.)

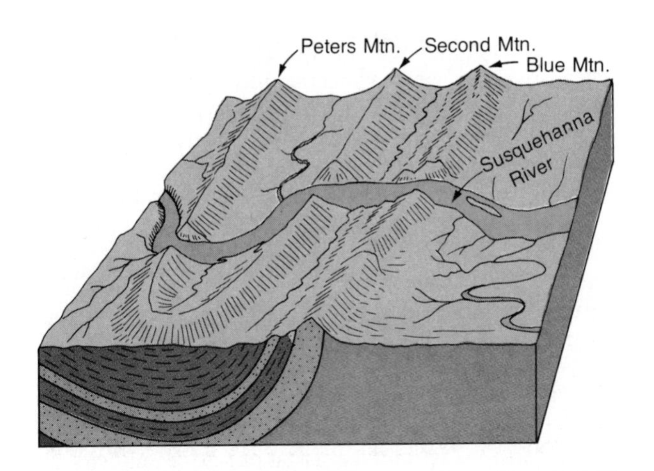

Figure 8–30 *The Susquehanna River has cut a series of watergaps through resistant ridges in the Appalachian Mountains of southeastern Pennsylvania.*

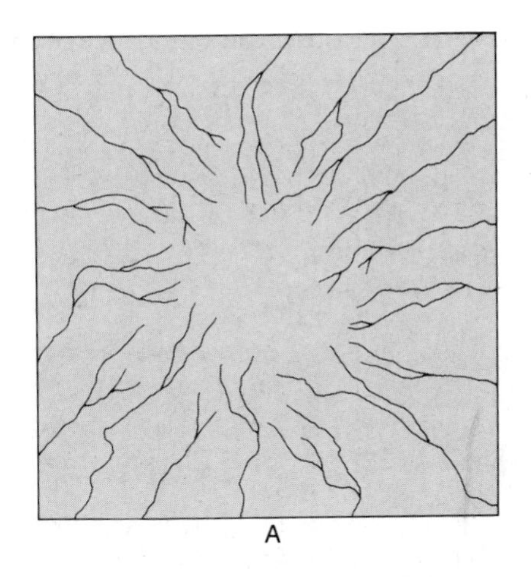

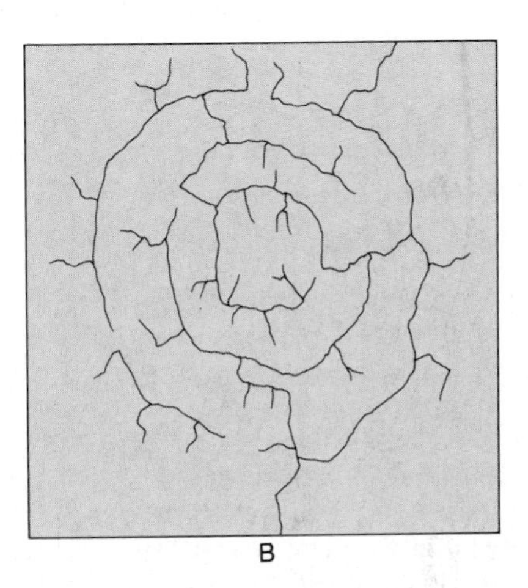

Figure 8–31 *(A) Radial drainage pattern. (B) Annular drainage pattern.*

A third type of drainage pattern develops on dome mountain uplifts or on the sides of volcanic cones. Because streams flow away from a high central area in all directions, the overall pattern is designated as **radial.** Drainage on exposed igneous stocks and laccoliths is also often radial in pattern (Fig. 8–31A).

Streams located around domal uplifts, where alternating layers of strata outcrop in a circular plan around the central mass, frequently develop an **annular drainage pattern** (Fig. 8–31B). As is the case with trellis drainage, the pattern of channels is structurally controlled. Streams flow across the outcrop of softer rocks and between stream divides composed of more resistant rock.

FLOODS

Nearly every month the news media report a disastrous flood somewhere in the world. A flood is any relatively high streamflow that spills out of its channel onto the valley floor (Fig. 8–32). Everyone knows that floods develop whenever there is heavy runoff due to the persistence or intensity of rain, rapid melting of snow, or alteration of land surfaces that increases runoff. The channel of the stream is shaped primarily by its usual amount of discharge, and therefore when it receives an extraordinary volume of water, the surplus can only spill over the banks and flow out onto the floodplain. Floodplains act as safety devices for streams in that they provide a place for the release of water the channel cannot hold. This simple fact is sometimes overlooked by overzealous developers of housing tracts.

A river will have only a moderate discharge on most days of the year. Perhaps a few days each year, there may be enough water to fill the channel bank full, but not sufficient to cause flooding. For most streams, actual flooding rarely occurs more frequently than about once every two or three years on the average. Streams have relatively high temporary water storage capacity, and it takes a heavy rainfall or sequence of rains to cause flooding. To plan for the exceptional floods, scientists examine river flow charts

Figure 8–32 Meramec River floods in 1973 inundating a small community on its floodplain. (Courtesy of C. Belt.)

for as many past years as there are records available. Using statistics, it is possible to calculate the probability of future flooding in specific streams. The method will not permit one to predict the precise data of a major flood, but it will indicate the probability of a flood of a certain size within a certain interval of years. For example, when geologists (and insurance adjusters) speak of a "10-year flood," they are referring to a flood having a certain discharge that has a *probability* of occurring once during 10 years. In general, two-year floods are rather modest. Ten-year floods can cause serious damage to buildings and utilities, and 50-year floods are usually major catastrophes. Fifty-year floods are capable of inundating an entire floodplain to the depth of the original stream channel. People living in areas that have just experienced a 50-year flood should not consider themselves safe from floods for the next 49 years. A major flood within a span of 50 years is merely a statistical probability and not a certainty. It would not be impossible for two so-called 50-year floods to occur in two successive years.

Certain engineering practices help to reduce the loss of property and lives from flooding. For example, artificial levees and dikes may be emplaced to help hold the water in its channel. Before such structures are emplaced, however, they should be carefully evaluated, for they may contribute to flooding upstream. Also, channels may be artificially straightened to speed up the flow and move water out quickly before it can spill over the banks. Here, too, there is a danger, for by speeding the flow of water, locations downstream may be flooded. In many flood-prone regions, dams have been constructed in order to store water and allow it to move downstream in a regulated manner. Dams, however, are not always as effective in flood control as we would like them to be. It is true that dams relieve flooding by containing some of the water, but they also collect the sediment that was carried by the river. The water released by the dam lacks a load of sediment and therefore quickly erodes the channel downstream from the dam and acquires a new load. Unfortunately, the discharge is relatively small because of the water retained behind the dam, and so the river drops its load as soon as there is a decrease in velocity. The deposition builds up the channel downstream, raising the level of the bed so that even modest flows may spill over the banks. Someone once remarked, "Nature can be perverse." The validity of that phrase seems evident whenever humans attempt to tamper with a complex natural system like a river.

STREAMS AND LANDSCAPE EVOLUTION

Geologists have long recognized that streams are fundamental to the development of landscapes. To describe the role of streams in landscape evolution, consider what would happen to an initially flat upland area having relatively frequent rainfall. The initial drainage system developed on such a terrain would consist of narrow stream valleys separated by relatively flat stretches of the original surface (Fig. 8–33A). Valley flats would be absent or sparse (B). At this stage, the landscape might be designated as *youthful.* As stream erosion and mass wasting progressed, valley flats would develop, drainage density would increase, divides would be rounded into slopes, and the major streams would attain the graded condition (C). Such a landscape might be termed *mature.* With continuing erosion (and provided there was no uplift), valley flats would be enormously expanded and divides gradually cut back until only a few isolated remnants remained (D). Swamps, lakes, and marshes would appear on the vast poorly drained lowlands of what might be called an *old age* landscape. The uplift of such a surface would very likely initiate renewed erosion and produce youthful topography once again. Thus, the sequence suggests a cycle, but actually it need not be cyclic, since tectonic events are likely to interrupt the orderly change from one kind of landscape to another at any time.

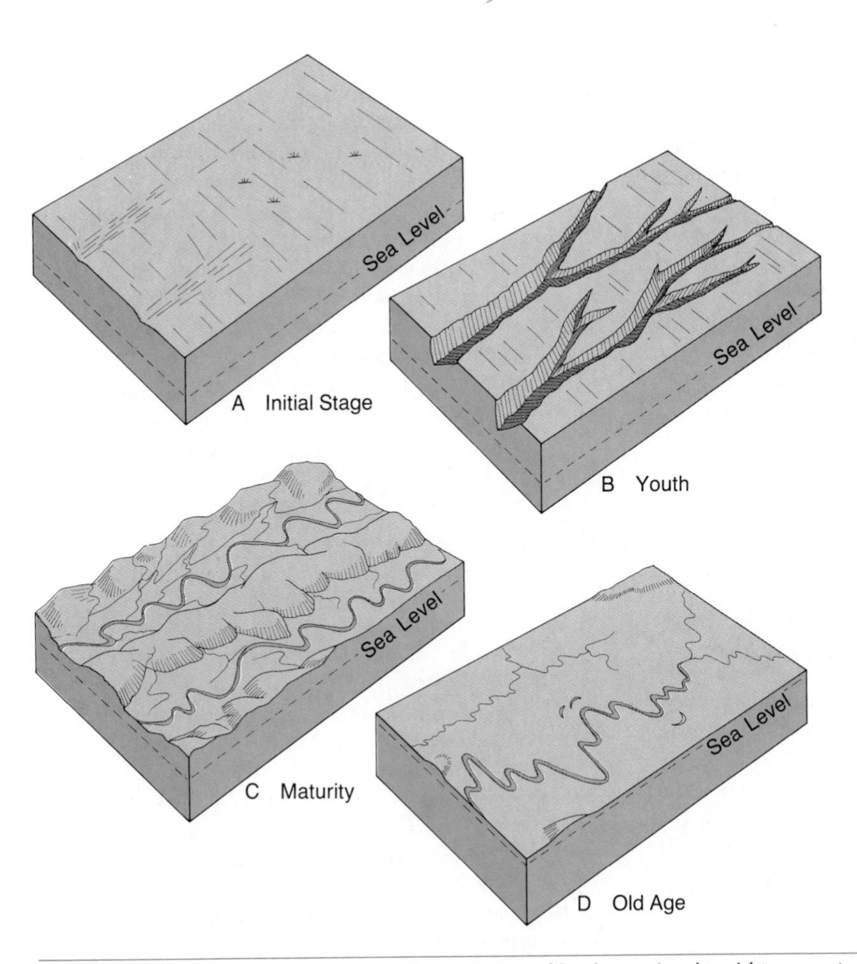

Figure 8–33 *Conceptual scheme for the evolution of landscape in a humid temperate climate.*

Summary

Stream erosion has done more to develop the landscapes of our continents than any other group of geologic processes. Everywhere we see the ravines, valleys, canyons, ridges, and floodplains that result from the activity of streams. There are times when that activity is inspirational, as when streams topple over resistant ledges to form waterfalls. At other times the activity of streams may be catastrophic, as when floods sweep away buildings and undermine bridges. But most of the work of streams goes on largely unnoticed, as the flowing water sculptures valleys and transports the material that once filled them to the sea.

As is the case with mass wasting, gravity is the fundamental force that causes water to flow and is thus indirectly responsible for the work of streams. The component of gravity acting along the slope over which the stream moves is one factor that influences the river's velocity. Velocity in turn influences the amount and nature of the erosional and transportational work of a stream. Velocity multiplied by the stream's cross-sectional area determines stream discharge (the amount of water passing through a given cross-section of the river in a given time).

The material transported by a stream is referred to as its load. Part of the load is dissolved material. Another part, consisting mostly of clay and fine silt, is carried in the body of moving water and is termed the suspended load. The saltation and tractive load consists of larger material rolling, sliding, and skipping along the bottom of the channel. The total load that a stream can carry constitutes its capacity. The companion term competency refers to the maximum-size particle the stream can move under a given set of conditions.

The gradient of a stream is the slope of its channel between any two specified points. It is readily obtained by dividing the horizontal distance between two points (in meters) into the vertical difference in elevation (also in meters) between the same two points. A graded stream is one that has sufficient gradient to transport the material delivered to it by its tributaries, its own bank erosion, and mass wasting. It is a stream in equilibrium, in which any change in load, channel morphology, or discharge will result in changes tending to absorb the effects of those altered conditions.

Streams are responsible for many familiar landscape features. In mountainous regions, for example, one finds alluvial fans developed at places where streams emerge from mountain fronts and deposit their load of silt, sand, and gravel. Deltas may develop where rivers enter oceans or lakes. Waterfalls and rapids are found frequently in relatively narrow valleys, whereas the prominent features of broad valleys are meanders, braided streams, natural levees, and stream terraces. The floodplain is perhaps the most readily apparent feature of most streams. Floodplains are formed primarily by lateral planation and the downstream migration of the stream's bends or meanders. In this migration, the stream acts somewhat like a horizontal saw cutting into its banks on the outside of bends and depositing sediment on the inside of meanders.

QUESTIONS FOR REVIEW

1 Calculate the discharge of a stream having a cross-sectional area of 120 square meters and a velocity of 2 meters per second. What would be the velocity of a stream with the discharge of 100 cubic meters per second and a cross-sectional area of 20 square meters?

2 Why is water velocity near the bed and banks of a stream generally lower than the velocity near the center of the stream?

3 What systematic changes in discharge, velocity, gradient, and channel shape occur in a river from its headwaters to its mouth?

4 How does a stream widen its valley?

5 Why is it that after a dam is constructed and the reservoir behind the dam is filled, the water that flows over the dam is clear, and this clear water tends to deepen the stream channel downstream from the dam?

6 What is the general configuration of the "longitudinal profile" of a stream? What is the cause of this configuration?

7 What geological inference might one make on observing a stream that has incised meanders? For a stream that cuts across the structural "grain" of a region (that is, across ridges of resistant rock)?

8 What type of drainage pattern usually develops on folded rocks? On horizontal rocks? What conditions are likely to result in a radial stream pattern?

9 What is a water gap? How might a water gap be converted into a wind gap?

10 What is the explanation for the fact that many streams continue to flow even though there has been no precipitation in their drainage basins for weeks or months?

SUPPLEMENTAL READINGS AND REFERENCES

Leopold, L.B., and Langbein, W.B., 1960. *A Primer on Water.* Washington, D.C., U.S. Department of the Interior, U.S. Government Printing Office.

Leopold, L.B., Wolman, M.G., and Miller, J.P., 1964. *Fluvial Processes in Geomorphology.* San Francisco, W.H. Freeman.

Morisawa, Marie, 1968. *Streams: Their Dynamics and Morphology.* New York, McGraw-Hill Book Co.

Ritter, D.F., 1974. *Process Geomorphology.* Dubuque, Iowa, Wm. C. Brown.

Schumm, S.A. (ed.), 1974. *River Morphology.* Stroudsburg, Pa., Dowden, Hutchinson & Ross.

Vitaliano, Dorothy B., 1973. *Legends of the Earth.* Bloomington, Indiana University Press.

Figure 8–34 *Kakabeka Falls in the Lake Superior region of Ontario, Canada. (Courtesy of National Film Board of Canada.)*

9 underground water

And Isaac's servants digged in the valley, and found there a well of springing water.
GENESIS 26:19

Prelude Of all the water that falls as rain or is derived from melting snow, about one third runs off at the surface as sheetwash and in streams. A similar amount evaporates or is used by plants, and the remainder infiltrates into the soil and rocks beneath, percolating downward until there are no longer any air-filled openings and then rising in level toward the surface as additional water is added. This underground water that nearly fills all openings in rock or sediment is called **groundwater** by geologists. It has been estimated that groundwater supplies about one fifth of the total water needs of the United States.

Several factors influence the amount of water that will become part of the groundwater reservoir for a particular region. The more important of these factors are the amount of precipitation, the volume of openings in the subsurface materials, the slope of the land surface, and the nature of the vegetative cover at the surface. Once the water has percolated into the subsurface, it is capable of movement toward such outlets as wells and springs, can accomplish certain kinds of geologic work, and can contribute to the hot waters and vapors emitted by geysers and hot springs. In this chapter, we will examine the movements and characteristics of groundwater.

Calcium carbonate, deposited from groundwater dripping from ceilings and walls of the "Big Room" in Carlsbad Caverns, New Mexico, has formed the dripstone features. (Courtesy of U.S. National Park Service.)

ROCK PROPERTIES AFFECTING GROUNDWATER SUPPLIES

Aquifers and aquicludes

A body of rock or sediment that can yield sufficient water for domestic or industrial use is termed an **aquifer.** Examples of aquifers include unconsolidated sands and gravels, sandstone strata, partly dissolved or cavernous limestones, and nearly any other rock type that is highly fractured (Fig. 9-1). Geologists employ the term **aquiclude** for rocks that do not permit passage of significant quantities of groundwater. By far the most common aquiclude is clay. Clay-rich rocks like shale may actually contain considerable amounts of water, but that water cannot flow freely through the microscopic pores between clay particles.

The quantity of water that can be obtained from wells drilled into aquifers depends in part upon the amount and characteristics of the void spaces in the water-bearing rock, as well as the manner in which these voids are interconnected. The hydrologic terms used to describe these two attributes of aquifers are porosity and permeability.

Porosity

Porosity is defined as the total volume of open spaces, pores, or voids in a rock or sediment. The spaces may be fractures, cavities dissolved in rock by water, or simply openings fortuitously formed between grains as they were deposited. In a clastic sediment or sedimentary rock, porosity will be markedly influenced by sorting. As noted in Chapter 3, sorting is a measure of the uniformity of the sizes of particles in sediment or rock. In well-sorted sandstones, for example, the grains are mostly about the same size or fall within the narrow range of sand-size particles. There are fewer of the tiniest clay and silt-sized particles to fill the spaces between the larger grains (Fig. 9-2). Poorly sorted sandstones contain a complete range of particles from clay-size to sand-size. Therefore, what might have been pore space in a rock having better sorting is occupied by fine particles in a poorly sorted rock. Therefore, poor sorting generally correlates with relatively lower porosity.

Another factor that influences the porosity of an aquifer is the manner in which the grains are packed together (Fig. 9-3). If that packing is relatively open, there will be a greater volume of space available for

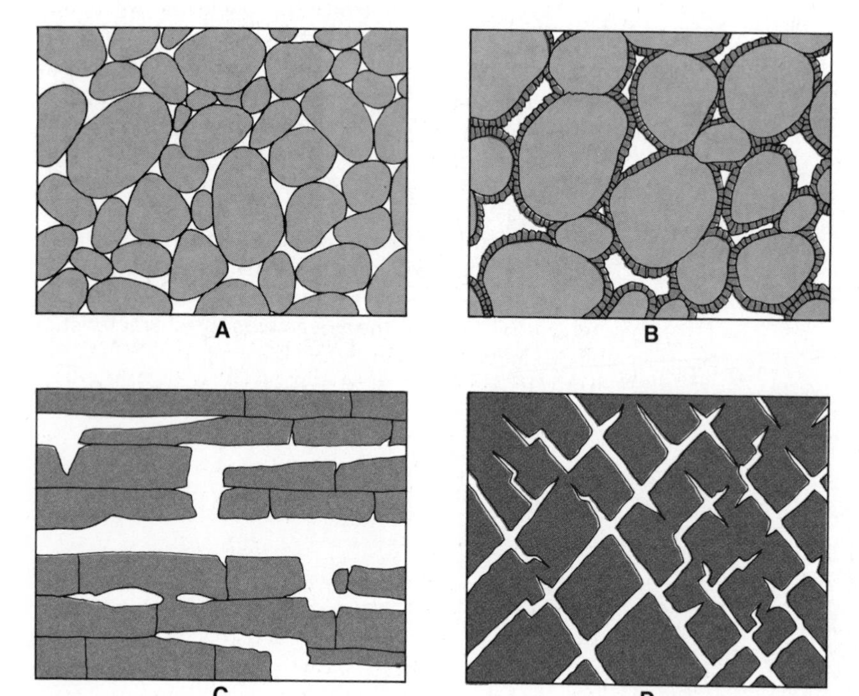

Figure 9-1 Differing modes of porosity. (A) Well-sorted sand having good porosity. (B) Porosity reduced by cement in sand converted to sandstone. (C) Rock rendered porous by dissolution. (D) Rock made porous by fracturing.

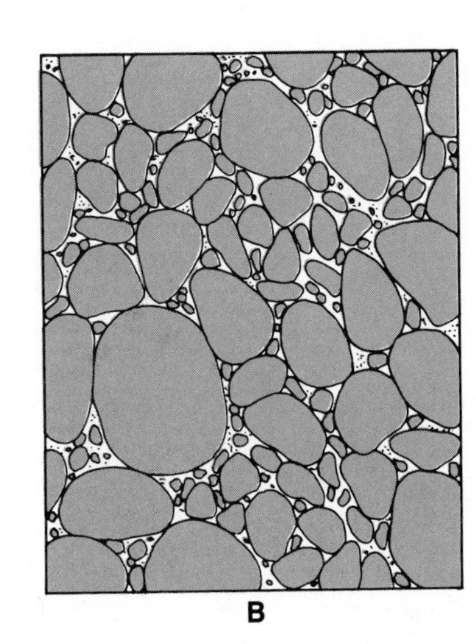

Figure 9–2 Because smaller particles occupy potential pore spaces, well-sorted sand (A) would generally have higher porosity than a sand having poor sorting (B).

water storage. The shape of grains and the presence of cementing materials like calcium carbonate or silica can also influence porosity. One factor that does not directly affect porosity is grain size. Given an aggregate of large, spherical grains packed in open arrangement, and a second aggregate of smaller but similarly shaped and packed grains, both would have identical porosities. A simple calculation shows this relation is valid. The volume of a spherical grain is

$$V_{sg} = \frac{\pi}{6} d^3$$

where V_{sg} is the volume of the spherical grain and d the diameter. Assume that one has arranged eight spherical grains as shown in Figure 9–4. The edge of the cube that circumscribes the eight spherical grains is twice the diameter of a single grain, and the volume of the cube is $(2d)^3$, or $8d^3$. The ratio of the volume of the eight uniform spherical grains to the volume of the circumscribed cube, V_c, is

$$\frac{V_s}{V_c} = \frac{8\frac{\pi}{6}d^3}{8d^3} = \frac{\pi}{6} \text{ or } \frac{3.14}{6} \text{ or } 0.52$$

Thus, the solid spherical grains occupy 52 percent of the volume of the circumscribed cube, and void spaces occupy the remaining 48 percent. Reducing the grain size will not diminish the porosity because grain size cancels out. If, however, the grains were packed more closely, porosity might be reduced to a value of 26 percent.

A rock, of course, is not composed of perfectly spherical grains arranged in the most open manner, and only rarely are sedimentary rocks found with porosity as high as 48 percent. Porosities from 5 to 15 percent are considered average in rocks, although higher porosities occur in unconsolidated sediments.

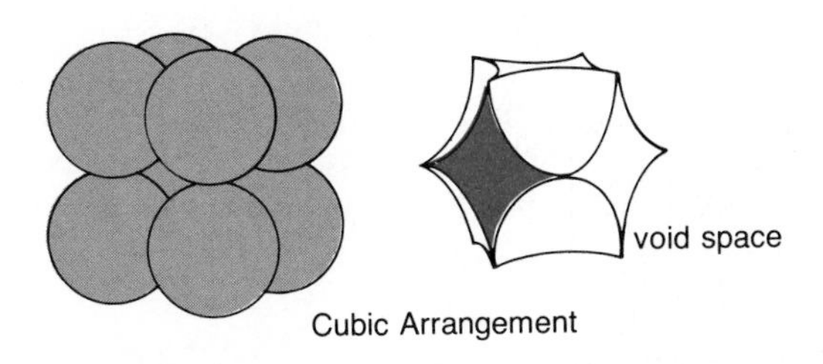

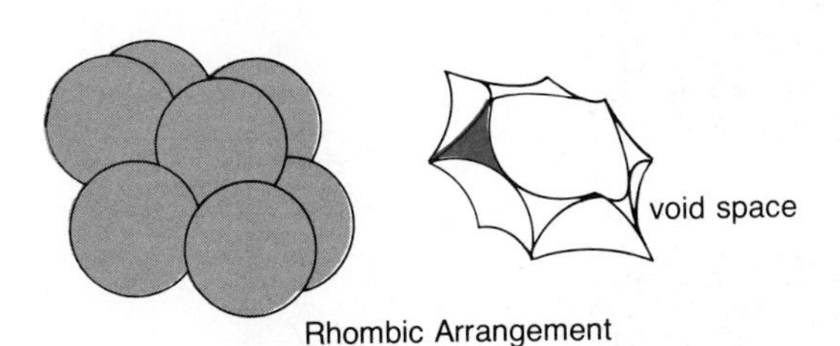

Figure 9–3 Illustration of the effect of packing on void space. The cubic arrangement of grains is the most open, and provides the greatest porosity. In addition, the minimum cross-sectional area of the pore (shown in orange) is greater in cubic packing. In natural materials like sand, the grains would not be perfect spheres, they would not be of identical size, nor would they be so precisely arranged. (From Graton, L.C., and Fraser, H.J., Systematic Packing of Spheres with Particular Relation to Porosity and Permeability, *Jour. Geol., 43:8, 1935.)*

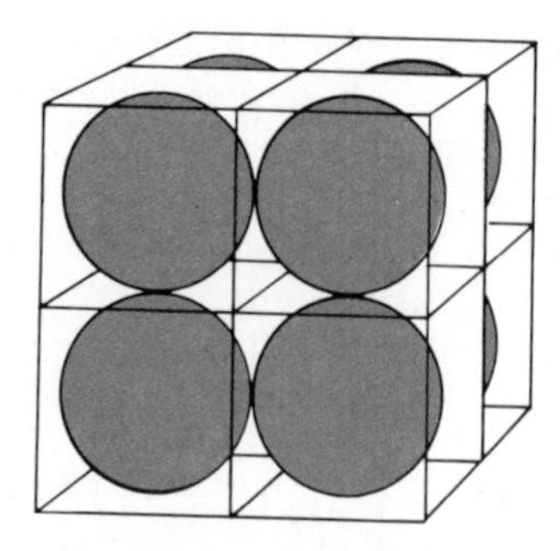

Figure 9–4 Each of the eight spheres packed in cubic arrangement occupies 52 per cent of the volume of the cube in which it rests. Porosity is therefore 48 percent. Because of variation in particle size, packing, and particle shape, natural materials will not attain so great a porosity.

Permeability

Permeability is a term used by geologists to express the relative ease with which water moves through the spaces or interstices in a rock or sediment. These spaces must be interconnected if a rock is to have permeability. It is possible to measure permeability by subjecting a piece of rock to a given water pressure and recording the amount of water that passes through the rock in a measured interval of time. When this is done for materials having a range of grain sizes, it is apparent that the flow rate is greatest in coarser materials. Unlike porosity, permeability increases as the grain size increases. The reason for this is that the size of voids or interstices also increases in a sediment composed of larger grains. Friction along the surfaces of grains retards flow, and there is far more total surface area to cause friction in fine-grained sediment than there is in coarse-grained material. In fine silts and clays the film of water adhering to each particle is in continuous contact with the film of adjacent grains, and there is less low-friction "throat area" for the free passage of fluid (Fig. 9–5). Rocks composed of clay may have considerably more porosity than those composed of sand, but because of their smaller interstices and the consequently higher frictional resistance to flow, clayey materials have much lower permeability than sands. Unconsolidated and well-sorted gravels, as well as cavernous limestones, are among the most permeable of earth materials.

THE OCCURRENCE OF WATER UNDERGROUND

Underground water zones

Unless trapped above impermeable layers of rock, meteoric water will percolate steadily downward into the earth under the pull of gravity. Eventually, however, the water will reach a depth at which the great pressures of overburden have closed most of the open

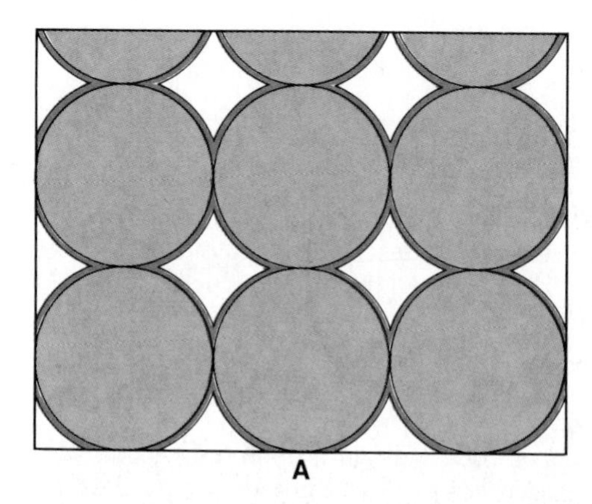

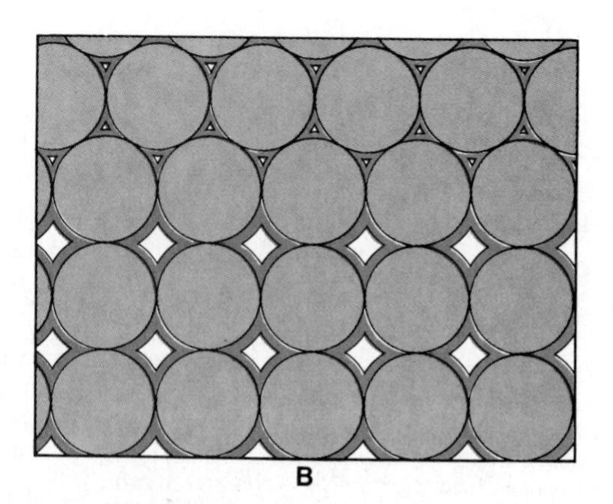

Figure 9–5 Conceptual example of the effect of grain size on permeability. Rocks composed of large grains (A) are likely also to have large openings between grains and less total surface area to cause frictional drag on fluids. In finer-grained material (B) the low-friction throat areas are smaller and surface area relatively greater than in coarser material.

Figure 9–6 A simple experiment that indicates the manner in which the zone of saturation *and* water table *develop.*

spaces and the rock has become impervious. Continued additions of water will then fill all pore spaces above the impervious zone, and the surface of the saturated zone will rise toward ground level as more water is added. The situation resembles what would occur if a tub were filled with gravel and then a shower of water added (Fig. 9–6). The water would percolate down to the bottom of the tub and then gradually rise as it filled up the spaces between the gravel. If the shower of water were stopped before the tub was filled, one could reach down through the gravel and find the level of the free water surface. In nature, that surface could be located by digging a hole into the ground until one reached a depth where water began to flow into the hole. After a period of time, the surface of the water in the well would stand at the level of the free water surface in the rock. The subsurface zone below that level is appropriately called the **zone of saturation** (Fig. 9–7). Above the zone of saturation is the vadose zone, or **zone of aeration,** so called because some of the pores are filled with air as the meteoric water infiltrates downward. If additional wells are dug nearby and the elevation of the free water surface is recorded in each well, then it would be possible to use that information to map the underground plane at the top of the zone of saturation. This upper surface of the saturated zone is called the **water table.**

Just above the water table, moisture may be pulled up into the vadose zone as a result of capillary attraction. This irregular zone of water is called the **capillary fringe.** In general, the capillary fringe is thicker in finer-grained materials than coarse clastic sediment. One can observe capillary attraction when a glass tube is inserted into a glass of water. Water will appear to oppose gravity and move upward in the tube (Fig. 9–8). Capillary attraction is the result of surface tension, or the tendency of a water surface to act like a stretched rubber membrane. Surface tension is caused by the mutual attraction of water molecules for one another. In a tubelike crevice or pore, water molecules

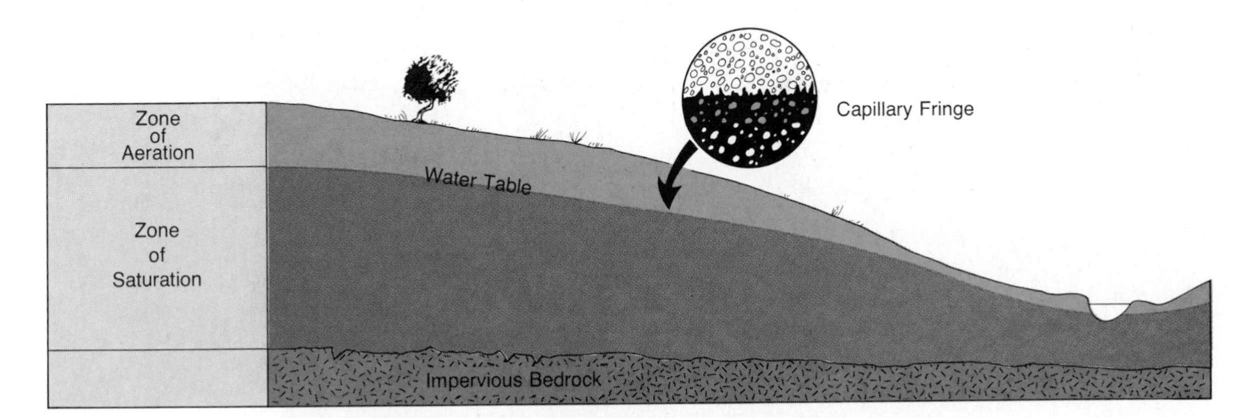

Figure 9–7 The water table in a humid region underlain by homogeneous sediment with impervious bedrock at depth. All openings below the water table are completely filled with water. In dry periods, the water table may drop, whereas in wet periods it is likely to rise.

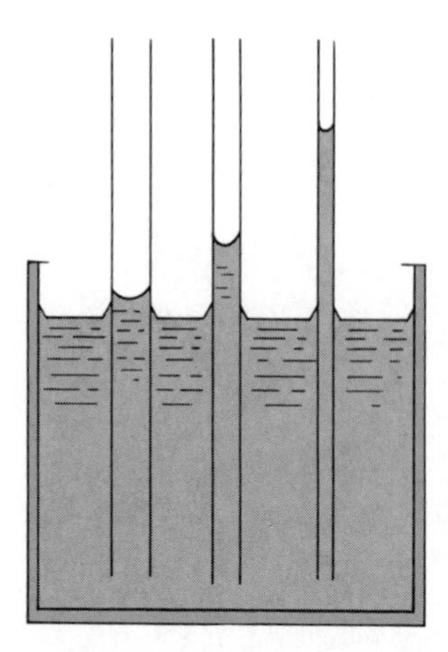

Figure 9–8 Capillarity can be demonstrated by suspending glass tubes in water. Water rises in the glass tube because the concave surface layer of molecules, in contracting, draws the water upward with it by molecular attraction. The height of rise varies inversely as the diameter of the tube. It is for this reason that the capillary fringe is thicker in fine sand than in a coarse gravel.

are attracted laterally to the minerals adjacent to the pore and downward to other molecules, so that a membrane-like concave surface is produced. The attempt by surface tension to flatten the concavity tends to draw the water up the tube. As soon as the weight of the raised column of water equals the force resulting from surface tension, water will cease to rise in the tube.

The configuration of the water table

The water table is not a static surface, but will vary in elevation and slope according to the configuration of the overlying topography, the permeability of the rocks and sediments, and the amount and frequency of rainfall. If a hilly terrain is underlain by homogeneous pervious earth materials, and if rainfall is relatively frequent, the configuration of the water table will roughly approximate the surface contours of hills and valleys (Fig. 9–9). In the event rain would fall continuously for a time, the zone of saturation would gradually rise toward the surface of the ground. On the other hand, if there were no rain for a long time, the water table would slowly flatten as the water in the saturated zone percolated downward and flowed laterally toward nearby stream valleys.

In humid regions, the time between periods of rainfall is usually not long enough to allow the water table to flatten appreciably. Also, water falling as rain over an entire hilly area does not infiltrate rapidly and rush to adjacent stream channels. In most aquifers, groundwater can only move slowly through the tortuous interstitial spaces, and before it can reach an outlet more rain falls, causing a "pile up" of water in the divides farthest from the streams. It is for this reason that the water table rises under hills and descends beneath valleys so as to take the form of a

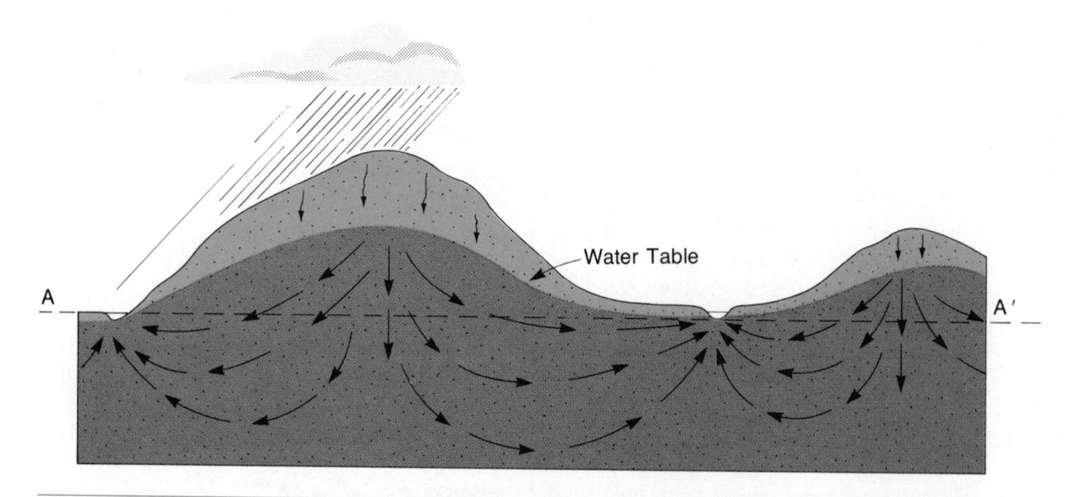

Figure 9–9 In a hilly area underlain by uniformly permeable material, the configuration of the water table will be a subdued reflection of surface topography, provided rainfall is frequent. The flow of groundwater through such permeable homogeneous material is indicated by the arrows.

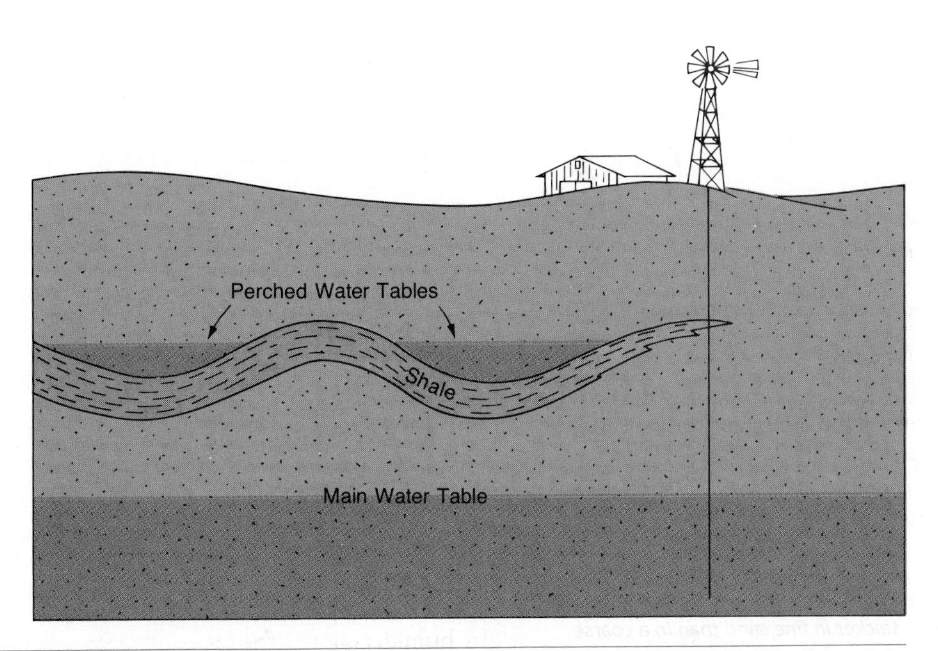

Figure 9–10 Perched water tables developed above a folded and pinched-out shale stratum.

subdued reflection of surface topography. It is a "subdued" reflection because the vertical distance from ground surface to the water table is greater along the divides than along the valleys.

Although most areas are underlain by a body of groundwater having only one water table, there are also some places where impermeable beds occur at a shallow depth below the surface and prevent the further downward percolation of water toward the main water table. Water may collect on top of the impermeable strata to produce a local secondary water table above the main one. Such features are appropriately called **perched water tables** (Fig. 9–10). The quantity of water available from the limited saturated zone beneath a perched water table may be considerably less than beneath the main water table. When large quantities of groundwater are needed, it is usually advisable to drill wells through the perched water table and penetrate the main water table.

GROUNDWATER MOVEMENT

Patterns of flow

The energy required for the movement of water molecules through rock or sediment is supplied by gravity. Water responds to the pull of gravity by attempting to seek its own level. This is why groundwater percolates from higher parts of the water table to lower parts and often out onto the surface as springs. Part of the groundwater beneath divides flows directly down the slope of the water table. An even greater part, however, moves slowly toward a nearby stream along curved paths of flow (see Fig. 9–9). Along these paths, water may be carried deep into bedrock before ascending toward the stream channel. These flow patterns replenish and replace the deeper realms of groundwater and prevent their stagnation. Water moving in this manner will dissolve and transport soluble matter and provide cements for formerly unconsolidated sediments.

The curved flow paths for groundwater are apparently the result of the fluid attempting to move toward points of least pressure. Along any horizontal line, such as A–A′ in Figure 9–9, the water is under greater pressure beneath a divide than beneath a stream, and so it moves toward the stream. The flow patterns of groundwater have been duplicated in laboratory models in which dye has been injected at various points so as to trace the path of water molecules. These experiments indicate that, in general, the rate of movement of water molecules along shallow paths is greater than in the deeper arcs.

Rates of flow

Two factors that influence the rate at which water in the zone of saturation will flow are the permeability of the aquifer and the slope of the water table. **Hydraulic gradient** is the term used to express the slope of the water table. Rather like the gradient of a stream, the hydraulic gradient is equal to the vertical distance between two points on the water table (h) divided by the horizontal distance between these two points (ℓ). The vertical measurement is termed the **hydraulic head.** One can quickly determine the hydraulic head by subtracting the elevation of the free-standing water surface in valley wells from similar measurements obtained in wells dug high on divides. For example, if water stood at an elevation of 550 meters above sea level in an upland well, and at 500 meters above sea level in a nearby valley well, then the hydraulic head (h) would be 50 meters. If the wells were 1000 meters apart, then the hydraulic gradient would be 50 meters divided by 1000 meters, or 0.05 (Fig. 9–11).

The relation among the rate of groundwater flow, hydraulic gradient, and permeability is expressed in the following equation formulated in 1865 by Henry Darcy, an engineer for the city of Dijon, France:

$$V = P\frac{h}{\ell}$$

In Darcy's equation V refers to the velocity of groundwater movement, P is a coefficient of permeability that depends upon the characteristics of the rock through which the water is moving, and $\frac{h}{\ell}$ is the hydraulic gradient. This famous equation has become known as Darcy's Law. For material of a certain permeability, Darcy's Law tells us that, as the slope of the water table (the hydraulic gradient) increases, so also will the velocity of groundwater movement. Applied to the pattern of groundwater movement beneath any tract of land, Darcy's Law indicates that the direction of maximum velocity of groundwater flow will be in the direction of the steepest slope of the water table; that is, in the direction of the maximum hydraulic gradient.

In most aquifers, the velocity of groundwater flow rarely exceeds 1 or 2 meters per day, although the range of velocities recorded has included rates as fast as 100 meters per day and as slow as only a few centimeters per year.

WELLS AND SPRINGS

It is always a delight and somehow surprising to see water flowing out from a ledge of rock or spouting from a pipe sunk deep into the earth. These are, however, rather ordinary occurrences of groundwater emerging at the earth's surface as springs and wells. In general, wells and springs are classified into two groups. Those that are fed by water flowing through unconfined aquifers are called ordinary wells and springs. If, however, the groundwater is confined to an aquifer by impermeable rock layers, the groundwater system is called artesian.

Ordinary wells

As indicated earlier, if a hole is dug to a depth below the water table and encounters permeable materials at

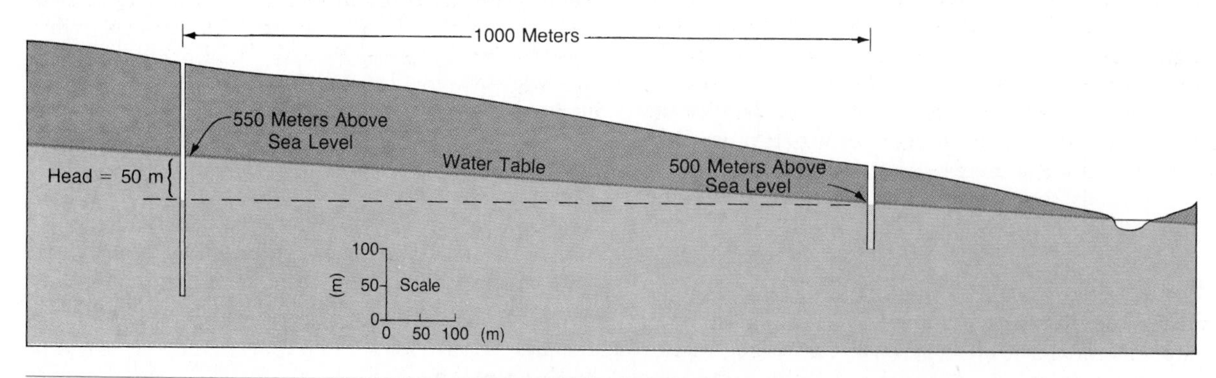

Figure 9–11 The hydraulic gradient between two points on the water table 1000 meters apart and having a 50-meter difference in elevation would be 50/1000 or 0.05.

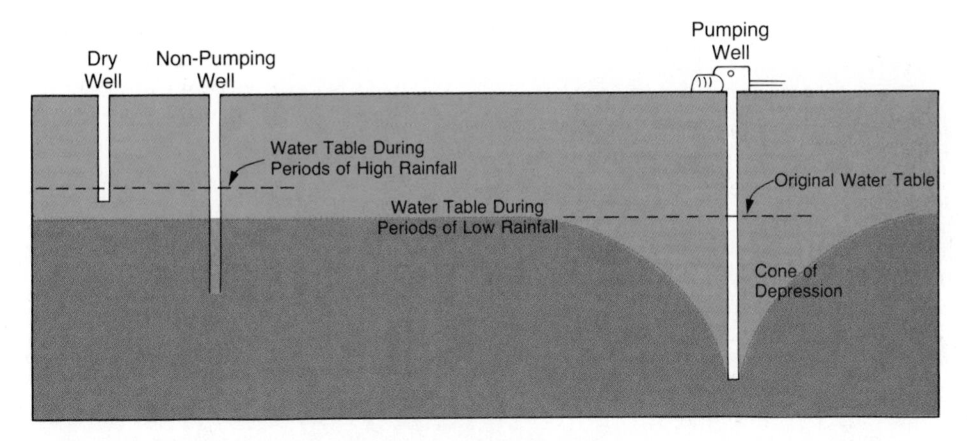

Figure 9–12 *Because the water table may subside during intervals of drought, wells should penetrate deeply into the zone of saturation. Note the basin-like effect or cone of depression developed in the water table around the pumping well.*

that depth, groundwater will flow into the hole.The hole has become a water well. The presence of permeable sediment or rock is important, for without good permeability water will not flow into the hole rapidly enough to replace that which is withdrawn. Also important is the necessity of digging a well deep enough. The water table rises during periods of high rainfall and falls during the drier season (or drought years). Clearly, the well should penetrate considerably below the elevation of the water table at its lowest predicted elevation. There is another reason for drilling a well to a depth appreciably below the water table. When water is pumped from a well, a conical depression in the water table forms in the area surrounding the bore hole. This **cone of depression** (Fig. 9–12) may eventually reach the intake level of the well, causing it to go dry. Much to the consternation of neighboring users of groundwater, the decline in the water table around large cones of depression may also cause nearby wells to go dry. Exceptionally wide and deep cones of depression develop around wells that supply the large quantities of water needed for irrigation and industry.

Ordinary or gravity springs

A spring is a natural flow of groundwater that has emerged onto the surface of the earth. The quantity of water emerging in this way may be only a trickle, or it may be a gushing flow that may supply water for a stream. As indicated in Figure 9–13, ordinary or gravity springs may develop at points where the water table intersects the ground surface along the sides of a valley. Springs may also be localized by impermeable strata, bedding planes, fractures, faults, or coverings of debris from slumping or landslides. Some springs are associated with perched water tables, and emerge above impermeable layers of rock exposed along hillsides.

Artesian wells and springs

In some wells, groundwater does not merely fill the borehole to the level of the water table but rises above that level and may even flow out at the surface. These wells, in which water rises under pressure above the level of the water table, tap confined aquifers and have been given the name **artesian** after the Old Roman province of Artesium in France. An artesian system (Fig. 9–14) is the result of the simultaneous occurrence of three geologic conditions. First, there must be an inclined aquifer. One or more dipping layers of sandstone or other coarse sediment, layers of gravel in old alluvial fans, or jointed limestones may be suitable inclined aquifers. Second, the aquifer must be overlain by an impervious material (an aquiclude), which serves to prevent the escape of water from the aquifer to the surface. As the third condition, the elevation of the intake area must be sufficiently high that enough hydraulic pressure will be developed to force the water in wells upward above the level of the aquifer. The pressure that forces the water to rise in the well is caused mainly by the weight of the water coming down

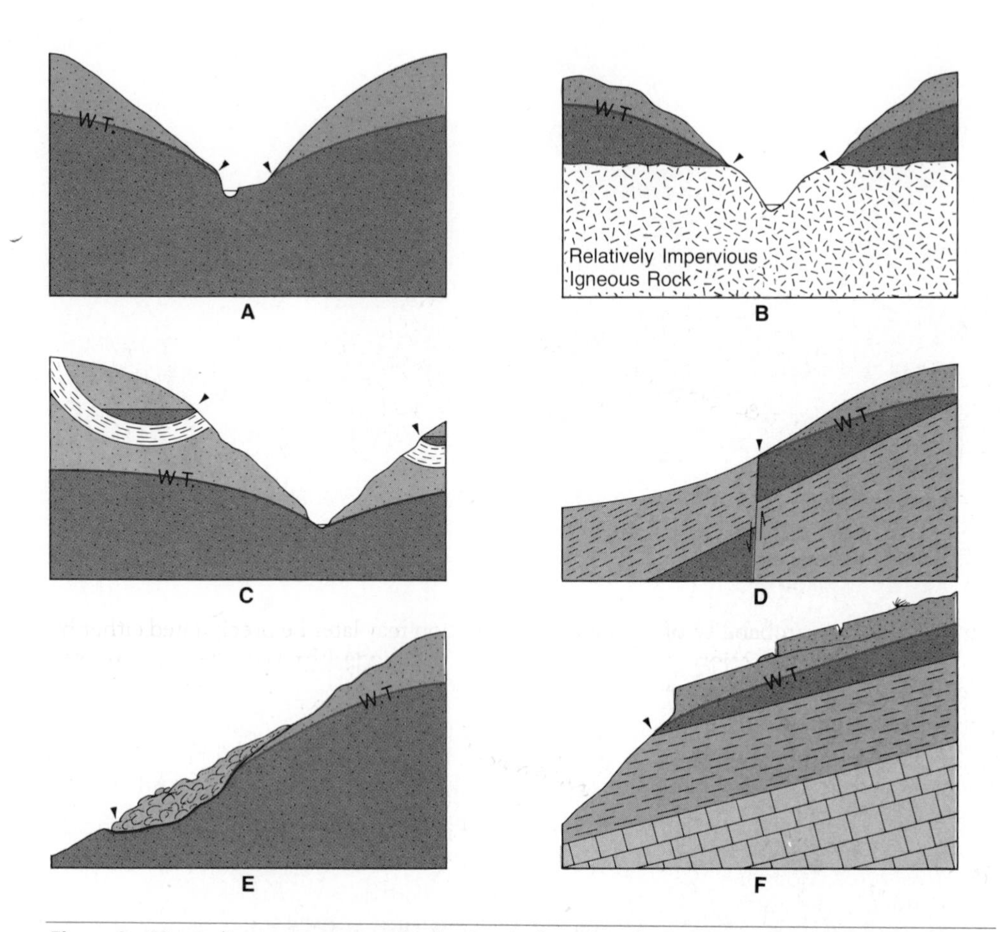

Figure 9–13 Ordinary springs may develop in a variety of geologic settings. (A) Springs developed at intersection of water table and valley slopes. (B) Springs formed at contact between impervious crystalline rocks and relatively permeable overlying sediment. (C) Perched springs. (D) Spring localized by faulting. (E) Spring developed at toe of a landslide. (F) Spring formed at contact between permeable sandstone stratum and impermeable shale stratum.

toward the well site from the higher collecting area. If the artesian system were free-flowing, then water would gush upward at the well site to the level of the water table in the upland intake area. This does not occur in natural artesian systems, however, because of the frictional loss of energy as the water percolates through the aquifer.

Perhaps the best known artesian system in the United States is that developed in the plains surrounding the Black Hills of South Dakota. Water enters the Dakota Sandstone where the formation is exposed in the outcrops that encircle the Black Hills. It then migrates through the sandstone and laterally outward beneath the plains. Overlying shale strata provide the impervious cover. Half a century ago, hydraulic pressure in this confined system was sufficient to cause water from wells to gush a few meters into the air and even turn waterwheels. There are now so many wells, however, that pumps must be employed on many wells that once flowed freely.

Where to dig a well

Drilling for water on a tract of land can be a risky business, and whether an abundant, barely sufficient, or insufficient supply of water will be obtained is cause for considerable anxiety on the part of a landowner. There are, however, reasonable and scientific methods

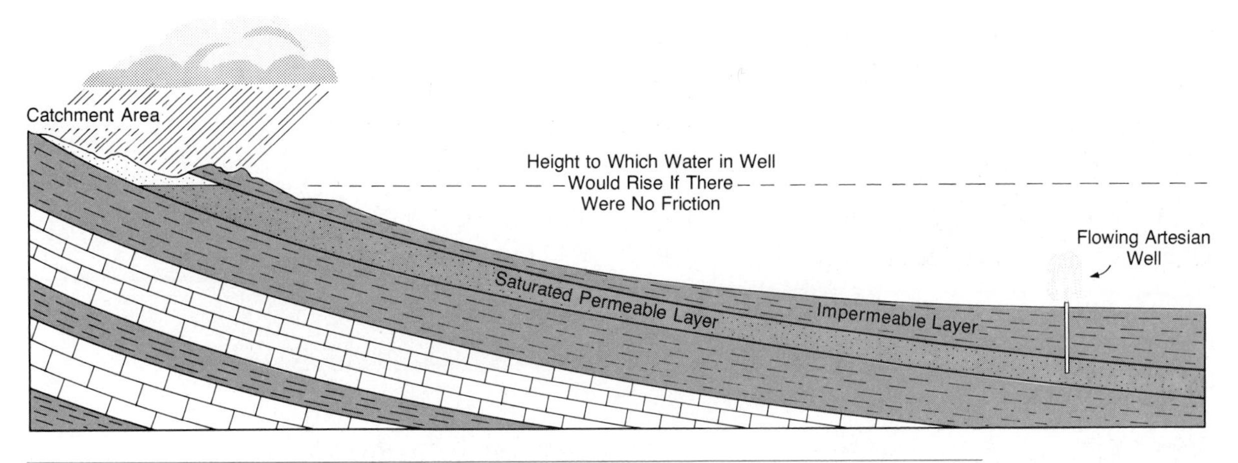

Figure 9–14 Typical conditions for the occurrence of an artesian well.

useful in finding a site having the best probability of yielding underground water. From the discussion of the water table, it is evident that one would be more likely to encounter water at shallower depths by drilling wells along the lower slopes and valley flats of a particular tract of land. Government geologists are usually helpful in providing groundwater maps and information about wells drilled in neighboring sites. This backlog of information can be useful in predicting the depth of known aquifers and in delimiting the vertical drop in the groundwater table during the drier seasons of the year. Geological reconnaissance can also provide information about the juxtaposition of potential aquifers and aquicludes, and their depth and dip.

GROUNDWATER IN SOLUBLE ROCKS

Except in those areas where water flows rapidly through caverns or lava tunnels, groundwater moves too slowly to cause appreciable mechanical erosion or to transport coarse particles. On a less perceptible scale, however, groundwater does accomplish geologic work by dissolving minerals, transporting the soluble matter, and precipitating that material as cements in granular sediments and as depositional features in caverns and other underground openings.

As described in the discussion of chemical weathering (Chapter 7), limestones, dolostones, marbles, and evaporites are particularly susceptible to solution by water that contains carbon dioxide. Limestone taken into solution may later be precipitated either by removing carbon dioxide (through agitation, photosynthesis or heating) or by evaporating part of the water. Evaporation would increase the concentration of dissolved matter to the point where there would be more in solution than the diminished volume of water could retain. The "surplus" calcium carbonate would come out of solution and be deposited. The basic equation for the relation among water, carbon dioxide, and limestone is as follows:

$$\underset{\text{calcium carbonate}}{CaCO_3} + \underset{\text{water}}{H_2O} + \underset{\text{carbon dioxide}}{CO_2} \rightleftharpoons \underset{\text{calcium bicarbonate}}{Ca(HCO_3)_2}$$

Notice that the addition of carbon dioxide to the left side of the equation enhances solution. A loss of carbon dioxide from the bicarbonate on the right side results in precipitation of calcium carbonate.

Solution of limestones occurs at the most rapid rate near the surface. Here, recently fallen rainwater still retains the carbon dioxide absorbed from the atmosphere, and there are further additions of the potent gas from organic materials in soil. Also, the water itself is not yet saturated with lime and is capable of further dissolution. The water flows along bedding planes, joints, and fractures, and begins to widen them by solution. Water may collect in low places on the ground and dissolve enough limestone to form depressions called **sinkholes** (Fig. 9–15). Sinkholes may also form where the ceiling of large solution cavities or parts of caverns collapse. Often sinkholes are connected to subterranean solution channels so that

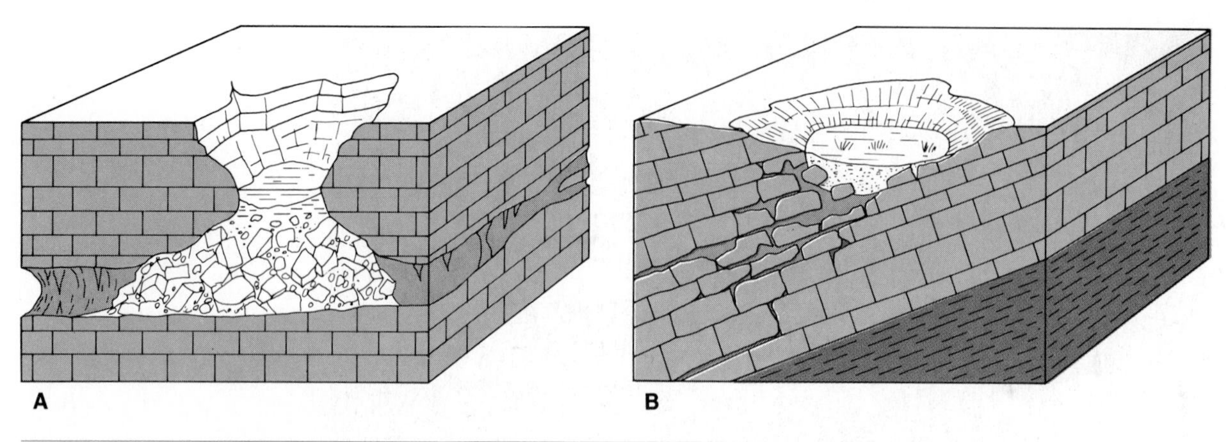

Figure 9–15 *(A) A collapse sinkhole resulting from failure of the roof of a cavern. The sinkhole has been plugged with clay and rock debris, so that water accumulates within the depression. Solution sinkholes (B) are due to surface dissolution around some favorable point like a joint intersection or topographic depression.*

water is drained from them. Many sinkholes, however, become plugged with clay and debris, so that they form lakes and ponds. Although some sinkholes are as much as 1000 meters in diameter, on the average they are less than about 30 meters across. The progressive collapse of the roof of a cavern may produce a linear series of sinkholes, and these may coalesce to form a steep-walled trough called a **solution valley.** Solution valleys are recognized by their peculiar right-angle bends, which tend to follow the original joint pattern in the limestone bedrock. Here and there along a solution valley, the stream may pass beneath a segment of the original cavern roof that has not yet collapsed, and thus form a natural bridge (Fig. 9–16). In other places, a considerable length of uncollapsed cavern may receive the flow of the surface stream. Such so-called disappearing streams may emerge elsewhere.

Caverns are, of course, the most fascinating of all solution features (Fig. 9–17). Each year thousands of people visit caverns to observe the rock formations

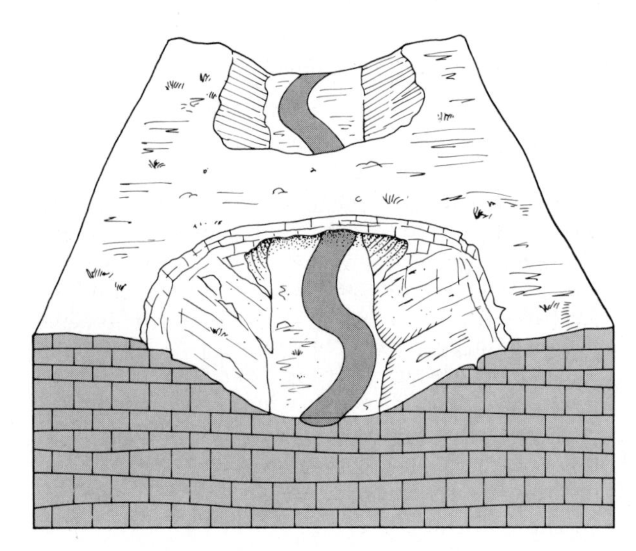

Figure 9–16 *Collapse of a significant portion of the roof of a cavern has produced this* natural bridge *and* solution valley.

Figure 9–17 Large symmetrical column (A); cluster of stalagmites (B); and stalactites (C) in Carlsbad Cavern. (Courtesy of U.S. National Park Service.)

and examine organisms strangely adapted to life in perpetual darkness. The development of caverns is almost inevitable in humid areas underlain by thick and extensive limestones. Caverns may be large or small, may have a simple or complex labyrinthic plan, may extend vertically as well as horizontally, and may have passages developed along multiple levels. Subterranean streams course through some caverns, whereas in other caverns streams are lacking. The rectilinear pattern of passages in some caverns indicates they developed along joint systems (Fig. 9–18), but not all caverns have such a relationship to joints.

The depositional features of caverns are well known. There are the downward-extending, icicle-like **stalactites,** as well as **stalagmites,** which grow in conical form upward from the floor of the cavern. The construction of a stalactite begins with a drop of lime-saturated water temporarily adhering to the cavern roof (Fig. 9–19). With the loss of CO_2, calcium carbonate comes out of solution to form a thin shell around the drop. As more water flows down through the center and along the sides of the original shell, the stalactite grows by successive additions of calcium carbonate. Stalagmites are formed by drops of water that

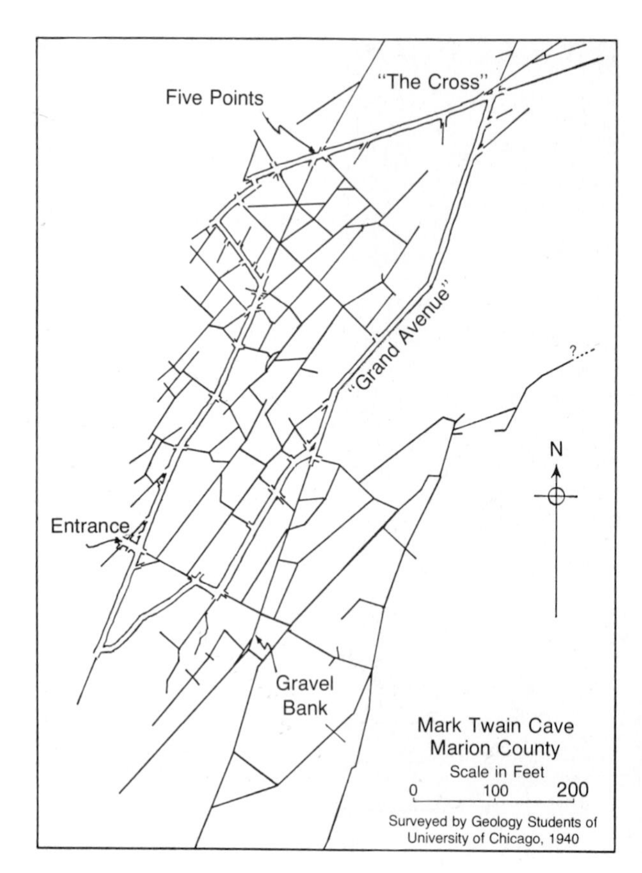

Figure 9–18 *Mark Twain Cave. This cave, made famous because of its description in the novel* Tom Sawyer, *is located in Hannibal, Missouri. It is easy to see how Tom Sawyer and Becky Thatcher may have become lost in the labyrinthine passages, the locations of which are determined by three or more intersecting sets of joints. (From Bretz, J.H.,* Caves of Missouri, *Missouri Geological Survey publication, vol. 39, 2nd sec., 1956.)*

Figure 9–19 *Stalactites in an early stage of development in Castleguard Cavern, Castleguard, Alberta, Canada. More precisely, these stalactites with twiglike lateral projections are called helictites. (From the film* Castleguard Caves, *Courtesy of the National Film Board of Canada.)*

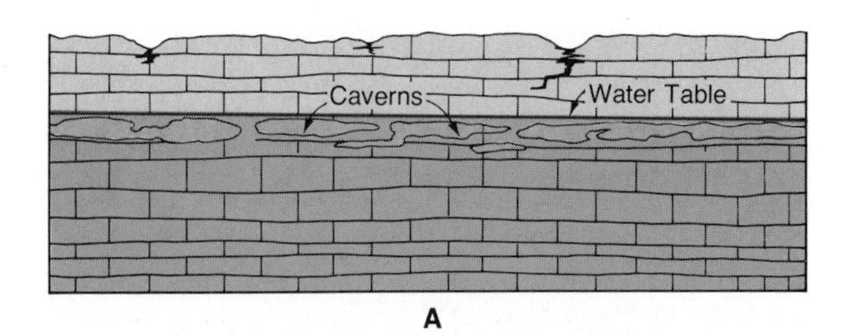

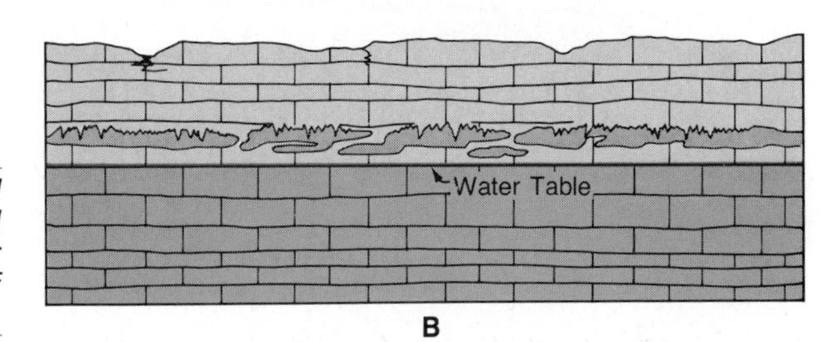

Figure 9–20 Caverns formed below the water table and completely filled with water (A) may subsequently be drained as the water table is lowered. Relative lowering of the water table may result from regional uplift, an extended period of diminished precipitation, or downcutting by local streams.

fall from the ceiling of the cavern and strike the floor, where both agitation and evaporation contribute to the buildup of calcium carbonate. Columns (see Fig. 9–17) develop when stalactites and stalagmites eventually grow together. Features such as stalactites, stalagmites, and columns are composed of a rather dense, banded variety of limestone called **cave travertine.**

Aside from their esthetic value, features like stalactites help geologists unravel the history of cavern development. Clearly, such features must have formed when the cavern was above the water table and filled with air. Yet, in some caverns, the walls and ceilings are lined with crystals that could only have developed beneath water, and hence beneath the water table. These observations indicate that the formation of most caverns began in the zone of saturation, but that the cavernous areas were subsequently drained as a result of a lowering of the water table (Fig. 9–20).

Karst topography

Karst is a term applied to topography formed in regions of limestone or dolomite bedrock by the vigorous solution work of groundwater. One recognizes karst topography by the presence of large numbers of sinkholes, solution valleys, disappearing streams, and caverns. Karst topography typically develops in well-jointed, thin-bedded carbonates (or evaporites) in regions having abundant rainfall, and sufficient relief to insure continuous movement of groundwater that will carry away dissolved matter. The term karst comes from a limestone plateau in Yugoslavia where solution features are well developed. Similar topography can be found in Kentucky, Tennessee, Indiana, northern Florida, and Puerto Rico.

HOT SPRINGS AND GEYSERS

To qualify as a **hot spring** (Fig. 9–21), water emerging from the subsurface must be at least as warm as the body temperature of a human. Most hot springs are found in regions where there has been geologically recent igneous activity. In such regions, rocks immediately below the surface may still retain part of their original heat. Indeed, that heat may be retained for centuries because of the insulating effect of overlying layers. Groundwater that comes in contact with the hot rocks is heated and sometimes even mixed with hot magmatic water before finding its way to the surface and emerging as a hot spring.

Geysers (Fig. 9–22) are special kinds of hot springs that discharge intermittently in the form of a tall column or fountain of water and steam. The essential component of a geyser is one or more tubular conduits extending from the surface into a zone of hot rocks. Water that percolates downward through the conduits is heated above the boiling point but is temporar-

Figure 9–21 The "Terrace Mount" of Minerva Hot Spring, Yellowstone National Park. (Courtesy of U.S. National Park Service.)

ily prevented from boiling because of the weight of the overlying water column. Eventually, water at some level in the tube reaches a temperature where it boils at the higher pressure. Steam is rapidly generated and lifts the water above it until it flows out at the surface. The deeper, superheated water, suddenly freed of the weight of overlying water, flashes into steam and blasts upward toward the surface. After the eruption has dissipated, water flows back down through the tube system and is heated for another eruptive cycle. The cycles may be irregular or regular, as is the case for Old Faithful in Yellowstone National Park. This famous geyser erupts, on average, every 65 minutes, but the actual time between eruptions ranges from 30 to 90 minutes.

In these times of concern over meeting our energy needs, increasing attention is being given to utilizing the steam produced in hot spring areas to generate power. To tap this resource, wells are drilled into moist hot rocks. The steam released into the well is fed into turbines that generate electricity. The energy produced in this way is relatively inexpensive, has no radiation hazards, and does not add pollutants to the atmosphere. Unfortunately, there are only a few sites around the world favorable for the development of geothermal power at reasonable costs. The best known in the United States is in Sonoma County, California, where 80 geothermal wells now generate enough power to serve the electricity demands of about a million people. In Reykjavik, Iceland, geothermal steam is not only used for generating electricity but is also pumped into buildings for heating and used to warm greenhouses where tropical fruits and vegetables are raised.

CONCRETIONS AND GEODES

Among the minor features associated with groundwater deposition are accumulations of mineral matter called **concretions.** Concretions may be ellipsoidal, discoidal, or variously shaped, and may range in size

Figure 9–22 Castle Geyser, Yellowstone National Park. (Courtesy of U.S. National Park Service, photograph by M.W. Williams.)

from a few millimeters to one or two meters. Most concretions are composed of calcite or silica, although one also frequently finds hematite, limonite, siderite (iron carbonate), and pyrite concretions. Concretions developed in a rock after it has been deposited are usually the result of localized chemical precipitations of dissolved substances from groundwater. Often precipitation starts around a mineral or fossil particle that differs chemically from the enclosing rock. Concretions formed within sedimentary rocks in this way may retain relic laminations and textural features of the original rock. Other concretions are merely local concentrations of the cementing material in a clastic sedimentary rock and therefore are similar in composition to the host rock. Concretions may differ in composition from the rock in which they have formed. Chert and flint nodules in limestone or dolostone are common examples.

Not all concretions develop in rocks after deposition. Some apparently form on the floors of lakes or seas as sediments are being deposited, as do the manganese nodules to be described in Chapter 11. Concretions that form in this way do not contain relic features of the rocks that enclose them.

Although the odd shapes of concretions are interesting, most people find geodes more interesting because of their beauty. **Geodes** (Fig. 9–23) are hollow,

Figure 9–23 Cavity in a rock which has become lined with quartz crystals. Such hollowed-out rocks are called geodes. (Courtesy of Wards National Science Establishment, Inc., Rochester, N.Y.)

crystal-lined, roughly spherical bodies varying from 1 or 2 cm to as much as 60 cm in diameter. Most are lined with quartz, but other minerals may occur as well. Quartz geodes are characteristic of certain limestone strata, although they are also known to occur in shales. In addition to their roughly globular shapes and hollow interiors, geodes usually have a thin clay film between their outer wall and the enclosing host rock. Beneath the layer of inwardly projecting crystals is a layer of banded silica (chalcedony).

THE USE AND MISUSE OF GROUNDWATER

As is the case with most geological resources, groundwater must be conserved and wisely managed if it is to serve present and future generations. For this reason, many state governments limit the number of water wells in any given area and also the rate at which water can be withdrawn from those wells. Along coastlines and small islands precautions must be taken to prevent saltwater intrusion. At such places, two kinds of water compete for space in permeable sediments and rocks. Meteoric water derived from rain and snow infiltrates in the manner already described and gradually builds a zone saturated with freshwater. Saltwater from the adjacent ocean, however, also enters permeable layers, but because of its greater density, the brine tends to form a wedgelike zone beneath the freshwater. The density differences between the two kinds of water are such that a balanced state develops in which there is about 40 meters of freshwater beneath sea level for every 1 meter of elevation of the water table above sea level (Fig. 9–24). Thus, if the water table is 5 meters above sea level at a given location, freshwater may extend to a depth of 200 meters below sea level. If the water table, however, is lowered by only 5 meters as a result of drawing too much water from wells, then the top of the saltwater zone may rise and contaminate wells.

Another consequence of heavy withdrawal of groundwater is subsidence of the ground above the earth materials serving as aquifers. Fluid withdrawals from loose fine-grained materials lead to volume reductions and, in artesian systems, reduce the pressure with which the aquifer formerly resisted compression. Venice, Italy, is a city whose problems with subsidence have drawn worldwide attention. Subsidence in Venice results both from groundwater withdrawal and from the compaction of loose sediments beneath heavy structures. The city has subsided more than three meters since it was founded 1500 years ago. Within the last few years, efforts to recharge the aquifers beneath the city and to regulate withdrawal of water from wells have reduced the rate of subsidence appreciably.

One problem concerning groundwater use that is not related to conservation is the problem of contami-

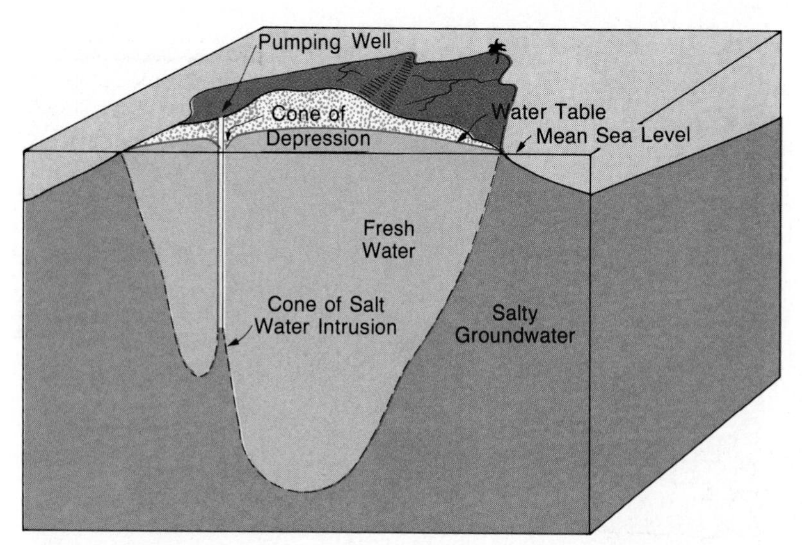

Figure 9–24 *Fresh groundwater floating on salt groundwater. Vertical scale of area below sea level is one-half that of area above sea level. Note that a cone of saltwater intrusion is developed by pumping a body of fresh water floating on saltwater.*

nation. Earlier we noted that groundwater derived from fine-grained sediment was often remarkably clean. This is because contaminating particles adhere to the surfaces of the myriads of mineral grains and become trapped in intersitial spaces. Of course, there are limits to the filtering capacity of even sandy materials, and the ultimate purity of water from a well may depend upon the volume of the pollutants being added to the underground reservoir and the distance over which the groundwater has traveled. Natural filtration is usually not effective for many dissolved chemical pollutants that may spoil groundwater. Also, not all aquifers are composed of granular materials. Large, open conduits, such as those in cavernous limestones, basalts containing lava tunnels, and dense rocks with large joints, provide little filtration.

Whenever sewage or chemical wastes are dumped into excavations or topographic depressions, there is a strong possibility that groundwater supplies may become contaminated. This is particularly true if the area for dumping is underlain by highly permeable materials (Fig. 9–25). Not infrequently one discovers places where refuse has been dumped into pits that lead directly into cavernous conduits that supply water for nearby towns. Salty water from oil fields and gasoline from the rusted storage tanks of neighborhood gasoline stations may similarly contaminate a community's primary source of water.

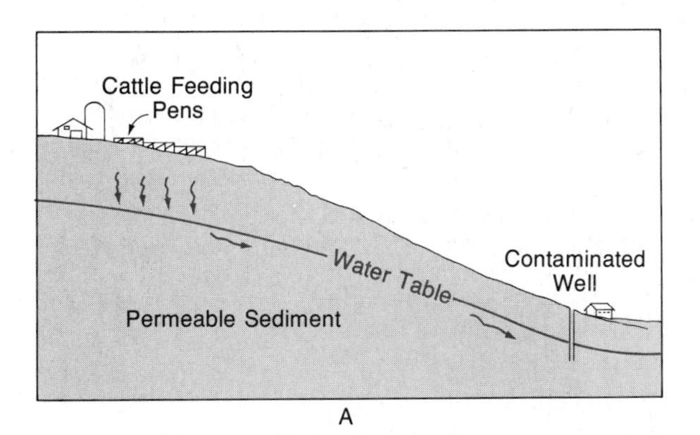

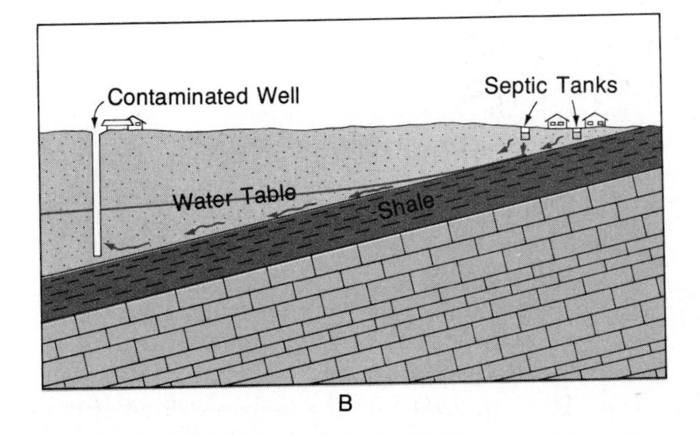

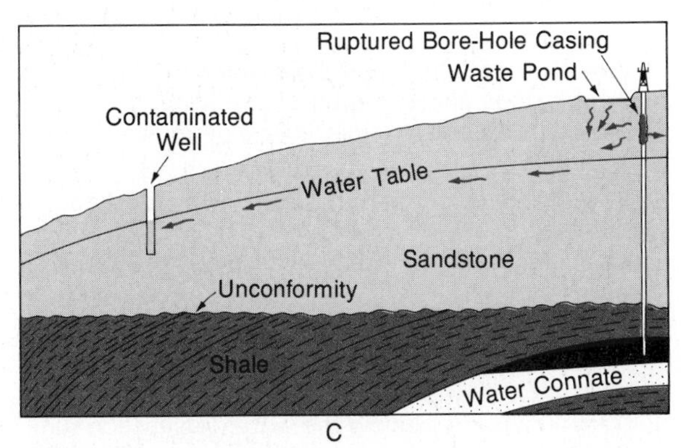

Figure 9–25 *Well pollution. (A) Seepage of pollutants from feeding pens results in contamination of nearby wells located downslope. (B) Septic tanks provide the contamination that moves toward a well because of the inclination of an impermeable shale stratum. (C) Contamination of groundwater from a waste pond and ruptured borehole casing of an oil well.*

Summary Approximately 20 percent of all the water used in the United States is drawn from the subsurface by means of wells. This underground water, which fills pores, cracks, and joints in rock and sediment, is called groundwater. Most groundwater is derived from that portion of rainfall or melting snow that does not immediately flow to streams but infiltrates into the ground. The downward infiltrating water eventually reaches rocks through which it cannot move and begins to fill all openings in overlying rocks so as to develop a zone of saturation. The upper surface of the zone of saturation is called the water table. Its configuration roughly parallels the land surface, and it will rise or fall according to the amount of precipitation that infiltrates from above. In general, wells must be deeper than the level of the water table if they are to provide an adequate supply of water. Where the water table intersects the sides of valleys, springs may be formed.

Rocks and sediments below the surface of the earth that are capable of producing water from wells are termed aquifers. The quantity of water that can be obtained from an aquifer depends not only on the amount of rainwater that percolates into it but also on porosity and permeability. Porosity is expressed as the percentage of the rock or sediment that consists of openings. Permeability refers to the ability of a rock to transmit groundwater or other fluids through pores and interconnected openings.

The movement of groundwater below the water table in the zone of saturation is principally a slow seepage downward and sideward toward streams and lakes. Normally, the water follows curved paths of flow and moves at rates determined by permeability of the aquifer and the slope of the water table; that is, the hydraulic gradient.

We see groundwater at or near the earth's surface when it emerges in well borings or as springs. An ordinary well is simply a drill hole that extends downward below the water table, thereby allowing water to seep into the cylindrical hole and be available for pumping. Such wells penetrate aquifers that are not confined by overlying, relatively impermeable beds. Where aquifers are confined and inclined so that there is sufficient hydraulic head, water may rise in the well above the level of the aquifer and flow without pumping. Such wells penetrate artesian groundwater systems.

The geologic work of groundwater consists primarily of solution of rocks and minerals, the transportation of the dissolved matter, and the precipitation of dissolved matter when conditions are suitable. The solution work of groundwater is particularly evident in areas underlain by limestone, although such rocks as dolostone and gypsum are also susceptible.

QUESTIONS FOR REVIEW

1. What is the water table? What would be the effect on the slope and elevation of the water table of a period of excessive rainfall? Of a period of marked aridity?
2. Rock A and rock B are found to have identical porosity. Rock B, however, is found to have greater permeability. What may be the cause for this observation?
3. In an obscure valley, Aaron Hatfield digs a well that penetrates two meters below the water table. In an adjacent valley, Jud McCoy has a well that penetrates ten meters below the water table. Each summer the Hatfield well goes dry (thus curtailing distilling operations). Why does Hatfield's well go dry?
4. What might you do to a sample of groundwater to prove that it contained substances in solution? What compound most frequently makes water "hard?"
5. Draw cross-sections illustrating the following:
   - **a** gravity spring
   - **b** artesian groundwater system
   - **c** zones of aeration and saturation
   - **d** perched water table
6. How does the functioning of a geyser differ from that of an artesian well?
7. What is the role of carbon dioxide in the solution of limestone and in the formation of such dripstone features as stalactites and stalagmites?
8. Why does the fountain produced by an uncapped artesian well not rise to the same elevation as the groundwater table in the catchment area?
9. What are the characteristics by which karst topography can be recognized?
10. Account for the observation that wells in coastal areas adjacent to the sea are often able to draw

freshwater from a depth well below sea level. What might be the effect of rapid withdrawal of freshwater from such wells?

11 A liquid dye is poured into a garbage dump at 1:00 P.M. on January 1st. The dye appears in a well one km away on January 6th at 1:00 P.M. What is the velocity of groundwater flow in meters per hour between the dump and the well?

12 The elevation of the water table at well A is 200 meters above sea level and at well B, 1000 meters away, the elevation of the water table is 180 meters above sea level. What are the head and hydraulic gradient between the two wells?

13 Describe the necessary natural conditions for an artesian groundwater system.

SUPPLEMENTAL READINGS AND REFERENCES

Anderson, J., 1974. *Cave Exploring*. New York, Association Press.

Bouwer, H., 1978. *Groundwater Hydrology*. New York, McGraw-Hill Book Co.

Freeze, R.A., and Cherry, J.A., 1979. *Groundwater*. Englewood Cliffs, N.J., Prentice-Hall, Inc.

Heath, R.C., and Trainer, F.W., 1968. *Introduction to Ground Water Hydrology*. New York, John Wiley & Sons.

Leopold, L.B., 1974. *Water, A Primer*. San Francisco, W.H. Freeman & Co.

Sweeting, M.M., 1973. *Karst Landforms*. New York, Columbia University Press.

Todd, D.K., 1959. *Ground Water Hydrology*. New York, John Wiley & Sons.

10
the work of ice and wind

The ground of Europe, previously covered with tropical vegetation and inhabited by herds of great elephants . . . became suddenly buried under a vast expanse of ice, covering plains, lakes, seas, and plateaus alike.

LOUIS AGASSIZ, 1840

Etudes de les glaciers

Prelude Although of paramount importance, running water is not the only geologic agent responsible for the sculpturing of our landscapes. Glacial ice and wind also work at the earth's surface to produce distinctive topographic features. In this chapter we will examine how wind and ice erode, transport, and deposit earth materials. We will begin with a description of those great flowing masses of ice and debris called glaciers. The origin and kinds of glaciers, the dynamics of their movement, the manner in which they carve our most magnificent scenery, and the reason for ice ages will be examined.

Compared with the moving ice of glaciers, moving air or wind is a far more subtle agent of erosion. Wind is capable of lifting and transporting only relatively small particles of rock and mineral matter. To accomplish even this relatively limited amount of work, certain conditions must exist. For example, the surface over which the wind blows has to contain loose, dry particles, preferably in sizes from silt to fine sand. There has to be little vegetative cover. Conditions such as these are found along some shorelines, on floodplains following periods of flooding, and in places where bare ground has been exposed by plow or bulldozer. Mostly, however, conditions favoring wind erosion occur in the warm deserts of the world.

Glacier in the Condoriri region of the Cordillera Real, north of La Paz, Bolivia. (Courtesy of Jonathan Turk, Telluride, Colorado.)

GLACIERS: THEIR MAGNITUDE AND ORIGIN

A **glacier** is a large mass of land ice derived from snow, having definite and recognizable limits, and moving slowly under the influence of gravity. At the present time, about 10 percent of the earth's land area is covered by glaciers, and in the recent geologic past that figure was about three times higher. Present-day coverage of glacial ice is sufficient to convince most geologists that we are now living in a warmer interval within an ice age whose end is not in sight. We would not have to persuade an Eskimo of the validity of that concept, but a sunbather in Florida might be more skeptical. Perhaps, that skepticism might be lessened on learning that glaciers cover about 14.5 million square km of our continents. Of this amount, 12.5 million square km are accounted for by the Antarctic ice sheet, and 1.7 million square km by glaciers in Greenland. The remaining glaciated areas consist mostly of smaller ice caps on polar islands, glaciers along the perimeter of Alaska and Scandinavia, and the mountain glaciers of the Yukon, Alps, and Himalayas.

Two conditions are necessary for the growth and development of glaciers. Temperatures must be below freezing at least part of the time, and there must be snowfall in sufficient amounts to exceed losses that normally occur through melting, evaporation, and sublimation. (**Sublimation** is a process whereby water molecules pass from the solid state to the vapor state or back to solid without passing through the liquid state.) These conditions are most prevalent in high latitudes, high elevations, or both. Glaciers will expand or contract depending upon either seasonal or long-term climatic changes. Certainly during the climatic changes of the recent Ice Age, glaciers covered lowlands more extensively, and extended much lower into mountain valleys than they do today.

Because of the impressive ice sheet that covers Antarctica, one might guess that temperatures well below freezing are conducive to glacial development. Actually, extremely low temperatures are not necessary. In fact, glaciers are more likely to form when temperatures are only slightly below freezing, because under such conditions air can hold more moisture than at lower temperatures. Although snowfall is meager in the frigid Antarctic, the ice has been accumulating there for millions of years.

If the 24 million cubic km of ice covering various parts of the earth were to melt, sea level around the world would rise by about 100 meters (328 feet). New York City, all of New Jersey, Florida, and Louisiana, and vast areas of our eastern states and interior California would be covered by shallow seas. Landlocked midwesterners might travel relatively short distances to enjoy the surf. Of course, one should not hurry to purchase "ocean front" property in inland places like Missouri, for such melting and marine encroachments would take place very slowly over a period of several thousand years.

ICE FROM SNOW

The first step in the formation of a glacier is the accumulation of snow as a permanent snow field. Flakes of freshly fallen snow are, of course, noted for their delicate hexagonal form and intricate growth patterns (Fig. 10-1). After the flakes lie at the surface for a period of time, or if they are buried by additional snow, they begin to lose their delicate form as a result of sublimation, partial melting, and refreezing of the water near the center of the flake. The conversion to ice is enhanced by the weight of overlying snow. Eventually, most of the water once distributed as ice throughout the snowflake is concentrated as a granule. The material has now ceased to be snow and is in

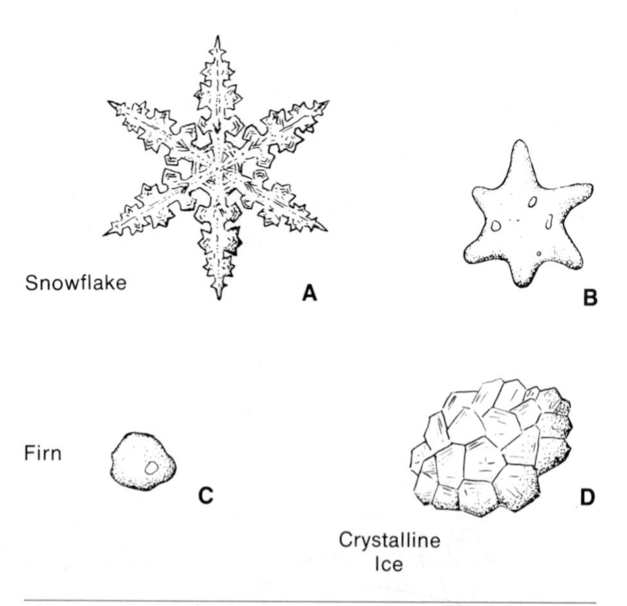

Figure 10-1 *Stages in the development of crystalline ice from snow. The snowflake is transformed into a pellet primarily by sublimation, although partial melting and refreezing also play a role.*

the form of granular ice called **firn.** The pellets of firn are really poorly formed hexagonal crystals of ice. Under the pressure of overlying ice and snow, these crystals begin a transformation into solid ice. Most of the air is gradually expelled from the mass as the ice pellets are packed tightly together. Melting occurs at contacts between grains (so-called pressure points), and the molecules of water are added elsewhere to growing ice crystals. As the process continues, adjacent crystals intergrow, and the mass becomes solid crystalline ice with a texture similar to but coarser than the texture of an intrusive igneous rock. If snowfall is rapid, the transformation from snow to solid ice may occur in as little time as five years.

KINDS OF GLACIERS

One way in which glaciers are classified is based upon their relation to surface topography. Great ice sheets, for example, may cover large parts of continents, burying mountains and plateaus indiscriminately. The size and shape of such glaciers is not dependent on the configuration of the ground surface. Glaciers that are confined by the mountain divides and valley sides constitute a second category of glaciers.

Glaciers not constrained by topography

Ice sheets and **ice caps** are great domal masses of ice that differ primarily in size. Ice caps are generally smaller than ice sheets and rarely exceed 50,000 square km. The ice in both types of glaciers flows outward from a central area of maximum ice accumulation much like thin pancake batter poured on a skillet. In cross-section, the maximum elevation is near the center of the ice sheet, and from that point the ice surface slopes away gently in all directions until it reaches the periphery where there is a marked steepening of slope.

Today ice caps are developed on Iceland, Baffin Island, and in the Canadian Arctic archipelago. As for the larger ice sheets, the only two that remain today cover most of Greenland and Antarctica. The Antarctic ice sheet attains a height of more than 4200 meters above sea level, and in places is more than 3000 meters thick. Buried beneath this thick blanket are hills, plateaus, and impressive mountain ranges (Fig. 10–2). Here and there one of the higher parts of the moun-

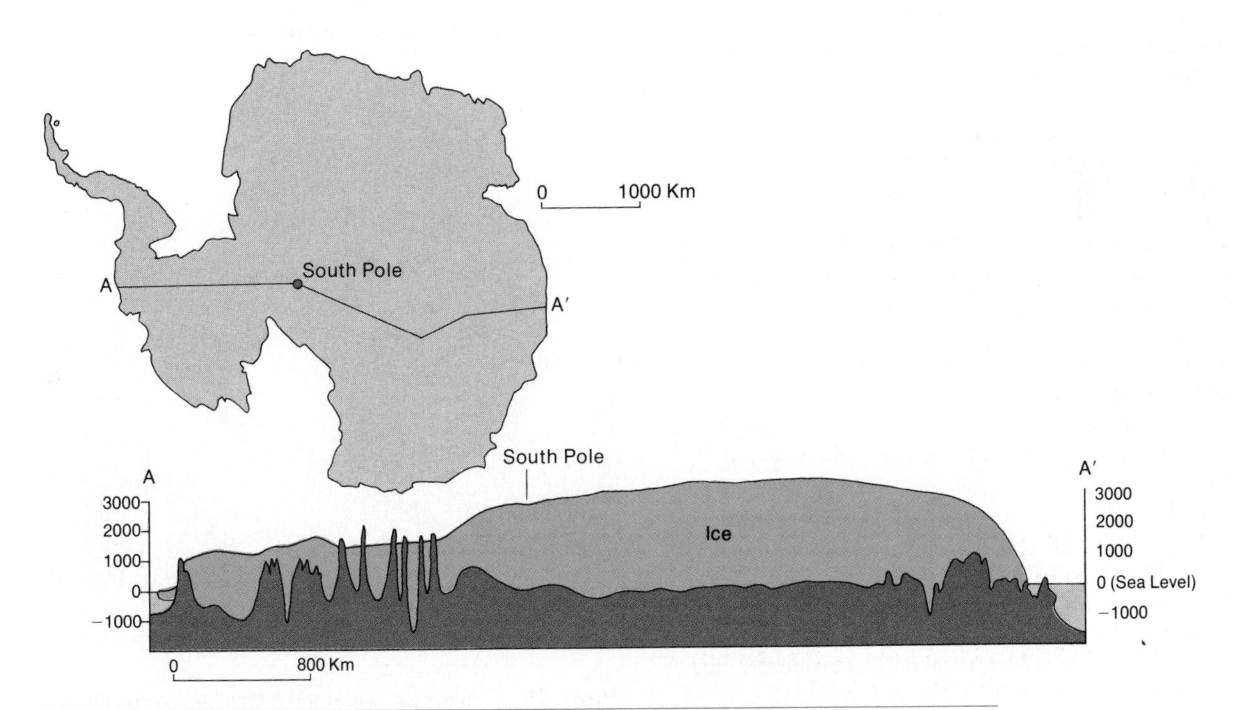

Figure 10–2 The Antarctic ice sheet in cross-section. Line of section A-A′ is indicated on index map. (Adapted from Twidale, C.R., Analysis of Landforms. *Sydney, Australia, and New York, John Wiley & Sons, Australasia, Pty. Ltd., 1976.)*

Figure 10–3 Snowfield at head of glacier in the Cordillera Real, north of La Paz, Bolivia. Each layer visible in the crevasse represents a year's accumulation of snow. (Courtesy of Jonathan Turk, Telluride, Colorado.)

tains protrudes through the glacial ice as an isolated peak called a **nunatak.** Around the margins of both Greenland and Antarctica, the moving ice reaches the ocean, where it may extend seaward as floating pack ice or break off into huge icebergs.

Glaciers constrained by topography

Glaciers that owe their size and shape to the characteristics of the landscape differ in many ways from ice caps and ice sheets. The most obvious difference is that they are smaller. In addition, they are not dome-shaped, and their direction of flow is not radial but down the existing slope of the valley. Typically, glaciers constrained by topography develop in high, mountainous regions. They are initiated in lofty catchment areas where snowfall is sufficiently heavy and frequent to accumulate as a snowfield (Fig. 10–3). The snowfield is the nourishment area for the ice that moves down the slope as a **valley glacier** (Fig. 10–4).

Figure 10–4 Yetna Glacier in Mount McKinley National Park, Alaska is famous for its well-developed banding. (Courtesy of U.S. National Park Service.)

Where valley glaciers emerge from a mountain range, they spread out across the unconfined mountain front as **piedmont glaciers.** Today, such glaciers are particularly numerous in Alaska. During the recent Ice Age, however, piedmont glaciers covered vast areas of the high plains along the eastern margin of the Rocky Mountains.

GLACIER DYNAMICS

The glacial balance

Glaciers that maintain their size and shape over a long period of time are said to be in balance. In such glaciers, the amount of frozen water that is added is balanced by the amount being lost through melting, sublimation, and evaporation. In a valley glacier, the area in which there is a net gain in snow and ice usually is found in the upper reaches of the valley, in what is designated the **accumulation zone** (Fig. 10–5). Accumulation generally decreases toward the snout of the glacier until one passes into the area where there is a net loss of ice due primarily to melting. This lower end is called the **ablation zone.** From the snout, ablation decreases up-glacier. An imaginary line called the *line of equilibrium* separates the ablation and accumulation zones and itself represents that part of the glacier in which gains approximately balance losses. The snout of a glacier will maintain itself at about the same position as long as the glacier is in equilibrium; that is, as long as net losses in the ablation zone are balanced by net gains in the accumulation zone. If, however, the glacier begins to acquire more ice than is lost through ablation, it will advance. Conversely, if the ablation rate exceeds the rate of accumulation, the glacier front will gradually recede. Even when a glacier is receding, however, the ice within it continues to move forward. Recession occurs when the terminus wastes away faster than the ice is advancing.

How glaciers move

We are accustomed to thinking of ice as rather brittle, unyielding material. When subjected to pressure of thick overlying layers of frozen waste, however, ice will deform or flow. In general, flow will begin when the ice mass reaches a thickness of about 60 meters. No matter what kind of glacier, the prime mover that causes flow is the pull exerted upon the ice by the force of gravity. Valley glaciers are drawn down the slope of the valley floor by a component of gravity directed parallel to the slope. In contrast, ice sheets move forward largely because the great weight near the centers

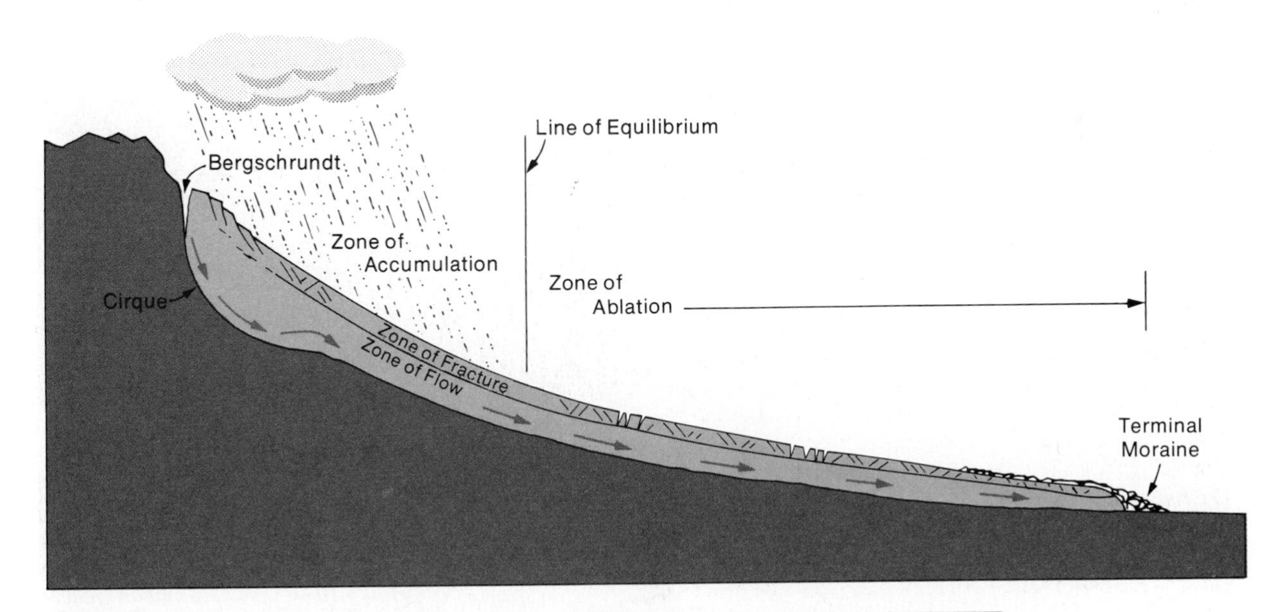

Figure 10–5 *The zones of accumulation, ablation, and line of equilibrium as ideally developed in a valley glacier.*

of accumulation exerts pressure on the underlying ice, causing it to flow outward.

As suggested by the even pattern of ribbons of dark debris seen on valley glaciers, ice flows laminarly, like slowly moving water. Because of friction with the bedrock floor, velocity of flow is least near the base of a valley glacier (Fig. 10–6) and increases from the base toward the top, except for the uppermost 60 meters. Ice in this top layer is under insufficient pressure to flow and is carried along as a brittle slab by the flowing ice below. This upper brittle portion of the glacier is designated the **zone of fracture** to distinguish it from the underlying **zone of flow.** As implied by its name, the zone of fracture is often characterized by chaotic fracturing and the development of deep crevasses. In plain view the center of a valley glacier moves along more rapidly than the flanks, where there is friction with the valley sides.

The actual flow of ice in a glacier represents an adjustment to the pressures acting upon the ice, and these pressures are in turn affected by the thickness of the mass and the slope of its surface. Meltwater trapped within the glacier and variations in temperature also influence how fast the ice moves. The most rapid movement measured in alpine glaciers is about 75 meters per year.

Several mechanisms are involved in the movement of glaciers. What appears as plastic flow is the result of mutual displacements of ice crystals relative to one another, and displacements within crystals along sheets of atoms. Such displacements produce ice crystals that on microscopic examination appear to have been bent.

Other mechanisms for movement include granulation of one portion of the ice mass as it moves against another. Partial and temporary melting may also enhance the movement of the glacier, provided temperatures within the ice are not exceptionally low. For example, ice flowing across bedrock obstacles will experience higher pressure on the "upstream" side of the obstacle than on the "downstream" side. As a result, partial melting will occur on the upstream side, and the meltwater will flow around the obstacle and freeze again on the downstream side. Elsewhere, water produced in this way (or by melting from other causes) may form thin layers of liquid that provide a low-friction horizon along which overlying ice may slide at increased speed.

Not all of the movement in a glacier is the result of melt-freeze or plastic flow. Whenever stress on the ice mass exceeds its ability to respond with displacements within and between crystals, rupture surfaces will develop in much the same way as do faults in rocks. Such rupturing and associated fracturing is mostly confined to the upper 30 to 60 meters of the glacier.

In addition to the movement of glaciers by actual flowage of the ice mass, glaciers progress by simply sliding along the underlying ground surface. Glaciologists refer to such movements as **basal slip.** Basal slip is most prevalent in valley glaciers of temperate regions.

The rates at which valley glaciers flow vary widely. Some move rapidly, others slowly, some erratically,

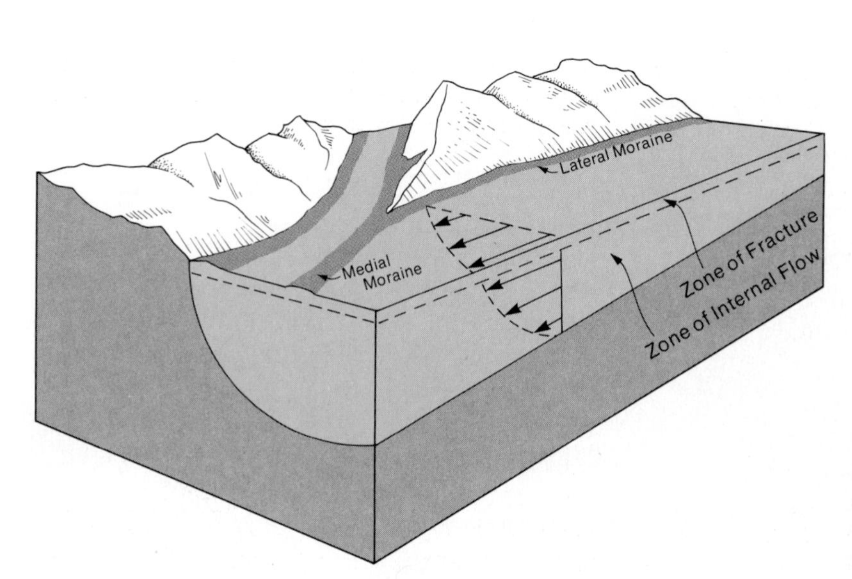

Figure 10–6 Forward motion in a glacier. The ice moves least rapidly near the contact with valley floor and sides because of friction.

and some with a constant rate throughout the year. A characteristic of many glaciers is a **surging behavior,** in which a glacier having a slow and uniform flow, periodically surges forward at higher velocity. Glaciers in the United States and the Himalayas are known to have sprinted forward on occasion at rates of more than 300 meters per day. The movement is usually uneven and jerky and is accompanied by a shivering motion of the ground and loud cracking noises.

THE WORK OF GLACIERS

Because most of us live far from regions undergoing active glaciation, we may fail to appreciate the awesome power of glaciers. Glacial sculpturing on a grand scale has enhanced the beauty of the Rocky Mountains, the Alps, the Himalayas, and coastal regions of Alaska, Greenland, and Scandinavia. Across northern Europe and North America, glaciers have scraped away topographic obstructions and molded hummocky lowlands. Vast amounts of fertile soils have been stripped from bedrock in some regions, whereas in others immense tonnages of rock debris have been spread across the land. The highly productive soils of the U.S. Corn Belt are partly composed of fertile soils transported into the region by glaciers.

Erosion by glaciers

Abrasion

Abrasion refers to the wearing, grinding, and rubbing away of solid rock by moving glacial ice armed with angular rock fragments. The debris within the ice acts somewhat like the teeth of a rasp in scratching and grooving the rock over which the glacier moves. The finer scratches are termed **glacial striations,** whereas coarser lineations are dubbed **glacial grooves** (Fig. 10-7). Boulders impinging on bedrock as the glacier inches forward may crack the bedrock into curved fractures called **chatter marks** (Fig. 10-8). The concave sides of these marks face the direction toward which the ice flows.

Glacial quarrying

Glacial quarrying (glacial plucking) is the process by which a glacier extracts and carries away pieces of the bedrock over which it is moving. Typically, quarrying occurs when meltwater beneath the glacier seeps into

Figure 10-7 *Fragment of limestone that has been scratched and grooved by glacial action. (Courtesy of the Institute of Geological Sciences, London.)*

fractures in the underlying bedrock and freezes. The expansion accompanying freezing opens the fractures more widely and may cause new breaks that free chunks of bedrock from the parent mass. As the glacier moves, these blocks of rock, solidly frozen into the ice, are pulled out and carried away. Also, the passing of the glacier over bedrock results in a pulling or tensional force on the ground surface which may open incipient fractures and facilitate lifting pieces of rock into the mass of moving ice.

Avalanching

As used by glaciologists, **avalanching** is the sudden and rapid movement of snow and rock debris down mountain slopes and onto the surface of a valley gla-

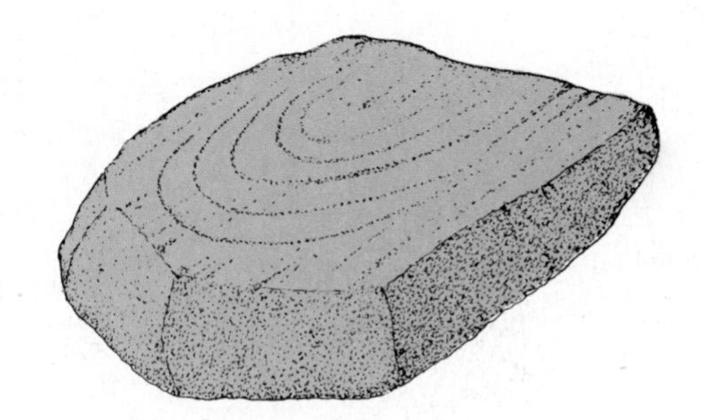

Figure 10-8 *Chatter marks. The glacier moved from the upper right to the lower left. Long dimension of rock is 40 cm.*

cier. The avalanching process is enhanced by the fact that abrasion and quarrying occur actively on the sides as well as the bottom of the glacier filling a valley. As a result, the valley sides are undercut, and this leads to loss of support for loose material lying on the valley slopes. Inevitably, the unconsolidated debris slides or falls onto the ice surface and is then moved along as if on a conveyer belt. From a distance, the debris appears in the form of graceful dark bands parallel to the valley sides (Fig. 10–9). These bands of dirt and rock flow with the ice and serve as indicators of direction of movement.

Erosional features

Anyone who has visited the higher ranges of the Sierra Nevada, the Rocky Mountains, or the Alps is certain to be impressed by their rugged grandeur. Indeed, the effects of glacial erosion are most spectacular in mountains. On every side one finds magnificent U-shaped valleys, elegant waterfalls cascading from hanging valleys, jagged peaks, and placid fjords.

Valley glaciers occupy valleys originally eroded by streams. In mountainous regions, stream valleys characteristically have a V-shape cross-section caused by the fact that the stream itself can only cut within its channel and depends upon mass wasting to establish the slope of the valley sides. In the glaciers that occupy those valleys, however, ice is in contact with, and can erode, rock at the base of the valley and well up onto the valley sides. As a result, glaciers produce valleys having parabolic cross-sections and often impressively steep walls.

Glaciers not only change the cross-sectional appearance of the valleys they occupy but also straighten the valley by wearing away the bends. The result is a

Figure 10–9 *Bands of dirt and ice and fracturing of the ice surface on a glacier in the northwestern corner of British Columbia, Canada. Tulsequah Lake at the center receives water from the glacier, and overflows catastrophically in late summer. (Courtesy of U.S. Geological Survey and A. Post.)*

straight, direct route of escape for the ice. The movement of ice down the valley truncates ridges that once existed between stream bends, so that steep cliffs, or **truncated spurs,** occur between tributary valleys.

Another distinctive feature of glaciated mountainous regions are **hanging valleys.** They are the result of the fact that tributary glaciers erode at a slower rate than does the main glacier. Thus, the floors of the tributary valleys meet the main valley high above the main valley's floor. After the glaciers have melted away, this difference in elevation between the tributary and main valley floor is dramatically exposed as a precipice over which waterfalls and cataracts may plummet (Fig. 10–10).

At the head of a glaciated valley, one frequently finds a steep-walled recess, shaped like a half bowl, and excavated primarily by glacial quarrying. These features are called **cirques** (Fig. 10–11). The walls of cirques may drop vertically more than 40 meters, and their floors are often gouged to a depth lower than the adjacent valley floor. A small lake called a **tarn** (Fig. 10–12) may occupy the depression within the floor of the cirque after the disappearance of ice it once contained. As the ice moves out of the cirque in association with the downslope movement of the entire glacier, an arcuate crevasse called a **bergschrund** develops next to the back wall. Meltwater flows into the bergschrund and freezes in bedrock fractures, thereby anchoring the ice to the rock. When the ice moves again, large pieces of rock are plucked from the mountainside, thus deepening and further sculpturing the cirque.

Characteristically in glaciated mountainous regions one finds places where adjacent valley glaciers have sharply reduced the width of divides, or where cirques on either side of a mountain are gradually eroded toward one another until only a knife-edge divide remains. Such sharp-edged precipitous ridges are known as **arêtes** (Fig. 10–13). A depression or notch in an arête may be produced by the convergence of the back walls of cirques. Alpine climbers refer to such depressions as **cols.** Where three or more cirques come together, the spectacularly steep projections of **horns** are sculptured. There are many horns in the Swiss Alps, but by far the most famous is the Matterhorn south of Zermatt (Fig. 10–14).

In any catalogue of inspirational landforms, **fjords** (Fig. 10–15) would certainly rival horns in scenic splendor. Fjords are long, deep, steep-sided glacial valleys into which the sea extends following the disappearance of the glacier. They are sculptured by thick glaciers with steeply sloping surfaces that flow down from the mountains into the sea and are able to erode their valleys to a depth below sea level. Today, the depth of water in some fjords is even deeper than when they were filled with ice because of the rise in sea level following the last Ice Age recession. In many of

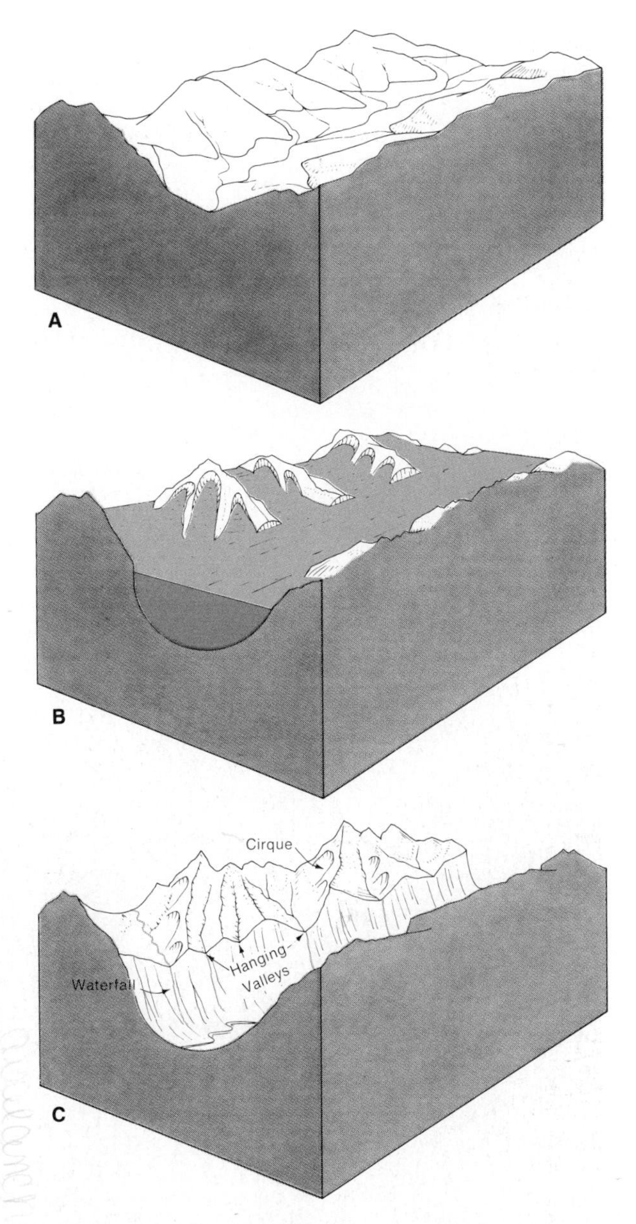

Figure 10–10 Landforms produced by glacial erosion in mountainous regions. (A) Before glaciation. (B) During glaciation. (C) After glaciation.

Figure 10–11 Cirque (upper left). *A meltwater stream flows in the foreground. Jasper National Park, Canada. (Courtesy of J.C. Brice.)*

the famous Norwegian fjords, water depths exceed 1200 meters.

Transportation by glaciers

The work potential of glaciers differs from that of streams in that there is practically no limit to the size or amount of material that can be carried by a glacier as suspended load. The competence and capacity of glaciers are truly enormous. Indeed, parts of some glaciers contain more rock debris than ice itself.

Ice sheets differ from valley glaciers in the distribution of their load. Very little of the load is carried on the ice surface because that surface is above hills and mountains, and hence receives no debris from avalanching. Also, ice sheets carry a smaller volume of

Figure 10–12 Lake of the Pines, a small tarn high on the south shoulder of Rock Creek, North Cascades National Park, Washington. (Courtesy of U.S. Geological Survey and M.H. Staatz.)

Figure 10–13 *Arêtes* and col *near Snowdon Summit, Carnarvon, northern Wales. (Courtesy of the Institute of Geological Sciences, London.)*

debris per unit volume of ice than do valley glaciers, and that load is carried primarily near the bottom of the sheet or pushed along at the perimeter.

In valley glaciers, avalanching and mass wasting contribute significantly to the glacier's load. The larger blocks of rock that fall on the surface of the ice work their way down so that the load becomes generally coarser near the base of the glacier. Debris from previous years may appear as superimposed layers in cross-sections of the glacier. Rock on the ice that does not work its way downward is transported along the upper brittle layer in conveyer-belt fashion and is eventually dumped at the snout of the glacier.

Figure 10–14 *The Matterhorn, Swiss Alps.*

Deposition by glaciers

Glacial sediment

Glaciers drop their load of coarse and fine sediment either directly from the moving ice mass or indirectly through meltwater. As one might expect, sediment deposited by streams of meltwater is likely to be better sorted than debris dropped directly from glacial ice. The general term for all deposits resulting from glaciers is **drift.** The term alludes to the erroneous idea that all glacial sediment "drifted" to present locations within floating ice during surging biblical floods. As used today, the term embraces debris carried and deposited by glaciers, glacial meltwater, or even icebergs derived from glaciers.

Because of differences in origin, it is useful to distinguish two kinds of drift; namely, stratified and unstratified drift. Unstratified drift is more appropriately called **till** (Fig. 10–16), a term originally used by nineteenth-century Scots to describe the rocky, clayey soils of northern Britain. Till is a poorly sorted jumble of rock and mineral fragments in a range of sizes and without a discernible arrangement of those particles. There is no trace of the kind of winnowing and sorting of particle sizes that characterize sediments deposited by water, and therefore till is believed to have been dropped directly from a glacier without appreciable reworking by meltwater.

Stratified drift is glacial sediment deposited and reworked by water and/or by wind. Stratified drift deposited by meltwater streams is described as outwash or **glaciofluvial** in recognition of its origin and its similarity to ordinary stream deposits. Glaciofluvial deposits are better sorted than till and often exhibit distinct sand and gravel layers. Cross-bedding may also be well developed.

Glaciolacustrine is a term used to designate stratified drift that accumulates in lakes. Many glaciolacustrine deposits have distinctive lamination. Sediment accumulating during the warmer summer months when water is stirred by waves forms a layer that is somewhat lighter and coarser than a darker, thinner layer deposited during the winter when the lake is frozen over. The regularly repeated pairs of layers, or **varves,** record the years during which a particular glacial lake existed.

Figure 10–15 A Norwegian fjord. (Courtesy of the Norwegian Information Service.)

Depositional features

Moraine is probably the term used most frequently in describing depositional features of glaciers. It is a term used by eighteenth-century Swiss peasants for ridges of rock debris found near the margins of glaciers. Today, geologists define moraines as accumulations of drift deposited directly by glaciers. The more easily recognizable types are ground moraines, terminal moraines, and lateral moraines.

Ground moraine (Fig. 10–17) is a broad blanket of till composed of detritus once carried in the bottom of the glacier, as well as debris that had melted downward during ablation. In contrast, **terminal moraines** are accumulations of till along the lobate front of the glacier. They are produced by the conveyer-belt action

Figure 10–16 Till deposit from an end moraine west of Madison, Wisconsin. (Courtesy of Professor Robert F. Black.)

of the moving ice, which brings rock fragments to the snout of the glacier even when it is in a state of equilibrium and not actively advancing. When the ice does move forward, however, it may push the terminal moraine ahead in much the same way as a bulldozer might. Glaciers that sporadically retreat may leave behind a succession of distinctive crescent-shaped terminal deposits called **recessional moraines** (Fig. 10–18). By contrast, when glaciers recede continuously, they deposit a layer of debris that is indistinguishable from ground moraine.

Rock debris that has fallen from mountainsides onto the margins of valley glaciers is carried along and eventually deposited as ribbons of till called **lateral moraines** (see Fig. 10–6). Long after glaciers have disappeared, these lateral moraines remain as ridges along margins of valleys. When glaciers from two adjoining valleys come together, neighboring lateral moraines coalesce and form a **medial moraine** (see Fig. 10–6). Out beyond the terminus of a glacier, one frequently finds an apron of large rock fragments, gravel, sand, and clayey materials. This irregular blanket of debris constitutes the **outwash plain** (see Fig. 10–18). Its constituent particles are transported outward from the glacier by streams of meltwater so choked with sediment that they frequently assume a braided pattern. In the outwash plain, material previously deposited as ground moraine may be reworked into deposits of stratified drift.

In regions once covered by ice sheets, one sometimes encounters long sinuous ridges of sand and gravel resembling raised railroad beds. These features

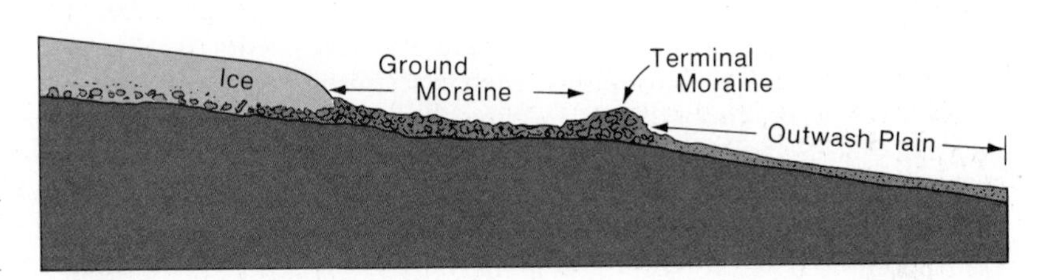

Figure 10–17 Terminal moraine, ground moraine, and outwash plain.

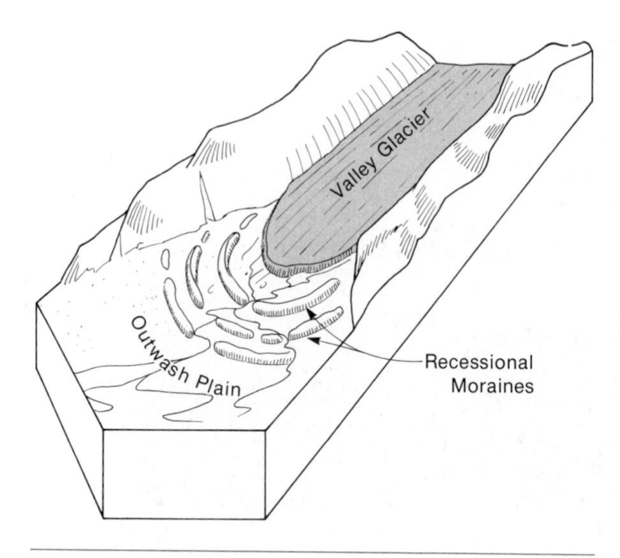

Figure 10–18 Recessional moraines.

never held rails, however, for they are ridges of stratified drift called **eskers.** Eskers are the deposits of former subglacial streams (Fig. 10–19).

The upper surface of a glacier is likely to hold numerous random depressions into which meltwater may wash sand and gravel. Small alluvial fans may also form at the margins of the glacier and extend out onto the ice. When the ice melts, the accumulation of sediment is lowered to the ground to form irregular mounds called **kames** (see Fig. 10–19). In valley glaciers, streams may develop along the margins of the glacier. The deposits of these streams when lowered to the ground by melting form raised embankments called **kame terraces** (see Fig. 10–19).

One kind of depositional feature that forms a depression rather than a terrace or mound is called a **kettle.** To understand how a kettle forms, imagine that the load of debris carried by a glacier is not uniformly distributed in the ice, so when the ice melts there is less sediment lowered to the ground in some places than in others. Where this occurs, there will be a small basin or kettle that may fill with water to form a kettle lake (Fig. 10–20). Kettles may also develop from the melting of large blocks of ice that were buried in till. Where excessive rock material existed in scattered parts of the glacier, **knobs** (Fig. 10–21) may form upon melting, thereby giving rise to kettle and knob topography.

Another feature particularly characteristic of regions once covered by continental glaciers are drumlins. **Drumlins** are low, elliptical, egg-shaped hills resembling inverted teaspoons and having their long axes oriented parallel to the direction of glacial movement (Fig. 10–22). They are rarely more than about 50 meters in height, generally occur in large numbers within a given area, and have steeper slopes on the sides from which the glaciers came. As for how they are formed, most offer evidence of a depositional origin followed by erosional modification of the deposited material. Clearly, their streamlined shape and parallel orientation of their long axes indicate that drumlins were shaped by moving ice.

THE QUATERNARY ICE AGE

The development of the glacial theory

If one were to leave the main highways and explore the back roads of the northern United States, Canada, and northern Europe, it would be possible to find large boulders of granitic rock lying at random about the countryside. One would surmise that these boulders were far from home, for they lie upon quite different rock, and exposures of similar igneous rocks are nowhere to be found nearby. Such displaced boulders and cobbles are called **erratics** (Fig. 10–23). They give us cause to wonder what gargantuan agency could have transported and dumped them onto the fields where they now rest.

Two centuries ago scientists would have explained erratics as having been carried by enormous currents of water and mud during the devastation accompanying the great biblical deluge. Indeed, these same scientists referred to the boulder deposits as *diluvium*, a Latin word meaning deluge. The eminent nineteenth-century British geologist Charles Lyell saw similarities between the deposits and rocky debris left by alpine glaciers. Unable to abandon the diluvian ideas, he surmised that boulder-laden icebergs had drifted across the inundated lands of Europe during the biblical flood, and as the ice began to melt, it dropped boulders and cobbles onto the submerged hills below.

An alternative theory was born in Switzerland, and its development began with studies made by a Swiss director of salt mines named Jean de Charpentier. Charpentier was fascinated by glaciers, and explored them whenever he had the opportunity. Initially, he was not concerned with an explanation of erratics, but rather used bouldery deposits as evidence that alpine glaciers once extended well beyond their present

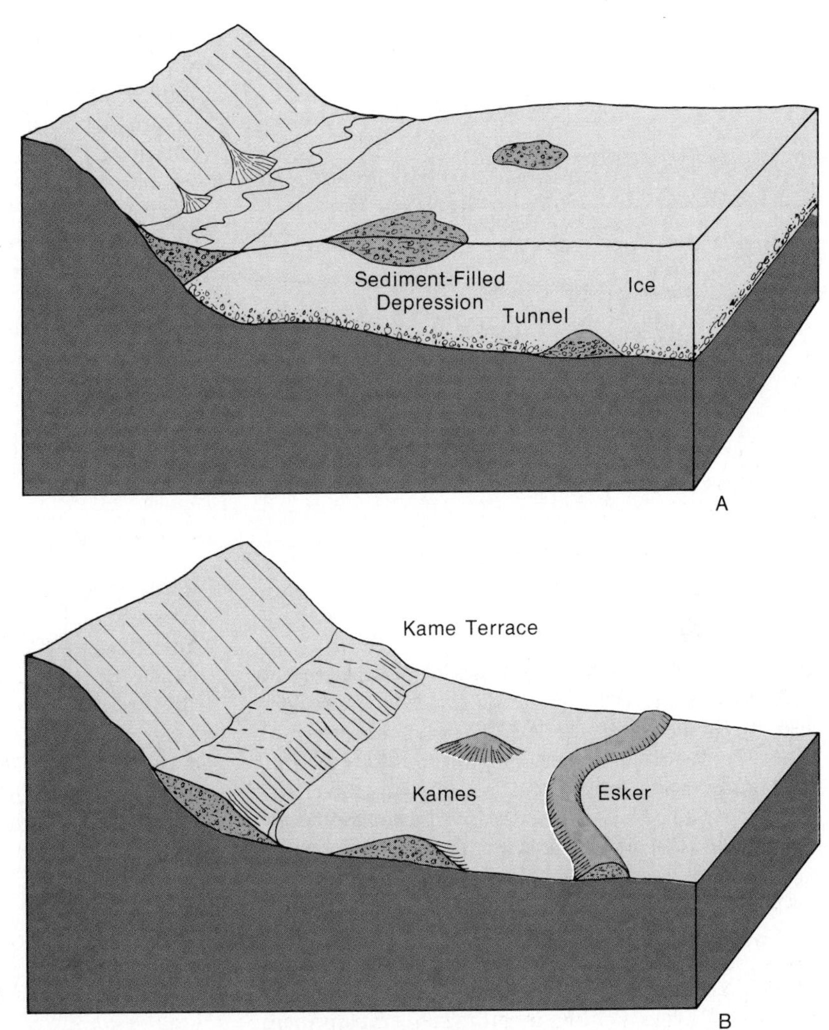

Figure 10–19 *The formation of kames and eskers.*

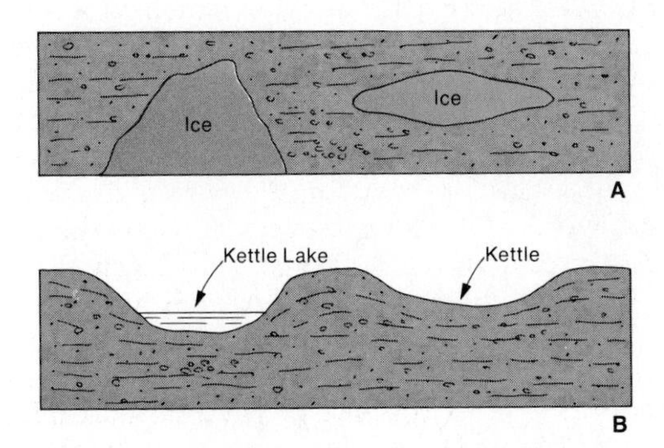

Figure 10–20 *Formation of kettles. (A) Cross-section of drift containing buried masses of ice. Upon melting (B) depressions are formed at the surface of the lost volume of ice.*

limits. In 1836, Charpentier brought a young friend and colleague named Louis Agassiz to his home in Bex, Switzerland, to show him the field evidence favoring his idea. Agassiz was a discerning scientist. On every side he saw indications of the way glaciers moved, witnessed the way they had altered the landscape, and observed their distinctive markings and deposits. He not only became convinced that Charpentier was correct about the former greater extent of alpine glaciers, but also was able to visualize how erratics were carried across the plains and lowlands of northern Europe. Erratics were transported, said Agassiz, by a blanket of ice so thick and so vast that nearly half of Europe was buried beneath it (Fig. 10–24).

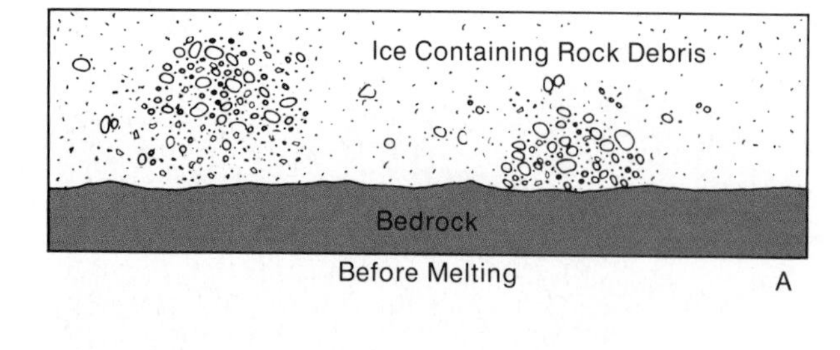

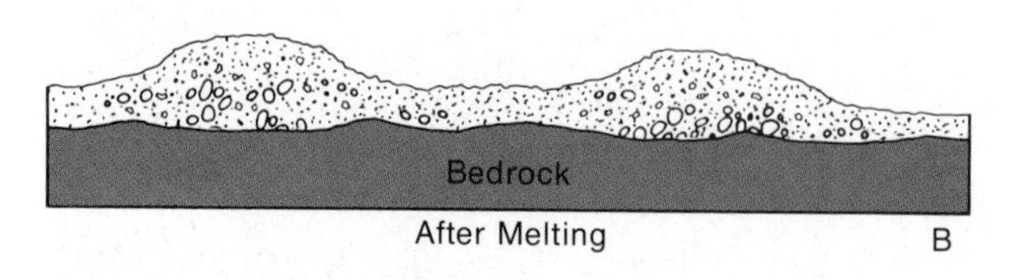

Figure 10–21 *Formation of knobs by melting of ice in which there are local concentrations of debris.*

It is not unusual in science for revolutionary new ideas to be greeted with skepticism. This is precisely what occurred when, in 1837, Agassiz presented his theory to the Swiss Society of Natural History. A few prominent members of the Society suggested that Agassiz would be best advised to give more attention to his study of fossil fishes and less to stories of gigantic ice sheets. Nevertheless, Agassiz's theory was based on objective and observable criteria. Soon others were able to see the evidence as well, and the ranks of believers in a Great Ice Age slowly began to swell. Even Charles Lyell eventually accepted Agassiz's ideas.

The development of vast continental ice sheets has occurred several times in the geologic past, but those ice sheets of the Quaternary Period are the most recent, and hence have the clearest geologic record. The Quaternary is believed to have begun about 2 million years ago and ended 8000 or 10,000 years ago with the onset of the Holocene, or Recent Epoch. The continental glaciers formed during the Quaternary ultimately covered about one fourth of the earth's land area with more than 40 million cubic miles of snow and ice. In North America alone, the ice sheet extended nearly 6400 km across the entire breadth of Canada (Fig. 10–25). Such an extensive coverage of ice had profound effects not only upon the glaciated terrains themselves but also upon lands far from the ice fronts. Climatic zones in the Northern Hemisphere shifted southward, and arctic conditions prevailed across northern Europe and the United States. Mountains and uplands in Eurasia and North America were sculptured by spectacular mountain glaciers. While the snow and ice were accumulating in higher latitudes, rainfall increased in the lower latitudes, with generally beneficial effects on plant and animal life. Even as recently as 10,000 years ago, currently arid

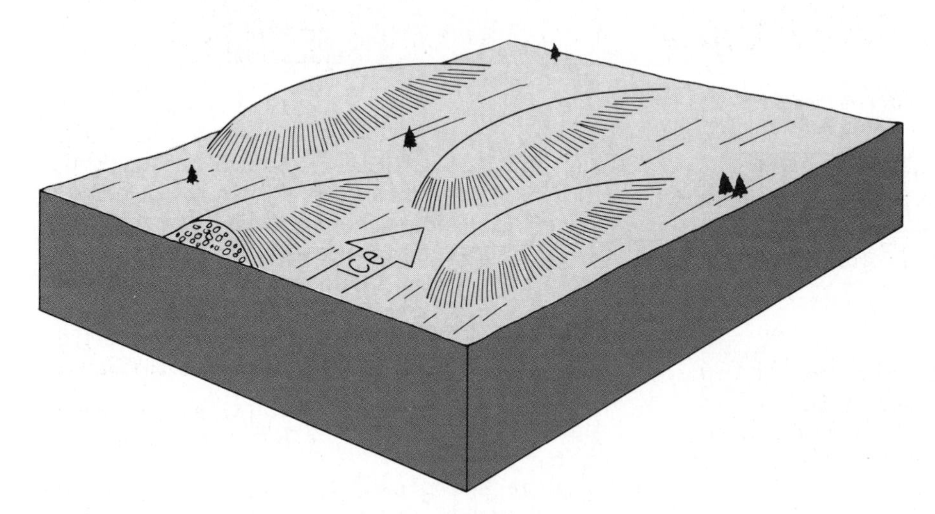

Figure 10–22 *Drumlins. The steeper end is the ice-approach side. These drumlins are about 600 meters long.*

Figure 10–23 *A glacial erratic has provided protection for underlying weak sedimentary rocks, while surrounding areas were eroded. As a result, the erratic is now perched on a pedestal of sandstone. An eagle has built a nest on top of the erratic. Chouteau County, Montana. (Courtesy of U.S. Geological Survey and W.G. Pierce.)*

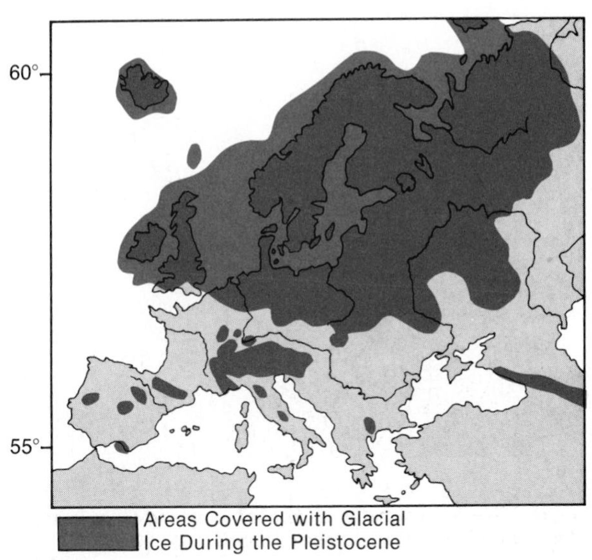

Figure 10–24 *Areal extent of major glaciers in Europe during the Pleistocene Epoch.*

Figure 10–25 *Maximum area of glacial coverage in North America. (From United States Geological Survey Pamphlet,* The Great Ice Age. *U.S. Government Printing Office, 0–357–128, 1969.)*

regions in north and eastern Africa were well watered, fertile, and populated by nomadic tribes.

During the Quaternary, there were at least four major advances of ice separated by three warmer interglacial intervals. The deposits of each of these time intervals is called a stage, and each glacial and interglacial stage is named after a locality where deposits of that age are well exposed for study (Fig. 10–26).

NORTH AMERICA	ALPINE REGION
WISCONSIN	Würm
Sangamon	Riss-Würm
ILLINOIAN	Riss
Yarmouth	Mindel-Riss
KANSAN	Mindel
Aftonian	Gunz-Mindel
NEBRASKAN	Gunz
Pre-Nebraskan	Pre-Gunz

A

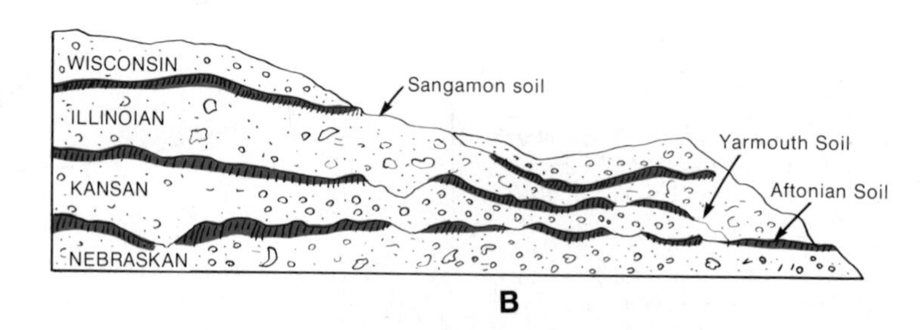

B

Figure 10–26 Standard Pleistocene nomenclature for glacial (all capital letters) and interglacial stages are shown in A. B is an idealized cross-section showing a succession of deposits of glacial stages and interglacially developed soils. (A section this complete would be a rare occurrence.)

In addition to these major glacial stages, geologists using carbon 14 techniques to study mountain moraines have discovered the occurrence of three significant glacial advances over only the past 5000 years. This recent episode of cold conditions has been dubbed the Little Ice Age. The most extensive glaciation of the Little Ice Age culminated within the last several centuries.

The effects of Pleistocene glaciations

The more than 40 million cubic kilometers of ice and snow that lay upon the continents during the Quaternary was equivalent to a tremendous amount of water. Removal of that water from the oceans had a multitude of effects on the environment. It has been estimated that sea level may have dropped at least 75 meters during maximum ice coverage. Extensive tracts of the present continental shelves became dry land and were covered with forests and grasslands. The British Isles were joined to Europe, and a land bridge stretched from Siberia to Alaska. During interglacial stages, marine waters returned to the low coastal areas, drowning the flora and forcing terrestrial animals inland.

The glaciers had a direct impact on the erosion of lands and the creation of glacial landforms. The great weight of the ice depressed the crust of the earth over large parts of the glaciated area, in some places to a level of 200 to 300 meters below the preglacial position. With the removal of the last ice sheet, the downwarped areas of the crust gradually began to return to their former positions. The rebound is dramatically apparent in parts of the Baltic, the Arctic, and the Great Lakes region of North America, where former coastal features are now elevated high above former levels (Fig. 10–27).

As the great continental glaciers advanced, they obliterated old drainage channels and caused streams to erode new channels. These dislocations are especially evident in the north-central United States. Prior to the Ice Age the northern segment of the Missouri River drained northward into Hudson Bay, and the northern part of the Ohio River flowed northeast-

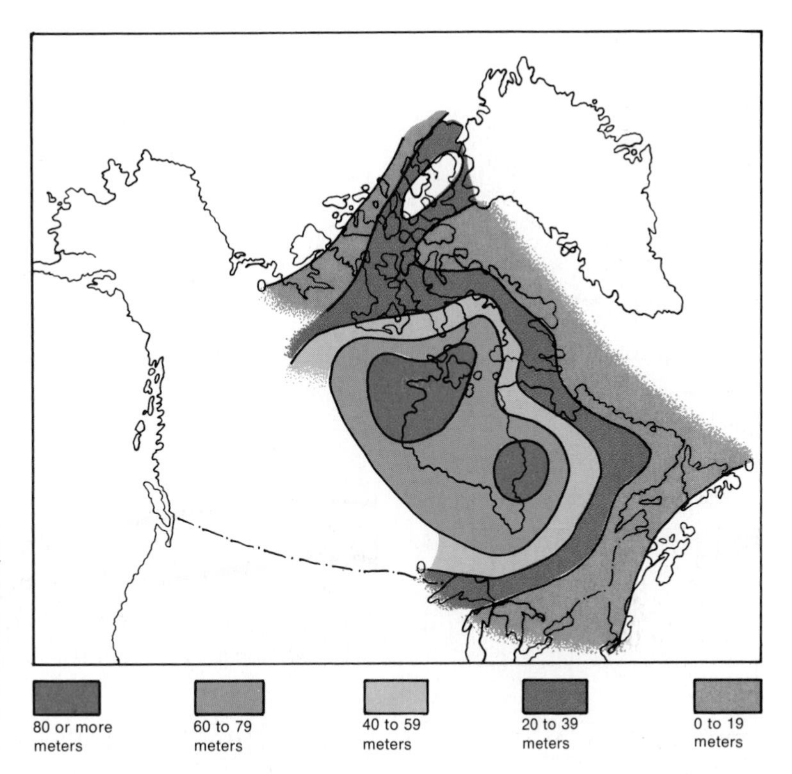

Figure 10–27 *Postglacial uplift of North America determined by measuring elevation of marine sediments 6000 years old. (Simplified and adapted from Andrews, J.T.,* The Pattern and Interpretation of Restrained, Postglacial and Residual Rebound in the Area of Hudson Bay. In *Earth Science Symposium on Hudson Bay. Ottawa, 1968: Canadian Geological Survey Paper 68–53, p. 53, 1969.)*

ward into the Gulf of St. Lawrence. The lower Ohio drained into a preglacial stream named the Teays River. Today, geologists recognize the former location of the Teays by a thick linear trend of sands and gravels. Those parts of the Missouri and Ohio rivers that once flowed toward the north were turned aside by the ice sheet and forced to flow along the fringe of the glacier until they found a southward outlet. The present trend of the Missouri and Ohio rivers approximates the margin of the most southerly advance of the ice. The equilibrium of streams was also affected, for with glacial advances there was a lowering of sea level and thus an increase in stream gradients near coastlines. A reverse effect occurred with wasting away of the ice sheets.

Prior to the Pleistocene, there were no Great Lakes in North America. The present floors of these water bodies were lowlands. Glaciers moved into these lowlands and scoured them deeper. As the glaciers retreated, their meltwaters collected in the vacated depressions (Fig. 10–28). Niagara Falls (see Fig. 8–26), between Lake Erie and Lake Ontario, came into existence when the retreating ice of the last glacial stage uncovered an escarpment formed by southwardly tilted resistant strata. Water from the Niagara River tumbled over the edge of the escarpment, which is supported by the resistant Lockport Dolostone. Weak shales beneath the dolostone are continuously undermined, causing southward retreat of the falls.

Another large system of ice-dammed lakes covered a vast area of North Dakota, Minnesota, Manitoba, and Saskatchewan. The largest of these lakes has been named Lake Agassiz. Today, rich wheatlands extend across what was once the floor of the lake. Other lakes developed during the Pleistocene, occupying basins that were not near ice sheets, but formed as a consequence of the greater precipitation falling on regions south of the glaciers. Such water bodies are called **pluvial lakes** (Latin *pluvia,* rain). Pluvial lakes were particularly numerous in the northern part of the basin and range province of North America. Lake Bonneville in Utah was such a lake. It once covered more than 50,000 square kilometers and was as deep as 400 meters in some places. Parts of Lake Bonneville persist today as Great Salt Lake, Utah Lake, and Sevier Lake.

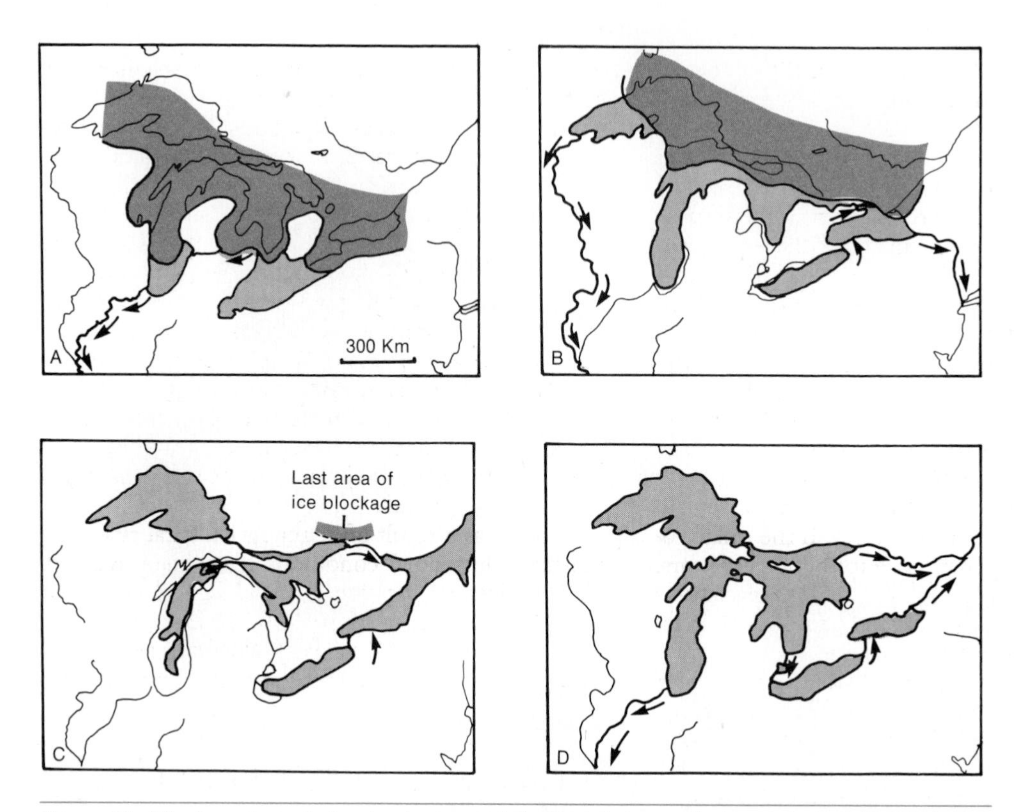

Figure 10–28 *Four stages in the development of the Great Lakes as the ice of the last glacial advance moved away. (After Hough, J.L.,* Geology of the Great Lakes. *Urbana, Ill., University of Illinois Press, 1958, Figs. 56, 69, 73, and 74.)*

A rather spectacular event associated with Pleistocene lake formation occurred in the northwestern corner of the United States. Lobes of the southwardly advancing ice sheet repeatedly blocked the Clark Fork River, and the impounded water formed a long, narrow lake extending diagonally across part of western Montana. The freshwater body, called Lake Missoula, contained an estimated 2000 cubic km of water. With the recession of the glacier, the ice dam broke and tremendous floods of water rushed out catastrophically across more than 38,000 square kilometers of what are now appropriately termed the channeled scablands.

Cause of Pleistocene climatic conditions

The results of oxygen isotope research indicate that world climates grew progressively cooler from the Middle Cenozoic to the Pleistocene. The culmination of this trend was not a single sudden plunge into frigidity but rather an oscillation of glacial and interglacial stages. Any hypothesis for the Ice Age must consider not only the cause of cooling but also the reason for the oscillations. Further, the hypothesis must include reasons for the ideal combination of temperature and precipitation required for the buildup of continental glaciers. Although geologists have been speculating about the cause of the Ice Age for more than a hundred years, no single agent has been found. Indeed, it appears likely that Pleistocene climatic conditions came about as a result of several simultaneously occurring factors.

One particularly interesting hypothesis of Ice Age origin was developed by M. Milankovitch, a Yugoslavian astronomer who convincingly calculated that irregularities of the earth's movement could cause climatic cycles. Milankovitch drew attention to the fact that the tilt of the earth's axis to the plane of orbit varies between about 22° and 24°. This tilt, of course, causes the seasons, and furthermore, the greater the tilt the greater the contrast between summer and winter temperatures at a particular latitude. The axis also has a wobbling motion called precession (see Fig. 17-10). In addition, the earth's orbit around the sun is an ellipse that varies in its eccentricity between known limits. According to the Milankovitch theory, a combination of these astronomical factors would periodically result in longer, colder winters and shorter, warmer summers at particular latitudes. The coldest point in the cycle would occur every 40,000 years or so. The timing of glaciations as predicted from the Milankovitch calculations corresponds rather well to indications of cooling obtained from oxygen-isotope and foraminiferal analyses of deep-sea cores. However, there is one problem with the theory. If the Milankovitch effect has been in existence for billions of years, why haven't there been Pleistocene-like glaciations down through geologic time? Apparently, other factors must also be involved.

One such factor might be a variation in the amount of solar energy reflected from the earth into space rather than being absorbed. The proportion of solar energy reflected into space is termed the earth's **albedo.** At the present time, it is about 33 percent. Theorists suggest that by the end of the Cenozoic, when continents were fully emergent and hence highly reflective, temperatures may have been lowered enough for the Milankovitch effect to begin to operate. This is not an unreasonable suggestion, for only a 1 percent lessening of retained solar energy could lead to as much as an 8°C drop in average surface temperatures. This would be sufficient to trigger a glacial advance if ample precipitation were available over continental areas. Still other geologists speculate that absorption of solar energy was hindered by cloud cover, volcanic ash and dust in the atmosphere, or fluctuations in carbon dioxide. A decrease in carbon dioxide content would cause a corresponding decrease in the warmth-gathering "greenhouse effect," for example. On the other hand, it is also possible that a buildup of CO_2 might trigger glaciation, for as warming occurred there might be more rapid evaporation and an increase in highly reflective cloud cover.

THE WORK OF WIND

The predominantly horizontal motion of air is called wind. In Chapter 15, the origin, dynamics, and global patterns of winds will be examined. At this point, however, our attention is directed to the ways in which wind accomplishes geologic work and to the distinctive landscape features produced by wind action.

The effectiveness of wind in lifting and transporting fine soil particles was dramatically demonstrated in the Western plains of the United States during the 1930s (Fig. 10-29). At that time, drought and crop failures had exposed millions of acres of loose soil to the ravages of gusting winds. A region from the Canadian prairie provinces into Texas became a huge dust bowl. Once-productive farms were stripped of their soils, and dust-weary farm families abandoned their homes and suffered the trauma of displacement so movingly described by John Steinbeck in his novel *The Grapes of Wrath.*

Today, as a result of better agricultural practices, full-scale dust bowl conditions have been avoided. Nevertheless, an occasional dust storm still occurs. It has been estimated that even in a rather average dust storm, more than 3000 metric tons of clay and silt are being carried in each cubic kilometer of moving air.

Transportation of sand, silt, and clay

The flow of air that we call wind is in some ways similar to the flow of water. Like water, air tends to flow in a laminar fashion when moving very slowly, but it flows turbulently at higher velocities. Air, like water, is retarded by friction with the surface over which it passes, and—as is the case with water in channels—its lowest velocity is immediately above the solid surface over which it flows. Of course, there are also differences between the behavior of flowing air and flowing water. Flowing water is confined to channels and is driven by a component of the force of gravity that acts parallel to the surface across which it moves. Winds, on the other hand, are impelled by differences in air pressure resulting from unequal heating of the earth's surface. It is also apparent that air lacks the specific gravity and viscosity of water and is thus less effective as an agent of erosion. Moving air does have one advantage over water, however, in that sand grains carried in wind are not impeded as much as in the denser liquid medium. As a result, a skipping sand grain on impact can dislodge another grain more than six times its own size.

As noted earlier, wind can only accomplish its work if fine-grained, dry, unconsolidated, unvegetated materials exist on the ground surface. Sand and silt are picked up most readily. Such particles are large

Figure 10–29 *This 1937 photograph of an abandoned farmstead in Oklahoma shows the disastrous effects of drought followed by wind erosion. (Courtesy of USDA.)*

enough to provide a surface against which the wind can push, and small enough that they are not too heavy to lift. The flattish clay particles tend to hug the ground and require extraordinary velocities before they are lifted into the air stream.

Once particles lying on the surface have been dislodged, they may be transported above the ground as part of the wind's suspended load or moved by saltation and rolling as the bed load (Fig. 10–30). Generally, particles in the suspended load are less than about 0.15 mm in diameter. Often the smaller particles of the suspended load are buffeted high into the atmosphere, where they are effectively segregated from more earthbound materials. This helps to account for the striking absence of such fine sediment in the sandy deposits of deserts.

A wind's bed load consists mostly of particles in the 0.15 to 2.0 mm size range. Such particles will not stay in suspension in air but rather move along the surface of the ground by rolling and saltation. As is the case with transportation of sedimentary particles

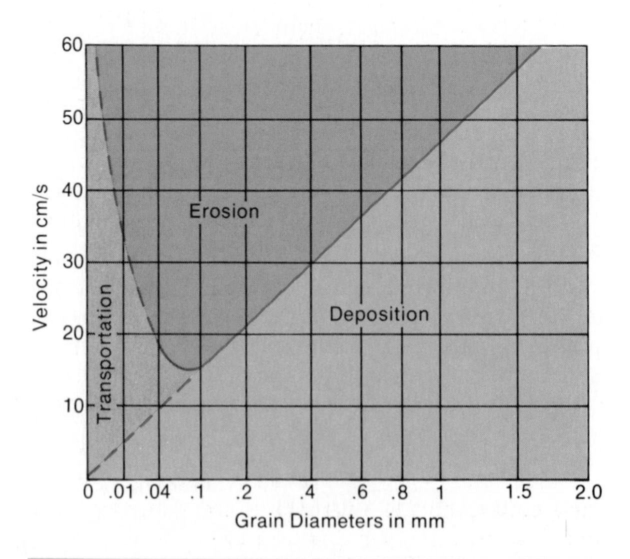

Figure 10–30 *Erosion and transportation curves for wind. (Adapted from Bagnold, R.A.,* The Physics of Blown Sand and Desert Dunes. *Methuen and Co., Ltd., London, 1941, p. 88.)*

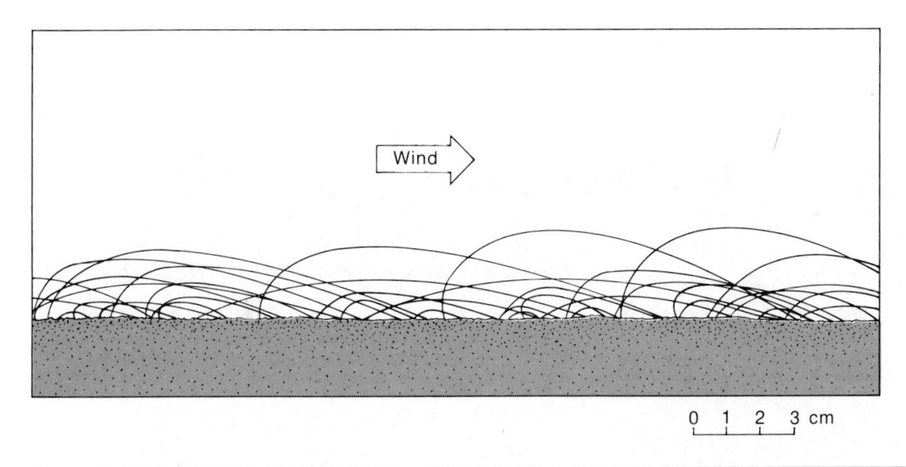

Figure 10–31 Saltation movement of sand particles over a surface composed of sand.

by streams, saltation is a pattern of movement whereby grains move forward in the direction of the air current by making low arcing jumps (Fig. 10–31). During saltation, grains may bounce off other grains on impact with the ground, and frequently they may also dislodge other grains so that they, too, begin the saltation movement downwind. Some additional transport occurs as myriads of impacting grains collide with grains on the ground, causing a slow forward "creep" of these larger particles.

Saltating sand grains frequently rise to a height of 100 to 150 cm above the ground. During a sand storm it is often possible to estimate the average maximum height of saltating grains by the clearly visible upper limit of the cloud of saltating grains. Even after the storms, the approximate height of saltation can be discerned on the sandblasted portion of the surface of telephone poles and other vertical objects. Only rarely do such sandblasted surfaces extend higher than about two meters.

Deflation and abrasion

The erosion of rock and soil surfaces by the wind is accomplished by the two processes of deflation and abrasion. Deflation refers to the removal of loose and dry sediment as the wind blows across an unprotected surface. Without deflation there would not be sand and dust storms. Deflation results in **desert pavements** (Fig. 10–32) of pebbles and cobbles by selectively blowing away finer particles. Wherever patches of ground are unprotected by vegetation, deflation may sweep away surface materials, producing shallow, roughly circular depressions called **blowouts** (Fig. 10–33). Erosion of this kind can often be prevented by planting suitable vegetative cover and building wind breaks.

Abrasion, the second means by which wind accomplishes erosion, is the wearing away of rock or soil masses as they are repeatedly struck by myriads of wind-borne sand and silt grains. The process resembles sandblasting. As the grains forcefully strike against rock surfaces, they break grains, weaken cements so that particles fall free of the rock, or actually dislodge grains from the matrix. The sand grains freed from the parent rock then become additional tools in causing further abrasion. When examined closely, sand grains of this kind display small impact pits on their surfaces, which give them an overall "frosted" appearance.

If a large cobble projects into the wind in an area where abrasion is occurring, the windward side will be sandblasted to a smooth sloping surface. Rocks modified by wind abrasion in this way are called **ventifacts** (Fig. 10–34), and when found in ancient sedimentary rocks, they are evidence for wind-blown, or eolian, deposition.

Deposition of wind-borne sediment

Sooner or later, the particles carried by wind are deposited. The finest materials are distributed widely over the earth. Some fall into the oceans, some onto forested areas, and some onto fields where they are

Figure 10–32 Desert pavement, Lake Mead National Recreation Area. (Courtesy of John V. Bezy and the Southwest Parks and Monuments Association.)

incorporated into soils. Rains wash the particles of dust from the air, and like other sediments, they are carried to the sea. The larger grains of wind-borne sediment are more readily visible to us, for they collect along shorelines and in deserts as dunes and blankets of sand. All wind-borne material, whether fine silt or sand, is described as **eolian,** a term derived from Aeolus, the Greek god who controlled the winds by releasing them at will from his cave.

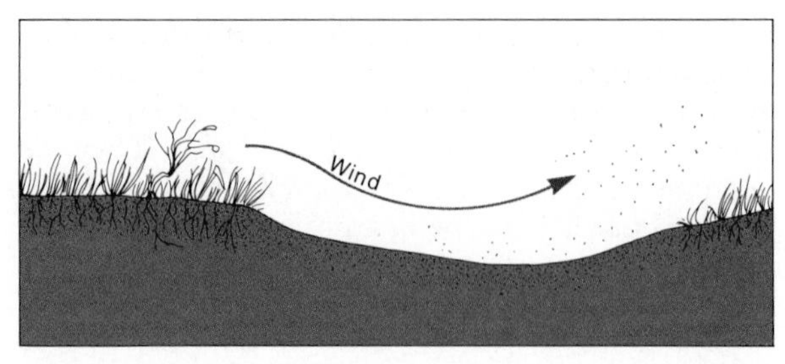

Figure 10–33 Blowout developing in area of patchy vegetative cover.

Loess

The term loess was first used to designate the tawny-colored, unconsolidated, silty deposits exposed along the bluffs overlooking the Rhine River in West Germany. Such deposits are also known in many other regions of the globe. In North America, loess deposits are several meters thick and blanket hundreds of square kilometers of terrain adjacent to the Missouri and Mississippi rivers. Vast blankets of loess also cover the plains of northern China.

Geologists define **loess** as a soft, homogeneous, porous, unconsolidated, and unstratified deposit consisting predominantly of silt but also containing lesser amounts of clay and very fine sand. Quartz is the most common mineral in loess. Nearly all deposits also contain 15 to 25 percent calcite, as well as minor amounts

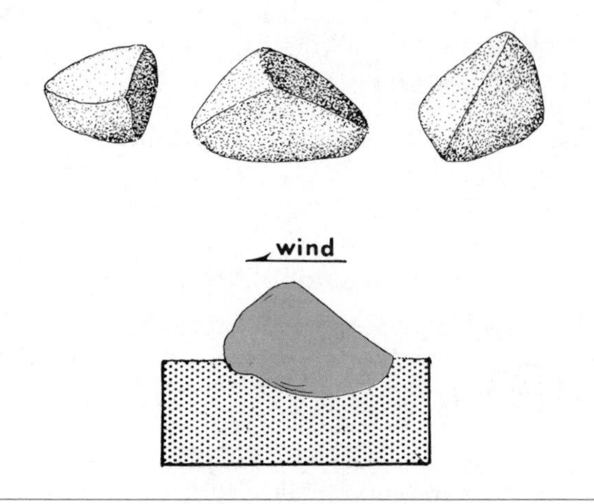

Figure 10–34 Ventifacts (top row). Pebble being shaped into ventifact by wind action (bottom).

of feldspar and mica. A striking characteristic of loess is the way in which exposures of the material break off in vertical slabs, leaving behind steep escarpments. Loess deposits develop into excellent soils. They are not only rich in the minerals required for healthy plant growth but also can be easily tilled and have good permeability.

Sand deposits

Contemporary dunes. Although eolian sand deposits may be blanket-like in form, they frequently accumulate as hills and ridges of sand called **dunes** (Fig. 10-35). Not all dunes are found in tropical deserts. They occur along the shorelines of oceans and large lakes and even on floodplains of streams in arid or semiarid regions. Even in arid regions, one finds dunes that derive their sand from the desert, but actually accumulate just beyond the desert's perimeter.

Dunes may form wherever there is a decrease in the velocity of wind that had been actively transporting sand grains. For example, anyone who has stood behind a tree during a windstorm is aware that the wind velocity in that protected area is much reduced. On any tract of land, one nearly always finds large and small obstacles that similarly offer a protected side for the accumulation of sand grains. The wind must sweep up and over such obstacles, and in doing so, an area of low-velocity eddying air is left behind on the lee side, as well as near the base on the windward side. These quieter areas are called **wind shadows,** and the boundary between the wind shadow and the overlying sweep of fast-moving air is termed a **surface of wind discontinuity** (Fig. 10-36). Sand grains that are carried into the lee wind shadow are deposited there, thus extending the obstacle downwind. Similarly, some of the grains swept over the top of the obstacle may fall through the surface of discontinuity and accumulate. Eddying air currents develop within the shadow zone

Figure 10-35 *The deserts of Tunisia contain vast dune fields which are exceptionally difficult to traverse. Here a bulldozer is moving into position to prepare an access road for a geophysical survey. Note the well-developed current ripple marks on the surfaces of the dunes. (Courtesy of Western Geophysical, photograph by Volker Vagt.)*

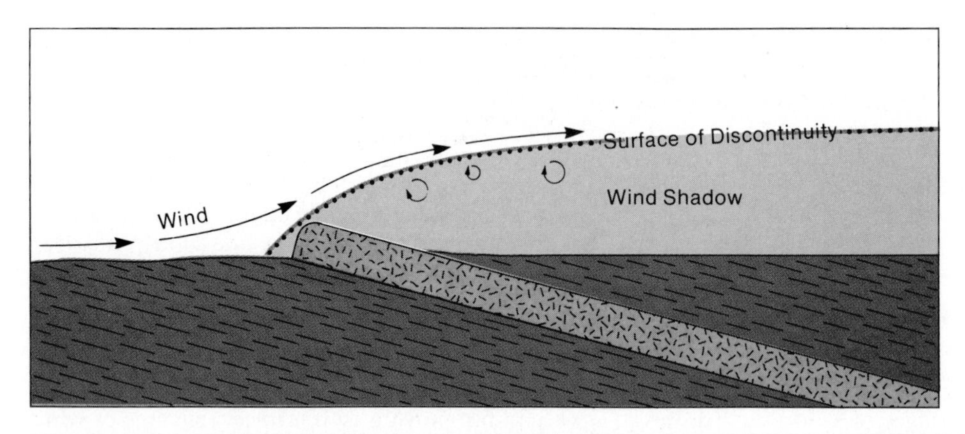

Figure 10–36 The air in front of an obstacle is divided by a surface of disconformity above which the air flows smoothly by, and below which the air moves forward in eddies, and has a lower average forward velocity than the wind above the surface of disconformity. (Adapted from Bagnold, R.A., The Physics of Blown Sand and Desert Dunes. *Methuen and Co., Ltd., London, 1941, p. 190.)*

and contribute to the process of dune formation by sweeping sand inward toward the central part of the wind shadow. Of course, once the dune begins to build, it provides an increasingly effective obstacle to wind. The steeper leeward slope of the dune is called the **slip face** because of the many small sand slides that occur there.

The windward slope of a dune is less steeply inclined than the lee slope (Fig. 10–37). Wind directed along the windward slope picks up and moves sand grains up the gentler incline toward the crest of the dune. Thus, the windward slope is continually being beveled by the wind and rarely acquires a slope of more than about 10°. On reaching the crest of the dune, sand grains that have been moved up the gentler windward slope are left to cascade down the leeward slope, where they assume a natural angle of repose of 30° to 40°.

The height to which sand will accumulate on dunes is limited by wind velocity. The current of wind blowing over a dune is compressed into an increasingly smaller space as it accelerates up the windward slope (Fig. 10–38). As more air is forced to rush through a smaller area, velocities increase to the point where grains blow off the top of the dune as fast as they are added. Naturally, the height attained depends upon the quantity of sand grains available for dune building, the overall wind speed, and the size of the sand grains being deposited. In the Sahara Desert, dunes as tall as 100 meters are commonplace.

From the explanation of how dunes are formed, it is evident that they are not stationary features but rather experience more or less continuous migration. In some parts of the world, the migration of dunes poses a serious threat to villages and farms. Wind-

Figure 10–37 Aerial view of transverse sand dunes in Imperial Valley, California. Wind is blowing from the upper left side to the lower right of the area. Notice the more gentle windward slope of each dune and the more steeply inclined leeward slope. (Courtesy of U.S. Geological Survey and J.R. Balsley.)

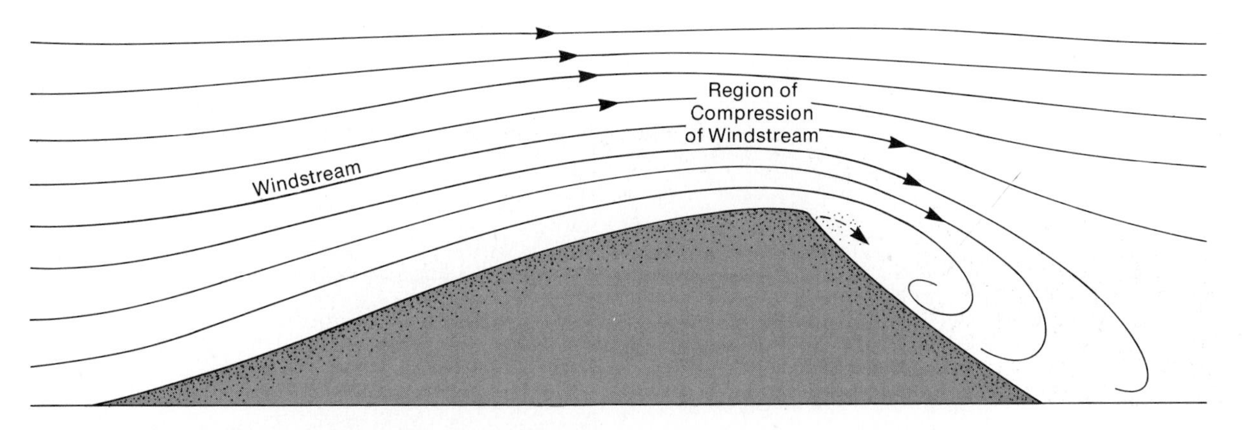

Figure 10–38 *As the wind blowing over a dune is compressed into a smaller space, it increases in velocity and in the power to erode. A point is reached where the wind is able to blow grains off the top of the dune as quickly as they are deposited.*

breaks are constructed, and vegetation is planted to protect fields and property from the encroachment of sand.

Dunes usually occur in groups, either in roughly parallel migrating series or as irregular migrating complexes. Depending upon the amount of sand available, the constancy of wind direction, and wind velocity, dunes may develop any of a variety of interesting and sometimes esthetically pleasing shapes. Perhaps the most interesting forms are the **barchans** (Fig. 10–39). These are crescent-shaped dunes in which the "horns" are directed downwind. Barchans characteristically develop in regions where wind direction is rather constant and the supply of sand relatively limited. In fact, the surfaces of the ground adjacent to barchans may consist of bare hard rock, with no sand cover whatsoever. The "horns" on the barchans develop because the wind sweeping around the sides of the accumulation of sand is less impeded and is thus able to move sand somewhat more efficiently. As a result, the long horns of sand are extended to the leeward. Crescent-shaped dunes may also form with their horns directed upwind. Such accumulations are designated **parabolic dunes** and nearly always represent an accumulation of sand around a blowout (Fig. 10–40).

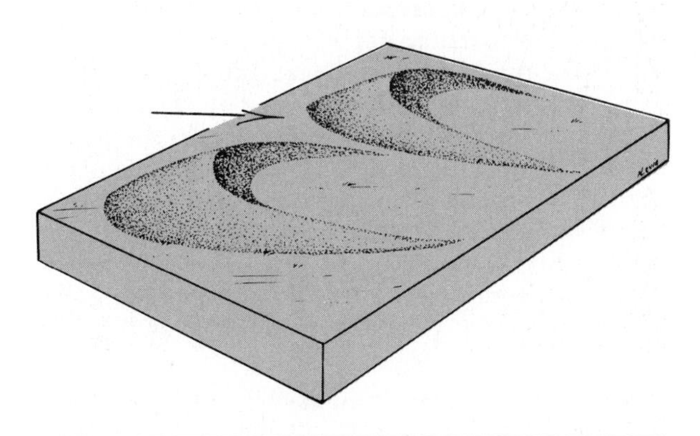

Figure 10–39 *Barchans are readily recognized by their crescent shape. "Horns" of crescent are directed downwind (direction of wind indicated by arrow).*

Where winds are strong and the supply of sand is abundant, **transverse dunes** (see Fig. 10–41) with crests at right angles to the wind may develop. If such elongate accumulations are aligned parallel to prevailing wind directions, they are referred to as **longitudinal dunes.** Longitudinal dunes are most prevalent in warm deserts where the wind blows strongly from a single direction and there is virtually no plant cover. In North Africa, such dunes are called *seifs,* an Arabic word meaning sword.

The sand in most dunes consists of quartz with lesser amounts of feldspar. At some localities, however, other minerals may predominate. For example, the dunes found in Bermuda are composed of calcite grains derived from the erosion of reefs and from ac-

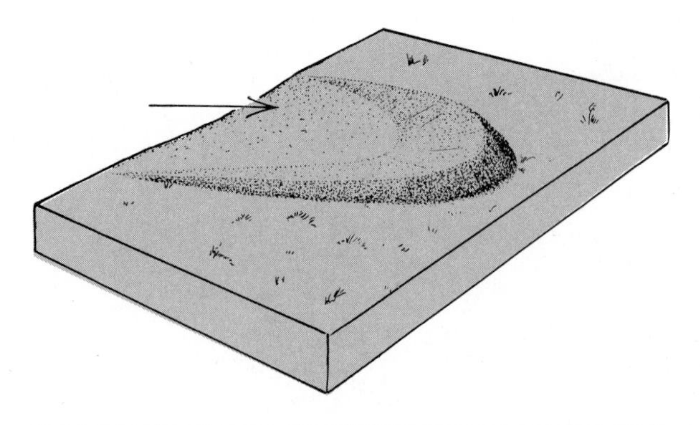

Figure 10–40 *Parabolic dune. Arrow indicates wind direction.*

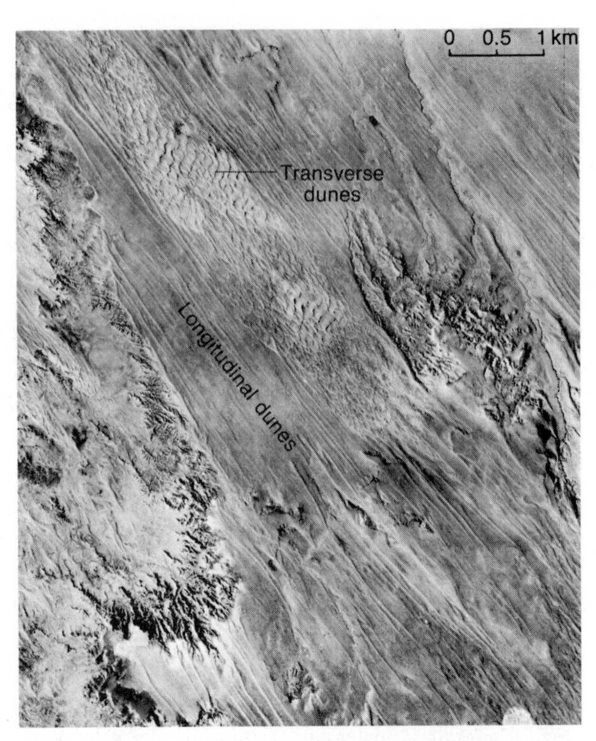

Figure 10–41 *Longitudinal dunes and transverse dunes, Coconino County, Arizona. Wind direction is toward southeast quadrant of photograph. (Courtesy of U.S. Geological Survey.)*

cumulation of the shells of marine invertebrates. The brilliant snow-white dunes of the White Sands area of New Mexico are composed of gypsum.

Ancient dunes. Geologists study modern dunes for the clues they hold to conditions long ago. In general, the sloping layers formed on the lee slopes of dunes are inclined in the downwind direction. Applying this observation to ancient eolian sandstones, one can sometimes determine the direction of prevailing winds that blew over land areas hundreds of millions of years ago. For example, measurements made of the 200-million-year-old New Red Sandstone of Great Britain indicate it was deposited by winds that blew mostly from the east and northeast. By comparison with the present global wind patterns, the measurements have bolstered a theory that suggests that, since the deposition of the New Red Sandstone, Britain and Europe have drifted northward and rotated clockwise by about 35°. These interpretations have recently been verified by paleomagnetic data.

Ripple marks. The surfaces of dunes and other sandy deposits, and also some of the bedding surfaces of eolian sandstones, are often characterized by subparallel ridges and hollows called **current ripple marks.** Current ripple marks may be formed by either water or wind. Those formed by water were described in Chapter 3, and therefore we need consider only eolian ripples here. Such wind ripples are a consequence of the saltation movement of sand grains. They will form spontaneously wherever there is an adequate supply of sand and the wind velocity is sufficient to erode and transport the sand-sized particles. After a few ripples have formed, additional ones develop quickly. Because of the average trajectory of incoming saltating grains, more sand will land on the windward side of the ripple than on the lee side (Fig. 10–42A). The windward side tends to be built up, while the lee side remains troughlike as a consequence of receiving too few grains. The ridge will steepen and the hollow deepen. A short distance downwind on the far edge of the trough, grains again impact with sufficient frequency to build a second ridge. This ridge in turn rises above the depression created by fewer grains impacting on its lee side. In this way ripple after ripple is formed. The spacing of ripples depends upon the average length of the saltating jumps (Fig. 10–42B), which is in turn determined by the size of the sand grains and wind velocity. Maximum height of each ripple crest is reached when deposited grains intercept winds moving so fast that grains are blown off the crest as quickly as they are added.

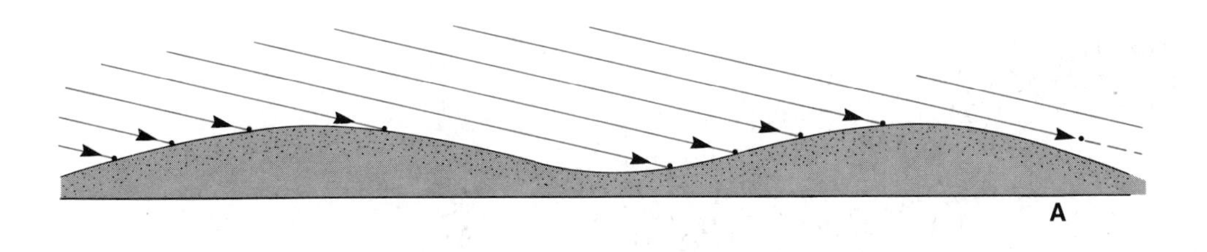

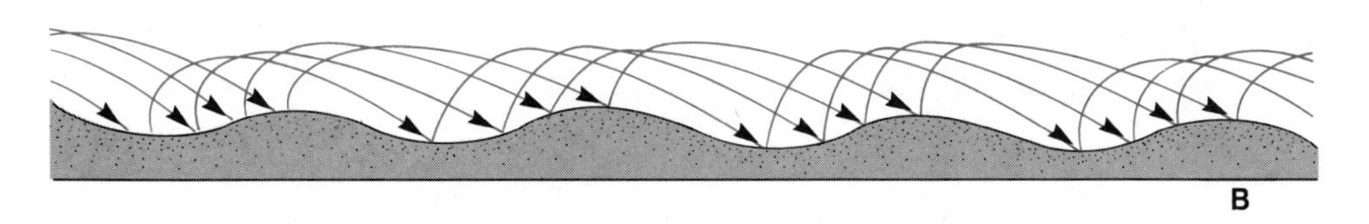

Figure 10–42 *From cross-section (A), it is apparent that the windward side of an uneven surface will receive a greater number of impacting grains than the lee side. Cross-section (B) suggests that there is a correspondence between the average distance between ripple crests and the average length of saltation paths. (After Bagnold, R.A.,* The Physics of Blown Sand and Desert Dunes. *Methuen and Co., Ltd., London, 1941, pp. 146, 150.)*

DESERTS

Characteristics of deserts

Because deserts generally have only sparse soil-holding vegetation, and also have dry soils, they are environments in which the effects of wind erosion are noticeable. Geologists study such areas assiduously, for by recognizing present-day desert features in ancient sedimentary rocks, they can infer where desert conditions existed millions of years ago. It may come as a surprise to devotees of Hollywood films depicting the French Foreign Legion, but not all deserts are vast areas of drifting sands and searing temperatures. Indeed, only about one fifth of the earth's total desert area is covered with sand. Also, not all deserts are hot, for deserts also exist in frigid polar regions.

What all deserts do have in common is a deficiency of vegetation so extreme that the region cannot support an appreciable human population. In polar regions, of course, the lack of vegetation results from the intense cold, but for most other deserts, the barren landscape is the result of insufficient rainfall. In these warm deserts, annual rainfall does not exceed 25 cm, and the potential rate of evaporation exceeds that amount. Because the amount of rainfall is so low and evaporation so great, streams in warm desert regions rarely extend to the oceans. The exception are streams like the Nile which have their headwaters in moist uplands. Vast floodplains are uncommon, and many streams are ephemeral. In addition to wind erosion, mass wasting, sheetwash, and rain pelting are important erosional processes in deserts. In fact, processes involving water are far more important than wind in the development of the major landforms of desert regions.

The majority of the world's great deserts are located in the low latitudes between about 15° and 30° (Fig. 10–43). Here one finds the famous Arabian and North African deserts, the Kalihari of southwestern Africa, and the immense Victoria Desert of Australia. Middle-latitude deserts are found beyond 30° to about 60° of latitude, and their aridity is usually enhanced by physiographic conditions. Thus, many middle-latitude deserts are located in the dry rain shadow of mountain ranges. Examples are the desert areas of the Basin and Range Province, the Gobi, and the small deserts that lie on the east flank of the Andes.

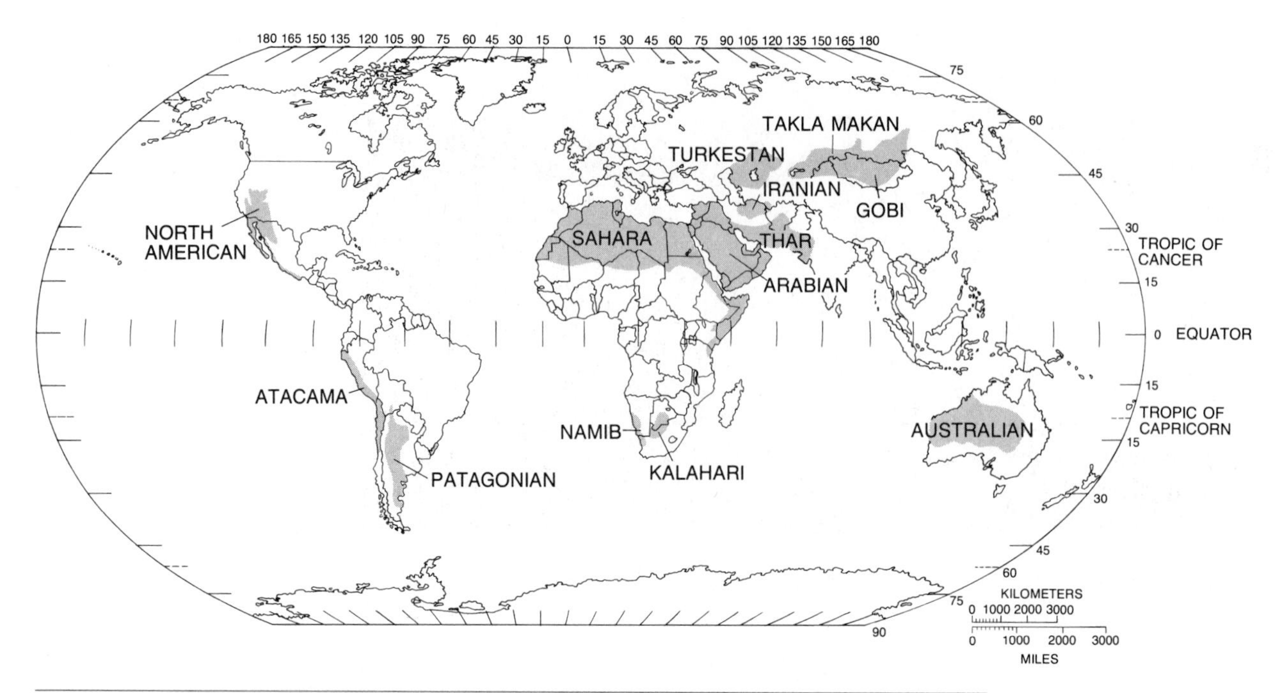

Figure 10–43 *Deserts of the world. (From Navarra, J.G.,* Atmosphere, Weather and Climate. *Philadelphia, W.B. Saunders Company, 1979.)*

Landforms of deserts

Deserts are often harsh, defiant regions that test the stamina of man and beast alike. Yet geologists freely relinquish studies of better-watered terrain in order to solve a problem in desert geology, for in such places rocks lie naked at the surface and can be traced across the land for scores of miles.

Because they are not covered with a heavy mantle of weathered material and dense vegetation, landscape features of deserts have a stark ruggedness and clarity (Fig. 10–44). On every side one sees evidence of insufficient water. As already noted, most desert streams are discontinuous and ephemeral. Channels may fill quickly after a sudden shower but are dry again in a few hours. For the most part, drainage systems are internal, and streams do not reach the ocean. The channels of most streams in deserts have rectangular cross-sections as a result of caving of dry banks along vertical cracks, the intermittent nature of the flow, and the high proportion of sediment flushed down the channel as bed load. The often dry, steep-walled channels are called **arroyos** in the United States, whereas in the Sahara, the Arabic term **wadi** is used.

Alluvial fans are often exceptionally well developed in desert regions that front on mountain ranges. As described in Chapter 8, alluvial fans are the result of rapid deposition of clastic sediment as a stream emerges from a mountain front and shifts radially back and forth across its former deposits. The alluvial fans of desert regions are noted for their symmetry. It is a characteristic of desert regions that as the fans along mountain fronts grow, they merge to form a large alluvial apron called a **bajada** (Fig. 10–45).

Alluvial fans are ubiquitous features in the desert regions of the western United States and Mexico, where topographic basins are often fault-controlled and more or less surrounded by mountains. The basins, called **bolsons,** are great collecting areas for the erosional debris shed from the encircling ranges. Streams flow inward toward the center of the bolson in what has been described as a centripedal drainage pattern.

During periods of rainfall, streams originating in the mountains may bring sufficient water into the center of the bolson, so that a temporary lake is formed (Fig. 10–46). Such lakes tend to lose their water rapidly as a result of evaporation and infiltration, and

Figure 10–44 Desert landscape in Zion National Park, Utah. (Courtesy of U.S. National Park Service.)

often only a salt-encrusted former lake bed or **playa** remains.

The mountains of arid regions are characteristically bordered by a sloping surface that extends downward toward the center of the bolson. The upper part of this slope is termed the **pediment** and is generally considered to have been formed primarily by erosion (see Fig. 10–47). Bare rock is often exposed along the surface of the pediment, but it may also be thinly covered with alluvium. Pediments are slopes of transportation across which streams, sheetwash, and mudflows carry sediment away from this mountain front. Although they are more characteristic of arid regions, pediments also are found in moister terrains. The history of pediment development begins with the uplift of a mountain range. Usually this uplift occurs along a fault system. For an initial period after uplift, alluvial fans form directly at the foot of the mountains. With time, the mountain front retreats and the smooth pediment begins to be leveled into the bedrock as indicated in Figure 10–47. Adjacent pediments may coalesce to form a pediplain.

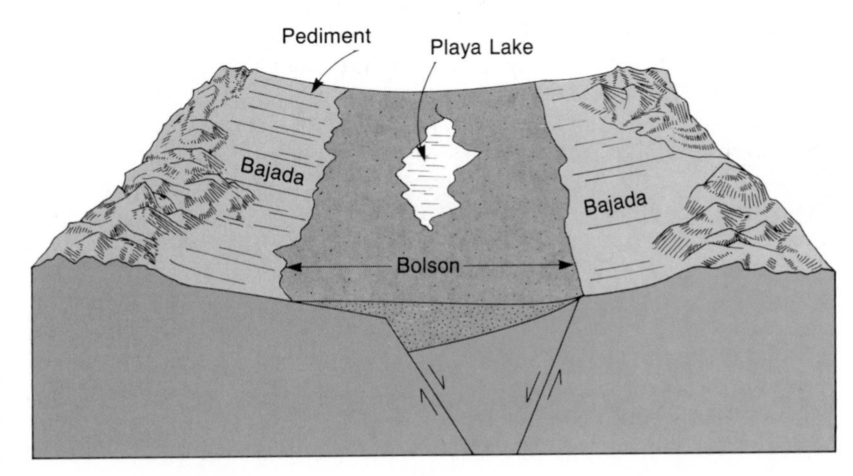

Figure 10–45 Playa Lake formed on the bolson between adjacent pediments.

Figure 10–46 Playa lake with alluvial fans in background, Death Valley, California.

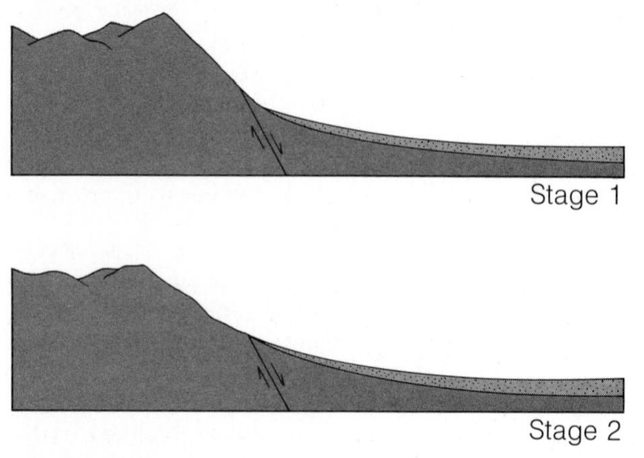

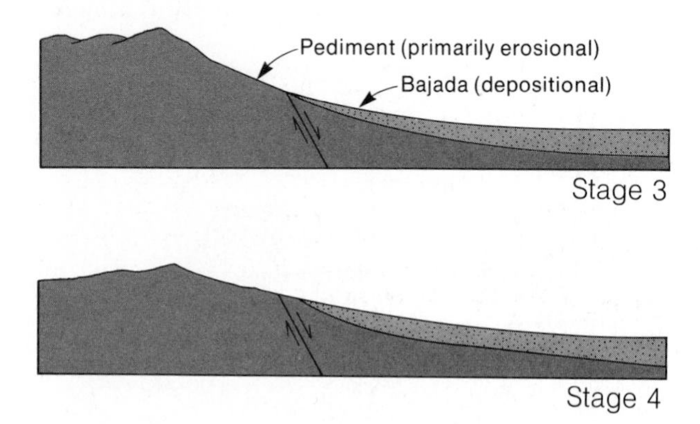

Figure 10–47 The development of a pediment and bajada.

Summary

Along with mass wasting and running water, glaciers and winds are important agents of erosion. Glaciers are flowing masses of land ice. The principal requirement for their development is the accumulation of snow in sufficient amounts to exceed losses that occur as a result of melting, evaporation, and sublimation. As a blanket of snow thickens because of continuing snowfall, solid crystalline ice is formed at depth and will begin to flow in response to gravity. At this point, a snowfield has been converted to a glacier. Some glaciers are constrained by topography and flow predominantly in one direction. Valley glaciers are of this type. Unconfined glaciers, called ice caps or ice sheets, move outward in all directions from a center of accumulation.

Rock and sediment transported and deposited by glaciers are termed *stratified drift* if they show evidence of sorting and crude layering, and *till* if they are unsorted and heterogenous. In general, most deposits of stratified drift have been deposited by meltwater, whereas till is the product of deposition directly by ice. Materials deposited by glaciers may accumulate in variously shaped masses called moraines.

The behavior of a glacier is determined by the state of balance between the rate of accumulation of snow and ice, the rate of flow, and the rate of ablation or wastage. When these factors are in perfect equilibrium, the terminus of the glacier will neither advance nor retreat. If accumulation increases, the snout of the glacier will advance, whereas a decrease in accumulation or increase in ablation will result in a glacial recession.

Glaciers flow as a result of plastic deformation caused primarily by displacements within and between ice crystals and by granulation. Movement is slowest adjacent to the bedrock surface beneath the ice and increases in velocity upward to a point about 60 meters from the ice surface. Ice above this level is under insufficient pressure to deform plastically and is carried by the glacier as a brittle unit. As the glacier moves along, it incorporates large and small pieces of rock that constitute the tools by which abrasion is accomplished. Further erosion is achieved by quarrying, in which bedrock frozen into the base of the glacier is lifted out as the glacier moves forward. The falling and sliding of rock debris from mountainsides onto the surface of the ice supplement the glacier's load further. Such action constitutes avalanching.

During the Quaternary Period of geologic time, ice sheets covered vast regions of Northern Hemisphere lands. The huge thicknesses of ice depressed the crust, caused changes in sea level, deranged drainage, altered topography, and scoured out basins that became lakes.

Winds are movements of air masses over the continents and oceans. They are caused by unequal heating of the various parts of the earth's surface. Geologists are interested in winds because of their effectiveness in transporting and depositing fine particles of sediment. Wind lacks the density and weight of either water or ice, and so it is limited in the amount of geologic work it can accomplish.

Bed load for wind is composed mostly of sand. These particles are transported by saltation or rolling along the ground. During saltation they follow low arcing trajectories that rarely exceed a meter or two in height. Upon striking the ground, the grain may temporarily remain, bounce back into the air, or dislodge another grain into the air stream.

Erosion by wind is accomplished by means of deflation and abrasion. Deflation is the actual lifting and blowing away of loose, dry, unprotected materials. Abrasion (corrasion) is a form of natural sandblasting. Sand particles in saltation forcibly strike against rock surfaces and, in doing so, weaken bonding materials, free grains from their matrices, and even fracture grains. Abrasion can produce ventifacts—stones with smooth, sandblasted surfaces.

Among the depositional features resulting from wind action are ripple marks and dunes. Ripple marks are formed more or less spontaneously as the result of bombardment of a sandy surface by saltating grains. Dunes result from the deposition of sand in an area of weaker eddying currents of air. On their windward side, dunes develop a low slope as high-velocity winds sweep over the accumulated sand. The lee or backslope normally assumes the steeper 35° angle of repose for dry sand.

The effects of wind action are most noticeable in the world's great deserts. The landscapes of desert regions are generally more angular and bleak than those of humid regions, where deep chemical weathering and a heavy covering of soil tend to round divides and reduce ruggedness. The drainage of deserts is internal, and streams flow into intermontane basins called bolsons. At the margins of bolsons one finds the coalesced alluvial fans of bajadas, and, at the front of the mountains, level surfaces called pediments.

QUESTIONS FOR REVIEW

1 Under what conditions does the front of a glacier remain stationary, moving neither forward nor backward?

2 What is the Milankovitch effect? Why is it unlikely that this effect alone could have been the cause of Pleistocene glaciations?

3 Why do glaciated valleys tend to have troughlike shapes (roughly semicircular in cross-section), whereas the valley of a mountain stream tends to be V-shaped?

4 What is the origin of the term *drift* and to what does it refer?

5 In general, how do the competence and capacity of a stream compare with the competence and capacity of a glacier?

6 In what way are the following features developed: a *cirque*, a *fjord*, an *end moraine*, an *esker?*

7 What evidence indicates that glaciers move? Describe the velocity distribution within the glacier. How does movement in the upper 30 meters of the glacier differ from the movement at greater depths?

8 What has been the effect on sea level of the advance and retreat of Pleistocene ice sheets? What role did continental glaciation play in the development of the Great Lakes?

9 Why is wind erosion more effective in arid regions than in humid regions?

10 Why do sand dunes rarely contain grains smaller than about 0.15 mm in diameter?

11 How do sand dunes form? Why do dunes have different shapes?

12 Why are the abrasional effects of winds limited to a height of a meter or so above the ground surface?

SUPPLEMENTAL READINGS AND REFERENCES

Bagnold, R.A., 1941. *The Physics of Blown Sand and Desert Dunes.* London, Methuen & Co. (Repr. 1965, Halsted Press, New York).

Cooke, R.U., and Warren, A., 1973. *Geomorphology in Deserts.* Berkeley, University of California Press.

Embleton, C., and King, C.A.M., 1975. *Glacial Geomorphology.* 2d ed. New York, John Wiley & Sons.

Flint, R.F., 1971. *Glacial and Quaternary Geology.* New York, John Wiley & Sons.

Glennie, K.W., 1970. *Desert Sedimentary Environments.* New York, Elsevier Publishing Co.

Kurten, B., 1972. *The Ice Age.* New York, G.P. Putnam's Sons.

Matsch, C.L., 1976. *North America and the Great Ice Age.* New York, McGraw-Hill Book Co.

11 the global ocean

Here in a protected environment covering almost three-quarters of the surface of the earth, the record of geologic events on and in the crust of the earth is likely to be preserved with minimum disturbance.

TJEERD H. VAN ANDEL
1968, *Science*

The Glomar Challenger, *an oceanographic research vessel designed for taking drill cores from the floor of the deep ocean. (Courtesy of Deep Sea Drilling Project, National Science Foundation.)*

Prelude There is a majesty about the sea. Poet and scientist alike are moved by its breadth and power, its ever-changing aspect, and its apparent timelessness. It is difficult not to be fascinated by the multitude of strange creatures that inhabit its depths, or awed by the deep chasms and spectacular mountains that exist in the darkness of its abyss. The ocean was the cradle of life, and in many ways it continues to nurture the inhabitants of this planet.

Programs to study the world ocean comprehensively are generally considered to have begun in 1872 with the scientific expedition of the naval ship *H.M.S. Challenger* (Fig. 11-1). The *Challenger,* a three-masted vessel with auxiliary steam power, was the research

Figure 11-1 H.M.S. Challenger. *(From the Report of the Scientific Results of the Exploring Voyage of* H.M.S. Challenger *during the years 1873-1876, Narrative, Part II, 1885.)*

tool of a team of scientists charged by the British government to chart the ocean depths, measure movements of water masses, describe the sea's many creatures, and examine the ocean's chemistry and bottom deposits. To accomplish this formidable assignment, *H.M.S. Challenger* sailed more than 110,000 km into all major ocean areas of the earth. During the three and a half years of the expedition, enough data were collected to fill 50 heavy volumes. Over a century has elapsed since the *Challenger* expedition, yet scientists still refer frequently to the information provided in those reports.

The scientists aboard *H.M.S. Challenger* would have been astonished by the sophisticated instrumentation now available for the study of the oceans. They did not have the automated echo-sounding devices that are now regularly used in studying the topography of the ocean floors. Nor were they able to photograph the sea bottom, precisely measure subtle variations in magnetic and gravitational properties, or actually drill into the ocean floor and obtain core samples. The ability to accomplish these feats became possible with the tools aboard a modern research vessel named the *Glomar Challenger* in an obvious tribute to its predecessor, *H.M.S. Challenger.* The *Glomar Challenger* (Fig. 11 – 2) was outfitted solely for the collection of fundamental

Figure 11 – 2 *The* Glomar Challenger. *Upper right, a view of the research vessel anchored at Okinawa; upper left, the drill assembly; lower left, micropaleontologist Dorothy J. Echols examines foraminifers in one of the ship's laboratories; lower right, the reentry cone, a device used to guide the drill pipe into the drill hole on the ocean floor. (Courtesy of D.J. Echols.)*

knowledge about the seas. It has permitted scientists to know the ages, composition, and relations of rocks and sediments on the sea floors, and it has provided the means to verify concepts of sea floor spreading. Much of this information was derived from the vessel's capability for drilling into the ocean floor. After 15 years of service, the *Glomar Challenger* was retired in November 1983. A new vessel, the SEDCO/BP 471, will permit scientists to continue their investigations of the sea floor.

OCEAN WATER

The chemistry of seawater

The waters of the oceans are remarkably uniform in the kinds and proportions of dissolved elements. An average sample of seawater consists of about 35 parts per thousand dissolved matter (Tables 11-1 and 11-2). Of that dissolved matter, nearly 86 percent consists of sodium and chlorine. The other major constituents in order of decreasing abundance are magnesium, sulfur, calcium, potassium, and bicarbonate. Minute quantities of almost all the other naturally occurring elements are also present.

Some of the elements in seawater are in the form of dissolved gases, the most important of which are carbon dioxide and oxygen. Carbon dioxide is essential to the growth of marine plants, a fact that is evident from the observation that the amount of this gas varies closely with the abundance of microscopic marine plants in any given area of the sea. By means of photosynthesis, plants replenish the supply of oxygen in seawater, and oxygen, of course, is essential to marine animal life.

The ocean has a great capacity for absorbing carbon dioxide from the air and this helps to regulate the amount of this gas in the earth's atmosphere. If the carbon dioxide content of the atmosphere rises, the rate at which it is dissolved in seawater also increases. Oxygen is also absorbed from the air above the oceans but in lesser amounts than carbon dioxide. Surface water tends to be richer in oxygen than deeper water because of its proximity to the air-water interface, and also because marine plants flourish in the upper layers of water, where there is abundant light for photosynthesis (Fig. 11-3). Beneath the surface layer, however, the amount of oxygen dissolved in the water is lower because it is consumed by animals and bacteria (Fig. 11-4) and used in the oxidation of organic waste as it sinks through the water column. Curiously, at depths greater than about 800 meters in many parts of the open ocean, oxygen concentrations increase somewhat due to the transport of oxygen by deep currents from high latitudes where cold oxygen-rich water sinks and flows generally toward the equatorial regions.

TABLE 11-1 **DISSOLVED SOLIDS IN OCEAN WATER***

Chemical constituent	*Content (parts per thousand)*
Calcium (Ca)	0.419
Magnesium (Mg)	1.304
Sodium (Na)	10.710
Potassium (K)	0.390
Bicarbonate (HCO_3)	0.146
Sulfate (SO_4)	2.690
Chloride (Cl)	19.350
Bromide (Br)	0.070
Total dissolved solids (salinity)	35.079

* From U.S. Geological Survey publication, *Why Is the Ocean Salty?*

TABLE 11-2 **COMPARISON OF PERCENT OF DISSOLVED SOLIDS IN OCEAN AND RIVER WATER***

Chemical constituent	*Percent of total salt content*	
	Ocean water	*River water*
Silica (SiO_2)	—	14.51
Iron (Fe)	—	0.74
Calcium (Ca)	1.19	16.62
Magnesium (Mg)	3.72	4.54
Sodium (Na)	30.53	6.98
Potassium (K)	1.11	2.55
Bicarbonate (HCO_3)	0.42	31.90
Sulfate (SO_4)	7.67	12.41
Chloride (Cl)	55.16	8.64
Nitrate (NO_3)	—	1.11
Bromide (Br)	0.20	—
TOTAL	100.00	100.00

* From U.S. Geological Survey publication, *Why Is the Ocean Salty?*

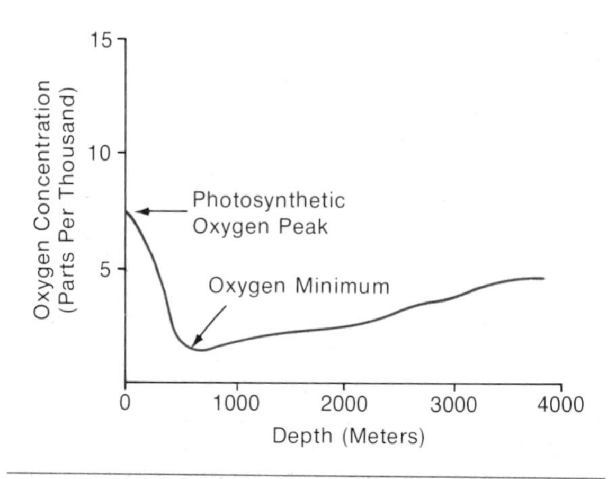

Figure 11–3 Variation of oxygen (O_2) with depth in the North Central Pacific. (After McCormick, J.M., and Thiruvathukal, J.V., Elements of Oceanography. *Philadelphia, W.B. Saunders Company, 1976.)*

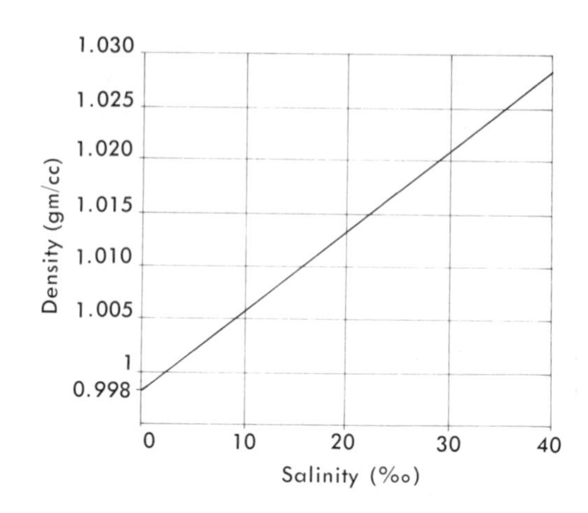

Figure 11–5 Variation in the density of ocean water with salinity, assuming constant temperature and atmospheric pressure. (From McCormick, J.M., and Thiruvathukal, J.V., Elements of Oceanography. *Philadelphia, W.B. Saunders Company, 1976.)*

Physical properties of seawater

Because of the dissolved salt in seawater, it differs from pure water in several physical properties. The density of pure water, for example, is one gram per cubic centimeter at 4°C. In contrast, the density of seawater is 2 to 3 percent greater than the density of fresh water, and increases with increasing salinity (Fig. 11–5).

The density of both sea and fresh water also changes slightly with variations in temperature. Once again, the changes are somewhat different in seawater. As fresh water is cooled from a temperature of 20°C to a temperature of 4°C, its density gradually increases. The dense water will sink to the bottom of the pond or lake and be continuously replaced by water from below until the entire water mass has reached 4°C. If fresh water is cooled further, it will expand and its density will decrease. The lighter water will rise to the surface where it freezes after the temperature has reached 0°C. By comparison, seawater of average salinity has a freezing temperature of about −2°C. Water much colder than −2°C is rarely encountered in even the high latitudes of the ocean. Unlike fresh water, the density of sea water increases all the way down to its freezing point (Fig. 11–6). It is evident from these relationships that the most dense water masses in the oceans are cold and have high salinity. Warm, less saline waters have lower density.

The temperature of the ocean varies with both latitude and depth and is strongly affected by oceanic circulation. Warmest water is found near the equator where temperatures average a balmy 28°C. Not unex-

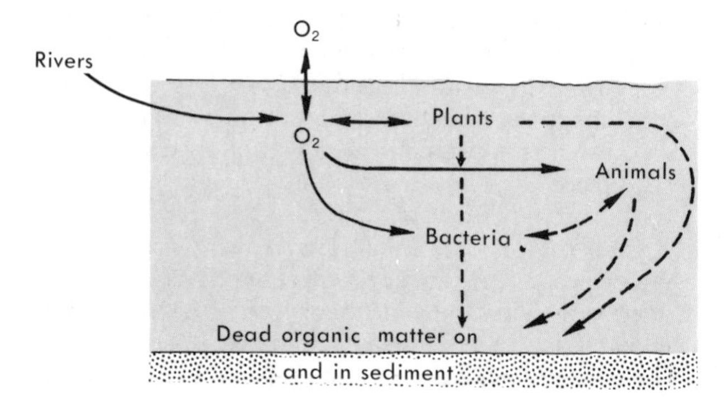

Figure 11–4 Oxygen cycle in the sea. Solid arrows indicate flow of elemental oxygen; dashed arrows indicate flow of oxygen in organic matter.

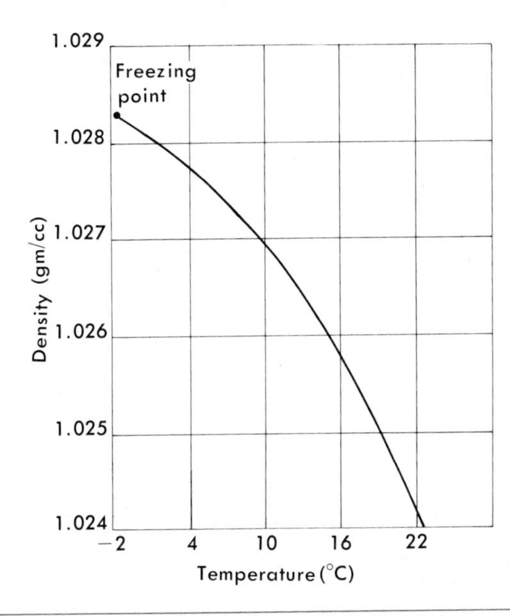

Figure 11–6 Variation in the density of ocean water with temperature at constant salinity and atmospheric pressure. (From McCormick, J.M., and Thiruvathukal, J.V., Elements of Oceanography. *Philadelphia, W.B. Saunders Company, 1976.)*

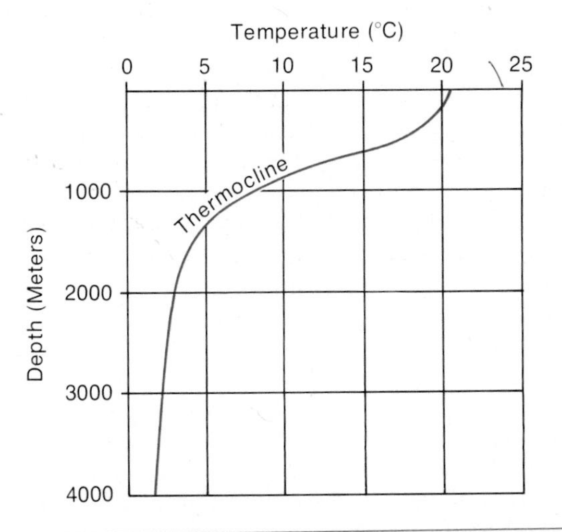

Figure 11–7 Warm water in the ocean is less dense than cold water and tends to stay at the surface, resulting in a layer of warm water over deeper cold water. The transition zone of rapid temperature change with depth is called the thermocline. (From McCormick, J.M., and Thiruvathukal, J.V., Elements of Oceanography. *Philadelphia, W.B. Saunders Company, 1976.)*

pectedly, the coldest waters are found in polar regions, where temperatures may reach −2°C. In the open ocean, the temperature of surface water changes very little with the seasons. This is because of the great volume of the oceans and the constant mixing that occurs. The ocean reacts very slowly to changes in air temperature and therefore serves as a heat regulator for the atmosphere. Of course, in shallow regions of the seas, water temperature is more markedly influenced by seasonal changes in air temperature.

In general, the temperature of ocean water decreases with depth. At moderate depths there is a layer of water in which the change of temperature with depth is at a maximum. This layer is called the **thermocline** (Fig. 11–7). Temperatures decrease only very slowly below the thermocline, until at a depth of 1500 meters the temperature remains virtually constant at 1° to 3°C.

THE DEPTH AND BREADTH OF THE OCEANS

As land dwellers, we sometimes need to be reminded that more than 71 percent of the earth's surface is covered by ocean. The ocean is not only broad but also deep. Although the average depth of the ocean is about 4000 meters, the floors of some deep-sea trenches are some 11,000 meters below sea level. The ocean depths are measured with the aid of **echo-sounding** devices. The method is based upon knowledge of the speed with which sound travels in water. A short, sharp signal or "ping" is emitted about every quarter of a second by the research vessel as it travels along. Sound waves from the signal move through the water column and, upon reaching the sea floor, are reflected back to the ship (Fig. 11–8). The time interval between the emission of the sound and the return of its echo from the sea floor is recorded, divided by two (because the sound makes a round trip), and multiplied by the speed of sound in ocean water (approximately 1500 meters per second). The results are automatically recorded on moving graph paper to provide a continuous profile of the sea floor.

The principle involved in echo-sounding is also used to assist in the study of the sediments that blanket the ocean floors. With a stronger signal, some of the energy can actually penetrate into the sea floor and is reflected off various layers of buried sediment and rock. This process is called **sub-bottom profiling.** To

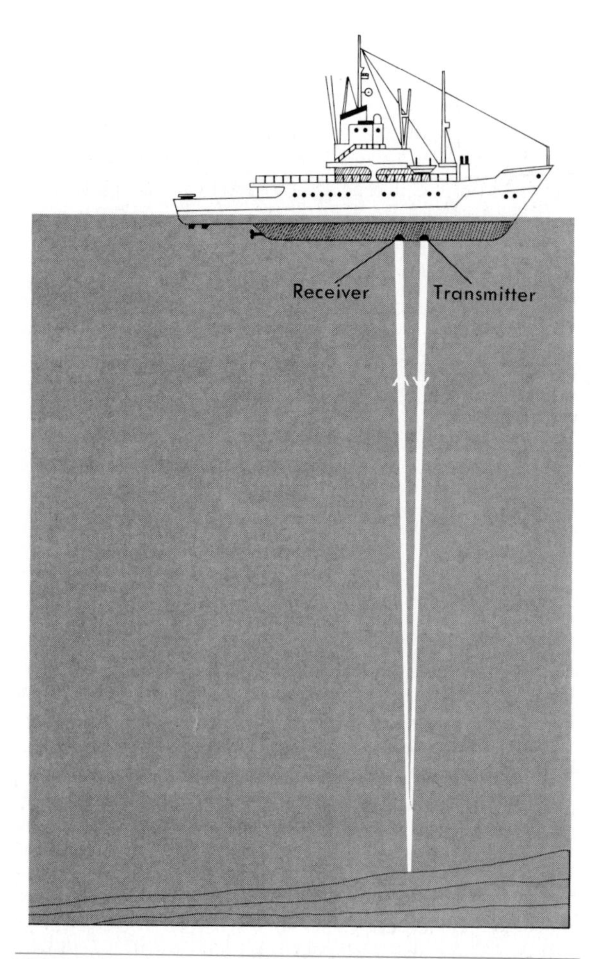

Figure 11–8 The principle of echo sounding. A transmitter sends a sound wave, which is reflected back to the surface by the ocean bottom and is picked up by a receiver. By knowing the total time involved and the speed of sound in ocean water (1500 m/sec), depth can be determined. (From McCormick, J.M., and Thiruvathukal, J.V., Elements of Oceanography. *Philadelphia, W.B. Saunders Company, 1976, p. 37.)*

produce a signal strong enough to penetrate deep into the ocean floor, ships utilize an air gun or electric spark. The returned energy, which may be reflected from horizons as deep as 10 km below the ocean floor, is then detected by hydrophones trailing behind the ship. Clearly, echo sounding and bottom profiling are splendid tools for oceanographic research. As the vessel steams along, the moving graph paper reveals not only submarine canyons and mountains but also an uninterrupted cross-section of what lies beneath the bottom of the sea (Fig. 11–9).

THE EARTH'S SURFACE BENEATH THE OCEANS

Eighteenth-century scientists had little knowledge of the topography of the ocean floors. They lived at a time when depth measurements were made by letting down a lead weight on the end of a rope. Not only was this method time consuming, but in the open ocean it was virtually impossible to prevent error from lateral drifting of the weight, the ship, or both. As a result of these problems, only a limited number of soundings were made, except in bays and offshore areas where such information was vital for safe navigation. Oceanographers interpreted the few measurements available as indicating that the ocean floors were monotonous flat plains. With the advent of continuous topographic profiles from echo-sounding devices, it was shown that the ocean floors are as irregular as the surface of the continents. Beneath the waves lay canyons deeper than the Grand Canyon, and mountain systems more magnificent than the Rockies.

The sea floors near the continents

Continental shelves

As one departs on an ocean voyage from New York City, the first major oceanic topographic feature above which the ship travels is the continental shelf. The **continental shelves** are very gently sloping (an average of of 0.1°), smooth surfaces that fringe the continents in widths that vary from only a few kilometers to about 300 kilometers, and depths that range from low tide to about 200 meters (Fig. 11–10). In a geologic sense, these shallow areas are not part of the oceanic crust, since they resemble the continents in their structure and composition. They are, in fact, the submerged edges of the continents, and have a readily apparent continuity with the coastal plains. The outer boundaries of the shelves are defined by a marked increase in slope.

For geologists specializing in the study of sedimentary rocks, the continental shelves hold great interest. All of the sediment eroded from the continents and carried to the sea in streams must ultimately cross the shelves or be deposited upon them. Study of depositional patterns on the shelves has provided valuable insights into the origin of features in ancient sedimentary rocks now found high and dry on the continents. The shelves also have enormous biological importance. Over most of their area, sunlight can penetrate

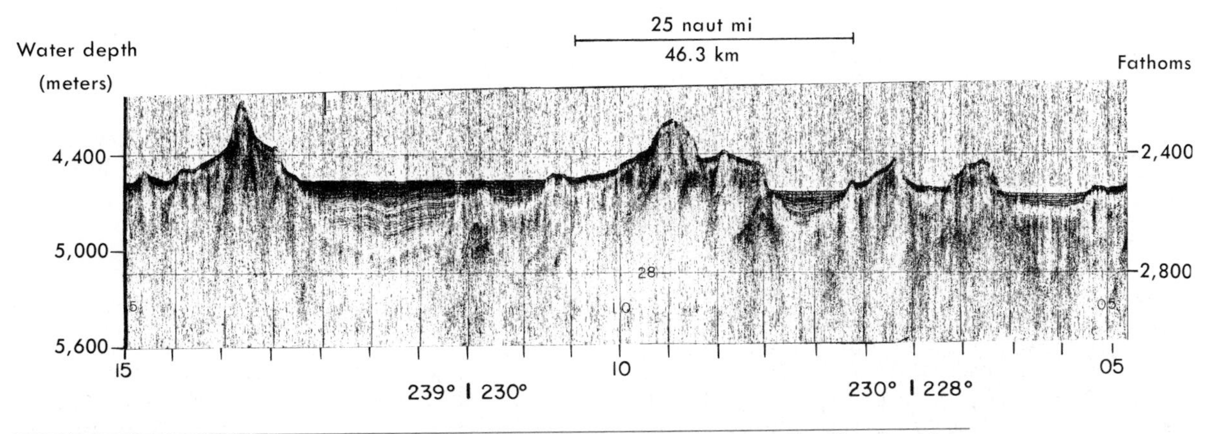

Figure 11–9 Continuous seismic profiles showing abyssal hills and abyssal plains along the edge of the Mid-Atlantic Ridge. (From Hayes, D.E., and Pimm, A.C., 1972 Initial Reports of Deep Sea Drilling Project, 14:341–376.)

all the way to the sea floor. Algae and other forms of plant life proliferate. Here one finds the "pastures of the sea" on which, directly or indirectly, a multitude of swimming and bottom-dwelling animals depend (Fig. 11-11).

Within the last few decades, we have become increasingly aware of the value of the continental shelves. Over 90 percent of the world's seafood is captured in the waters of the shelves. Photographs of enormous offshore drilling rigs remind us daily of the oil and gas resources that lie beneath this part of the sea floor.

Marginal seas and plateaus

The continental shelves are not the only marginal features of the oceans. For example, depressed areas of the continents near the continental borders are sometimes inundated to form **epicontinental marginal seas.** Examples are the Gulf of St. Lawrence and the Gulf of Maine. Such areas are distinguished from the shelves by their greater topographic irregularity and depths. **Marginal plateaus** are also present here and there along the periphery of continents. The plateaus are shelflike features that generally lie at greater depths

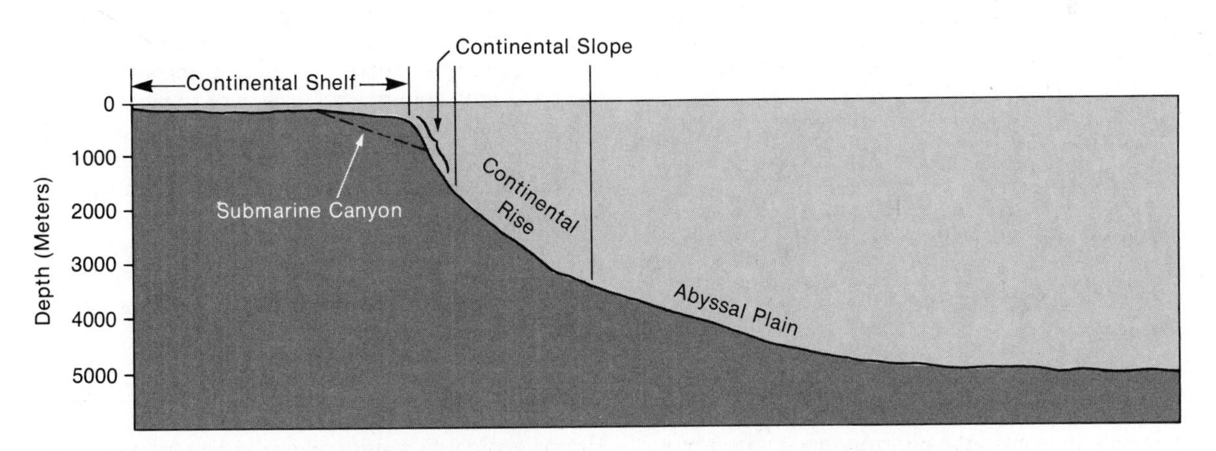

Figure 11–10 The continental shelf is a relatively shallow flat portion of the sea floor which extends from the coastline to a line along which there is an abrupt increase in slope. (Vertical exaggeration × 150.)

Figure 11–11 Marine life flourishes in many areas of the continental shelves, for there is light necessary for marine plants, and plants are the basic components of the food chains that support marine animals such as the staghorn corals and fish shown here. (Courtesy of U.S. National Park Service.)

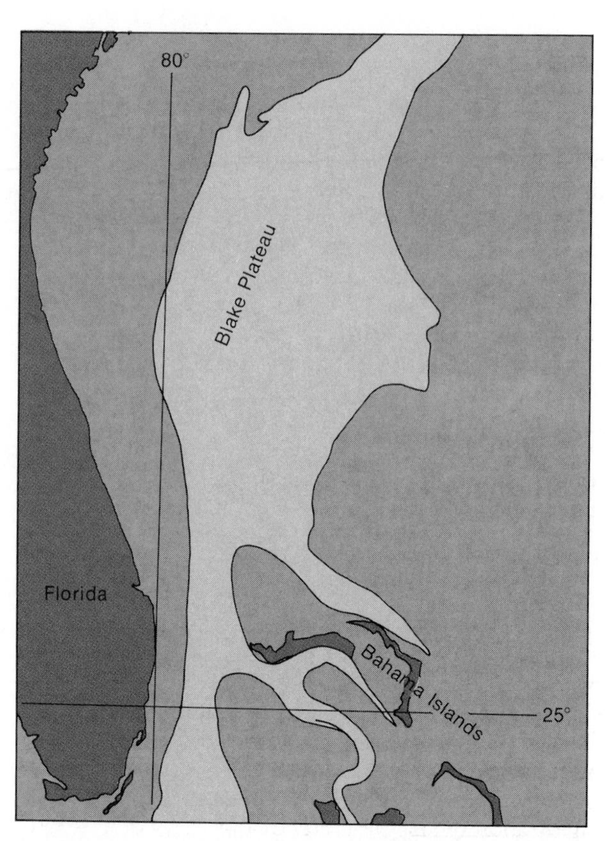

Figure 11–12 Location of the submarine Blake Plateau off the east coast of Florida.

than the shelves and are separated from shelves by a slight slope. The Blake Plateau east of Florida is a particularly fine example of a marginal plateau (Fig. 11–12).

The continental slopes

Those areas of the ocean floor that extend from the seaward edge of the continental shelves down to the ocean deeps are named the **continental slopes.** Physiographic diagrams of the ocean floor are usually drawn with a large amount of vertical exaggeration, so that the continental slopes appear as steep escarpments. Actually, the inclination of the surface is only 3° to 6°. From the sharply defined upper boundary of the continental slope, the surface of the ocean floor drops to depths of 1400 to 3200 meters. At these depths the inclination of the ocean floor becomes gentler. The less pronounced slopes are the **continental rises.** Most of these areas of lesser slope and low relief appear to be wedgelike accumulations of sediment that have been transported across the shelves and deposited in the deep ocean at the base of the continental slopes.

Submarine canyons and valleys

Marine geologists conducting submarine topographic surveys of the continental shelves have long been aware of the large canyons and deep valleys on the continental shelves and slopes. Some of these features, like the Congo Canyon at the mouth of the Congo River and the submarine canyon seaward of the mouth of the Hudson River, are clearly delineated on profiles obtained by echo-sounding. Other canyon-like features are now filled with sediment and have been detected by drilling and sub-bottom seismic profiling. The majority of such features on the continental shelves appear to have been formed by rivers at times when sea level was lower and broad tracts of the shelves were exposed to subaerial erosion. The best evidence for their fluvial origin is their continuity with valleys on land. Indeed, about 80 percent of submarine valleys appear to be extensions of land valleys. Those that are not associated with land valleys are most often found at the outer edge of the shelf and on the continental slope. These canyons were eroded by submarine processes, such as sand slides, slumping, and currents of sediment-laden water called turbidity currents.

Water is said to be turbid when it contains an abundance of suspended particles, and a **turbidity current** is the movement of this dense mixture down a slope beneath clear or less turbid water (Fig. 11-13). The sediment in suspension causes the moving mass to have greater density than less turbid water. The heavy water mass flows close to the sloping ocean floor, picking up additional loose material as it moves along and developing a characteristic turbulent motion. Turbidity currents can develop astonishing speed and are potent agents for the transport of sediment. A turbidity current initiated by an earthquake-triggered landslide off the Grand Banks of Newfoundland in 1929 broke a series of trans-Atlantic telegraph cables. The time each cable was broken was automatically recorded, indicating that the troublesome turbidity current was traveling 100 km per hour over the steeper parts of the continental slope, and it continued into the Atlantic for well over 600 km.

Submarine canyons formed by submarine processes are different from canyons and valleys formed subaerially on land. In most cases, their gradients are steeper, the canyon floors deepen consistently seaward, and the walls are more precipitous. Indeed, the height of the walls of submarine canyons may far exceed that of even the most magnificent of land canyons. The Grand Canyon of Colorado has a north wall that is about 1676 meters high, whereas the sides of the Great Bahama Submarine Canyon reach an astonishing 4400 meters above the canyon floor. Submarine canyons vary in length from only a few kilometers to hundreds of kilometers. The Bering Canyon north of the Bering Islands, for example, is 442 km long.

In those places where submarine canyons emerge onto the sea floor, the sediment carried down the canyons by turbidity currents may be deposited as **deep-sea fans** that resemble the alluvial fans found at the foot of mountains on land.

The floors of the ocean basins

The ocean basins are located between the continental rises. In general, they lie in the depth range from 4600 to 5500 meters. Unlike the crust of the continents, the ocean basins are underlain by igneous rocks of basaltic composition. The basalts are characteristically rather deficient in alkali elements like sodium and potassium and have been given the name **tholeiites** (Fig. 11-14). There are a variety of topographic forms developed on the floors of the ocean basins, but the most distinctive are abyssal plains, oceanic rises, and seamounts (see Figure I-4).

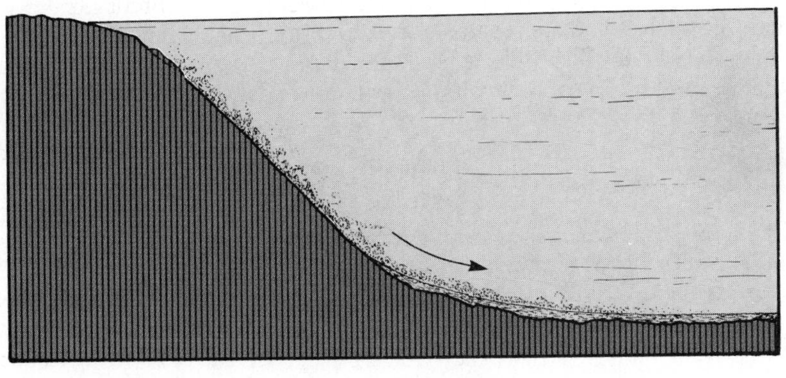

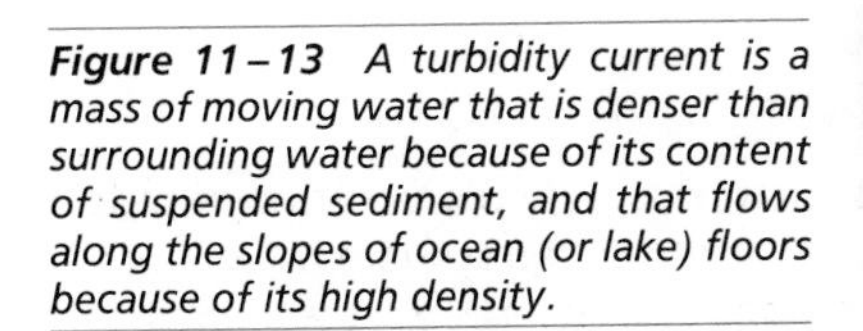

Figure 11-13 A turbidity current is a mass of moving water that is denser than surrounding water because of its content of suspended sediment, and that flows along the slopes of ocean (or lake) floors because of its high density.

Figure 11–14 Pillow of tholeiitic lava from the summit area of a small oceanic volcano near the East Pacific Rise at about 12°N Lat. The structure is termed a "trapdoor" lava pillow because liquid magma is withdrawn during growth through an opening and central cavity. The central cavity is visible on the top broken surface. (Courtesy of Prof. Rodey Batiza, Washington University.)

Abyssal plains

The **abyssal plains** are the vast flattish regions that extend seaward from the base of the continental rises. Their flatness can be attributed to sedimentation that has buried many of the irregularities that once existed on the ocean floor. Most of the sediment consists either of very fine mineral particles derived from land or of the microscopic remains of marine organisms. In addition, an appreciable amount of the sediment on the abyssal plains appears to be fine particles from nearly spent turbidity currents. Such currents do spread into thin sheetlike masses on reaching the deep-sea floor and would be effective agents in transporting and spreading sediment on the abyssal plains.

Oceanic rises, seamounts, and guyots

Oceanic rises (see Figure 11–10) are elevated tracts of the sea floor that are hundreds of square miles in area, stand a few hundred meters above the surrounding abyssal plains, and are distinctly separated from continental margins or ocean ridges. The surface of a rise may be either relatively smooth or quite irregular. Rises are found around Bermuda (the Bermuda Rise), as well as in the Pacific, Indian, and South Atlantic oceans. They are thought to develop as gentle upwarpings of the ocean floor. Oceanic rises should not be confused with features like the East Pacific Rise, which is part of the ocean ridge system and not a true oceanic rise.

Among the most remarkable features of the sea floor are rather symmetrical isolated volcanic peaks having a height of 900 meters or more. They are called **seamounts** (Fig. 11–15) if they are conical and **guyots** if, in addition, they have flat tops. Seamounts are known on the continental rises but are much more numerous on the ocean basin floors. Indeed, oceanographers estimate that more than 10,000 seamounts and guyots occur in the Pacific Ocean basin alone. Many seamounts appear to have developed along fracture zones in the crust where lavas were able to issue forth and accumulate as undersea volcanoes. Others may have been formed in linear sequences as the ocean crust moved over "hot spots." Where the lava is built up above sea level, seamounts form islands like those in the Hawaiian and Society islands volcanic chains (see Figure 2–26).

Guyots, named for the Swiss geographer Arnold Guyot (1807–1884), are of special interest to geologists, because of their remarkable flat tops. The flat tops are either drowned wave-beveled surfaces or coral reefs that had been built upon the summits of barely submerged volcanoes. In support of these ideas, the top surfaces of many guyots are encrusted with the remains of ancient coral reefs; a few others have yielded dredge samples of cobbles and pebbles that appear to have been shaped by wave action. Waves and coral reefs, however, are features that characterize the uppermost level of the sea. How then did the guyots reach the great depths at which they are now found? Either sea level rose to cover them, or the guyots were carried downward by subsidence of the oceanic crust. Because some guyots lie 2000 to 3000 meters below sea level, it is unlikely that eustatic rise in sea level is the answer. A more likely explanation is that some guyots reached their present depths as a result of the gradual subsidence of oceanic crust when it cools and is conveyed off the elevated flanks of the midocean ridges (Fig. 11–16).

Midocean ridges

Because of their impressive topography, distinctive structure, areal extent, and importance in global tectonics, the midocean ridges deserve a prominent place

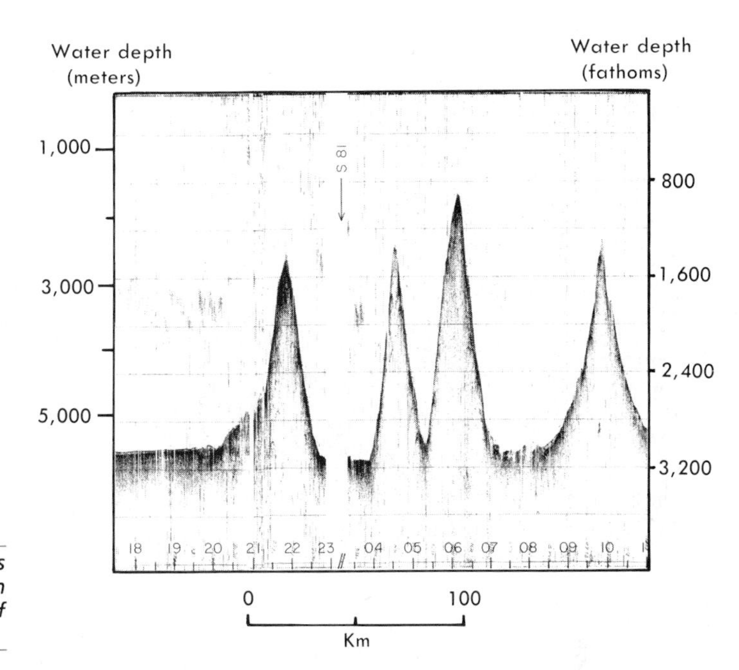

Figure 11-15 Continuous seismic profile across seamounts near the Mid-Atlantic Ridge. (From Haynes, D.E., and Pimm, A.C., 1972 Initial Reports of the Deep Sea Drilling Project, 14:341-376.)

in any classification of this planet's major geologic features (Fig. 11-17). Midocean ridges extend across 64,000 kilometers of the sea floor and nearly equal the continents in total area. This world-encircling system can be traced down the middle of the Greenland Sea and the North and South Atlantic oceans. It then swings eastward into the Indian Ocean basin and continues along between Australia and Antarctica to enter the Pacific Ocean. Once in the Pacific, the ridge is continued northward along the eastern side of the ocean basin where it is called the East Pacific Rise.

The Mid-Atlantic Ridge is a widely studied segment of the ridge system and illustrates well many of the characteristics of midocean ridges. In its southward trend down the Atlantic Ocean basin it parallels the margins of Africa and South America. Altogether, the ridge occupies about a third of the Atlantic Ocean basin (see Figure I-4) and rises more than 3 km above

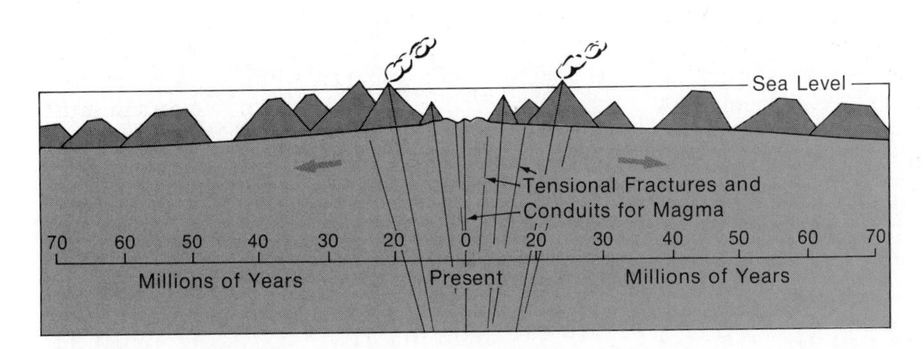

Figure 11-16 The history of guyots begins with volcanoes that originate near the axis of a midocean ridge or at a hot-spot, and progressively move away from their original location as they are conveyed by lithospheric plates. As the volcanoes are conveyed down the flank of the ridge and/or subside isostatically, their tops may be flattened by wave action and they ultimately sink to great depths beneath the ocean surface.

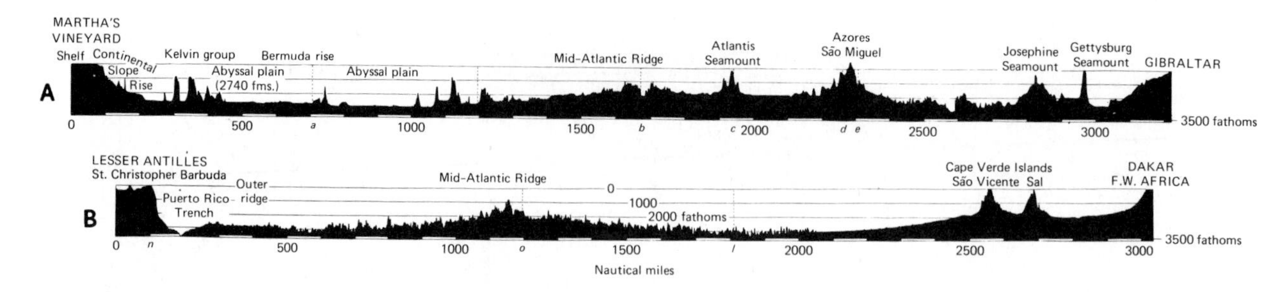

Figure 11–17 Bathymetric profiles of the Atlantic Ocean from: (A) Martha's Vineyard, Massachusetts to Gibraltar; and (B) the island of Barbuda (about 100 km east of Puerto Rico) to Dakar, Africa. (From Heezen, B.C., et al., The Floors of the Ocean, Pt. I, The North Atlantic, *Geol. Soc. Amer. Special Paper 65, 1959.)*

the basin floors. Here and there some of its highest peaks project above sea level to form islands, such as those of Iceland and the Azores (see Figure 11–17).

In 1973 and 1974, the French-American Mid-Ocean Undersea Study (FAMOUS) undertook detailed mapping of the Mid-Atlantic Ridge southwest of the Azores. Included in the results of this study were cross-sectional profiles across the ridge that clearly showed the presence of a central rift valley resembling a huge graben. This central rift was bounded on either side by normal faults, along which basaltic melts rose from deep zones to form dikes, sills, and submarine lava flows. Similar rift zones have been inferred from echo-sounding profiles along many other segments of the world ridge system. They appear to mark the location of divergent plate boundaries. As lithospheric plates move apart, the crust is subjected to tension, normal faults develop, and basaltic lava is able to rise and become incorporated into the trailing edges of the diverging plates. From this lava, the midocean ridges are constructed (see Fig. 11–16).

Study of the world ridge system is not necessarily confined to the ocean floor. In places like the west coast of North America, eastern Africa, and the Red Sea, the ridges run into the continents. As is the case for the Mid-Atlantic Ridge, seismic refraction studies indicate that the rift occupied by the Red Sea is also characterized by normal faulting.

Deep-sea trenches

The deepest parts of the ocean are elongate, often arcuate troughs called **deep-sea trenches.** The floors of these great deeps lie as much as 11,000 meters below sea level. There is no light in the trenches, and pressures may exceed that at sea level by some 1000 times. Most deep-sea trenches are only 40 to 120 km in width, but a few are thousands of kilometers in length. Trenches lie either along the oceanward side of volcanic island arcs, as does the Aleutian Trench, or adjacent to coastal ranges of continents, as is the case with the Peru-Chile Trench off the west coast of South America. The Mariana Trench (Fig. 11–18) in the North Pacific is particularly noteworthy for its record depth of 11,022 meters.

Deep-sea trenches and associated island arcs are vibrant with geologic activity. Volcanic eruptions and earthquakes are frequent events. When plotted on cross-sections of the trench and adjacent tracts, the foci of the earthquakes define a zone that slopes downward at an angle of 30° to 50° under the arcs and toward continental margins. Earlier we defined these trends as Benioff Seismic Zones. We also noted that strong negative gravity anomalies occur over deep-sea trenches. The Benioff Zones and negative gravity anomalies indicate that, along the deep-sea trenches, the oceanic crust is bent and carried downward into the upper mantle. Thus, deep-sea trenches are ocean floor manifestations of *subduction zones.* They are one of several major global features that form along the *leading edges* of tectonic plates.

DEEP-SEA SEDIMENTS

Scientists aboard the H.M.S. *Challenger* found that there were two abundant kinds of sediment on the basin floors. In the deepest regions, the *Challenger* crew brought up samples of clay, which they called "red clay." Actually, abyssal clays are more of a brownish color, and so the term *brown clay* is now preferred.

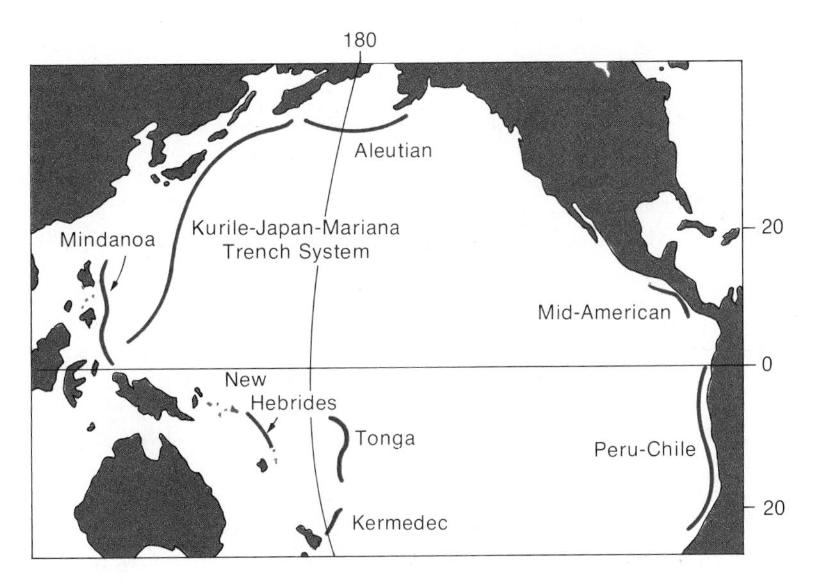

Figure 11–18 *Major trenches of the Pacific Ocean.*

In the shallower regions of the ocean basins, various kinds of very fine organic deposits were raised onto the deck of the *Challenger*. Because of their fine texture and slippery feel while still wet, these materials were named *oozes*.

Inorganic deep-sea deposits

The deep-sea sediment called brown clay consists of clay minerals and tiny particles of other silicates. For the most part, these particles originated on land and have been carried by winds and currents into the open oceans. Along with lesser quantities of volcanic and cosmic dust, these materials accumulate at the nearly imperceptible rate of a millimeter per thousand years. Centuries elapse before a particle reaches the ocean floor, and so there is ample time for iron in the particles to be oxidized by even meager amounts of dissolved oxygen in the water column. Not all of the components of abyssal clays are derived from a distant source. Manganese nodules (Fig. 11–19) may grow in place on the sea floor. These curious spherical concretions are chemically precipitated from seawater as coatings of manganese dioxide (as well as iron and other elements) around some "seed object," such as a fish tooth or rock fragment.

Occasionally, while examining the long cylindrical cores of deep-sea sediments (Fig. 11–20), geologists are startled to discover a layer of sand or silt-sized particles. In most cases, the coarser material was brought to the site of deposition by turbidity currents. Relatively large pebbles and cobbles are also sometimes found in deep-sea tracts near the poles. This kind of coarse debris has been dropped from drifting masses of melting glacial ice. It is especially prevalent on the floor of the ocean around Antarctica.

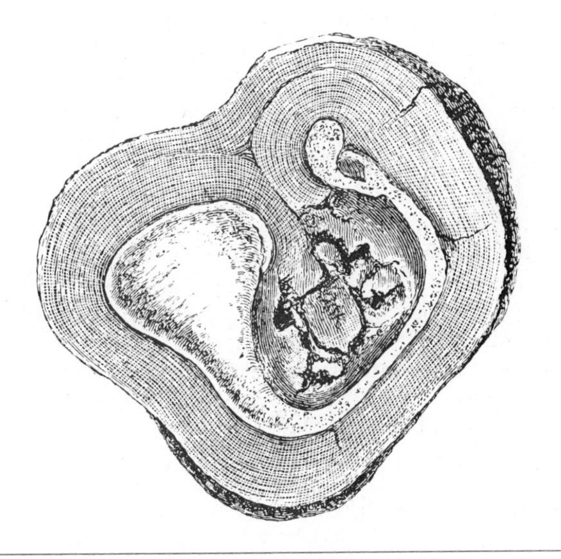

Figure 11–19 *Manganese nodule recovered at depth of 1400 meters by scientists of the* H.M.S. Challenger *expedition. The nodule has formed around the ear bone of a whale. (From the Report of the Scientific Results of the Exploring Voyage of* H.M.S. Challenger, *1873, Narrative, Part 2.)*

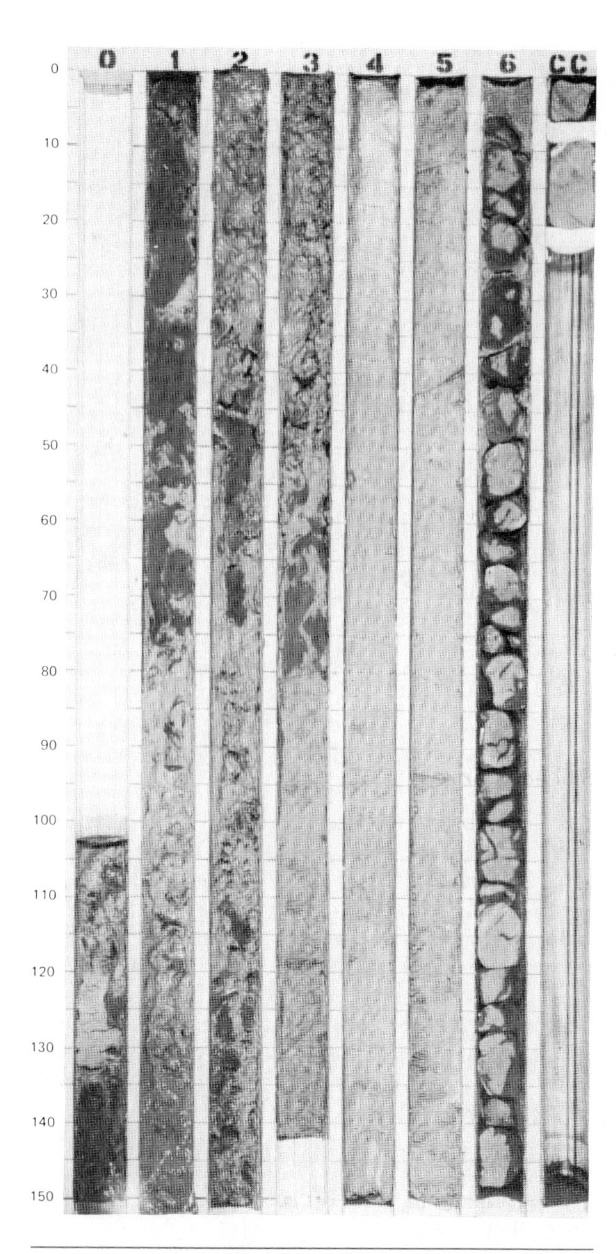

Figure 11–20 *Core recovered from a location just east of the Mid-Atlantic ridge, about 23°N, 43°W, by the* Glomar Challenger *as part of the Deep Sea Drilling Project. Sections 1–3 are disturbed sediment and oozes, sections 4–6 are largely calcareous oozes. Basalt begins at top of section 6. Scale is in centimeters. (Courtesy of U.S. National Science Foundation, Initial Reports of Deep Sea Drilling Project, volume XLV, Washington, D.C.)*

Organic deep-sea deposits

There are science students who have been so impressed by the beauty of the skeletal remains found in oozes that they have devoted their lives to their study. The principal contributors to deep-sea oozes are protozoans, tiny mollusks, and algae. Depending on the composition of their hard parts, these organisms contribute to the formation of either calcareous or siliceous oozes. By definition, an ooze must be at least 30 percent organically derived material. Therefore, oozes are not found in nearshore areas: The influx of terrigenous (land-derived) sediment is so great that the organic contribution does not reach the 30 percent level.

Calcareous oozes

The principal constituents of calcareous oozes are the calcium carbonate coverings of foraminifers, coccoliths, and pteropods. Foraminifers (Fig. 11–21) are marine protozoans, whereas coccoliths (Fig. 11–22) are tiny calcite structures that encase the living cell of members of the algal family Coccolithophoridae. The term **Globigerina ooze** is frequently employed for a calcareous ooze composed of a mixture of coccoliths and tests (shells) of foraminifers. The "forams" in Globigerina ooze are species that, like *Globigerina*, are adapted to a floating (planktonic) existence. The tests are regularly vacated as part of the reproductive cycle in these organisms, and the abandoned shells settle to the sea floor. Only a very small percentage of the remains in oozes represent fatalities. The pteropods (Fig. 11–23) are a group of tiny marine gastropods (snails) that are adapted to a floating life. Their delicate shells are frequently found in Globigerina ooze, and locally they may be so abundant as to justify the name **pteropod ooze.**

Surface waters of the oceans are usually supersaturated with calcium carbonate. Foraminifers, coccolithophores, and other organisms extract this material in quantities that exceed by six times the amount of calcium carbonate contributed each year to the oceans from rivers. Initially, one might surmise that the biological consumption of calcium carbonate would have depleted the supply long ago. There is, however, a recycling mechanism that helps to maintain the supply. If the tiny shells settle through a column of water no deeper than about 3000 meters, they will accumulate intact and form calcareous ooze (Fig. 11–24). Below that depth, however, the colder water holds

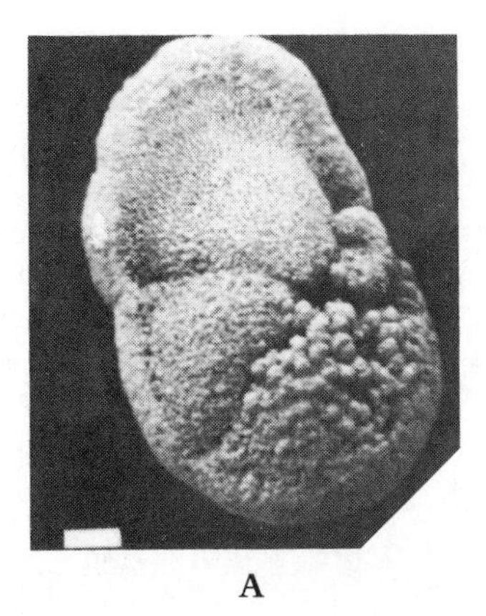

A

B

C

Figure 11–21 *Present-day planktonic foraminifers. (A)* Globorotalia tumidia; *(B)* Globigerinoides conglobatus; *(C)* Globigerinoides rubra. *(Courtesy of C.G. Adelseck, Jr., and W.H. Berger, Scripps Institute of Oceanography.)*

more carbon dioxide. This increases its acidity and causes the dissolution of calcium carbonate. These deep waters are corrosive to the fragile shells of foraminifers and coccolithophores. At a depth of about 4600 meters, the supply of shell materials from above is approximately balanced by the amount being dissolved. As a result, calcareous oozes cannot accumulate. The dissolved calcium carbonate is returned to the oceans to be used again by floating organisms near the surface. It is a well-balanced system, for if the organisms that utilize calcium carbonate in their shells increase in number, the depleted ocean water would respond by dissolving still more calcareous material on the ocean floor.

The ability of cold, deep water to dissolve calcareous materials accounts for the fact that calcareous oozes are most extensive on the shallower regions of the ocean basins. Tall submarine mountains may have their summits, which reach shallower levels, capped by calcareous ooze. They resemble snow-capped peaks, with "snow lines" at about 4600 meters below sea level.

Siliceous oozes

In order to be termed a siliceous ooze, a sediment must be at least 30 percent organic silica. The two principal contributors to siliceous oozes are the unicellular diatoms (Fig. 11–25), which are algae, and the microscopic protozoans known as radiolaria (Fig. 11–26). Because the ocean is undersaturated at all depths with silica, microscopic siliceous shells are often dissolved before reaching the ocean floor. The rate of solution depends on the thickness of the skeletal structure and whether they have protective coatings of organic compounds. Siliceous oozes do accumulate in colder or deeper oceanic regions where other sediments are lacking, and in regions where an abundance of nutrients promotes high productivity of siliceous organisms (Fig. 11–27).

Climatic history from deep-sea sediments

A core of deep-sea sediment (see Figure 11–20) can yield a bonanza of geologic information. Superposition tells us that the lower layers of sediment are older and the higher layers consecutively younger. Using radioactive dating methods, the actual period of time during which the sediments in the core accumulated can be

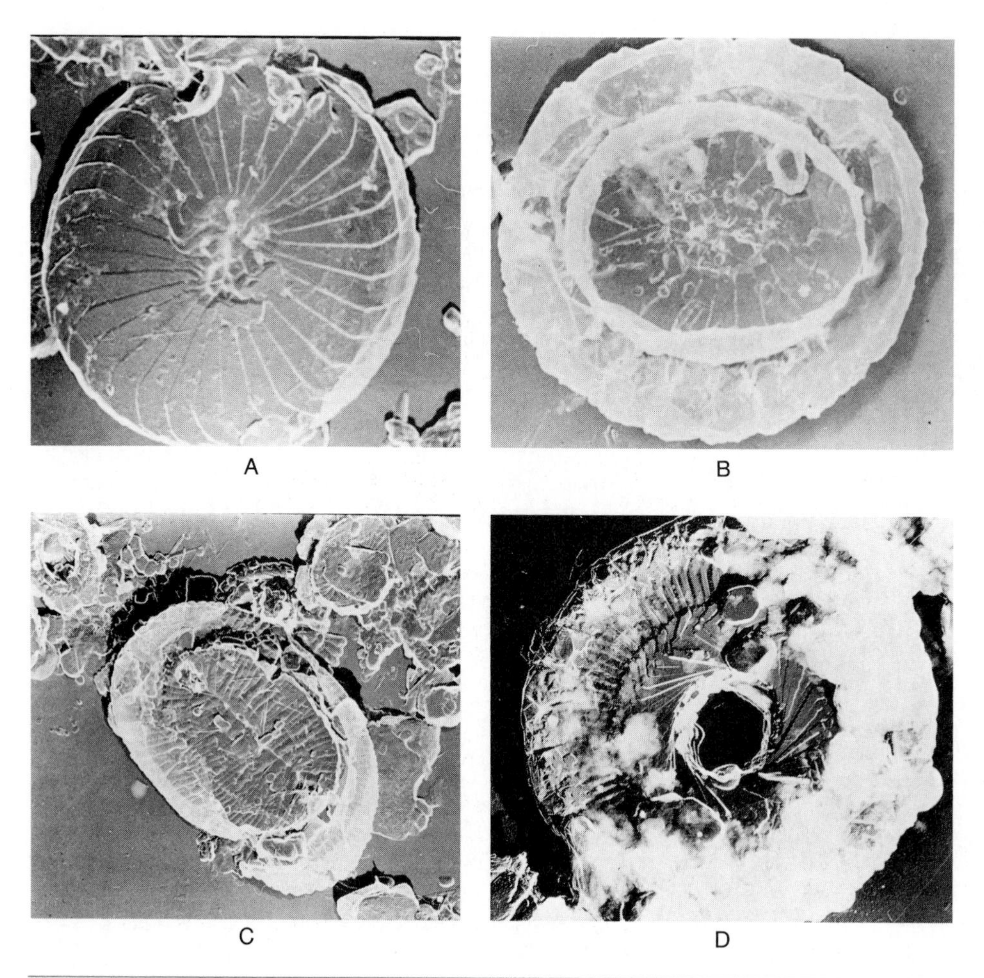

Figure 11–22 *(A)* Coccolithus leptoporus *(distal side). (B)* Coccolithus leptoporous *(proximal side). (C)* Helicosphaera carteri. *(D)* Cyclococcolithus *sp.*

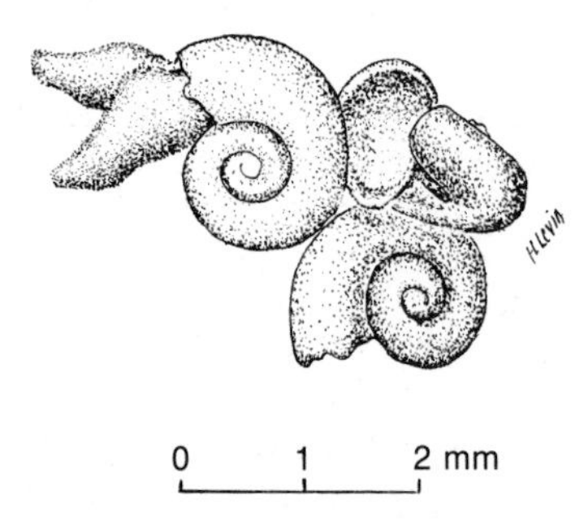

Figure 11–23 Limacina, *a tiny, swimming marine snail or pteropod with foot modified into a pair of winglike fins for swimming (shown in specimen at left). At the right are two empty conchs.*

determined. Next, the foraminifers, radiolaria, diatoms, and coccolithophores are carefully identified at each successively higher level in the core. Because many species of these organisms are able to live only within certain limits of water temperature, their fossil remains are indicators of ancient temperature. The layers might, for example, suggest an ancient cooling trend if warm-water species in the lower parts of the core gave way to cold-water species in higher layers. Even where good temperature indicator organisms are absent, the calcareous shells may provide chemical clues to ancient temperatures. Scientists have discovered that the ratio of the isotopes oxygen-16 and oxygen-18 dissolved in ocean water varies according to water temperature. Similarly, the ratio of these iso-

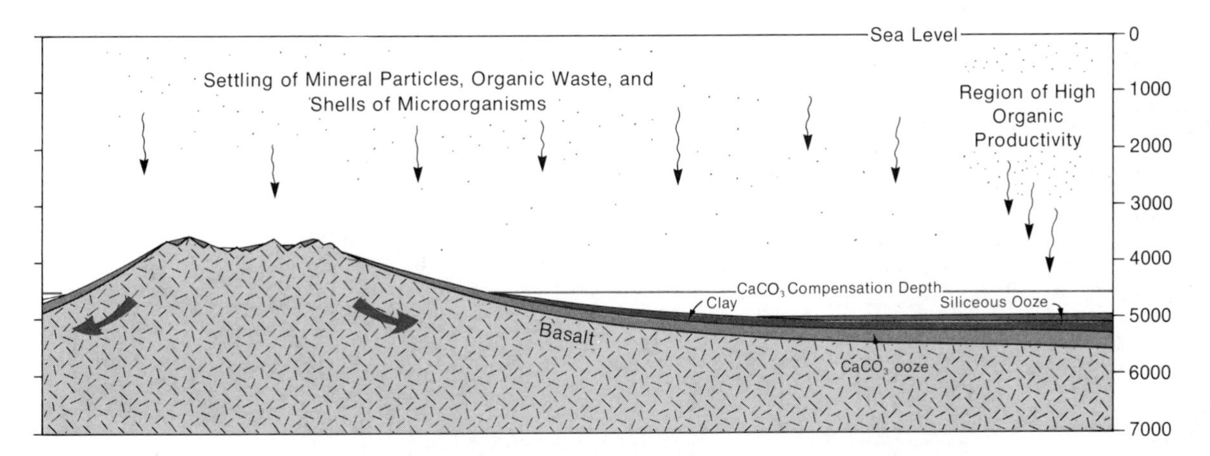

Figure 11–24 *The calcium carbonate compensation depth is the depth below which particles of calcium carbonate from microorganisms are dissolved. The compensation level varies at different locations in the world ocean. In this conceptual drawing, note that the calcium carbonate collects along parts of the midocean ridges that are above the compensation depth. The accumulated layer of calcareous ooze is then carried laterally as lithospheric plates diverge. When a given region of the ocean floor has reached depths in excess of the compensation depth, calcium carbonate is no longer deposited on the ocean floor, whereas clay particles and/or siliceous remains of radiolaria and diatoms may accumulate.*

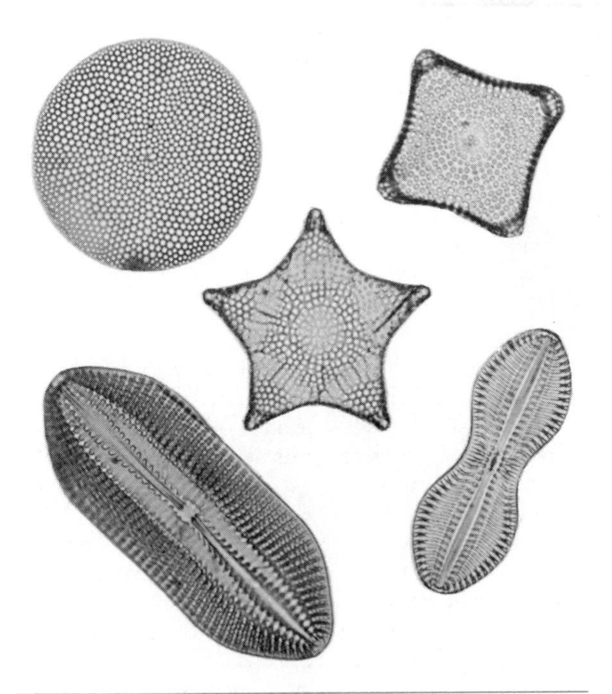

Figure 11–25 *Diatoms. These modern diatoms were collected off the coast of Crete. (Photographs courtesy of Naja Mikkelsen, Scripps Institute of Oceanography.)*

topes in the calcium carbonate shells will reflect the enclosing water's temperature at the time the carbonate was secreted. Oxygen isotope studies and paleontology can thus provide information about climatic conditions on earth long before the birth of our own species.

CORAL REEFS AND ATOLLS

Coral reefs are distinctive features of the marine environment and are common in the shallower warmer oceanic regions. The coral reefs growing today consist of a rigid framework composed of calcium carbonate structures built by vast colonies of tiny, tentacled animals called *coral polyps* (Fig. 11–28). The coral reef, however, is not composed of coral structures alone, for encrusting calcareous algae and a variety of lime-secreting marine animals attach themselves to the framework of the reef and contribute to its strength and mass. Reefs are classified according to their proximity to shore. A **fringing reef** forms a narrow apron directly around the perimeter of an island. If there is a

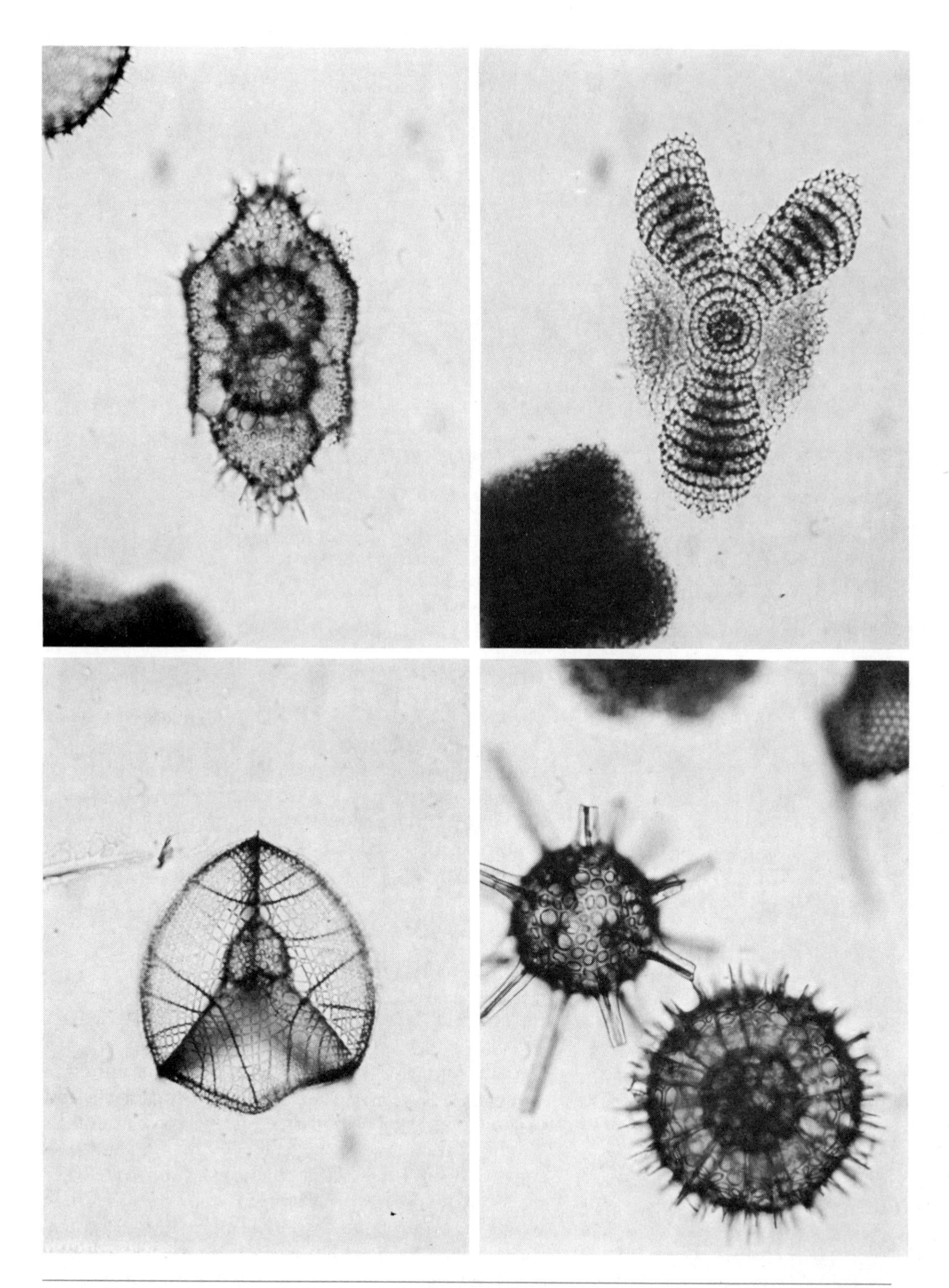

Figure 11–26 *Quaternary radiolarians from the tropical Pacific Ocean. (Courtesy of Annika Sanfilippo, Scripps Institute of Oceanography.)*

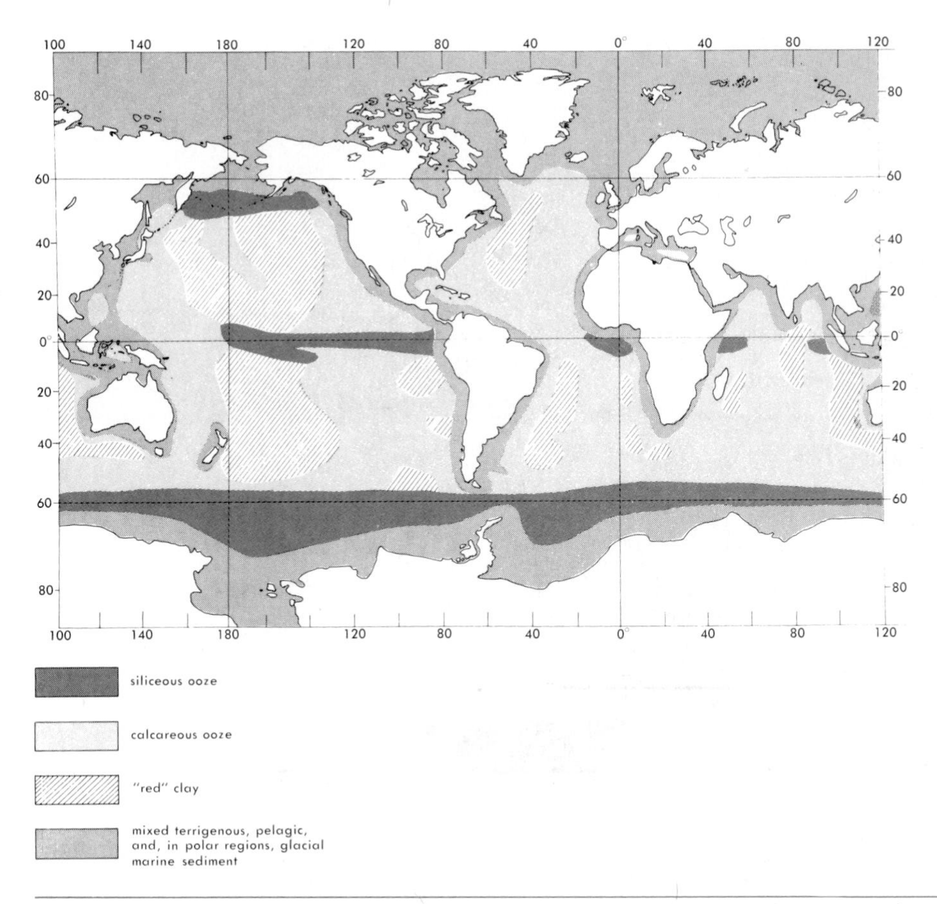

Figure 11–27 *Distribution of sea floor sediment. (After Heezen, B.C., and Hollister, C.D.,* The Face of the Deep. *New York, Oxford University Press, 1971.)*

lagoon between the inner edge of the reef and the shoreline, the structure is termed a **barrier reef.** Finally, if there is no central island but merely a circle of coral reefs and coralline islands surrounding a lagoon, the feature is called an **atoll.** All of these kinds of reef structures are related in origin. This fact was perceived long ago by Charles Darwin during his historic service aboard the naval survey ship *H.M.S. Beagle* in 1832.

Darwin had ample evidence that the coral reefs of the Pacific grew around the perimeter of volcanic islands. At a meeting of the Geological Society of London during the summer of 1837, he proposed that volcanic islands tend to sink slowly because of the inability of the ocean floor to support their great weight. Before this begins to happen, however, free-swimming larvae of coral animals reach newly formed volcanic islands, attach themselves to the shallow submerged areas around the island, and begin their development into adult polyps. Gradually, the coralline framework is constructed and other organisms add their own skeletal structures to the reef. The corals, however, are essential to the vitality of the reef. Darwin was aware that corals are rather fussy creatures in that they thrive only in shallow, well-lighted, relatively warm ocean waters. Depth of water is critical to the health of coral polyps. Thus, with the sinking of their volcanic island, the myriads of polyps might be doomed were it not for their astonishing efficiency in building their colonies upward so as to maintain an optimum habitat near the ocean surface. As the proc-

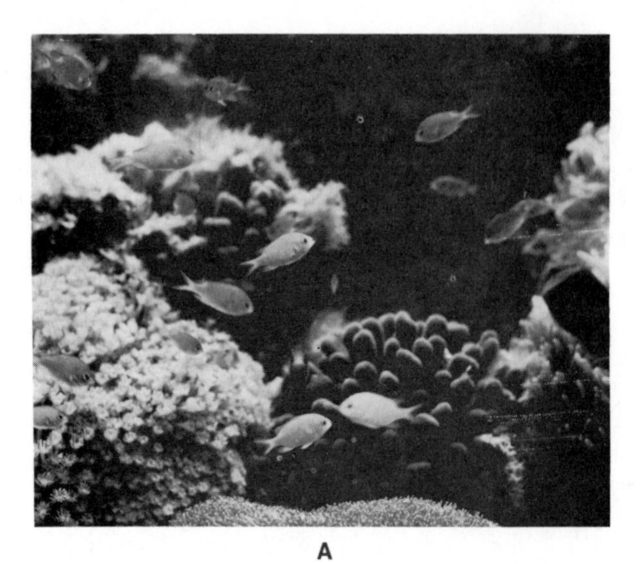

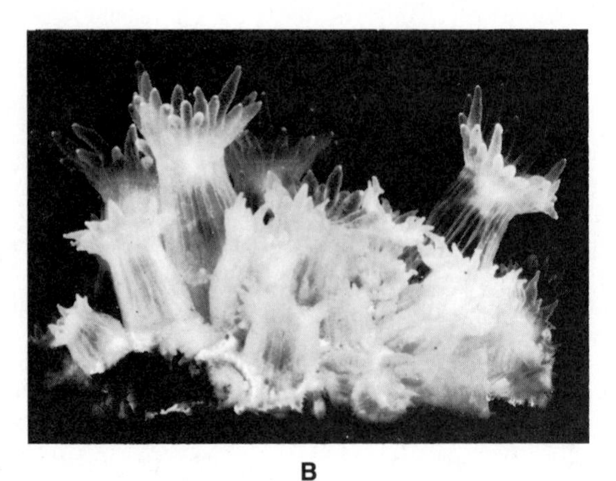

Figure 11–28 *(A) Australian coral reef. (B) Polyps of the so-called star coral. (A, Courtesy of Australian Tourist Commission and Quantas Airlines; B, Courtesy of American Museum of Natural History.)*

ess of subsidence and growth continues, a barrier reef would develop as shown in Figure 11-29. Eventually, the island would sink beneath the waves, leaving only a circle of coralline islands—the atoll. In Darwin's own words, ". . . as the land with the attached reefs subsides very gradually from the action of subterranean causes, the coral building polypi soon raise again their solid masses to the level of the water: but not so the land; every inch of which is irreclaimably gone: as the whole gradually sinks, the water gains foot by foot upon the shore until the last and highest peak is finally submerged."

Several deep borings into atolls have validated Darwin's theory. A boring at Eniwetok in the Pacific encountered the basaltic summit of the volcano at depths of about 1200 meters. Eocene reef structures were directly above these igneous rocks.

There is an alternate possibility for atoll formation. It is possible that corals might build **fringing reefs** around the perimeter of volcanic islands that had been erosionally truncated during an interglacial period of low sea level. If the sea level subsequently rose (as it did when the great Pleistocene ice sheets began to melt), then the corals would strive to build the reef upward in order to stay at their optimum living depth. It seems not unreasonable that particular atolls may be primarily the result of subsidence of the host island, and others may have formed in response to a rise in sea level.

LIFE IN THE OCEAN

Biological realms of the sea

One need only view a Jacques Cousteau film on television or pause to examine the creatures in a tidal pool to appreciate the fact that the ocean teems with life. The multitude of marine creatures are not distributed uniformly through the oceans; rather, individual groups inhabit particular parts of the oceans according to their mobility and needs (Fig. 11-30). Such variables as temperature, pressure, currents, light penetration, and nutrients limit the distribution of certain organisms. In addition, the different sediments that blanket the sea floor influence the distribution of organisms that live on, or burrow into, the ocean bottoms.

So that one may effectively communicate informa-

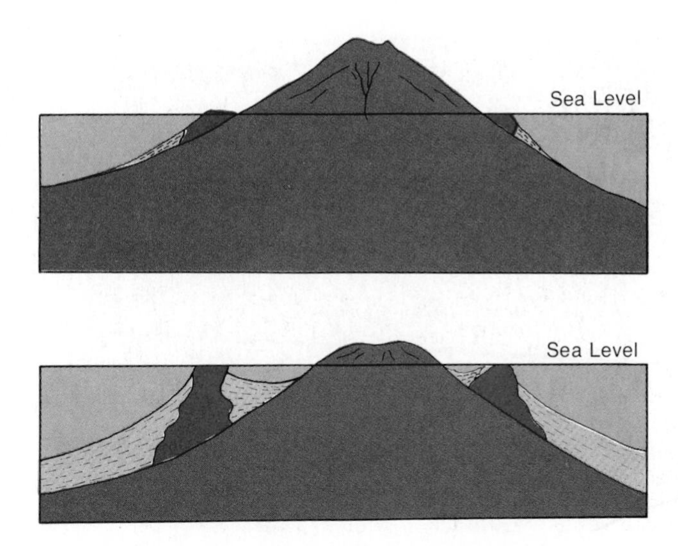

Figure 11–29 Three stages in the development of an atoll. In initial stage (top) a fringing reef develops around the shoreline of a volcanic island. The island begins to subside isostatically (center), and corals build upward in order to stay in their optimum shallow-water life zone. The result is the development of a barrier reef backed by lagoons. As subsidence continues (bottom), the original land area has become inundated, and a circle of reefs and coralline islands remain.

tion about life in the oceans, scientists have developed a widely accepted classification of marine environments. In this classification, the entire ocean is divided into two broad environments: the **pelagic realm,** or water environment, and the **benthic** or bottom environment. The pelagic realm can in turn be divided into the neritic environment, which overlies the continental shelves (Fig. 11–31), and the oceanic environment, which extends seaward beyond the shelves.

Bottom environments

The benthic environments are of particular importance to geologists because of the multitude of shell-building organisms that live there. In the past, these creatures have become the fossils that are now widely used in correlating strata and determining their relative ages. The benthic environment actually begins with a narrow zone above high tide, called the **supratidal zone.** A few algae, other plants, and insects have adapted to this unique marine environment, in which spray from waves or wetting during storms provides necessary moisture. Seaward from this supratidal belt is the intertidal or **littoral zone.** This zone is perhaps the most rigorous of marine environments, for its inhabitants must be able to tolerate the alternation of wet and dry conditions. Some creatures of the littoral zone are sufficiently mobile to move back and forth with the tides, whereas others will burrow into the moist sand to escape dessication during low tide. The littoral zone is the marine environment that we can all readily observe and on which we often play as well.

The greatest multitude of bottom-dwelling animals and plants in the ocean live on the relatively shallow, continuously submerged **sublittoral zone.**

A

B

Figure 11–30 (A) A sublittoral benthic environment dominated by sea stars, brittle stars, and sea lilies (crinoids). (B) An abyssal benthic environment in the southern Indian Ocean. The creature with the spines is a sea cucumber. At left one can see the trail of feces left by a marine worm. (Courtesy of Smithsonian Institution, Oceanographic Sorting Center; from McCormick, J.M., and Thiruvathukal, J.V., Elements of Oceanography. *Philadelphia, W.B. Saunders Company, 1976.)*

This zone corresponds rather well to the continental shelf. The boundary of the sublittoral zone is drawn at the base of the zone of light penetration. Light does not penetrate as deeply in the littoral zone as in the open ocean because of bottom turbulence and turbidity. In the benthic littoral zone one finds a host of blue-green, green, brown, and red algae, as well as abundant protozoans (foraminifers), sponges, corals and sea anemones, worms, mollusks, crustaceans, and various echinoderms (starfish, sand dollars, sea urchins). All of these organisms that live on the ocean floor are **benthos.** Those that crawl about constitute the vagrant benthos, and those that are attached to the sea floor are called sessile benthos.

Beyond the continental shelves, the benthic environment is subjected to colder temperatures, little or no light penetration, and high pressures. Plants are unable to live at these depths. One encounters the **bathyal** environment from the edge of the shelf to a depth of about 4000 meters. Still deeper levels constitute the **abyssal** environment. The term *hadal* is reserved for the extreme depths found in oceanic

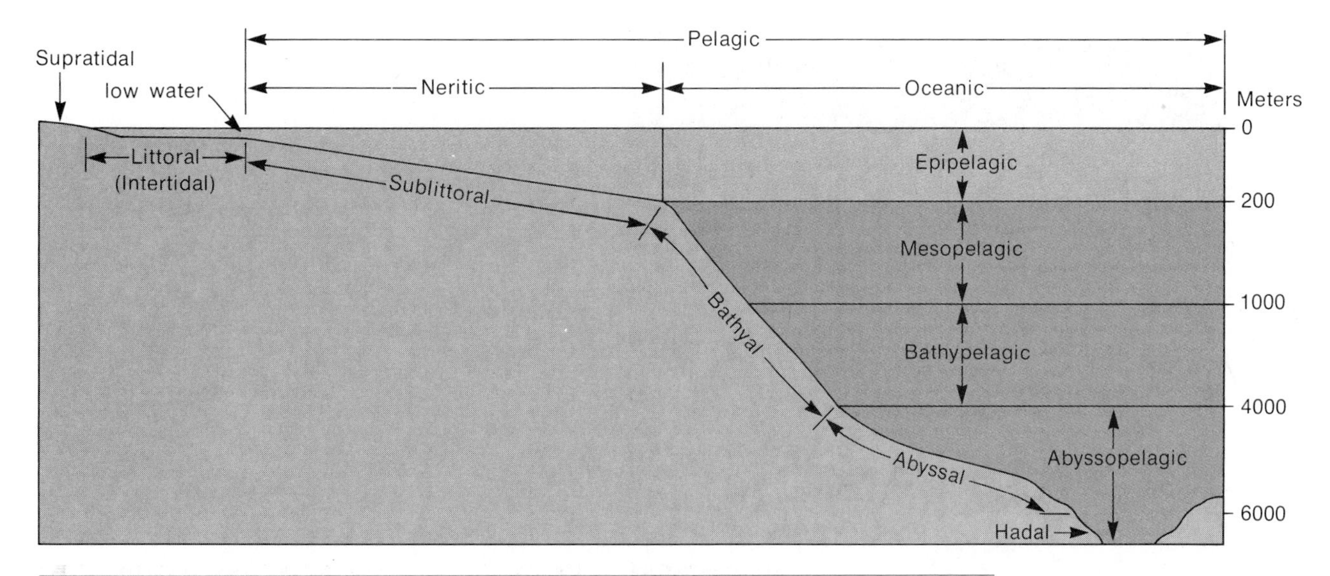

Figure 11–31 *Classification of marine environments (After Hedgpeth, J.W. (ed.),* Treatise on Marine Ecology and Paleoecology, *Geol. Soc. Amer. Memoir 67, 1:18, 1957.)*

trenches. Animals are less abundant in the abyssal and hadal realms. Mostly, they are carnivores and scavengers that depend upon the slow fall of food from higher levels.

The drifters

The primarily microscopic drifting and floating organisms in the ocean are called plankton. Some do have feeble swimming abilities, but not sufficient to permit them to migrate long distances against the pull of oceanic currents. As we have seen, both plants and animals are included in this category. The plants are called **phytoplankton** and include the Coccolithophoridae and diatoms. A group of phytoplankton not previously mentioned are the dinoflagellates (Fig. 11–32). These creatures have whiplike flagella that they use to move themselves feebly through the water.

The foraminifers and radiolaria are true protozoan animals (Fig. 11–33), and thus are designated **zooplankton.** However, many other phyla of invertebrate animals are zooplankton either as adults or in their

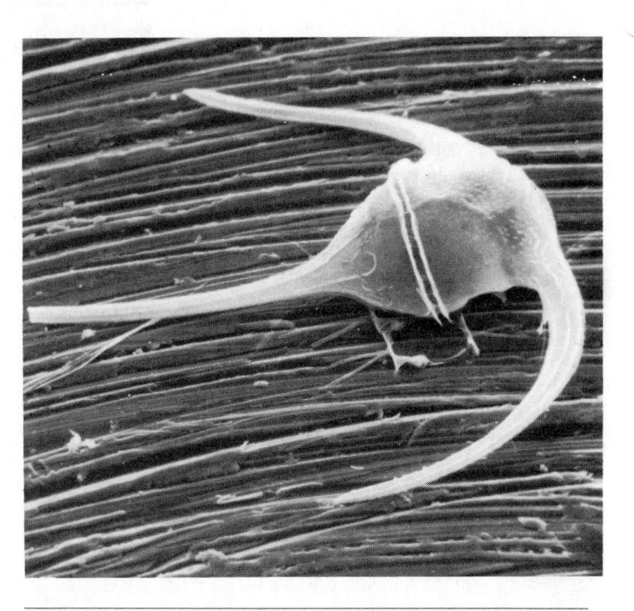

Figure 11–32 Ceratium, *a common dinoflagellate of coastal waters. (From McCormick, J.M., and Thiruvathukal, J.V.,* Elements of Oceanography. *Philadelphia, W.B. Saunders Company, 1976; photograph courtesy of P.E. Hargraves, University of Rhode Island.)*

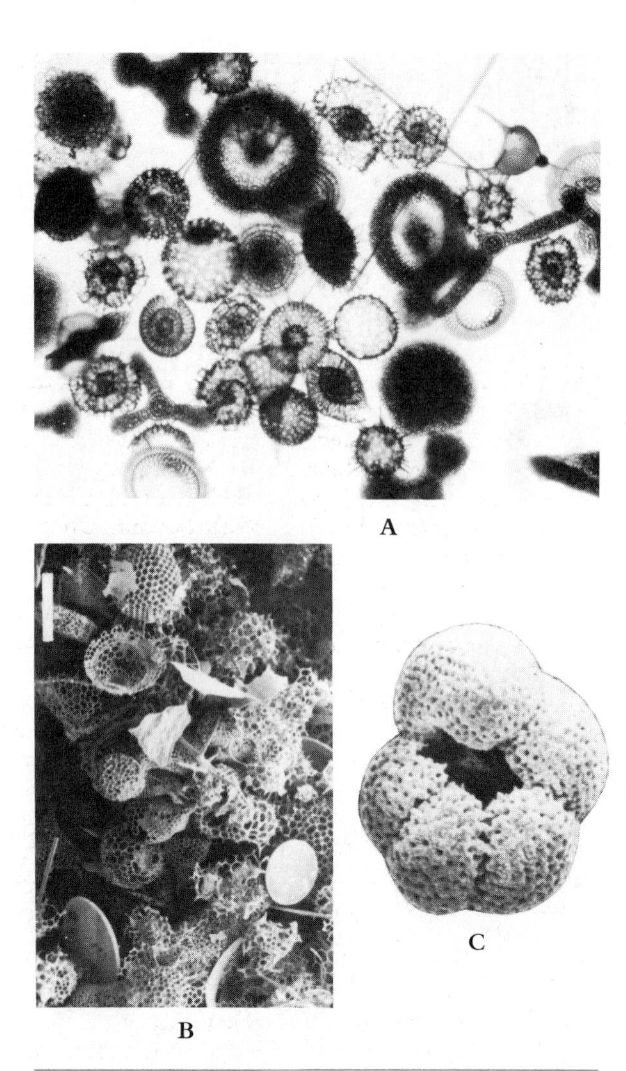

Figure 11-33 *(A) Radiolarians from the tropical Pacific Ocean viewed through ordinary microscope (×70). (B) Radiolarian and diatom remains from surface sediment, eastern tropical Pacific Ocean (white bar scale = 100 microns). (C) The foraminifera* Globoquadrina *(actual size 0.5 mm). This specimen settled below the calcium carbonate compensation depth, and as a result part of its central area has been dissolved. (A, courtesy of A. Sanfilipo; B and C courtesy of C.G. Adelseck, Jr., and W.H. Berger; all of Scripps Institute of Oceanography.)*

larval stages. Jellyfish of the Phylum Coelenterata, for example, are zooplankton, as are the larvae of certain segmented worms, the previously mentioned pteropods, the larvae of echinoderms (sea urchins, sea lilies), and many small crustaceans.

The swimmers

Almost all organisms in the oceanic realm are the prey of some other organism (Fig. 11-34). Phytoplankton are consumed by zooplankton, which also consume one another, and may ultimately become the food of **nekton.** Nekton are true swimming animals. They are able to move where they choose under their own power, and this is a very advantageous adaptation. A

Figure 11-34 *In the oceans plants capture the sun's energy and store it as organic matter. The plants are eaten by animal herbivores, and the herbivores are in turn eaten by primary carnivores such as echinoderms and most fish. The primary carnivores are in turn eaten by larger carnivores and so on. (From McCormick, J.M., and Thiruvathukal, J.V.,* Elements of Oceanography. *Philadelphia, W.B. Saunders Company, 1976.)*

swimming creature can search for its food and does not have to depend upon a drifting current to bring a chance morsel. It can escape predators at times, and can move to more favorable areas when conditions become difficult. The nekton are a diverse group. For the most part, they are members of the Chordata (animals with backbones), and include a vast array of fish, and fewer numbers of reptiles and mammals. There are, however, nektonic nonchordates as well, the most numerous of which are squid, shrimp, and krill.

OCEAN CURRENTS

Because of the importance of ocean currents in navigation, they have been closely observed since the days of sailing ships. Benjamin Franklin, whose lively intelligence and interest in natural phenomena are well known, made one of the earliest studies of the Gulf Stream. Using ships' logs and notes provided by ships' captains, he mapped the Gulf Stream from the southeast coast of the United States northward to Cape Hatteras and thence eastward toward the northwest coast of Europe. Franklin's chart was useful in shortening the number of days required to sail from Boston to London. By avoiding the Gulf Stream on the return trip, these vessels also achieved a faster passage.

There are two major types of movements of ocean water. One kind is caused by differences in density. Where the density differences are the result of variations in salinity and temperature, these currents are termed **thermohaline.** People tend to be more aware of the second kind of ocean movements, which are the wind-driven, or surface, currents.

The role of wind

Surface currents are slow drifts of ocean water set in motion by winds that blow over the surface of the water. Energy is transferred from the moving air (wind) to the water. Because the energy transfer occurs at the surface, the topmost level of the ocean water moves the fastest, and successively lower levels have decreasing velocity. At depths of about 100 meters, the current can hardly be detected.

If oceanic currents derive their energy largely from wind, then one should be able to see similarities between the global current pattern and the pattern of atmospheric circulation. In a very general way, such correlations do exist (Figs. 11-35, 11-36). In the low latitudes, trade winds blow from the east diagonally toward the equator. To the north and south of the trades are the westerlies, which blow from the west diagonally away from the equator in the middle latitudes. The result is a system of large, rather elliptical whirls of water circulation called **gyres.**

Influence of land masses

Wind is not the only factor that determines the global pattern of ocean currents. Land masses and the earth's rotation also play important roles. In considering the influence of land masses, one need only follow the path of currents indicated in Figure 11-36. The North Equatorial Current, for example, is deflected by the land areas of the Caribbean and North America and moves northeastward as the Gulf Stream. On approaching Europe, it is again deflected and divides into the northward-flowing North Atlantic Current and a southward-moving Canary Current. The Canary Current joins the North Equatorial Current to complete the gyre.

In addition to such factors as wind direction and locations of continental barriers, oceanic currents are strongly influenced by the movement of the earth itself. The earth rotates from west to east, and as the solid earth moves eastward under the liquid oceans, water is more or less pushed toward the land mass on the west side of the oceans. Thus, the center of the gyres is usually offset to the west.

Influence of earth's rotation

An important effect resulting from the earth's rotation is the tendency of particles of matter in motion on the earth's surface to be deflected toward the right (clockwise) in the Northern Hemisphere and toward the left in the Southern Hemisphere regardless of the direction in which the body may be moving. This phenomenon has been named the **Coriolis effect** after the nineteenth-century French mathematician G.G. Coriolis who first studied it. Because of the Coriolis effect, the subtropical gyres rotate in a clockwise sense in the northern half of our planet, and counterclockwise in the Southern Hemisphere.

The Ekman spiral

As mentioned above, the major current systems of the oceans are generated by the frictional drag of the wind on the ocean surface. Because of the rotation of the earth, however, the water does not move in the direc-

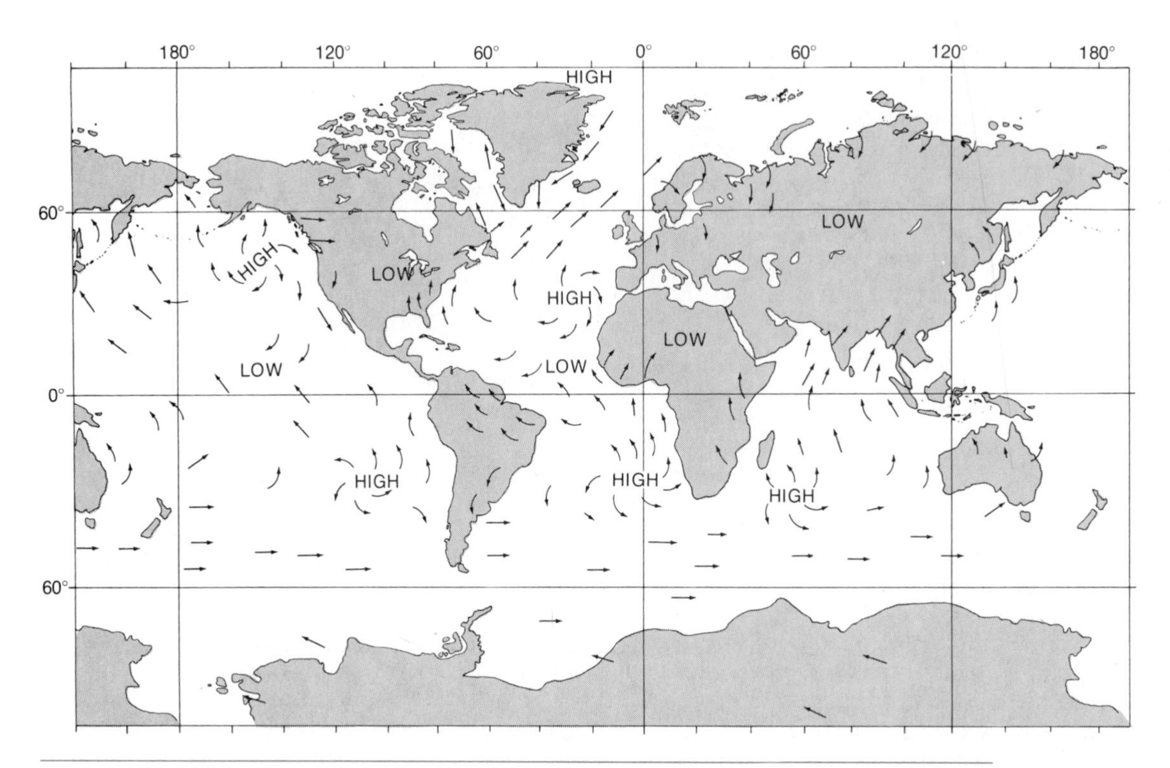

Figure 11–35 Wind systems of the world in July. (From Turk, J., and Turk, A., Physical Science. *Philadelphia, W.B. Saunders Company, 1977.)*

tion of the wind, but rather the surface water moves 45° to the right of the wind direction in the Northern Hemisphere and 45° to the left in the Southern Hemisphere. In addition, as depth increases, each "layer" of water undergoes a similar deviation, while simultaneously friction causes a decrease in velocity with depth. The successive shifts in direction and decrease in speed can be described as a spiral (Fig. 11–37), which has been named the **Ekman spiral** after V.W. Ekman, who described the phenomenon in 1902. An important result of the Ekman phenomenon is that net transport of the water column moved by the wind is at right angles to the wind direction. This movement along the trends of the major current gyres causes a pile-up of low-density water near the centers of gyres. These "hills," as they are sometimes called, provide a slight slope along which water tends to move because of gravity. Once again, because of the Coriolis effect, the downward moving water is deflected at right angles (Fig. 11–38).

THE ORIGIN OF THE OCEAN AND ATMOSPHERE

The histories of the ocean and atmosphere are closely related. The initial chapter of that history began about 4.6 billion years ago when the earth had accumulated most of its mass and was undergoing the internal changes involved in the formation of a core, mantle, and primordial crust. During this differentiation, large quantities of magma saturated with dissolved gases moved toward the surface of the earth and expelled their volatiles as they began to crystallize. The gases escaped in the course of volcanic eruptions or more slowly by rising through holes or vents above magmatic bodies. Such igneous activity would produce a mixture of gases that very likely changed somewhat as differentiation proceeded. For example, before iron in the planet had separated and descended to the core, this metal would have been more abundantly dissemi-

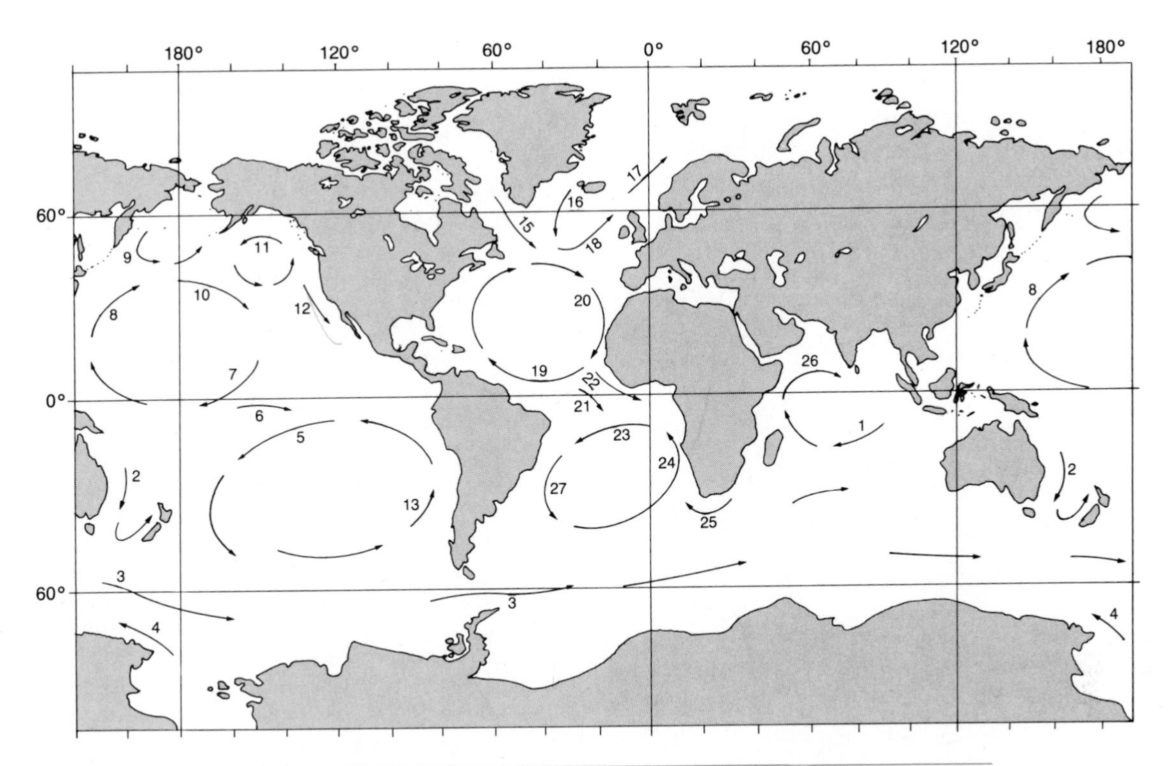

Figure 11–36 *Surface currents of the world in July. (Key: 1, Indian South Equatorial Current. 2, East Australia Current. 3, West Wind Drift. 4, East Wind Drift. 5, Pacific South Equatorial Current. 6, Pacific South Equatorial Counter Current. 7, Pacific North Equatorial Current. 8, Kuroshio. 9, Oyashio. 10, North Pacific Current. 11, Alaska Gyre. 12, California Current. 13, Peru Current. 14, Gulf Stream. 15, Labrador Current. 16, East Greenland Current. 17, Norway Current. 18, North Atlantic Current. 19, Atlantic North Equatorial Current. 20, Canary Current. 21, Atlantic Equatorial Counter Current. 22, Guinea Current. 23, Atlantic South Equatorial Current. 24, Benguela Current. 25, Agulhas. 26, Indian Equatorial Counter Current. 27, Brazil Current.) (After Naval Oceanographic Office, SP-68.)*

nated near the surface and would probably have combined with oxygen and removed oxygen from early volcanic exhalations. Gases rich in hydrogen (H_2), water, carbon dioxide, carbon monoxide, and hydrogen sulfide were emitted. Later, when the iron had melted and percolated toward the planet's center, the gases given off by volcanoes came to resemble those emitted during present-day eruptions. These include water vapor, carbon dioxide, sulfur dioxide, nitrogen, and small amounts of hydrogen chloride and carbon monoxide.

Such a water- and carbon dioxide–rich atmosphere was suitable for the chain of chemical reactions required to produce life. However, it was an atmosphere quite different from that existing today in that it lacked abundant free (uncombined) oxygen.

Oxygen-rich atmosphere

The change from an oxygen-poor to an oxygen-rich atmosphere began to take place about 1.8 billion years ago. The amount of oxygen in the atmosphere slowly increased and reached 3 to 10 percent of present levels about 0.6 billion years ago. Oxygen was added to the atmosphere as a result of two processes. The first was dissociation of water molecules into hydrogen and oxygen. Termed **photochemical dissociation,** this process occurs in the upper atmosphere when water molecules are split by high-energy beams of ultraviolet light from the sun. Photochemical dissociation, however, does not produce free oxygen at a rate sufficient to balance loss of the gas by dissipation into space. Another far more important oxygen-generating mecha-

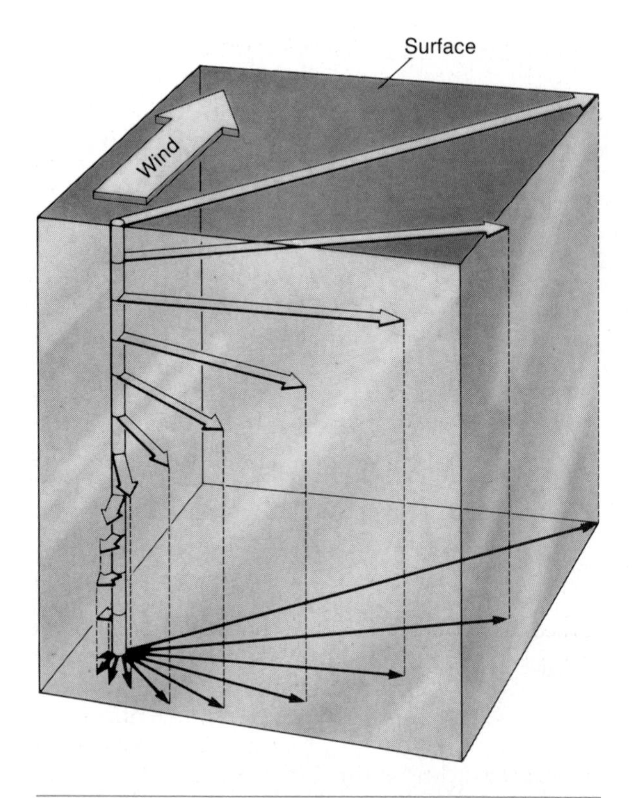

Figure 11–37 *The Ekman spiral as developed in the Northern Hemisphere. (From McCormick, J.M., and Thiruvathukal, J.V.,* Elements of Oceanography. *Philadelphia, W.B. Saunders Company, 1976.)*

nism came with the advent of life on earth and, more specifically, with life that had evolved the remarkable capability of dissociating carbon dioxide into carbon and free oxygen. We now know this process as **photosynthesis.** It has made the earth a truly unique planet in our solar system.

Geologic clues to the nature of ancient atmospheres

The oldest rocks that tell us something about the nature of earth's early atmosphere are 3.5-billion-year-old metamorphosed sediments from Africa. These rocks have retained evidence of having been deposited in water, and their component mineral grains have been formed or modified by weathering processes. They could not have developed in the absence of an atmosphere and hydrosphere. In these ancient formations, one can also find detrital particles of pyrite and uraninite. Such minerals do not form in an oxygen-rich environment because the metals would combine to form oxides and quite different minerals. Carbonate rocks, such as dolostone or limestone, are also absent. This indicates that the atmosphere was rich in carbon dioxide and probably somewhat acidic. In such an environment, chert could form, but carbonate rocks could not. Indeed, the inferred abundance of carbon dioxide and water vapor in the atmosphere probably contributed to the formation of carbonic acids, which would have prevented the precipitation of carbonate minerals.

Iron in crustal rocks provides rather direct evidence of the advent of an oxygen-rich atmosphere. In the oldest rocks for which we have reliable radiometric dates, the iron is not deeply oxidized. Beginning in rocks that are about 3 billion years old, one observes evidence of periodic abundance of oxygen. Such formations exhibit an alternation of rusty red and gray bands. The former are colored by ferric iron oxide (Fe_2O_3) and indicate abundant free oxygen in the environment. The grayish layers contain oxygen-deficient iron compounds and suggest a lesser availability of oxygen. The alternation of color bands might have been caused by a fluctuating and perhaps season-related supply of oxygen. In the iron mining district north of Lake Superior, the ores above the banded

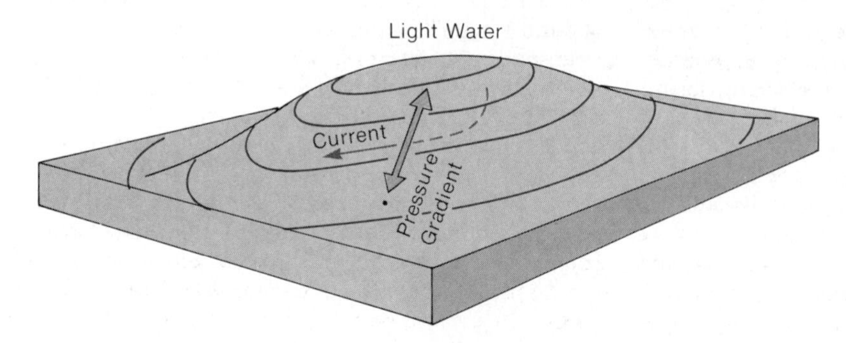

Figure 11–38 *Major gyres have a tendency to pile up low density water at their centers (here exaggerated). Water particles tend to move directly down the slope due to gravity, but are deflected by the Coriolis Effect.*

iron deposits are uniformly oxidized. These rocks are about 1.8 billion years old and are evidence of the persistent presence of an oxygenic atmosphere by that time. In other parts of the world, one finds extensive outcrops of dolostone and limestone of about the same age. If carbon dioxide had still been abundant in the atmosphere, these carbonate rocks would not have been deposited.

THE SOURCE OF OCEAN WATER

If the water in the atmosphere was outgassed from the interior, then the bodies of water that accumulated on the surface of the earth were derived from the atmosphere after it had cooled sufficiently to condense and allow rain to fall. Gradually, the great depressions in the primordial crust began to fill and oceans began to take form. One cannot help but wonder whether the enormous volume of oceanic water could have all come from the interior. Calculations provide a very strong "yes" answer to this question. Vast amounts of water were locked within hydrous silicate minerals that were ubiquitous within the primordial earth. These waters, sweated from the earth's interior and precipitated onto the uplands, began immediately to dissolve soluble minerals and carry the solutes to the sea. In this way, the oceans quickly acquired the saltiness that is their most obvious compositional characteristic. They have maintained a relatively consistent composition by precipitating surplus solutes at about the same rate as they are supplied. Of course, because of its high solubility, sodium chloride remains in seawater more readily than other common elements. However, the fossil record of marine organisms suggests that even this element has not varied appreciably in seawater for at least the past 600 million years.

The earth probably outgassed its present quantity of water rather early in its history and has been partially recycling it ever since. The heating that was required to melt the core, as well as possible stresses induced in the crust by the nearby moon, might have so disturbed the crust as to promote outgassing at a rapid rate. Soon thereafter, the oceans settled into a sort of steady-state system and probably have not gained appreciably in volume. Water is continuously recirculated by evaporation and precipitation—processes powered by the sun and gravity. Some of the water in the oceans is temporarily lost by being incorporated into hydrous clay minerals that settle to the ocean floors. However, even this water has not permanently vanished, for the sediments may be moved to orogenic belts and melted into magmas that return the water to the surface in the course of volcanic eruptions.

Summary

The world ocean covers 71 percent of the earth's surface. This vast body of water has received careful scientific study for well over a century, beginning with the *Challenger* expedition of 1872. Scientists examining the chemistry of the oceans have found that ocean water contains about 3.5 percent dissolved matter, of which sodium chloride is the most abundant. In general, the greater the amount of salt dissolved in seawater and the colder its temperature, the more dense it becomes.

The depths and configuration of the ocean floors are measured primarily by echo-sounding devices. These instruments determine depth by measuring the time required for a sonic signal to travel from a source of energy at the surface to the sea floor and back.

Major features of the ocean floors include the continental shelves, continental slopes, and continental rises that exist around the edges of land masses. Submarine canyons are found incised into the continental shelves and slopes. Other features of the ocean basins include seamounts and guyots. The most majestic elements of the ocean basins, however, are the midocean ridges and deep-sea trenches. Midocean ridges are submarine volcanic mountain systems formed along zones of crustal tension. They are spreading centers between tectonic plates. Deep-sea trenches are thought to be zones of subduction.

Based on their origin, the sediments of the ocean basins may be considered as either inorganic or organic. Brown clays, consisting of clay and other clay-size particles derived from the continents, are the predominant inorganic sediments of the deep sea. The organic materials consist of siliceous oozes, composed of the skeletal structures of diatoms and radiolaria, and calcareous oozes that consist of myriads of tests of foraminifers and coverings of tiny plants of the family Coccolithophoridae.

Because they contribute to the formation of new land areas, coral reefs are also important oceanic features. Most coral reefs require the proliferation of colonial coelenterates, which build the general structure of the reef, thereby providing a framework on which invertebrates and calcareous algae can attach themselves. Depending on their relationship to the host island or seamount, reefs are classified as fringing reefs, barrier reefs, or atolls. Barrier reefs and atolls form as a result of subsidence of the island around which coral growth is occurring. Certain atolls and barrier reefs may also have developed as a result of worldwide rise in sea level.

The environments of the sea include a pelagic realm, which consists of all the water lying above the ocean floor, and a benthic realm, which encompasses the ocean bottom itself. The pelagic zone is divisible into a neritic zone, which overlies the continental shelves, and an oceanic zone, which extends seaward. The neritic zone is important because of the abundance of marine creatures that live there.

Movements of ocean water may be wind-driven or thermohaline. Surface currents are directly influenced by atmospheric circulation, as well as by the locations of land masses and the effects of the rotation of the earth. Thermohaline currents are slow-moving density currents resulting from difference in the temperature and salinity of various water masses.

The water that fills the ocean basins is of great antiquity. During an early stage in earth history, this water was outgassed as vapor from the interior, mostly from volcanoes. The water vapor ultimately cooled sufficiently to condense and fall as rain. Gradually the larger depressions of the earth's primordial surface began to fill and the first oceans began to take form.

QUESTIONS FOR REVIEW

1. Analyses of the water from major streams entering the oceans indicate a far greater calcium content than sodium (see Table 11 – 2). Why then are most of the dissolved salts in ocean water sodium chloride rather than calcium carbonate?
2. What is the effect on the density of ocean water if it is cooled? If it is evaporated? If it is mixed with rainwater?
3. How do scientists utilize their knowledge of the speed of sound in ocean water to determine submarine topography?
4. In what way may such features as midocean ridges and deep-sea trenches be related to plate tectonics? With which of these features are Benioff Zones associated?
5. What differences in composition and texture are likely to exist between sediments recovered from the floor of the continental shelf in the Gulf of Mexico and sediments recovered from the abyssal plains of the ocean basins? What is the geologic explanation for these differences?
6. Account for the observation that the number and variety of benthic organisms decrease markedly below a depth of about 200 meters.
7. What is the origin of atolls? guyots? submarine canyons?
8. What characteristic of the living coral animal is important to any theory for the origin of atolls?
9. A "pinger" on an oceanic research vessel transmits a signal that is reflected off the ocean floor and returned to the ship in 1.8 seconds. What is the depth of the ocean floor beneath the ship?
10. What organisms are largely responsible for the formation of siliceous and calcareous oozes, respectively?

SUPPLEMENTAL READINGS AND REFERENCES

Andel, T. van, 1977. *Tales of an Old Ocean*. New York, W.W. Norton Co., Inc.

Anikouchine, W.A., and Sternberg, R.W., 1973. *The World Ocean*. Englewood Cliffs, N.J., Prentice-Hall, Inc.

Corliss, W.R., 1970. *Mysteries Beneath the Sea*. New York, Thomas Crowell Co.

Davis, R.A., Jr., 1977. *Principles of Oceanography*. Reading, Mass., Addison-Wesley Co.

Heezen, B.C., and Hollister, C.D., 1971. *The Face of the Deep*. New York, Oxford Press.

McCormick, J.M., and Thiruvathukal, J.V., 1981. *Elements of Oceanography*. 2d ed. Philadelphia, W.B. Saunders Co.

Turekian, Karl K., 1976. *Oceans*. 2d ed. Englewood Cliffs, N.J., Prentice-Hall, Inc.

12
tides, waves, and shorelines

We stand on a rugged coast and watch the waves strike blow after blow with the relentless persistence of a trip hammer. The display of vast power is impressive, and some disintegration of the rocky walls proceeds before our eyes.

SIR CHARLES LYELL
Principles of Geology, 1860

Prelude Only a few fortunate people have had an opportunity to see the ocean's depths or chart its currents. Most of us, however, have at least been able to visit the shore, where we can observe the advance and retreat of the tides and the rhythmic breaking of waves along a stretch of beach. When we are in a meditative state of mind, the shore seems to evoke questions. What mighty force drives the tides? Why do they vary in size at different times and places? How do waves, which occur only at the surface of the sea, assist in the sculpture of tall sea cliffs? Why are there sandy beaches along some coastlines, but only bare bedrock along others? And why are some coastlines straight, whereas others are punctuated with numerous bays and jutting promontories? These are a few of the questions we will address in this chapter.

Ruby Beach at Olympic National Park in northwestern Washington. (Courtesy of U.S. National Park Service.)

TIDES

Anyone who has spent a few hours at the seashore knows that the level of the ocean changes in a regular and predictable way twice in each span of approximately 24 hours. These changes are called tides. Their cause has fascinated curious people for thousands of years. Herodotus in 450 B.C. wrote about the tides he observed on the shores of the Mediterranean. Because both tides and moon followed a similar cyclical pattern, he concluded that they were in some way related. It was not until 1696, however, that the relationship of tides to the orbit of the moon could be adequately understood. In that year, Sir Isaac Newton presented his Law of Universal Gravitation.

The forces that cause tides

Newton's **Law of Universal Gravitation** states that every body in the universe attracts every other body, and that the attraction is proportional to the product of the masses of the bodies and inversely proportional to the square of the distance between their centers of mass. The law may be stated in mathematical terms as follows:

$$F = G\frac{M_1M_2}{d^2}$$

In this equation, F is the gravitational force between the two objects, G the gravitational constant, and d the distance between the centers of the objects. The mass of each of the two bodies is represented by M_1 and M_2. Newton's law is indeed universal, for it applies to rocks lying side by side in a field, to two people on opposite sides of a room, or to bodies like the earth and its moon.

The only two planetary bodies in our solar system to exert a significant gravitational force on the earth are the moon and the sun. The sun is far more massive than the moon and thus has an enormous inherent gravitational pull. The sun, however, is 390 times farther away from the earth than is the moon, and that distance is squared in Newton's formula. As a result, the sun's influence is small relative to that of the nearby moon (Fig. 12-1).

Gravitational attraction is not the only factor involved in the production of tides. There is also an effect caused by the earth's rotation. As the earth spins on its axis, objects not at the poles tend to be thrown off. This tendency is called centrifugal force, even though it is only an apparent force caused by our rotating frame of reference. The centrifugal tendency serves to counteract the force of gravity everywhere except at the poles. However, its influence is small, never amounting to more than about 0.33 percent of gravity. Nevertheless, if there were no centrifugal tendency, the ocean would bulge slightly outward only on the side of the earth facing the moon. There is, however, always a second tidal bulge on the opposite side of the earth. This is because the gravitational attraction of the earth and moon is almost completely balanced by centrifugal forces resulting from slight motions of the earth around the center of gravity of

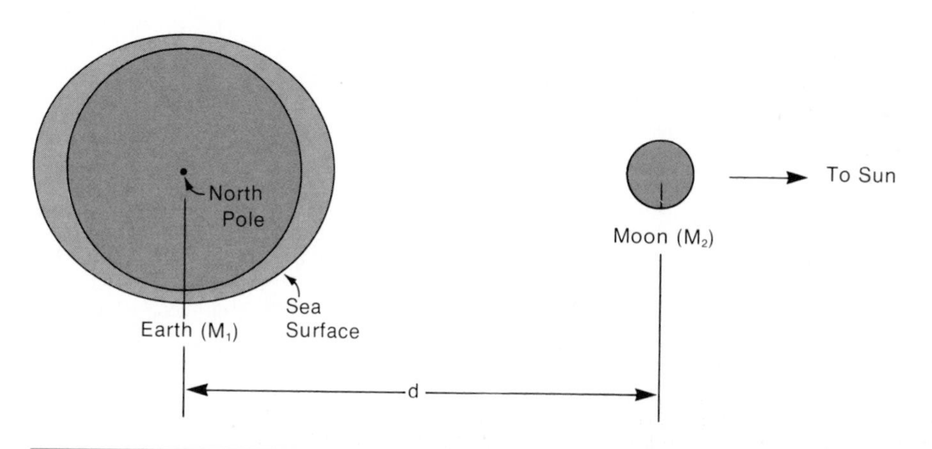

Figure 12-1 The tide-producing effect of the moon. Tidal bulges are exaggerated.

the earth-moon system. At the earth's surface there is an imbalance between these gravitational and centrifugal forces. On the side of the earth facing the moon, the attractive force of gravity dominates, whereas centrifugal force dominates on the opposite side of the earth. These forces which tug at the earth in opposite directions result in two opposing bulges of water (Fig. 12–2). If the earth were completely covered by water, the global tidal pattern would resemble two giant swells, one on either side of the planet. Even the solid earth behaves in a rather similar manner, but because of its rigidity, the bulges are extremely small.

Behavior of tides

As the earth rotates from west to east, tidal highs and lows appear to move generally westward. Many coastal locations will experience two high tides every 24 hours and 50 minutes, or one high tide every 12 hours and 25 minutes. The reason the tides are not exactly 12 hours apart is that during the earth's daily rotation, the moon has been moving forward in its orbit. The earth, therefore, must turn for an extra 50 minutes to reach its previous day's position relative to the moon.

The two high tides experienced at certain coastal

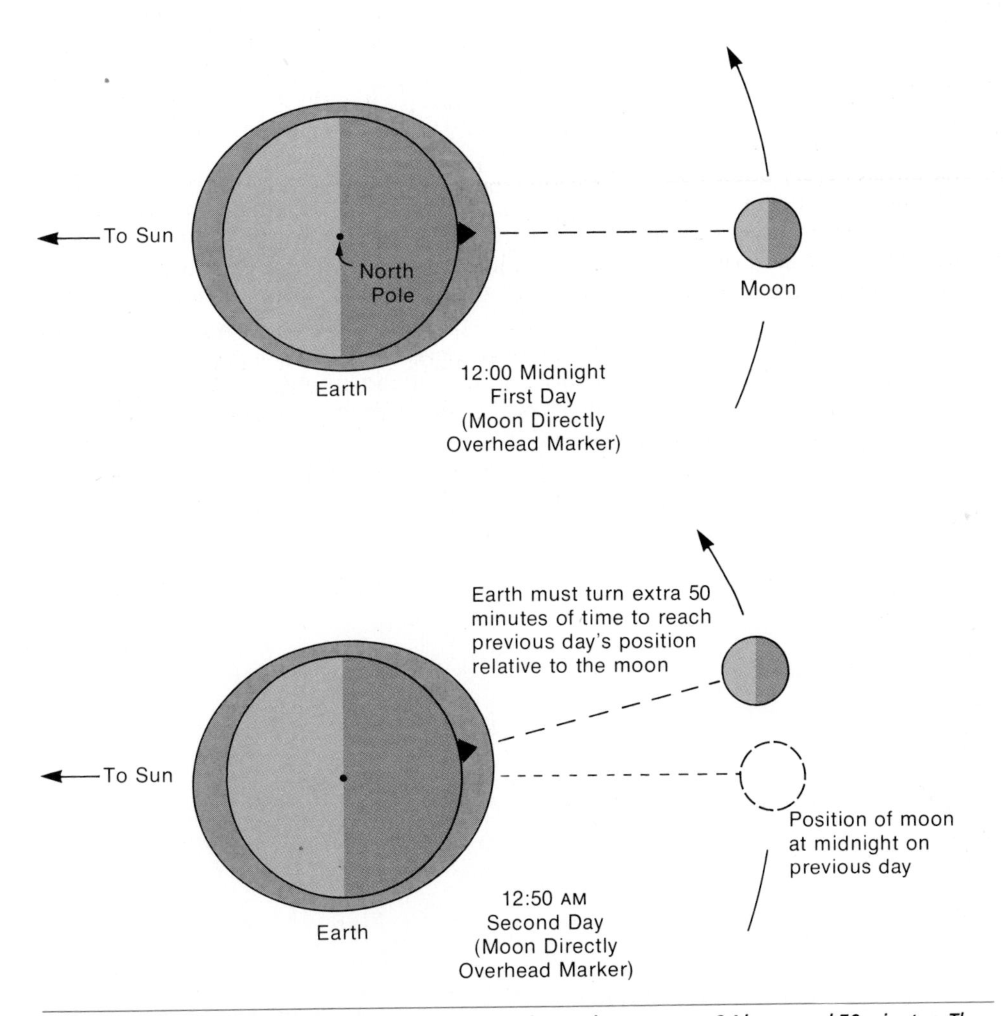

Figure 12–2 The moon passes a given location on the earth once every 24 hours and 50 minutes. Thus, the earth, after having completed its 24-hour rotation (the solar day), must continue to turn for 50 additional minutes to arrive at its previous day's position relative to the moon.

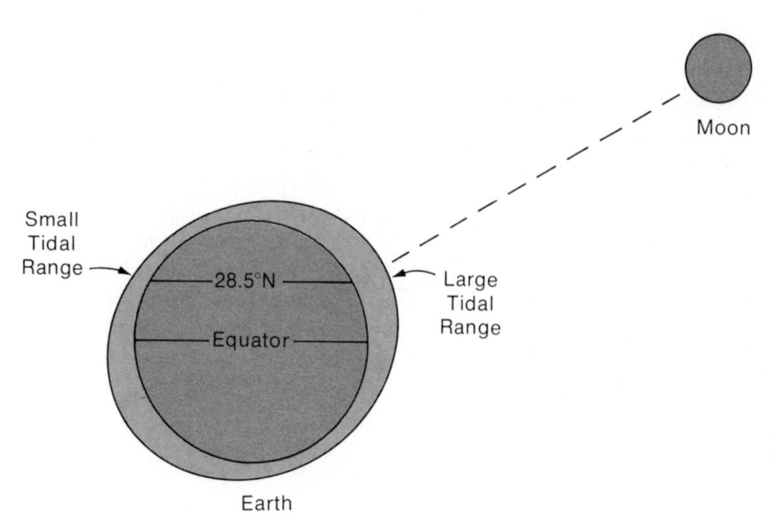

Figure 12–3 Because the center of the tidal bulges may lie at any angle from the equator to a maximum of 28.5°, tidal ranges along latitudes are not the same on opposite sides of the earth.

locations are usually of different magnitude. This is because the orbit of the moon is at an angle to the plane of the earth's equator. Indeed, the moon's orbital path carries our satellite as high as 28.5° north and south of the equator. The tidal bulges occur on a line passing from the earth's center to the moon's center. As the earth rotates, a point along the coastline at any given latitude will experience first and second high tides of different magnitude except when the moon happens to be in the equatorial plane (Fig. 12–3).

Although the sun is not as important in causing tides as the moon, it nevertheless produces a tide that is about half the magnitude of a lunar tide. At new and full moon, when the sun and moon are aligned with the earth (Fig. 12–4), the gravitational attraction of the sun reinforces that of the moon, and tides are at their maximum height. These unusually high tides that occur twice a month are termed **spring tides** (although they have no relationship to the season). At quarter moon, the sun and moon pull at right angles to each other so as to produce semi-monthly weak tides called **neap tides.**

There are many other factors that complicate the earth's tidal patterns (Fig. 12–5). The movement of

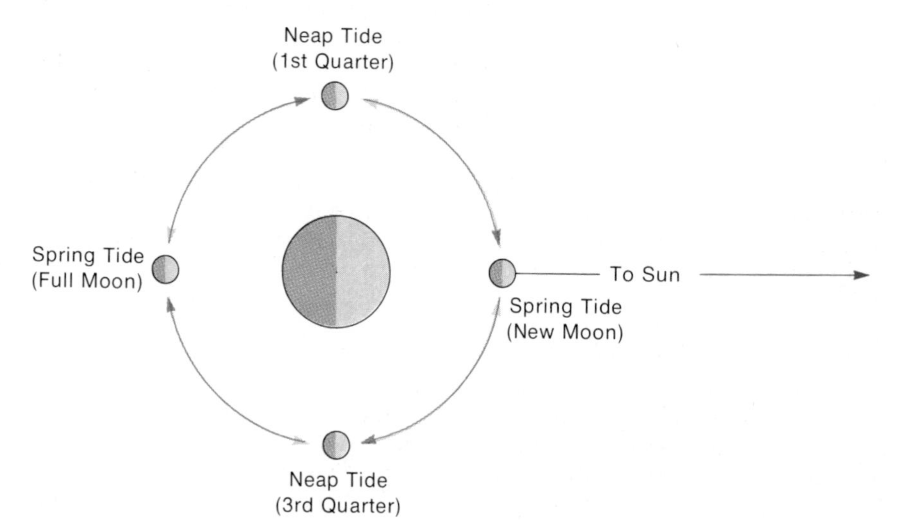

Figure 12–4 Position of the sun, earth, and moon at times of neap and spring tides.

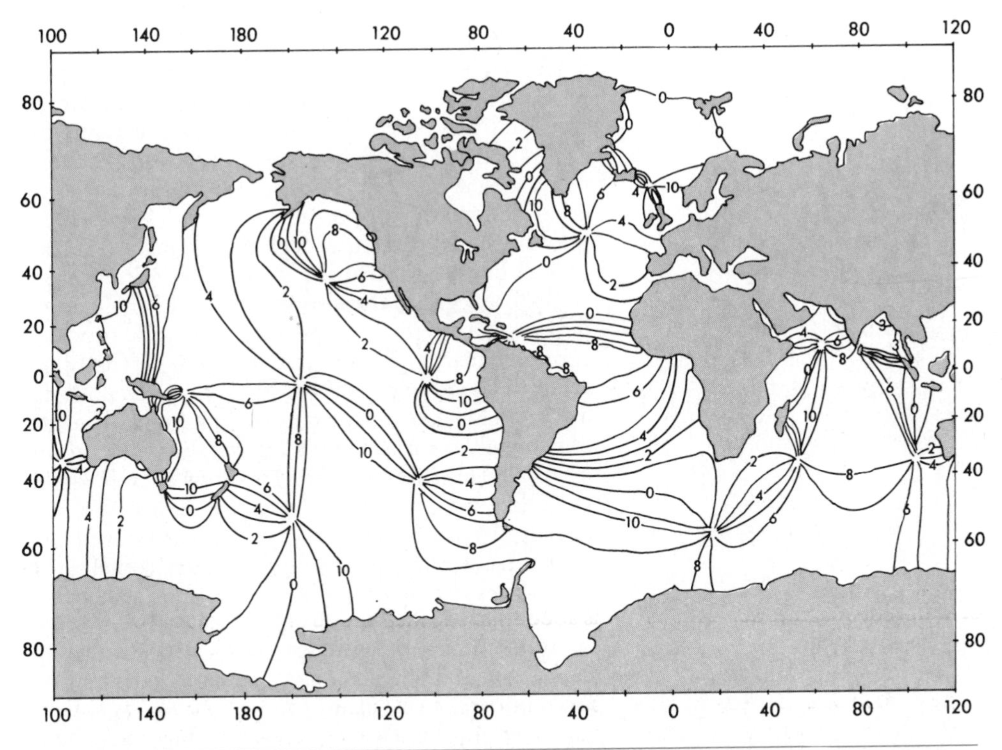

Figure 12–5 Tidal map of the ocean for an assumed position of the moon directly above the equator at 0° longitude. The lines are drawn through points connecting simultaneous high tides for each hour after the moon had crossed the 0° meridian. As is evident here, tides move around central points counterclockwise in the Northern Hemisphere and clockwise in the Southern Hemisphere. (After Cartright, Science Journal, *5:60–67, 1969.)*

tidal bulges around the earth is strongly influenced by the shape, depth, and interconnections of ocean basins. Tidal bulges may be diverted by continents and are subject to the Coriolis effect. As a result, tidal predictions for particular coastal locations require actual tidal observations (Fig. 12–6) over long periods of time.

It is difficult to measure the magnitude of tides in the open ocean, but observation stations on islands in the Pacific show that the water level seldom varies by more than about 1 meter. Along the irregular shorelines of continents, however, water movements may become concentrated by the configuration of the land and friction with the ocean bottom. The result is exceptionally high tides at some coastal locations. For example, in Nova Scotia and along the Brittany coast of France, tides frequently exceed 15 meters. The maximum tidal range for most coastlines, however, is less than 3 meters.

Tidal currents are only minor agents of transportation and erosion. Nevertheless, in some shallow areas they have sufficient strength to transport sand and scour and erode the sea floor near the entrance to inlets and bays. At the mouths of some rivers, the advance of the tidal bulge may take the form of a rapidly flowing turbulent wave, which moves up the river for miles before reversing its flow. Such tide-induced river waves are called **bores.** Along the Amazon River, the tidal bore is often 5 meters high and travels upstream at 22 km per hour.

Tidal friction

As the solid earth rotates, there is a tendency for the two tidal bulges to be carried around with it. However, the moon's attraction prevents the earth from dragging the bulges very far. Thus, the two tidal crests tend to act as rather inefficient brake bands clamped on

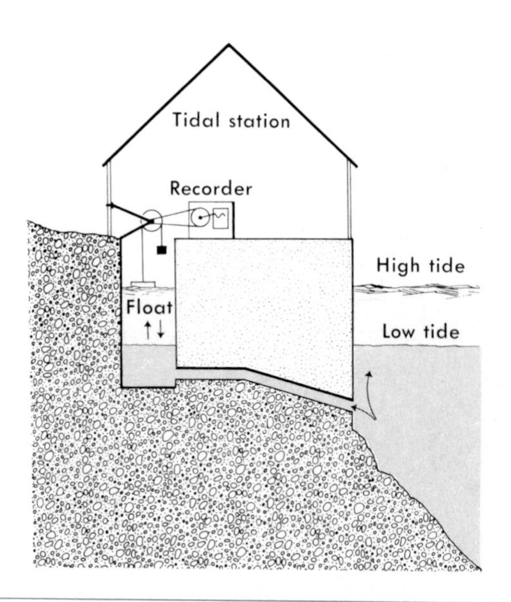

Figure 12–6 A tide gauge station. (From McCormick, J.M., and Thiruvathukal, J.V., Elements of Oceanography, *2nd ed., Philadelphia, Saunders College Publishing, 1981.)*

either side of the rotating earth, and dragging back on the planet as it rotates. The bulges experience friction along shallower areas of the ocean bottom and are also retarded by the continents that stand in their way. These effects all contribute to a slight slowing of the earth's rotation. The slowdown means that days have been increasing in length through geologic time, and the number of days in the year have been decreasing. The rate of slowing, however, is very small. It has been calculated that the average length of a day has been extended by only one thousandth of a second in the past 100 years. The change is imperceptible in a human lifespan, but over hundreds of millions of years of earth history, that amount of slowing adds up. Earlier in geologic time, the planet probably rotated faster, and the moon was closer to the earth. Under these conditions, tides were of greater magnitude, and the interval of time between tides shorter.

WAVES

The shores of the continents would not be as esthetically interesting to us were it not for waves (Fig. 12–7). Waves contribute significantly to the enjoyment of bathers and make possible the sport of surfing. For geologists, waves are important because they accomplish more geologic work than either tides or surface currents. Waves are constantly at work changing the configurations of our shorelines. They transport sediment themselves and contribute to the formation of other sediment-transporting currents. Waves are a great erosional mill for reducing large masses of rock to finer debris. From an environmental point of view, waves breaking at the sea surface bring vital carbon dioxide and oxygen into the water mass, and provide a constant mixing of gases and nutrients in the upper layers of water. Some day waves may have importance in producing electricity for use in lighting and heating our homes, for it is already technically (but not economically) feasible to generate this power from waves.

There are many kinds of water waves. We have already discussed waves generated by earthquakes and volcanic eruptions (tsunami). Tides are actually also true waves characterized by great distances between crests and very low amplitude. The most important waves from a geologic point of view, however, are waves generated by wind. Wind waves occur in many sizes and shapes, ranging from a few centimeters in height to those with crests towering 30 meters above the wave trough. They are formed whenever the wind blows over the surface of the ocean and thereby transfers part of its energy through friction and pressure fluctuations to the water surface. Three factors control the size of waves. The first is the velocity of the wind, the second is the **fetch** or distance over which the wind has blown, and the third is the constancy or duration of time the wind has moved over the water. Waves frequently continue to travel hundreds of miles beyond the area in which they were generated. Such waves are termed **swells.**

Wave terminology

Waves are described in terms of such properties as length, period, frequency, velocity, and height (Fig. 12–8). **Wavelength** is simply the distance between two successive crests or troughs. The **period** of a wave is the time required for one full wavelength to pass a fixed point. Thus, if it takes 2 seconds for two crests to pass a fixed buoy, the wave period is 2 seconds. The **frequency** of a wave is the number of waves passing an observation point in a given interval of time. **Wave velocity** is the speed at which a crest or trough travels and is equal to the wavelength divided by the period. **Wave height,** of course, is the vertical distance between the top of a crest and bottom of the trough.

Figure 12–7 *Waves crashing against coastline of granitic rocks, Acadia National Park, Maine. (Courtesy of U.S. National Park Service, photograph by Richard Frear.)*

Waves in the open ocean

In deep water, the motion of waves is not affected by the sea floor, and so open ocean waves have different characteristics than waves in shallow water. A wave is called a deep water wave if the ratio of water depth to wave length, that is $\frac{d}{L}$, is at least one half. The motion of waves in deep water is a bit more complex than an uninformed person might surmise. For example, although the wave *form* appears to travel along, there is very little real transport of water. This seemingly con-

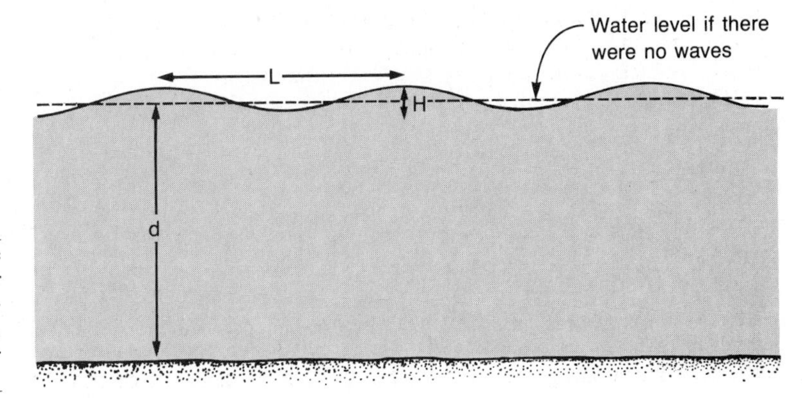

Figure 12–8 *Ideal ocean waves. L = wavelength, H = wave height, D = water depth. (From McCormick, J.M., and Thiruvathukal, J.V.,* Elements of Oceanography, *2nd ed., Philadelphia, Saunders College Publishing, 1981.)*

tradictory fact is apparent to an observer watching a piece of driftwood floating on the ocean. The wood does not move appreciably forward with the wave, but stays in approximately the same place as it rides up on the wave crests and down on the troughs.

While the wave *form* is moving across the ocean surface, water particles within the wave travel through a circular orbit and return approximately to their original positions. On the crest of a wave, all water particles are moving forward in the direction of wave travel. Particles move backward in the trough. In this manner, any given water particle makes one complete circuit in each wave period.

As indicated in Figure 12-9, the particle orbits exist not only at the surface but also at shallow depths. As we have seen, energy is imparted to a wave by wind at the surface, and this energy is progressively lost with depth. For this reason, the diameter of the particle orbits decreases with depth. At a depth of about half the wavelength, motion is negligible. This depth is referred to as **wave base.** Scuba divers are usually not disturbed by small or medium-size waves because wave base is relatively near the ocean surface.

If it were somehow possible to "freeze" a wave at any given instant and then draw lines between corresponding water particles in orbits from the surface downward as shown in Figure 12-9, then one would notice that the lines bend toward the crests and away from the troughs. In effect, the traveling wave is a series of convergences and divergences of water particles. Water, being incompressible, does not respond to these compressions and expansions by changing volume, as air might, but rather is forced to rise as a crest under converging water and fall as a trough in divergent areas.

Far out to sea, waves usually do not exhibit a symmetrical pattern of successive elongate crests and troughs (Fig. 12-10). Rather, the crests and troughs tend to have an irregularly elliptical form and uneven spacing. This irregularity is the result of waves from different source areas interfering with one another, as well as fluctuations in the velocity and directions of the wind that generates the waves.

Waves in shallow water

As waves enter the shallow zone near the shoreline, their movement and form are significantly changed. Near a depth of half the wavelength, the deepest orbits below the surface come into contact with the shallow sea bottom (Fig. 12-11). Here, actual lateral movement of water is initiated for the first time. Friction with the bottom causes the wave to be slowed. The still rapidly moving waves that follow tend to catch up with

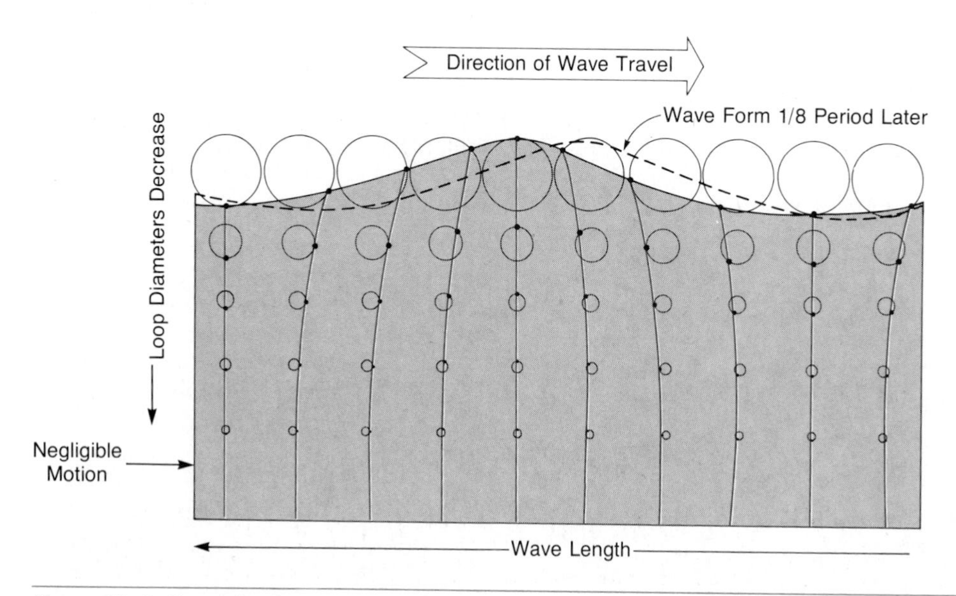

Figure 12-9 Generalized cross-section of wave showing orbital motion of water particles (black dots). Orange lines connect corresponding particles. (From Kuenen, Ph. H., Marine Geology. *New York, John Wiley, 1950, p. 70.)*

Figure 12–10 *The complex wave pattern often developed on the open sea. (From McCormick, J.M., and Thiruvathukal, J.V.,* Elements of Oceanography, *2nd ed., Philadelphia, Saunders College Publishing, 1981; photograph courtesy of U.S. Coast Guard.)*

the wave that has been slowed, and in this way wavelength is shortened. The effect rather reminds one of a pile-up of automobiles that must slow down at a highway toll gate. Also, in an attempt to accommodate the volume of water in the incoming waves, wave height increases. As these changes are taking place, the shape of the water particle orbits are distorted from circular orbits to more oval configurations. The orbits near the bottom are eventually flattened, so that water particles move back and forth parallel to the sea floor. Near the surface, the water particle orbits are less distorted. As the wave enters still shallower water, the water particles at the interface between the ocean and the atmosphere are moving faster than the wave itself, with the result that the wave steepens and collapses as a line of breakers. The motion can be seen particularly well in **plunging breakers** (Fig. 12–12B), which have a curling crest because water particles have outrun the wave and so have nothing beneath them to support their motion. Such plunging breakers usually form on steeply sloping areas of the sea floor where the energy of the wave is extracted rather abruptly by friction. On sea floors that slope more gently seaward, the more commonly observed **spilling breakers** occur (Fig. 12–12A). Spilling breakers give surfers a longer but perhaps less exciting ride than do plunging breakers. Incoming waves may also dissipate as surging breakers, which do not break at all but simply surge up on the slope of steep beaches or sea walls (Fig. 12–12C).

Wave refraction

There are many similarities in the behavior of water waves and light waves. Light waves, for example, are bent or refracted as they pass from one medium to another, as from air to water. Surface water waves are also bent, and this bending or refraction occurs when one part of a wave reaches shallow water before another part (Fig. 12–13). If waves were approaching a straight and uniformly sloping shoreline with crests parallel to the shore, then refraction would not be possible. Such conditions, however, occur only rarely. Because of the earth's spherical shape, as well as irregularities in coastlines and in the topography of the near-

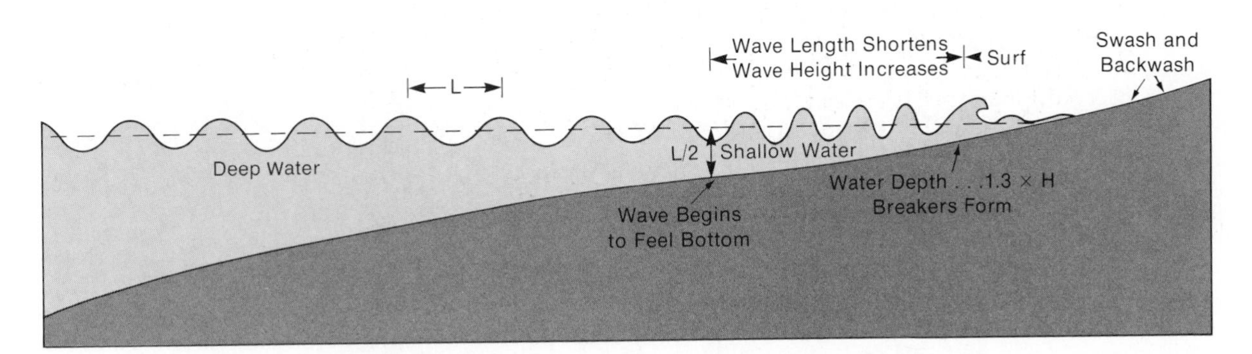

Figure 12–11 *Profile of the nearshore zone indicating change in wave form as waves travel from deep water to shallow water to shore.*

Figure 12–12 *(A) Spilling breaker. (B) Plunging breaker. (C) Surging breaker. (From McCormick, J.M., and Thiruvathukal, J.V.,* Elements of Oceanography, *2nd ed., Philadelphia, Saunders College Publishing, 1981; photograph by U.S. Army Coastal Engineering Research Center.)*

shore sea bottom, waves nearly always approach a shoreline from an angle and are then refracted. As waves approach the shore obliquely, one end of the wave will begin to touch bottom before the other. The part of the wave that touches bottom first is slowed by friction, while the still unaffected more seaward part of the wave proceeds along at a faster speed. The result is that the wave crests and troughs swing around until they are within about 5° of being parallel to the shoreline. One would observe the crest-to-crest distance diminishing, and the crests forming a pattern of convex-landward curves.

If we were to take a helicopter flight over a very irregular shoreline, we would observe the refraction of wave crests into a pattern of curves that are concave landward around jutting headlands and convex landward in open deep bays. This pattern is a manifestation

Figure 12–13 *An incoming wave is refracted as it approaches the beach at an angle so that the right side is slowed first when encountering shallow water.*

of the fact that approaching waves touch bottom first adjacent to headlands where shallow water is initially encountered. Those parts of the wave front opposite bays do not encounter shallow water for an additional distance.

One of the effects of wave refraction along an irregular shoreline is to straighten that shoreline by vigorously eroding the headlands and filling the coves with transported sand and gravel. The headlands receive the first attack of the waves, and these wave segments have not dragged the bottom for any appreciable distance. Meanwhile, the remainder of the wave crests sweep into an arc, so their energy is dissipated over a wider area. The relationships are illustrated in Figure 12-14, which shows wave patterns as well as lines, called **orthogonals,** that are drawn perpendicular to wave fronts. Orthogonals are drawn with uniform spacing and are intended to indicate that equal amounts of energy are contained in the segments of the wave that are between two adjacent orthogonals. Thus, the amount of energy between any pair of orthogonals is constant regardless of refraction. Wherever orthogonals converge (as around headlands), there is a convergence of energy and a vigorous erosional attack. In the bays, orthogonals diverge, and the wave energy is spread widely. With lesser energy available, erosion is minimal, and waves are primarily at work distributing and depositing sand and silt.

Along some stretches of straight coasts, one finds a pattern of concave and convex curves of waves somewhat resembling the refractions along more uneven coasts. Usually such a pattern is the result of irregularities in bottom topography. A submerged elevated area would retard waves and produce wave fronts with concave-landward crests (Fig. 12-15).

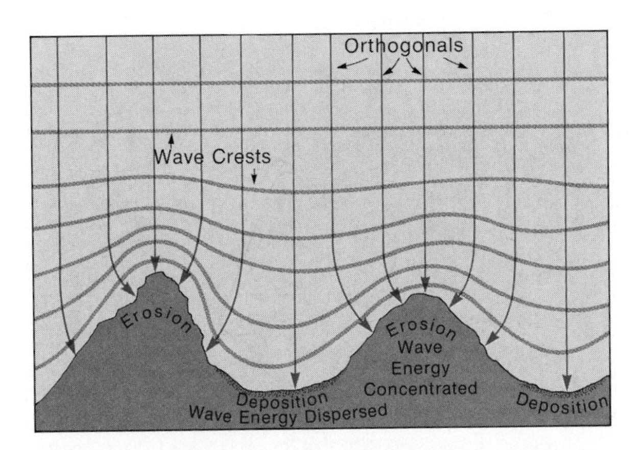

Figure 12-14 Wave refraction pattern for an irregular shoreline with a uniformly sloping bottom. Sea floor contours approximately parallel wave crests.

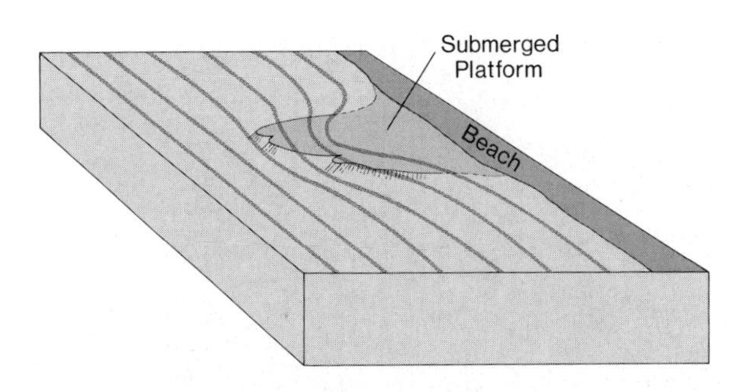

Figure 12-15 Wave refraction pattern for a straight coastline but with irregularity in bottom topography.

NEARSHORE CURRENTS

Anyone who has floated placidly on the sea on a balmy day has noticed how quickly one's position changes relative to some fixed point on the shore. Usually, we are being carried along by either longshore or rip currents. **Longshore currents** are nearshore drifts of water that parallel the shoreline. They are developed when waves approach a shoreline at an angle, so that there is a component of energy that causes water to move parallel to the beach and away from the acute angle made by the incoming waves and the shoreline (Fig. 12-16). The velocity of these currents generally increases with increases in the angle made by incoming waves and the beach, as well as increases in the slope angle of the beach, wave height, and wave period. Longshore currents are important agents in transporting sediment along a coastline.

Lifeguards at beaches sometimes caution vacationers about jetlike currents that flow out to sea (Fig. 12-17). Such narrow currents are called **rip currents.** They are formed when water brought to the shore by breakers escapes seaward. Often these currents are localized by depressions in the sea floor. Signs posted on beaches that warn swimmers of "dangerous undertow" are actually referring to these rip currents. Seen from above, they may be visible as long streams of foamy, turbid water extending at approximately right

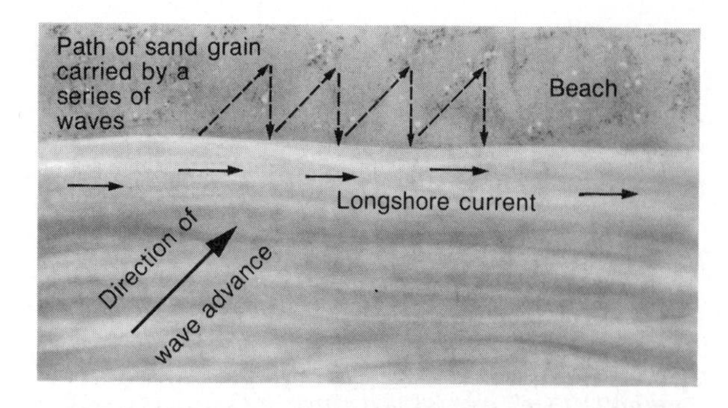

Figure 12–16 Longshore drifting (also called beach drifting) and longshore currents produced by waves approaching the coast obliquely. (From McCormick, J.M., and Thiruvathukal, J.V., Elements of Oceanography, *2nd ed., Philadelphia, Saunders College Publishing, 1981.)*

angles to the shore. If you should find yourself transported out to sea on a rip current, it would be unwise to exhaust yourself by attempting to swim back to shore against the current. Rather, swim parallel to the shore until out of the rip, and then swim directly to the beach.

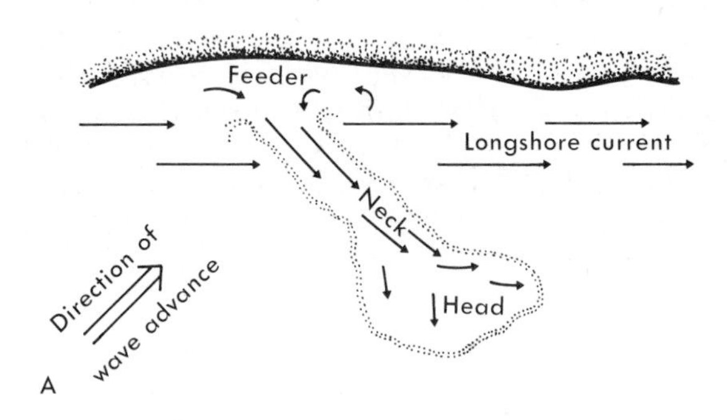

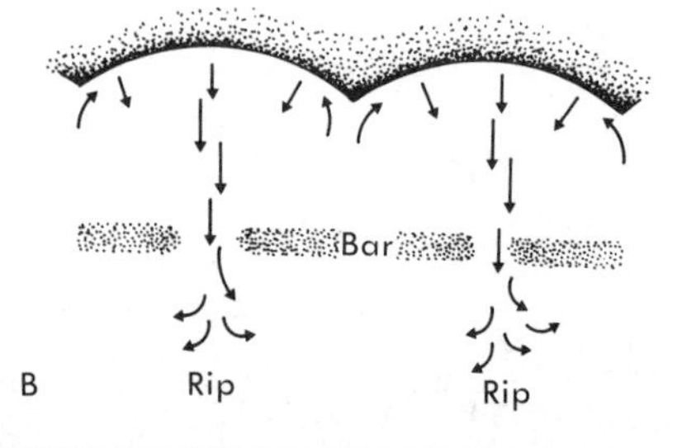

Figure 12–17 (A) Rip current formed along a coast by escape of water along a submarine topographic low. (B) Rip current caused by water piling up behind nearshore sandbars. (From McCormick, J.M., and Thiruvathukal, J.V., Elements of Oceanography, *2nd ed., Philadelphia, Saunders College Publishing, 1981.)*

TRANSPORT OF SEDIMENT ALONG THE SHORE

The transportation of sediment along a coastline is accomplished by longshore currents and by an interesting process known as **longshore drifting.** We may define longshore drifting as the irregular movement of sand and gravel along the shore that results from the combined effects of incoming waves and the return flow of water from the beach. As indicated earlier, waves nearly always strike the shore at an angle and then continue along the same direction as they push up onto the beach. This incoming mass of turbulent water, or **swash,** picks up particles of sand and carries them obliquely up the beach. The water then flows directly back under the influence of gravity, and this **backwash** similarly lifts and transports grains of sand. The result is a kind of sawtooth pattern of sand grain transport along the shoreline (see Figure 12–16).

The direction of drifting may vary from time to time, but for most shorelines there is a cumulative movement determined by the prevailing wind direction. Longshore drifting and the carrying power of longshore currents are responsible for transport of sand over hundreds of miles of coastline. Because of these processes, beaches may exist even where there is no local supply of sand.

THE BEACH

Millions of people who visit the beaches of the world will agree that beaches are indeed enjoyable and fascinating places. A **beach** can be rather simply defined as the zone of sediment that accumulates between the average low-water level and a landward change in topography (such as a sea cliff). Depending upon the tectonic characteristics of a shoreline, beaches may extend continuously for hundreds of miles along a coast or be limited to protected bays and coves (Fig. 12–18). The materials that go into the building of beaches may be derived from either land or sea (e.g., reworked shell debris), but the far greater part is contributed by streams that drop their transported load of

Figure 12–18 Beach in protected cove north of Brookings, Oregon. (Courtesy of J.C. Brice.)

Figure 12–19 A beach composed of gravel (called a shingle beach) at East Sussex, England. (Courtesy of the Institute of Geological Sciences, London.)

sediment when they reach the sea. Although many beaches are composed of gravel (Fig. 12-19), the majority consist of sand, and the most abundant mineral in most beach sands is quartz. Along granitic coastlines, feldspars also contribute to the composition of beach sand, whereas in semitropical or tropical regions, sand composed of broken and worn shell fragments may be prevalent. Grains of black basalt and green olivine are common in the beaches of volcanic islands.

Once deposited on a beach, sediment may be subjected to repeated redistribution. During seasons when storms are frequent, the thickness of sand on a beach may be significantly reduced. Beaches generally tend to grow in width and thickness during periods of calm seas.

As seen in profile (Fig. 12-20), a beach can be divided into several subdivisions, although each of the subdivisions may not be present on every beach. Most beaches have a nearly flat upper part formed by deposition of material by waves. This level feature is called the **berm.** A sloping lower part is called the beach face. The **beach face** is the area across which waves swash upslope across the beach and return as backwash. The

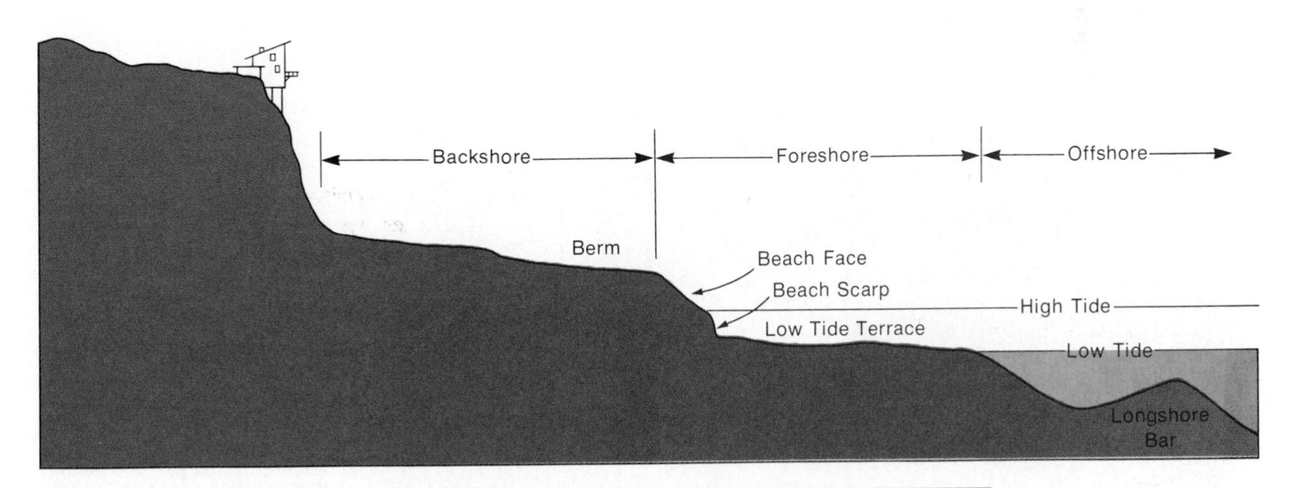

Figure 12–20 Major divisions of the beach and nearshore area. (From McCormick, J.M., and Thiruvathukal, J.V., Elements of Oceanography. *Philadelphia, W.B. Saunders Company, 1976.)*

slope of the beach face is largely determined by sediment particle size. In general, the larger the grain size, the steeper the slope.

EROSION OF SHORELINES

Waves are important agents in causing the erosion of coastlines (Fig. 12–21). As one might expect, large waves, such as those generated during storms, are exceptionally effective and may cause more change than smaller waves operating over much longer periods of time. Aside from the size of the waves themselves, the rate at which erosion proceeds is also influenced by the durability of rock along the shoreline, how that rock is jointed, the openness of the coast to attack, the slope of the sea floor adjacent to the coast, and the abundance and size of sedimentary particles that can be hurled by waves against the shore (Fig. 12–22). The abrasive action of sand, gravel, and pebbles in surging water effectively reduces rock particles to sizes that can be carried seaward in rip currents.

Rugged coastlines provide an impressive show of the effectiveness of wave impact in eroding rock masses (Fig. 12–23). The crash of incoming waves may exert pressures exceeding 6000 pounds per square foot. Air in joints and fractures is suddenly compressed and acts as a wedge in driving rocks apart. In the next moment, the water recedes, and the compressed air expands, often with explosive force. These mechanical stresses on the rocks are capable of reducing solid granitic headlands to masses of boulders and rubble.

There are many features of coastlines that are formed primarily by erosional activity. Among these features are wave-cut cliffs, wave-cut benches, and an assortment of erosional remnants termed stacks, sea caves, and sea arches. **Wave-cut cliffs** are coastline escarpments resulting from the erosive powers of water that surges landward with each breaking wave. The action of the surf reminds one of a horizontal saw that repeatedly bites into the base of the cliff, sometimes chiseling out an indentation called a **notch** (Fig. 12–24). Undermining by wave action causes periodic collapse of the face of the cliff, and the downfallen rubble is then further diminished by the surf. As the wave-cut cliff retreats before the onslaught of the waves, remnants may be left behind as **stacks** (see Figure 12–22). Elsewhere, waves differentially erode along the more readily erodible joints and sculpture **sea caves** and **sea arches** (Fig. 12–25). Extending seaward from the base of the cliff along shorelines where erosion predominates, a platform is produced by wave erosion called a **wave-cut bench.**

Figure 12–21 Wave erosion at Sea Arch Overlook, Volcanoes National Park, Hawaii. This coastline is open to attack, and the jointed solidified lavas are readily eroded to form sea arches. (Courtesy of U.S. National Park Service.)

Figure 12–22 Beach composed largely of chert nodules derived from the sea cliffs in the background. The chert nodules are used as tools by storm waves to aid in the erosion of this coastline of the Isle of Wight, England. The isolated masses of rock are erosional features called stacks. (Courtesy of the Institute of Geological Sciences, London.)

DEPOSITIONAL FEATURES OF SHORELINES

In addition to beaches, there are several other depositional features that develop along coasts as a result of the activity of nearshore currents and waves. Some of these features are developed below sea level, and others are built above sea level. Among the submerged features are wave-built terraces and longshore bars. **Wave-built terraces** (see Figure 12–24) are simply accumulations of unconsolidated sediment swept back to sea from the beach environment or deposited by longshore currents. In most locations, the boundary between the beach and terrace is transitional. **Longshore bars** (see Figure 12–20) are elongate barely submerged ridges of sand that parallel the shoreline. They are particularly characteristic of coastal areas with a very gently sloping ocean floor and an abundant supply of sand. Longshore bars migrate shoreward and increase in height during winter seasons when larger waves are prevalent. These trends are reversed during the less stormy summer season.

Figure 12–23 Breaking waves are hurled against granitic rocks along the coast of Mount Desert Island, Maine. Air trapped in joints and fractures is alternately compressed and allowed to expand by incoming waves. This hydraulic action aids in the disintegration of rocks along the shore. (Courtesy of J.C. Brice.)

Several different coastal topographic features are built above sea level. They are distinguished by their form and position. For example, **spits** are ridges of sediment connected at one end to land and extending into open water at the other end (Fig. 12–26). Where spits partially extend across the mouth of a bay, wave action may cause them to curve landward. The result is a feature called a **recurved spit** or hook. Spits may build completely across the bay opening so as to close off the harbor. In this way, they are transformed into **bay barriers.** When a bar built of sand or gravel ex-

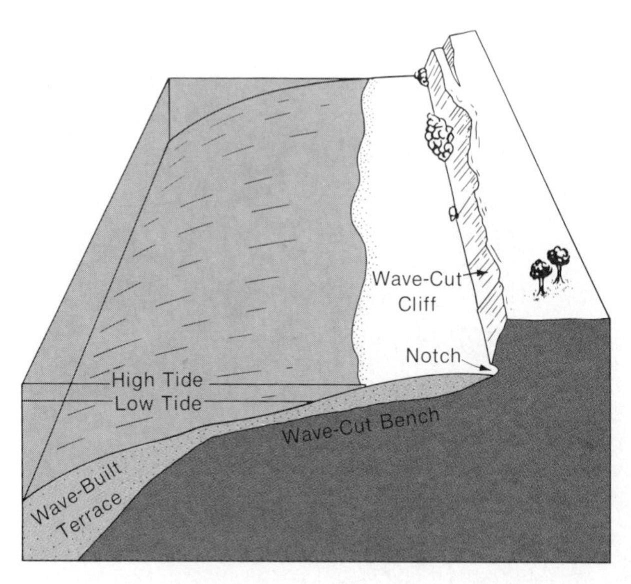

Figure 12–24 Erosional features of the shore profile are the wave-cut cliff, wave-cut bench, and notch. Depositional features include the beach and wave-built terrace.

tends from the mainland out to a nearby island or connects two islands, the resulting feature is termed a **tombolo.**

Barrier islands (Fig. 12–27) are elongate islands of sand extending parallel to a coast. These features are rather continuous along the Atlantic coast of the United States as well as around Florida and along the Gulf of Mexico. Barrier islands may attain widths of several kilometers and extend more than 100 km in length. Coney Island off the New York coast, Miami Beach, and Padre Island off Texas are good examples of barrier islands. Barrier islands may develop in more than one way. Some appear to have been formed by the continuous extension of a spit along a coastline. The shoreward migration of offshore bars may also be responsible for having formed some barrier islands. One favored theory for the barrier islands along our eastern seaboard is that they formed during an increase in sea level with the last melting of glaciers about 10,000 years ago. The rise in sea level would have drowned low tracts of land behind coastal dunes and isolated the former dune areas as barrier islands offshore (Fig. 12–28).

Figure 12–25 Arch, stacks, and sea cliffs eroded in chalk strata between Swanage and Ballard Point, Dorset, England. (Courtesy of J.C. Brice.)

KINDS OF COASTS

The marvelous variety of topographic features along coastlines is, as we have noted, the result of many interacting variables. Coasts are dynamic geographic elements that are constantly changing as a result of erosion, deposition, and tectonic movements. Because of the changes and the variables, classifications of shorelines are difficult to formulate and should be considered as generalizations to which there are nearly always exceptions.

Drowned coasts

The majority of the world's coastlines appear to have experienced varying amounts of submergence in rela-

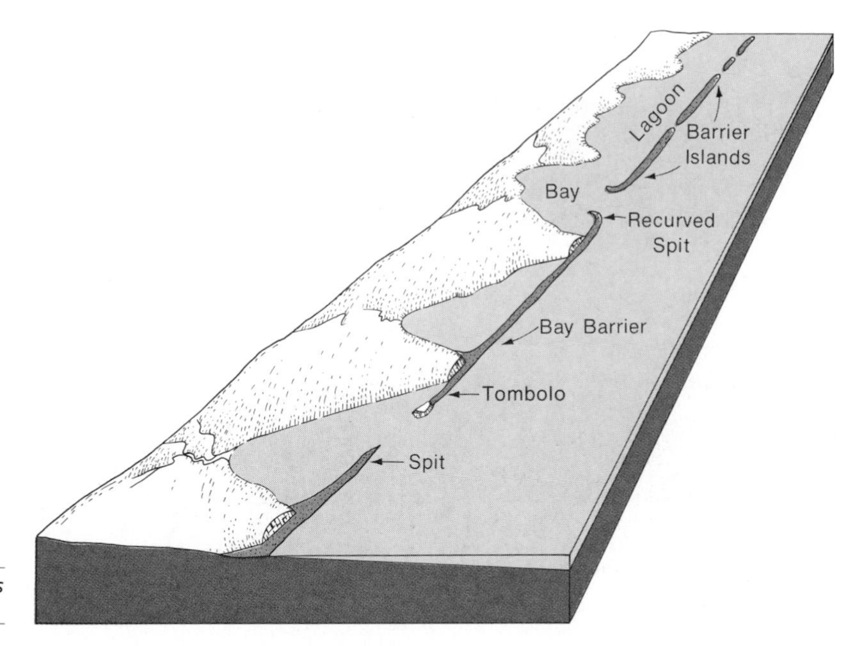

Figure 12–26 *Depositional features along a coastline.*

tively recent geologic time. This does not necessarily mean that all of the inundated coasts subsided, for inundation can also result from rising sea level. Indeed, rising sea level appears to be the reason for most submerged coastlines today. A global rise in sea level began about 10,000 years ago as meltwater from continental ice sheets was poured back into the ocean. Whether the result of subsiding land or rising sea level, drowned coastlines exhibit highly irregular outlines. The configuration results from the inundation of low-lying areas as the sea advances into river valleys and tributaries (Fig. 12–29). The distinctive branching

Figure 12–27 *The barrier island that is Miami Beach has been the location of extensive urbanization. (Courtesy of U.S. Geological Survey.)*

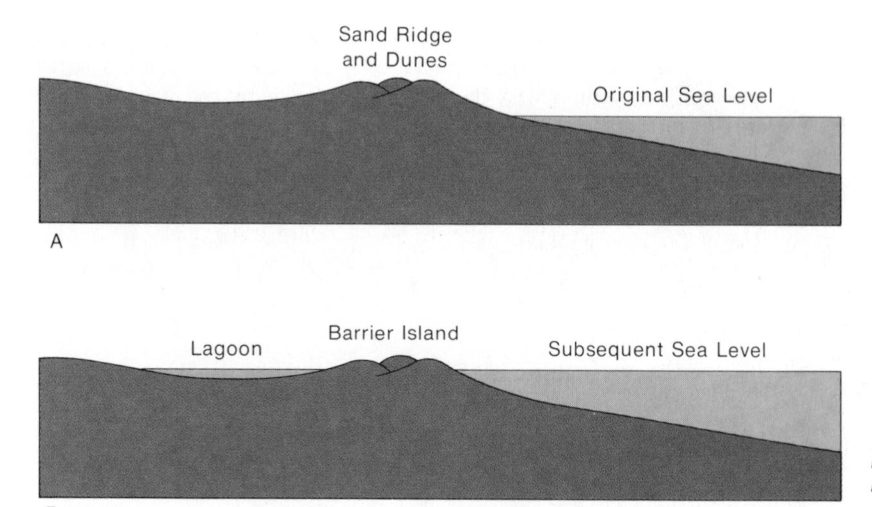

Figure 12–28 Development of a barrier island as a result of a rise in sea level.

Figure 12–29 Earth Resource Technology Satellite photograph of Chesapeake Bay. The geologically recent increase in the volume of water in the ocean with the melting of glaciers has caused an inundation of river valleys to produce bays such as this one. The city of Washington, D.C. is located at the "W" on the Potomac River. (Courtesy of the National Aeronautics and Space Administration.)

pattern is well developed in such areas as Delaware Bay (Delaware and New Jersey), Pamlico Sound (North Carolina), and Chesapeake Bay (Maryland and Virginia).

Submerged coastlines in glaciated areas may exhibit narrow, steep-sided inlets. These features are called **fjords** (Fig. 12–30). They were developed by valley glaciers that excavated the lower ends of the valleys to depths well below present sea level. The great depth of water in some fjords is the result of not only glacial action but also the previously mentioned eustatic increase in sea level following the Ice Age. Fjord coasts have splendid protected harbors and some of the most spectacular scenery on earth. Although the fjords of Norway are probably the most famous, these features are also found along the coast of New Zealand, Greenland, Alaska, and British Columbia.

Deltaic coasts

Deltaic coasts are those built seaward as a result of the deposition of sediment brought to the ocean by rivers. For a delta to form, the river must provide sediment at a rate faster than waves and nearshore currents can redistribute it. For this reason only rather large rivers are able to build and maintain deltas. If the coastline at

Figure 12–30 *Two views of the Norwegian fjord known as the Geirangerfiord. (Courtesy of Mineralogisk Geologisk Museum, Oslo, Norway; photograph by B. Elgvad.)*

the mouth of a river is relatively protected, and if tidal action is minimal (as in the Mediterranean and the Gulf of Mexico), the removal of sediment by wave action will be slowed and the opportunities for the development of deltas improved. Deltas vary considerably in overall form. Some are truly in the form of the Greek letter delta (Δ), others are more broadly rounded, whereas others are called bird's-foot deltas in reference to their finger-like extensions into the ocean (Fig. 12–31).

In the great marine deltas of the earth, the littoral zone has its maximum width. Here one finds shallow-water lagoons and saltwater marshes alternating with patches of land over a belt that may be 50 km wide. Not only is there a mixture of environments of sedimentation across the surface of the delta, but vertical sections are characterized by an intertonguing of marine and nonmarine deposits caused by repeated shifting of the shoreline. Nearly always, sedimentation in deltas involves transportation of sands far out into the marine basins, where they may become enclosed in organic-rich muds. The sands are potential reservoir

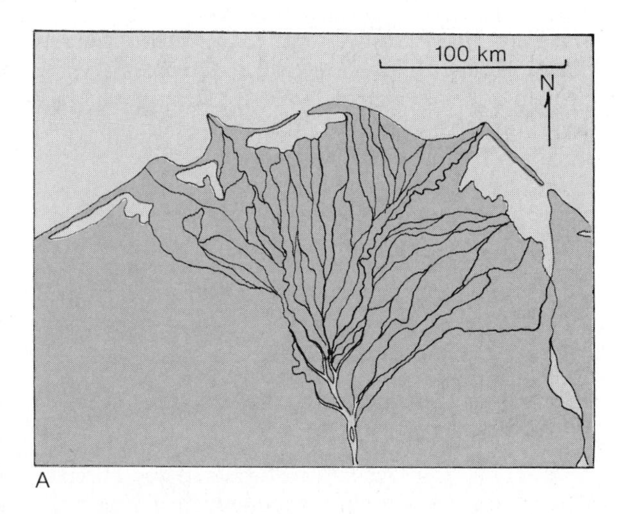

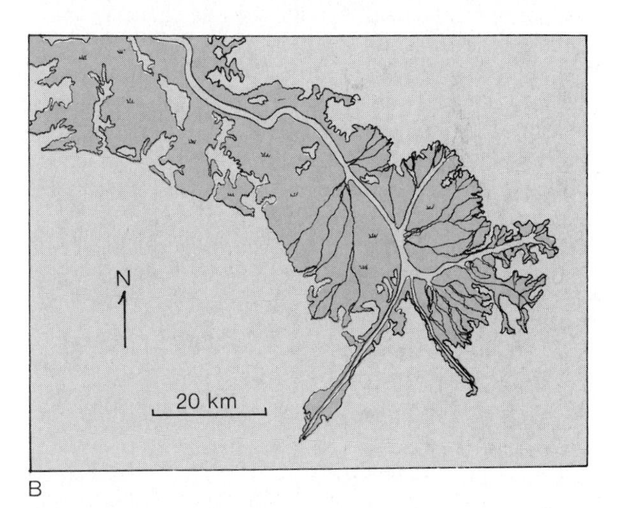

Figure 12–31 *(A) The Nile Delta. The branching streams of this delta begin about 160 km inland at Cairo and fan out seaward over a great triangular area. (B) The most southern subdelta of the Mississippi Delta exhibits the bird's-foot pattern.*

rocks for oil and gas, whereas the muds may provide the hydrocarbons and serve as impermeable barriers needed for oil entrapment. Clearly, ancient deltas are important because of the store of fossil fuels they may contain. They are no less important, however, because of the record they preserve of past conditions on earth.

Tectonically active erosional coasts

Some of the most beautiful coasts in the world are those that have been subjected to periodic tectonic uplift and then vigorously eroded by marine agencies. Raised wave-cut cliffs and stairlike sequences of wave-cut benches (see Fig. 5-1) are common along such coasts, and these features attest to the periodic uplift experienced by the unstable margin of the continent. Often, very deep water exists immediately offshore from the mountainous coastal belt. Because of the rapid steepening of the sea floor, offshore bars are rarely developed. Tectonically active erosional coasts are particularly prevalent around the Pacific Ocean.

Coastlines along which deposition predominates

Another prevalent type of coastline is characterized by a broad, shallow, flat offshore zone lying adjacent to a low-lying coastal plain. Such coasts also have an abundant supply of sandy sediment. In such an environment, barrier islands and beaches are common and extensive. Some of the barrier islands may attain widths of several kilometers and extend for great distances. Galveston, Atlantic City, and Miami Beach are built upon barrier islands. Usually the barriers are separated from the mainland by shallow lagoons. It is not uncommon for such lagoons to gradually silt up and thereby increase the size of continents.

Coasts modified by organisms

Yet another kind of coastline is shaped largely by the activity of organisms. As indicated in the previous chapter, coral reefs and calcareous algae abound in tropical areas and can add significantly to the dimensions of a coastal area. The Florida Keys are coral reef coasts formed during an interglacial stage of the Ice Age when sea level was about six meters higher than at the present. Similar organic coasts are prevalent around the islands of the southwestern Pacific. Besides corals, a host of other organisms, including oysters, tube-secreting worms, calcareous algae, and mangrove plants, contribute to the formation of coastlines modified by plants and animals.

MINERAL RESOURCES FROM THE OCEAN

Humans have extracted salt from seawater for thousands of years. Yet only within the last few decades have we begun to appreciate the importance of other mineral resources that occur on and beneath the ocean floors. As we exhaust the resources on land we shall turn increasingly to the oceans as a potential source for many of our future mineral needs.

Petroleum and natural gas

Oil and gas are now produced on continental shelves adjacent to the coastlines of 25 nations (Fig. 12-32). These hydrocarbons account for more than 90 percent of the total value of resources extracted from beneath the ocean floor and currently constitute approximately 20 percent of the world's production. Currently, most offshore petroleum comes from water depths of less than 100 meters and from areas within 120 km of the coastline. Such areas of the continental shelves are sufficiently shallow that drilling costs are not prohibitive. Further, the shallow shelves are commonly underlain by a thick sequence of sedimentary layers that include source beds rich in organic material as well as permeable and porous sandstone "reservoir rocks." The same kinds of oil-trapping structures found beneath the nearby coastal plains also occur beneath the ocean floor in offshore oil fields.

Saline minerals

Although nearly every chemical element has been detected in seawater, only common salt (sodium chloride), magnesium, and bromine are now being commercially extracted. Of these three substances, magnesium is by far the most valuable, with annual production worth about 75 million dollars. The ocean is the source of 70 percent of the world's production of bromine, an element now used primarily as an antiknock compound for gasoline. Sediments beneath the sea floor in such places as the Gulf of Mexico and Persian Gulf include thick beds of gypsum (calcium sulfate), potassium salts, and common salt. These deposits were formed long ago by evaporation of seawater in basins having restricted circulation. Important magnesium salts were also deposited in such basins. Under the heavy pressure of a thick overburden of strata, rock salt and gypsum tend to flow plastically. In many instances in the Gulf Coast region of the United States and the Persian Gulf, the salt has squeezed upward along a zone of relative weakness to form salt domes (see Fig. 3-37). Overlying younger sediments are faulted and folded, thereby producing favorable structures for petroleum. Elemental sulfur is also usually present in the cap rock of salt domes and can be extracted by the Frasch process, in which sulfur is melted by the injection of hot water into drill holes.

Figure 12-32 An offshore drilling rig. (Courtesy of Standard Oil Co. of California, photograph by John Larrecq.)

Manganese oxide nodules

In the previous chapter, we noted the existence of manganese oxide nodules on the deep ocean floor. These concretions also occur near land in such areas as the Blake Plateau east of Florida and off the west coast of Baja California. The nodules are of interest not only because of their manganese content but also because they contain nickel, copper, and cobalt. Unfortunately, dredging for nodules is very costly, and it is not always economically feasible to gather them. They are nevertheless an important resource that may be more widely used within a decade or two.

Summary Tides are caused primarily by the gravitational attraction of the moon, and to a lesser extent, by the sun's gravitational pull. The earth develops two tidal bulges simultaneously. One is on the side facing the moon, and the other is on the opposite side of the earth. The bulge opposite the moon results primarily from the strong attraction of the moon for the water mass that is closest to it. Because gravity decreases with distance, the force acting on the body of the earth itself is greater than that acting on the ocean of the opposite hemisphere. Thus, the moon tends to pull the entire earth away from the opposite hemisphere's ocean. Also, on the far side of the earth, centrifugal force dominates over the gravitational pull of the moon, thus enhancing the tidal effect. Because of those two opposite tidal bulges, as the planet rotates any given section of coastline will experience two high tides and two low tides each day.

Waves are caused by friction of wind as it passes over the water surface. The size of waves is determined by such factors as wind velocity, its duration, and the distance over which the wind has blown. Waves are described according to their wavelength (distance from crest to crest), wave height (vertical distance between crest and trough) and wave period (the time necessary for a crest to travel one wavelength). Water particles in the wave move in circular orbits that diminish with depth of water until little or no motion is evident at a depth equivalent to half the wavelength. As waves approach a coastline, they undergo a marked change in shape because of friction with the sea bottom. The waves become higher and more closely spaced, but the wave period does not change significantly. Eventually the unstable crest topples over and forms a line of breakers.

Most waves approach land obliquely. They "feel" the drag of the bottom at that part of the wave that first attains a depth of half a wavelength. Because of the lessening of wave velocity at the part of the wave that first touches bottom, the wave bends by refraction and breaks against the shore at an angle. As sheets of water (swash) move obliquely upward on the beach, they carry along grains of sediment, which are subsequently carried directly back down the beach as the water returns in the backwash. This pattern of swash and backwash results in a net transport of sediment along the beach. The transport system is called longshore drifting.

Waves erode shorelines predominantly by hydraulic action and abrasion, both of which help to produce such erosional features as wave-cut benches, sea cliffs, stacks, sea caves, and sea arches. Refraction plays an important role in wave erosion by concentrating the energy of waves on headlands.

Waves are indirectly responsible for such inshore water movements as longshore currents and rip currents. Longshore currents are produced as waves strike a shore obliquely. The coast acts as a barrier, and water is deflected so as to flow parallel to the shore. Rip currents represent a return flow of water brought to the nearshore area by waves.

Sediment carried along a coastline by longshore currents contributes to the formation of beaches, as well as longshore bars and troughs, wave-built terraces, spits, bay barriers, tombolos, and barrier islands. Longshore bars are submerged features composed of sand and gravel that extend approximately parallel to the coast. Sediment carried seaward beyond the beach may collect as a wave-built terrace. Spits are sandy ridges extending for a distance from shore into open water. When spits completely shut off a harbor, they become bay barriers, and when they grow from a headland to a nearby island, they are called tombolos. Barrier islands are exposed depositional features built parallel to the continental margins from which they may be separated by a shallow-water lagoon. The physical appearance of any particular coastal area is determined by the size of waves, openness of the coast to wave action, nature of coastal bedrock, and changes in sea level of either a tectonic or a eustatic nature. Coastlines having an irregular profile with many shallow inlets and estuaries are termed drowned coastlines. Deltaic coastlines often have a lobate or fanlike pattern and develop where streams enter the sea and deposit their load, thereby extending the shoreline seaward. Tectonically active coasts are those in which recent tectonic uplifts have raised wave-formed features well above present sea level. Such coastlines are known for their rugged beauty. Along marine depositional coasts, features formed by the deposition of sand and gravel predominate, such as broad beaches, bay barriers, spits, and barrier islands. Some coasts are built largely by the activity of calcium carbonate–secreting organisms. Coral reef coasts are the most important in this category.

The ocean contains quantities of mineral wealth. Some of this wealth is now being tapped by drilling and mining on the shallower sea floor in nearshore environments. Petroleum and natural gas are currently being produced on the continental shelves of many countries and are clearly the most valuable resources currently obtained from beneath the sea floor. Sulfur is being produced from salt domes beneath the ocean bed, and common salt, bromine, and magnesium from the water itself.

QUESTIONS FOR REVIEW

1 What is the relationship between the motion of water particles within a wave and the external form of a wave?

2 What is the relationship between wavelength and the water depth at which waves begin to slow down and form breakers?

3 What is the cause of wave refraction? What is the role of wave refraction in causing a general straightening of a formerly rugged and irregular coastline?

4 How might a coastline that has experienced recent uplift differ in appearance from one that has experienced recent subsidence?

5 Sketch an imaginary length of shoreline that exhibits beaches, spits, bay barriers, barrier islands, and tombolos. Label each feature.

6 Explain, using a diagram, the reason for neap and spring tides.

7 The velocity of waves in the open ocean can be

determined from the formula $V = \frac{L}{T}$ (where V is velocity, L is wavelength, and T is wave period). What is the velocity of a wave having a wavelength of 10 meters and a period of 10 seconds? What is the period of a wave having a wavelength of 12 meters and a velocity of 2 meters per second?

8 Some stretches of shoreline do not have beaches. What might be a reason for a shoreline that lacks beaches?

9 What factors determine the distance between the breaker (surf) zone and the beach?

10 What is the most valuable resource currently being extracted from sedimentary rocks beneath the ocean floor?

SUPPLEMENTAL READINGS AND REFERENCES

Bascom, W., 1980. *Waves and Beaches*. 2d ed. Garden City, N.Y., Doubleday & Co.

Clancy, E.P., 1968. *The Tides, Pulses of the Earth*. Garden City, N.Y., Doubleday & Co.

Davis, R.A., Jr., 1972. *Principles of Oceanography*. Reading, Mass., Addison-Wesley Co.

Emery, K.O., 1969. The continental shelves. *Sci. Amer.* 221:3, 106–122.

King, C.A.M., 1972. *Beaches and Coasts*. 2d ed. New York, St. Martin's Press.

Newell, N.D., 1972. The evolution of reefs. *Sci. Amer.* 226:6, 54–64.

Pirie, R.G. (ed.), 1973. *Oceanography: Contemporary Readings in Ocean Sciences*. New York, Oxford Press.

Shepard, F.P., and Wanless, H.R., 1971. *Our Changing Coastlines*. New York, McGraw-Hill Book Co.

Smith, F.G.W., 1973. *The Seas in Motion*. New York, Thomas Crowell.

13

the nature of the atmosphere

Men judge by the complexion of the sky
The state and inclination of the day.
SHAKESPEARE
Richard II

Prelude Rather like creatures on the floor of the sea, we humans look up from below at a virtual ocean of gases that we call the atmosphere. We are so accustomed to this great ocean of air that we often forget how essential it is to us. Besides being necessary for life itself, the atmosphere dominates our daily activities. It influences the way we dress, the foods we produce, our moods, and our physical well-being. The atmosphere serves as a blanket to protect us from lethal radiation, from excessively high temperatures during the day, and from too great a loss of heat at night. Without this beneficial canopy of gases, there would be no blue sky or crimson sunsets, no wind to drive the ocean waves, no clouds, and no streams to sculpture our landscapes. Clearly, the atmosphere merits our study.

Cumulus congestus. These clouds are distinctive because of their massive, dense appearance and sharp outlines. They develop vertically and often form magnificant domes and towers. (Photograph courtesy of Stephen D. Levin.)

THE CONTENT AND STRUCTURE OF THE ATMOSPHERE

The composition of air

The earth's atmosphere is composed of a mixture of gases in motion that we call air. Normally, the mixture is colorless and odorless, but these pleasing qualities may be altered locally by smoke, gases from decaying organic material, or any one of a number of emissions from factories and automobiles. In its relatively pure dry state, air is composed primarily of four gases—nitrogen, oxygen, argon, and carbon dioxide (Table 13-1). Of these, nitrogen (N_2) accounts for about 78 percent of the total volume, and oxygen about 21 percent. The greater part of the remaining 1 percent consists of argon (0.93%), and neon, helium, krypton, hydrogen, xenon, ozone, radon, and minuscule amounts of other gases.

Nitrogen, a relatively inert component of the atmosphere, is nevertheless important because it dilutes the oxygen we breathe and thereby beneficially protects us from too rapid a rate of oxidation. In addition, nitrogen from the atmosphere is converted by soil bacteria into the nitrates essential for protein synthesis in plants.

In contrast to nitrogen, *oxygen* is a very active element and combines readily with a variety of other elements at the earth's surface. It is, of course, essential for respiration and the conversion of food into energy. Most of the oxygen in the atmosphere is produced by plants. For example, carbon dioxide exhaled by animals is taken up by plants during photosynthesis, and its oxygen constituent is later released. The oxygen used by animal life is returned to the atmosphere by plant life in approximately equal amounts, and thus the percentage of the vital gas in the atmosphere remains relatively uniform.

TABLE 13-1 **PRESENT COMPOSITION OF THE LOWER ATMOSPHERE**

Gas	Symbol or formula	Percent by volume
Nitrogen	N_2	78.08
Oxygen	O_2	20.94
Argon	Ar	0.934
Carbon dioxide	CO_2	0.033
Neon	Ne	0.00182
Helium	He	0.00052
Methane	CH_4	0.00015
Krypton	Kr	0.00011
Hydrogen	H_2	0.00005
Nitrous oxide	N_2O	0.00005
Xenon	Xe	0.000009

The small proportion of *argon* in the atmosphere is generated by decay of radioactive potassium in minerals of the earth's crust. As a noble gas, argon is not involved in the chemical reactions of weathering, nor in the chemical reactions critical to living organisms.

Ozone (O_3) is a form of oxygen in which three oxygen atoms are bound together into a single molecule (in contrast to the two-atom linkage of the more common O_2 molecule). Although the proportion of this gas in dry air is only 0.000007 percent by volume, it is of vital importance because of its ability to absorb lethal ultraviolet radiation from the sun before these rays can reach the earth's surface. Scientists are deeply concerned that exhaust gases from jet aircraft, freon propellants used in aerosol cans, and refrigerants carelessly released from air-conditioners and refrigerators may destroy part of the protective ozone layer in the atmosphere. The layer having the greatest concentration of ozone lies between 20 and 40 km above the earth's surface. It is currently being carefully studied to determine whether this vital shield is being destroyed faster than it is replenished by natural processes.

Water vapor and *carbon dioxide* are two vital compounds whose proportions in the atmosphere are more variable than the proportions of nitrogen and oxygen. Both water vapor and carbon dioxide are capable of absorbing part of the infrared (heat) radiation given off from the earth's surface. Water vapor is more abundant in the lower levels of the atmosphere, where it may be as much as 2 to 7 percent by volume (average air contains about 1.4 percent water vapor). As will be explained later, it has an important effect on air temperature, density, and humidity. Water vapor is added to the atmosphere by evaporation from various bodies of water (or from damp ground), steam from industrial complexes, and transpiration from plants. On the other hand, it is extracted from the air by condensation to form clouds and dew. Thus it is being actively added to the atmosphere at one location and extracted at another. As a result, its proportion in air is variable.

Measurements of the carbon dioxide content of the atmosphere reveal that it varies not only spatially (its concentration is higher near cities and industrial complexes) but also in time. For example, its average percentage by volume in the atmosphere increased from about 0.029 percent in the early 1900s to 0.033 at the beginning of the 1980s. This increase is of great concern to scientists because of the effect it may have in increasing the earth's average temperature. Such an increase might cause melting of polar ice caps, which in turn would raise global sea level and inundate many densely inhabited coastal areas.

Although carbon dioxide constitutes only a small proportion of air, it is a vital gas. Without it, most plants could not survive, and plants are ultimately essential to animal life. Combustion, animal respiration, and volcanic eruptions are among the natural sources of atmospheric carbon dioxide. Its proportion in the atmosphere is influenced not only by rates of natural and artificial emission but also by photosynthetic activity and the ability of ocean water to absorb and release the gas. For example, when carbon dioxide in the air increases, it is absorbed by ocean water. Conversely, when the amount of atmospheric carbon dioxide diminishes, some of the gas dissolved in seawater is released.

There are still other components of the atmosphere that vary in concentration both spatially and from day to day. These include a host of tiny particles, such as dust, bacteria, pollen grains, spores, and microscopic pieces of salt from ocean spray. Many of these particles serve as tiny "nuclei" around which water may condense to form droplets of rain and fog. Variable components of air also include such pollutants as hydrocarbons, oxides of nitrogen and sulfur, carbon monoxide, and a variety of particles resulting from industrial processes.

The physical properties of air

Although air is a mixture of gases, it acts in most ways like a single gas and exhibits similar relationships among its volume (the space it occupies), temperature (a measure of the kinetic energy of its molecules), and pressure (the impact force of its molecules on a given surface area). Simply stated, these relationships are as follows:

1 If the temperature of air is kept constant, the volume of a given quantity of air will vary inversely as the pressure, a larger pressure resulting in a smaller volume (Fig. 13-1).

2 At constant volume, the pressure of a gas increases for each Celsius degree rise in temperature by about $\frac{1}{273}$ of its pressure at 0°C.

3 If pressure is held constant, the volume of a gas increases for every Celsius degree rise in temperature by about $\frac{1}{273}$ of its volume at 0°C. This relationship and the one preceding constitute **Charles' and Gay-Lussac's Law.**

4 If allowed to expand, air will become cooled, and conversely, if compressed, it will be heated.

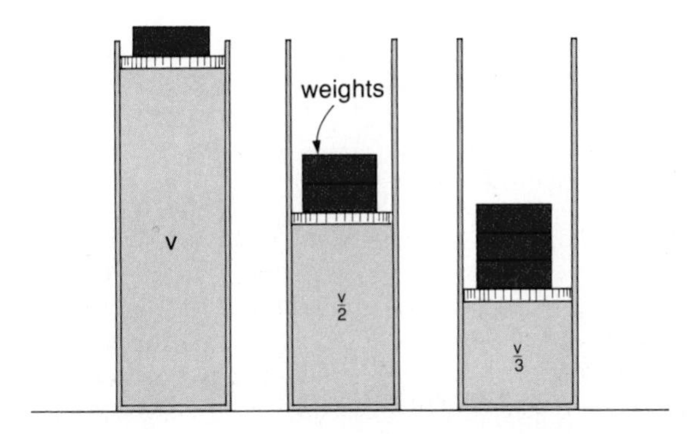

Figure 13-1 *The volume of a gas varies inversely as the pressure.*

Any sample of a single gas exerts a certain pressure that is the result of rapid bombardment of a unit area by billions of actively moving molecules. In the mixture of gases we call air, each component gas exerts the same pressure that it would exert if it alone were present and occupied the same volume as the whole mixture. That part of the total pressure that each gas exerts is called the **partial pressure.** When water vapor escapes into the atmosphere and mixes with the other gases of air, it similarly exerts its partial pressure in all directions. This partial pressure is more commonly known as the *vapor pressure of the air*. The sum of all the partial pressures in air constitutes the air's **total pressure.** Water vapor may produce small but significant differences in total pressure.

Vertical changes in atmospheric properties

The expression "light as air" might imply that the atmosphere has no weight. This, of course, is not true. The atmosphere does have weight, and because the weight of the upper layers of air bears down upon the lower layers, air closest to the earth's surface is under the greatest pressure. The force per unit area of the atmosphere on the surface is called **atmospheric pressure.** Atmospheric pressure can be measured with an instrument called a **barometer.** Some instruments link the barometer to a recording device called a **barograph** (Fig. 13-2) and thereby attain a continu-

Figure 13–2 A barograph. This device provides a continuous record of changes in air pressure. (Courtesy of National Oceanic and Atmospheric Administration.)

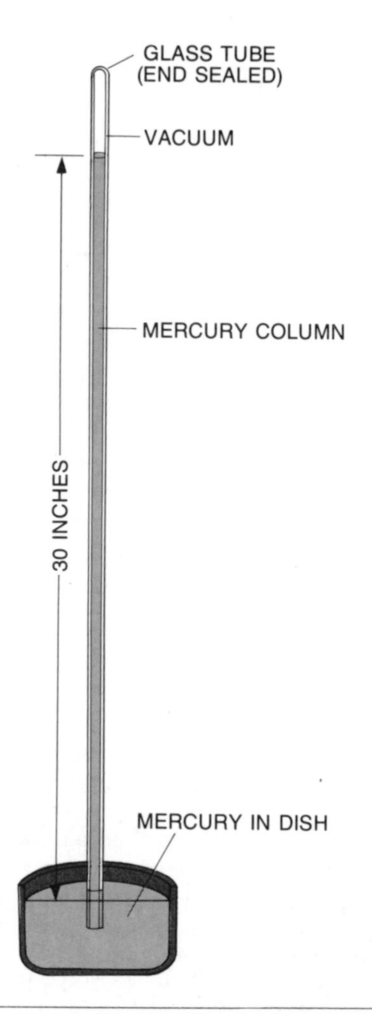

Figure 13–3 The Torricelli barometer is simple in design, but it was used effectively to establish some very important scientific concepts. (From Navarra, J.G., Atmosphere, Weather, and Climate. Philadelphia, Saunders College Publishing, 1979.)

ous record of changes in atmospheric pressure. In the mercury barometer, the weight of a column of mercury is balanced against the weight of the column of air (Fig. 13–3). Pressure is indicated by the height to which the mercury rises in an evacuated tube in order to achieve balance. At sea level, for example, atmospheric pressure would support 760 mm, or 29.92 inches, of mercury, and atmospheric pressure would be similarly expressed as inches or millimeters of mercury. It seems rather unsuitable, however, to express pressure by means of units of length, and therefore the preferred modern unit is a true measure of pressure called the *millibar*. A millibar is equivalent to about 0.03 inches of mercury. As indicated on Table 13–2, atmospheric pressure of 29.92 inches of mercury is equivalent to 1013.25 millibars.

If you can imagine a vertical container having a base one inch square and standing vertically from sea level to the top of the atmosphere, the weight of the column of air on the base would be about 14.7 pounds (Fig. 13–4). Thus, the average atmospheric pressure at sea level is about 14.7 pounds per square inch (1033 g/cm^2). If we were then to examine successive levels in our long column of air, we would observe further that there was a systematic decrease in pressure with each increment of ascent above the earth's surface. Indeed, at an altitude of 30 km, atmospheric pressure would be only 1/1000 of the sea level value (Fig. 13–5).

The *density* of the atmosphere (its mass per unit volume) also decreases with increasing height above the earth's surface. The density of a gas is often expressed in kilograms per cubic meter (kg/m^3). At sea level, the density of the atmosphere is about 1.2 kg/m^3. At an altitude of only 6 km, the density of air has already fallen to only 0.6 kg/m^3, or half the density reading obtained at sea level. The decrease continues with each increase in altitude until at 30 km the density of air is a small fraction of the sea level value (Table 13–3). Clearly, the bulk of gas molecules in the atmosphere are found close to the earth's surface.

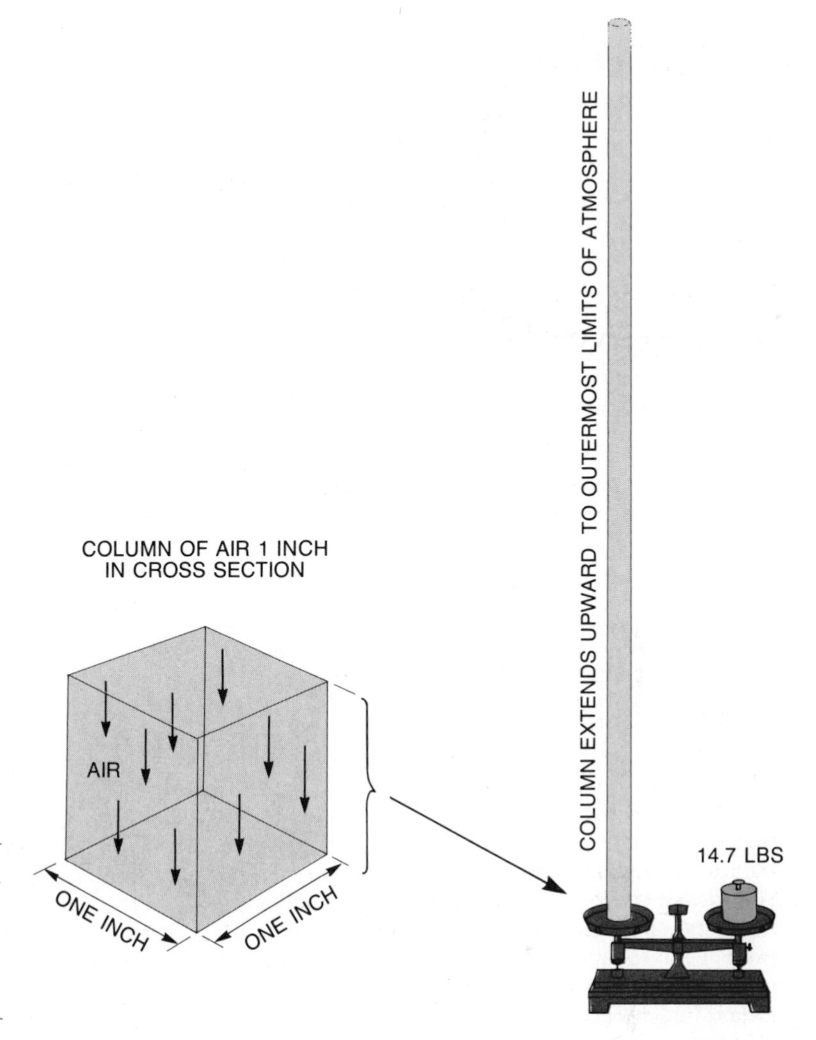

Figure 13–4 *The weight of a vertical column of air with a 1-inch square base that extends to the outermost limits of the atmosphere is 14.7 pounds. (From Navarra, J.G.,* Atmosphere, Weather, and Climate. *Philadelphia, Saunders College Publishing, 1979.)*

If one were to plot the change in temperature of the air with increasing altitude, one would not obtain a smooth curve such as that shown for atmospheric pressure. With ascent upward from the earth's surface, one would observe a general cooling trend followed by gradual warming. At still higher altitudes, cooling would resume and then revert again to pronounced warming in the highest levels of the atmosphere.

The beginning of the upward trend of cooling temperatures occurs within a layer of air closest to the earth's surface. This layer, from ground level to about 12 km, is called the **troposphere** (Fig. 13–6). More than 80 percent of the total mass of the atmosphere is concentrated in the troposphere. It is characterized by strong vertical overturning, mixing, and weather phenomena. Temperatures decline with increasing altitude in the troposphere because this blanket of air is warmed *from below* by conduction and heat radiated from the rocks, soil, and water bodies of the earth's surface. The rate at which temperature declines with increasing height is called the **lapse rate,** and the lapse rate for the troposphere is about 6.5°C per kilometer.

The land and water bodies that warm the troposphere derive their heat from solar radiation. That radiation includes visible light, ultraviolet, and infrared components, but because most of the ultraviolet and infrared rays are filtered out before reaching the

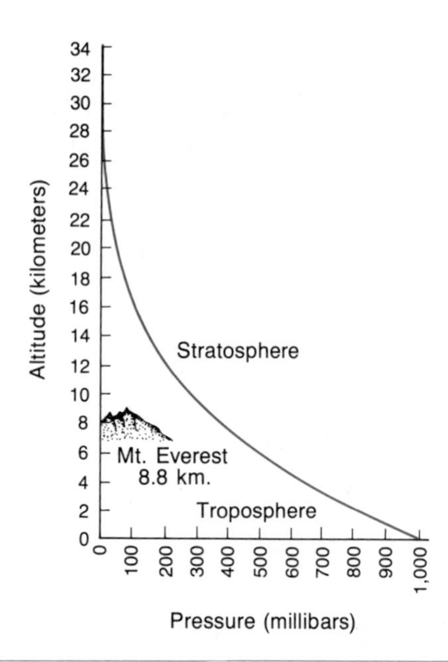

Figure 13–5 Decrease of atmospheric pressure with altitude. (From Turk, J., and Turk, A., Physical Science. *Philadelphia, Saunders College Publishing, 1977.)*

TABLE 13–2 **ATMOSPHERIC PRESSURE EQUIVALENTS**

Inches of mercury	*Millimeters of mercury*	*Millibars*
27.00	658.8	914.3
28.00	711.2	948.2
28.50	723.9	965.1
29.00	736.6	982.1
29.50	749.3	999.0
29.75	755.6	1007.5
29.92	760.0	1013.2
30.00	762.0	1015.9
30.25	768.3	1024.4

TABLE 13–3 **DENSITY CHANGES WITH ALTITUDE**

Altitude in km	*Density in kg*
0	1.23
2	1.01
4	0.82
6	0.66
8	0.53
10	0.41
15	0.19
20	0.09
30	0.018
40	0.004
50	0.001
60	0.0003
70	0.00009
80	0.00002
90	0.000003
100	0.0000005

earth's surface, the radiation that actually reaches the surface consists mostly of energy in the visible range of the electromagnetic spectrum (see Fig. 18–4). The ground is a very good absorber of this visible radiation, and almost all of it is absorbed. As the ground heats up, it also begins to radiate. The ground, however, radiates in the infrared rather than the visible range, and this infrared radiation is readily absorbed by water vapor, carbon dioxide, and ozone in the troposphere. Were it not for this heat-absorbing capacity of the troposphere, the loss of heat from the ground would be exceptionally rapid, especially at night, when heat loss would not be compensated by heat gain from the sun.

The upper limit of the troposphere (where the decline in temperature ceases) is called the **tropopause.** As indicated in Figure 13–6, the tropopause lies at an average altitude of about 11.5 km. Its altitude, however, varies with the seasons and with location. For example, the tropopause is higher at the equator than at the poles. There is also variation in tropopause air temperature. Over the equator, for example, the average temperature along the tropopause is about −80°C, whereas it is a somewhat less frigid −55°C over polar regions.

At an altitude of about 12.5 km, the steady decline in temperature with increasing height ceases rather abruptly. This change marks the passage into the next atmospheric layer, called the **stratosphere.** The stratosphere extends from the tropopause to an altitude of about 50 km. Its uppermost boundary is termed the **stratopause** (see Figure 13–6).

Unlike conditions in the underlying troposphere, there is an absence of vigorous overturning and verti-

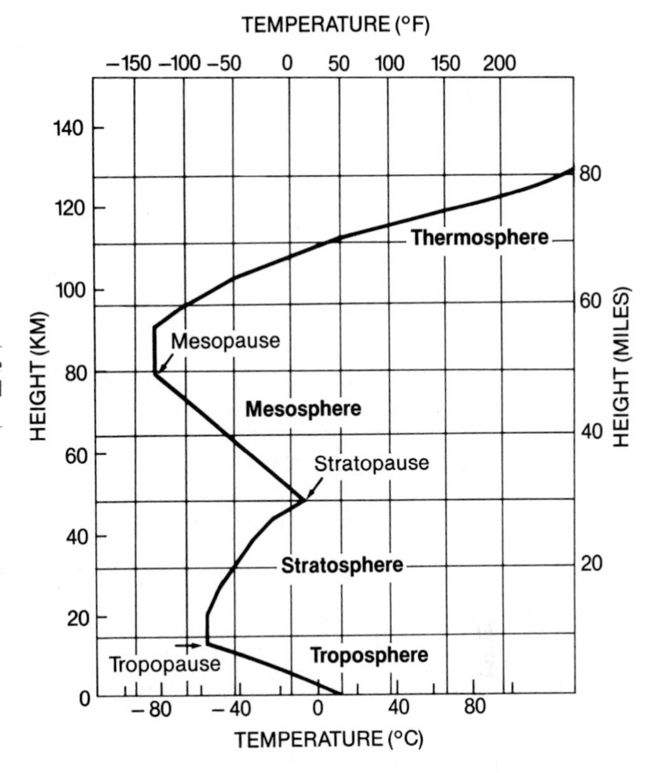

Figure 13–6 *The four major temperature regions of the atmosphere. (From Navarra, J.G.,* Atmosphere, Weather, and Climate. *Philadelphia, Saunders College Publishing, 1979.)*

cal mixing in the stratosphere. Within this second atmospheric zone, temperatures increase with increasing altitude, eventually reaching a maximum at the stratopause. The stratopause coincides with the top of the ozone layer, and its high temperature is the result of absorption of ultraviolet solar radiation. Unlike the troposphere, which is heated from below, the stratosphere is heated from above. As a result, temperatures decrease downward until encountering the reverse trend of the underlying troposphere.

The region directly above the stratosphere is dubbed the **mesosphere,** which means the middle sphere. The mesosphere extends from about 50 to 85 km. It is a zone in which temperature diminishes with height, finally dropping to −140°C at its upper boundary or mesopause. As in the troposphere, there is considerable overturning and turbulence in the mesosphere.

Above the mesosphere, the temperature increases with height once again. This hot layer is called the **thermosphere.** The thermosphere extends upward from an altitude of 85 km and attains temperatures as high as 700°C at an altitude of 200 km. The high temperatures in the thermosphere result from absorption of very short wavelength ultraviolet radiation by oxygen (O_2) and nitrogen (N_2) molecules. In absorption, air molecules take up some of the radiant energy coming from the sun and convert it into internal energy, which is expressed as an increase in temperature.

The thermosphere includes most of a region of the atmosphere known as the **ionosphere.** In this region, very short wavelength ultraviolet radiation, x-rays, and gamma rays cause atoms of nitrogen molecules to dissociate and expel electrons. Thus they are transformed into the ions for which the "ionosphere" is named. The ejected electrons are set free to form conducting layers, which reflect radio waves back to earth.

Above an altitude of about 600 km, the abundance of atoms and molecules has considerably diminished. Indeed, there are so few atoms in this highest level of the atmosphere that any given atom of oxygen is not likely to encounter another one for some 160 km. This zone, called the **exosphere,** represents the transition from the earth's atmosphere to the thin interplanetary gases of outer space.

THE DISTRIBUTION AND CONTROLS OF GLOBAL TEMPERATURES

The earth's heat budget

Although there is evidence that the earth has experienced significant episodes of cooler or warmer conditions in its long geologic history, for the last few centuries our planet's average temperature has remained essentially constant. This suggests a balance between energy received as radiation from the sun and the amount of that energy the earth returns back to space. The balance of energy is sometimes called the earth's heat budget (Figure 13-7).

Only a small percentage of the total energy output of the sun reaches the outer limits of our atmosphere. A proportion of that is reflected back into space and is thus not involved in warming the earth. Of the portion that does enter the atmosphere, a small amount is absorbed by ozone, carbon dioxide, and water vapor. The still unabsorbed radiation continues its descent, and some of it is reflected from clouds and the earth's surface or reflected from dust particles and gas molecules. The remaining amount of solar radiation, constituting as little as half that at the top of the atmosphere, is absorbed by land and water masses and constitutes the "credit side" of the earth's heat budget.

At the surface of the earth, energy is radiated back into the troposphere. As noted earlier, the warm ground emits energy in the form of infrared radiation. Infrared is efficiently absorbed by water vapor and CO_2 in the troposphere and by ozone in the stratopause. That same water vapor and CO_2 not only absorb energy well but also are capable of releasing it

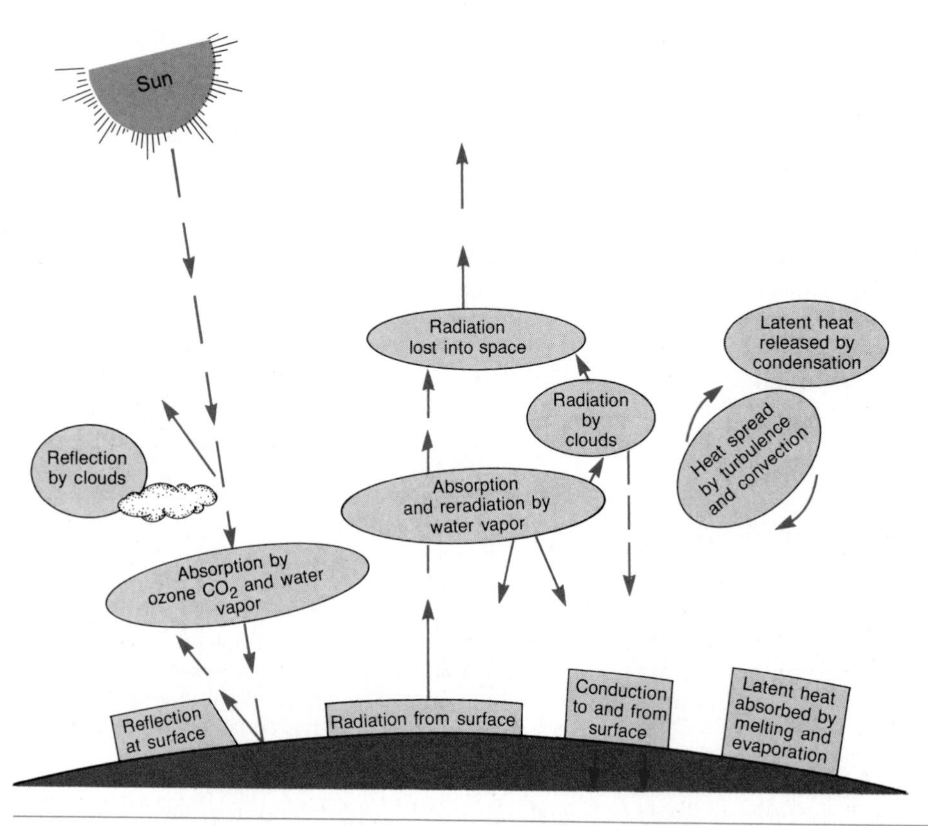

Figure 13-7 The earth's energy budget.

back into the atmosphere. Thus, heat is in part returned to space and in part directed back toward the earth. The recycling serves to keep the air close to the earth's surface at a worldwide average temperature of about 15°C. As will be described shortly, wind, conduction, convection, evaporation, and condensation aid in the recycling of energy.

In considering the earth's balance of energy, it is necessary to remember that the word "balance" refers to the earth as a whole. Specific geographic regions are characteristically in a state of imbalance between radiative energy received and lost. One result of the imbalance is wind, which helps to distribute surpluses of heat and cold around the planet.

Heat and temperature

To better understand the way energy is transferred in the atmosphere, it is useful to distinguish between the terms heat and temperature. The gas molecules of which the atmosphere is composed are in constant motion. They ceaselessly collide with neighboring molecules, rebound, and vibrate. Thus, they possess **kinetic energy,** a term that refers to the energy a body has because of the motion of its molecules. **Heat** can be described as the total kinetic energy of the random motions of a quantity of molecules in a body or a given volume of air. Adding heat to a given sample of air really means causing its molecules to move about faster. Conversely, cooling means making the molecules slow down.

Unlike heat, temperature does not depend upon the quantity of molecules in a given body but is simply a measurement of the degree of hotness or coldness. A large mass of air may have a low temperature but still contain a large amount of heat because of the enormous quantity of molecules contained within the air mass.

The measurement of temperature

Temperature is measured by use of that familiar device, the thermometer. One kind of thermometer consists of a gas-free capillary tube or stem of uniform diameter that terminates in a bulb containing a fluid such as mercury or alcohol. When the temperature rises, the fluid in the bulb expands and rises into the stem, which has been calibrated so that the length of the column of fluid will correspond to increasing increments of temperature.

The temperature scales in most common use are the **Celsius** and **Fahrenheit.** On the Celsius scale, the point at which water freezes at sea level is arbitrarily placed at zero degrees (written 0°C), and the point at which it boils is placed at 100°C. The freezing temperature on the Fahrenheit scale is 32°F and the boiling point 212°F (Fig. 13–8).

Frequently, we need to convert a Celsius reading to the corresponding Fahrenheit reading or vice-versa. As indicated in Figure 13–8, the 180 divisions on the Fahrenheit scale are equal to 100 divisions on the Celsius. Thus, each Fahrenheit degree is $^{100}/_{180}$ or $^{5}/_{9}$ as large as a Celsius degree. We also know that 0°C is equal to 32°F. These relations are the basis for the following equations for converting between the two scales:

$$F = \tfrac{9}{5}\,C + 32$$

$$C = \tfrac{5}{9}\,(F - 32)$$

Although the Celsius scale is widely used for scientific purposes, there is another scale that scientists

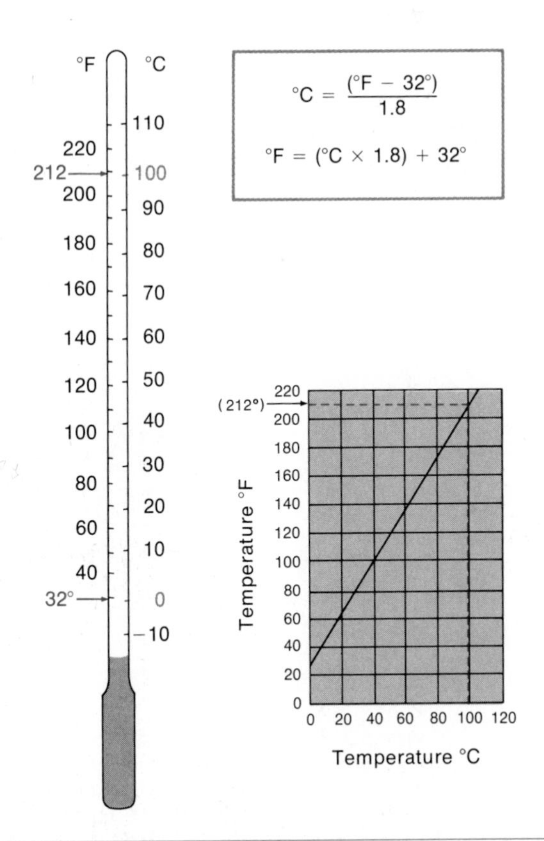

Figure 13–8 Relation between the Fahrenheit and Centigrade temperature scales.

employ. It is called the **Kelvin,** or **absolute, scale,** named after the eminent nineteenth-century physicist Lord Kelvin. The lowest point on the Kelvin scale starts with the coldest temperature that can possibly exist. It is the temperature at which the molecules cease their motion so that the body containing those molecules has no heat. This lowest possible temperature corresponds to −273°C and is called *absolute zero* or *zero degrees Kelvin* (0°K). The Kelvin scale uses degrees of the same size as in the Celsius scale. Therefore, by adding 273 to any Celsius temperature we obtain the corresponding reading on the Kelvin scale (Kelvin temperature = Celsius temperature + 273).

Temperature readings, usually in the Celsius scale, are taken at regular intervals at weather stations around the world. In North America, thermometers and temperature-recording devices are mounted in shelters of standard construction consisting of a white box about half a meter on each side with a sloping double roof and latticed sides. The shelter shields the thermometers from direct sunlight and at the same time allows for the free circulation of air. One of the instruments within the shelter is a **thermograph** (Fig. 13-9), which provides a continuous record of temperature changes. Other important devices are the maximum and minimum thermometers, which record the highest and lowest temperatures for a given period of time and thereby provide the temperature range.

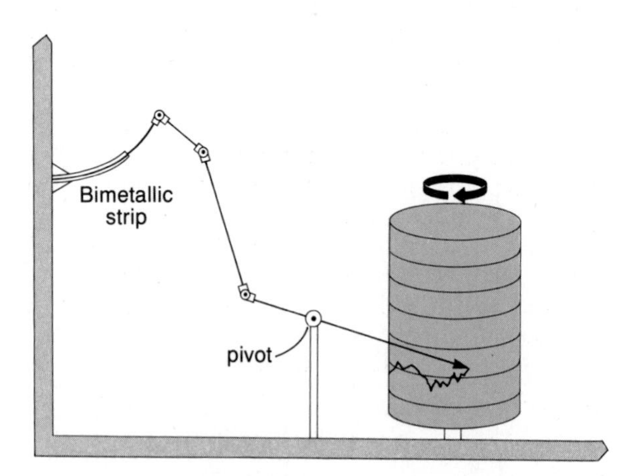

Figure 13-9 Principal of the thermograph. A strip consisting of two different kinds of metals bonded together serves as a thermometer. One of the metals experiences greater expansion with heat than the other, causing the strip to warp. The motion caused by temperature change is transmitted to a pen, which records temperatures on a graph wrapped on a clock-driven drum.

From the records of instruments in the thermometer shelters, meteorologists can calculate the daily mean temperature, the annual mean, and the annual temperature range. For example, the *daily mean* is obtained by adding the maximum and minimum temperature and dividing by 2. By adding the daily means for a month and dividing by the number of days in the month, the *monthly mean* is obtained. The *annual mean* is the average of 12 monthly means. Finally, the *annual temperature range* is the difference between the warmest and coldest monthly means.

The seasons

The solar radiation received at the earth's surface is termed **insolation.** Variations in the amount of insolation arriving at any latitude on the earth's surface are caused by seasonal changes in the earth's position relative to the sun and result in differences in temperature conditions between equatorial and polar regions. The two most important factors influencing insolation at any given location are the angle of the sun's rays at that latitude and the length of the period of daylight.

To understand the importance of the angle of the sun's rays incident on the earth's surface, consider what would happen if you were to shine a beam of light directly at the equator of a globe and then slowly shift the beam of light northward toward the North Pole as shown in Figure 13-10. At the equator, the beam illuminates a relatively small area. The light energy is concentrated there. As the light is moved northward, it arrives at a slant and illuminates a larger area. As a result, the light intensity per unit area is diminished. Similarly, on the real earth, the sun's radiation is spread over a wider area at high latitudes. This accounts in part for colder polar conditions. In addition, when the sun is low on the horizon, its rays pass through a greater distance of atmosphere, and insolation is diminished as a result of the greater scattering and absorption of incoming radiation.

Length of day, the second factor influencing insolation, is governed by the seasons. The most important reason for the seasons is the 23.5° tilt of the earth's equatorial plane relative to the plane of the earth's orbit (the *plane of the ecliptic*). Throughout each of its revolutions around the sun, the earth maintains its 23.5° angle, and the orientation of the axis of rotation does not change (except over thousands of years by precession, as will be discussed in Chapter 17). This places the North Pole within the circle of the sun's illumination during half of the earth's revolution and in darkness during the other half (see Fig. 17-13).

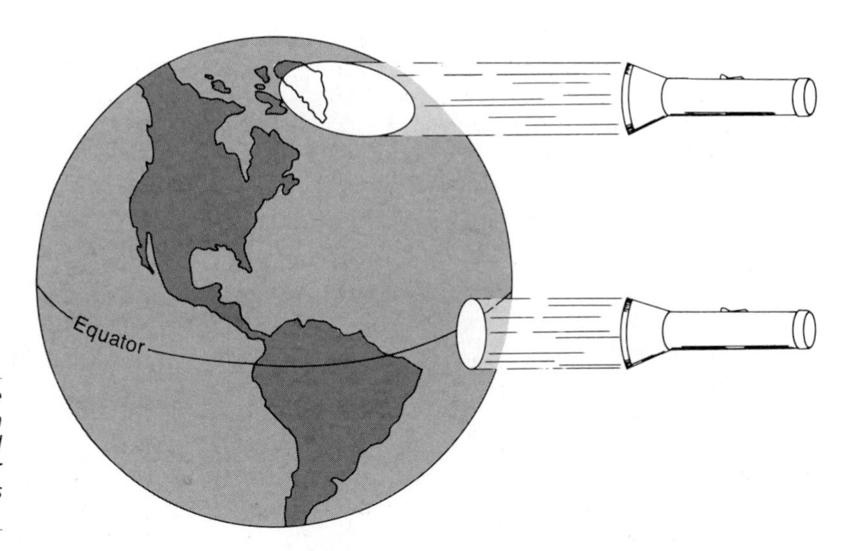

Figure 13–10 *Experiment to illustrate how the sun's rays are spread across a wider area at high latitudes as compared to the equator. This contributes to lower temperatures for points near the poles than for points at or near the equator.*

In December the earth has arrived at a position in its orbit such that the North Pole is directed away from the sun and is in darkness (Fig. 13–11). At this time, other parts of the Northern Hemisphere also receive less than the annual mean amount of insolation and therefore experience cooler temperatures. People living in Europe, Canada, and the United States notice that the sun appears low in the sky even at noon and that the days are shorter than those in summer. It is winter in the Northern Hemisphere. December 22d, the **winter solstice** and shortest day of the year, marks the beginning of the winter season. On this date, one would have to travel to a location on the Tropic of Capricorn, 23.5°S latitude, to see the sun directly overhead at noon.

In June, the earth has moved to a position where its North Pole is tilted toward the sun and the Northern Hemisphere experiences summer. June 22d is the **summer solstice,** the longest day of the year. On this date the sun appears directly overhead at noon along the Tropic of Cancer (23.5°N latitude), and north of that latitude it appears at its highest angle above the horizon (Fig. 13–12). It is warmer in the United States not only because a given quantity of sunlight is concentrated in a smaller area of the continental surface, but also because the sun is above the horizon longer than at any other time of the year. Indeed, at a latitude of 45°N there will be about 16 hours of sunlight. All of these conditions, of course, are reversed south of the equator, where the winter season has begun.

In its progression around the sun, the earth will reach two positions on opposite sides of the orbital path where the sun's rays fall vertically along the equator and every point on the earth's surface at any latitude north and south of the equator receives 12 hours of daylight. The first of these positions, reached on about March 21st, is called the vernal or **spring equinox.** The earth reaches the second position on about September 23rd. This date is the **autumnal equinox.** These are very appropriate terms, for equinox means equal night.

Because the earth revolves around the sun in an elliptical path, there are times when the planet is closer to or farther from the sun. This small variation in distance has some minor effect on weather and climate, but it is not sufficient to alter the fact that the seasons are primarily the result of the inclination of the earth's equatorial plane relative to the plane of its orbit. As evidence of this, it is interesting to note that when the earth is closest to the sun (currently January 4th), the Northern Hemisphere is experiencing winter.

Temperature variation

The amount of insolation received at the earth's surface is not totally governed by seasonal shifts in incoming rays from the sun. Also important is the depletion of incoming radiation by reflection, scattering, and

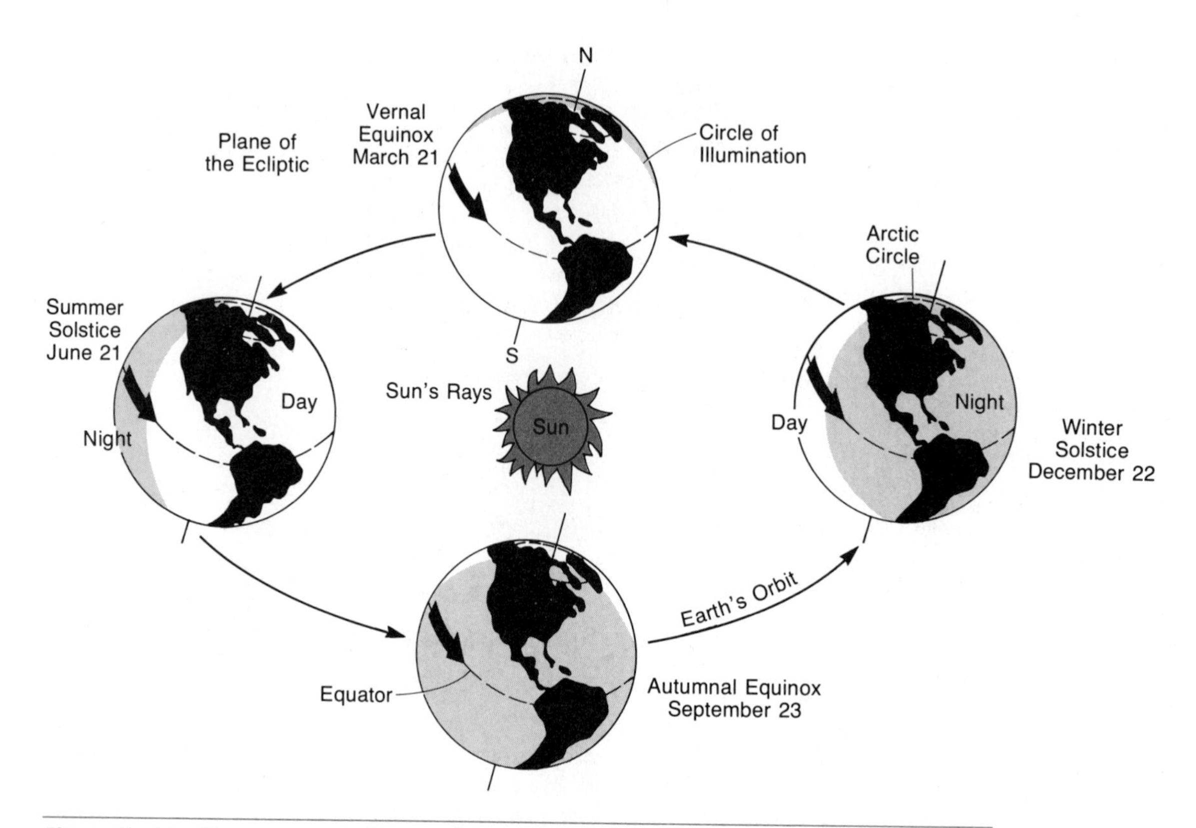

Figure 13–11 *The seasons occur because the earth's axis is tilted relative to the plane of its revolution around the sun. When the Northern Hemisphere is tilted toward the sun, it has summertime. Winter occurs at the same time in the Southern Hemisphere.*

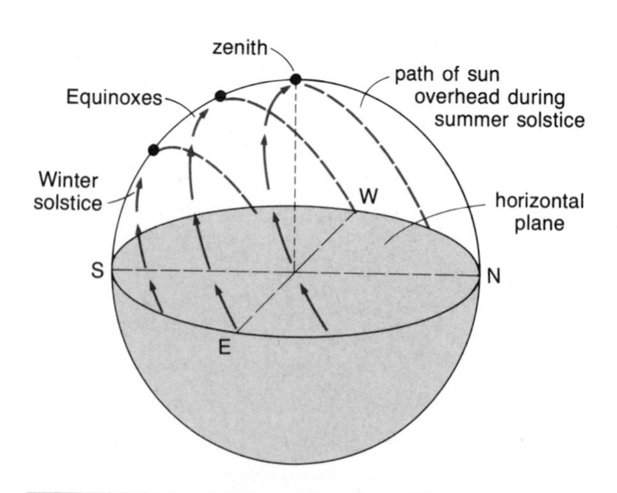

Figure 13–12 *The seasonal variation in the path of the sun across the sky at the Tropic of Cancer. Note that from a location along the Tropic of Cancer, the sun would be directly overhead at noon during the summer solstice.*

absorption. Earlier it was noted that only about half of the radiation received at the outer limits of the atmosphere actually strikes the earth's surface. Even that proportion will vary according to the concentration of water vapor in the air, as well as the thickness of atmosphere through which the rays must travel. The greater that distance, the greater will be the amount of reflection, scattering, and absorption.

Another important factor that results in temperature differences at different locations is the size, shape, and location of larger bodies of land and water. We have noted how land and water bodies absorb solar radiation and then become radiating bodies themselves. In the process, they heat the overlying air. The rate at which the absorption of solar energy occurs, however, is different for land than for water. Land areas heat more rapidly and to a higher temperature than water, and also cool more rapidly and to a lower temperature than water. As a result, variations in temperature are more extreme over continents than over oceans. The moderating influence of the ocean is evident in many coastal cities, which have milder climates than inland cities. The moderating effect of the oceans is quite apparent in San Francisco, where the annual temperature range is only 6°C. Towns at the same latitude in Missouri have a temperature range about four times as great. New York City does not experience pronounced oceanic moderation because it is affected by winds that blow predominantly from the interior of the continent rather than from the ocean.

There are three basic reasons for the differences between land and ocean temperature conditions. The first results from circulation. In the ocean, surface water warmed by the sun is continuously mixed and replaced by cooler water from below. This circulation pattern slows down the surface warming process. Circulation also causes slowness in cooling, because water which has been chilled at the surface becomes denser, sinks, and is replaced by ascending warmer water. It is interesting to note that ocean water having a certain temperature may be carried by major oceanic surface currents into a region having a different temperature. Air blowing over such water tends to warm or cool adjacent land areas. For example, the climate of the British Isles, which lie above 50°N latitude, would be far more severe were it not for the moderating effect of the Gulf Stream, which brings relatively warm water (and air warmed by that warm water) from the southeast coast of the United States northeastward across the Atlantic.

The second reason for differences in temperature conditions between land and sea results from the different capacity with which ground, in contrast to water surfaces, absorbs and liberates heat. Water must absorb more heat than land to raise its temperature any given number of degrees. For this reason, water warms up more slowly than land. Also, because it has absorbed more heat to attain a stated temperature, it has more to give up and will cool more slowly than land.

The third factor that accounts for the unequal heating and cooling conditions of land and ocean is related to evaporation. A considerable part of the sun's radiation energy striking the ocean surface is used in evaporation, which is a heat-absorbing process. Evaporation also occurs on land, but to a far lesser degree. Thus, evaporation does not cool the land on a hot day as effectively as it does the ocean.

Another kind of temperature variation is related to altitude. Earlier, we noted how temperatures generally decrease with increasing altitude in the troposphere. As a result, high mountain summits, even in the otherwise warm equatorial belt, may be frigid and snow covered.

Mapping global temperatures

The distribution of temperatures over the earth is depicted on maps by means of isotherms. **Isotherms** are lines connecting points of equal temperature. As shown in Figure 13-13, isotherms do not correspond closely to parallels of latitude as they might if the earth were a completely uniform body. This observation provides us with further evidence that the angle of the sun's rays above any given latitude is not the only factor that governs temperature conditions around

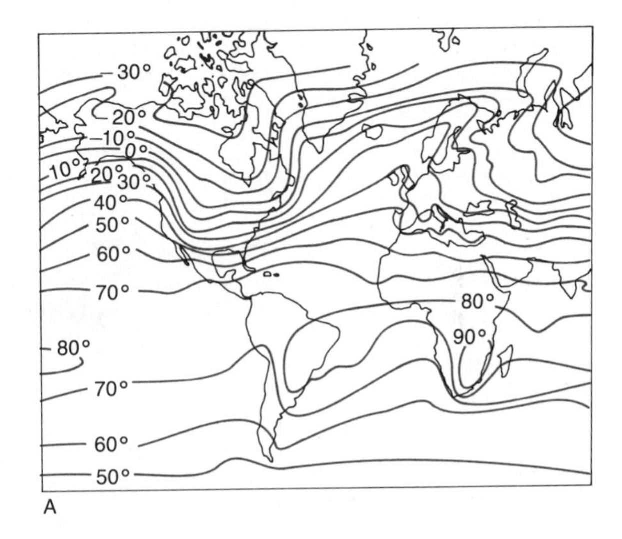

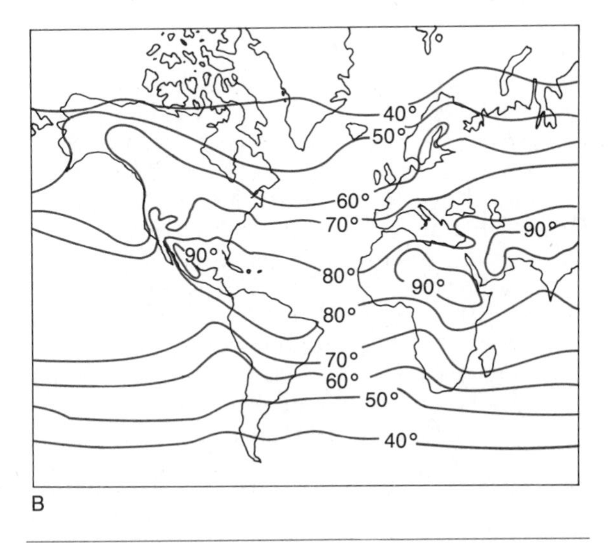

Figure 13-13 Isotherms of average temperature for January (A) and July (B). The differences are less pronounced in the Southern Hemisphere because of the moderating effect of the more extensive ocean.

the world. The spatial distribution of lands and seas is also important. For example, note how isotherms for the Northern Hemisphere shift northward over the continents in July (lands become relatively warmer) and southward in January (lands become relatively cooler). The maps also reveal the north and south shift of isotherms with the seasons. On comparing the summer and winter isotherm maps, it is apparent that temperature values in the Southern Hemisphere are more uniform than in the Northern Hemisphere. This is because more than 80 percent of the Southern Hemisphere is covered with water. The moderating marine influence is evident at the Cape of Good Hope on the southern tip of Africa, where the average January temperature is only 20°C (69°F) and the average July temperature is a balmy 17°C (53°F).

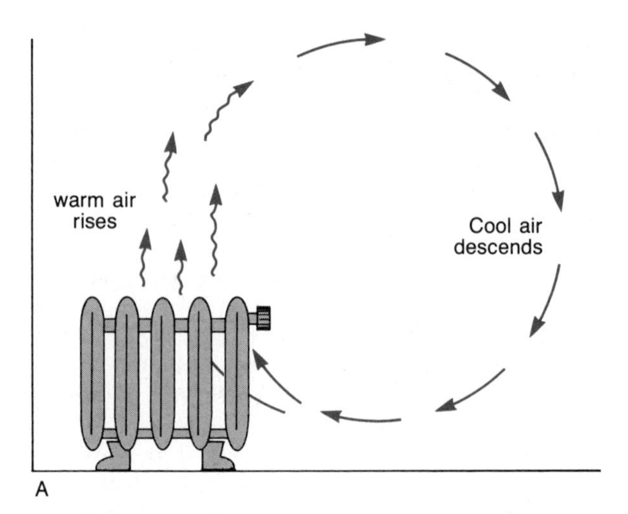

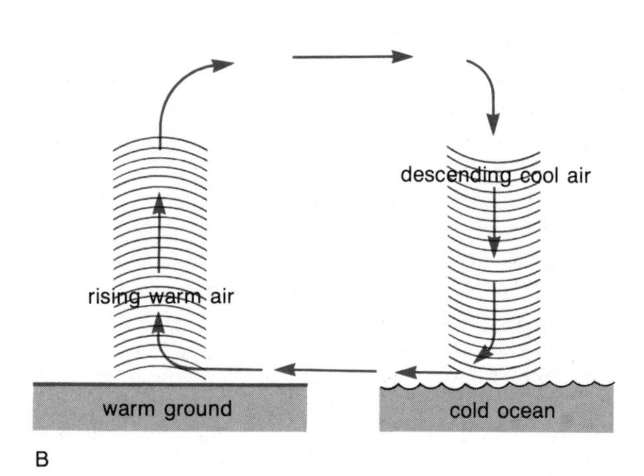

Figure 13–14 Convection currents in room heated by a radiator (A), and at earth's surface where warm ground heats the air above it (B). In either situation, the warm air expands, and since its density is lower than surrounding air, it rises and is replaced by an inflow of cooler air.

Transferring the sun's energy

The heat that exists in the atmosphere as a result of solar radiation is transferred from place to place largely by conduction, convection, and radiation. **Conduction** is accomplished by the transfer of heat from molecule to molecule (without the molecules themselves moving from one place to another). During conduction, a molecule that has absorbed some energy vibrates rapidly and collides with neighboring molecules, causing them to move faster. The process continues until the increased motion has been transmitted to all the molecules in a given body, thus making it warmer. Some solid substances, especially metals, are good conductors. If one end of a metal bar is held in a fire, the other end becomes hot. Molecules are relatively far apart in liquids and gases, which accounts for their poor conductivity. The atmosphere, being gaseous, is also a poor conductor.

Convection is far more effective than conduction in transmitting heat within the atmosphere. **Convection,** the transfer of heat by means of currents (which move groups of energized molecules from place to place), is characteristic of substances like air and water in which the molecules are free to move about. Motion of the air is easily begun, and therefore convection is by far the major way by which heat is transported in the atmosphere. The heating of a room by a radiator provides an example of convection. The air in contact with the radiator is warmed, expands, and rises, and a continuous convectional circulation is established (Fig. 13–14).

Whereas both conduction and convection involve the exchange of heat through some sort of material medium (a solid, liquid, or gas), **radiation** is the transfer of heat energy through space that can be empty of matter. The space between the earth and the sun is largely devoid of matter. It is a near vacuum. Thus, radiation is the major sun-to-earth heat transfer mechanism.

We can think of radiation as waves that resemble the ocean waves described in Chapter 12. Like water waves, radiation may differ in wavelength, which you will recall is the distance between the crests of two successive waves. Indeed, the differences in wavelength are the basis for recognizing the different kinds of radiation and organizing them into a system called the **electromagnetic spectrum** (see Fig. 18–4). In order of long to short, the electromagnetic spectrum includes the ultralong radio waves (ranging to tens of kilometers), infrared, visible light, ultraviolet, x-rays, and the minuscule gamma rays with wavelengths as short as a billionth of a millimeter. Of these various forms of radiation, we are able to directly see only visible light. The human eye is not sensitive to radiation of a wavelength much shorter than 0.000035 cm or longer than 0.00007 cm.

The invisible radiations, however, have fundamental similarities to visible light. All forms of radiation, whether ordinary light, radio waves, or ultraviolet, travel at a speed of 186,282 miles per second (300,000 km per second). Called the speed of light, this speed is considered to be the greatest possible in the universe. Also, as radiation travels from its source, its speed remains constant, but it diminishes in intensity by the square of the distance from its source. Thus, the measurement of a source of radiation at two kilometers will be one fourth as intense as at one kilometer.

Scientists recognize that both hot and cold objects emit radiation, but hot bodies emit more total energy for a given area than cold objects. Thus, radiation is being emitted today even from cold Antarctic ice sheets. Although both cold and warm bodies emit radiation, they do not emit radiant energy in the same wavelengths. Very hot bodies like the sun emit radiation in the shorter wavelengths (in the visible part of the electromagnetic spectrum), whereas cooler bodies like the earth emit longer-wavelength radiation.

It is important to note that the relatively small part of the electromagnetic spectrum that encompasses visible light includes all the colors of the rainbow. Color is really radiation in the visible light range that has a specific wavelength. Red light, for example, has a wavelength of about 0.000076 cm. This is 1½ times the wavelength of blue light. The wavelength of yellow lies between red and blue.

Most of the radiation received by the sun in the visible light range is transmitted through the atmosphere to the earth's surface. The earth's surface then returns this energy in the form of longer wavelength infrared radiation (heat). Whereas atmospheric gases such as water vapor, ozone, and carbon dioxide pose no obstacle to incoming short-wavelength radiation, they do block the infrared rays emanating from the earth's surface (Fig. 13–15). The resulting entrapment of heat has been called the greenhouse effect, although the term is not appropriate. At one time it was thought that most of the warmth in a greenhouse accumulated because the glass allowed solar radiation to enter but blocked the escape of the radiated longer wavelengths (infrared radiation). Thus, greenhouse glass was thought to serve the same function as water vapor, carbon dioxide, and ozone in the atmosphere. The more important factor in maintaining warmth within the greenhouse, however, is its enclosed condition, which prevents the mixing of inside air with outside air.

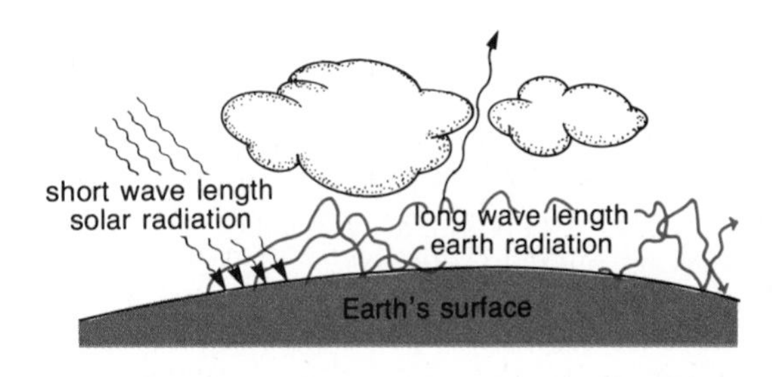

Figure 13–15 The greenhouse effect. Incoming solar radiation that is absorbed and re-radiated cannot escape back into space because the earth's atmosphere is not transparent to the re-radiated infrared radiation.

Two additional factors related to the effectiveness of the sun's rays in warming the earth are reflection and scattering. A significant quantity of solar radiation is reflected into space by clouds and solid particles of dust and ice in the atmosphere, as well as by snow and other materials on the earth's surface. The fraction of solar energy reflected into space is termed the earth's **albedo.** At the present time, the earth's total albedo has been estimated at about 33 percent, most of which is due to reflection from clouds. **Scattering** is the term used to describe the deflection and dispersion of incoming radiation by various particles in the atmosphere, including water droplets, dust, smoke, ice crystals, salt, and gas molecules. Scattering occurs only when the particles that act as obstacles to rays are of the same size or smaller than the ray's wavelength. Furthermore, it has been observed that the amount of scattering caused by particles the size of gas molecules is much greater for short-wavelength radiation (such as blue) than for long-wavelength radiation (reds). This explains the blue color of the sky, which is the result of scattering primarily by the gaseous constituents of the atmosphere.

POLLUTANTS IN THE ATMOSPHERE

An adult man or woman requires about 13 kilograms of air each day. One would hope that this essential requirement for life would be relatively free of contamination, but this is not the case. The air we breathe contains both natural pollutants (such as spores and pollen from plants, sulfur dioxide from marsh environments, hydrocarbons from forested areas, ash and gas from volcanoes, salt from the sea, and smoke from forest fires) and the human-controlled emissions from engines, factories (Fig. 13 – 16), and power plants. The pollutants caused by industry and automobiles are the most persistently hazardous to our health (Table 13 – 4). Today, more than 45 million Americans live in cities described by the federal government as having periodic episodes of major air pollution. Lung cancer and emphysema occur in these urban areas at twice the rates recorded in less industrialized rural areas. Clearly, the cleanliness of the air we breathe is important to our physical vitality. It is no less important to the esthetic appreciation of our environment and our psychological well-being.

Principal human-controlled pollutants

Among the most hazardous and offensive of the human-controlled contaminants to our air supply are carbon monoxide, various sulfur compounds, oxides of nitrogen, hydrocarbons from the combustion of fossil fuels, and industrial particulates. The first of these, *carbon monoxide*, is a colorless, odorless gas that results primarily from the incomplete burning of carbon or carbon compounds in automobile engines. Natural carbon monoxide in the atmosphere rarely exceeds 0.5 parts per million (ppm). In contrast, the carbon monoxide level inside your car when you drive through heavy traffic ranges between 25 and 50 ppm. For many people this concentration is sufficient to cause dizziness, headache, and fatigue. In extreme cases, carbon monoxide can cause death, as when people remain in a closed garage while an automobile engine is running.

Sulfur dioxide (SO_2) is another pollutant that is particularly widespread and difficult to control. This gas is produced by the burning of oil or coal that contains sulfur, or when metal sulfides are "roasted" to extract the pure metal. Sulfur dioxide may cause serious respiratory ailments by irritating delicate membranes in the throat and lungs. *Sulfur trioxide* (SO_3), which is produced by the oxidation of SO_2 in sunlight, causes similar respiratory problems. Meteorologists usually designate pollutants like SO_2, which are emitted directly from an industrial process, as **primary pollutants.** Noxious gases that result from the reaction of the primary pollutants with air and sunlight are called **secondary air pollutants.** Sulfur trioxide is a secondary pollutant. When it combines with droplets of water in the air, it forms a mist of sulfuric acid that is capable of destroying animal and plant cells, corroding metals, and chemically weathering buildings and statues (see Fig. 7 – 12). Similarly, rain washes such pollutants as sulfur and nitrogen oxides from the air, producing sulfuric and nitric acids. The resulting **acid rain** may be 200 times more acidic than normal. Without doubt, it is harmful to fish and other forms of wildlife in streams and lakes. It also has a deleterious effect on vegetation and soils, reducing agricultural yields and forest growth.

The compounds of nitrogen that pollute the atmosphere are primarily *nitrous oxide* (NO) and *nitrogen dioxide* (NO_2). Nitrous oxide results from the oxidation of nitrogen (N_2) in engines and industrial furnaces ($N_2 + O_2 \xrightarrow{\text{heat}} 2NO$). Once released into the air, nitrous oxide reacts with oxygen to produce nitrogen dioxide, a reddish-brown gas having a pungent odor that can be detected even in concentrations as low as 0.1 ppm. As noted above, nitrogen dioxide in turn combines with water to form nitric acid (HNO_3) and nitrous acid (HNO_2), and these compounds also have a deleterious effect on the environment when they are

Figure 13 – 16 An undesirable byproduct of our industrial society is the pollution of air from smokestacks of factories and smelters. (Courtesy NOAA.)

TABLE 13–4 **EMISSION OF AIR POLLUTANTS BY SOURCE IN THE UNITED STATES IN 1974, IN MILLIONS OF TONS PER YEAR***

Source	*Sulfur oxides*	*Nitrogen oxides*	*Hydrocarbons*	*Carbon dioxide*	*Particles*
Transportation	0.8	10.7	12.8	73.5	1.3
Stationary fuel consumption	24.3	11.0	11.7	0.9	5.9
Industrial	6.2	0.6	3.1	12.7	11.0
Solid waste disposal	0.0	0.1	0.6	2.4	0.5
Miscellaneous	0.1	0.1	12.2	5.1	0.8
TOTAL	31.4	22.5	40.4	94.6	19.5

* Sixth Annual Report, Council on Environmental Quality, 1975

precipitated as acid rain. Both of these acids can attack the tissues of the respiratory tract.

Hydrocarbon pollutants are released into the atmosphere primarily from the inefficient burning of fossil fuels. Although unhealthy pollutants in themselves, hydrocarbons pose a greater danger because of the even more toxic secondary pollutants formed from them in the atmosphere. Among these secondary pollutants is *ozone* (O_3). As we have already mentioned, ozone in the upper atmosphere protects life from lethal ultraviolet radiation. Ozone, however, is also very reactive chemically and would be fatal to humans in concentrations greater than a few parts per million. The maximum allowable concentration for factory workers in an eight-hour working day is only 0.1 ppm. Contrary to popular belief, ozone produced from home "purification" devices is too dilute (around 0.1 ppm) to serve as an antigermicidal agent, and in fact, such appliances pollute rather than purify the air.

Aerosols and particulates

More than 150 million tons of soot, dust, smoke, and liquid droplets were discharged into the atmosphere of the United States in 1983, principally by automobiles and factories. These materials constitute components of atmospheric pollution termed **aerosols** and **particulates.** Aerosols are less than 1 micron in diameter; particulates are larger. Both kinds of pollutants serve as carriers for chemically active SO_2, nitrogen oxides, acids, ozone, and lead compounds derived from antiknock agents in gasoline. These substances interfere with healthy respiration in humans, and some are considered carcinogenic. The increasing use of unleaded gasoline has reduced the lead hazard for city dwellers, but the danger persists for those working in, or living near, lead smelters.

Smog

The original use of the term **smog** was to describe a hazy air pollution composed primarily of smoke and fog (Fig. 13–17). Currently, however, the term is used for any type of air pollution. Two kinds of smog are generally recognized. One is the chemically reducing type that comes largely from the burning of coal and oil. It contains sulfur dioxide, soot, fly ash, and partially decomposed organic compounds. This is the "London type" of smog. A second kind of smog, sometimes dubbed the "Los Angeles type," is referred to as **photochemical smog** because sunlight is needed to produce its high content of secondary pollutants. Los Angeles smog contains relatively little sulfur dioxide but is rich in ozone, oxides of nitrogen, organic peroxides, and hydrocarbons.

Air pollution and temperature inversions

Earlier in this chapter the troposphere was described as an atmospheric zone characterized by active vertical mixing and overturning of air. This mixing is entirely beneficial, for it helps to dispose of pollutants. There are times, however, when air circulation is poor. When this occurs, the usual decrease in temperature with increasing altitude may be upset and **temperature inversions** occur. One kind of temperature inversion can be produced on a cold, clear night when the rapid loss of heat from the ground surface quickly renders that surface cold. The air immediately above the ground also becomes chilled, and for a certain

Figure 13–17 *Chicago under smog. (Courtesy of Chicago Tribune.)*

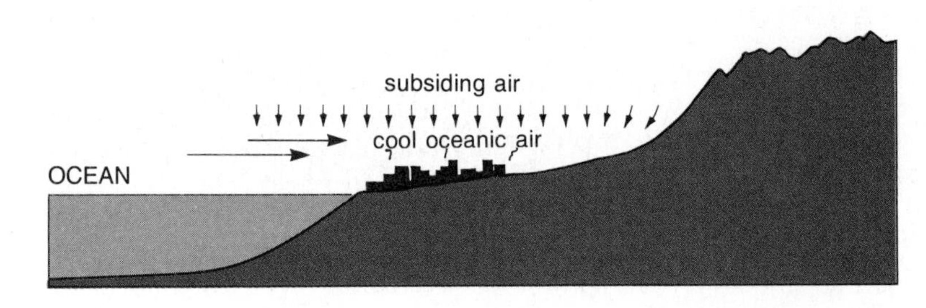

Figure 13–18 *Topographic barriers, descending air currents, and onshore breezes combine on many occasions to trap smog in the Los Angeles basin.*

additional vertical distance, temperatures actually increase with altitude. This situation is just the reverse of the normal condition in the troposphere as a whole, and hence the term temperature inversion is appropriate. Ground fogs and heavy morning dew occur within the low-lying blanket of dense cool air.

After sunrise, the temperature inversion may disappear as the ground again begins to absorb solar radiation. Some of the heavy cold air, however, may persist in valleys. Pollutants become entrapped in the dense cool air and may pose a health hazard to communities in low-lying areas.

Temperature inversions may also occur when the velocity of cool high-altitude winds diminishes, resulting in a tendency for this cold air to sink and compress the underlying air layer. The compression heats the underlying air, which then forms a relatively warmer layer sandwiched between subsiding cool air above and colder air adjacent to the ground below. This phenomenon is termed **subsidence inversion.** Subsidence inversions, combined with geographic conditions, contribute to the persistence of the smog that afflicts the Los Angeles area. In the Los Angeles basin, cool air from the ocean moves in over the city and is simultaneously covered by a layer of subsiding warm air. The resulting inversion tends to persist, for the surrounding mountains inhibit eastward dispersal of the stagnant air (Fig. 13–18).

Summary

The envelope of gases around the earth constitutes the atmosphere. It is composed of such gases as nitrogen (N_2), oxygen (O_2), and argon (Ar) as well as gases whose proportion in air is more variable in time and place, such as water vapor and carbon dioxide. An important lesser constituent of the upper atmosphere is ozone (O_3), which protects life on earth by absorbing harmful ultraviolet radiation from the sun.

Although the atmosphere is a mixture of gases, in physical properties it resembles a single gas. Like a single gas, the atmosphere has weight and therefore can exert atmospheric pressure on the earth's surface. The atmospheric pressure at any given location is the force exerted by the weight of the overlying air. The bulk of atmospheric gases remains close to the earth's surface because of gravity, and the quantity of gas molecules diminishes with altitude and gradually merges into the near emptiness of space. Thus, both atmospheric pressure and density decrease with increasing altitude.

The atmosphere is divided vertically into four major zones, primarily on the basis of temperature changes. The lowermost layer is the troposphere, where air is heated from below and therefore decreases in temperature with increasing height. The layer above the troposphere, characterized by constant or increasing temperature with height, is called the stratosphere. The warmth of the stratosphere results from the direct absorption of ultraviolet radiation by ozone. The mesosphere is a zone in which temperature decreases rapidly with altitude. It is overlain by a hot layer known as the thermosphere.

The sun is the ultimate source of the energy that affects atmospheric properties. Some of this energy is directly absorbed by the atmosphere, some is reflected back into space, and some is absorbed at the earth's surface. Energy is constantly being transferred between the earth's surface and the atmosphere, and between different parts of the atmosphere by radiation, conduction, evaporation, and convection.

The energy that a body has because of the motion of its molecules is called heat. How much heat a body or volume of air contains is governed by the quantity of gas molecules present and their motion. Temperature is the measure of the average hotness or coldness of a body. Because of the immense number of molecules in a large volume of air, it may contain considerable heat even though its temperature is relatively low.

Seasonal variations result from the tilt of the earth's equatorial plane. The earth's equatorial plane is tilted 23.5° from the plane of the earth's revolution around the sun. That orientation remains fixed as the earth revolves around the sun, so that the sun appears directly overhead at noon on latitude 23.5°N (Tropic of Cancer) on June 21 (summer solstice), directly overhead at noon on latitude 23.5°S (Tropic of Capricorn) on December 22d (winter solstice), and directly overhead at noon on the equator during the spring and autumnal equinoxes. The angle of incoming rays and the length of the day govern the temperatures associated with the seasons.

The size, shape, and arrangement of land relative to ocean bodies influence the climate at particular locations. The ocean cools and warms more slowly than land areas and thus may moderate temperatures at favorably situated locations. The distribution of temperatures can be depicted on maps by means of lines connecting points of equal temperature, called isotherms.

Pollutants in the atmosphere have both natural and human-controlled sources. The principal human-controlled pollutants are carbon monoxide, carbon dioxide, hydrocarbons, sulfur compounds, oxides of nitrogen, water vapor, and particulate matter. Temperature inversions occur when the usual tropospheric condition of decreasing temperature with height is inverted. Cool dense air may serve to hold pollutants close to the ground and thereby pose a health problem to all people living in the affected area.

QUESTIONS FOR REVIEW

1 What are the most abundant components of the earth's atmosphere? Why is the proportion of water vapor so variable from place to place in the atmosphere? What conditions may be causing a buildup of carbon dioxide in the atmosphere, and how may that increase relate to change in the earth's average surface temperature?

2 What is ozone? At approximately what altitude is ozone concentrated in the atmosphere? Why is an ozone layer in the atmosphere necessary for the well-being of plants and animals?

3 Define the terms troposphere, stratosphere, mesosphere, and thermosphere. What are the characteristics of each?

4 Why does the atmosphere exert pressure? How does the pressure change with altitude?

5 Imagine that the earth's axis of rotation was exactly perpendicular to the plane of the ecliptic. Would there be seasons? At what location(s) would insolation be minimal? Would there still be latitudinal changes in mean annual temperature?

6 Why does the amount of insolation received at any given location depend upon the obliqueness of the sun's rays?

7 Distinguish between the terms heat and temperature. What are the three temperature scales and how are the fixed points (freezing point, boiling point) numbered in each scale?

8 Distinguish between the terms partial pressure and total pressure. What is the instrument used for measuring the total atmospheric pressure?

9 New York City and San Francisco are both coastal cities located between 37°N and 42°N latitude, yet the annual temperature range in New York is considerably greater than in San Francisco. What is the reason for this difference? The cities of London in England and Winnipeg in Canada are located at about the same latitude. Why does Winnipeg have harsher climates and greater annual temperature range than London?

10 In which of the major atmospheric layers is convection most pronounced? In a mixture of gases like air, is convection or conduction more effective in heat transfer? Why?

11 What is the meaning of the term albedo? Would albedo be greater in central Antarctica or central Kansas? Would it be greater in pure clean air or clouds?

12 What is meant by a temperature inversion? Under what conditions do temperature inversions occur, and how are they related to periodic episodes of intense local air pollution?

SUPPLEMENTAL READINGS AND REFERENCES

Anthes, R.A., Panofsky, H.A., and Cahir, J.J., 1983. *The Atmosphere*. 3rd ed. Columbus, Ohio, Charles E. Merrill.

Bach, W., 1972. *Atmospheric Pollution*. New York, McGraw-Hill Book Co.

Battan, L.J., 1979. *Fundamentals of Meteorology*. Englewood Cliffs, N.J., Prentice-Hall, Inc.

Cole, F.W., 1975. *Introduction to Meteorology*. 2d ed. New York, John Wiley & Sons.

Navarra, J.G., 1979. *Atmosphere, Weather, and Climate*. Philadelphia, W.B. Saunders Co.

14

moisture in the atmosphere

The fog comes in
on little cat feet.

It sits looking
over harbor and city
on silent haunches
and then moves on.
CARL SANDBURG
"Fog"

Prelude The air around us contains a quantity of invisible moisture called water vapor. It is supplied to the atmosphere by evaporation from the ocean, the earth's smaller water bodies, and the moist surface of continents. Some is also derived from solid snow and ice through sublimation, a process by which a solid (ice) is converted directly to a gas (water vapor). Plants and animals also give off moisture as part of their life processes.

Once in the air, water vapor and the energy that resides within it contribute significantly to those continuously changing conditions in the lower atmosphere that we call weather. The importance of water vapor to weather is apparent from the frequency with which we hear such terms as clouds, rain, snow, sleet, hail, humidity, frost, and dew on the radio weather reports. All of these terms tell us something about what is happening to water vapor in the atmosphere.

Moisture in the atmosphere in the form of fog obscures part of the Golden Gate Bridge, San Francisco.

RELATION OF WATER VAPOR TO HEAT AND PRESSURE

Water vapor is present in the atmosphere in amounts that rarely exceed 3 percent by volume. This seems a small quantity, yet its importance is far greater than that modest percentage might suggest. Water vapor is the source of our fresh water supplies. It is an essential component in the vital, unending circuit of evaporation and precipitation called the hydrologic cycle (see Fig. 8-1). As an important repository of heat, water vapor helps retain warmth close to the earth's surface even at night, when radiation from the sun is not received. It also affects atmospheric pressure. Water vapor is only about three fifths as heavy as dry air. Thus, evaporating water into the atmosphere is rather like adding helium to it. The moist air becomes more buoyant and tends to float over dry air.

Latent heat

The evaporation of water vapor from bodies of liquid water and the sublimation of water vapor from ice require that water molecules break away from their neighbors and enter the overlying atmosphere. This breaking of the adhesion that holds water molecules together requires an expenditure of heat energy, which is taken from the adjacent environment. Evaporation, therefore, is a cooling process. We recognize the cooling effect of evaporation whenever wind blows over skin that is moist with perspiration. The heat absorbed by water molecules during evaporation is one reason why the temperature of the ocean rises so little, even when bathed in the sun's rays. An enormous amount of the energy in those rays is utilized in evaporation, rather than in heating the ocean. Another example of the cooling effect of evaporation is experienced when temperatures drop after a summer shower. The air becomes cooler partly because of the heat extracted from it by evaporation of falling rain.

The heat needed to convert liquid water to water vapor is called **latent heat of vaporization.** The term "latent" conveys the notion that the heat has accompanied the departing water molecules during evaporation and resides within them during the vapor phase. During subsequent condensation, the latent heat is released into the environment as heat, warming the air around us.

The term calorie describes the amount of heat involved in evaporation or condensation of water. A **calorie** is the amount of heat needed to raise the temperature of 1 gram of water by 1°C. The number of calories needed to convert liquid water to vapor (the heat of vaporization) will vary with the temperature of the water. For example, 540 calories per gram of boiling water (100°C water) is required for vaporization, whereas if the water is at the melting point of ice (0°C), 596 calories per gram are needed. Thus, the colder the water, the more latent heat is taken up during evaporation.

When ice changes to water, latent heat is also taken up by the liquid water molecules. At 0°C, the latent heat needed to melt ice is 80 calories per gram. If the ice or snow changes directly to vapor without melting first, then the latent heat required is equivalent to the latent heat of melting plus the latent heat of vaporization, or a total of 676 calories per gram ($80 + 596 = 676$).

Factors affecting water vapor in the atmosphere

The amount of moisture in the air at any time on earth depends upon a variety of factors, including temperature, the nature of the surface from which evaporation is taking place, the amount of water vapor already present in the atmosphere, wind velocity, and atmospheric pressure. With regard to different kinds of surfaces, densely forested areas give up over 30 percent more water than does the surface of a lake under similar temperature and pressure conditions. A single, ordinary tree loses water to the atmosphere at an average rate of about four U.S. gallons per hour. Such water loss from the leaves of plants is termed **transpiration.** Often scientists combine the loss of water by evaporation with the loss through transpiration and call the combined process **evapotranspiration.**

The amount of water vapor already present in air influences the amount that can be added. For example, the more water molecules in the air, the slower the evaporation. Initial humidity, however, may be low and evaporation may still proceed slowly unless wind and temperatures are favorable.

Humidity

The word **humidity** refers to the concentration of water vapor in the atmosphere at any given time. As noted earlier, the percentage of water vapor in the air is small and limited by temperature. This is apparent to anyone who has attempted to evaporate water in a confined space like a jar. After a small amount has been evaporated, any further evaporation produces an equal amount of condensation, so the amount of water vapor in the air within the jar thereafter remains constant. The air in the jar is said to be saturated.

If one plots the amount of water vapor in air against the air's temperature, it becomes apparent that the colder the air, the less moisture it can hold. At 5°C, for example, the water vapor content at sea level pressure is only 6.8 grams per cubic meter (6.8 gm/m^3), whereas at 25°C the vapor content reaches 22.8 gm/m^3 (Fig. 14-1). The everyday significance of this relationship is that warm summer air can, and usually does, hold more moisture than cold winter air. The seemingly greater humidity of winter weather in some regions can be explained by the slower rate of evaporation during cold winter days, causing streets to remain wet for longer periods of time. Also, cool winter air is usually closer to saturation than warm summer air, even though it doesn't hold very much water.

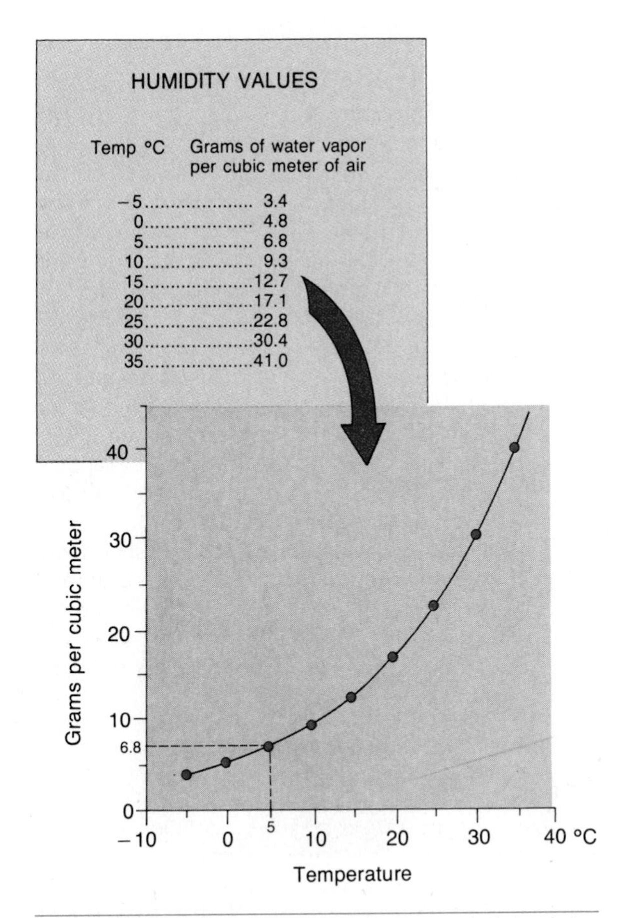

Figure 14-1 *Humidity values plotted on a graph show that the atmosphere's capacity for retaining water vapor increases at higher temperatures.*

Three terms commonly used to express humidity are absolute humidity, vapor pressure, and relative humidity. **Absolute humidity** is a measure of the mass of water vapor in a unit volume of air. It is expressed in grams of water vapor per cubic meter. *Vapor pressure* refers to that part of the total air pressure exerted solely by the water vapor. It is therefore the partial pressure of the water vapor. Vapor pressure is usually expressed in units of absolute pressure, such as millibars. **Relative humidity** is defined as the ratio of the actual amount of water vapor present (absolute humidity) to the amount that could be present if the air at the given temperature were saturated. It is a way of expressing how near the air is to being saturated. Air at 21°C, for example, can hold 18 gm/m^3. If it actually contains only 12 gm/m^3, then the air has $12/18$ or $2/3$ of its total possible capacity at that temperature. Because relative humidity is expressed as a percentage, $2/3$ becomes 66.6 percent. Note that if either the temperature or the amount of water vapor changes, so also will the relative humidity. Relative humidity is measured by an instrument called a **hygrometer.** This device employs a fiber (sometimes a human hair) that lengthens with increasing humidity and shortens with decreasing humidity.

A pair of specially mounted thermometers may also be used to measure relative humidity. The instrument, called a **psychrometer** (Fig. 14-2), consists of two identical thermometers attached to a frame that can be whirled about a handle or ventilated by a fan. A wet, tight-fitting muslin cloth is wrapped around the bulb of one of the thermometers, which is therefore designated the wet-bulb thermometer. Evaporation from the cloth cools the "wet bulb" and thereby causes a difference in the readings of the two thermometers. The basic principle involved is that the greater the humidity, the less evaporation and thus the smaller the difference between the two temperatures. Psychrometer readings are taken several times, and the device is whirled between each reading to ensure a supply of fresh air to the thermometers. By referring to tabulations such as that given in Table 14-1, psychrometer readings may be used to determine relative humidity.

There are more sophisticated instruments for determining humidity than the hygrometer and psychrometer. A device incorporated into the radiosonde (Fig. 14-3), for example, determines relative humidity by measuring changes in the conductivity of

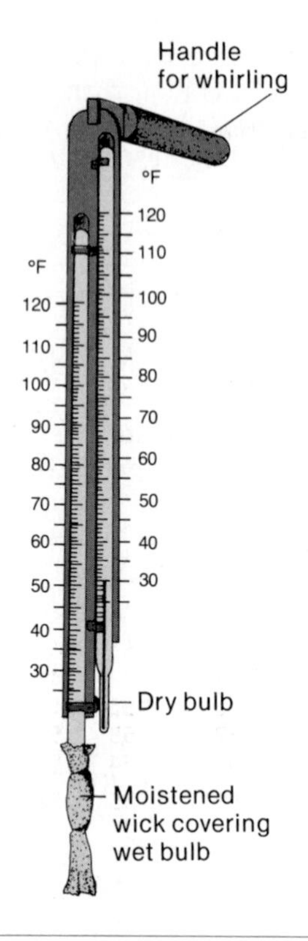

Figure 14–2 *A sling psychrometer. (From Navarra, J.G.,* Atmosphere, Weather, and Climate. *Philadelphia, Saunders College Publishing, 1979.)*

a metal plate that has been coated with a moisture-absorbing chemical compound.

Another term is useful in studies associated with humidity. If the temperature of air containing moisture is gradually lowered, a temperature will be reached where the air is saturated and can therefore hold no more water vapor. That temperature is the **dew point.** Reducing the temperature beyond the dew point results in condensation in the form of dew, frost, fog, clouds, or precipitation. Dew point can be measured with an instrument called a dew point hygrometer. This device cools air until dew appears on a metal surface, thereby indicating the dew point temperature.

Condensation

Condensation is the opposite of evaporation. It is the passage of water vapor, an invisible gas, into visible liquid water. As described in the previous section, evaporation of water absorbs heat from the environment and thereby produces a cooling affect. The absorbed heat is the previously defined latent heat of vaporization. When water vapor condenses, the latent heat formerly absorbed is liberated, thus warming the adjacent environment.

Because the capacity of air to hold water in the form of vapor decreases with temperature, cooling of air is the usual way by which first saturation and then condensation occur. Such cooling occurs when ascending air expands, when relatively warm air is mixed with cold air, or when air comes into contact with a very cold area of the earth's surface. In the latter case, either dew or frost may form. Frost is particularly interesting because its formation involves the previously mentioned process of sublimation. Frost, a solid composed of ice crystals, forms directly from a gas (water

Figure 14–3 *A radiosonde suspended from a balloon and parachute will be carried aloft and send back information about relative humidity, winds, and temperatures high in the atmosphere.*

TABLE 14–1 **RELATIVE HUMIDITY IN PERCENTAGES**

Dry-bulb reading (°F)	*Differences between dry-bulb and wet-bulb readings (°F)*															
	1°	*2°*	*3°*	*4°*	*5°*	*6°*	*7°*	*8°*	*9°*	*10°*	*11°*	*12°*	*13°*	*14°*	*15°*	*16°*
50°	93	87	80	74	67	61	55	49	43	38	32	27	21	16	10	5
52°	94	87	81	75	69	63	57	51	46	40	35	29	24	19	14	9
54°	94	88	82	76	70	64	59	53	48	42	37	32	27	22	17	12
56°	94	88	82	76	71	65	60	55	50	44	39	34	30	25	20	16
58°	94	88	83	77	72	66	61	56	51	46	41	37	32	27	23	18
60°	94	89	83	78	73	68	63	58	53	48	43	39	34	30	26	21
62°	94	89	84	79	74	69	64	59	54	50	45	41	36	32	28	24
64°	95	90	84	79	74	70	65	60	56	51	47	43	38	34	30	26
66°	95	90	85	80	75	71	66	61	57	53	48	44	40	36	32	29
68°	95	90	85	80	76	71	67	62	58	54	50	46	42	38	34	31
70°	95	90	86	81	77	72	68	64	59	55	51	48	44	40	36	33
72°	95	91	86	82	77	73	69	65	61	57	53	49	45	42	38	34
74°	95	91	86	82	78	74	69	65	61	58	54	50	47	43	39	36
76°	96	91	87	82	78	74	70	66	62	59	55	51	48	44	41	38
78°	96	91	87	83	79	75	71	67	63	60	56	53	49	46	43	39
80°	96	91	87	83	79	75	72	68	64	61	57	54	50	47	44	41
82°	96	92	88	84	80	76	72	69	65	61	58	55	51	48	45	42
84°	96	92	88	84	80	76	73	69	66	62	59	56	52	49	46	43
86°	96	92	88	84	81	77	73	70	66	63	60	57	53	50	47	44
88°	96	92	88	85	81	77	74	70	67	64	61	57	54	51	48	46
90°	96	92	89	85	81	78	74	71	68	65	61	58	55	52	49	47
92°	96	92	89	85	82	78	75	72	68	65	62	59	56	53	50	48
94°	96	93	89	85	82	79	75	72	69	66	63	60	57	54	51	49
96°	96	93	89	86	82	79	76	73	69	66	63	61	58	55	52	50

vapor) when the dew point of air is below 0°C. Thus, frost is not frozen dew (which forms from liquid water). Sublimation of water vapor to form frost involves the release of heat in an amount equal to the sum of the heat otherwise released during condensation (from water vapor to liquid water) and freezing (from liquid water to solid ice).

One might surmise that as soon as the concentration of water vapor in the atmosphere exceeded the saturation level, condensation would quickly follow. In nature, however, mere saturation is not enough to cause condensation. There must be surfaces on which condensation can take place. Such condensates as dew and frost form readily on solid surfaces whenever air temperature reaches the dew point. In the open air, however, there are no extensive solid surfaces. If the air were pure, then condensation would depend upon the chance collision of hundreds of individual water molecules to form even the smallest droplet of liquid water. Because of the larger space between molecules, the probability for chance collisions would be low. Even a relative humidity of 400 percent would not be sufficient to cause the condensation necessary to produce a cloud. Clouds commonly form, however, at much lower humidity levels, largely because the atmosphere contains myriads of particles of ice, salt, dust, and solid particles that provide the necessary surfaces for condensation. These particles are called **condensation nuclei.** On them, water molecules become attached and, by further additions, grow into water droplets. Most condensation nuclei are less than about one micron (1/1000 mm) in diameter. Because of their small size, they are able to remain suspended in the atmosphere for long periods of time. Some particles resist condensation and are called *hygrophobic.* The majority, including salt, ice, and products of combustion, are *hygroscopic* and have a definite affinity for water. In recent years, meteorologists have attempted to bring about the formation of rain and snow by dumping small quantities of hygroscopic particles like dry ice (frozen CO_2) and silver iodide into clouds. The process is called cloud seeding, and although it is popularly thought to be a rain-making process, it really makes ice. The dry ice particles produce millions of tiny ice crystals by cooling the air to about −40°C. Silver iodide, which has a six-sided crystal form similar to ice, also serves as a template or nucleus for ice crystal growth. The effectiveness of cloud seeding is still being evaluated. In the American Southwest it has had questionable success in increasing the snowpack in the mountains.

CLOUDS

Cloud formation

Clouds are visible, mistlike masses of water or ice particles suspended in the atmosphere above or at the earth's surface. They develop as a result of condensation throughout a large volume of air. The condensation, in turn, results from cooling of the same air. If the water vapor content of the air is high, less cooling is needed to induce condensation than if the air contains relatively little water vapor. The cooling necessary to bring about condensation and cloud formation results from such processes as convective cooling, cooling by mixing of unlike air masses, advective cooling, and forceful lifting.

Convective cooling refers to the lowering of the temperature of a mass of air as it expands and rises. When a body of air rises into a region of lower atmospheric pressure, it expands, and some of the air's internal energy is expended in doing the work of expanding. As a result, the parcel of air, having expended part of its internal energy, becomes cooler. Conversely, just as air is cooled by expansion, it is heated by compression. Temperature changes resulting from the expansion or compression of air that do not involve the addition or subtraction of heat from the surroundings are termed **adiabatic temperature changes.** We have witnessed adiabatic temperature changes when a large volume of air is pumped into the small space of a bicycle tire. The air in the tire becomes warmer because the work we have done in compressing the air has been converted to heat. When the air is released from the tire, it uses up its heat in the work needed for expansion and becomes noticeably cooler.

Meteorologists agree that adiabatic cooling associated with rising convective air currents is the most important factor in forming most kinds of clouds. The adiabatic temperature changes involved in cloud formation take place at definite rates. Specifically, as long as a given parcel of air remains undersaturated, it cools everywhere at the fixed rate of about 1°C per 100 meters of ascent (Fig. 14-4). This rate is called the **dry adiabatic lapse rate.** (Descending air, whether saturated or unsaturated, heats at the same uniform rate of 1°C for every 100 meters of descent). Usually, how-

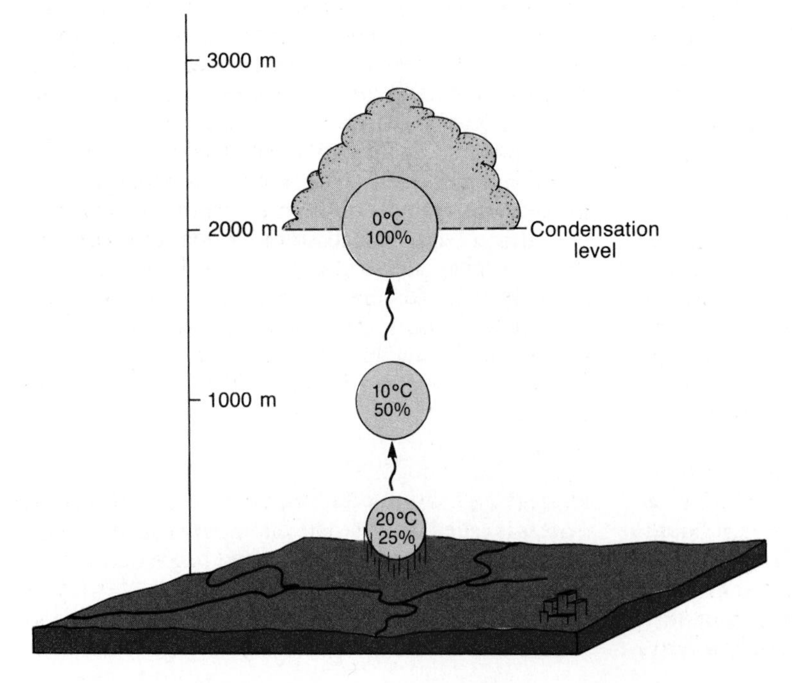

Figure 14-4 A rising parcel of air cools at the dry adiabatic lapse rate of 1°C per 100 meters (10° per 1000 km). When the air reaches a level where temperature overtakes the dew point, cloud development begins. Percentages refer to relative humidity.

ever, air contains moisture, and this has a decided impact on the adiabatic rate. As water vapor condenses in the rising air, it gives off heat, and this slows the rate of cooling. The moist-air rate of cooling is called the **moist adiabatic lapse rate.** It varies with the amount of moisture in the air, but an average rate would be about 0.7°C per 100 meters.

It is important to distinguish between the general rate of decrease in temperature with increasing altitude in the troposphere (called the *measured lapse rate*) and the adiabatic lapse rate. The former has nothing to do with rising or cooling air and merely expresses the temperature distribution in the troposphere as measured directly by thermometers in weather balloons or airplanes. The adiabatic lapse rate refers only to the temperature change experienced by a parcel of air ascending or descending through the atmosphere.

As the temperature of a parcel of rising air decreases, the relative humidity increases. Upon reaching a level where temperature becomes less than the dew point, the air parcel experiences condensation and cloud development. That level is called the **condensation level** (see Fig. 14–4). It is marked by the base of the clouds, and clouds continue to grow above that level by further condensation.

If the parcel of ascending air described above was especially humid before it began its ascent, it will not have to rise as high as a dry-air parcel before cloud formation begins. Thus, the higher the humidity near ground level, the lower will be the condensation level. This condition is apparent in the relatively lower level of summer cloud bases over humid Gulf Coastal states compared with higher cloud bases in the arid regions of the Southwest.

Convective cooling resulting from adiabatic processes is the most important but not the only mechanism conducive to cloud development. For example, the mixing of two parcels of moist air having different temperatures may induce saturation in the mixture and cause cloud formation. Such mixing occurs along weather fronts where large masses of air having different temperature and humidity come into contact. Clouds like those shown in Figure 14–5 may form by this mixing process.

Clouds may also result from **advective cooling.** Advection is merely a horizontal convective movement of air, such as wind. When wind conveys warm moist air horizontally across a cold land or ocean surface, there is a tendency for cloud or fog formation. If the wind velocity is low, advective cooling favors fog, but with stronger winds the moisture may lift to form a low-lying cloud layer (Fig. 14–6). Advective clouds are particularly prevalent in cold oceanic regions.

Yet another way in which air may be cooled so as to induce cloud formation is by **forceful lifting.** A moving mass of air may encounter a sloping terrain, such as a mountain range, which forces air to ascend, form clouds, and produce precipitation on the windward side of the range (Fig. 14–7). By the time the air has reached the far or lee side of the mountains, much of its water vapor has been lost. Furthermore, if it descends, it warms and would be more likely to take on moisture than to yield it as rain. Thus, the dry lee sides of highland areas are said to be in the **rain shadow.** The term used to describe the forced upslope movement of air because of topography is **orographic lifting.** Orographic lifting is an important producer of clouds and precipitation along windward sides of such mountains as the Andes of South America and Pacific Coastal ranges of North America. The Patagonia Desert of South America and the Great Basin Desert of the United States are examples of rain shadow regions.

The windward sides of coastal mountains are often wet not only because of simple orographic action but also because they form barriers that confine or retard air masses. In addition, the rugged irregularity of mountainous terrain provides many sites for differential heating, channelization of air currents, and thermal irregularities.

The slope required for upslope movement of air need not always be a land surface. A mass of cool air may also act as a sloping barrier (Fig. 14–8) over which warmer, less dense air may rise, cool, and undergo condensation. This kind of interface between two air masses can be a major cause of precipitation.

Finally, air may be forcefully lifted by convergence of similar masses of air. The Florida peninsula provides an excellent example of this mechanism. Winds from both the east and west coasts of Florida converge over the peninsula. Being forced into a smaller space by the convergence, and prevented from moving downward by the solid ground, the air is shunted upward. The process is enhanced by the strong solar heating that characterizes the Florida peninsula.

Atmospheric stability

Meteorologists recognize that particular bodies of air may be either stable or unstable and that their general condition of stability governs the development of clouds and the nature of precipitation. Stability refers to the degree to which the atmosphere is susceptible to vertical movements and convection. In brief, air is

Figure 14–5 *Altocumulus cloud formed by the mixing of two parcels of moist air having different temperatures along a weather front. (Courtesy of NOAA.)*

Figure 14–6 *Low stratus clouds formed as a result of advective cooling partially obscure the towers of San Francisco's Golden Gate Bridge. (Courtesy NOAA.)*

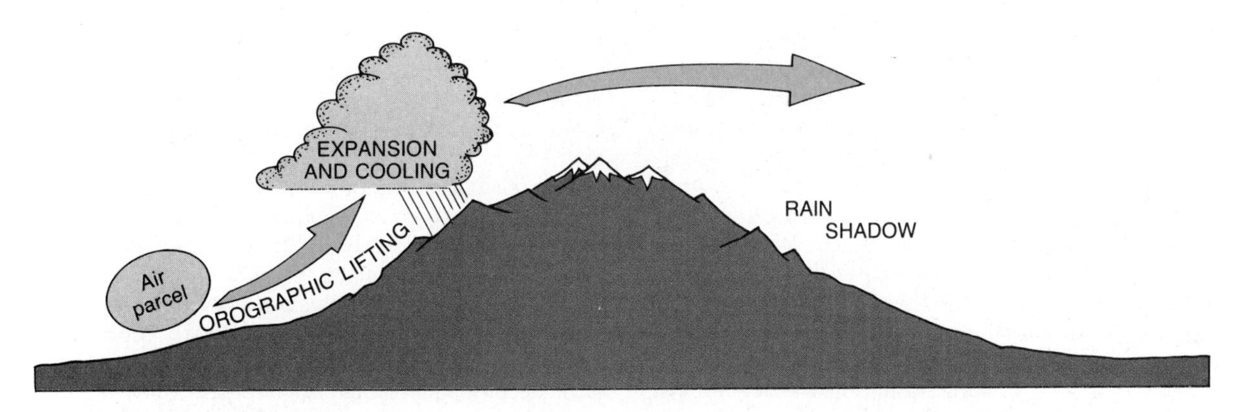

Figure 14–7 Orographic lifting so as to form clouds.

considered stable if such movements are hindered and unstable if they are occurring or have the potential to occur.

The degree of stability of an air column is determined by the relation between the measured lapse rate (the general temperature drop with increased altitude in the troposphere) and the adiabatic lapse rate. **Absolute stability,** for example, occurs when the measured lapse rate is less than the moist adiabatic lapse rate (that is, less than 0.5°C per 100 meters). Thus, the air tends to be cooler and heavier than surrounding air (Fig. 14–9), usually requires forceful lifting, and has a tendency to return to its original position. (Conversely, if

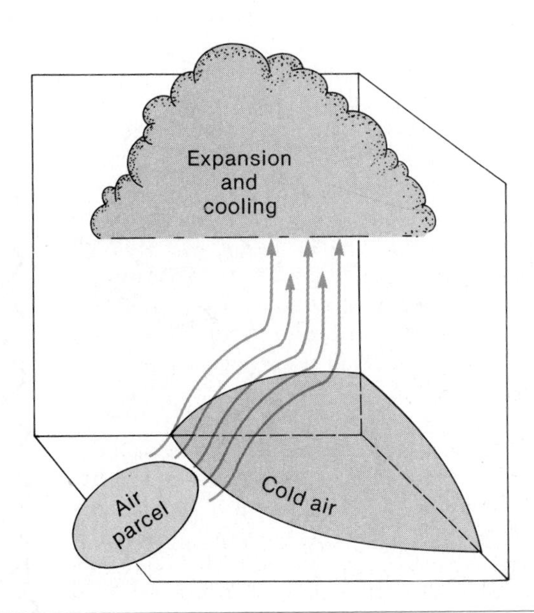

Figure 14–8 Relatively warm air may rise, cool, and condense to form clouds when moving up and over a mass of cooler air.

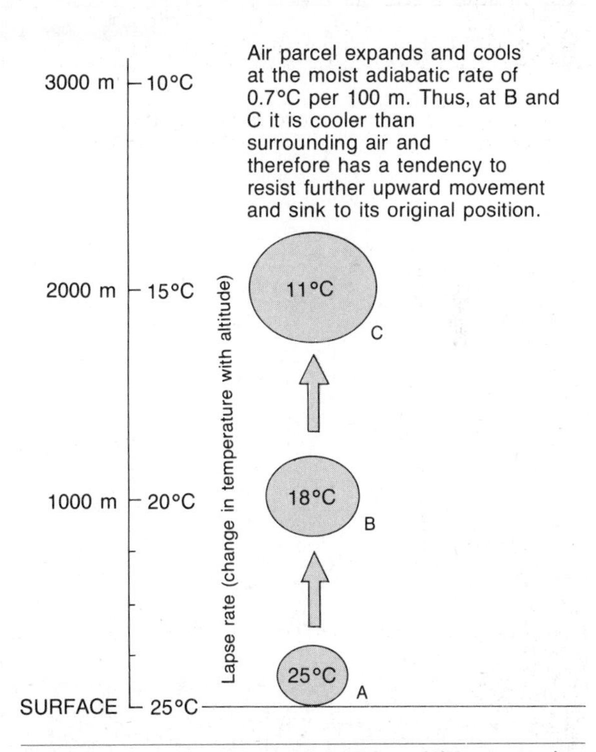

Figure 14–9 The condition of absolute stability occurs when the lapse rate (decrease in temperature with altitude) is less than the moist adiabatic rate, so that the rising air parcel quickly becomes cooler and heavier than the surrounding air.

displaced downward, it would rise to its original position.) Under such conditions, convection currents tend to be suppressed and inversions favored. A relatively thin but widespread blanket of low clouds accompanied by only light precipitation may characterize this condition of stability.

A condition of instability known as **absolute instability** occurs when the air's measured lapse rate exceeds the dry adiabatic rate of cooling (that is, the lapse rate exceeds 1 °C per 100 meters). If such an unstable parcel is forced aloft (by heating at the earth's surface, orographic lifting, or other processes), it will be warmer than the surrounding air (Fig. 14-10) and will continue to rise and cool adiabatically. When the saturation temperature of the air is reached, condensation with resulting clouds and precipitation may occur.

A third situation with regard to the behavior of air is conditional on the presence of considerable water vapor and is therefore called **conditional instability.**

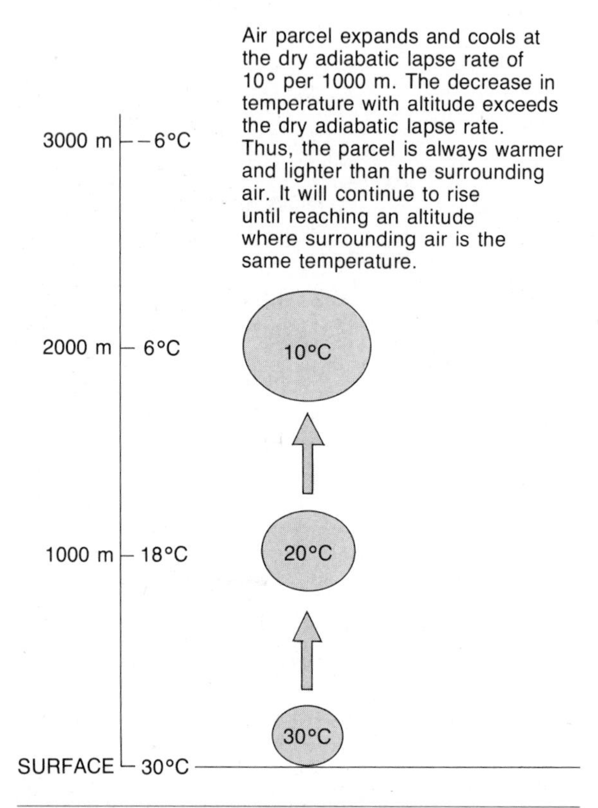

Figure 14–10 An example of absolute instability using a measured lapse rate of 12 °C per 1000 meters. The dry adiabatic lapse rate is 10 °C per 1000 meters.

It often happens that an air column is stable while it rises and cools at the dry rate (without condensation) but becomes unstable when condensation begins. Consider a parcel of moderately stable air being forced aloft. As it ascends, it cools at the dry adiabatic rate until it reaches saturation. On reaching saturation, the release of latent heat of condensation provides warmth for further lifting, and the air subsequently cools at the slower, moist adiabatic rate. What was initially a stable condition has become unstable above the condensation level. Clouds associated with such conditional instability tend to be massive and towering, and are often accompanied by thunderstorms.

Cloud types

Clouds have always held some degree of fascination for people because they are simultaneously the product of weather and the harbingers of weather yet to come. Proverbs whose origins are lost in antiquity tell us, "Dark clouds in the west, stay indoors and rest," or "When high clouds and low clouds do not march together, prepare for a blow and a change in the weather." For serious weather watchers, clouds provide clues to temperature and moisture conditions in the atmosphere and indicate the direction of winds aloft. On a global scale, satellite pictures of cloud patterns help us monitor the progress of storms and other weather conditions.

Clouds are classified and named according to their appearance and height. With regard to appearance, the descriptive root terms *cirrus* (meaning curly, wispy, or feathery), *stratus* (meaning layered), and *cumulus* (meaning heaped up) help us remember the basic forms. The recognized categories for cloud height are simply *high, middle,* and *low*. The actual altitudes for these levels vary with location as well as the seasons. In middle latitudes, *low* clouds range from near the earth's surface to about 2000 meters. At the same latitude, *middle* clouds occur from 2000 meters to 5500 meters, and *high* clouds occur from 5500 to 14,000 meters. The 10 basic cloud types within these three divisions are called **cloud genera** (Table 14-2).

Among *low* clouds, the typical genera are *stratus, stratocumulus,* and *nimbostratus*. **Stratus** clouds (Fig. 14-11) occur as broad sheets that spread across most or all of the sky. They generally form by condensation within layered air that is not actively disturbed by strong vertical movements. **Stratocumulus** clouds are often seen as long parallel bands or regularly patterned masses that also cover nearly all of the sky. The root

TABLE 14–2 **CLOUD TYPES AND ASSOCIATED CONDITIONS**

Altitude over middle latitude (in feet)	*Name, abbreviation, and symbol*	*Description*	*Composition*	*Possible weather changes*
High Clouds 18,000 to 45,000	Cirrus (Ci)	Marestails Wispy	Ice crystals	May indicate a storm, showery weather close by
	Cirrostratus (Cs)	High veil Halo cloud	Ice crystals	Storm may be approaching
	Cirrocumulus (Cc)	Mackerel sky	Ice crystals	Mixed significance, indication of turbulence aloft, possible storm
Middle Clouds 6500 to 18,000	Altocumulus (Ac)	Widespread, cotton ball	Ice and water	Steady rain or snow
	Altostratus (As)	Thick to thin overcast; high, no halos	Water and ice	Impending rain or snow
Low Clouds Sea Level to 6500	Stratocumulus (Sc)	Heavy rolls, low, widespread Wavy base of even height	Water	Rain may be possible
	Stratus (St)	Hazy cloud layer, like high fog Somewhat uniform base	Water	May produce drizzle
	Nimbostratus (Ns)	Low, dark gray	Water, or ice crystals	Continuous rain or snow
Vertical Clouds Few hundred to 65,000	Cumulus (Cu)	Fluffy, billowy clouds Flat base, cotton ball top	Water	Fair weather
	Cumulonimbus (Cb)	Thunderhead Flat bottom and lofty top Anvil at top	Ice (upper levels) Water (lower levels)	Violent winds, rain, hail are possible Thunderstorms

From J.G. Navarra, *Atmosphere, Weather, and Climate,* Philadelphia, Saunders College Publishing, 1979.

nimb- means rain, and **nimbostratus** clouds are dark, shapeless masses from which rain or snow is falling.

The two principal middle-group clouds bear the prefix *alto-*. They are named *altostratus* and *altocumulus*. The uniform grayish blanket of altostratus clouds is often thick enough to obscure the sun, or at least to make the sun appear as it might when viewed through ground glass. **Altostratus** consist primarily of water droplets and form when warm air ascends over denser and colder air. **Altocumulus** (Fig. 14–12) clouds consist of thin, flakey patches or globular, billowy masses arranged in lines.

Cirrus, cirrocumulus, and *cirrostratus* are the basic categories of high clouds. In their most usual form, **cirrus** clouds are delicate, wispy, and filamentous (Fig. 14–13). Because they develop entirely above the freezing level, cirrus clouds are composed of ice particles. If cirrus-like clouds occur as a thin continuous veil of more or less uniform texture (often producing a halo around the moon) they are called **cirrostratus** (Fig.

Figure 14–11 Characteristically layered clouds include stratus *(upper left),* altocumulus *(upper right),* stratocumulus *(lower left), and* nimbostratus *(lower right). (Courtesy of NOAA.)*

14–14). The first indication that bad weather is imminent may be the advance of a thin, even sheet of cirrostratus. **Cirrocumulus** clouds occur as grainy, white, rippled patches. Their rippled aspect imparts a distinctive appearance that is called a "mackerel sky" because of its resemblance to the pattern of bluish and silvery scales on a mackerel.

There are two genera of clouds that are not readily placed in the high, middle, or low categories because they range vertically through these levels from near the earth's surface to about 20,000 meters. These are classified as vertical clouds in Table 14–2. Included here are the familiar *cumulus* and *cumulonimbus* cloud types. **Cumulus** clouds are readily recognized by their puffy, heaped-up, cauliflower-like appearance (Fig. 14–15). Frequently, they exhibit flat bases that indicate the position of the condensation level. Cumulus clouds are the result of condensation within rising unstable air parcels. Mostly, these clouds are associated with fair weather, but with time they may be-

Figure 14–12 *Altocumulus clouds. (Courtesy NOAA.)*

come overdeveloped and grow into majestic thunderheads, the **cumulonimbus** clouds (Fig. 14–16) associated with thunderstorms.

Fog

When air near the surface of the earth is cooled sufficiently, it becomes saturated and condenses to form the microscopically small droplets of water that comprise fog. The cooling required for the formation of fog may occur in a variety of ways giving rise to a classification of fog types into radiation fog, advective fog, upslope fog, frontal fog, and ice fog.

Radiation fog, also called ground fog, develops as the earth's surface cools by loss of radiation to space. Most often radiation fog develops at night, when air in contact with the cold ground is cooled below the dew point. The cool dense air is sufficiently heavy to stay close to the ground and may flow into valleys forming thick pockets of fog. Radiation fog may be particularly

Figure 14–13 *Cirrus and cirrostratus clouds occur at high altitudes. (Courtesy of NOAA.)*

Figure 14–14 Cirrostratus clouds often produce a halo around the moon. (Courtesy of NOAA.)

Figure 14–16 Cumulonimbus clouds bring thunderstorms.

thick around cities because of the abundance of smoke and dust particles that serve as condensation nuclei.

Advective fog is generated by warm moist wind blowing over the cold surface of the ground (or cold air blowing over a warm moist surface). Such fogs form along seacoasts where breezes carry masses of warm air over regions of cold water. The dense fogs off the Grand Banks of Newfoundland originate in this way. Cold air moving over a warm surface may also produce advective fog of a type sometimes called steam fog (also evaporation fog and sea smoke). In the early morning when the air is still chilled, this fog does indeed resemble steam as it rises off the surface of lakes and ponds (Fig. 14–17).

Figure 14–15 Cumulus clouds are characterized by a billowy appearance and flat base.

Upslope fog is formed as air sweeps up rising land slopes. As the air ascends diagonally to a level where atmospheric pressure is less, it is cooled by expansion and begins to condense. Because of the manner in which they develop, upslope fogs are most prevalent on the windward slopes of highland areas, as well as along the Pacific and Gulf coasts of the United States.

Frontal fogs are developed along the boundaries (fronts) of two great bodies of air that differ in temperature and moisture characteristics. These large bodies of air, called air masses, will be described more fully in Chapter 16. If the air mass is warm and is advancing against and replacing cold air, the boundary is called a warm front. Along a warm front, warm air tends to climb up and over the cooler air. Rain falling from the overriding warm air passes through cold air along the front, evaporating as it does so. The evaporated moisture causes saturation or supersaturation in the cold air mass, with resulting development of fog.

The fifth kind of fog, **ice fog,** occurs at temperatures below about −30°C. Ice fog consists of myriads of tiny ice crystals formed from water droplets in the atmosphere. The airport in Fairbanks, Alaska, where temperatures are frequently as cold as −45°C, is regularly blanketed with ice fog generated by idling aircraft engines.

Figure 14–17 *Advective fog rises from the surface of a lake. (Courtesy of NOAA.)*

PRECIPITATION

Mechanisms for precipitation

If there are sufficient condensation nuclei in the atmosphere, saturated air will condense on these nuclei and form the myriads of tiny water droplets and ice crystals that are the essential ingredients for the development of precipitation. **Precipitation** is any form of liquid or solid water particles that fall from the atmosphere to the ground. For precipitation to occur, it is necessary to bring together atmospheric water to form water drops or ice particles that are large enough not only to escape evaporation before they reach the ground but also to overcome air friction and updrafts that might otherwise keep them from falling. The reason that the size of water drops is important becomes apparent when we realize that the average water droplet in a cloud is less than 10 microns in diameter and would require two days to fall the relatively short distance of 1000 meters. The drop would certainly be evaporated before reaching the ground. About 100,000 droplets must join to form drops large enough to sustain themselves on the downward journey.

There are two processes that are important in producing water drops or ice crystals sufficiently large to become precipitation. The first is called the **collision-coalescence** or **warm cloud process,** and the second is called the **Bergeron** or **cold cloud process.**

The collision-coalescence process is influenced by the unequal size of water droplets in a cloud. Because of differences in the nature of condensation nuclei and the rates at which condensation proceeds in various parts of a cloud, the big droplets may be several times larger than the smallest. As the air in the cloud swirls about, the larger droplets, because of their larger mass and greater inertia, follow different paths than the smaller droplets. Because they do not move in unison, they frequently collide and coalesce with the smaller droplets. Once the droplets have acquired sufficient mass to begin falling freely through the cloud, their growth is further promoted as they overtake and collect smaller droplets that happen to be in their way, or as they encounter droplets in the updraft. The pattern of air flow (Fig. 14–18) in the wake of the falling larger droplet causes additional smaller droplets to be added at the rear. (Similar patterns of air flow bring dust into the open rear window of a station wagon traveling on a dirt road.) Repeated additions of water may cause a falling drop to become so large that it shatters into smaller drops, and each of these begins to collect smaller droplets in a sort of chain reaction.

The Bergeron or cold cloud process requires clouds containing supercooled water droplets (i.e., below 0°C), ice crystals, and water vapor. Such clouds

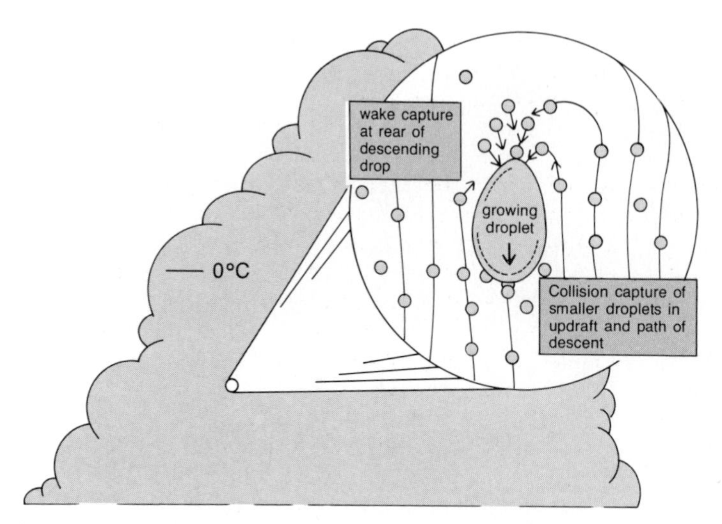

Figure 14–18 Development of precipitation by the collision-coalescence process.

are the usual kind in the middle latitudes. The temperature of water droplets in these clouds is often as low as −10°C, yet the water does not freeze because of the relative sparseness of the nuclei needed to initiate freezing. Below temperatures of −10°C, clouds are composed mostly of ice crystals.

The Bergeron process begins with the formation of a cloud composed of both the required supercooled water droplets and ice crystals. In such a cloud, the ice crystals will grow at the expense of the water droplets. This is because the vapor pressure of the liquid water droplets is much greater than the vapor pressure of the ice crystals. The molecular explanation is that it is much easier for water molecules to escape from a liquid, where they have considerable mobility, than from solid ice, where they are held rather firmly in place. Thus, water is evaporated readily from the droplets, whereupon they cause saturation, and the resulting vapor is deposited on the ice crystals (Fig. 14–19). The crystals then proceed to grow, and after acquiring sufficient mass, they begin to fall as snowflakes. As indicated above, because they are solid, snowflakes are less prone to evaporation during their fall than are water drops. Should the snow encounter warm air during the descent (as frequently occurs), the snow is converted to rain.

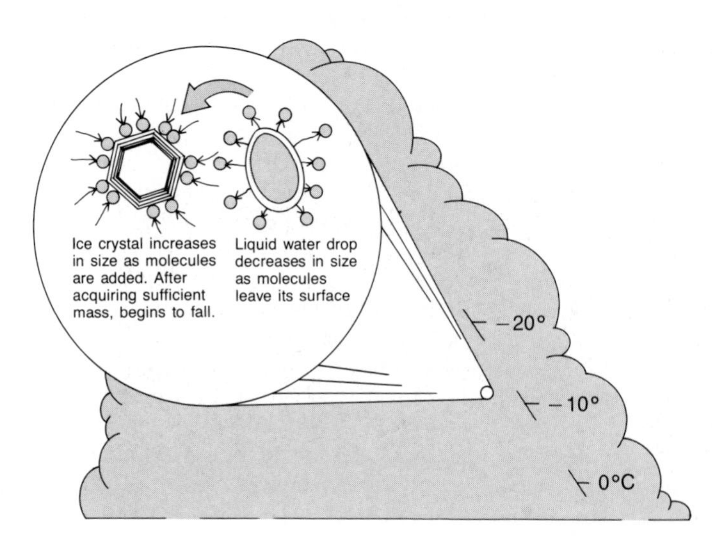

Figure 14–19 Development of precipitation by the Bergeron or cold cloud process.

It is now apparent why the Bergeron process is also termed the cold cloud process, for it requires supercooled water droplets and ice crystals. It is the more prevalent of the two processes and operates in temperate and cool regions. In contrast, the collision-coalescence mechanism operates predominantly in tropical regions where temperatures within clouds do not always exceed the freezing level.

Types of precipitation

Rain, of course, is the most common kind of precipitation of water in liquid form. Raindrops range in size from 0.5 mm to about 5 mm in diameter. There is a natural limit to their size, for drops larger than 5 mm are unstable and break up into smaller ones as they fall. Steady, continuous rains usually fall from stratus clouds, whereas showers that stop and start or vary in intensity are more characteristic of convective clouds.

Water drops less than 0.5 mm in diameter constitute a wet mist called **drizzle.** Like steady rain, drizzle is a product of stratus clouds and is normally a warm cloud process. It rarely accumulates any faster than about 1 mm per hour.

The most common type of solid precipitation is **snow.** Snow develops from water vapor that is deposited directly as a solid. (In this process, called sublimation, the liquid state is bypassed.) The initial snowflake is a tiny hexagonal crystal that grows more rapidly at the ends of the six rays because these areas are more openly exposed to the supply of water vapor molecules (Fig. 14-20).

One form of solid precipitation rarely seen except in high mountains and polar regions is **ice needles,** or "diamond dust." The tiny needles are so light they appear to float on air and often sparkle as they reflect sunlight. Ice needles may originate in stratus clouds or even appear from a sky that is clear.

Snow pellets are opaque white spherical particles that sometimes fall from convective clouds in mountainous areas. Sometimes also called *graupel,* they develop when supercooled water droplets freeze on ice crystals. **Ice pellets,** such as hail, differ from snow pellets in that they are harder, have a greater resistance to crushing, and are transparent to translucent. Most ice pellets in sleet originate as rain but freeze as they fall through a cold layer near the ground. Occasionally, the subfreezing zone of air near the earth's surface is not sufficiently deep to freeze the drops. The supercooled raindrops, however, do freeze on striking the cold ground. This kind of precipitation is called **freezing rain** and results in the thin, slick ice coating called **glaze.** Glaze makes driving hazardous and often collapses tree branches and power lines (Fig. 14-21).

Hail (Fig. 14-22) is a solid form of precipitation formed when strong, rising, convective currents (as in a cumulonimbus cloud) carry raindrops to high levels where they freeze. The frozen drops then fall again and take on one or more additional coatings of ice as they pass through levels containing supercooled water droplets. Repeated ascent and descent of the ice particle produces the concentric layers of ice in the hailstone. Hailstones are roughly spherical to irregular in shape and range between 5 and 50 mm in diameter. A distinctive characteristic of many hailstones is their alternating concentric layers of clear and opaque ice. The clear layers form when the growing hailstones encounter particularly moist air so that liquid water covers the ice and subsequently freezes. The opaque layers derive their opacity from numerous air bubbles that become entrapped when coalescing cloud droplets freeze immediately on contact with the hailstone.

All forms of precipitation can occasionally cause problems. Excessive rainfall may result in flooding. Too much snow can slow or even stop automobile traffic and curtail a host of everyday activities. Heavy

Figure 14-20 The branched hexagonal plates of snow crystals. (Courtesy NOAA.)

Figure 14–21 *The result of an ice storm is a heavy glaze on streets, trees, and power lines. (From Navarra, J.G.,* Atmosphere, Weather, and Climate. *Philadelphia, Saunders College Publishing, 1979. Courtesy of ESSA.)*

Figure 14–22 *Hailstones, all larger than the golf ball. (From Navarra, J.G.,* Atmosphere, Weather, and Climate. *Philadelphia, Saunders College Publishing, 1979. Courtesy of ESSA.)*

snow accumulation may also cause dangerous avalanching in mountainous areas. Glaze makes driving hazardous and may result in extensive damage to power lines and trees. Hailstones can dent the metal of automobiles, shatter windows, and devastate crops. Small wonder that many of us listen faithfully to the evening weather report so as to prepare for what the atmosphere may have in store for us tomorrow.

Summary An important component of the atmosphere is moisture in the form of water vapor. Water vapor is supplied to the atmosphere by the evaporation of liquid water, sublimation of ice, and transpiration by plants. The capacity of the atmosphere for absorption and retention of water is governed primarily by temperature. In general, the higher the temperature, the greater the amount of water vapor that can be absorbed and retained.

Evaporation of water or sublimation of ice requires energy so that water molecules can escape from the main body of the liquid and enter the air. This energy is in the form of heat, which is taken from the neighboring environment, thus cooling it. The heat needed to convert liquid water to water vapor is called latent heat of vaporization.

Humidity refers to the amount of water vapor in air. Absolute humidity is the mass of water vapor in a unit volume of air, whereas relative humidity refers to the ratio of the water vapor in the air to its potential capacity for holding water vapor at any given temperature. Dew point is the temperature to which a given parcel of air must be cooled to reach 100 percent relative humidity and begin to give up its moisture to condensation.

When water vapor is steadily added to air, the air will eventually reach a condition of saturation at a particular temperature. If still more water is added beyond that point, or if the air is cooled, the water vapor will condense as liquid or be deposited as ice. When condensation occurs, the latent heat taken in during evaporation is released. For condensation to occur in the atmosphere, there must be surfaces to which the water molecules can attach themselves. Such surfaces are provided in the atmosphere by salt, particles derived from combustion, dust, and ice—all of which constitute condensation nuclei.

Among the familiar atmospheric manifestations of condensation are clouds, fogs, and such forms of precipitation as snow, rain, and hail. Clouds are formed mostly by upward movements of air, which are then cooled by expansion. As air is cooled beyond saturation, condensation begins around condensation nuclei. The cooling of air by expansion is called adiabatic cooling. Adiabatic cooling occurs at rates of 1°C per 100 m of ascent for dry air and about 0.7°C per 100 m for moist air. Clouds also develop from advective cooling, as when warm air is blown across cold surfaces, and from various kinds of forced lifting, as when a warm air mass is shunted above a colder dense mass.

The stability of the atmosphere refers to the degree to which it is susceptible to vertical movements and convection. In general, air is considered stable if vertical movements are impeded and unstable if they are occurring. Clouds associated with unstable air tend to be massive, towering, and capable of yielding heavy precipitation. Clouds of stable air tend to be blanket-like and provide little or no precipitation.

Precipitation is water that falls from the atmosphere to the earth's surface. For precipitation to occur, cloud droplets or ice crystals must grow to sufficient size to permit a rapid rate of fall and to withstand complete evaporation during the descent. The Bergeron and collision-coalescence processes are both important in the origin of precipitation. The former provides for growth of large ice crystals in a cloud by deposition of moisture derived from evaporating water droplets; the latter depends upon random collision to gather small droplets into rain-size drops.

QUESTIONS FOR REVIEW

1 What is the effect of water vapor on atmospheric pressure?

2 What is latent heat? What happens to latent heat during condensation? Why is more latent heat required for sublimation than for evaporation?

3 Distinguish between absolute humidity and relative humidity. Air at 20°C can hold a maximum of 14 grams of water per kilogram. What is the relative humidity when the temperature is 20°C but the air holds only 7 grams of water per kilogram?

4 What is an adiabatic temperature change? What is the dry adiabatic lapse rate? Why is it greater than the moist adiabatic lapse rate?

5 If the reading given by a dry bulb thermometer is 22°C and the wet bulb reads 16°C, what is the relative humidity? (Use Table 14–1.)

6 Given the same conditions of temperature and humidity, why does condensation occur more readily on the ground than in the atmosphere? Why is silver iodide used in cloud seeding?

7 Describe two ways in which forceful lifting of a parcel of air may occur.

8 What is meant by atmospheric stability? Distinguish between absolute stability and absolute instability. Under which conditions would thunderstorms be most likely to occur?

9 Describe and contrast cirrus, cumulus, and stratus clouds and indicate the nature of precipitation that each might bring.

10 Distinguish between clouds and fogs. What is the cause of radiation fog, advective fog, and frontal fog?

11 Describe the formation of precipitation according to the Bergeron and collision-coalescence processes. What is the difference between precipitation and condensation?

12 Why are the lee sides of coastal mountain ranges often arid? What is the term that describes this condition?

SUPPLEMENTAL READINGS AND REFERENCES

Bentley, W.A., and Humphreys, W.J., 1962. *Snow Crystals.* New York, Dover Publications.

Eagleman, Joe R., 1980. *Meteorology.* New York, D. Van Nostrand Co.

Lutgens, F.K., and Tarbuck, E.J., 1982. *The Atmosphere.* Englewood Cliffs, N.J., Prentice-Hall Inc.

Mason, B.J., 1979. The growth of snow crystals, in *The Physics of Everyday Phenomena,* Sci. Amer., San Francisco, W.H. Freeman & Co.

McDonald, J.E., 1979. The shape of raindrops, in *The Physics of Everyday Phenomena,* Sci. Amer., San Francisco, W.H. Freeman & Co.

Schaefer, V.J., and Day, J., 1981. *A Field Guide to the Atmosphere.* Boston, Houghton Mifflin.

15

the restless atmosphere

Observe which way the hedgehog builds her nest,
To front the north or south, or east or west;
For 'tis true what common people say,
The wind will blow the quite contrary way.
Poor Richard's Almanack, 1733

Prelude One of the most noticeable characteristics of air is that it is almost constantly in motion. At times the motion is so gentle as to be barely discernible, but at other times air may move with the gusty violence of a gale or hurricane. Whether fierce or moderate, the movement of air is of vital importance. It serves to distribute the sun's heat around the earth, to transport moisture from the seas to the lands, and to bring the combination of temperature and pressure conditions that characterizes our weather and climate.

The movements of air, on both a global and a local scale, are ultimately powered by the sun. Temperature and pressure variations that result from variations in the solar energy received in different areas are the immediate cause of air movements.

A seemingly anachronistic windmill at an oil refinery conserves fossil fuel by using the energy from wind to pump water for nearby lawns and gardens. (Courtesy of E.I. du Pont de Nemours & Co.)

AIR IN MOTION

When one thinks of moving air, the word wind comes immediately to mind. All air movements, however, do not qualify as wind. **Wind** is a predominantly *horizontal* motion of air. Such terms as updraft and downdraft are employed for vertical motions of air like those in local convective air circulation.

Wind and wind measurement

Winds are named according to the direction from which they come. Thus, if a wind comes out of the west and moves toward the east, it is called a *west wind*. One determines wind direction from that most venerable of all wind instruments, the wind vane (Fig. 15-1). There are two conventional schemes for recording wind direction. The older and less commonly used convention refers direction to 1 of 16 points on the compass (Fig. 15-2A), whereas the newer system employs numbers to indicate wind direction to the nearest tens of degrees (Fig. 15-2B). Thus, 09 represents 90° or an east wind, 36 refers to 360° or a north wind, 18 indicates 180° or a south wind, and 00 represents a dead calm.

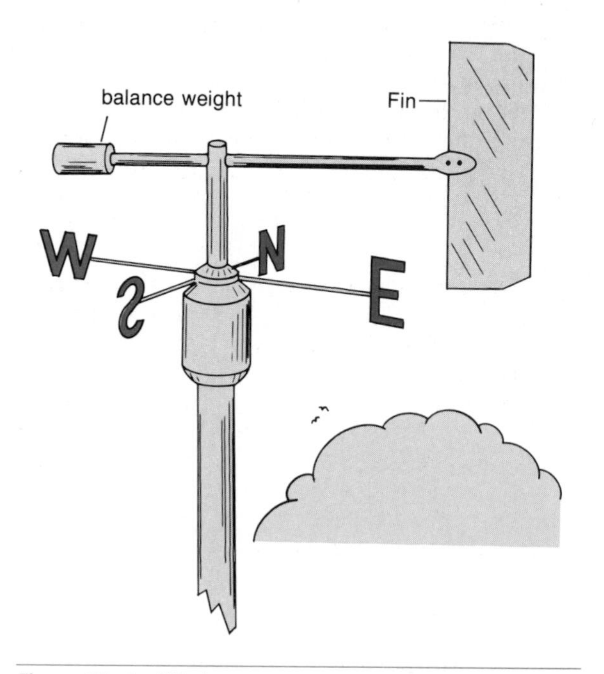

Figure 15-1 *Wind vane.*

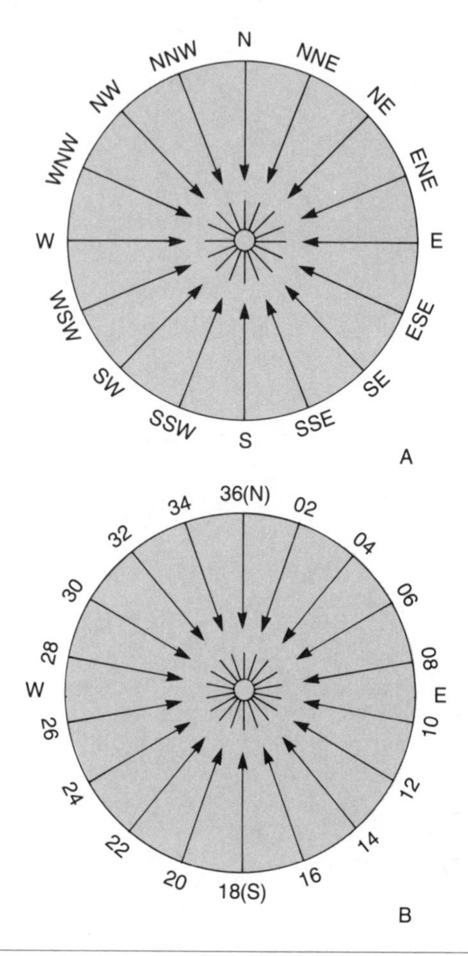

Figure 15-2 *Two systems for recording wind direction. In A, wind direction is based on points on the compass. Where greater precision is needed, directions are given in tens of degrees, as in B.*

A useful way to graphically show the distribution of wind direction experienced at a particular location over a considerable period is the **wind rose** (Fig. 15-3). In the wind rose, lines radiate from a central point to each of the eight compass directions. The lengths of the lines are scaled to represent the monthly or annual percentage of time that winds blew from each directional segment. For example, from the wind rose illustrated in Figure 15-3, one knows that the wind blew 6 percent from the north, 14 percent from the northeast, 10 percent from the east, 15 percent from the southeast, 21 percent from the south, and so on. The numeral 4 in the center of the rose indicates that calm prevailed 4 percent of the time. Wind roses

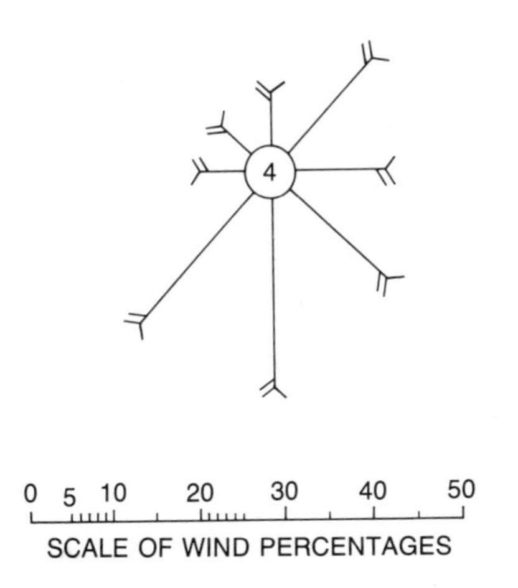

Figure 15–3 *Wind rose for a location in South Carolina. (From Navarra, J.G.,* Atmosphere, Weather, and Climate, *Philadelphia, Saunders College Publishing, 1979.)*

Figure 15–4 *A three-cup anemometer. (Courtesy of NOAA.)*

plotted on maps can be of considerable value in studying air pollution, laying out airport runways, and locating wind and snow screens.

No less important than a wind's direction is its speed. Wind speed is measured by one of several varieties of **anemometers** (Fig. 15–4). It is standard practice to express speed in **knots,** one knot being one nautical mile per hour and equivalent to a speed of 1.15 miles per hour (1.85 km/hr). One can estimate wind speed without instruments by observing the effect of wind on trees, leaves, waves, and other objects. The scale used in making these estimates is called the Beaufort scale (Table 15–1), after the British admiral who devised the scheme in 1806.

The origin of the wind

The energy needed to produce wind is derived from the sun. When the sun warms one part of the earth more than another, the warm air expands and rises. This warm rising air is replaced by cooler (hence denser) air, which moves in as wind at relatively low altitude from adjacent regions (Fig. 15–5). As described earlier, because air molecules are closer together in cool air, a column of cool air will weigh more than a column of warm air of the same volume. As a result, a mass of cool air above the ground usually constitutes a **high-pressure area,** whereas warm air masses often constitute **low-pressure areas.** Air moves as wind from high-pressure areas to low-pressure areas. The rate of change of pressure across a given horizontal distance is termed the **pressure gradient** and is usually expressed in millibars per 100 miles (160 km).

One may construct a map showing the variation in atmospheric pressure across an area at any given time by connecting points of equal pressure with lines called **isobars.** Suppose the isobars on such a map extend north-south, as shown in Figure 15–6. The high pressure on the east will cause air to initially flow toward the west. The pressure gradient on the east is about two units of pressure per 100 km, whereas that on the west side of the map area is four units per 100 km. Because of the steeper gradient in the west, those winds are likely to be stronger. Thus, closely spaced isobars indicate marked pressure differences and relatively high rates of air flow (wind speed), whereas widely spaced isobars indicate a gentle pressure gradient and relatively light winds. When pressure differences over a region are negligible, calm prevails.

Factors affecting air in motion

As noted earlier, winds blow because of differences in pressure. The greater the pressure difference, the

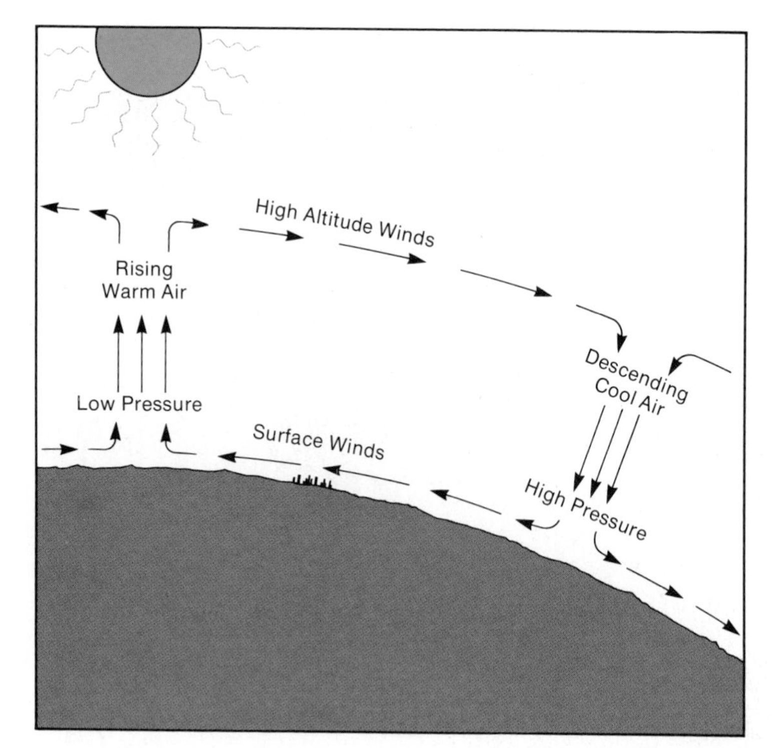

Figure 15–5 The generation of wind as a consequence of unequal heating.

greater the wind's speed. The force that drives the air from a high-pressure area to a low-pressure area across the pressure gradient is called the **pressure-gradient force.** This force is clearly the most important factor in producing wind. After air has been set in motion by the pressure-gradient force, however, other factors come into play that modify the direction and speed of wind. Among these modifying factors, the Coriolis effect and friction are particularly important.

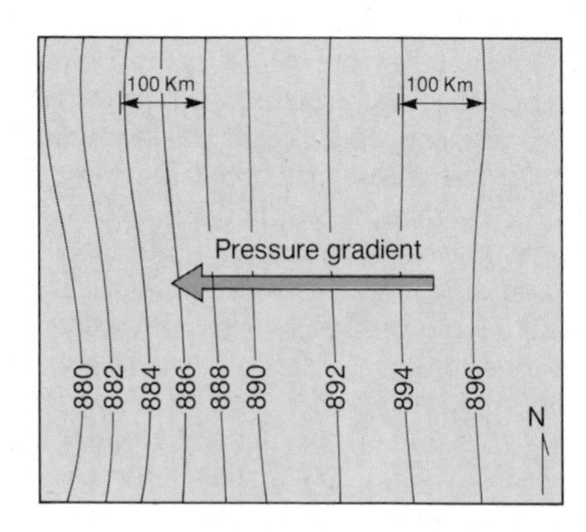

Figure 15–6 Map view of pressure gradient from a high-pressure area to a low-pressure area. The lines connecting points of equal pressure are called isobars. *In general, winds are relatively stronger where isobars are more closely spaced.*

If the earth did not rotate, wind would blow in the same direction as the pressure-gradient force. The earth, of course, does rotate. At the equator, the earth's surface is moving eastward with a velocity of about 1600 km/hr, whereas areas near the poles move very slowly. The rotation causes winds to be deflected toward the right in the Northern Hemisphere and toward the left in the Southern Hemisphere. This deflecting action is the result of the Coriolis effect. To understand the reason for the Coriolis effect, imagine you are an observer far out in space watching a rocket that has been fired toward the south from the North Pole. From space, the rocket would indeed be seen to move in a straight line. If you were to plot its course across the surface of the earth, however, it would appear to curve gradually westward. This is because, the earth is turning from west to east beneath the rocket as it speeds southward. Air movement is subject to this very same effect. For example, if the pressure force

TABLE 15–1 **SIMPLIFIED BEAUFORT SCALE**

Beaufort number	*Explanatory titles*	*Effect of wind* *On land*	*At sea*	*Miles per hour*	*Knots*
0	Calm	Smoke rises vertically	Sea completely smooth	Less than 1	0–0.9
1	Light air	Smoke drifts	Small ripples	1–3	1–3
2	Light breeze	Leaves rustle	Waves are short and pronounced	4–7	4–6
3	Gentle breeze	Leaves and twigs in constant motion	Crests begin to break	8–12	7–10
4	Moderate breeze	Raises dust, loose paper; small branches are moved	Long waves—many whitecaps	13–18	11–16
5	Fresh breeze	Small trees begin to sway	White foaming crests everywhere	19–24	17–21
6	Strong breeze	Large branches in motion overhead begin to whistle; umbrellas used with difficulty	Larger waves form; crests more extensive	25–31	22–27
7	Moderate gale	Whole trees in motion; difficult to walk	Foam blows in streaks	32–38	28–33
8	Fresh gale	Twigs break off trees	↓	39–46	34–40
9	Strong gale	Slight damage to roof and homes	↓	47–54	41–47
10	Whole gale	Trees uprooted	High waves, great foam patches	55–63	48–55
11	Storm	Widespread damage	Ships hidden in troughs of waves	64–73	56–63
12	Hurricane	Devastation		Above 74	Above 64

From J.G. Navarra, *Atmosphere, Weather, and Climate,* Philadelphia, Saunders College Publishing, 1979.

causes an air mass located in the Northern Hemisphere to move southward, the earth's rotation beneath the moving air mass will cause it to turn toward the west (that is, to the right). Conversely, an air mass moving northward in the Northern Hemisphere (like the prevailing southwesterlies in Figure 15–7) will be deflected eastward. The deflection caused by the Coriolis effect increases with increasing wind speed. It also varies with latitude, being zero at the equator and increasing poleward.

When generated by the pressure gradient alone, wind moves perpendicular to the isobars. The Coriolis effect, however, causes winds to turn gradually until they blow approximately parallel to isobars (Fig. 15–8). When this occurs, the pressure gradient and Coriolis effect are acting in opposite directions, and one is balanced against the other. A wind that blows parallel to isobars because of this state of balance is termed **geostrophic.** Winds far enough aloft (one kilometer or so) that they are unaffected by frictional drag with the earth's surface are nearly always geostrophic.

A second major factor that modifies wind patterns is friction. The frictional drag resulting when air moves across the earth's surface tends to slow winds. The effect is more pronounced for air moving across rugged terrain than level terrain, and more pronounced for air moving over land than expanses of ocean. Friction not only lessens wind speed, but also weakens the Coriolis effect. You will recall that the Coriolis effect lessens as wind speed decreases. Thus, as friction weakens the Coriolis effect by reducing wind speed, the wind becomes more responsive to the tug of the pressure-gradient force. Air therefore moves diagonally across isobars toward low pressure in a direction that is a compromise between the parallel-to-isobar Coriolis flow and the perpendicular-to-isobar pressure-gradient flow (Fig. 15–9). In this way friction not only reduces wind speed but also is an important influence on the direction in which near-surface winds blow. Under normal conditions, frictional forces rarely affect winds above an altitude of about 900 meters. The lower altitude movements are sometimes called **friction-layer** winds.

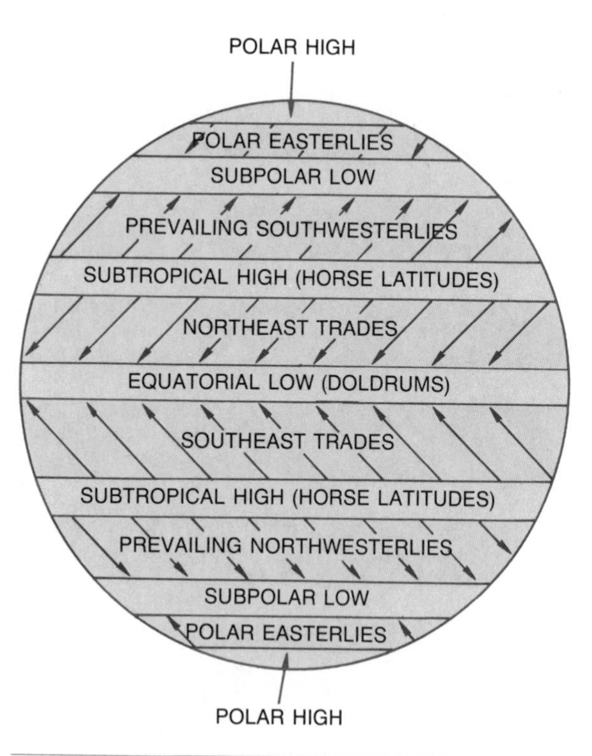

Figure 15–7 Prevailing surface winds and pressure patterns. (From Navarra, J.G., Contemporary Physical Geography. *Philadelphia, Saunders College Publishing, 1981.)*

The effect of friction in changing wind movement is not always horizontal. Friction can induce turbulence and generate strong vertical movements as well. Dust devils and waterspouts are evidence of such turbulence.

CYCLONES AND ANTICYCLONES

A **cyclone** is an atmospheric pressure distribution in which there is a low central atmospheric pressure relative to surrounding areas. It is shown graphically by a system of isobars enclosing the central area of low pressure so as to form a roughly circular or oval pattern (see Fig. 15–10). As winds blow inward from all sides toward the center of cyclones, the Coriolis effect causes the winds to be deflected toward the right in the Northern Hemisphere (and to the left in the Southern Hemisphere). At the same time, the warm air in the cyclone's center rises to higher and cooler areas where clouds form and precipitation may ensue. Indeed, cyclones are characteristically associated with poor weather. As an extreme example, hurricanes (Fig. 15–11) are severe tropical cyclones.

A pressure distribution in which there is high central pressure relative to the surroundings is called an anticyclone. Circulation of air in an anticyclone is opposite to that in a cyclone. In the Northern Hemisphere, air moves downward through the center of the anticyclone and spirals outward with a clockwise deflection (see Fig. 15–10). The descending air of anticyclones tends to be cool and dry, and therefore anticyclones or **highs** are generally associated with clear weather. Even the winds associated with anticyclones are gentler, largely because the more widely spaced isobars result in lower pressure gradients.

In the middle latitudes, cyclones and anticyclones are carried along from west to east by prevailing westerly winds. Normally, they require about a week to complete a transcontinental journey across North America. Television weather broadcasters depict their

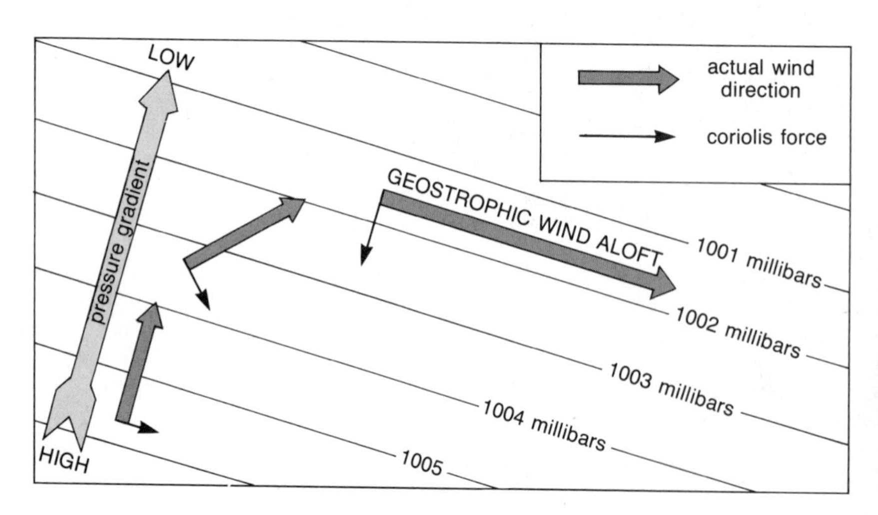

Figure 15–8 At about 900 meters above the earth's surface, frictional forces on wind are negligible. If the pressure gradient is constant, wind direction will deviate until it is moving along the isobars and perpendicular to the pressure gradient. Thus, surface isobars indicate wind direction at the 900-meter level.

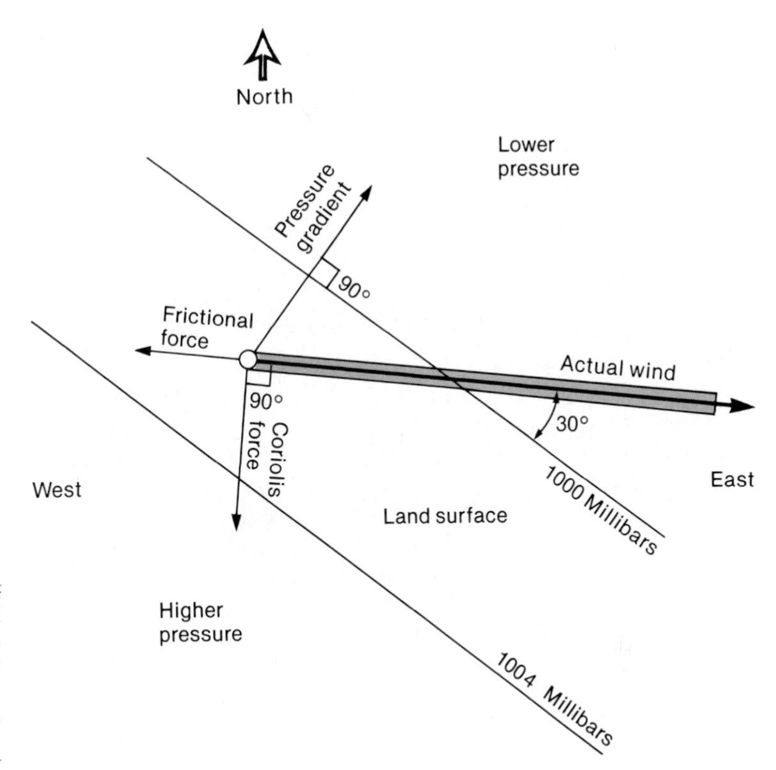

Figure 15–9 *Near the surface, the effect of friction on wind is to deflect it diagonally across surface isobars. In this example, the wind cuts across a surface isobar at a 30° angle. (From Navarra, J.G.,* Atmosphere, Weather, and Climate. *Philadelphia, Saunders College Publishing, 1979.)*

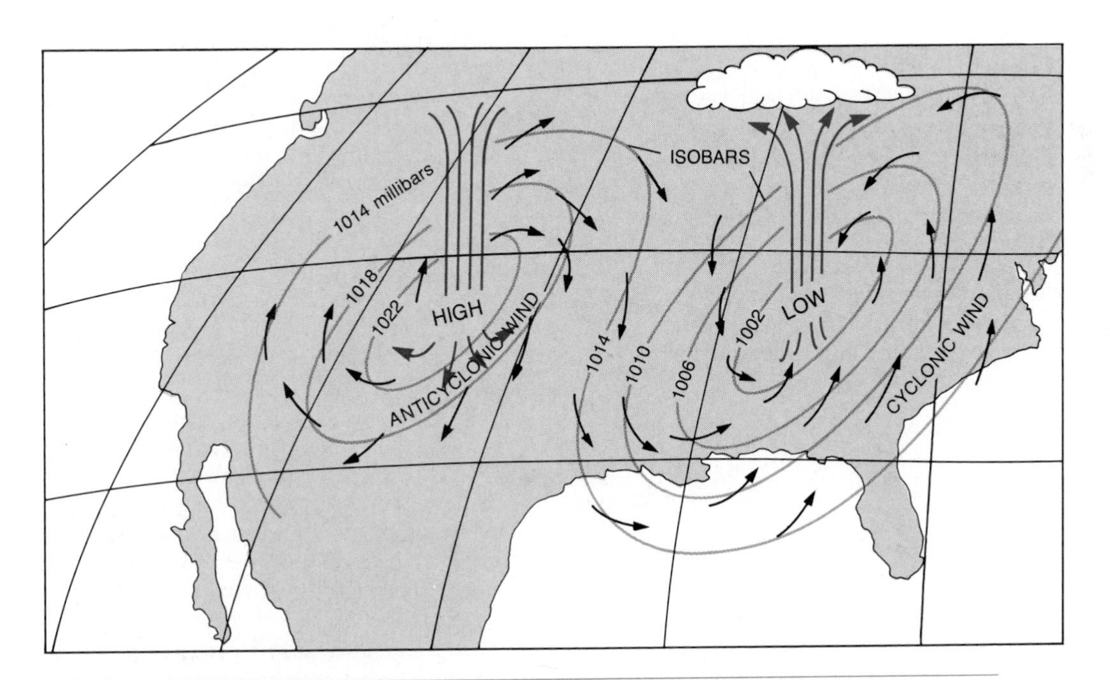

Figure 15–10 *Pattern of winds around a low-pressure* cyclone *and a high-pressure* anticyclone.

Figure 15–11 National Oceanographic and Atmospheric Administration satellite photograph of the United States on September 12, 1979. Hurricane Frederick is clearly visible in the Gulf Coast region.

positions on evening news programs. It becomes apparent from these broadcasts that the **lows,** or cyclones, are the real weather rascals, for they often bring unpleasant weather. In the United States, the broad belts over which these lows travel are referred to as **storm tracks.** The west to east migration of lows explains why weather forecasters in the middle latitudes base their predictions on conditions in the west, for these storms are approaching.

Cyclones and anticyclones are large-scale features often covering more than a million square kilometers. In general, they are less intense and travel eastward at lower speeds in the summer than in the winter. They also change locations somewhat with the seasons, shifting northward in the summer and southward in the winter (in the Northern Hemisphere).

THE GLOBAL CIRCULATION OF AIR

A simplified model

The great wind systems operating across the surface of the earth are components of huge convection cells in the atmosphere. Global wind patterns are quite complex, for they represent responses not only to latitudinal temperature changes but also to the earth's rotation around an axis, the unequal distribution of lands and seas, and the wide variations in the heights of physiographic features. To aid our understanding of the global wind system, consider for a moment what would happen if the earth did not rotate on its axis, was covered entirely by water, and experienced heating along the equator and cooling at the poles. On such a hypothetical earth, air warmed at the equator would expand and rise. As this occurred, cooler air would move in from the polar regions to fill the space vacated by warm air, and a relatively simple pattern of convection cells would be established resembling those shown in Figure 15-12. The pressure differences from high pressure at the poles and low pressure at the equator would generate two large air-mass movements. In the Northern Hemisphere, the predominant surface winds would blow southward, and in the upper atmosphere, warm air would flow toward the North Pole. The situation would be similar but with reversed directions in the Southern Hemisphere.

Of course, because the earth rotates, the simple atmospheric circulation described above does not resemble the more complex actual pattern, which is composed of many smaller convection cells. Warm air

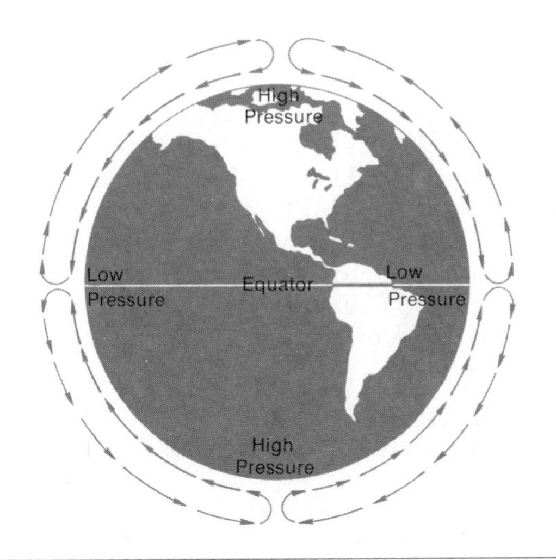

Figure 15–12 Wind systems on a hypothetical nonrotating earth. (From Turk, J., and Turk, A., Physical Science. *Philadelphia, W.B. Saunders Company, 1977.)*

rising at the equator first divides and then begins a high-altitude journey toward the poles. It does not, however, move directly north or south, but in response to the Coriolis effect veers toward the east (Fig. 15–13). When the air has reached about 30° north and south latitude, it is traveling directly eastward, has cooled, and begins to descend, producing the subtropical high-pressure areas (see Fig. 15–7) known also as the **horse latitudes.** The near-surface winds in the horse latitudes are light and variable, rainfall is slight, and skies are generally clear. In this region, bygone sailing ships were often becalmed. As water and food aboard the ships became scarce, horses carried as cargo died and were thrown in the ocean. The many floating carcasses of horses provided the odd name for these high-pressure areas. On land, such deserts as those in North Africa and the southwestern United States characterize the horse latitudes.

The air masses that descend at the horse latitudes return to the equator at low altitudes. Because of the Coriolis effect, they shift and blow from the northeast in the Northern Hemisphere and from the southeast in the Southern Hemisphere. These reliable winds are called **trade winds,** because they once provided trade routes to the west for merchant sailing vessels. Winds like the trade winds, which blow from one direction more frequently than any other, are termed **prevailing winds.**

Poleward from the subtropical high-pressure regions are the wind belts known as the **westerlies.** In these regions, the prevailing wind directions are from the southwest in the Northern Hemisphere and from

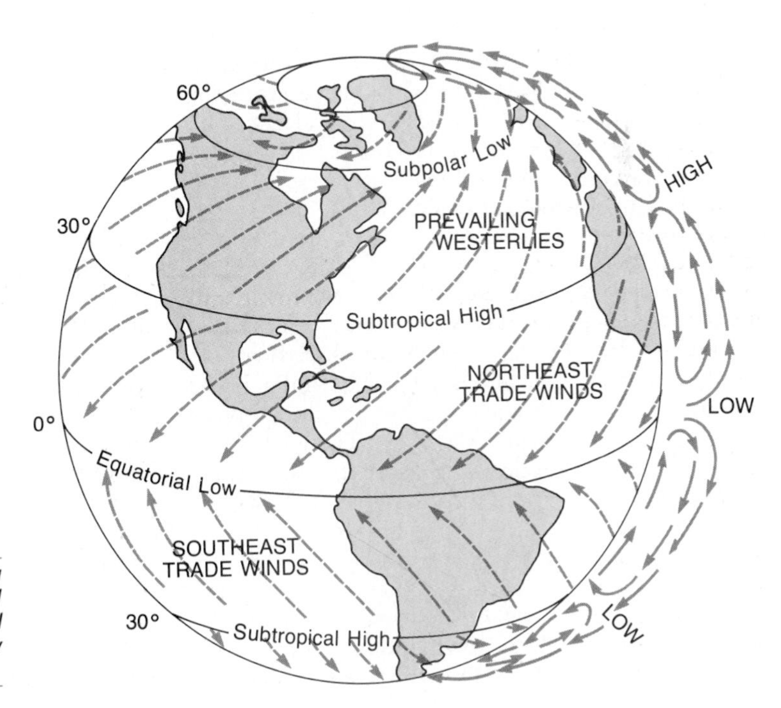

Figure 15–13 Simplified model of prevailing surface winds and location of major high- and low-pressure bands. (After Ramsey, W.L., and Burckley, R.A., Modern Earth Science. *New York, Holt, Rinehart, and Winston, 1961.)*

the northwest in Southern Hemisphere. Subpolar low-pressure belts are developed at latitudes of about 60° north and south. Poleward of these areas, the air is cooled and compressed to form high-pressure zones. Cold air in this region sinks and moves generally southward in the Northern Hemisphere and northward in the Southern Hemisphere. The winds are deflected and are appropriately designated the **polar easterlies.** At latitudes of about 40° to 60°, the polar easterlies encounter the northern margin of the belt of westerlies. The warmer westerlies are forced above the colder polar air. The interaction of the warm and cool winds produces a stormy belt called the **polar front.** The temperature contrast along the polar front gives rise to cyclonic storms that are carried along in the westerlies. These cyclones strongly influence weather in the middle latitudes.

Because the sun's vertical rays shift north and south of the equator during each year, the pressure belts and wind systems also change positions. The shift in pressure belts, however, is less than the 23.5° shift of the sun's vertical rays. In fact, the average change is only about 6°. Even this seemingly small amount, however, can subject some locations to different wind zones during different times of the year.

From this brief survey of atmospheric circulation, we see that three factors determine the horizontal and vertical movements of air masses. Of predominant importance is the sun, which provides the energy needed to drive the system. A second related factor is the change in temperature from the equator to the poles. Finally, currents of air generated by temperature differences are influenced by the Coriolis effect. The entire pattern of circulation acts to moderate the temperature differences on earth and carry "excess" heat from the equator toward the poles.

In this simplified view of the global circulation of air, it is apparent that the atmosphere is characterized by four distinct pressure zones. There are two highs in each hemisphere (the subtropical and polar highs) where dry air descends and flows outward at the earth's surface, producing the prevailing winds. There are also two lows (the equatorial low and the subpolar low located at about 60° of latitude). These low-pressure zones, like the cyclones described earlier, are associated with converging and ascending airflow that brings precipitation.

The globe-encircling low-pressure area that lies near the equator is called the intertropical convergence zone, or ITCZ. Along this band of low pressure, the trade winds from the two hemispheres converge. In general, the air from the two source regions has similar moisture and temperature characteristics. Pressure gradients, therefore, are usually weak and produce mostly light and variable winds. If it has passed over an expanse of the ocean, the air is laden with moisture, and this causes the cloudiness and frequent rainfall within the ITCZ. Another characteristic is the slight seasonal displacement of the ITCZ to the north and south during the year. This is to be expected, for the ITCZ depends upon solar radiation, whose intensity also shifts seasonally.

Disruption caused by distribution of continents

The model of pressure and wind systems described above assumes a homogeneous surface for the earth. In the real world, however, pressure belts and their associated wind systems are modified by the geographic distribution and differences in the thermal properties of continents and ocean. Especially in the Northern Hemisphere, where there is more land, the simple zonal model of pressure belts is severely disrupted (Figs. 15-14, 15-15). Continents heat and cool more rapidly than ocean surfaces. The temperature range is more extreme on the continents, and the seasonal displacement of the ITCZ is also greater.

As already noted, areas of high temperature have a tendency to form low-pressure cells. Thus, low pressure tends to characterize continents during the summer as the land surface warms and heats the overlying air. The warmed air rises and flows away from the continents, accumulating over adjacent cooler oceanic expanses, where it results in broad regions of high pressure. At low levels near the earth's surface, air from the ocean, where pressure is higher, flows into the continental interiors, where low pressure exists. In winter, these conditions are reversed: High pressure develops on the cold continents, and lower pressure over the relatively warmer oceans.

The thermal differences between continents and oceans (with their associated highs and lows) are not the only factors responsible for displacing pressure belts and wind systems. The general pattern is further disrupted by semipermanent cells of high pressure located in the subtropics (which give us the trade winds and westerlies), areas of persistent low pressure over subpolar regions, and the jet streams (to be discussed in the next section). Regardless of these influences, however, the general pattern of global air circulation is produced by semipermanent cells of high and low pressure over the ocean, and the effect of these large

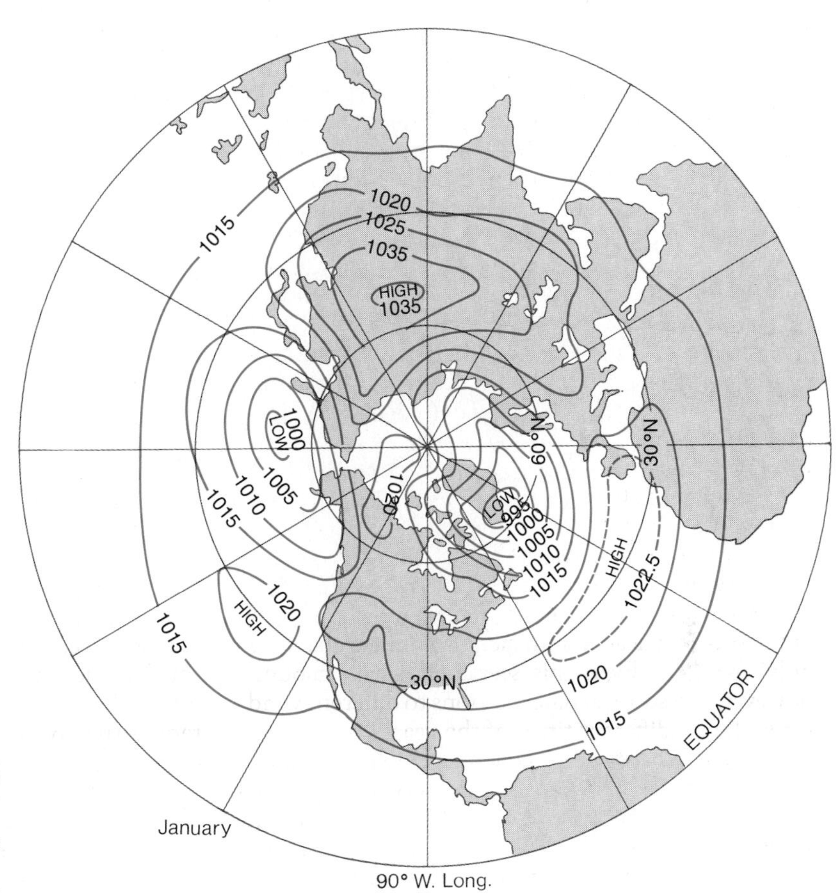

Figure 15–14 Average atmospheric pressure at sea-level in millibars for January in the Northern Hemisphere. (Modified from U.S. Department of Commerce, National Weather Service.)

cells is modified by temperature-induced pressure changes over the continents.

Global circulation and precipitation

The simplified model of general atmospheric circulation shown in Figure 15–13 provides some interesting corollaries to the general patterns of average precipitation and cloudiness around the globe. As already mentioned, the earth's heaviest precipitation occurs along the ITCZ, where there are rising currents of moist air (Fig. 15–16). The next-heaviest precipitation occurs at about 45° of latitude in association with the prevailing westerlies. The instability of the westerlies produces cyclonic eddies and stormy weather with accompanying clouds and precipitation. Between these two zones of abundant precipitation are the regions where air flow is descending (see Fig. 15–13), and therefore precipitation is relatively light.

Except for certain persistent areas of nonprecipitating low clouds in the eastern regions of oceans, at high latitudes, and in the tropics, the general pattern of cloudiness around the world resembles that of precipitation. Along the ITCZ, where rainfall is heavy, clouds prevail. Cloudiness also occurs at the latitudes of the prevailing westerlies, where the poleward flow of air converges with the equatorial flow coming from the poles. As expected, relatively clear skies predominate in the zone of descending air flow along the subtropical highs.

Monsoons

Monsoons are winds that result from seasonal temperature differences on continents relative to adjacent oceanic areas. Because of their enormous capacity for retaining heat and their continuous circulation, oceans tend to maintain temperatures within relatively

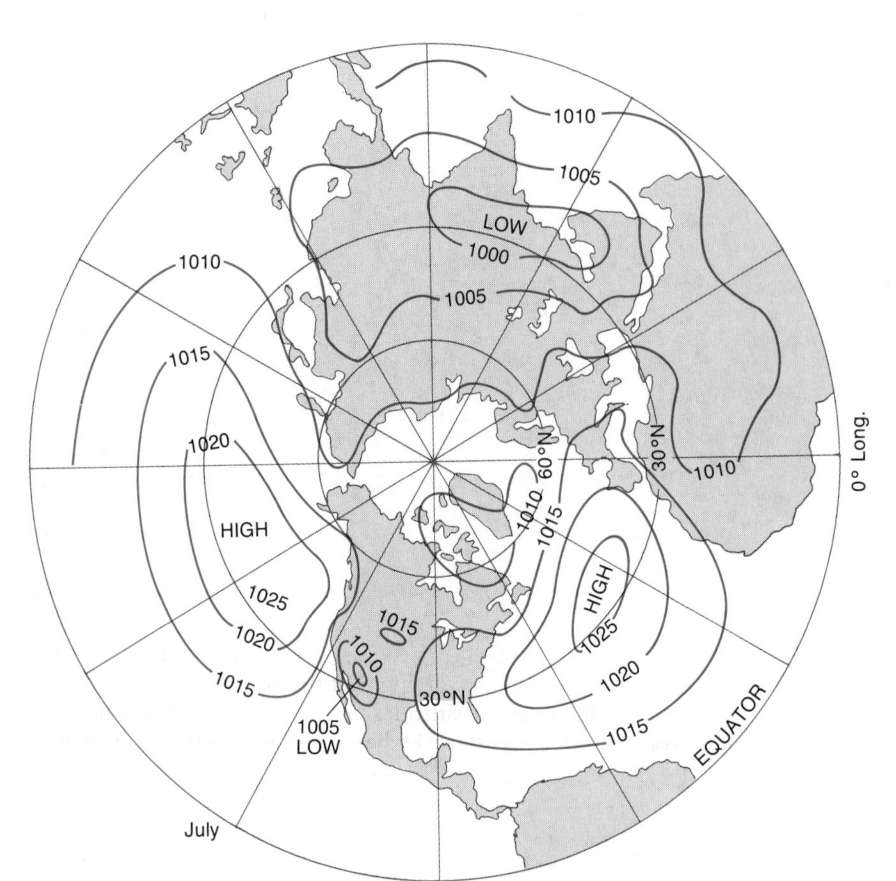

Figure 15–15 *Average atmospheric pressure at sea-level in millibars for July in the Northern Hemisphere. (Modified from U.S. Department of Commerce, National Weather Service.)*

narrow limits. As described above, continental areas are not capable of such heat retention and become cooler than nearby water bodies during the winter. At such times, there is a flow of cooler, denser air from the continents toward the ocean (Fig. 15–17). The resulting winds are called winter monsoons. In the summer, land areas may become warmer than the adjacent ocean, and summer monsoons will blow inward from the oceans (Fig. 15–18). Monsoons are particularly well developed in southeastern Asia, where there is a complete reversal in wind direction between the winter and summer seasons. Southeastern Asia is situated relatively close to the ITCZ, and its monsoonal conditions are influenced by the annual migration of that zone.

Because of their intensity, India's monsoons are famous. In summer, the moist southerly winds from the northern Indian Ocean pass across India and drop up to 450 inches of rain along the southern slopes of the Himalayan Mountains. This rainfall is of great importance to agriculture in the region. In winter, cool air from the north is brought in by prevailing northeast winds and there is a corresponding dry season.

Although the term monsoon is usually applied to winds in southeastern Asia, similar conditions occur at other locations where there are large temperature contrasts between oceans and continents. Monsoons occur in Australia, Central and South Africa, and South America. They are present also in higher latitudes, but in such locations they are overshadowed by the major global winds.

JET STREAMS

The **jet streams** are important components of high-level circulation. They consist of narrow bands of

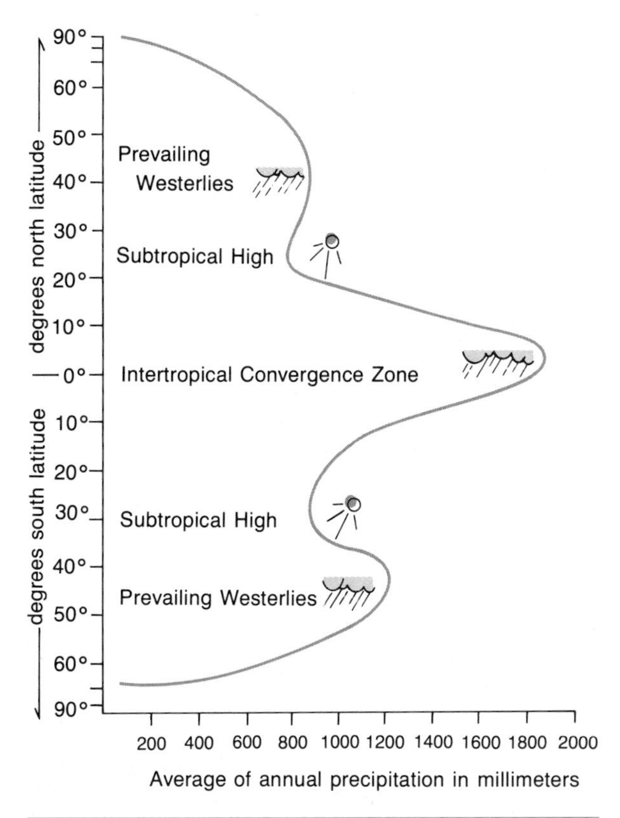

Figure 15–16 *Variation of average global annual precipitation with latitude.*

high-velocity winds that meander like great rivers around each hemisphere at elevations of 4 or 5 km to just above the tropopause. There are two major jet stream systems (Fig. 15–19). The more northerly one, called the polar front jet stream, is developed along the boundary between the relatively warm westerlies and the cold polar easterlies. The second system is called the subtropical jet stream. It makes its way around the globe between 20°N and 30°N latitude (Fig. 15–20).

The presence of the polar front jet stream became dramatically evident during World War II when bombers making high-altitude runs to Japan encountered jet stream headwinds roughly equivalent to the air-speed capabilities of their aircraft. Indeed, the velocity of the polar front jet stream is remarkable. It can flow at 150 knots and on occasion rushes along at 250 knots. The higher velocities are attained in the winter. The path taken by the polar jet varies seasonally but does not extend beyond the area bounded by 25°N and 60°N latitude. The area of strongest winds (e.g., the core) is usually about 100 km wide and 2 to 3 km thick.

In addition to the major jet streams described above, two lesser jets are known. One of these, the polar night jet, occurs during the Northern Hemisphere winter at high latitudes. The other has been identified at latitudes of about 13°N and is called the tropical-easterly jet because of its movement from east to west. A tropical-easterly jet occurs only during Northern Hemisphere summer.

The polar front jet stream is strongly affected by the distribution of solar radiation in each hemisphere. The boundary between the cold polar air and warmer middle-latitude air shifts with the seasons, and thus the jet stream also shifts southward during the winter and northward during the summer. The seasonal shift in solar heating also explains why the speed of the

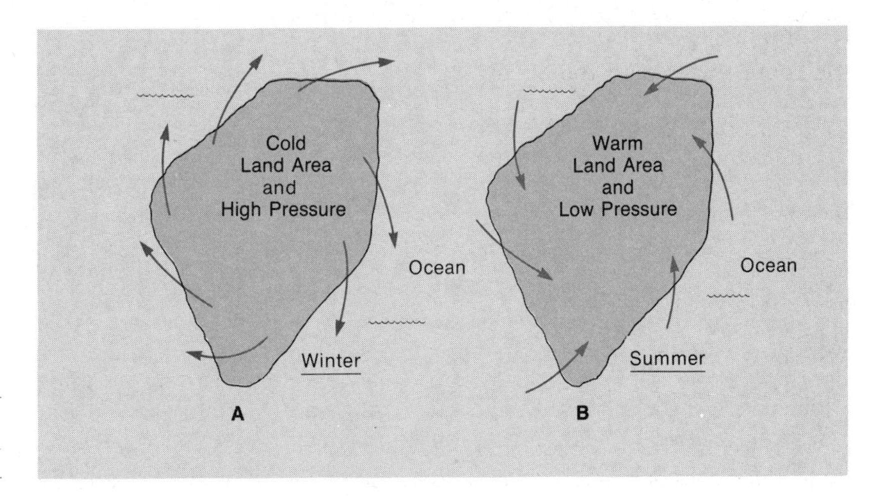

Figure 15–17 *Idealized monsoon effect on a hypothetical continent in winter (A) and summer (B).*

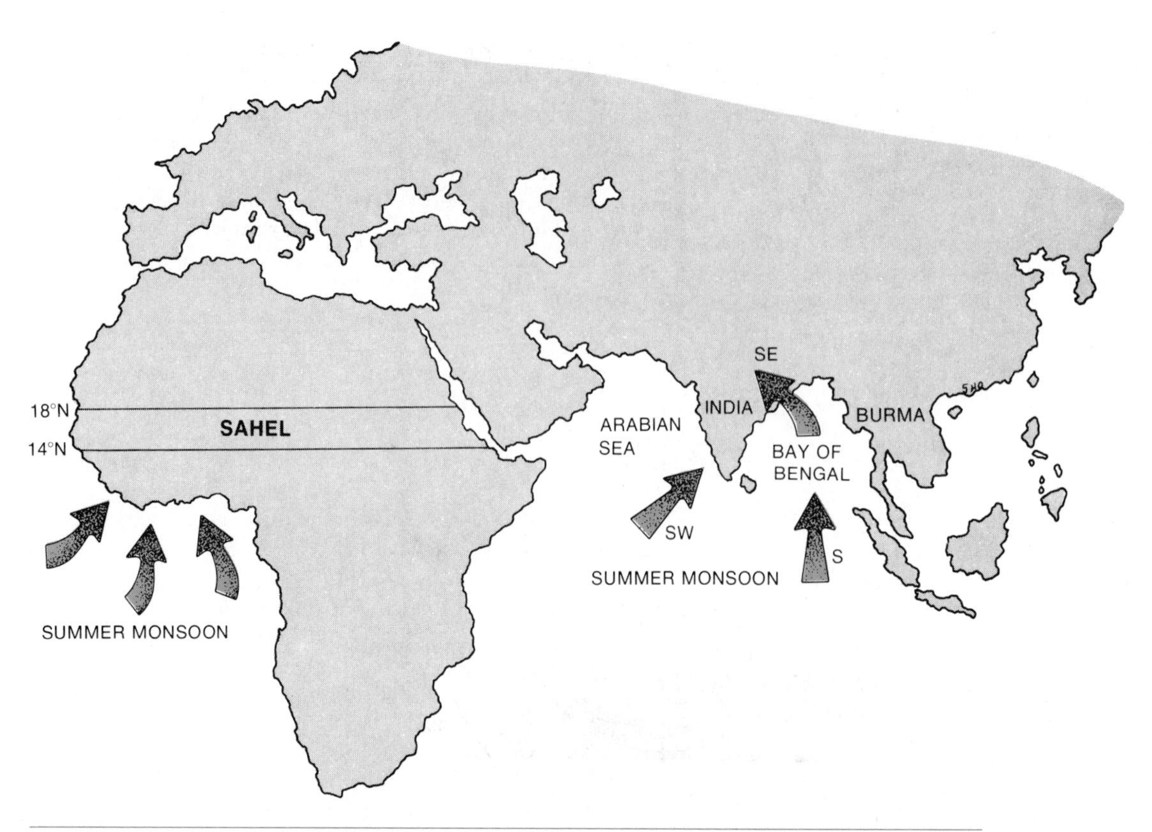

Figure 15–18 Locations well known for summer monsoons. The Sahel (Arabic meaning fringe) is a sub-Sahara region of Africa which is dependent upon rainfall brought inland by monsoons. Famine threatens the Sahel whenever the monsoons fail to bring adequate moisture.

polar front jet stream is greater in the winter than in the summer. In the winter, the North Pole is tilted away from the sun, and therefore the polar region receives minimal solar heat. This produces a marked contrast between temperatures at the poles and the appreciably warmer air in the more equatorial regions. These temperature contrasts result in density differences that translate into pressure differences and wind. In the summer, all these contrasts are lessened, and consequently the jet stream has less speed.

Once a jet stream is established, it has an important influence on surface weather. The large cyclonic bends in the jet stream generate midlatitude cyclones, and these in turn produce smaller storms, such as thunderstorms and tornadoes. In fact, tornadoes are more common in the spring before the poles have warmed because the jet stream still has full power and transfers its energy to the thunderstorms from which twisters arise. When, during winter, a meander of the jet stream stays farther northward than usual, middle-latitude locations may experience higher-than-usual temperatures. Conversely, when the jet stream is displaced far to the south of its normal winter position, colder-than-usual weather can be expected. Such an event occurred in December 1983 when a meander of the jet stream shifted far south of its normal track, bringing subfreezing temperatures to Florida and southern Texas. The Florida citrus crop was largely destroyed, and billions of fish in the Gulf of Mexico died of the cold. For the United States, it was the coldest December in 50 years.

A jet stream characteristically progresses through a cycle of three stages. In the initial stage it forms a relatively straight band of wind that flows uniformly from west to east. During the second stage, bends (meanders or waves) in the stream become more pro-

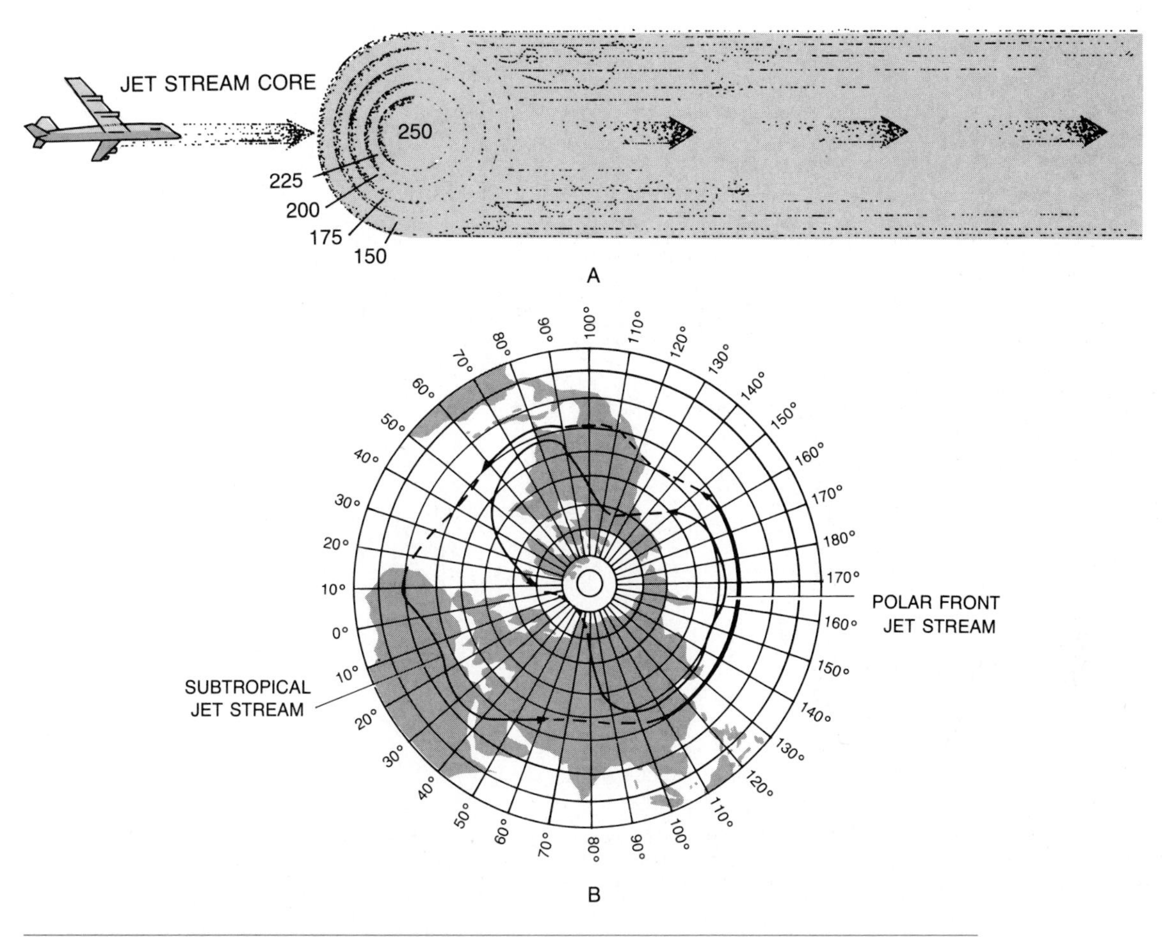

Figure 15–19 *The jet stream core (A) and the two major jet streams (B) in the upper atmosphere at altitudes in excess of 10,000 meters. (From Navarra, J.G.,* Atmosphere, Weather, and Climate. *Philadelphia, Saunders College Publishing, 1979.)*

nounced (Fig. 15–21). Some of the bends will have a cyclonic curvature, and these alternate with bends having anticyclonic curvature. In stage three, the bends break loose, forming pockets of cold air in the south and warm air in the north, both of which markedly affect local weather.

LOCAL WINDS

In addition to the earth's prevailing winds, local winds may develop in particular areas whenever or wherever a pressure gradient is established. For example, people living along coastlines are familiar with the so-called **land and sea breezes** resulting from the unequal heat capacities of land and water. (Earlier, we noted that water has one of the highest known heat capacities.) For this reason, seawater warms and cools much more slowly than solid ground. The land adjacent to a coastline heats quickly during the day under the action of the sun's rays and also cools rapidly during the night. In the daytime, air heated by contact with the warm soil rises and is replaced by cooler air flowing in from the sea. At night, these conditions are reversed. The land areas cool more quickly than the sea, and air flows from the cool land surface toward the water, which has retained its warmth (Fig. 15–22).

Similar pressure gradients are found in mountainous regions. On warm clear days, mountain slopes

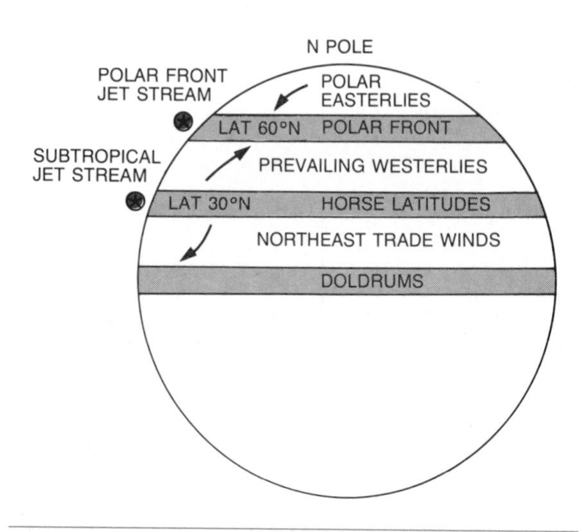

Figure 15–20 Location of the polar front and subtropical jet streams relative to the polar front and horse latitudes. Both systems flow from west to east. (From Navarra, J.G., Atmosphere, Weather, and Climate. *Philadelphia, Saunders College Publishing, 1979.)*

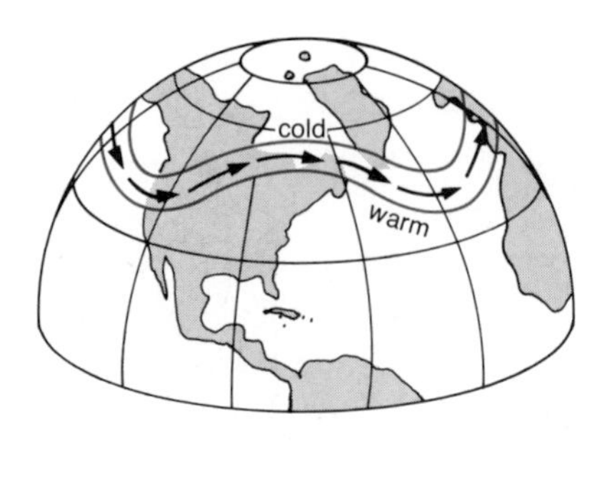

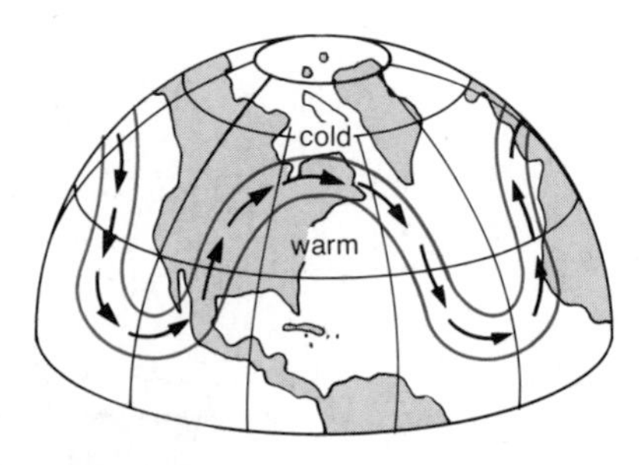

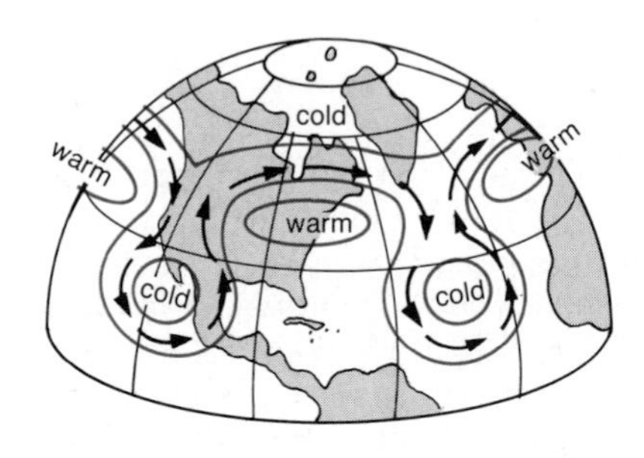

Figure 15–21 The zig-zag pattern of the jet stream creates pockets of warm and cold air far to the north and south of their usual locations. (After Ramsey, W.L., and Burckley, R.A., Modern Earth Science. *New York, Holt, Rinehart, and Winston, 1961.)*

exposed to the sun's rays are heated. The air in contact with the slopes is warmed, expands, and flows up the flanks of the mountains. Because of the relatively low heat capacity of rock and soil, at night the mountain walls cool rapidly. Air in contact with the cooled surfaces is chilled, becomes denser, and flows down the mountains into the lowlands as a **mountain breeze.**

Another type of local wind is referred to as **chinook** in the area east of the Rockies and **foehn** in the Alps. These names apply to warm dry winds that flow downward along the lee side of mountain ranges. The development of these winds begins as air ascends on the windward side of a range, is cooled, and drops its moisture as condensation. Warming due to compression during the descent on the leeward side raises the air temperature, giving the chinook its warm, dry characteristics. Chinook winds are named after an Indian territory in Montana. The Indians in this area called these warm, dry winds "snow eaters" because of the speed with which they melted and evaporated snow. Chinook-type winds often attain high speeds and may even cause damage to buildings.

There are many examples around the world of cold air that descends from highland areas primarily under the influence of gravity. For example, the **mistral** is a northerly wind blowing from the Alps down the Rhine Valley of France, and the **bora** sporadically brings cold air down the steep slopes of Yugoslav mountains to the otherwise warm Adriatic Sea. Cold air coming off glaciers may similarly flow downward along valleys and fjords, as is often observed in Alaska, Greenland, and Norway. Although there is some heating due to compression in these moving bodies of air, they were so cold to begin with that they are still chilled when they reach lower elevations.

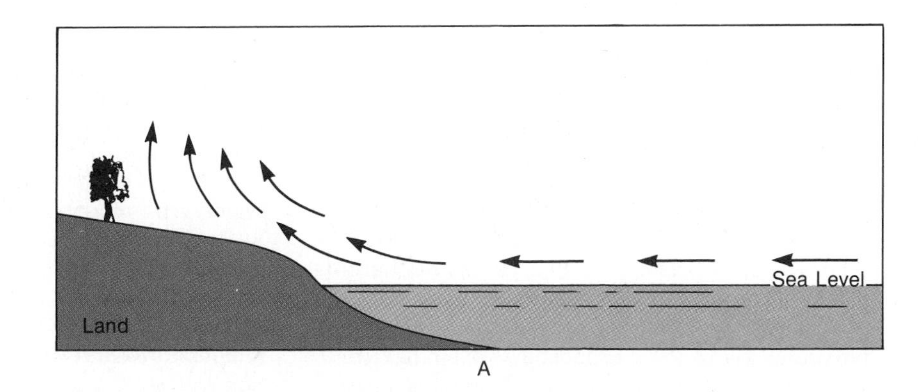

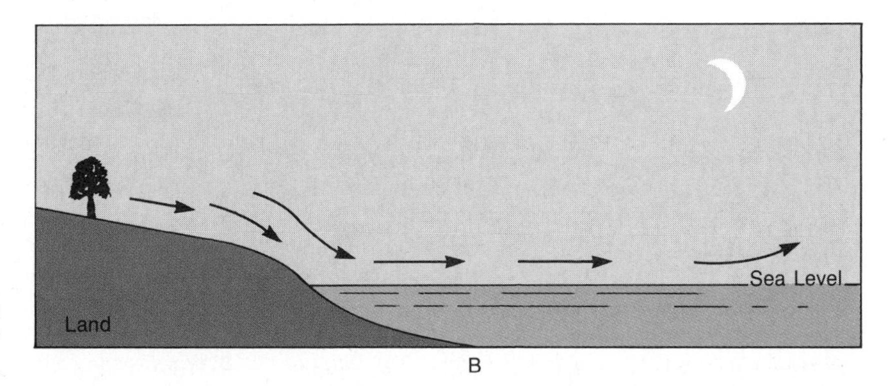

***Figure 15–22** Land and sea breezes result from unequal heating of land and water.*

Summary The ultimate source of energy needed to produce wind is the sun. Unequal solar heating of the earth's surface results in temperature differences, which in turn cause pressure differences. From these pressure differences winds and air currents arise. Air moves as wind from areas of high atmospheric pressure to areas of low atmospheric pressure, and the rate of pressure change across a given distance is termed the pressure gradient. The more rapid the rate of change, the greater the speed of the resulting wind.

Lines on a map that connect points of equal pressure are called isobars. Winds are termed geostrophic if the effect of the pressure gradient equals that of the Coriolis effect, so that the air flows parallel to isobars. Winds are influenced not only by the Coriolis effect but also by friction with the earth's surface. Because of frictional drag, wind is slowed and the Coriolis effect correspondingly lessened. As a result, winds affected by friction often blow at an angle to isobars.

A simplified model of the earth's air circulation begins along the equator, where warm air ascends and is shunted toward the poles at high levels. This high-level flow is deflected to the right in the Northern Hemisphere, and to the left in the Southern Hemisphere by the Coriolis effect. At about 30° north and south latitude, some of the air, now cooler and drier, descends for a return trip to the tropics. It does not flow directly toward the equator but again is influenced by the Coriolis effect and blows from the northeast in the Northern Hemisphere and the southeast in the Southern Hemisphere. These are the trade winds. Some of the air that descends at about 30° north and south latitude does not move to lower latitudes but instead flows poleward at low levels. These winds also experience a Coriolis deflection and become the prevailing westerlies. Eventually (at a latitude of 40° to 60°), the warm winds encounter cold polar air along a boundary called the polar front. Here the warm air is forced over the chilled, dense polar air. At high levels some of it continues toward lower latitudes, is deflected by the Coriolis effect, and becomes the frigid polar easterlies.

Cyclones are circular areas of low pressure having counterclockwise air circulation in the Northern Hemisphere. Friction-induced changes in wind direction cause air to move inward in the direction of low pressure. The opposite of a cyclone is an anticyclone, or high-pressure area. In anticyclones, winds in the Northern Hemisphere blow in a clockwise direction, and friction turns the surface winds outward in a diverging pattern. Cyclones are often associated with bad weather, whereas anticyclones are providers of fair weather.

Monsoons are a special category of winds resulting from seasonal temperature differences on land relative to the adjacent ocean. In winter, continents become colder than oceans, and there is a flow of cool, dense air from the continents toward the sea. The offshore wind constitutes the winter monsoon. In the summer, continents become warm and develop low pressure, causing cooler air to blow inward from the ocean into the continent as summer monsoons.

In addition to the monsoons and large-scale global winds, there are several kinds of local winds that result from variations in temperature and pressure over small areas of the earth. These local winds include land and sea breezes, mountain breezes, chinooks, and mistrals.

The jet stream is a sinuous band of high-velocity, high-altitude wind that flows eastward around the Northern Hemisphere between about 25° and 60° of latitude. It is associated with zones of strong horizontal temperature contrasts and therefore oscillates in position with the seasons.

QUESTIONS FOR REVIEW

1 What is the general relationship between atmospheric pressure and the temperature of air? Between atmospheric pressure and wind?

2 What factors influence the direction and speed of winds?

3 In general, what is the effect of atmospheric moisture on air temperature?

4 Explain how the rotation of the earth causes deflection of the major wind systems.

5 Explain how the frictional drag of wind in contact with the earth's surface weakens the Coriolis effect. What factors other than friction affect the direction of horizontal wind?

6 What kinds of pressure and weather conditions accompany cyclones? What are the usual weather and pressure conditions for anticyclones? Contrast the circulation of air in cyclones and anticyclones.

7 Each hemisphere is encircled by four major pressure belts. Describe each and identify its location by latitude. What is the effect of the continents in the Northern Hemisphere on the location of these belts? What is the reason for this effect?

8 What is the cause of monsoons?

9 Where in the atmosphere is one likely to encounter the polar front jet stream? Why is the jet stream more powerful in the winter than in the summer? Why does it flow farther toward the equator during the winter?

10 In what way is the mechanism by which land and sea breezes originate rather like that for monsoons?

REFERENCES AND SUPPLEMENTAL READINGS

Anthes, R.A., Panofsky, H.A., Cahir, J.J., and Rango, A., 1978. *The Atmosphere.* Columbus, Ohio, Charles E. Merrill.

Edinger, J.G., 1967. *Watching the Winds.* Garden City, N.Y., Doubleday & Co.

Neiburger, M., Edinger, J.G., and Bonner, W.D., 1982. *Understanding our Atmospheric Environment.* 2d ed. San Francisco, W.H. Freeman & Co.

Wallace, J.M., and Hobbs, P.V., 1977. *Atmospheric Science: An Introductory Survey,* New York, Academic Press.

16 weather and climate

The rain it raineth every day
Upon the just and unjust fella
But mostly on the just because
The unjust hath the just's umbrella.
ANONYMOUS

Prelude Weather and climate are related terms, both concerned with the condition of the atmosphere. **Weather,** however, involves the more immediate or daily state of various atmospheric elements at a given place, whereas climate refers to the general characteristics of the atmosphere over a long period of time in a broad region. We speak of the weather in Seattle on such and such a day, and may describe it as cloudy with rain, or clear and fair. When we describe Seattle's climate, however, we refer to the generally cool, moist, and often cloudy conditions that prevail through most of the year.

Weather involves all the elements of the atmosphere, including temperature, wind, humidity, clouds, and precipitation. These conditions change constantly, and this variability is a major characteristic of weather. Changes in weather commonly originate when cold air moves down from the poles and pushes in under warm air of lower latitudes. At the same time, warmer air from equatorial regions is constantly being routed toward the poles. When such different air masses come into contact, they form the storm centers that generate much of the weather experienced in the middle latitudes of North America and Europe.

Result of heavy snowfall, Hamburg, New York, 1981. (Courtesy of NOAA.)

AIR MASSES

Characteristics and origin of air masses

A large body of air having more or less uniform temperature and moisture characteristics at any particular altitude is called an **air mass.** Air masses may extend across thousands of square kilometers (Fig. 16-1). When we are in the central region of an air mass, we recognize its presence by fairly uniform weather and the occurrence of a relatively constant wind from a given direction. When, however, one air mass comes into contact with another, the confrontation produces marked atmospheric disturbance and changeable weather.

Air masses form over large areas of water or land that are fairly uniform in temperature and free of strong winds. These areas are the source regions from which the air mass develops by acquiring temperature characteristics from the underlying surface. Most air masses originate within the anticyclonic flow of subtropical and polar high-pressure belts. Air moving around the semipermanent Atlantic anticyclone, for example, assumes the moisture and warmth of the Atlantic Ocean, the Caribbean and the Gulf of Mexico. Similarly, air located above the frigid terrain of northern Canada slowly forms cold dense air masses. Once formed, the earth's large-scale circulation moves the air mass for distances that may exceed thousands of kilometers.

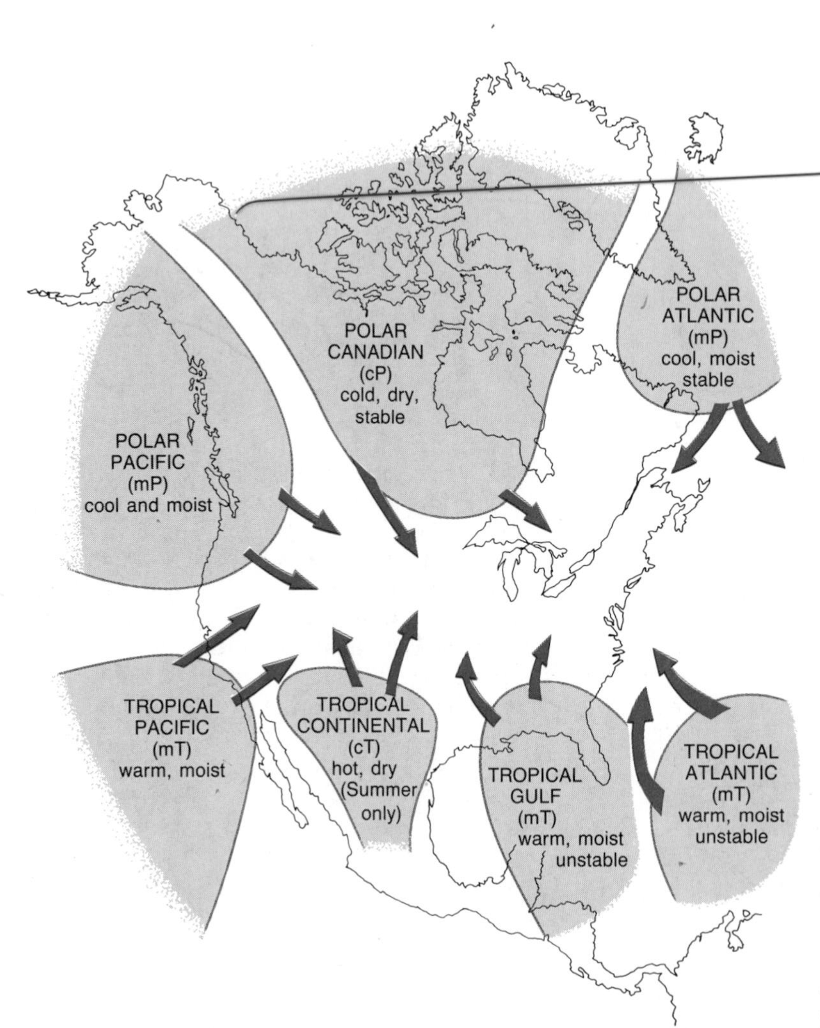

Figure 16-1 Principal air masses that influence weather in North America. Arrows indicate general direction of movement of the air masses. The letters c *for continental and* m *for maritime are indications of moisture content. General temperature conditions are indicated by* P *for polar and* T *for tropical. (After Ramsey, W.L., and Burckley, R.A.,* Modern Earth Science. *New York, Holt, Rinehart, and Winston, 1961.)*

Source regions of air masses

Air masses are divided into categories according to source region. The two largest categories are *tropical* (designated by *T*) and *polar (P)*, with the former applying to air masses having their source in low latitudes, and the latter air masses originating in high latitudes. Each of these categories is then divided into maritime *(m)* or continental *(c)*, depending upon whether they formed over water or land. Thus, the four principal types of air masses are designated *mP*, *mT*, *cP*, and *cT*. For extremes of cold dry air, the expression *cA* (continental Arctic) is used.

As soon as air masses move away from their source regions, modification of their temperature and moisture characteristics begins. They are, after all, moving across regions unlike the ones in which they were produced. Cold air masses moving over warm land or sea may be warmed from below, and this is usually sufficient to cause instability. If a dry tropical continental air mass should move to a position over the ocean, it is likely to be converted to a moist maritime air mass.

There are seven source regions for air masses that directly affect weather in North America. These are depicted in Figure 16-1. The continental polar *(cP)* Canadian air masses, the maritime polar *(mP)* Pacific air masses, and the maritime tropical *(mT)* Gulf of Mexico and tropical Atlantic air masses all have an important effect on North America's climate and weather. The Canadian continental polar air masses originate over snow- and ice-covered tracts of Canada and move in a general southeasterly direction across Canada and the northern United States. They cause some of the intensely cold weather that affects the northern states, and sometimes even states as far south as the Gulf Coast. In the summer, the Canadian *cP* may bring cool, dry relief from the heat.

Weather along the West Coast of the United States is governed in winter by maritime polar *(mP)* Pacific air sweeping out of the northern Pacific Ocean and southwestern Alaska. These air masses, although cool and moist, are not extremely cold. In summer they produce cool, often foggy weather along the coast; in winter they may bring rain and snow as the air is lifted over the coast ranges. The *mT* air from the Pacific provides mild winter conditions in the southwestern United States, and in the summer gives California a Mediterranean-like climate.

The maritime tropical Gulf and tropical Atlantic air masses (both also designated *mT*) develop over warm oceanic tracts of the Gulf of Mexico and South Atlantic and then move northward into the eastern United States. They bring hot, humid weather with thunderstorms in the summer, and mild, often cloudy and rainy weather in winter.

The tropical continental *(cT)* air masses affecting North America form over Mexico and the southwestern United States. They have a significant influence on climate in this region, and usually they bring clear, dry, and very hot weather.

The air masses formed over the northern Atlantic are maritime polar in character. They move generally eastward toward Europe, but also occasionally bring cold, overcast, weather to New England and the maritime provinces of Canada.

FRONTS

The boundary between two air masses is called a **weather front.** When cold and warm air masses come into contact, the colder air, which is heavier, tends to push beneath and lift the warmer air. This movement produced an inclined boundary in which the slope is always up and over the cooler air mass. Fronts vary in width from about 15 to 200 km, depending upon their steepness (the steeper the narrower). On a weather map, fronts are represented by bold lines.

Cold fronts

When cold air encroaches upon warm air and slides beneath the warm air mass, the front is designated a **cold front.** Cold fronts have two important characteristics. They tend to travel relatively rapidly because the air is heavy and has a tendency to flow outward from the center of mass. In addition, the slope of a cold front is relatively steep, resembling a blunt wedge, as the cold, heavier air rushes under the adjacent warm, lighter air. The average speed of a cold front is about 3 km an hour, and the slope may be as great as 1 meter vertically per 100 meters horizontally. In winter, the movement of the cold front is more rapid than in summer because upper-level winds are stronger.

Along a rapidly moving cold front, warm air is quickly lifted within a rather narrow zone (Fig. 16-2). Provided there is sufficient moisture in the warm air, heavy towering clouds and thunderstorms will develop. Although precipitation can be heavy, it is usually of short duration. If, however, the cold front is advancing rather slowly, it will bring more diffuse

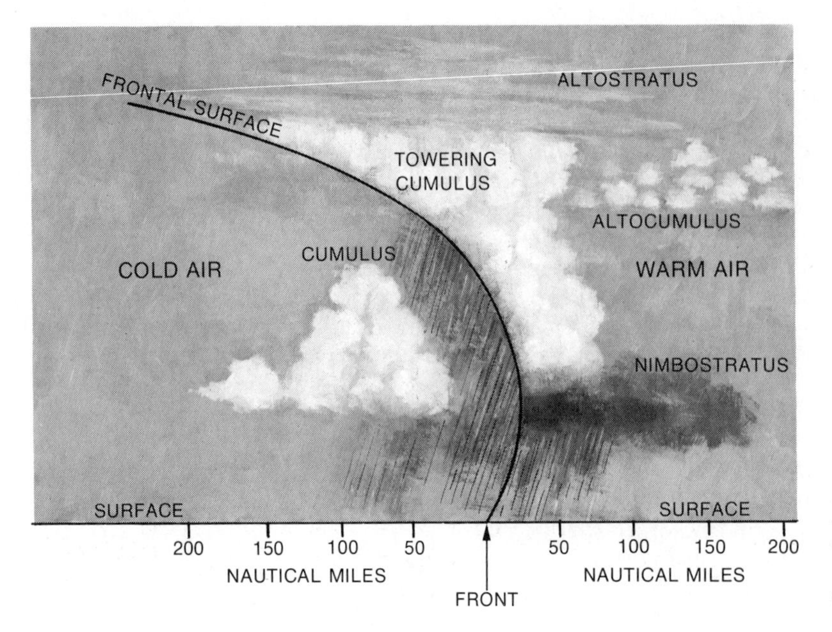

Figure 16–2 Vertical view of air masses at a cold front (From Navarra, J.G., Atmosphere, Weather, and Climate. *Philadelphia, Saunders College Publishing, 1979.)*

cloudiness and precipitation. Once the cold front has passed, one can expect fair weather. On the weather map, the cold front is indicated by a line having triangles directed into the warmer air mass (Fig. 16–3).

Warm fronts

Whereas a cold front develops when a cold air mass encroaches on a warm air mass, a **warm front** is produced when a mass of warm air overtakes the trailing edge of a cold air mass. On a weather map, the position of such a front is indicated by a line having half circles extending into the colder air (see Fig. 16–3). Warm fronts regularly move into the central and eastern United States from the Gulf of Mexico or the Southwest and encounter cold air masses as they progress northeastward at rates of about 25 km per hour.

The slope angle of a warm air mass is far less than in a cold front. As a result, clouds that form in the more broadly lifted warm air may extend for hundreds of miles ahead of the base of the front (Fig. 16–4). Whereas a fast-moving cold front often brings towering cumulus clouds, a warm front is accompanied by extensive stratus clouds. Precipitation is likely to be relatively steady and persistent. An exception is when warm air is forced over very moist cold air. Under such conditions, stormy weather can be expected. In contrast, if the warm air is dry, the warm front may produce few clouds and little or no precipitation.

Stationary and occluded fronts

Occasionally, a situation occurs in which there is little or no forward motion along the boundary of cold and warm fronts. Meteorologists refer to such boundaries as **stationary fronts.** They note their presence on weather maps by a line having half circles on the cold side and triangles on the warm side (see Fig. 16–3). In general, the weather along stationary fronts resembles that along warm fronts, often with continuous rain, drizzle, and fog.

Because a cold front normally moves more rapidly than a warm front, it is not unusual for a cold front to overtake the warm air in its path and converge on the trailing edge of the cold air that precedes it. The result is an **occluded front** (Fig. 16–5), the symbol for which is a line containing alternate half circles and triangles on the side toward which the front is moving. When the rear cold air mass has made contact with the forward air mass, the intervening warm air is lifted completely off the ground (Fig. 16–6). Abundant rain or snow may be released if the elevated warm air is sufficiently moist.

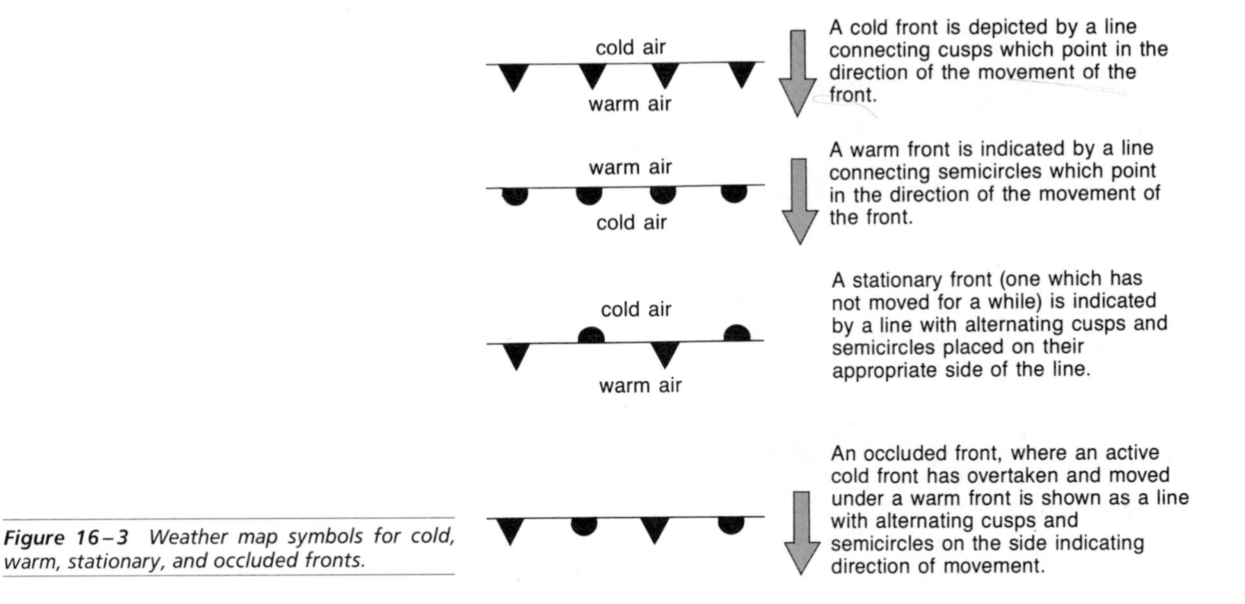

Figure 16–3 *Weather map symbols for cold, warm, stationary, and occluded fronts.*

It is possible to distinguish two kinds of occluded fronts. If the rear cold air mass has a lower temperature than the forward cold air mass, the rear mass will displace the forward mass along a boundary called a **cold front occlusion.** If, on the other hand, the forward air mass is the cooler of the two, the encroaching air to the rear will be lifted, forming a **warm front occlusion** (see Fig. 16–6). An important characteristic of both types of occluded fronts is that they are relatively slow moving and may therefore bring weather episodes of relatively lengthy duration.

It should be recognized that weather fronts are complex features, and one cannot always expect a certain kind of cloud formation or precipitation from any

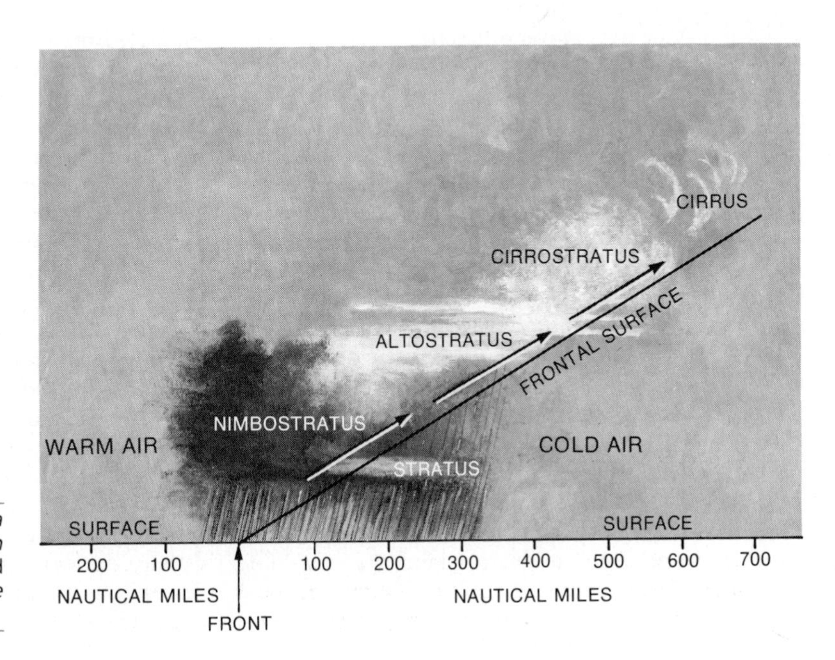

Figure 16–4 *Typical conditions along a warm front having stable warm air. (From Navarra, J.G.,* Atmosphere, Weather, and Climate. *Philadelphia, Saunders College Publishing, 1979.)*

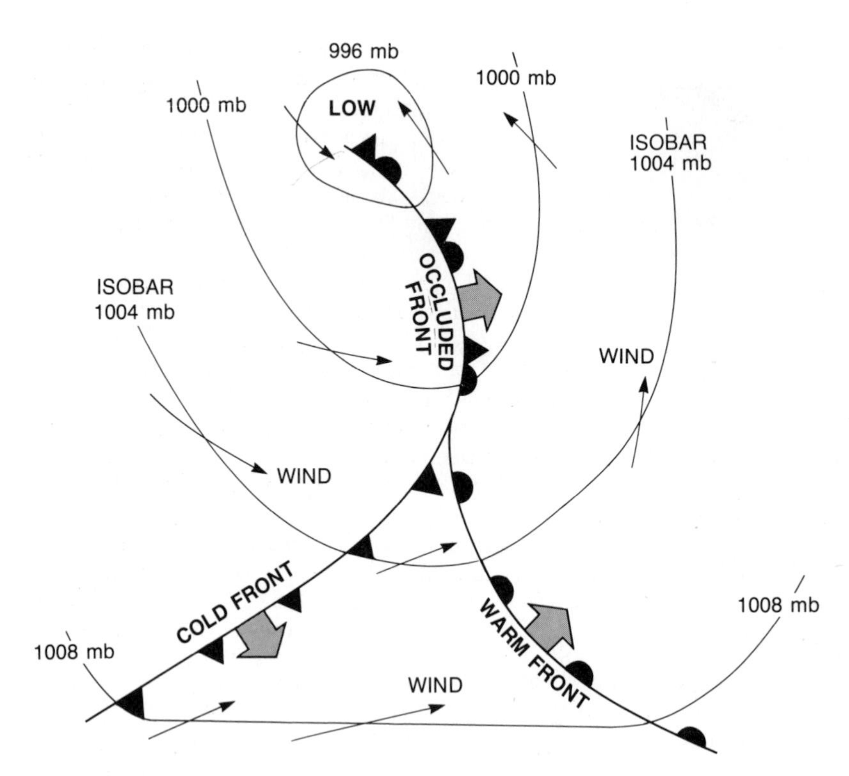

Figure 16–5 Map view of a developing occluded front. (From Navarra, J.G., Atmosphere, Weather, and Climate. *Philadelphia, Saunders College Publishing, 1979.)*

given type of front. Thunderstorms and gusty winds may accompany a strong cold front, but if it is weak, its passage may be almost unnoticeable.

Cyclones formed along fronts

All of the fronts just described pervasively influence weather in the middle latitudes. In particular, cold fronts and stationary fronts tend to develop large undulations called wave cyclones. **Wave cyclones** are really low-pressure systems that travel across the continents at rates of about 800 to 1600 kilometers per day. These wave cyclones have a characteristic life cycle. They are born when a cold or stationary front develops a bulge of cold air that advances slightly ahead of the rest of the front (Fig. 16–7B). As the bulge of cold air pushes toward the equator, warm air moves poleward along an adjacent warm front. The junction of the two fronts coincides with the center of a low (Fig. 16–7C). As the infant cyclone develops to a youthful stage, pressure decreases in its center. The fast cold front and warm front pivot counterclockwise (in the Northern Hemisphere) around the center of low pressure.

The wave cyclone's mature stage is reached when the more rapidly moving cold front overtakes the warm front and, by doing so, forms an occlusion (Fig. 16–7D). The wave cyclone is now in its most vigorous state. The strong lifting action of the warm air brings stormy weather, the pressure gradient steepens, and winds move at high speed around the low-pressure area at the center of the disturbance. Eventually, the system expends its energy resources, and the cyclone dies (Fig. 16–7E).

Frontal weather

We have all experienced the changes in weather that occur when fronts move past, and therefore a review of the usual sequence of events will seem familiar. Imagine that it is spring (when there is greater temperature contrast between air masses), that your home is in the middle latitudes, and that a wave cyclone is approaching your location from the west (as is normally the case). From your location east of the approaching warm front (see Fig. 16–4), the first sign of the approaching cyclone will be the appearance of thickening cirrostratus clouds. Because of the gentle

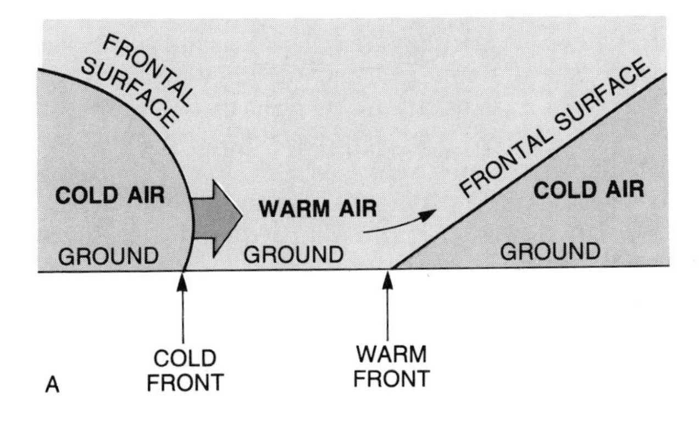

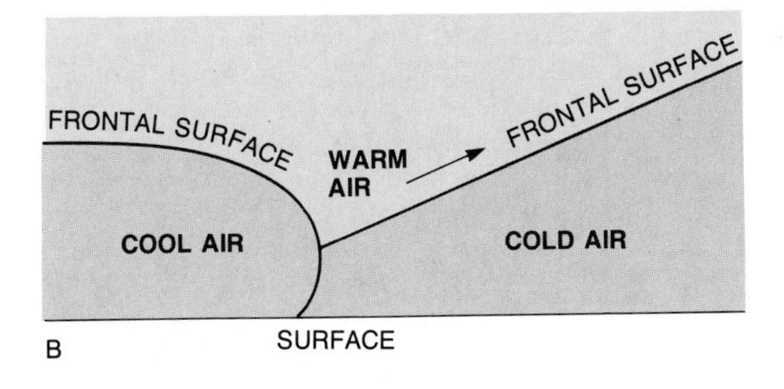

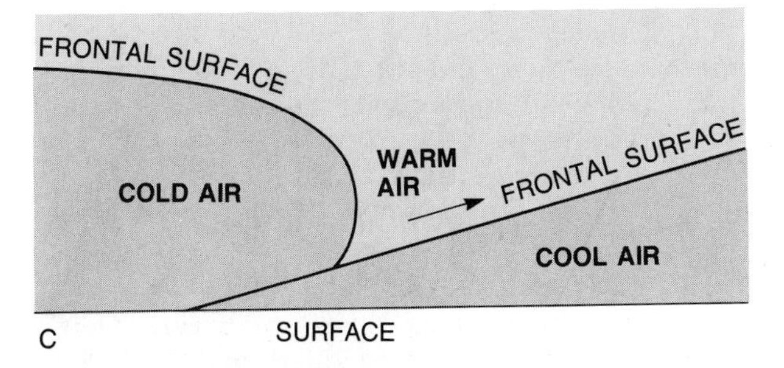

Figure 16–6 *The development of occluded fronts. (A) Vertical cross-section of a cold front approaching a warm front; (B) Cold-front type occlusion; (C) Warm-front type occlusion. (From Navarra, J.G.,* Atmosphere, Weather and Climate. *Philadelphia, Saunders College Publishing, 1979.)*

slope of the warm front, these lofty clouds may precede the ground-level position of the front by 1000 km or more. As the cyclone continues its eastward journey, there is a gradual lowering of atmospheric pressure, and the clouds become lower and thicker, changing to the lower and more continuous altostratus variety. The clouds become still lower, heavier, and yield rain as the warm front passes overhead. Next you may experience warmer temperatures, south winds and reasonably clear skies because you are beneath the central part of the warm air mass. Then the cold front approaches. The kind of weather accompanying the passage of the cold front depends upon the intensity of temperature contrast, the forward speed of the front, and the stability of the air being lifted. In the spring, the approach of the cold front is often marked by a wall of heavy dark clouds, gusty winds, and severe squalls. After the cold front passes, however, wind direction shifts abruptly to the north or northwest, temperatures drop, and atmospheric pressure rises. These conditions produce the typically clear, good weather of high pressure systems. Normally, a wave cyclone requires between one and two days to form, with further development occurring over an additional two or three days.

As is evident from the preceding description of a typical middle-latitude wave cyclone, the weather in the area between the warm and the cold fronts (called the warm sector), as well as the weather behind the cold front, is determined by the temperature, pressure, and the moisture content of the air masses themselves. The highs and lows need not always develop as closed, roughly circular systems, but may assume the form of troughs of low pressure or tongues of high pressure. Also, the weather along the occluded front may vary from the previous description. Occluded areas are slow moving and tend to bend northwestward and remain over a given location longer. Characteristically, occluded fronts bring the snow and ice of fierce winter storms. Precipitation may be heavy and bad weather persistent as moist air of the warm sector is squeezed upward and over adjacent colder air masses.

DESTRUCTIVE STORMS

The common cyclonic storms associated with the development of fronts occur across broad regions. There are, however, storms that develop locally, and because of their concentration of energy over a relatively small area, they are characteristically violent and destructive. The most important of these local storms are hurricanes, tornadoes, and thunderstorms.

Hurricanes

Tropical cyclones in the Atlantic Ocean having winds that reach 120 km per hour are called **hurricanes.** In

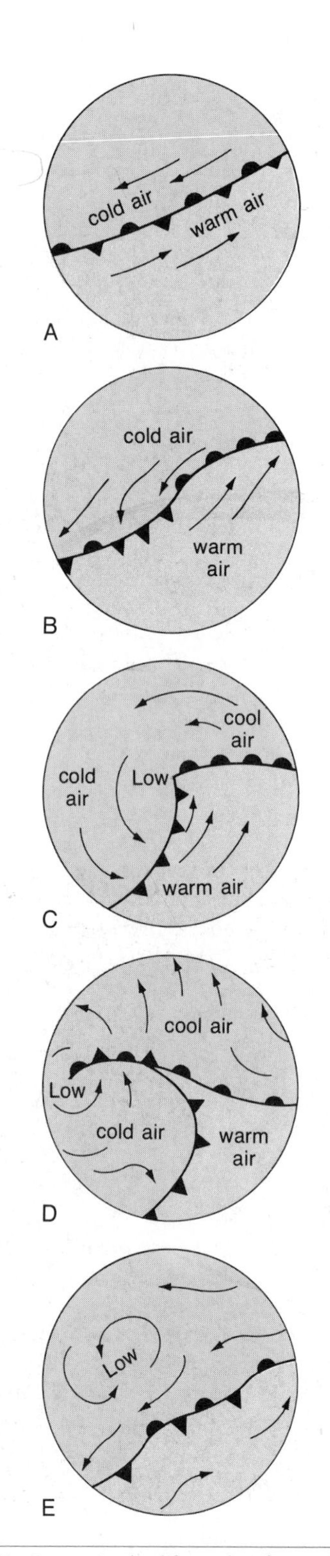

Figure 16–7 *Stages in the life cycle of a typical wave cyclone.*

the western Pacific they are called *typhoons,* and in the South Pacific and the Indian Ocean they are known as *cyclones* (not to be confused with the usual meaning of cyclone, a broadly dispersed mass of low-pressure air moving in a counterclockwise whirl in the Northern Hemisphere). Regardless of the name, these storms originate in the same way and have similar structure and behavior.

A hurricane is a large convectional system that derives its energy from heat released during the condensation of moisture-laden air of tropical oceans. As air flows into the base of the hurricane and is swept upward and around the central area, 580 calories of heat is released for each cubic centimeter of condensed water (in the form of cloud droplets).

Hurricanes (Fig. 16–8) may begin their development from wavelike motions of the trade winds. The wave motion establishes a rotational whirl (counterclockwise in the Northern Hemisphere), which may evolve into the tight spiral of a hurricane after a day or two. Such an event occurs only over tropical ocean expanses during those months when surface water temperature is relatively high. Hurricane formation appears to require an ocean surface temperature of greater than 26.5°C. Such temperatures are prevalent during August, September, and October in the region between 5° and 20° North Latitude. Hurricanes tend not to develop precisely at the equator. The Coriolis effect is zero there, and hence there is no mechanism for initiating the required spiral motion. Once formed, a hurricane travels westward with the trade winds, and then veers to the north in the Northern Hemisphere or to the south in the Southern Hemisphere. A typical hurricane is about 500 km in diameter.

Imagine that you are an observer located directly in the path of an oncoming hurricane and that you are able to record the sequential changes that occur as it approaches and passes. The first sign that a hurricane is on the way would be the appearance of high cirrus clouds. Often these are 400 to 500 km in advance of the stormy portion of the hurricane. One might next notice that the barometer shows a slow decline in atmospheric pressure. Winds increase in intensity. By the time a point located about 300 km from the hurricane's center is passing overhead, winds have reached gale force (about 50 km per hour) and subsequently increase steadily in their intensity. At the same time, pressure begins a more rapid decline. When the center of the hurricane is only about 150 km away, winds in excess of 80 km per hour are recorded. Pressure declines still further while gray, unruly clouds threaten overhead. As the ring-shaped inner region of the hurricane passes, torrential rains fall, driven by winds in

Figure 16–8 Hurricane Allen forms a characteristic spiral over the western Gulf Coast region in August 1980. (Courtesy of NOAA.)

excess of 160 km per hour. Then with surprising abruptness, the wind slows to less than 30 km per hour, and the clouds become so thin as to allow the sun to occasionally shine through. The *eye* (Fig. 16–9) of the hurricane would be passing overhead.

Around the eye of a hurricane, clouds are piled in nearly vertical walls (Fig. 16–10) that extend upward from a few thousand meters to 12 or 14 km. Atmospheric pressure in the eye of the hurricane is at its lowest. The air here is subsiding, and yet it is warmer by several degrees than the air in surrounding regions. The warmth results from compressional heating in the descending air.

As the remaining half of the hurricane begins its traverse across the observation point, the calm associated with the eye changes quickly to turbulence. The conditions experienced previously occur again but in reverse order. Thus, the strong winds and heavy rains give way to lessening rain and wind, followed by a return of cirrus clouds and conditions nearly normal for the region.

Those people living in hurricane-prone places like the Gulf and Atlantic Coasts of the United States are well aware of the property damage and possibility for injury from hurricanes. The danger begins as the hurricane comes within about 160 km of the coast. Often

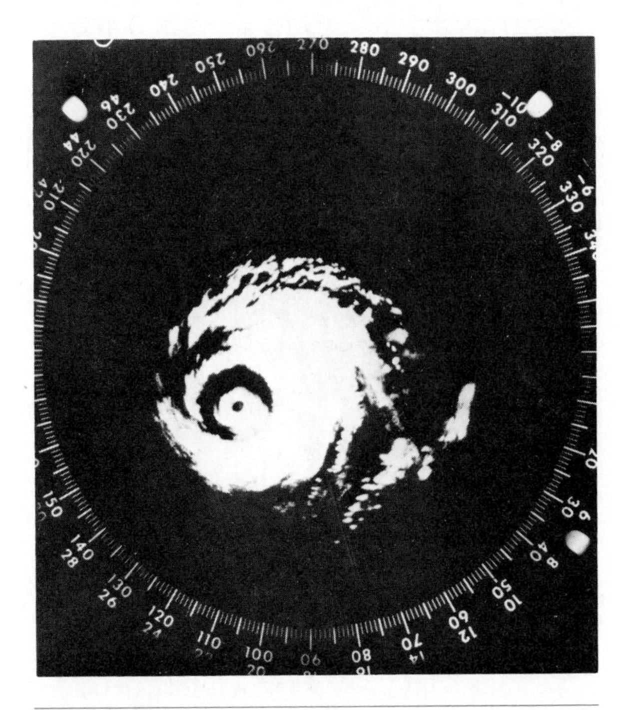

Figure 16–9 The eye of Hurricane Beulah over Brownsville, Texas, as it appeared on the radarscope on September 12, 1967.

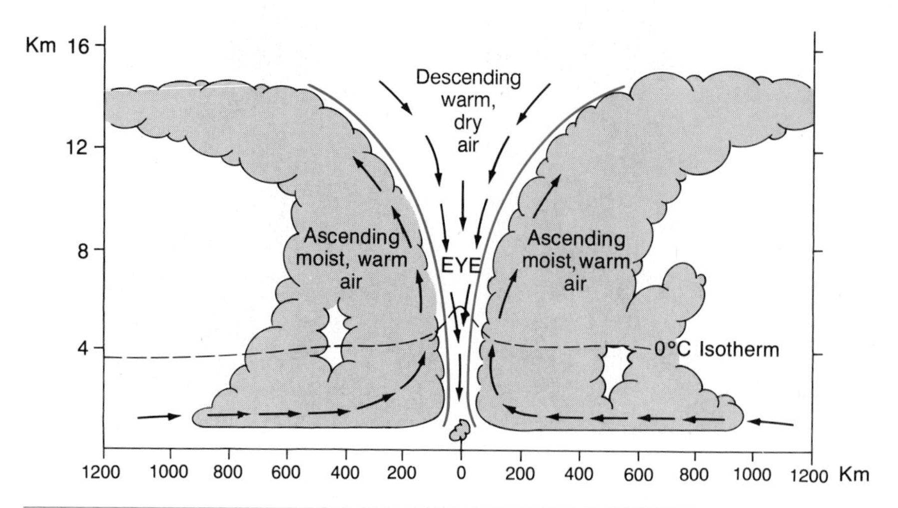

Figure 16–10 *Conceptual vertical cross-section of a hurricane showing the location of the eye and directions of major air movements.*

the greatest damage is caused by flooding of the coastal areas by ocean surges and high waves. The surges are "hills" of water driven against the coast by high winds. The shoreline itself may act as a temporary barrier against the water, which may then pile up to more than 5 meters above normal sea level. Added to this are waves more than 6 meters high. These rush inland and inundate large areas. Gale-force winds also contribute their share of havoc by breaking power lines, hurling debris against windows, and even tearing the roofs from buildings. Fortunately, these misfortunes lessen as the hurricane progresses inland, for friction with the surface of the land retards the wind, and there is no longer warm humid air to provide energy.

It is apparent from the above description that hurricanes are different from the middle latitude cyclones discussed earlier. They do not contain fronts, they are more symmetrical, and they occur within a single air mass (tropical maritime). In addition, hurricanes produce far more rain and far more violent winds.

Tornadoes

A **tornado** is an intense, rapidly rotating column of air having a funnel-shaped neck of whirling air hanging beneath a cumulonimbus cloud (Fig. 16–11). Winds race around the tubular funnel at speeds of 200 to 500 km per hour, and on rarer occasions up to 800 km per hour. For these speeds to develop there must be extremely low pressure inside the funnel and a correspondingly enormous steep pressure gradient. In fact, the drop in pressure between the outer perimeter and the center of a tornado is usually at least 25 millibars, and decreases of as much as 200 millibars have been recorded.

Tornadoes occur most frequently in the central and Gulf Coastal regions of the United States (Fig. 16–12) during late spring and early summer, and usually take place in the afternoon and early evening. Generally, they develop in the vicinity of intense cold fronts marked by moving bands of thunderstorms. Such locations have a combination of great vertical instability and strong shearing action from adjacent air currents. The shearing action imparts the whirling motion characteristic of tornadoes. Once that motion has been established, it is maintained by convective action until dissipated by friction.

Although not as broadly destructive as hurricanes, tornadoes are nevertheless powerful and often damaging storms (Fig. 16–13). The strong winds pick up and hurl heavy objects, topple television towers, and carry away roofs and cars. Pressure differences inside and outside of buildings in the path of the tornado cause them to virtually explode. Entire neighborhoods can be destroyed in minutes, leaving stark rows of foundations and stunned survivors.

Fortunately, tornadoes are of relatively small size, and the damage they cause is somewhat restricted in area. Rarely do they sweep a path more than about 500 meters wide. The average path length is about 6 kilometers. Usually, twisters dissipate after about 6 or 8 minutes, but they can persist for more than

Figure 16–11 *A tornado forms (Stage 1), touches down (Stage 2), and grows while whipping up dust and debris (Stage 3). (Courtesy of NOAA.)*

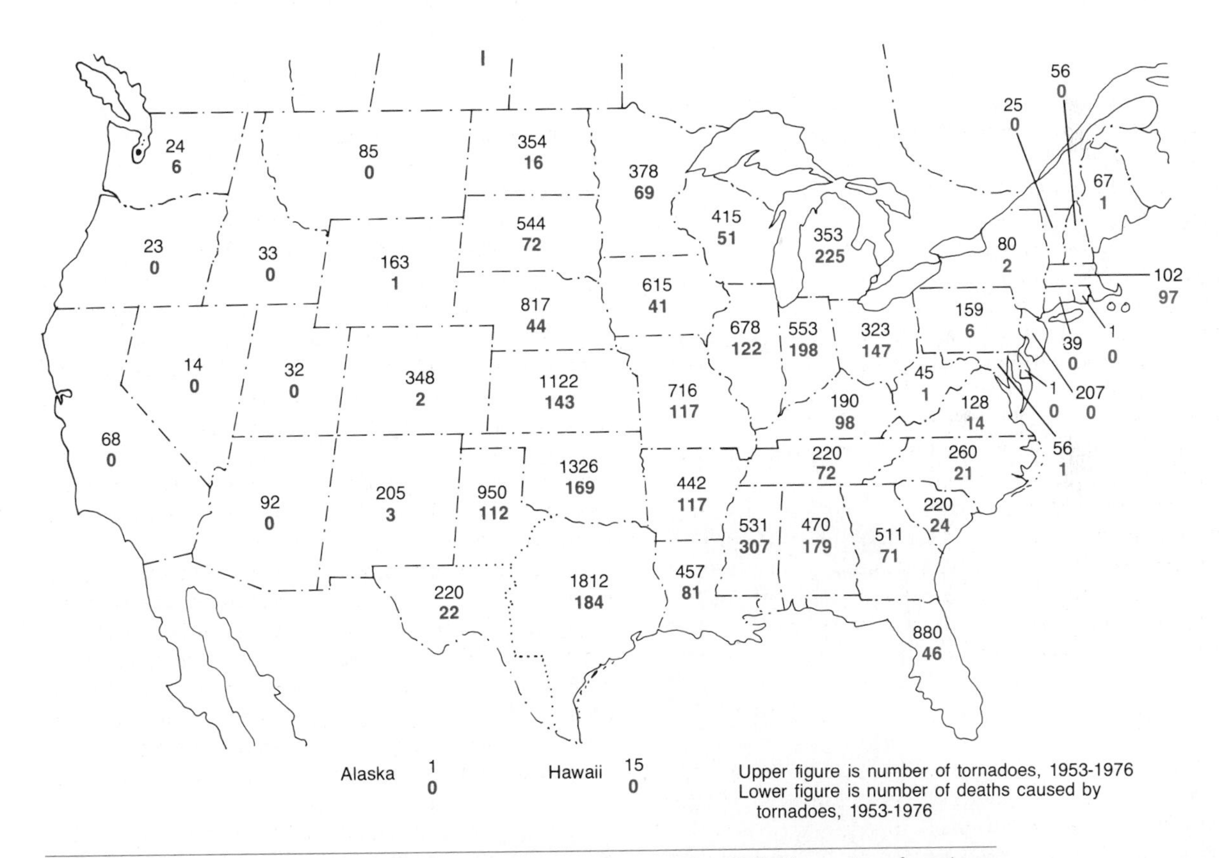

Figure 16–12 *Tornado incidence and deaths from tornadoes, 1953–1976. The region of maximum threat lies east of the Continental Divide. Note that the greatest potential for casualties from tornadoes is not necessarily where the greatest number of tornadoes occur, but where there is a high tornado incidence, a dense population, and many mobile homes or poorly constructed wood frame houses without basements. (Data courtesy of NOAA.)*

Figure 16–13 Damage to homes caused by hurricane-force winds in Gulfport, Mississippi, in 1969.

Figure 16–15 Waterspouts photographed off the Grand Bahama Islands. (Courtesy of NOAA.)

20 minutes. The funnel of the tornado whips back and forth and sporadically lifts off the ground, taking an erratic and unpredictable track. Forward motion is normally about 50 to 60 km per hour.

So that people are informed about the possibility of a tornado in their community, the National Weather Service (Fig. 16–14) issues tornado watches and tornado warnings to commercial radio and television stations for transmission to the general public. A tornado *watch* means tornadoes are possible. During a watch, people should be attentive to their local radio and television for more information. They should also watch for threatening weather, and if such bad weather arrives, they should take shelter immediately. Tornado *warnings* are issued when a tornado has actually been sighted or is indicated by radar. When a warning is issued, people close to the storm should immediately take cover. Go to the basement or part of the interior at the lowest level. Stay beneath something sturdy. Interior halls, closets, and bathrooms may have the best support. Open windows to equalize pressure, but then stay away from windows, doors, and outside walls. Leave mobile homes for sturdier shelter. If you are outside and there is no shelter nearby, lie flat in a ditch. Whether inside or out, protect your head.

Figure 16–14 National Weather Service meteorologist operating computerized communication system used to speed data handling.

Occasionally, the conditions of extreme atmospheric instability that produce tornadoes occur over bodies of water. If the neck of the funnel cloud extends to the water surface, a **waterspout** (Fig. 16–15) is formed. The waterspout is a whirling cloud of spray, rather than dust and debris as is the case in a tornado. Like tornadoes, waterspouts descend from cumulonimbus clouds, but many have been observed in growing cumulus clouds as well.

Thunderstorms

A **thunderstorm** is a local storm, usually produced by a cumulonimbus cloud, accompanied by lightning, thunder, strong gusts of wind, and often hail. These storms have the potential for causing extraordinary

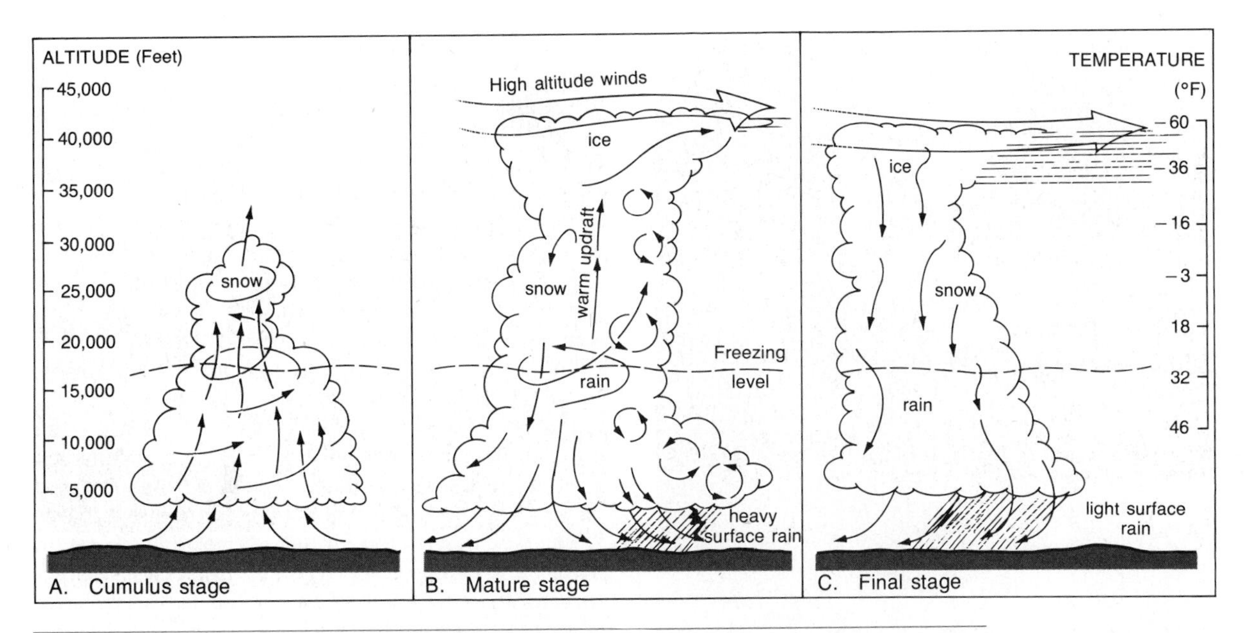

Figure 16–16 Stages in the development of a thunderstorm. (After National Oceanic and Atmospheric Administration, National Weather Service pamphlet, Thunderstorms, *1977.)*

damage. For this reason, thunderstorms and the conditions that cause them are continuously monitored at the National Severe Storms Weather Center in Kansas City. As with tornado watches, thunderstorm watches and warnings issued from this facility are reported through the National Weather Service to local radio and television stations. A thunderstorm *watch* means severe thunderstorms are possible; a *warning* means severe thunderstorms have been sighted or are indicated by radar.

Although the development of a thunderstorm is a continuous process, there are several readily recognized stages in the storm's history. The initial or *cumulus stage* begins with the convective upward expansion of moist air to a level where the water vapor within it condenses and forms a cumulus cloud. The condensation releases heat, which adds power and lift to the updraft. Still more impetus is given to the updraft by air entering from the sides and by winds aloft, which produce a chimney effect in drawing the air upward.

If conditions are suitable, the growing cumulus cloud will evolve into a cumulonimbus cloud, in which ice crystals, water drops, and hailstones begin to form. When the updraft is no longer strong enough to prevent the fall of these bodies, the *mature stage* of thunderstorm has been reached. Heavy showers now begin to fall from the cloud. The rain, hail, and ice particles, however, tend to fall through the less vigorous currents of the peripheral region, where they contribute to the formation of cool downdrafts. At the same time, however, a warm updraft persists in the core (Fig. 16–16B). When the cool downdrafts leave the cloud base and strike the earth's surface, they spread out and generate the familiar gusty winds that blow outward from the area of rainfall. Within the cloud, repeated updrafts and downdrafts provide the conditions needed for the formation of hailstones.

The *final stage* in the history of a thunderstorm begins when the upward air currents are reduced and the supply of moisture being carried by them diminishes. With the moisture supply cut off, condensation is reduced and the power of the thunderstorm subsides. The rain, thunder, and lightning fade away, and temperatures within the cloud change slowly to that of the surrounding air.

The familiar companion to thunderstorms is lightning. **Lightning** (Fig. 16–17) is a huge electrical discharge that passes from cloud to cloud, from clouds to the earth's surface, or from two points within a cloud. The kind of lightning that accompanies a thunderstorm results from a concentration of unlike charges in various parts of the cloud. For example, positive charges tend to accumulate in the higher levels of

Figure 16–17 *Lightning forms a spectacular display in the night sky. (Courtesy of NOAA.)*

cumulonimbus clouds, and negative charges near their lower levels. This distribution of charges is believed to arise when electrons are forcibly torn off water and ice particles as they experience the friction of updrafts and turbulence that accompany storms. The electrons removed in this way accumulate near the base of the cloud, giving it a negative charge. Having lost electrons, the moisture particles ascending in the updrafts are positively charged and impart this positive charge to the upper reaches of the cloud. When the electrical difference between the higher and the lower parts of the cloud becomes sufficient, a great spark (lightning) travels between the two parts.

Lightning may also pass from a cloud to the ground. Normally, the ground is negatively charged relative to the atmosphere. When a thunderstorm passes over the ground, however, the negative charge in the base of the cloud induces a positive charge on the ground below. The positive ground charge follows the storm like a shadow and builds as the negative charge in the cloud becomes stronger. When the difference between the positive and the negative charges (the electrical potential) becomes great enough to overcome the resistance of the insulating air, lightning travels from the cloud to the ground (Fig. 16–18).

Lightning provides not only spectacular visual displays but also dramatic claps of thunder. Thunder

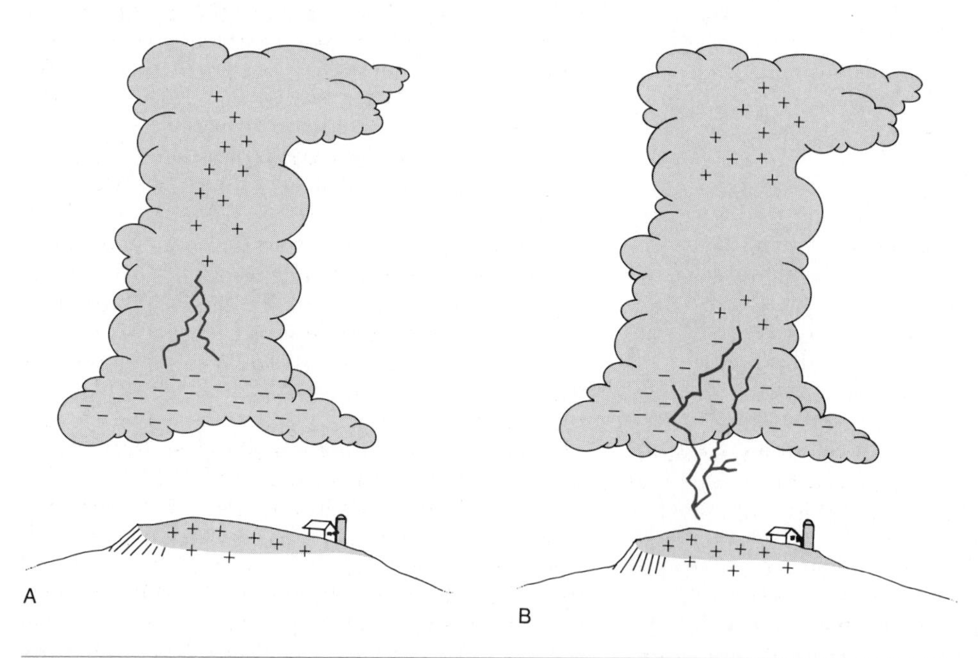

Figure 16–18 *Development of lightning within a cloud (A) and from cloud to ground (B).*

results from the explosive expansion and collapse of air that has been heated by a stroke of lightning. When lightning is close by, the sound of thunder will resemble a sharp, window-rattling crack, whereas more distant lightning produces the familiar rumbling noises that result from the refraction of sound within the turbulent air of the thunderstorm. Because light travels much faster than sound, we see lightning before we hear it. This makes it possible to obtain a rough estimate of the distance between an observer and the lightning. That distance in kilometers is approximately equal to the number of seconds between lightning and thunder divided by three.

WEATHER FORECASTING

Because weather influences so many of our activities, it is important to be able to anticipate the kind of weather we will experience each day, whether it be on the way to work, before a crop is harvested, or prior to a flight to a distant city. We depend upon the meteorologists to provide the predictions we need. They must weigh and interpret past and present atmospheric conditions, estimate the direction and speed of fronts, and anticipate where and when a weather system will dump snow or discharge rain. Because of the many variables affecting atmospheric conditions, weather forecasting is a very risky business. The risk, however, is reduced by the care taken in each of the three steps that lead ultimately to the weather forecast. These steps are *observation of present weather, collection and mapping of data,* and *analysis of data.*

Observation of present weather

One must have basic data to make a forecast, and therefore weather prediction necessarily begins with the observation of atmospheric conditions at many localities and speedy transmission of those observations to meteorological centers. Cooperation on a global scale is clearly required. To foster such cooperation, a World Meteorological Organization has established international standards for the weather items to be observed, the symbols employed in representing meteorologic data, and the time during which the data are to be reported. Indeed, all of the direct observations are reported at the same time by more than 8000 meteorological stations around the world four times a day. Each report includes observations of clouds, temperature, humidity, wind speed and direction, atmospheric pressure, and precipitation. The observations are transmitted through a three-level communication system. At the first level is a main trunk circuit connecting the *world meteorological centers* in Washington, D.C., Melbourne, and Moscow. With help from high-speed computers and advanced communication devices, the world center relays data to the second level, or *regional meteorological centers.* Here data from around the earth are summarized graphically and revised three or four times each day. The data are then forwarded to the *national meteorological centers.*

Collection and mapping of data

The second stage in the development of a weather forecast consists of collecting all data and displaying it in the form of readily interpretable maps and charts. Thanks to advanced telecommunication devices mounted on satellites, the collection of data from stations on land and at sea is accomplished with extraordinary efficiency. Not only are data from surface observations collected, but also upper-atmosphere observations recorded by radiosondes (see Fig. 14-3) and rawinosondes carried aloft in helium balloons. As noted in Chapter 14, radiosondes record information on temperature, humidity, and atmospheric pressure. The rawinosonde adds information on wind velocities. Finally, surface and upper-atmosphere data are supplemented by information from weather satellites (Fig. 16-19). These marvelous space-age devices are increasingly used to determine the distribution of clouds (Fig. 16-20) and the size of storm systems. Satellite measurements are particularly valuable over the oceans, where observation stations are few and far between.

The data collected at surface stations, from instrument packages suspended by balloons, and from satellites are used in developing current weather maps. A portion of one of these maps and a diagram explaining its symbols are depicted in Figure 16-21. Note that a circle marks the location of the weather observation station. The circle can be viewed as the head of an arrow from which a shaft extends in the direction from which the wind is blowing. Each full barb on the shaft represents a wind speed of 10 knots, with a half barb indicating 5 knots. A percentage of the circle may be colored black to show the extent of cloud cover. The station model depicted in Figure 16-21B indicates a wind speed of 20 knots blowing from the northwest. It is snowing at this station, and there is total cloud cover. Visibility is ¾ mile, dew point is 30°F, amount of precipitation, which began about four hours previously, is 0.45 inch, past weather was rain, and barometric pressure reduced to sea level is 1024.7 millibars.

Figure 16–19 The geostationary orbiting environmental satellite turns through the same arc distance in the same time as the earth. As a result, it is always above the same point on the earth's surface. This satellite can transmit essentially continuous photo coverage of a large region of the earth. (Courtesy of NOAA.)

Analysis of data

After the basic information has been placed on weather maps, meteorologists begin the task of analysis. Isobars (lines of equal pressure) are sketched onto the weather maps and centers of high and low pressure located. Determinations of pressure gradients are made and used to verify wind speed observations. From temperature data, maps showing isotherms (lines of equal temperature) are constructed. Shaded areas representing current rainfall are applied to the basic weather maps. In addition, upper-air maps are developed which show atmospheric conditions at or near the level of the jet streams. As noted earlier, these high-level conditions affect our weather at the surface.

The forecast

Having completed the collection, display, and analysis of atmospheric conditions, the final step of actually forecasting the weather can be attempted. Many methods are combined to make the prediction. Initially, the meteorologist scrutinizes the weather maps and develops a forecast based on recent weather history. The assumption is made that present conditions will persist for a short period of time. The history of an existing cold front, for example, may reveal that it is moving southeastward at 35 knots. Assuming it will continue to move in the same direction and approximate speed, one can predict its arrival at specific communities in its path. But atmospheric conditions may change, and since change is not considered, some predictions may fail. For this reason, wary forecasters seldom use this method for periods longer than 6 to 12 hours.

To forecast days into the future, computer-based numerical methods are used. Here one starts with the known initial state of such atmospheric variables as wind, temperature, and pressure, and calculates the changes likely to occur in each of these variables within a short period, routinely 10 minutes. These

Figure 16–20 Satellite photographic panorama of the United States for November 17, 1981. Layered clouds are wrapped around a storm off the North Carolina coast. High-level clouds are visible over Kansas and Oklahoma, whereas low-level clouds cover parts of Texas and form a band from the Ohio Valley to Minnesota. (Courtesy of NOAA.)

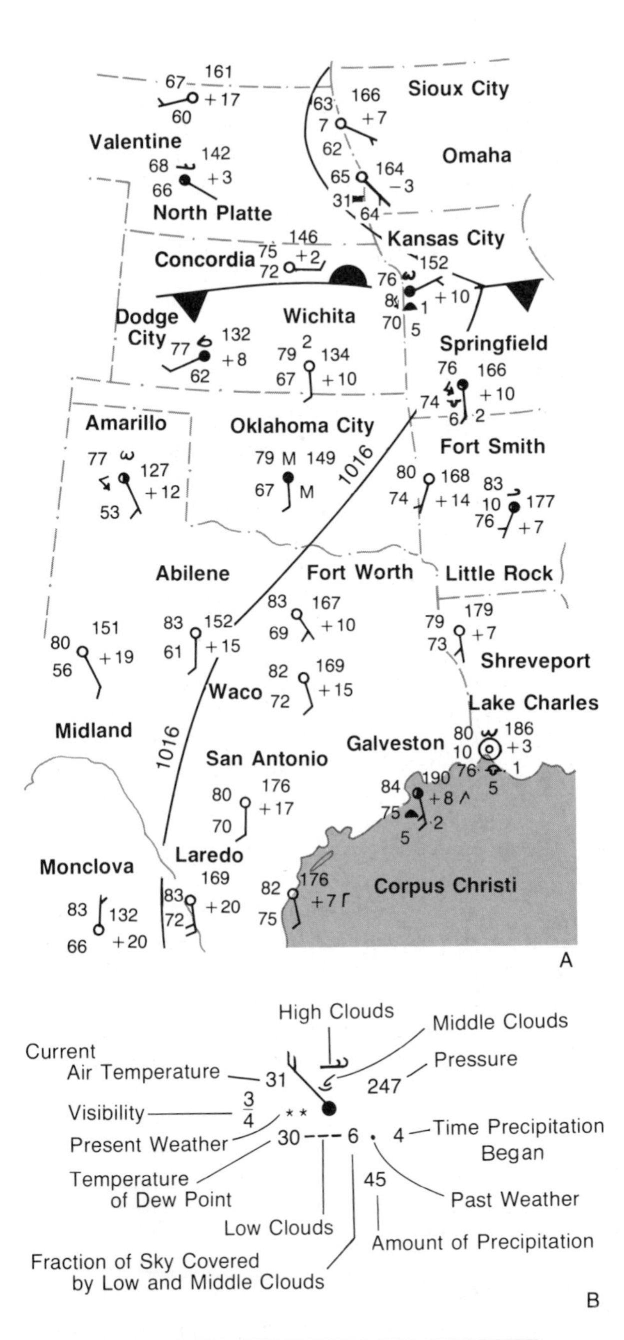

Figure 16–21 Simplified portion of a weather map of the U.S. is shown in A. An example of a station notation and associated symbols is depicted in B. (From Navarra, J.G., Atmosphere, Weather, and Climate. *Philadelphia, Saunders College Publishing, 1979.)*

calculations yield a new set of conditions that are used to calculate further changes over succeeding short intervals for as many hours or days in the future as feasible. Statistical procedures are also used. Meteorologists may statistically examine past data to discern which relationships are predictive. They can then apply what has been learned about currently known weather conditions to arrive at a prediction of future conditions. In the United States and most developed countries, the weather services combine statistical methods with numerical forecasting and the constant evaluation of data on weather maps to arrive at the daily forecasts.

CLIMATE

The role of insolation

Climate and weather are closely related phenomena. We have already noted that climate is the average weather of a particular area, as well as its extremes of variation. Thus, the same factors (insolation, wind, and moisture) that control weather govern climate as well. Insolation—the amount of heat received from the sun—is the most fundamental of these factors. For any given location on the earth, insolation is controlled by the angle at which the sun's rays strike the surface and by the duration of daylight. Both conditions are in turn related to latitude and the tilt of the earth's axis. In general, insolation decreases as one proceeds away from the equator. At the equator itself, variations in insolation are minimal. The sun along the equator is either directly overhead or close to that position each day. As a result, there is little variation in temperature throughout the year. This condition is reflected on a world map by the location of the so-called **heat equator,** a line connecting points of highest average temperature for any given time interval (Fig. 16–22). This line is slightly to the north of, but roughly parallel to, the geographic equator. Poleward from the heat equator, climates become generally cooler, largely because of the increasing slant of incoming solar radiation. There are, however, many variations in this trend resulting from the influences of wind, ocean currents, altitude, and the juxtaposition of continents and oceanic tracts.

Wind and precipitation

Wind—its usual direction and its force—is most certainly an important element of climate. In general, the

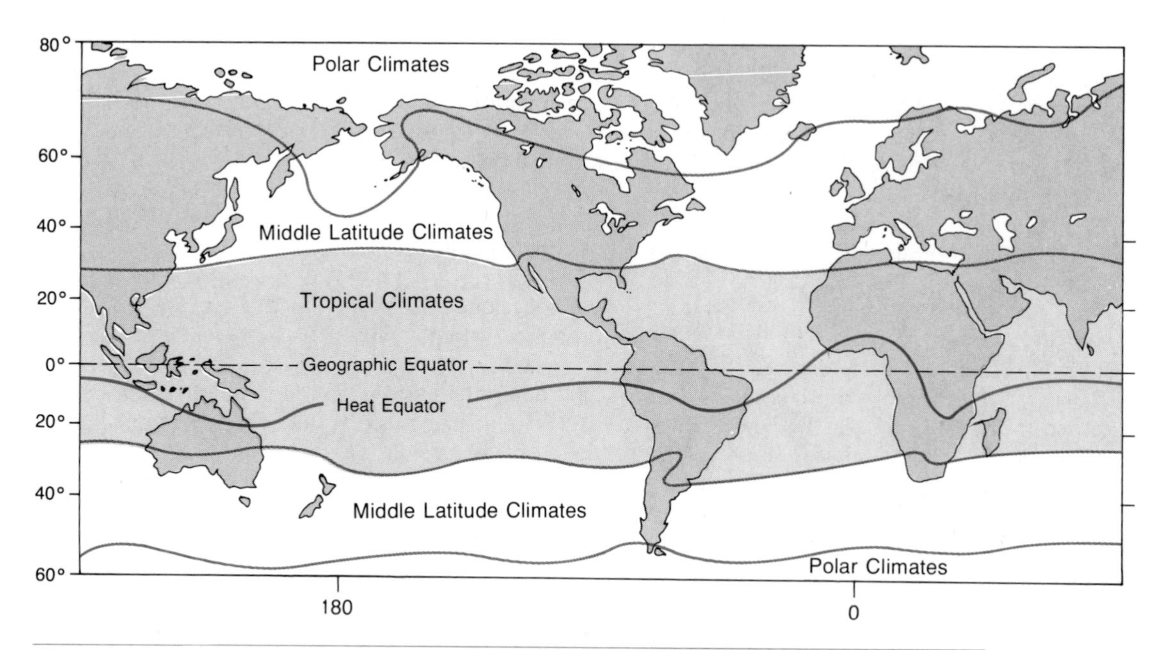

Figure 16–22 *The heat equator for January. The heat equator does not follow the geographic equator perfectly because of differences in heating between land and water. With the same amount of insolation, land heats faster and to a higher temperature than water. Wind and ocean currents also cause minor deflections in the heat equator. The map also depicts the generalized boundaries of major climatic zones.*

wind blowing across a particular location tends to parallel the direction of movement within a larger major wind belt. There are, of course, exceptions to this generalization. The seasonal monsoons, caused by the difference in heating between continent and ocean, regularly provide winds that have no relation to wind belts and blow in one direction during the summer and the opposite direction in winter. Storms and local weather also provide variations within the larger trend of the wind belt.

The amount and nature of precipitation in an area is another important element of climate. One can make certain generalizations about precipitation. Regions of the earth that are persistently covered by low-pressure belts, for example, experience generally heavy precipitation. This is especially true around the equatorial low, where moist, warm air is rising and rainfall attains a yearly average of about 1900 mm. This glut of precipitation decreases away from the equator, and in belts of descending air along the horse latitudes, average annual rainfall is only about 800 mm. Still farther toward the poles precipitation again increases, reaching as much as 1200 mm per year along the subpolar lows. At still higher latitudes, the amount of precipitation decreases rapidly to less than about 50 mm. (Actual precipitation in these high latitudes is in the form of snow, but for comparison, the snowfall is converted here to millimeters of rainfall.)

Because the direct rays of the sun shift as the earth makes its annual orbital journey, so also is there a seasonal shift in pressure belts and associated zones of precipitation. Thus, the band encircling the globe from about 10° to 20° latitude is likely to have more precipitation in summer than in winter. The equator, however, has heavy precipitation throughout the year. Because of the continuous interplay between cold polar and warm equator air masses, the subpolar lows have moderately abundant annual precipitation.

Pressure belts are certainly a dominant influence on the major zones of high or low precipitation. This does not mean, however, that all locations within a given zone will have the same quantity of precipitation, for the direction of prevailing winds relative to continents and oceans may result in wide variations at particular localities. For example, in the ordinarily dry zones between 20° and 30° latitude, easterly trade winds blowing across oceanic tracts bring moist air to the eastern borders of continents that would otherwise be dry. In fact, further inland and along the west coasts of these latitudes, drier conditions usually prevail. In contrast, continental masses in the belt of the prevailing westerlies (about 40° to 60° latitude) generally have ample rainfall on their west coasts. The air, having lost its moisture along the coast, is dry as it moves farther inland. East coasts may escape aridity when there are landward-blowing winds. Finally, we have already noted how mountain belts may receive precipitation from air forced up windward slopes. This leaves the region to the lee of the mountains with minimal precipitation.

Summary The condition of the atmosphere close to the ground at any given time is called weather. The advent of a particular kind of weather and subsequent changes in weather are usually related to the passage of large, horizontally homogeneous bodies of air called air masses. Air masses tend to take on the temperature and moisture characteristics of the surfaces above which they formed. There are two main categories of air masses, tropical and polar. The former develops over warm equatorial regions, and the latter in the frigid high latitudes. Each of these major air mass types has a continental and maritime subdivision, depending on whether the air mass originated over land or ocean.

When an air mass having particular temperature and moisture characteristics converges on another having dissimilar characteristics, the boundary between the two bodies is termed a front. The passage of a front nearly always brings a change in the weather. The principal kinds of fronts are cold fronts, warm fronts, stationary fronts, and occluded fronts. A cold front is the advancing edge of a cold air mass. As the dense, cold air advances, it remains close to the ground, pushing itself under adjacent bodies of warm air and forcing the warm air aloft. Gusty winds and heavy precipitation may result from this movement.

A warm front occurs when the advancing edge of warm air displaces colder air at the surface. Because of the lesser density, the warm air cannot penetrate the cold air mass from below, and so it overrides it. In general, weather events associated with the advance of a warm air front are more extensive and of longer duration than along a cold front. When there is little or no forward motion along a front, it is called a stationary front. If a cold front overtakes and slides beneath a warm front, lifting the warm air off the surface, the result is an occluded front.

Cold and stationary fronts in the middle latitudes tend to develop large undulations, which are called wave cyclones. Wave cyclones are really giant low-pressure systems that form along the boundary between polar and tropical air masses. The two air masses have different densities, and air flows in opposite directions along each side of the front they produce. Soon the front assumes a wavelike form and the counterclockwise circulation of a wave cyclone is established. The wave cyclone begins to dissipate when the cold front overtakes the warm front so as to form an occluded front.

Occasional storms also characterize our atmosphere. Hurricanes are severe storms that develop in the North Atlantic, Caribbean Sea, Gulf of Mexico, and the eastern North Pacific off the coast of Mexico. Exceptionally strong winds and heavy rains accompany hurricanes. Tornadoes develop mostly over continents during hot humid days. Although they affect only a relatively small area and have short duration, they are extraordinarily violent. Less violent but nevertheless dangerous are thunderstorms, which can bring damaging high winds, brief torrential rains, lightning, and hail.

Various regions of the earth tend to have a characteristic assemblage of meteorologic elements that together constitute climate. Unlike weather, which comprises the condition and behavior of the atmosphere at a given time, climate implies the totality of day-to-day weather conditions over an extended period. Climate is not just average weather; it also includes weather extremes, such as greatest and least rainfall and highest and lowest temperatures. Climate, like weather, is largely controlled by insolation, wind, and precipitation. Other factors, like position relative to continents and oceans, altitude, and local geographic features, secondarily influence climate.

QUESTIONS FOR REVIEW

1 What distinction is usually made between the terms weather and climate?

2 What is an air mass? How does an air mass originate? What are the source regions for air masses affecting North America?

3 What is a weather front? Contrast the weather

likely to occur along cold fronts and warm fronts during the spring in the middle latitudes. How does the weather along a cold front differ from weather within the central area of the cold air mass?

4 Why is the boundary along a weather front inclined? What factors influence the severity of storms that may occur along a cold front?

5 How might *cP* and *mT* air masses differ in moisture content and temperature?

6 What is an occluded front, and under what circumstances does it develop?

7 How do hurricanes differ from middle-latitude cyclones? Where do most of the hurricanes that affect the United States originate?

8 Describe the three developmental stages that characterize wave cyclones. Describe those for thunderstorms.

9 What is lightning? How does it originate? What is the cause of thunder? Why does one often see lightning before hearing thunder?

10 Why do climates become cooler poleward from the heat equator?

11 In general, why does the region around the equatorial lows have greater rainfall than the region that straddles the horse latitudes?

SUPPLEMENTAL READINGS AND REFERENCES

Battan, L.J., 1983. *Weather in Your Life.* San Francisco, W.H. Freeman & Co.

Chalmers, J.A., 1968. *Atmospheric Electricity.* New York, Pergamon Press.

Eagleman, J.R., 1976. *The Visualization of Climate.* Lexington, Mass., D.C. Heath & Co.

Eagleman, J.R., Muirhead, V.U., and Willems, N., 1975. *Thunderstorms, Tornadoes, and Building Damage.* Lexington, Mass., D.C. Heath & Co.

Few, A.A., 1979. Thunder, in *The Physics of Everyday Phenomena.* Readings from *Sci. Amer.*, San Francisco, W.H. Freeman & Co.

17
planet earth

The theory of the Earth's motion is admittedly difficult to comprehend, for it runs counter to appearances and all tradition. But if God wills, I shall in this book make it clearer than the sun, at least to mathematicians.
COPERNICUS, 1507

The earth rising over the lunar landscape. (NASA photograph by Apollo 17 astronauts in lunar orbit.)

Prelude Imagine that you have taken a journey by spacecraft to a location high above our solar system. From your vantage point in space (and provided you had suitable equipment) you might look downward and study the entire solar system. You would see at its center a rather ordinary star, our sun, circled by nine planets (Fig. 17–1). On close examination you would discern that the four planets nearest the sun were relatively small (Fig. 17–2), and that the next four were bigger and spaced more widely apart. The last one, Pluto, would be seen at a distance from the sun of more than 40 times that which separates the earth from the sun. Between Mars and Jupiter one might detect a belt of asteroids.

Even from so distant an observation point, you might discern some of the characteristics of the third planet, the earth. It is the largest of the inner group of rocky or terrestrial planets. You might detect its blue and white color and notice that its moon, illuminated on one side by the sun, is larger relative to the size of the earth than the moon of any other planet except that of distant Pluto. The earth differs also in that it contains more water than any other planet. It occupies a fortunate place in space, for it is neither so close to the sun as to be seared by intense heat nor so far away as to be locked in intense cold. The swirling cloud patterns visible from space are evidence of its atmosphere and rotation. Patches of green suggest that life has developed and has been hospitably sustained.

Given sufficient time, an observer in space would be able to detect not only rotation but also the passage of the earth and other planets around the sun. It would become evident that the paths of the planets lie in approximately the same plane. Thus, the solar system is rather like a disc with the sun at its center. Looking down on the solar system from a location above the sun's north pole, one would recognize an interesting uniformity of movements. The revolution of the earth

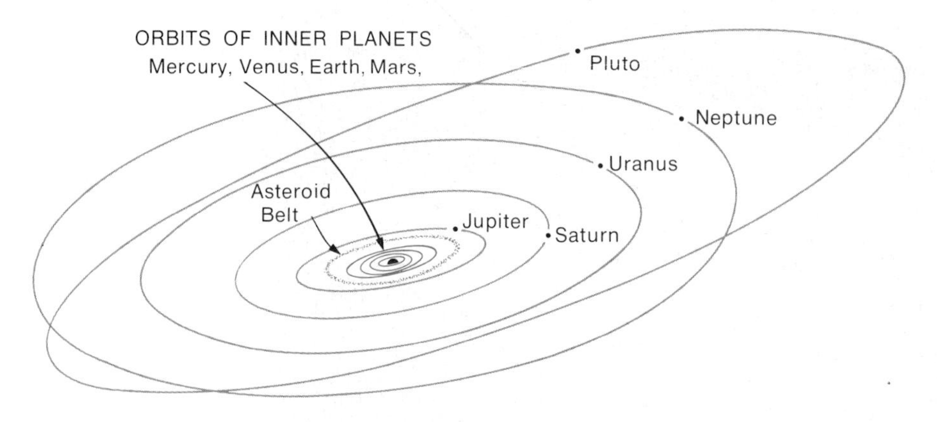

Figure 17–1 *Orbits of the nine planets in the solar system.*

about the sun, of the moon about the earth, and the rotation of all three bodies on their individual axes are all in the same counterclockwise direction. For the earth, these movements relative to the sun provide a means for keeping track of time. They give us day and night, seasons, and tides.

As our spacecraft departs from the solar system and enters the farther realms of space, we might look back to see that out system is only a small part of an enormously larger aggregate of stars, planets, dust, and gases called a galaxy. Our galaxy is called the Milky Way because as we look in a direction parallel to the plane of its disclike form, we see a milky haze of light emanating from the many stars located along that plane (Fig. 17–3). The huge spiral of the Milky Way Galaxy may be more than 100,000 light years in diameter. Our sun is located about two thirds of the distance outward from its center.

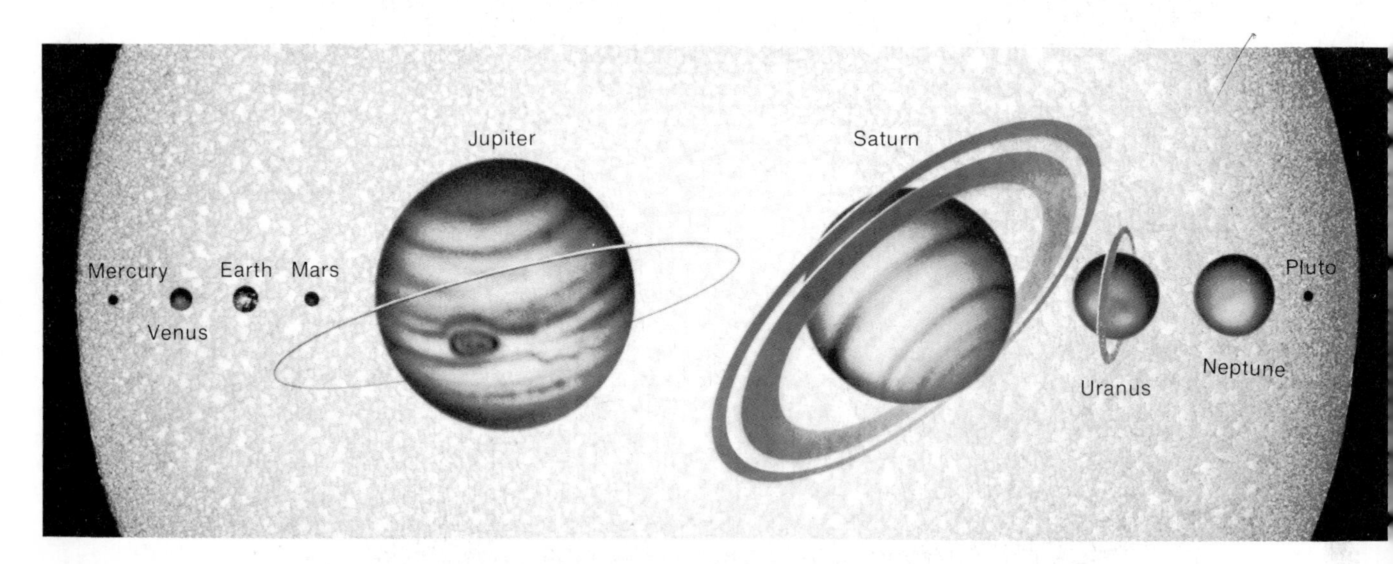

Figure 17–2 *Diagram of the planets arranged along the equator of the sun to show their relative sizes. (Modified from Pasachoff, Jay M.,* Contemporary Astronomy, *3rd ed., Philadelphia, Saunders College Publishing, 1985.)*

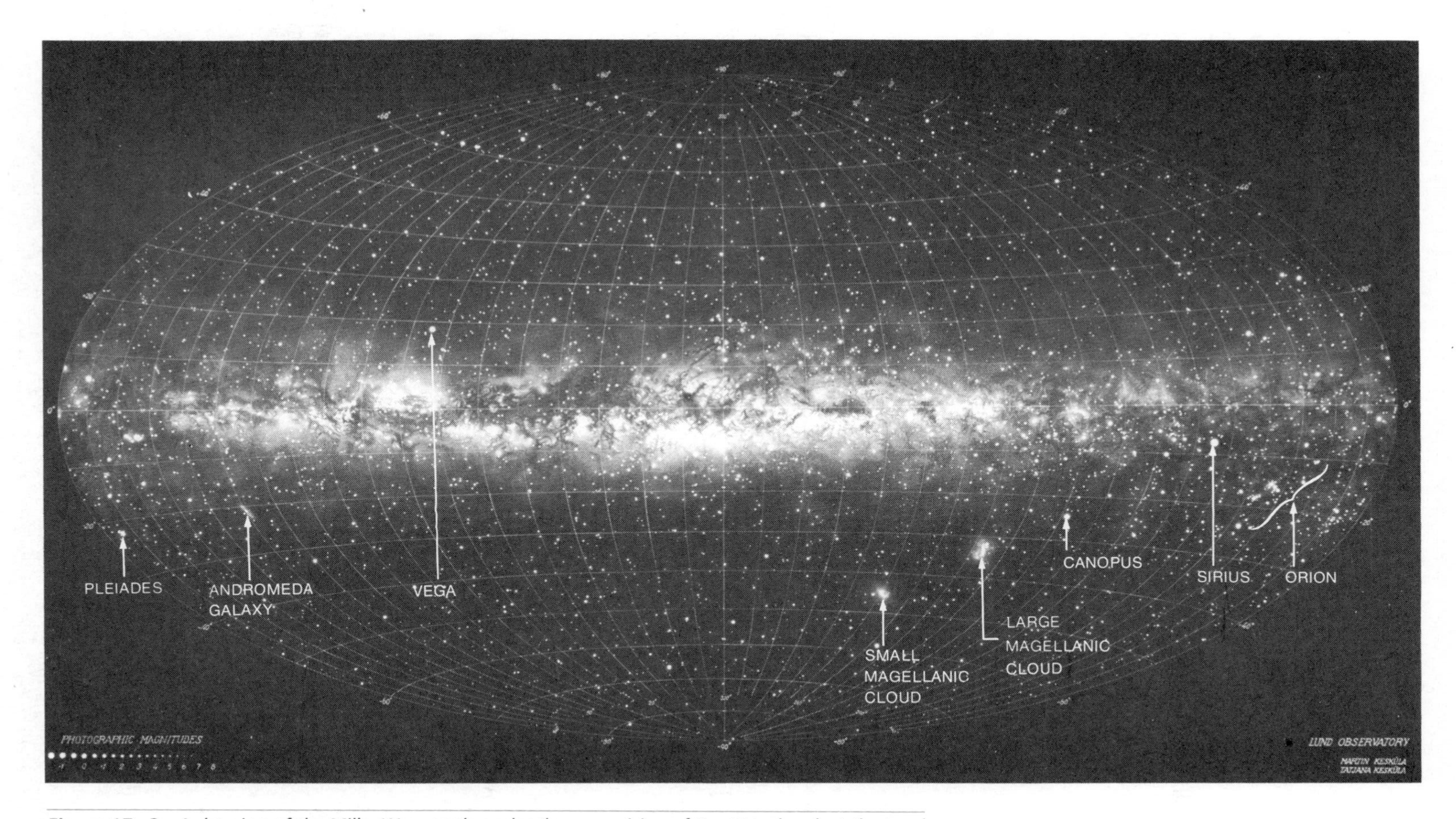

Figure 17–3 *A drawing of the Milky Way, made under the supervision of Knut Lundmark at the Lund Observatory in Sweden. 7000 stars plus the Milky Way are shown in this panorama, which is in coordinates such that the Milky Way falls along the equator. (Courtesy of Lund Observatory, Sweden; from Pasachoff, J.M.,* Contemporary Astronomy, *3rd ed. Philadelphia, Saunders College Publishing, 1985.)*

MOVEMENTS

The two major movements of the earth are rotation and revolution. **Rotation** refers to the spinning of a planet or satellite on its axis, whereas **revolution** is the orbiting of one body around another. Thus, the earth *rotates* on its axis and *revolves* around the sun.

Rotation

Only about 600 years ago, it was widely accepted that the stars, the planets, and the sun moved around the earth, which stood still at the center of the universe. This seemed a logical premise to anyone who casually observed celestial objects move across the sky. From the time of Copernicus, however, indisputable evidence has accumulated indicating that the apparent motion of stars across the sky is the result of the motion of the earth.

The earth rotates counterclockwise when viewed from above the North Pole, and therefore each of us is carried eastward beneath the celestial scenery. At the equator, the velocity of that movement is more than 1600 km per hour. Although the earth's rotation has recently been observed directly from spacecraft, it is interesting to recall an experiment that demonstrated the rotation of the earth before the modern space age. The experiment was performed in 1851 by the French physicist Foucault. His experimental apparatus consisted of a heavy lead ball suspended from a high ceiling by a wire. The wire was more than 60 meters long, and the ball at the end of the wire was free to swing as a pendulum in all directions. Foucault placed a smooth layer of sand beneath the ball so that a point attached to its bottom would mark a path in the sand. When he started the pendulum swinging by drawing it to one side and carefully releasing it, he found that the ball did not swing across the room and return to its exact starting point. Instead, there was a slight deflection with each sweep of the pendulum. Indeed, the plane of oscillation appeared to be slowly shifting clockwise (Fig. 17-4A).

If the earth were stationary, then the ball, released at point A in Figure 17-4 would have swung to B and returned to A. The earth's rotation beneath the freely swinging pendulum, however, caused the pendulum to *appear* to change its direction, and it followed the curved path to C. On the return swing, it was deflected to the right of its new motion and swung to D rather than E. The pendulum was actually maintaining its direction of swing as the earth moved beneath it. As the earth rotated counterclockwise, the plane in which the pendulum oscillated was rotated in a clockwise sense (in the Northern Hemisphere). Foucault found the shift to be 11° per hour in Paris. Later it was noted that pendulums placed at locations farther away from the equator had a greater rate of shift, until at the poles the rate reached the maximum of 15° per hour. It is easy to see the effect of Foucault's experiment if one imagines the pendulum swinging over the polar region, as indicated in Figure 17-4B.

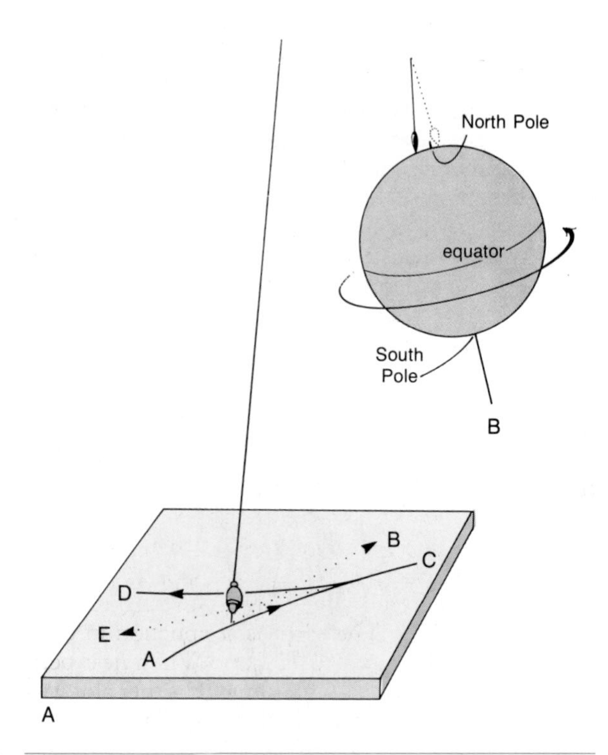

Figure 17-4 (A) The motion of the Foucault pendulum. The deflection is exaggerated. The deflection is more easily understood if one imagines that the pendulum is suspended over the North Pole, as in B.

Time and rotation

One source for our measurement of time is the nearly constant rotation of the earth. This rotation accounts for the *apparent* motion of stars across the sky, which is the basis for **sidereal time** (star time), and for the apparent motion of the sun, which is the basis for **solar time.** It is solar time that we use in our everyday activities and that is recorded by ordinary clocks.

The length of the day is based upon the rotational period of the earth. This is the time required for the earth to make one complete rotation in relation to the sun's position. Because the earth rotates toward the east, the sun seems to rise in the east, travel across the sky, and set in the west. Noon on any specific day is defined as the moment the sun reaches its highest point in the sky. The time that elapses between two successive noons, called the **apparent solar day,** is not exactly 24 hours, and it varies slightly through the year. The variation in apparent solar day occurs because the earth's orbital motion is somewhat faster when the planet is closer to the sun, as well as because the sun's path does not run exactly east and west in the sky but is inclined to the equator by 23.5°. As illustrated in Figure 17-5, the earth moves along an elliptical orbital path and therefore approaches the sun more closely on one side of its orbit than the other. When the earth is at the point nearest the sun, it is said to be at **perihelion,** whereas **aphelion** describes its most distant point. The earth is at aphelion in early July and perihelion in early January. When near perihelion, our planet advances faster and farther along its orbit than when it swings by the aphelion position. While it is making more rapid progress around the sun during the same interval of rotation, the sun appears to move more slowly across the sky than at other times of the year.

The second cause of the slight variations in the duration of an apparent solar day results from the 23.5° tilt of the earth's equator from the plane of the earth's orbit (called the **plane of the ecliptic**). As a result, the sun appears to move during the year along a path inclined with respect to the equator, and this causes a yearly periodic variation in the duration of an apparent solar day.

Variations in the length of the apparent solar day would not exist if the earth had a perfectly circular orbit and no axial tilt. Under such imaginary circumstances, the length of each day would be the same. Because they vary slightly in duration, it would be bothersome and difficult to keep track of the length of apparent solar days. During one time of the year a day would encompass 24 hours and 2 seconds, and during another 23 hours, 59 minutes and 52.2 seconds. Clearly, for keeping track of time we require a day of unvarying length. Such a day would have as its duration the *average* of all apparent solar days of the year. That average is precisely 24 hours and is named a **mean solar day.**

Star time

As mentioned in the previous section, the length of an apparent solar day can be considered as the interval between two successive appearances of the sun at its highest point in the sky. The **sidereal day** can be measured by two successive appearances of a particular star at an identical position in the sky. The sidereal day is about four minutes shorter than the 24-hour mean solar day, or about 23 hours and 56 minutes. The reason for the difference between the two can be easily understood from Figure 17-6. When the earth has traveled around the sun from position A to B, it has completed one sidereal day. Before the earth has completed one rotation relative to the sun, however, it must rotate one more degree at the end of the sidereal

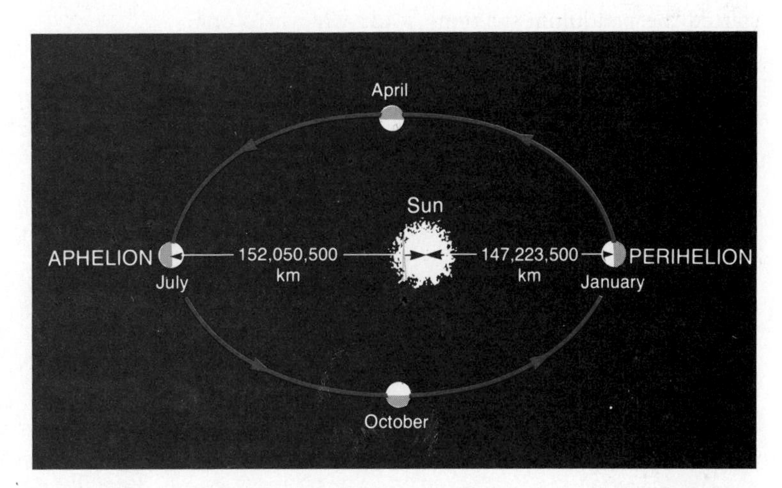

Figure 17-5 The earth's orbit as seen from above so as to show perihelion and aphelion. Orbital eccentricity is exaggerated. The mean distance of the earth to the sun is about 149,504,000 km (92,897,000 miles), a distance equivalent to 1 astronomic unit (1 A.U.).

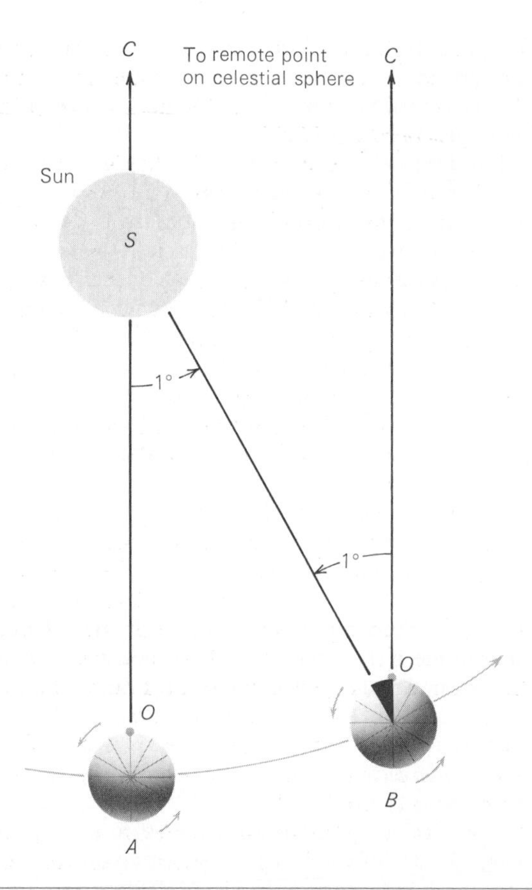

Figure 17–6 The difference between sidereal and solar days. Point O on the earth makes one complete revolution with respect to the stars as the earth moves from A to B. It will take almost 4 minutes more to complete one revolution relative to the sun. (From Abell, G.O., Realm of the Universe, *3rd ed. Philadelphia, Saunders College Publishing, 1984.)*

day. To accomplish this, it is necessary for the planet to move in its orbit for another four minutes.

Standard time and longitude

An understanding of differences in time around the planet is made simpler if one is familiar with the grid system by which we describe locations on earth. The east-west lines that encircle a globe parallel to the equator are called **parallels of latitude.** Zero degree of latitude corresponds to the equator, and 90°N to the North Pole. There are also north-south lines on the globe, which converge at the poles and are called **meridians of longitude.** By international agreement, the zero line of longitude is placed so as to cross the Royal Observatory in Greenwich, England (Fig. 17–7).

Figure 17–7 The Royal Observatory, Greenwich, England. By straddling a brass marker in the courtyard of the observatory, one can place one foot in the Eastern Hemisphere and one in the Western Hemisphere.

Longitude is measured in degrees east or west of the zero meridian, which is also called the **prime meridian** (Fig. 17–8). Meridians are usually drawn every 15° around the globe to the east and west of the prime meridian, to the 180° meridian. Locations to the east of the prime meridian are a certain number of degrees of east longitude, whereas those to the west have west longitude.

Meridians of longitude are the basis for the modern system of standard time. Many of us have had the experience of traveling east or west across the conti-

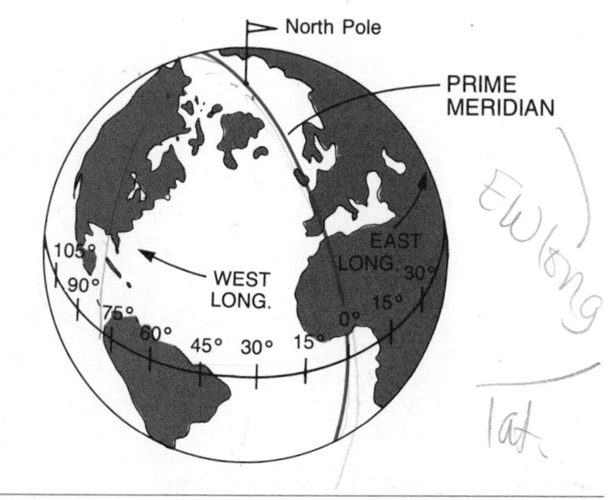

Figure 17–8 The prime meridian, from which east and west longitude are measured.

nent and having to periodically adjust our clocks as we pass into new time zones. The ultimate reason for this rather bothersome procedure is the earth's rotation. Because of rotation, the sun appears to rise in the east, and people living in the eastern part of the continent see the sun rise before people far to the west. For example, when sunrise occurs at 6 A.M. in Chicago, it is an hour past 6 A.M. in New York.

The earth rotates through 360° in 24 hours, or 15° each hour (360° ÷ 24 = 15°). Hence, for every 15° east in longitude, time is one hour later, and for every 15° west in longitude, time is one hour earlier. For each 1° the earth turns, four minutes elapse. Imagine two towns, 1° of longitude apart. Time in the eastern town would be about 4 minutes earlier than in the other town. If it were 11 A.M. in the first town, mean solar time in the second town would be 10:56 A.M. Imagine the confusion if each town used its own local mean solar time. For the United States, this problem was solved in 1883 when the railroads established four standard time zones (named Eastern, Central, Mountain and Pacific time). Within each zone, the time everywhere was to be that of an important city in that zone. Four standard meridians (75°W, 90°W, 105°W, and 120°W) pass through the approximate center of each zone (Fig. 17–9).

Shortly after the establishment of the U.S. system of time zones, an international scheme was developed in which the globe is marked off into 24 standard time zones. These are based upon meridians of longitude 15° apart, beginning with the prime meridian. Standard time in each zone is taken as the mean solar time in the center of that zone. The scheme, however, is not rigidly practiced worldwide.

For the world traveler, there is a further complication. Suppose you leave from New York on a Friday on a westward flight around the world. As you passed over the Central time zone, you would set your wristwatch back one hour. You would reset the watch again over the Mountain and Pacific zones and continue turning it back one hour for each 15° of longitude traversed. After a flight completely around the earth you would have set your watch back 24 times. On landing in New York you would think the day to be Friday, but you would discover the day was actually Saturday.

To avoid such confusion, an imaginary line near (but not always on) the 180th meridian has been designated the **international date line** (Fig. 17–10). In your imaginary trip westward, when you crossed this line you would change not your watch but the calendar. For example, if the international date line were crossed going from east to west at 6 P.M. on Friday, the time would become 6 P.M. on Saturday. This would compensate for the 24 times that you set your watch back one hour. When traveling eastward, a day is repeated as the line is crossed.

For scientific purposes, it is convenient to have a universal system of time that does not depend on location. Universal time, abbreviated UT, is commonly

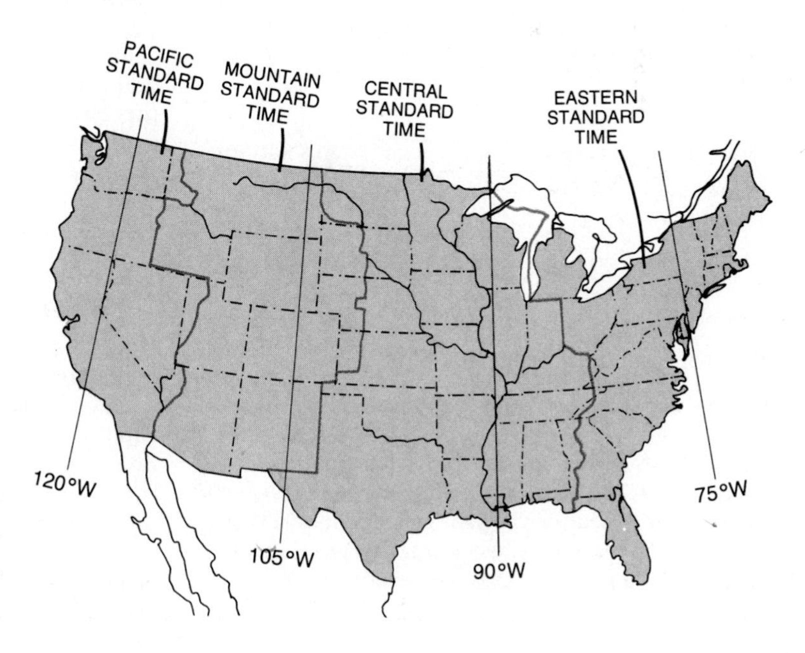

Figure 17–9 Standard time zones in the United States. The boundaries of time zones are not always parallel to meridians.

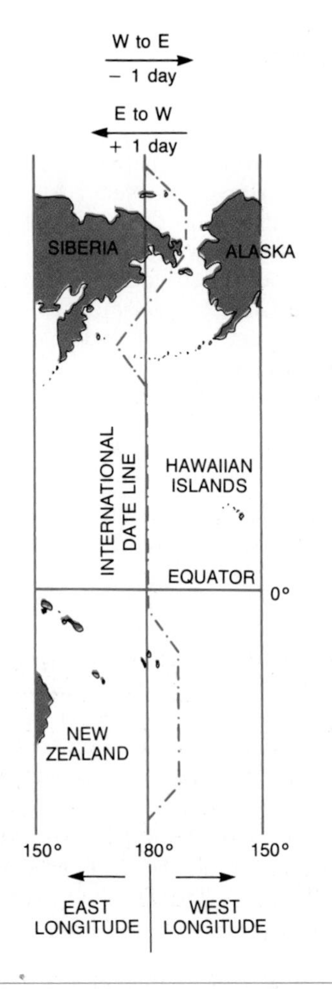

Figure 17–10 The international date line.

used for this purpose. UT is the local mean solar time of the meridian running through Greenwich, England.

Revolution of the earth around the sun

The tropical and the sidereal year

Having examined some of the effects of the earth's rotation, we are now ready to consider the second most important motion of the earth, its revolution. As noted earlier, the term revolution refers to the orbital passage of one body (such as the earth) around a larger body (such as the sun). The earth completes one revolution around the sun, measured relative to the stars, in a period of time called the sidereal year. In our ordinary activities, we do not employ the sidereal year but use the tropical year, which is 20 minutes shorter. To understand the basis for the tropical year, it is necessary to recall that the earth's axis of rotation is tilted at an angle of 23.5° from the perpendicular to the plane of the ecliptic, and except for a very slow conical motion, remains in the same orientation as the earth revolves around the sun (see Fig. 17-13). Two positions along the earth's orbit are designated equinoxes. At equinox, which means "equal night," the lengths of the day and the night are equal, and the sun appears directly overhead at noon on the equator. During the autumnal equinox, the sun appears to cross the equator from north to south, and during the vernal equinox, the sun crosses the equator from south to north. Knowing this, we can define the *tropical year* as the time interval between two successive vernal equinoxes.

The tropical year is shorter than the sidereal year because equinoxes slowly shift their positions in the earth's orbit as a result of the previously mentioned slow conical motion of the earth's axis. That motion is called **precession** (Fig. 17-11). An analogy to axial precession is seen in an ordinary spinning top (Fig. 17-12) in which the upper end slowly moves in a circle, while the top itself rotates rapidly. Precession results from the gravitational attraction of the sun and moon for the earth. The sun pulls the equatorial bulge toward the plane of the ecliptic while simultaneously the moon pulls the earth's bulge toward its own orbital plane. What all of this means is that the orientation of the earth's axis shifts steadily through a circular path, completing a precession once in about every 26,000 years. When viewed from above the North Pole, the precessional movement is clockwise. Currently, the earth's axis points toward the star Polaris, but about 12,000 years hence it will be pointing toward the star Vega in the constellation of Lyra.

The obliquity of the ecliptic

While the earth makes its annual trip around the sun, it rotates about 365½ times on its axis. The earth's equatorial plane during this revolution remains tilted about 23.5° relative to the plane of the earth's orbit (the plane of the ecliptic). That 23.5° angle between the earth's equatorial plane and the plane of the ecliptic is dubbed the **obliquity of the ecliptic.** It is because of the obliquity of the ecliptic that we experience seasons.

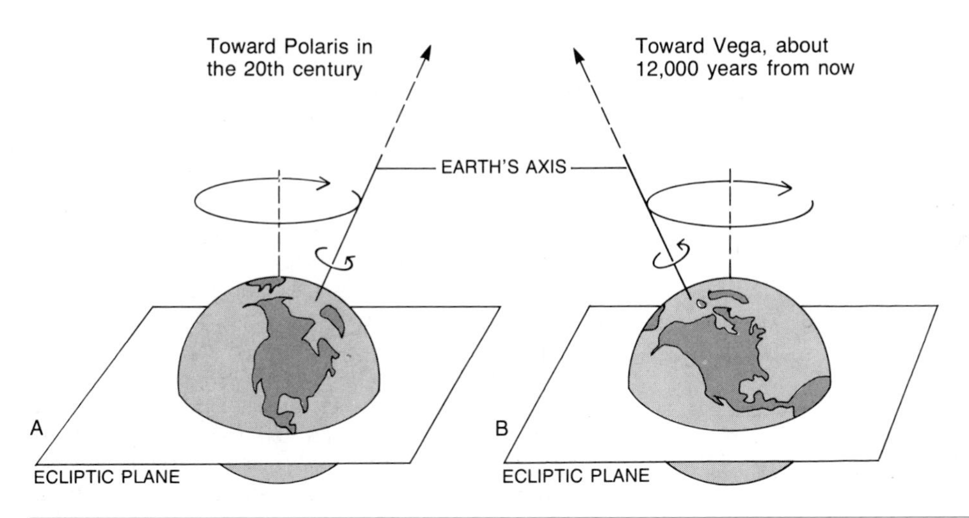

Figure 17–11 Two positions in the precession of the earth's axis of rotation. (Modified from Pasachoff, J. M., Contemporary Astronomy, *3rd ed. Philadelphia, Saunders College Publishing, 1985.)*

As the earth revolves around the sun, the axis maintains its orientation, so the North Pole is tilted first toward the sun and then away from it. As indicated in Figure 17-13, the north end of the axis is tilted away from the sun during December, and so the Northern Hemisphere receives relatively less heat. During this time, the Northern Hemisphere experiences winter. (It would be summer in the Southern Hemisphere). The sun on December 22d would have reached the most southerly point in its annual migration and would stand directly overhead at noon on the Tropic of Capricorn (23.5°S latitude). At this time, both the elevation of the sun above the horizon and the length of the day are at their lowest values in the Northern Hemisphere. At the North Pole, the sun does not rise at all.

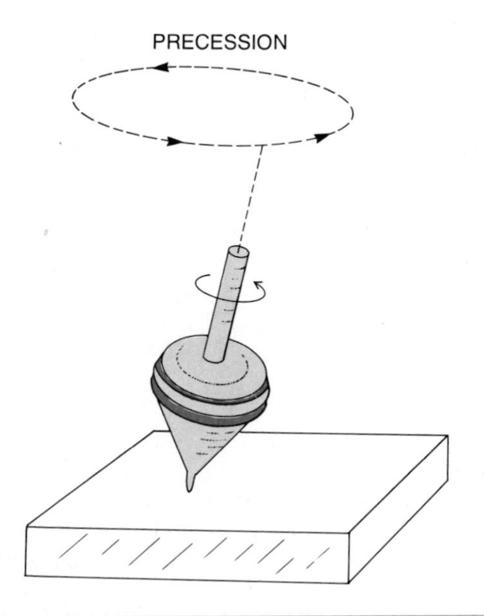

Figure 17–12 Precession of the earth's axis resembles the slow wobble of a rapidly rotating top.

In June, the North Pole is tilted toward the sun and the Northern Hemisphere experiences summer. At this time, both the elevation of the sun above the horizon and the length of day are at their maximum.

THE EARTH'S SIZE AND BULK

Dimensions

From classical times, it has been known that the earth is roughly spherical in shape. Aristotle (384-322 B.C.) knew this because of the curved form of the shadow cast by the earth onto the moon during a lunar eclipse. Somewhat later, Eratosthenes arrived at a similar conclusion and also calculated the earth's circumference with surprising accuracy. Eratosthenes noted that at noon on June 21 at Aswan, Egypt, a plumb line cast no

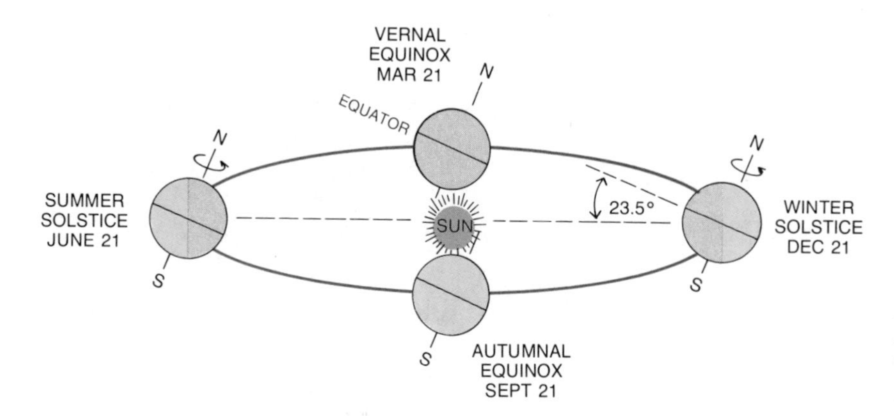

Figure 17–13 *The cause of seasons on earth. Note that the earth's axis always points in the same direction, toward the north star Polaris.*

shadow. At Alexandria, located 925 km north of Aswan, the plumb line did cast a shadow. The angle between the outer edge of the shadow and the plumb line was about 7° (Fig. 17–14). By geometry, 7° would equal the angle between lines extending from the two plumb lines to the earth's center. The 7° angle is 1/50 of 360°, and therefore the approximate circumference would be 50 × 925 km, or 46,000 km. The earth's average true circumference is 40,010 km.

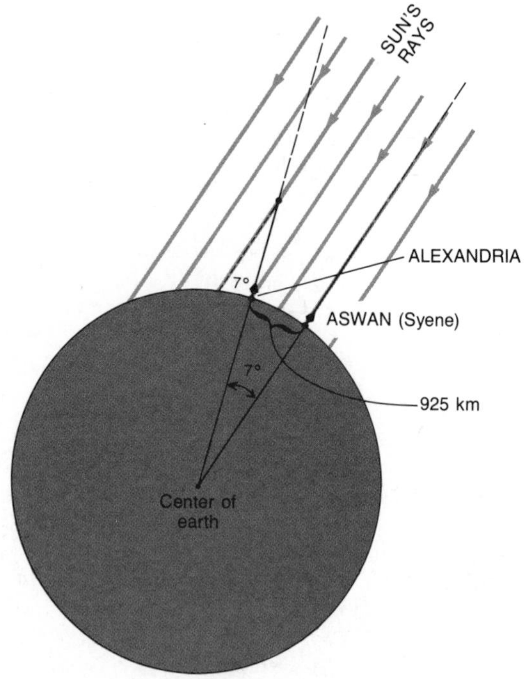

Figure 17–14 *The method used by Eratosthenes to measure the circumference of the earth.*

Today, we are fortunate in having a very direct way of determining the earth's dimensions. We simply make measurements from photographs transmitted to earth from artificial satellites and spacecraft. In the opening illustration for this chapter the sphericity of the earth is clearly evident. One can only imagine the delight Eratosthenes would have experienced if he could have had a similar view of the earth through the window of the Apollo spacecraft.

Eratosthenes thought of the earth as being perfectly spherical, but actually the planet is slightly oblate, like a somewhat flattened ball. The spinning of the earth on its axis causes polar regions to flatten slightly and equatorial zones to bulge. Thus, the earth's equatorial radius (6378 km) is 21 km longer than its polar radius. This is a very small variation, and on a model of the earth 760 centimeters in diameter, it would amount to less than 3 centimeters.

The earth's mass, volume, and density

The term *mass* refers to the quantity of matter an object contains. The earth's mass, calculated from the gravitational force it exerts, has been found to be 6×10^{24} kilograms. The best value we have for the total volume of the earth is 1.08×10^{12} cubic kilometers. Once the mass and volume of the earth are known, it is possible to determine its density. The *density* of any object is defined as its mass divided by its volume $\left(D = \frac{m}{V}\right)$. The earth has a density of 5.5 grams per cubic centimeter. This density is greater than that of any other planet in our solar system, but not appreciably different from that of Mercury, Venus, and Mars.

Because the density of surface rocks on earth is only about 2.7 g/cm^3, the material lying deep in the earth's interior must have density values well in excess of 5.5 g/cm^3.

EARTH-MOON RELATIONSHIPS

Lunar motions

The moon, which is an average distance of 384,393 km from the earth, is our closest neighbor in the solar system. Because of the moon's proximity and relatively large size, the moon and earth are a sort of double planet. In fact, although the moon appears to revolve around the earth as the earth revolves around the sun, actually the earth and moon revolve around their common center of mass. This center of mass is called the **barycenter** (Fig. 17–15). It lies on a line that joins the center of the earth and moon, and it moves around the sun in an elliptical orbit.

Whereas the earth travels around the sun in a year, the moon encircles the earth once each month, or more precisely, in about 27.33 days. This period of time is called a **sidereal month.** The sidereal month is measured relative to the stars. As shown in Figure 17–16, because the earth is itself moving around the sun, the moon must travel farther than one sidereal revolution to complete a revolution relative to the sun.

When the sun is used as a reference point, the moon revolves around the earth in about 29.5 days. This is called the **synodic month,** and is the interval of time from one new moon to the next.

The moon's orbit is an ellipse that is inclined to the orbital plane of the earth's orbit by an average of 5.1°. As a result of this difference, the moon does not ordinarily block the light coming to the earth from the sun, nor does the earth interfere with light reaching the moon. The exceptions are the eclipses to be described subsequently.

As the moon revolves around the earth, it simultaneously turns on its axis, completing one rotation in precisely the same time it takes to go around the earth. It is for this reason that the moon always presents the same face to the earth.

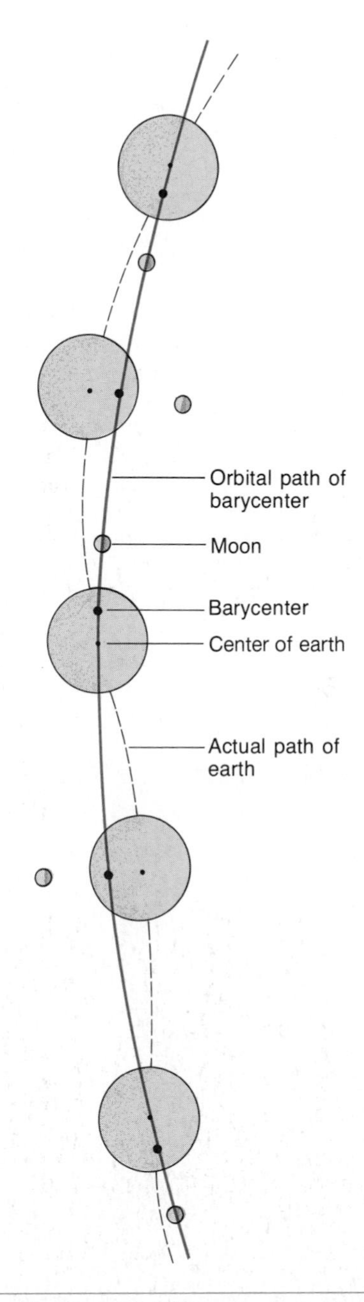

Figure 17–15 The center of mass for the earth-moon system is called the barycenter. *It is actually the barycenter rather than the center of the earth that follows an elliptical path around the sun. The earth's path has a slight wobble, exaggerated here.*

Phases of the moon

The most striking thing the ordinary observer notices about the moon is the change in area that appears to be illuminated when we view the moon from earth. These changes are caused by the orbital motion of our satellite and are called the **phases of the moon.** Because the moon is without illumination of its own, its brightness results from reflected sunlight. Thus, the

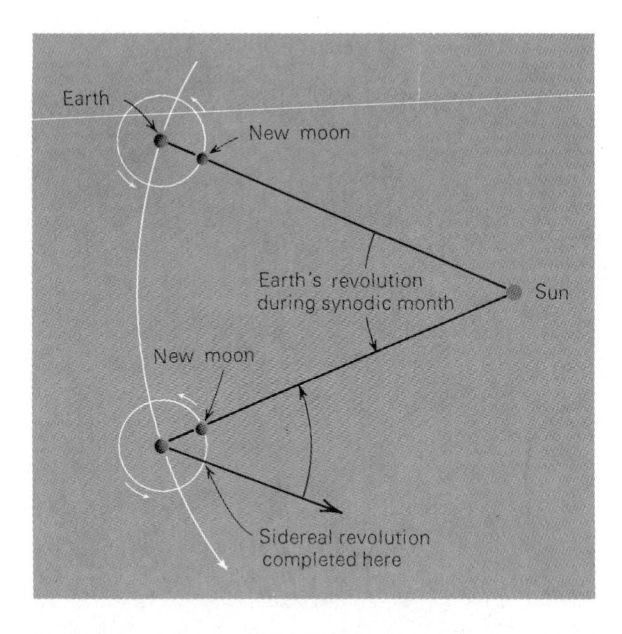

Figure 17–16 The difference between the sidereal and synodic months. (From Abell, G.O., Realm of the Universe, *3rd ed. Philadelphia, Saunders College Publishing, 1984.)*

side of the moon that faces the sun is brilliantly lighted and the opposite side is dark.

When the moon is between the earth and the sun, it appears dark (Fig. 17–17). This phase is called the **new moon** and is usually thought of as starting the synodic month. The new moon rises and sets with the sun. Within two or three days, the motion of the moon carries it to a position where a small waxing crescent of illumination can be seen from earth. About a week after the time of the new moon, it has traveled about a quarter of its path around the earth and is now about 90° east of the sun. The moon now appears to have half of its face illuminated and is described as being in its first quarter. This first-quarter moon rises at about noon and sets near midnight. Thereafter the moon continues eastward, causing the half-moon appearance to pass through what is called the waxing gibbous phase. About 15 days after new moon, the full moon phase is reached. Because the moon is now opposite the sun as seen from earth, it rises just as the sun sets and is up all night long. After full moon, the phases are reversed, so that the moon passes from full, to waning gibbous, to third quarter, to waning crescent and back to new moon.

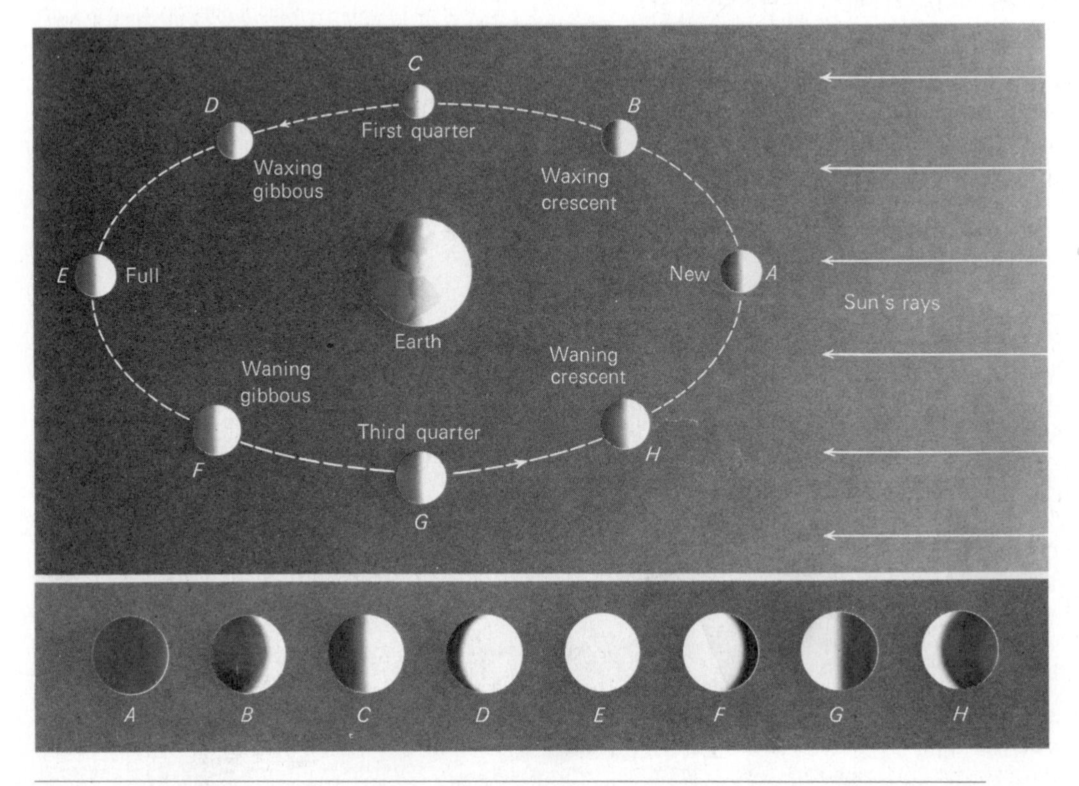

Figure 17–17 The phases of the moon. In the upper drawing, the moon's orbit is viewed obliquely. (From Abell, G.O., Realm of the Universe, *3rd ed. Saunders College Publishing, 1984.)*

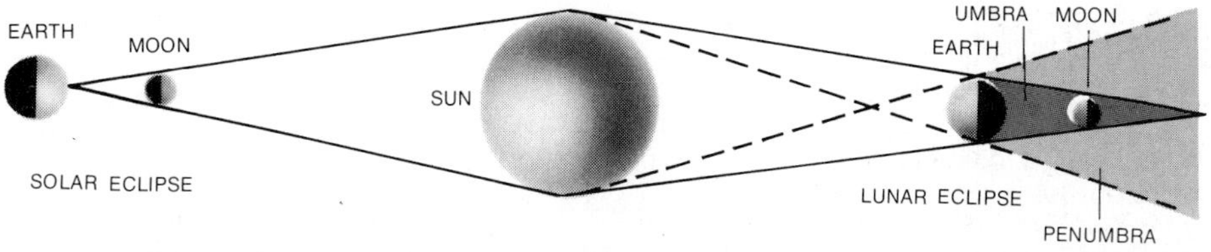

Figure 17–18 *An eclipse of the sun occurs when the moon is between the earth and the sun. When the moon is on the far side of the earth from the sun, we see a lunar eclipse. That part of the earth's shadow from which the sun is only partially shielded from the moon's view is called the* penumbra, *whereas the part of the earth's shadow from which the sun is entirely shielded from the moon's view is called the* umbra. *(Drawing not to scale. From Pasachoff, J.M.,* Astronomy: From the Earth to the Universe, *2nd ed. Philadelphia, Saunders College Publishing, 1983.)*

Eclipses of the moon and sun

About twice a year as the moon makes its monthly journey around the earth, it passes through the shadow of the earth. The result is an eclipse of the moon. For the moon to get into the earth's shadow, the alignment of the earth, moon, and sun must be nearly perfect. When the earth lies precisely between the moon and the sun at a time of full moon, the shadow of the earth is cast on the moon and a **lunar eclipse** occurs (Fig. 17–18). The maximum duration of the eclipse is around 3 hours and 45 minutes. Should full moon occur when the moon and sun are near but not exactly aligned, the eclipse will be incomplete (or partial) and of shorter duration.

A **solar eclipse** occurs if, during a time of new moon, the moon moves exactly between the earth and the sun. Under these circumstances, the shadow of the moon is cast on the earth. The duration of the total solar eclipse can be up to about 7½ minutes. During a total eclipse, the moon casts a dark shadow called the **umbra.** People not in the shadow of the umbra will not see a *total eclipse* because only a part of the sun will be blocked out, and they will witness a *partial eclipse.* The partial eclipse is seen within the area of weaker shadow surrounding the umbra that is called the **penumbra.**

Summary

The daily passage of day into night and the annual progression of seasons are the result of the earth's two major motions: rotation and revolution. The earth rotates, or spins, about an imaginary line extending from pole to pole. The axis of rotation is tilted 23.5° from a line perpendicular to the earth's orbital plane. The axial tilt results in a corresponding 23½° angle between the earth's orbital plane and its equatorial plane, and this relationship is responsible for the seasons. During rotation, the earth turns through 360° in 24 hours or 15° each hour. Thus, there is a time change of one hour for each 15° of longitude.

Revolution, the second basic earth movement, refers to the travel of the earth around the sun. The earth's orbit is slightly elliptical, and the planet completes one orbital trip every 365 days, 5 hours, and 48 minutes. The plane of the earth's orbit is called the plane of the ecliptic.

Because of a slight equatorial bulge resulting from its rotation, the earth is not a perfect sphere but rather an oblate spheroid. The equatorial radius is about 6378 km, and the planet as a whole has a density of 5.5 grams per cubic centimeter.

The moon follows an elliptical path around the earth. Its average distance from the earth is about 384,393 km, and it requires 29½ days to make the revolution as measured relative to the sun. The revolution of our satellite around the earth causes the new, first-quarter, full, and third-quarter phases of the moon. A lunar eclipse occurs when the moon passes into the earth's shadow. A solar eclipse occurs when the moon passes between the earth and the sun and causes the moon's shadow to fall upon the earth.

QUESTIONS FOR REVIEW

1. Imagine a planet like the earth in every way except that its axis is perpendicular to the plane of the ecliptic. What characteristics of the earth would be lacking on such a planet?
2. Using the Eratosthenes' method, determine the circumference of a fictional planet if the distance between the two cities corresponding to Aswan and Alexandria is 1000 kilometers and the angle between the shadow and plumb bob at the city corresponding to Alexandria is 12°.
3. Why is it that we do not see an eclipse of the sun every month?
4. For a person living in the Northern Hemisphere, which season occurs when the noon sun is directly overhead at 23.5°S latitude?
5. Describe the motion of a Foucault Pendulum suspended at the North Pole. How many degrees would the pendulum shift in one hour? Why?
6. What are meridians? Why are they normally drawn every 15° on a globe? How many of these 15° meridians are there on the globe? What is the maximum longitude? What is the latitude of your home town or city? What is the maximum latitude?
7. Where is the international date line located? If you crossed this line when traveling from east to west, would you set your clock forward or backward, and by what amount?
8. What is the difference between the sidereal and synodic month? What is the reason for this difference?
9. Why is the same side of the moon always facing the earth?

SUPPLEMENTAL READINGS AND REFERENCES

Gamow, G., 1963. *A Planet Called Earth*. New York, Viking Press.

Glass, B.P., 1982. *Introduction to Planetary Geology*. New York, Cambridge University Press.

Hartmann, W.K., 1983. *Moons and Planets*. 2d ed. Belmont, Calif., Wadsworth Publishing Co., Inc.

Spar, J., 1962. *Earth, Sea, and Air*. Reading, Mass., Addison-Wesley Publishing Co., Inc.

Skinner, B.J. (ed.), 1981. The solar system and its strange objects. Readings from *American Scientist*. Los Altos, Calif., Wm. Kaufmann, Inc.

Wood, J.A., 1979. *The Solar System*. Englewood Cliffs, N.J., Prentice-Hall, Inc.

York, D., 1975. *Planet Earth*. New York, McGraw-Hill Book Co.

18
planets and moons

Give me matter and motion and I will construct the universe.
DESCARTES, 1640

This image of Europa, smallest of Jupiter's four Galilean satellites, was acquired by Voyager 2 on July 9, 1979, from a range of 241,000 km. Europa, the brightest of the four large inner satellites, has a density that suggests a substantial content of water. Scientists have speculated that the water must have cooled from the interior and formed a mantle of ice perhaps 100 km thick. The complex patterns on its surface suggest that the icy surface was fractured, and the cracks filled with dark material from below. (Courtesy of the National Aeronautics and Space Administration, and R.E. Arvidson.)

Prelude Throughout history, people have explored and developed new frontiers. Yet today on our widely traveled planet, only a few frontiers remain. One essentially boundless frontier does still exist, however. It is the frontier of space.

Space begins some 80 to 160 km above the earth's surface where the atmosphere has thinned to a vacuum. The environment beyond this level is an incredibly harsh one for living things. Our bodies, for example, would explode in the vacuum of space. To prevent this, and to protect against lethal radiation and extremes of temperature, complex spacecraft and cumbersome space suits are required. The spacecraft must have the power to escape the earth's mighty gravitational pull, as well as special shielding to resist the intense heat generated on reentering the earth's atmosphere. Time was needed to overcome these many obstacles, so it was not until the late 1950s that space became a directly explorable frontier. Today, a multitude of artificial satellites orbit the earth providing storm warnings, military intelligence, observations of distant galaxies, crop forecasting, global communications, and much more.

Rather like the opening of the American West, the space frontier will steadily shift farther away from its places of origin. The electronic eyes of earlier artificial satellites were mostly directed down toward the earth. The next step was to look outward at the moon. Having solved the problems of manned space travel, in 1969 humans actually landed on the moon (Fig. 18–1) and explored its surface on foot and in motorized vehicles. Since that time, there have been robot missions to Venus and Mars, probes of Jupiter and Saturn and the development of spacecraft that can enter space and return to the earth repeatedly. These incredible machines have given us a flood of new information about

Figure 18–1 Geologist-astronaut H.H. Schmitt samples the moon's regolith with a specially designed rake. (Courtesy of the National Aeronautics and Space Administration.)

the planets, moons, asteroids, and meteorites that occupy the part of the space frontier we call the solar system (Fig. 18–2). Our neighboring planets are of keen interest. We cannot help but wonder how they originated, of what materials they are made, and how they compare with our own planet. What does Venus look like at ground level beneath its gaseous shroud? How were the channels of Mars developed? Why does Jupiter have a great red spot? What is the nature of the rings encircling Saturn, as well as Jupiter and Uranus? These are some of the questions to be considered in this chapter.

Figure 18–2 The solar system. (From Pasachoff, J.M., Contemporary Astronomy, *3rd ed. Philadelphia, Saunders College Publishing, 1985.)*

THE SUN

The largest and for many reasons the most important member of our solar system is the sun. Although only a modest star compared with others in our galaxy, it nevertheless has a mass 333,000 times that of the earth and is 109 times larger in diameter. It is composed mostly of hydrogen (about 73 percent) and helium (about 25 percent). Remaining heavier elements exist as gases in the hot interior of the star, where temperatures exceed about 15 million degrees Celsius.

That part of the sun we are able to observe directly is called the solar atmosphere. There are three general regions of the solar atmosphere, each having rather transitional boundaries with adjacent regions. The uppermost region is called the **photosphere** (Fig. 18–3). The light that reaches us here on earth emanates from the photosphere. Immediately above the photosphere is the **chromosphere.** One can see the hot hydrogen that forms the chromosphere as a reddish glow visible during a solar eclipse. The chromosphere merges gradually into the outermost atmospheric region known as the **corona.** This seething, tenuous shroud of fiery gases extends millions of kilometers outward from the sun.

Although an enormous amount of solar energy is intercepted by the earth, our planet is neither frozen nor roasted but is able to maintain a range of temperatures roughly between $-50°$ and $+50°C$. The maintenance of this vital temperature range is made possible by three factors. First, because of rotation, the earth receives energy from the sun on one hemisphere at a time, while it returns heat to space over its entire surface. Second, some of the incoming radiation is reflected off the atmosphere and clouds and is directed back into space without ever reaching ground level. Finally, a part of the intercepted radiation is absorbed by the atmosphere and radiated back into space without warming the earth's surface.

There are other forms of radiation besides visible sunlight that are intercepted by the earth (Fig. 18–4). Three percent of the incoming rays are ultraviolet. Because of their destructive effects on life, it is fortunate that most of this short-wavelength radiation is absorbed in the ozone layer of the atmosphere. X-rays and gamma rays are also absorbed in the upper atmosphere. Much of the infrared radiation is absorbed by water vapor and carbon dioxide, causing the lower zones of the atmosphere to be warmed.

The energy that maintains the sun as a great glowing sphere of gases is derived from a continuous thermonuclear reaction called fusion. In the fusion process, four hydrogen nuclei combine in a series of steps to form helium. Helium has slightly less mass than the hydrogen nuclei from which it was derived, and that "missing mass" is converted into energy in the form of heat (Fig. 18–5). Each second, the sun transmits an amount of energy equivalent to that which would result from the burning of 25×10^{18} pounds of coal.

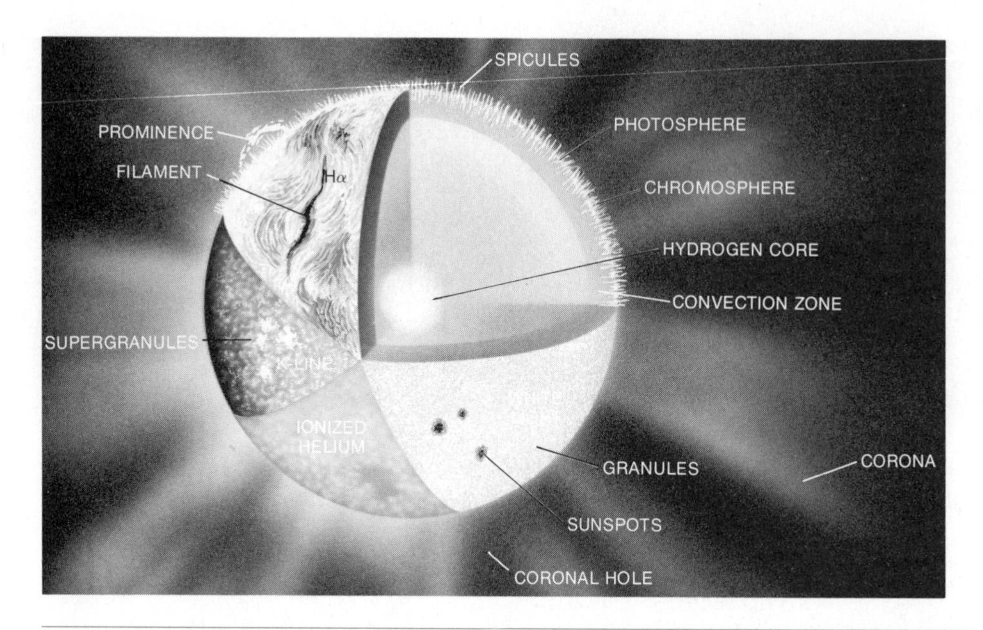

Figure 18–3 *Parts of the solar atmosphere and interior. (From Pasachoff, J.M.,* Contemporary Astronomy, *2nd ed. Philadelphia, Saunders College Publishing, 1981.)*

This energy from the sun is essential for life on earth and is the ultimate force behind the many geologic processes that sculpture the earth's surface. For example, the sun's rays aid in the evaporation of surface waters that in turn results in clouds that provide the precipitation for erosion. Along with the earth's rotation, the sun's radiation results in winds and ocean currents. Some scientists believe that protracted variations in the heat received from the sun may trigger episodes of continental glaciation or may reduce lush forests to barren wastelands. Even primitive people knew the sun's importance and recognized it as the fountainhead and sustainer of life.

THE BIRTH OF THE SOLAR SYSTEM

Scientists now have fairly reliable estimates of when the earth originated. The question of how it originated

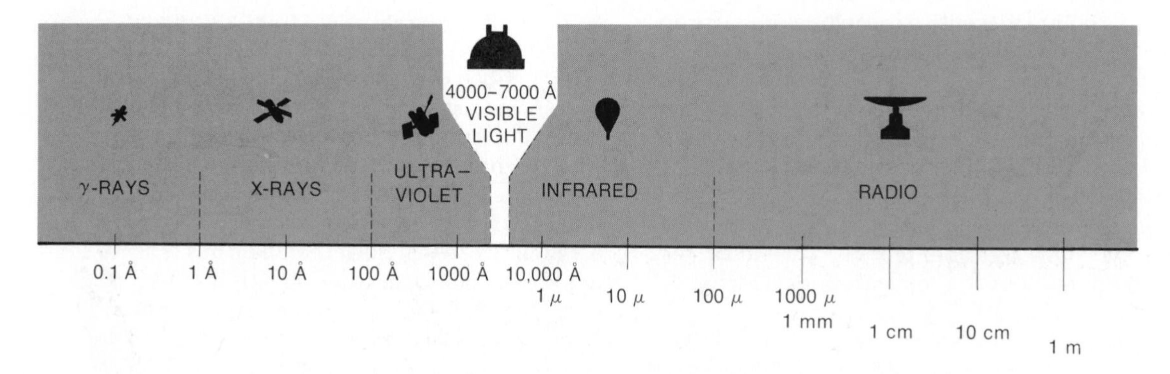

Figure 18–4 *The electromagnetic spectrum. In the scale, an angstrom (Å) is 10^{-8} cm (0.000 000 01 cm) and a micron (μ) is 10^{-4} cm (0.000 1). (From Pasachoff, J.M.,* Contemporary Astronomy, *2nd ed. Philadelphia, Saunders College Publishing, 1981.)*

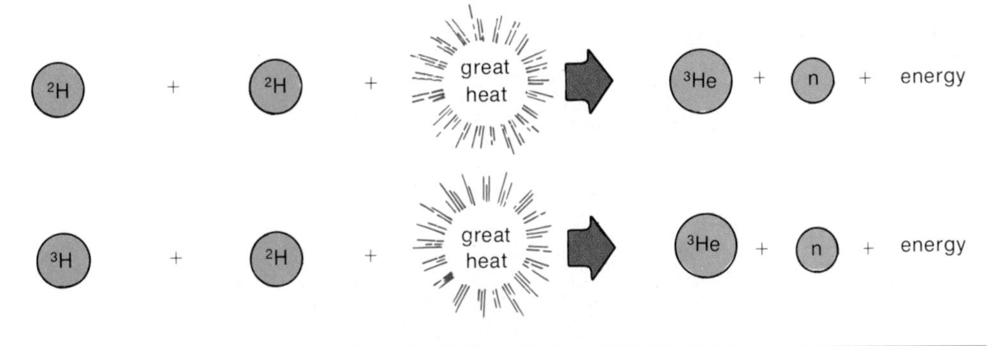

Figure 18–5 Examples of two fusion reactions (n = neutron).

is a subject of continuing investigation. Current programs in space exploration, including the landing of spacecraft on the moon, Venus, and Mars and the fly-by missions to Mercury, Saturn, and Jupiter, have vastly improved our ability to speculate with some validity about the earth's origin. The new observations have added many refinements to older theories, all of which were obliged to account for the following characteristics of the solar system.

1 The orbital paths of planets occur within approximately the same plane, so the solar system has the shape of a flat disc.
2 The orbits are nearly circular and all lie near the plane of the sun's rotation (Fig. 18–6).
3 The direction of planetary revolution is counterclockwise (called prograde) when viewed from a hypothetical point high in space above the earth's North Pole. The direction of rotation of the planets (with the exception of Venus and Uranus) is also counterclockwise.
4 With few exceptions, the angle between the axis of rotation and the pole or orbit of each planet is small (Fig. 18–7). This angle of inclination is 23.5° for the earth.
5 The sun is overwhelmingly the most massive body in the solar system. It contains 99.9 percent of all the mass in the system. However, the sun is rotating so slowly that it has only 2 percent of the total angular momentum in the solar system. (The momentum of a body is the product of its mass and its velocity. Angular momentum is the momentum multiplied by the radial distance of the body from the point about which it is rotating.)

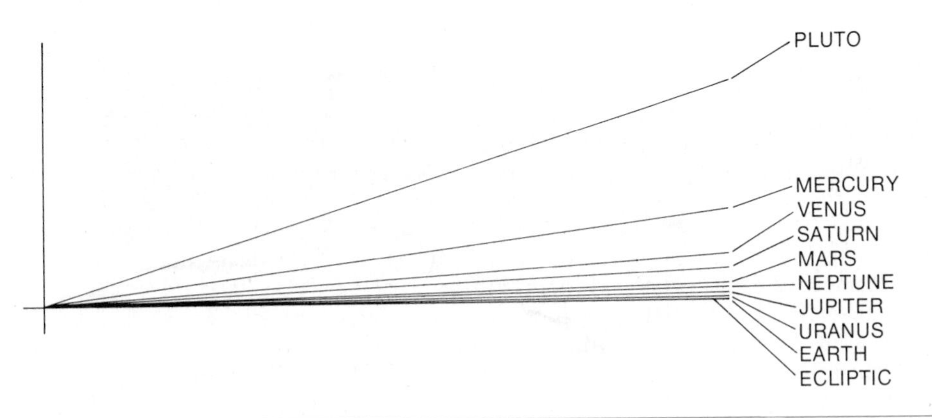

Figure 18–6 The orbits of the planets, with the exception of Pluto, have only small inclinations to the ecliptic plane. (From Pasachoff, J.M. Contemporary Astronomy, *2nd ed. Philadelphia, Saunders College Publishing, 1981.)*

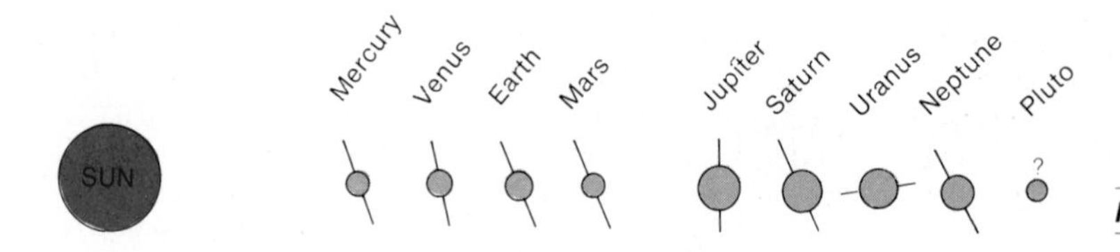

Figure 18–7 *Planetary angles of inclination.*

6 Most planetary distances from the sun can be described by a relationship called **Bode's Law.** Bode, a German astronomer, noted that all planets (except Pluto) have distances from the sun that fit a single mathematical sequence (provided the asteroid belt is counted as a missing fifth planet). The sequence is obtained by writing down the numbers 0, 3, 6, 12, etc., so that each number after the first is double the preceding one. If 4 is added to each number and then divided by 10, the resulting numbers are the approximate mean distances of the planets from the sun in astronomic units (one astronomic unit being roughly the mean distance from the earth to the sun).

7 With few exceptions, satellite systems tend to mimic the movements of the larger planetary bodies.

8 As indicated by their differing densities, as well as by moon, earth, and meteorite samples, the planets have different compositions. Mercury, Venus, Earth, and Mars have mean densities of 5.4, 5.2, 5.5, and 3.9, respectively. They constitute the small, dense, inner or terrestrial planets. Jupiter, Saturn, Uranus, and Neptune have lesser densities of 1.34, 0.69, 1.27, and 1.58, respectively. They are termed the outer, or **Jovian planets.**

9 Hydrogen and helium constitute about 98 percent of the sun's mass. In contrast, the terrestrial planets have large percentages of such elements as oxygen, silicon, and iron. Once again, the Jovian planets are observed to contain an enormous proportion of hydrogen and helium.

The nebular hypothesis

One of the earliest scientific attempts to explain the origin of the solar system was made in 1755 by the German philosopher Immanuel Kant. About 40 years later, the brilliant French mathematician Pierre de Laplace independently developed a theory much like that of Kant, but he described it in a more comprehensive and scientific manner. This early account of the earth's origin was titled the **nebular hypothesis.** Kant and Laplace speculated that the solar system was once a great cloud of slowly rotating hot, gaseous material. As it cooled, it shrank, and as it shrank, it spun ever more rapidly, much as an ice skater spins faster when he pulls in his arms. The spinning motion would cause the nebula to flatten to a disclike shape. Eventually, centrifugal force around the margin of the disc would become sufficient to cause a ring of material to separate and be left behind as the rest of the nebula continued to contract. Subsequently, as the shrinking parent disc continued to spin faster, smaller rings would separate. In the next stage, matter in each ring would begin to condense, particles would collect into large aggregates, and aggregates would eventually collide and collect into planets and satellites. Laplace had neatly accounted for the flattened shape of the solar system, the circular orbits of the planets, and their spacing outward from the sun. However, the theory had a serious flaw. It predicted that the sun should have most of the angular momentum, and as we have noted, this is not the case.

The planetesimal hypothesis

Recognition of the inability of the nebular hypothesis to account for the slow rotation of the sun led to other theories that were based on a chance encounter between the sun and another body. One such theory, called the **planetesimal hypothesis,** was proposed in 1900 by Forest R. Moulton, an astronomer, and Thomas C. Chamberlin, a geologist. These two University of Chicago professors proposed that another star passed so near the sun that it raised enormous tides and pulled outward great gaseous filaments of matter. The material of the filaments condensed, formed into gases and particles, and the particles in turn accreted into larger masses (planetesimals) that ultimately evolved into planets and satellites. The passing star was considered to have imparted a strong oblique thrust to the ejected materials. Thus, the sun was not caused to spin rapidly, but considerable momentum was transferred to the outer material that was to become the planets. In the two decades that followed presentation of the planetesimal hypothesis, several modifications of the basic encounter concept were published. The best known of these was developed by James H. Jeans and Harold Jeffries. Their theory was based on the grazing collision of a passing star with the sun and did not depend on tidal disturbances.

The encounter hypotheses were not without weaknesses, and they soon came to be viewed with skepticism. Collisions of stars appear to be rare events in our galaxy, and calculations suggested that the filamentous matter torn from the sun would simply dissipate into space rather than be drawn together into planets.

The protoplanet hypothesis

From turbulent eddies to protoplanets

As a result of dissatisfaction with the encounter theories, astronomers began to reevaluate the old nebular concepts in the light of new information about the characteristics of the sun. The revised theory, sometimes called the **protoplanet hypothesis,** was first proposed by C.F. Von Weizacker and later modified by Gerard P. Kuiper. The hypothesis begins with a cold, rarefied cloud of dust particles and gases. The cloud gradually contracts and flattens as it takes on a counterclockwise rotation. About 90 percent of its mass remains concentrated in the thicker central part of the disclike cloud. While rotation and contraction are occurring, the phenomenon of turbulence (not considered in the earlier nebular hypothesis) begins to affect the cloud system. Any departure from the smooth flow of a gas or fluid constitutes turbulence. In the dust cloud it consisted of chaotic swirling and churning movements that were superimposed on the grander primary motion of the entire cloud. Each turbulent eddy was affected by the movement of adjacent eddies, and some served as collecting sites for matter brought to their location by neighboring swirls. When the disc had shrunk to a size somewhat larger than the present solar system, its now denser condition permitted condensation of the concentrated knots of matter. Smaller particles merged to build chunks up to meters in diameter, and the dense swarms swept up finer particles within their orbital paths, thereby growing in size. The swarms of solids and gases were designated **protoplanets** by Kuiper. They were enormously larger than present-day planets. Each rotated somewhat like a miniature dust cloud, and each eventually swept away most of the debris in its orbital path and was able to revolve around the central mass without collision with other protoplanets (Fig. 18–8). Their

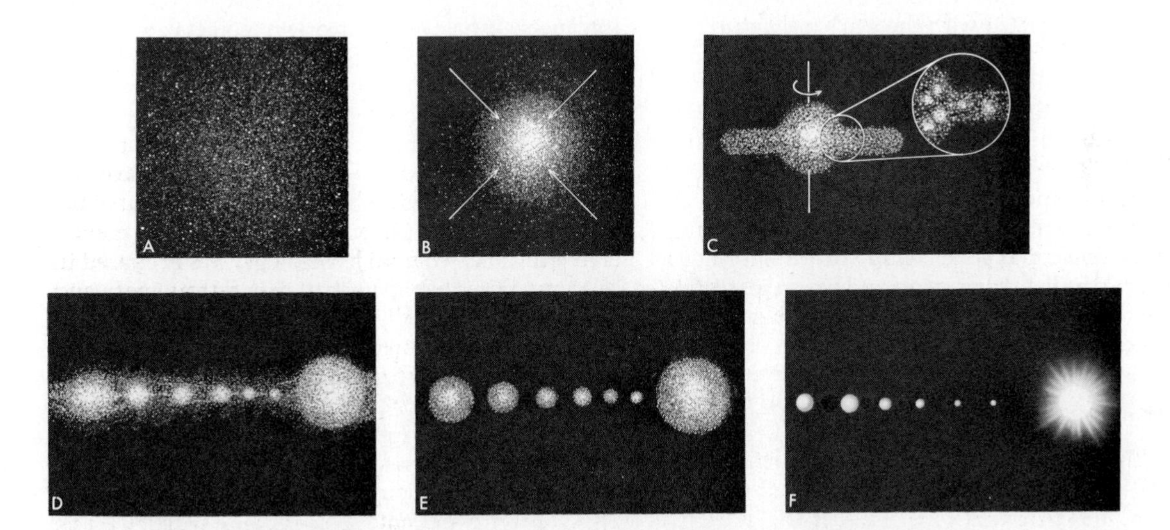

Figure 18–8 *An illustration of the protoplanet hypothesis. A and B show the dust cloud taking form and beginning to rotate counterclockwise as it contracts and flattens. In C the nebula is distinctly disklike, with a central bulge representing the protosun. Materials in the thin portion of the nebula condense and accrete to form protoplanets (D, E, and F), which in turn contract to become planets. (From Pasachoff, J.M.,* Contemporary Astronomy, *2nd ed. Philadelphia, Saunders College Publishing, 1981.)*

formation from the original dark, cold cloud required an estimated 10 million years.

Within the protoplanets, condensation produced solid particles, and accretion of the particles into larger chunks led eventually from protoplanets to true planetary bodies. The process is presumed to have involved gravitational settling of the heavier particles (like iron) toward the center of the protoplanet, contraction of the initially rarefied cloud, and with that contraction, an increase in the rate at which the mass rotated.

The birth of the sun and planets

While the protoplanets were in the process of condensation and accretion, material that had been pulled into the central region of the nebula condensed, shrank, and was heated to several million degrees by gravitational compression. Later, when pressures and temperatures within the core of the sun became sufficiently high, thermonuclear reactions began, pouring out additional energy derived from matter by the nuclear fusion of hydrogen atoms into atoms of helium. (Today, the sun converts about 596 million metric tons of hydrogen into 532 million metric tons of helium each second. The difference represents matter that has been converted to energy.)

As the sun began its thermal history, its radiation ionized surrounding gases in the nebula. These ionized gases in turn interacted with the sun's magnetic lines of force, causing its magnetic field to adhere to the surrounding nebula and drag it around during rotation. At the same time, the nebula caused a drag on the magnetic field and slowed down the sun's rotation. In this way, the old problem of why the sun rotates more slowly was explained.

The stream of radiation and particles from the sun, sometimes called the **solar wind,** drove enormous quantities of the lighter elements and frozen gases outward into space. (This solar force is what causes a comet's tail to show wavy streaming as illustrated in Figure 18-9, or to be bent away from the sun.) As would be expected, the planets nearest the sun lost enormous quantities of lighter matter and thus have smaller masses but greater densities than the outer planets (Table 18-1). They are composed primarily of rock and metal that could not be moved away by the solar wind. The Jovian planets were the least affected and retained their considerable volumes of hydrogen and helium.

Asteroids and comets

Proponents of the protoplanet hypothesis assume that a major protoplanet failed to develop in the belt now occupied by the asteroids. Possibly a number of small eddies evolved in this zone and eventually produced the larger asteroids. Some of these may have collided to produce smaller asteroids and meteoroids. The origin of comets is far more speculative. Astronomers examining the spectra of comets find indications that they are composed of frozen methane, ammonia, and water in which particles of heavier elements may be embedded, somewhat like sand grains in a snowball. These comets may be composed in part of the flux driven spaceward by solar wind, or they may be made of matter that once was too far out on the periphery of the nebula to have been drawn into the evolving protoplanets. Whatever their origin, comets are visitors from interstellar space. Some comets approach sufficiently close to earth to be almost within range of our present space technology. Plans are currently under way for space scientists to send a probe to comets, analyze them chemically, and thereby obtain new insights into solar system history.

The protoplanet hypothesis should not be considered the final word in cosmology. Many questions relating to the role of gravitation and the distribution

Figure 18-9 *Photograph of the Comet Mrkos. The wavy streaming in the tail was caused by solar wind. (Hale Observatory photograph; from Pasachoff, J.M.,* Contemporary Astronomy, *2nd ed. Philadelphia, Saunders College Publishing, 1981.)*

TABLE 18–1 **PHYSICAL AND ROTATIONAL CHARACTERISTICS OF THE PLANETS**

Name	*Equatorial radius* *km*	*÷earth's*	*Mass ÷earth's*	*Mean density (gm/cm³)*	*Oblate-ness*	*Surface gravity (Earth = 1)*	*Sidereal rotation period**	*Inclination of equator to orbit*	*Apparent magni-tude at opposition*
Mercury	2,439	0.3824	0.0553	5.44	0.0	0.378	59^d	<28°	−1.8
Venus	6,052	0.9489	0.8150	5.24	0.0	0.894	244.3^d R	177°	−4.3
Earth	6,378.140	1	1	5.497	0.0034	1	$23^h56^m04.1^s$	23°27′	—
Mars	3,397.2	0.5326	0.1074	3.9	0.009	0.379	$24^h37^m22.6^s$	23°59′	−2.01
Jupiter	71,398	11.194	317.89	1.3	0.063	2.54	9^h50^m to $>9^h55^m$	3°05′	−2.55
Saturn	60,000	9.41	95.17	0.7	0.098	1.07	10^h14^m to $>10^h38^m$	26°44′	+0.67
Uranus	27,900	4.4	14.56	1.0	0.01	0.8	12^h R	97°55′	+5.52
Neptune	24,300	3.8	17.24	1.7	0.02	1.2	15^h48^m	28°48′	+7.84
Pluto	2,500	0.4	0.11?	?	?	?	$6^d9^h17^m$	?	+14.9

* R signifies retrograde rotation.
The masses and diameters are the values recommended by the International Astronomical Union in 1976, except for a 1977 value for the radius of Uranus. Densities and surface gravities were calculated from these values.
Source: Pasachoff, J.M., *Contemporary Astronomy*. 2d ed., Philadelphia, Saunders College Publishing, 1981.

of certain of the heavy elements remain to be answered. Even if the hypothesis continues to be accepted, it can be expected to undergo considerable change in the years ahead. As the concept now stands, it is probably the most widely favored of hypotheses for solar system origin. Large telescopes have revealed the existence of true nebulae between stars, and some of these great swirls of gas and dust appear to be forming new stars. Planetary systems like our own may be forming even now. The hypothesis also accounts for the known spacing of planets, their directions of rotation and revolution, and the distribution of angular momentum within the system.

For the earth scientist, the protoplanet hypothesis presents a foundation on which to build the history of the earth. One may begin with a solid earth that originated in the cold state by the collection of particles that were derived from a cloud of interstellar matter, and view the solar system as having formed from rather ordinary cosmic processes. The formation of the earth was probably not a freakish occurrence.

METEORITES AS SAMPLES OF THE SOLAR SYSTEM

Because meteorites are masses of mineral or rock that have reached the earth from space, they are of great importance to scientists interested in the origin and history of planets. They are the only objects from the universe beyond the moon and the earth that can currently be held in hand and scrutinized in the laboratory. Besides revealing something of the composition of the solar system, they provide important clues to the earth's age.

Meteors (Fig. 18–10) are familiarly known as "shooting stars," a misleading term because they are not stars at all. They are flashes of light produced by rocky objects originally left in space during the formation of the solar system (or formed by asteroid collision), later captured by the earth's gravity, and then completely vaporized before reaching the earth's surface. If a portion of the object survives its fall through the atmosphere and crashes into the earth, it is then termed a **meteorite.** A related term, **meteoroid,** refers to a meteoritic particle in space before any encounter with the earth. Most of those meteoroids that enter the earth's atmosphere are vaporized before they strike the earth's surface, but about 500 meteorites as large as or larger than a baseball reach the earth's surface each year. Weathering and erosion on the earth have erased most of the craters made by meteorite impacts, but there remain about 70 partially preserved craters or clusters of craters. In some of these, all that remains is the root zone of shattered and shocked rocks. Geologists refer to these remnants as astroblems. A large number of astroblems are known from the area of ancient rocks that surround Hudson's Bay in Canada.

Figure 18–10 A meteor crossing the field of view of the Palomar telescope while it was taking a 15-minute exposure of the Comet Kobayashi-Berger-Milon, August 1975. (Hale Observatory photograph/ John Huchra; from Pasachoff, J.M., Contemporary Astronomy, *2nd ed. Philadelphia, Saunders College Publishing, 1981.)*

Craters caused by meteorite bombardment are clearly evident on the moon. The moon is also affected by **micrometeorites** (less than 1 mm in diameter). These smaller falling particles are considered a major cause of lunar erosion. Most micrometeorites come from comets, whereas the larger meteorites are probably derived from collisions in the asteroid belt.

Upon entering the atmosphere, meteoroids are heated to incandescence by the friction created during collision with air molecules, and for a few moments they appear as glowing orbs with bright, gaseous tails. At this point, they can be photographed and their spectra analyzed. These analyses indicate that the larger meteoroids are composed chiefly of iron, calcium, silicon, aluminum, and sodium. Meteors are frequent events. Over the whole earth, many thousands can be seen each night. More meteors are seen at certain times than at others. When the number of sightings is above average, the earth is experiencing a **meteor shower.** Showers tend to occur at the same time each year and represent movement of the earth through the orbits of defunct comets. Also, the rate at which meteors are seen increases after midnight because that side of the earth is facing the oncoming interplanetary debris.

Meteorites can be classified according to their composition as ordinary chondrites, carbonaceous chondrites, achondrites, irons, and stony-irons (Fig. 18–11). The most abundant of these types are the **ordinary chondrites** (Fig. 18–12), which are crystalline stony bodies composed of high-temperature ferromagnesian minerals. Their mineralogy indicates they formed in a closed system at temperatures of about 500° to 1000°C. Ordinary chondrites can be dated by the uranium-lead, strontium-rubidium, and potassium-argon methods, and they are found to be about 4600 million years old. Many ordinary and carbonaceous chondrites contain spherical bodies called **chondrules** (Fig. 18–13). Their spherical shape sug-

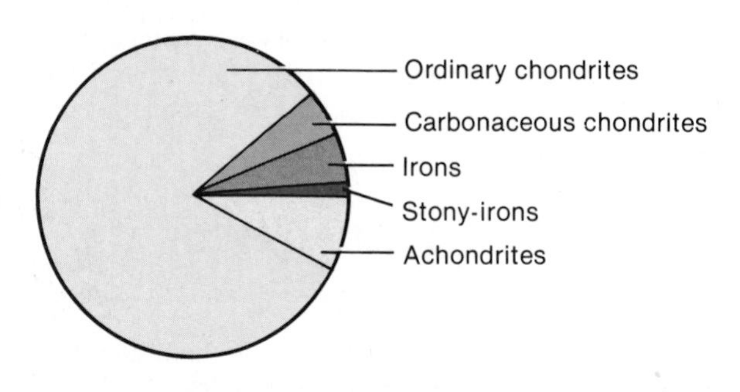

Figure 18–11 The major categories and proportions of meteorites.

Figure 18–12 Stony meteorite of the type termed a chondrite. Chondrites are stony meteorites that contain BB-sized round spherules (chondrules) composed of olivine and certain pyroxenes. (Courtesy of the Institute of Meteoritics of the University of New Mexico.)

gests that chondrules may have formed as droplets of molten material that was then rapidly chilled to preserve the rounded form of the original droplet. Perhaps the melting occurred during the early heating of the sun, and the chondrules solidified as they moved outward toward cooler regions of space.

Less abundant than the ordinary chondrites are the **carbonaceous chondrites.** In these meteorites one finds the same abundance of metallic elements as in ordinary chondrites, with the addition of hydrous minerals, more volatile trace elements, and about 5 percent organic compounds, including inorganically produced amino acids. Suspended in the blackish, earthy matrix of carbonaceous chondrites are both chondrules and irregular pieces of crystalline, high-temperature minerals that may have condensed from a cooling vapor. These bodies are referred to as calcium-aluminum–rich inclusions.

The matrix material of most carbonaceous chondrites consists of relatively low temperature minerals, whereas the chondrules and calcium-aluminum–rich inclusions are composed of minerals that form at higher temperatures. It appears that these two components originated under different conditions and were subsequently brought together.

One of the most interesting findings about carbonaceous chondrites is that their overall composition is very similar to that of the sun. Indeed, if it were possible to extract some solar material, cool it down, and condense it, the condensate would be chemically

Figure 18–13 Chondrules in a carbonaceous chondrite. (Photograph courtesy of the Institute of Meteoritics of the University of New Mexico.)

Figure 18–14 *Iron meteorites. The upper specimen has been cut and polished, then etched with acid to show the interesting pattern of interlocking crystals called the "Widmanstätten Pattern." (Courtesy of the Institute of Meteoritics of the University of New Mexico.)*

similar to carbonaceous chondrites. This similarly suggests that carbonaceous chondrites are samples of primitive planetary material that formed when the sun formed and since that time have never experienced melting and the kind of mineral fractionation that characterizes igneous rocks.

A small percentage of stony meteorites do not contain chondrules and therefore are termed **achondrites.** Most of these have compositions similar to terrestrial basalts, and many exhibit angular, broken grains. These fragmental textures indicate that achondrites may be the products of collisions of larger bodies. One can imagine, for example, the collision of two asteroids having metallic cores enveloped in silicates. The fragments resulting from the collision would consist of a range of compositions from metallic to stony.

The **iron meteorites** (Fig. 18–14) account for about 6 percent of all meteorites found on earth. Most iron meteorites are intergrowths of two varieties of nickel-iron alloy. The large size of the crystals and metallurgic calculations indicate that some iron-nickel meteorites cooled as slowly as 1°C per million years. Such a slow rate of cooling would be possible only in objects at least as large as asteroids. Thus the history of the iron meteorites probably involved a long episode of slow crystallization within a parent body sufficiently large to provide insulation for the hot interior, followed by violent disruption of that body by a collision to produce the meteorites.

The least abundant of all the categories of meteorites are the **stony-irons** (Fig. 18–15). As indicated by their name, they are composed of silicate minerals and iron-nickel metal in about equal amounts. The stony-iron meteorites are generally considered to represent fragments from the interfaces between the silicate and metal portions of asteroidal bodies. In some cases, they may have originated at the boundary between an iron core and the silicate mantle of a planetary body destroyed by collision.

Figure 18–15 *Stony-iron meteorite. The white areas are composed of iron-nickel, and the gray areas are mostly olivine. (Courtesy of the Institute of Meteoritics of the University of New Mexico.)*

THE MOON

Moon features

As satellites go, our moon is large relative to its parent planet. Its diameter is more than a fourth that of the earth, and its density of about 3.3 g/cm^3 is the same as that of the upper part of the earth's mantle. Gravity on the moon is insufficient to maintain an atmosphere. The moon and the earth revolve around their mutual center of gravity as a sort of double planet. The moon orbits the earth and rotates on its axis at the same rate. As a result, we always see the same side of the moon and must depend on transmissions from space vehicles for images of the far side (Fig. 18-16). These photographs have revealed a densely cratered surface interrupted only by few and relatively small level areas.

Telescopic observation of the near side of the moon was first made in 1609 by a rather irascible professor of mathematics at the University of Padua. The mathematician was Galileo Galilei. With the use of his primitive telescope, Galileo was able to see that the surface of the moon was not smooth, but "uneven, rough, full of cavities and prominences. . . . " He recognized "small spots," which we now know are

Figure 18-16 The far side of the moon as photographed by Apollo 16. (Courtesy of the National Aeronautics and Space Administration.)

craters, and dark patches that children imagine to be the eyes and mouth of the "man in the moon." These dark patches lie within large basins. Galileo named them maria ("seas") and incorrectly suggested they were filled with water. Today, we know the dark areas have been flooded not with water but with basaltic lavas. In many maria, the lava has been contained within depressions that are hundreds of kilometers across and 10 to 20 km deep. These depressions are termed **mare basins** (Fig. 18–17). Mare basins had to have developed prior to the extrusion of the dark lavas that form the maria, for these lavas are contained within the basins.

Galileo also provided the name for the lighter-hued, rougher terrains of the moon. He called these regions terrae, although modern space scientists refer to them as **lunar highlands.** The highlands are the oldest parts of the moon, having formed long before

Figure 18–17 In order to show the whole moon, but still show the detail that does not show up well at full moon, the Lick Observatory has put together this composite of first and third quarters. Note the dark maria and the lighter, heavily cratered highlands. Two young craters, Copernicus and Kepler, can be seen to have rays of light (that is, darker than the background) material emanating from them. (Lick Observatory photograph; from Pasachoff, J.M., Contemporary Astronomy. *Philadelphia, W. B. Saunders Co., 1977.)*

Figure 18–18 This view from the orbiting Apollo 15 Command Module shows a smaller crater, Krieger B, superimposed on a larger crater, Krieger. Obviously, the smaller crater is younger than the larger one. Several rilles (clefts along the lunar surface that can be hundreds of kilometers in length) and ridges are also visible. (Courtesy of the National Aeronautics and Space Administration.)

the earliest mare lavas appeared. The highlands are heavily cratered regions reminiscent of a battlefield following a heavy artillery barrage. They provide stark evidence of the moon's early episodes of intense meteoritic bombardment.

On the moon, as on the earth, craters are roughly circular, steep-sided basins normally caused by either volcanic activity or meteoritic impact (Fig. 18-18). Although a few craters in the moon show questionable evidence of volcanic origin, the great majority are clearly the result of meteoroid impact. Unlike the earth, the moon has insufficient atmosphere to burn up approaching meteorites, and thus the frequency of impact is high. Lunar geologists are able to determine the relative ages of certain areas on the moon's surface by the density of craters. Younger areas have fewer craters than do older ones. Another aid to recognizing differences in the age of lunar features is provided by the rays of material splashed radially outward from the crater at the time of impact. Younger rays and other impact features cross and partially cover older features. Light-colored impact rays composed of finely crushed rock are exceptionally well developed in the large lunar crater known as Copernicus (Fig. 18-19). This familiar feature is about 90 km across and rimmed by concentric ridges of hummocky material blasted out by the impact of the crater-forming body. The center of the crater floor is raised and thought to represent an upward adjustment following impact. Not all craters have such spectacular features. Impact rays are sometimes obliterated by the rather continuous rain of

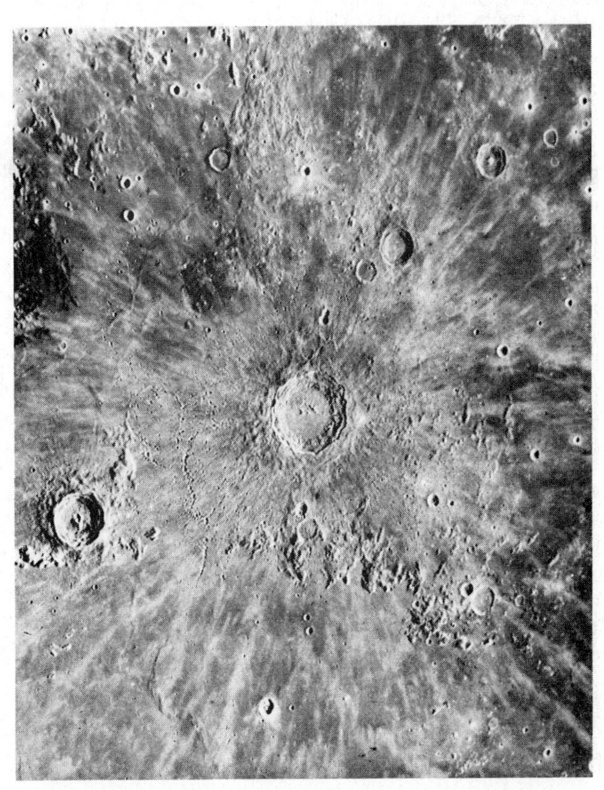

Figure 18–19 The crater Copernicus, as seen from earth-based telescope, has rays of lighter colored material emanating from it. This light material was ejected radially when the meteorite that formed the crater struck the moon's surface. (From Pasachoff, J.M., Contemporary Astronomy, *2nd ed. Philadelphia, Saunders College Publishing, 1981.)*

tiny meteorites that strike the moon. For this reason, younger craters may still be adorned with impact rays, whereas older craters have lost these features.

There is ample evidence of former igneous activity on the moon. We have already noted the extensive lava flooding of mare basins. In addition to these flood basalts, a few dome-shaped volcanoes measuring about 10 km in diameter at the base have been located. About 1 percent of the lunar craters may have resulted from volcanic activity. Meandering channelways called *sinuous rilles* occur on the surface of our satellite. Most of these are thought to have been produced by the turbulent flow of lava or by the collapse of lava tunnels, although the evidence is not conclusive. Sinuous rilles are distinctly different from straight linear rilles, which probably result from faulting. In the maria, astronomers have also observed features named wrinkle ridges, also thought to develop by the extrusion of lava. Tall volcanic peaks are not found on the moon. Perhaps this is because lunar lavas, like lavas associated with shield volcanoes on the earth, are highly fluid and spread laterally rather than piling up around a vent.

Are lunar volcanoes still active today? In 1958, Russian astronomers observed evidence of gas emission from the crater Alphonsus, and on two occasions in 1963, observers at the Lowell Observatory saw glowing red spots in and near the crater Aristarchus. These observations suggest that gas emissions or possibly small ventings of ash may occur occasionally on the moon. Significant lava eruptions seem unlikely.

Moon rocks

Direct examination of samples returned from the moon, as well as data gathered by orbiting spacecraft, indicates that there are three main kinds of rocks on the moon. One of these forms the floor of the smooth mare regions, and the other two are characteristic of the lunar highlands. The mare rocks resemble *basalt* (Fig. 18-20), a finely crystalline, dark-colored igneous rock that solidifies from lava and occurs abundantly on earth as well as the moon. Plagioclase, pyroxene, olivine, and ilmenite are the principal minerals in basalt. These minerals are described in Chapter 4.

The two rock types that occur in the lunar highlands differ from the mare basalts in both texture and composition. Texturally, they are decidedly coarser. Most of them are aggregates of crushed rock mixed with material that appears to have been molten. Compositionally, they resemble the family of rocks known on earth as gabbros, but one is able to recognize two distinct groups. One group is given the nickname **ANT** rocks (Fig. 18-21), an easily remembered acronym derived from the words *a*northosite, *n*orite, and *t*roctolite — three gabbroic rocks known from earth. Calcium-rich plagioclase comprises over 75% of ANT rocks.

Figure 18-20 A basalt collected from the lunar maria by Apollo 15 astronauts. This lava solidified so rapidly that bubbles formed by escaping gas were trapped, forming vesicles. Specimen is 14 inches across. (Courtesy of the National Aeronautics and Space Administration.)

The second category of lunar highland rocks has been dubbed **KREEP** by geologists at the Johnson Space Center in Houston. The name indicates that compared with other moon rocks, KREEP rocks contain relatively more potassium (chemical symbol K), rare earth elements (REE) and phosphorus (chemical symbol P). Mineralogically, KREEP rocks contain more pyroxene and less feldspar than ANT rocks. They differ from mare basalts in having a much lower percentage of ilmenite (Fig. 18-22).

The samples brought to earth from the moon were all obtained from the blanket of rock fragments and dust that covers the moon's surface. Within this loose material, called **regolith,** it is possible to find fragments of rocks of all sizes, ranging from microscopic grains to huge boulders. Some of the boulders and cobbles consist of aggregates of smaller rock fragments welded together as lunar breccias.

At Tranquillity Base, site of the first moon landing, the regolith was estimated to be three to four meters thick. It is believed to have formed as a result of meteorite impacts, each of which would dislodge a mass of debris many hundred times greater than its own mass.

Figure 18–21 ANT rock from the lunar highlands. As indicated by the larger crystals, these rocks may be among the earliest to solidify when the moon formed. The specimen shown here is 4.6 billion years old. It is composed primarily of olivine and feldspar. (Courtesy of the National Aeronautics and Space Administration.)

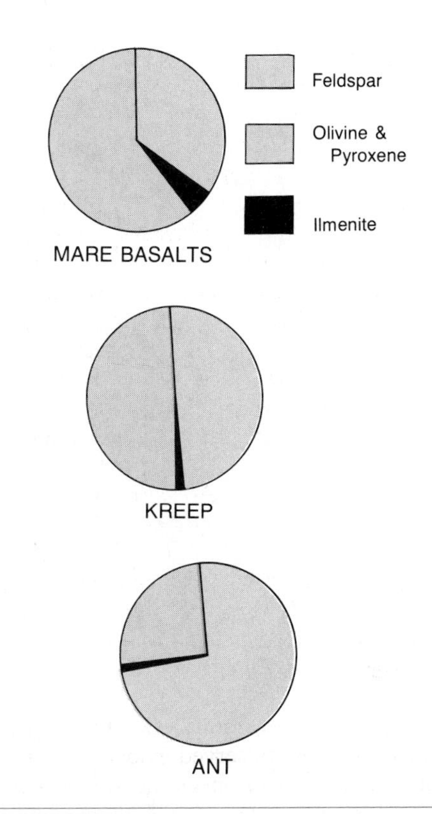

Figure 18–22 The percentages of minerals in mare basalts as compared to those in ANT and KREEP rocks.

Calculations indicate that meteorite impacts amply account for the loose regolith on the moon.

Radiometric dating of lunar rocks has been of great importance in understanding the sequence of events in the moon's history. Most mare basalts are between 3.2 and 3.8 billion years old and were extruded during the great lava floods that produced the mare basins (Fig. 18–23). For the most part, the lunar breccias were formed during impacts that occurred prior to these floodings. The oldest rocks on the moon were taken from the lunar highlands, and they have been found to be about 4.6 billion years old. This age represents the time that meteoritic impacts formed the rugged highland terrain and crushed and melted the KREEP and ANT rocks.

Samples of moon rocks now being studied in the United States have been found to retain weak magnetism imparted to the rocks at the time they solidified. Lunar geologists have suggested that this so-called *remanent magnetism* indicates the existence of a once-liquid but now solidified core and that the moon had a significant magnetic field about 4.2 to 3.2 billion years ago. The moon has no general magnetic field today.

Mascons

Prior to sending a manned spacecraft to the moon, the United States had launched five spacecraft designed to circle the moon and transmit high-resolution pictures of the moon's surface back to earth. The speed and location of these craft (named orbiters) were precisely monitored by radio-tracking advices. Much to the surprise of the scientists monitoring the orbiters, their speed was not constant but increased as the craft approached one of the dark, circular maria and decreased as they passed over the mare. Evidently, there were mass concentrations, dubbed **mascons,** beneath the mare. Gravitational attraction of the mascon caused approaching orbiters to speed up and then slowed the orbiter down as it passed overhead. Very likely, mascons are caused by dense layers of basalt within mare

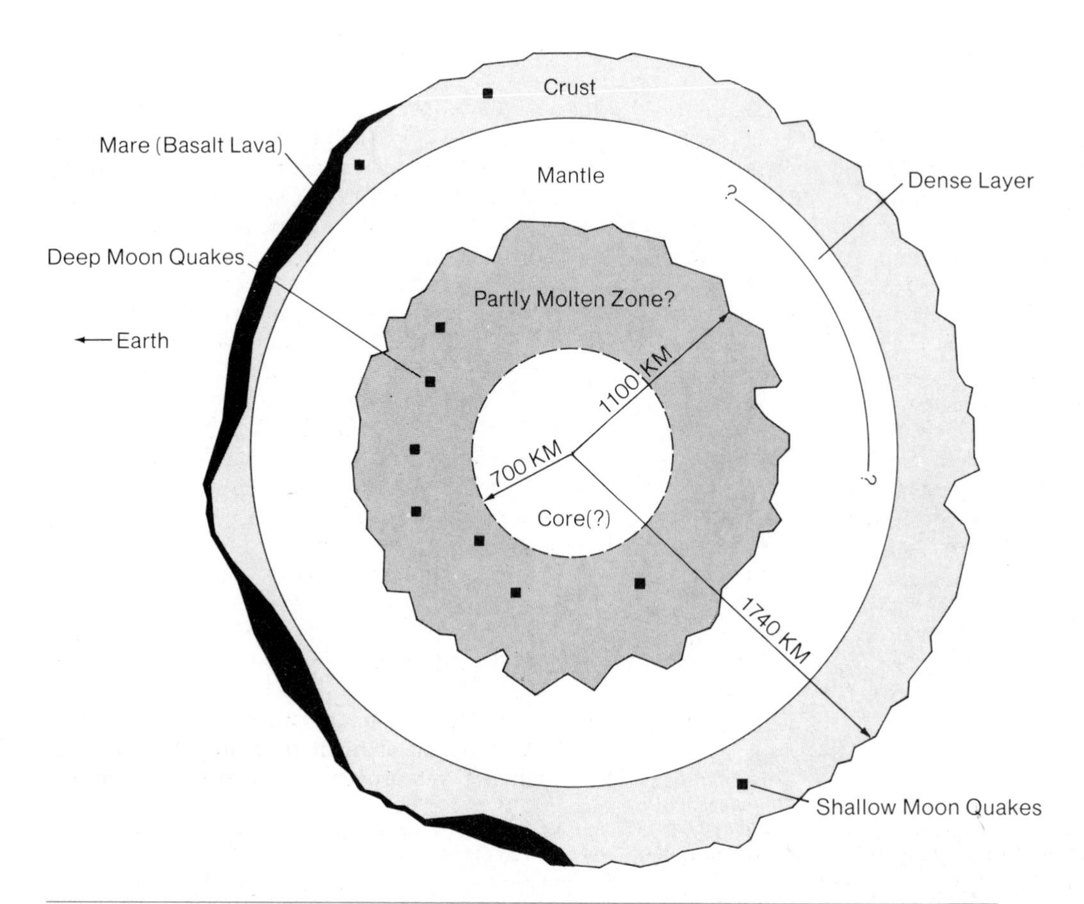

Figure 18–23 Cross-section through the moon showing an outer crust, an inner mantle, and an innermost zone that may still be partly molten. Within the innermost zone there may be a small metallic core (dashed circle). The earth-facing side is relatively smooth and shows larger accumulations of the mare basin lavas. The far side is more rugged and has almost no lava. The drawing is not to scale and the ruggedness of the far side is exaggerated. Data for the drawing were provided by A.M. Dainty, N.R. Goins, and M.N. Toksoz. (Courtesy of the National Aeronautics and Space Administration. From French, B.M., What's New on the Moon?*)*

basins. Calculations indicate that a layer 5 to 10 km thick and about 10 percent denser than lunar highland rocks would produce the effect noticed in the orbiters. Scientists are puzzled, however, that these concentrations of mass have not sunk into the lunar interior. This indicates that the inside of the moon is not weak and plastic but rather is sufficiently rigid to support mascons for billions of years.

Lunar processes

In Chapter 5 we described earth processes capable of raising mountain ranges, and we have considered such opposing processes as erosion and weathering that are constantly at work wearing away the lands. Geologic processes also operate on the moon, but they are slower and of a different kind. The moon is a desolate world in which there is no air or surface water. Whenever gases or liquids escape to the lunar surface, they are dissipated into space because the moon lacks sufficient mass, and hence gravitational attraction, to hold them. There can be no stream-cut canyons and valleys, no glacial deposits, and no accumulations of wave-sorted sand. Rocks brought back to earth from the moon seem bright and newly formed and lack the discoloration common to weathered earth rocks. Yet erosion does occur on the moon, primarily through the continual bombardment of the lunar surface by large and small meteorites. The larger meteorites, of course, produce craters and rough terrain, but it is the rain of small meteorites and micrometeorites (those with diameters of less than a millimeter) that subtly alter the moon's surface and are important in the formation of the lunar soil. The upper or exposed surfaces of lunar rocks brought back to earth from the moon were riddled with tiny, glass-lined pits produced by the impact of fast-moving micrometeorites. Geologists call these tiny depressions microcraters, or **zap pits.** Most are less than 0.0001 mm in diameter (Fig. 18–24). The micrometeorite barrage continues today, as was evident on examination of the Apollo spacecraft after its return to earth. Ten tiny microcraters were chipped into its windows.

Bombardment of the lunar surface by objects from space is largely responsible for the development of lunar soil. Beginning with a newly solidified lava, incoming particles blast out a multitude of small craters. As if struck by bullets, large particles are pulverized to produce fine fragments that pile up around the new crater and cover older ones. The impacts of larger meteorites are capable of scattering debris over a wide area.

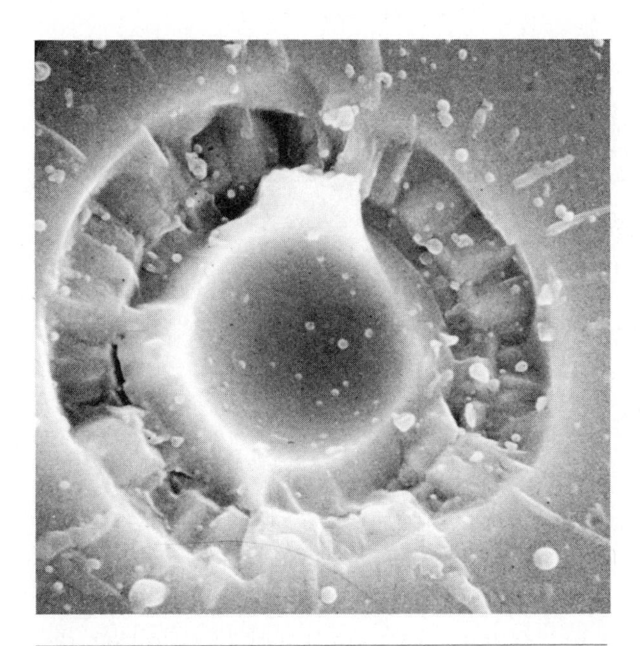

Figure 18–24 Photograph of a zap pit *or microcrater. The tiny pit is only 20 microns in diameter and was photographed with a scanning electron microscope. (Photograph by F. Horz, courtesy of G.J. Taylor.)*

Cosmic radiation is another process that plays a less vigorous but persistent role in weathering the lunar surface. Cosmic rays are fast-moving atomic particles, consisting mostly of protons and helium nuclei. On earth, we are largely protected from cosmic rays by the atmosphere, but the moon is subjected to a continuous onslaught. Cosmic rays penetrate moon minerals and damage their atomic structure. If these minerals are dipped in a corrosive chemical bath, the chemical will enlarge the damage paths made by the passage of cosmic particles. On examination of these minerals with a microscope, they are frequently found to be riddled with cosmic ray particle tracks. Even the plastic space helmets worn by the Apollo II astronauts contained tiny cosmic ray damage tracks. When closing their eyes, the astronauts could "see" bright streaks of light that were actually produced by the flight of a cosmic ray particle through the eyeball. Fortunately, the astronauts were not exposed to this radiation long enough to endanger their health.

Temperature variations on the moon may also contribute to the production of loose surface materials. Lunar day and night are each about two weeks long. During daytime, temperatures at the surface rise to 134°C, whereas at night they drop to −170°C. Although not yet proven, the alternate expansion with heating and contraction with cooling weakens grain boundaries and could result in fracturing of rock.

Although gravity on the moon is only a sixth that on earth, it is nevertheless an effective agent in moving lunar materials from high to low places. Whenever loose material on slopes is unable to resist the pull of gravity, it will break away and produce masses of slumped debris. Steplike slump masses are recognized as a common feature along the rims of some lunar craters. Elsewhere, photographs reveal tracks in lunar soils that record the rolling and sliding of large boulders down slopes.

The moon's early history

Even with the explosion of scientific data from the space program, scientists are unable to support confidently a single theory for the origin of the moon. As a result, there are three hypotheses now being actively examined. The first of these stipulates that the moon accreted along with the earth as an integrated two-planet system. The second concept suggests that each body accreted independently and that the moon was then captured by the earth. The third hypothesis proposes that the moon broke off from the earth before either had completely solidified.

If both the earth and the moon formed together from the same part of the solar nebula, then they should have similar composition (as indicated in part by specific gravity). However, the average specific gravity of the earth is about 5.5, whereas that of the moon is only 3.3. This difference, together with the absence of a dipolar magnetic field on the moon, indicates that it lacks a large iron core. These considerations have caused many scientists to view with skepticism the double-planet concept for the origin of the moon.

The compositional difficulties seem to be resolved in the second hypothesis, in which the moon is considered to have formed in a distant part of the solar system but was subsequently captured when it approached the earth's gravitational field. Unfortunately, there are some severe problems relating to the mechanics of such a capture, and long-term measurements of the moon's movements required for verification of the concept are lacking. However, the capture hypothesis will be periodically reexamined in the light of new, highly accurate measurements made with the aid of laser reflectors left behind on the moon's surface.

In 1898, George Darwin, the son of Charles Darwin, speculated that the moon was composed of a large mass of material that spun off the earth at an early time when its rotation was much faster than now. The earth's centrifugal effect, together with the sun's tide-raising ability, first caused a huge tidal bulge on the earth. The bulge then separated to form the moon. When the moon's composition was found to be much like that of the earth's mantle, scientists speculated that the development of the earth had first proceeded to the point at which the earth's core of iron and other heavy elements had formed. After these heavy elements (which are substantially depleted in the moon's composition) had settled inward, a great bulge of mantle material formed, then pinched off, and began to orbit in widening circles around the earth. This interesting theory will continue to be examined over the next several years. It cannot be accepted as yet, largely because of calculations indicating that if the moon spun off the mantle, it would subsequently be drawn back again. Geophysicists are also uncertain that there would be sufficient force available to lift the mass of the moon from the earth and throw it into orbit.

As some scientists continue to debate the possibility that the moon was derived from the earth's mantle, a large number of investigators have come to favor the hypothesis that the moon, along with the other solar system bodies, was born about 4.6 billion years ago by the accumulation of matter from the original dust cloud. Theoretically, the infall of small objects released so much energy that the outer part of the moon became molten to a depth of several hundred kilometers. At this time, a melt of lighter minerals (such as feldspars) floated to the surface of this vast ocean of molten rock and formed a crust of ANT rocks. Heavier minerals like pyroxene, olivine, and ilmenite crystallized from a melt that settled to form an underlying layer about three times thicker than the crust. Then, about 200 or 300 million years later, great quantities of molten rock rose through fractures in the crust and hardened on the surface as KREEP.

The ANT and KREEP rocks of the moon's primordial surface were next subjected to a meteoritic barrage lasting nearly 1.5 billion years that produced the intricately scarred and cratered landscapes of the lunar highlands. Then internal heat generated by the steady decay of such radioactive minerals as uranium and thorium began to melt the moon again, especially at rather shallow depths of 100 to 250 km. The results of this melting event were massive extrusions of basaltic lava that spread out over parts of the lunar surface about 3.8 to 3.1 billion years ago. The moon has been quieter since the last of these eruptions. Meteorite impacts continued to sculpture its surface, sometimes forming spectacular craters like Copernicus (see Fig. 18–19).

THE EARTH'S NEIGHBORING INNER PLANETS

Mercury

Mercury has the distinction of being the smallest and swiftest of the inner planets. It revolves rapidly, making a complete journey around the sun every 88 earth days. Mercury makes a complete rotation on its axis every 59.65 days, so the planet rotates three times while encircling the sun twice.

Mariner 10, launched in 1973, provided our first close look at Mercury (Fig. 18–25). Following its launching, Mariner 10 flew by Venus and, using the gravity of Venus to deflect its trajectory, went to Mercury. Indeed, the spacecraft's orbit permitted visits over Mercury on three occasions. Photographs transmitted to earth revealed a planet with a moonlike surface. Densely cratered terrains as well as smooth areas resembling maria were well developed. Some of the

Figure 18–25 *Photomosaic of Mercury as photographed by the Mariner spacecraft. (Courtesy of the National Aeronautics and Space Administration.)*

craters exhibited rays of light-colored impact debris, just as do craters on the moon. Long, linear scarps were discerned, and these are thought to be fracture zones along which crustal adjustments occurred. Perhaps tidal forces, or shrinkage of the planet as it cooled, produced these elongate scarps.

Mercury has a weak magnetic field, a lightweight crust covered with fine dust, and an iron core that constitutes nearly 75 percent of its radius. The existence of the iron core is inferred from Mercury's high density, its magnetic field, and other evidence suggesting the planet experienced melting and differentiation.

Venus

Although Venus is similar to the earth in size and shape, its surface conditions are different from those of earth. Indeed, our planet is a paradise compared with Venus. Maximum surface temperature on our nearest planetary neighbor is a searing 475°C. The planet is shrouded in a thick layer of clouds that extend 40 km above the surface (Fig. 18–26). Carbon dioxide is the principal gas in the atmosphere of Venus. The large amount of carbon dioxide serves as an insulating blanket in trapping solar energy by the greenhouse effect (the heating of an atmosphere by the absorption of infrared energy remitted by the planet as it receives light energy in the visible range from the sun). Surface atmospheric pressure on Venus is 90 times greater than it is on earth. The yellowish appearance of the planet as seen from the earth is the result of droplets of sulfuric acid in the planet's clouds.

Figure 18–26 Photomosaic of Venus, made in ultraviolet light from the Mariner 10 spacecraft in 1974. The photograph shows the general circulation pattern of Venus's upper atmosphere. (Courtesy of the National Aeronautics and Space Administration.)

Venus rotates through one complete turn each 243 days. Its direction of rotation is retrograde with respect to the stars, that is, opposite to the paths of revolution of the planets around the sun. Its period of revolution around the sun is 225 days. From its size and density, planetary scientists infer an interior rather like that of the earth. Unlike the earth, however, Venus has no oceans. This is probably because it is too close to the sun for water vapor to condense and form oceans. Without oceans in which CO_2 can be combined with calcium and magnesium to form carbonate rocks, there is no apparent mechanism to remove CO_2 from the atmosphere. The gas persists, locked to its planet by gravity.

Because of the dense cloud cover, direct photography of Venus's hard surface from orbiting spacecraft is not possible. The American Pioneer Venus Orbiter, however, was equipped with cloud-piercing ground-scanning radar and with this device was able to map the entire surface of the planet. These maps reveal that more than 60 percent of Venus consists of gently rolling hills. Most of this terrain varies about 1000 m between high and low points. There are however, far more dramatic features on Venus. Mountains as high as Mount Everest have been identified, as well as spectacular rift (fault) valleys. One of these valleys extends more than 1,400 km and is 250 km wide and nearly 5 km deep. Another striking topographic feature is a magnificent plateau (named Ishtar Terra after the Assyrian goddess of love and war) that is higher than the earth's Tibetan Plateau and more than twice as large in area (Fig. 18–27).

The first close look at the actual surface of Venus was provided in 1975 when the Russian robot spacecraft Venera 9 and Venera 10 succeeded in landing on the planet. Despite the crushing atmospheric pressure and scorching temperatures (sufficient to melt tin), these spacecraft survived for about an hour. Even more successful, however, were the landings of Venera 13 and 14 in March 1982. Like their predecessors, the two robots separated from their mother ships and drifted under parachute through the corrosive clouds to touch down on a mountainous region called Phoebe just below the Venusian equator. Venera 13 lasted two full hours; its sister craft lasted about half as long. The electronic eyes on both landers began working immediately and transmitted to earth remarkable panoramic photographs of a landscape strewn with rust-colored

Figure 18–27 *Artist's conception of Ishtar Terra on Venus. Ishtar is a high plateau, carrying several mountain ranges, and is about the size of the continental United States. (Courtesy of the National Aeronautics and Space Administration.)*

slabby rocks and pebbly rubble (Fig. 18–28). In addition to accomplishing their photographic mission, the landers were able to drill out a few centimeters of Venusian soil and determine its composition. Analysis indicated the samples were basalt. Indeed, they had an uncanny resemblance to the basalts extruded from fissures along midocean ridges on earth. Many scientists believe that this discovery, along with the observation of great rift valleys, indicates that Venus, like the earth, is still a very dynamic planet.

Mars

General characteristics

The planet Mars has always been of special interest to humans because of speculation that some form of life, however humble and microscopic, may now exist or at one time have existed there. Mars is only a little farther from the sun than is our planet; it has an atmosphere that includes clouds, it has developed white polar caps, and it has seasons and a richly varied landscape. That

Figure 18–28 *A view of the surface of Venus taken by Venera 14. The rocks are interpreted to be flow-layered basalts. The photograph was given by the U.S.S.R. to scientists at the United States Geological Survey to help in choosing the landing sites for Venera 13 and 14.*

landscape (Figs. 18–29, 18–30) has been splendidly revealed to us by the Mariner and Viking missions. It includes magnificent craters, colossal volcanic peaks, deep gorges, sinuous channels, and an extensive fracture system. Evidence of wind erosion is clearly seen on photographs of the Martian surface, and there are indications of the work of ice as well. The planet has a diameter only about 50 percent that of the earth and has only 10 percent the mass of the earth.

A year on Mars is 687 earth days long, and its axial tilt is similar to that of the earth. Carbon dioxide is the principal gas in the Martian atmosphere, although small amounts of nitrogen, oxygen, and carbon monoxide are also present. The atmosphere is nearly 150 times thinner than the earth's atmosphere. The planet has virtually no greenhouse effect, and temperatures vary at the equator from about +21° to −85°C.

Mars has two moons (Fig. 18–31). Their names are *Phobos* (meaning fear) and *Deimos* (meaning panic). They are large chunks of rock about 27 and 15 km across, respectively.

The somewhat lower density of Mars, compared with other inner planets, and its lack of a detectable magnetic field suggest that its iron may be largely scattered throughout the planet. This interpretation is strengthened by the color photographs transmitted from the Viking Landers. Those photographs show the rusty red and orange hues typically developed on earth by limonite and hematite. Indeed, the sky often takes on a pinkish hue, probably because of the rust-colored dust. Apparently, iron at the Martian surface has reacted with water vapor and oxygen to provide the color that so fascinated early planetary observers.

Martian landscape features

As a result of the work accomplished by the Mariner and Viking Orbiters, nearly the entire surface of Mars

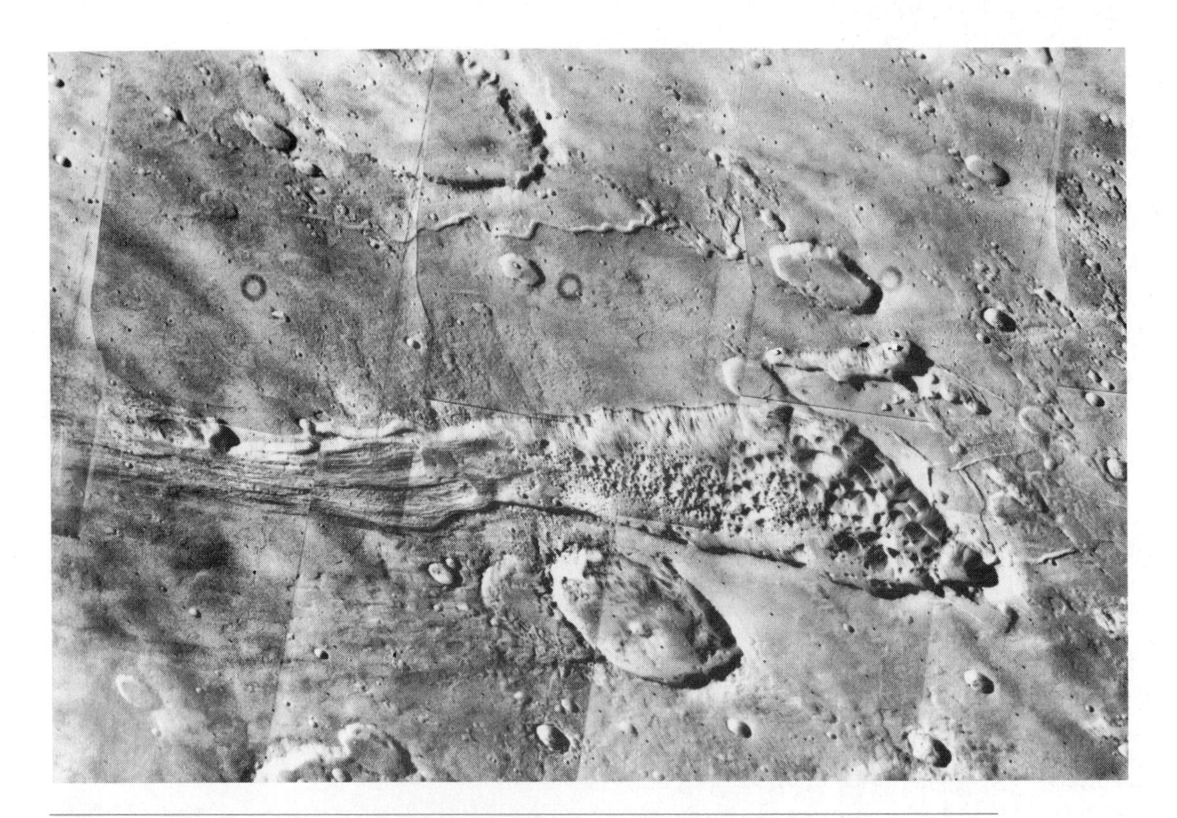

Figure 18–29 *Martian surface as revealed in a mosaic of photographs taken by Viking 1 on July 3, 1976. The large valley in the center was probably caused by downfaulting, possibly in association with the melting of subsurface ice. The hummocky topography on the valley floor may have also resulted from this process. (Courtesy of the National Aeronautics and Space Administration.)*

Figure 18–30 *Viking 1 view of the great canyon on Mars named Valles Marineris. The canyon is about 500 km long, about 6 km deep, and about 120 km wide. In this view, sections of the canyon walls have collapsed to form huge landslides that have slid down and across the canyon floor. (Courtesy of the National Aeronautics and Space Administration.)*

had been photographed by 1976. Crisp, high-resolution photographs revealed craters, dunes, volcanoes, and canyons in unsurpassed clarity. On a grand scale, the planet's physiography is divisible into two halves. The southern hemisphere is mostly heavily cratered rough terrain. This surface is inferred to be older than the smoother, only lightly cratered surface of the northern hemisphere.

The Viking spacecraft that reached Mars in the summer of 1976 each contained two parts, a lander and an orbiter. The landers provided scientists with the first ground view of Mars. The photographs transmitted from the landers to the orbiters and thence to the earth show a sandy-looking terrain littered with cobbles and boulders (Fig. 18–32). Here and there, undisturbed bedrock was visible. Many of the rocks strewn about were clearly of volcanic origin. Some had spherical holes made by gas bubbles, whereas others were finely crystalline like many igneous rocks on earth. The devices on the lander that analyzed the soil indicated that the fine, loose material was rich in iron, magnesium, and calcium. Very likely, it was derived from parent rocks rich in ferromagnesian minerals.

Craters and dunes

The density of craters in the southern half of Mars is similar to the crater density in the lunar highlands. Most of these craters were probably excavated between 4 billion and 4.5 billion years ago by the same

Figure 18–31 *Photomosaic of Mar's inner satellite, Phobos. The photographs incorporated into this mosaic were obtained when Viking Orbiter 1 approached to within 480 km of Phobos in 1977. Phobos is about 21 km across and 19 km from top to bottom. North is at the top. (Courtesy of the National Aeronautics and Space Administration.)*

torrential barrage of meteorites that pelted the lunar highlands.

Photographs clearly show row upon row of dunes on the floors of some of the larger craters. If the thinness of the Martian atmosphere is taken into account, then the size and spacing of these dunes can be used to estimate the velocity of the winds that formed them. Calculations indicate that the dune-forming winds blew at speeds up to 100 meters per second. Dunes and other evidence of the work of wind are especially common near the poles. The incorporation of dust in the polar ice itself is inferred from the stratified appearance, interpreted to be alternate layers of dust and ice (Fig. 18–33). Dust storms on Mars are prodigious events. As Mariner 9 began encircling the planet, a dust storm was in progress, and useful photographs of the Martian surface could not be transmitted until the dust subsided. Viking orbiters encountered similar problems. The dust storms are not driven by excessively violent winds, however. One of the Viking landers recorded a gust of 120 km per hour, but average wind velocities were much lower.

Volcanic features

Although most of the craters on Mars have been formed by the impact of meteorites, a fewer number

Figure 18–32 *Photograph of Martian dune field remarkably similar to dunes seen in deserts on earth. Photographs taken by camera on Viking Lander on August 3, 1976. The sharp dune crests indicate the most recent wind storms transported sediment particles from upper left to lower right. Large boulder at left is about 3 meters long. (Courtesy of the National Aeronautics and Space Administration.)*

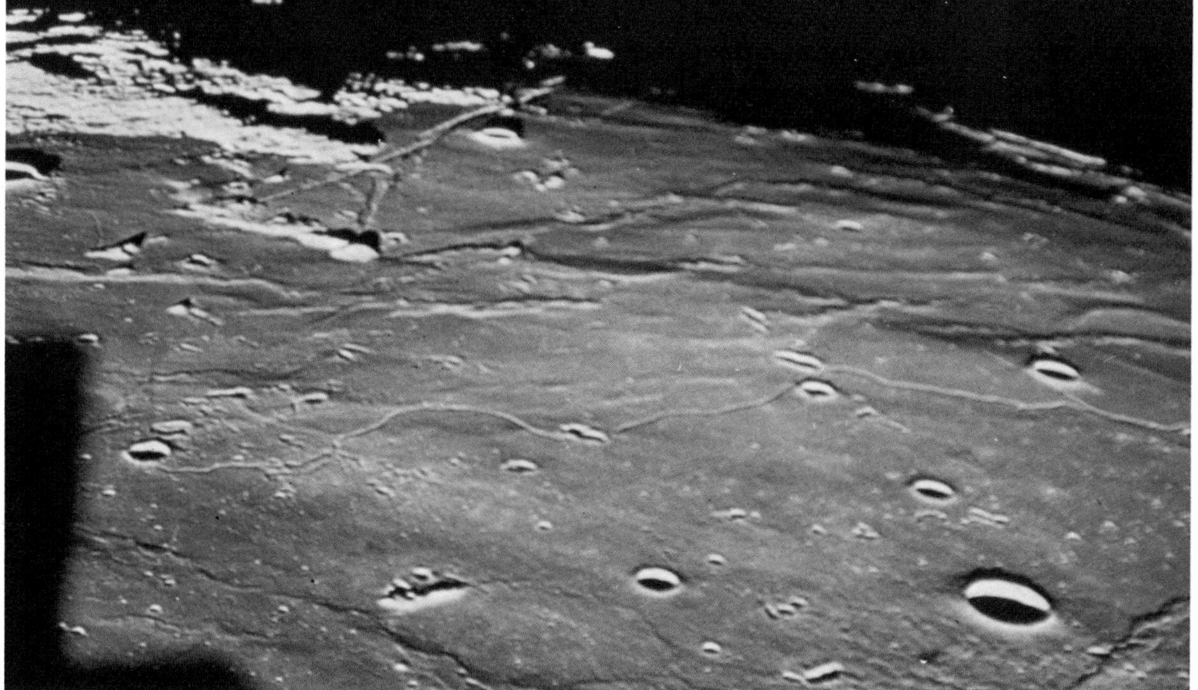

The surface of the earth's moon. The heavily cratered terrain on the far side of the moon is revealed in the upper photograph; the lower photograph shows the less cratered Sea of Tranquility, where Apollo 11 made its landing.

(NASA)

The gigantic Martian volcano Olympus Mons, photographed by Viking Orbiter in July 1976. The original black and white photograph has been colored in by the artist.

(NASA/JPL) (From Abell, G.O., Realm of the Universe, *3rd ed. Philadelphia, Saunders College Publishing, 1983.)*

Figure 18–33 The frosty Martian North Pole. In this view one can see the perennial ice cap that overlies stratified deposits thought to be alternating layers of dust and ice. (Courtesy of the National Aeronautics and Space Administration and Michael Botts.)

are certainly volcanic. Easily the most spectacular of the volcanic craters is *Olympus Mons* (see Fig. 2–29). This gigantic feature is more than 500 km across at its base and rises 30 km above the surrounding terrain. It is equal in volume to the total mass of all the lava extruded in the entire Hawaiian Island chain. Around its summit are lava flows and the narrow rilles interpreted to be old lava channels or possibly collapsed lava tubes. Flood lavas are also evident, not only in the smoother northern hemisphere but also among the craters of the rugged southern half of Mars.

Channels and canyons

There is so very little water today in the Martian atmosphere that if all of it were to be condensed in one place, it would probably provide only enough water to fill a community swimming pool. Yet the surface of the planet is dissected with channels and canyons that grow wider and deeper in the downslope direction. Some of these channels have braided patterns, sinuous form, and tributary branches such as are characteristically developed by streams on earth (Fig. 18–34). These features suggest that flowing water was an important geologic agent at some earlier time in Martian history. There may once have been a considerable reserve of water frozen in the subsurface as a kind of permafrost. Permafrost is ground that remains below the freezing temperature and contains ice throughout the year. Many investigators theorize that the larger channels, some of which are more than 5 km wide and 500 km long, were excavated by water gushing out onto the surface as the subsurface layer melted. Large tongues of debris closely resembling mudflows may also have received their water from such a process.

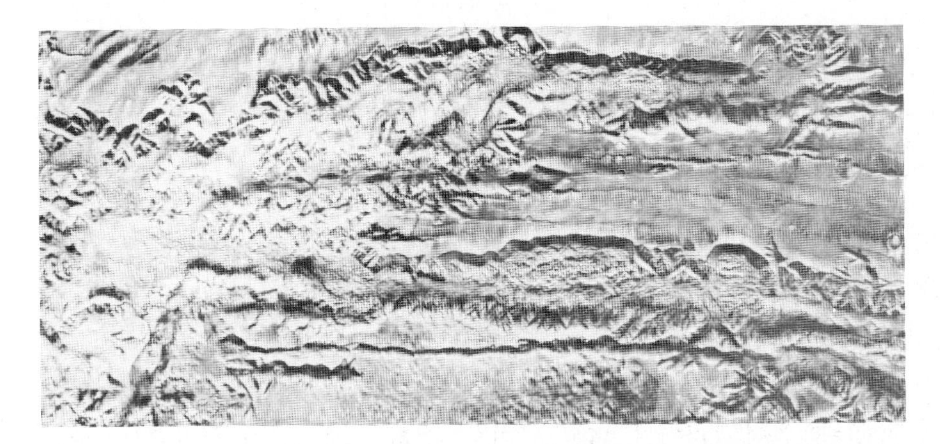

Figure 18–34 The immense Martian canyon system named Valles Marineris as photographed by Viking orbiters. It has been suggested that these canyons are in part the result of slumping and fluvial erosion as ground ice melted and gushed onto the surface. (Courtesy of the National Aeronautics and Space Administration.)

Many channels on Mars dwarf our own Grand Canyon in size and, in order to form, would have required torrential floods so spectacular as to be hard to visualize by earth standards.

Ice caps

Long before spacecraft landed on Mars, observers using telescopes noticed the planet had ice caps that grew and shrank with the seasons. Initially, scientists inferred that the ice was frozen carbon dioxide (rather like commercial dry ice). Data transmitted by Viking Orbiters, however, indicated that the residual northern ice cap was actually made of water ice (although the seasonal ice that forms in the fall and dissipates in the spring is confirmed as carbon dioxide). At the Martian north pole, temperatures rarely exceed 205°K (−68°C) in the summer, but this temperature is not low enough to maintain a permanent cap of carbon dioxide ice under the planet's atmospheric pressure of only 6 millibars. Because of dust storms, it is more difficult to interpret data from the southern cap, but it is likely that it, too, is composed mostly of water ice. In addition to the water locked in the ice caps, there may be subsurface ice on Mars in the form of permafrost.

Martian history

Like the earth, Mars was transformed from a protoplanet by accretion of a multitude of smaller objects. This accretion was followed by an episode of differentiation during which the once-homogeneous body became partitioned into masses of different chemical composition and physical properties. Very likely the differentiation process involved partial or complete melting, so fluids moved from one region of the planet to another.

During the initial billion years or so of Martian history, the planet experienced heavy bombardment of meteorites and asteroids, with the result that the exposed crust became densely cratered. The craters have been modified by relatively recent wind erosion. The crust was disturbed not only by cratering but also by the faulting and fracturing that accompanied volume changes in the mantle.

It has been postulated that a dense atmosphere formed on Mars as a result of volcanic outgassing (removal of embedded gas by heating) and that the production of water and carbon dioxide far exceeded its loss into space. The planet was warmer at that time, perhaps because of a greenhouse effect. For a time, running water was an important geologic agent, as evidenced by the sinuous furrows incised throughout the cratered terrain. In certain regions of the planet, a small decline in surface temperatures may have caused the water in surface materials to freeze, thus trapping still-liquid subsurface water in pockets and channels. With local melting, the subsurface water may have gushed upward and eroded large channels, simultaneously causing entire areas of surficial material to collapse and form mudflows.

Eventually, the supply of atmospheric water and carbon dioxide became so depleted that the planet began to cool. Remaining carbon dioxide and water migrated to polar caps and to subsurface reservoirs, where it has remained.

Because Mars was once warmer, had more water, and possessed the elemental raw materials thought to be required for the development of organisms, scientists have long speculated on the possibility of finding primitive forms of life on the planet. For this reason, experiments aboard the Viking lander were designed to search for signs of present or former life. The results were disappointing. Not a single microorganism or even a trace of organic molecules could be detected in the Martian soil at the landing site.

THE OUTER PLANETS

Beyond the orbit of Mars lies the ring of asteroids, and beyond the asteroids are the orbits of Jupiter, Saturn, Uranus, Neptune, and Pluto. The first four of these planets are also called the **Jovian** (for Jupiter) **planets.** They are similar to one another in having large size and relatively low densities. Pluto is not a Jovian planet. It is so far away and small that it is difficult to observe. As a result, there is considerable uncertainty about its origin.

Jupiter

Jupiter, named for the leader of the ancient Roman gods, is the largest planet in our solar system. The diameter of this huge planet is 11 times greater than the earth's diameter, and its volume exceeds that of all the other planets combined. Its mass is 318 times that of the earth yet it is only a quarter as dense. Jupiter spins rapidly on its axis, making one full rotation in slightly less than 10 earth hours. This rapid rotation results in the formation of the encircling colored atmospheric bands for which the planet is famous (Fig. 18-35). Hydrogen, helium, and lesser quantities of

Figure 18–35 Jupiter. Voyager 1 took this photograph of Jupiter in 1979 at a range of 32.7 million km.

ammonia and methane are the predominant gases in Jupiter's atmosphere. On the basis of experiments with these gases, some investigators believe that the reddish and orange colors in the atmosphere are derived from sulfur- or phosphorous-based compounds. It is impossible to discern the surface of the solid interior of Jupiter with optical telescopes. The atmosphere, which is several hundred kilometers thick, passes gradually to liquid and eventually solid matter. The interior of the giant planet may consist of highly compressed hydrogen, possibly surrounding a rocky core. Jupiter has at least 16 satellites. With its many moons, this large planet rather resembles a miniature solar system.

We have learned about Jupiter from earth-based telescopes and from close encounters by unmanned spacecraft. The most successful of the robot explorations occurred in 1979 when the two Voyager spacecraft swept past Jupiter and transmitted magnificently clear images of the planet's brightly colored atmosphere, the tempestuous Great Red Spot (Fig. 18–36), five of Jupiter's moons (Fig. 18–37), and a ring of debris less than 0.6 km thick that circles the planet about 55,000 km above the tops of the clouds (Fig. 18–38). The discovery of the ring around Jupiter makes that planet the third in the solar system known to possess such a feature (Saturn and Uranus are the other planets known to have rings).

Jupiter's Great Red Spot, which is twice the size of the earth, is an interesting feature known to astronomers for centuries. As clearly revealed by the Voyager mission, this many-hued disturbance is a gar-

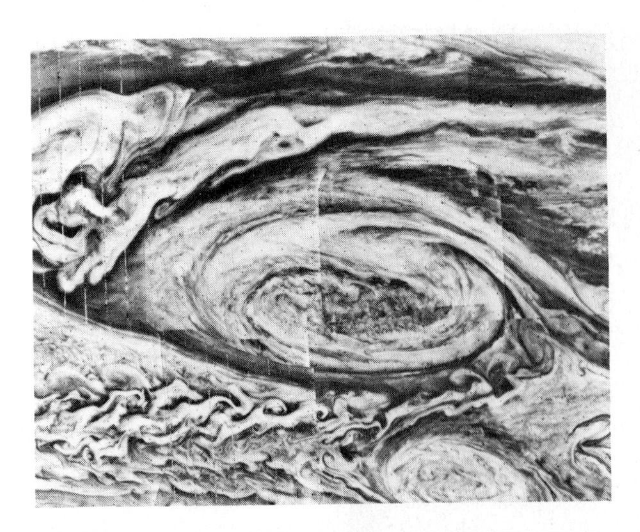

Figure 18–36 *Photomosaic of Jupiter's Great Red Spot, assembled from pictures taken by Voyager 1 in 1979 at a distance of 1.8 million km. The smallest clouds visible in this mosaic are 33 km across. (Courtesy of the National Aeronautics and Space Administration.)*

Figure 18–38 *Jupiter's ring. This image of the ring of Jupiter was assembled from photographs taken by Voyager 2. The thickness of the ring has been estimated at less than 1 km. (Courtesy of the National Aeronautics and Space Administration.)*

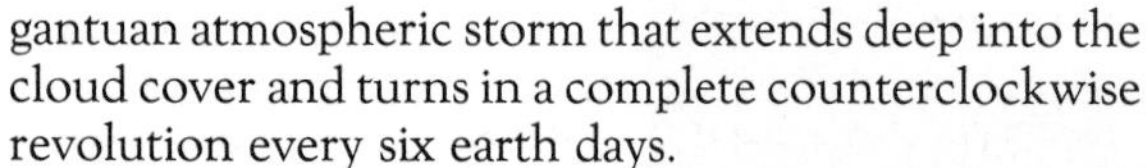

gantuan atmospheric storm that extends deep into the cloud cover and turns in a complete counterclockwise revolution every six earth days.

In addition to the Great Red Spot, the computer-controlled devices aboard the Voyager spacecraft radioed back the distinct images of great bolts of lightning, whose occurrence had only been suspected before the flyby. Another discovery was an auroral display far brighter than any northern lights ever seen on earth. Voyager also confirmed the presence of an immense magnetic field surrounding Jupiter (Fig. 18–39). First detected by the Pioneer 10 spacecraft in 1973, the magnetic field extends outward from the planet for 16 million km, encompasses all of the larger satellites, and traps enormous quantities of charged particles from the solar wind. The presence of the field indicates that Jupiter has a magnetic core resulting from fluid movements within its spinning interior. The two Voyager spacecraft were not launched at the same time. Voyager 2 reached Jupiter four months after its companion spacecraft and was able, therefore, to fill in many of the gaps in information. Voyager 2 also pro-

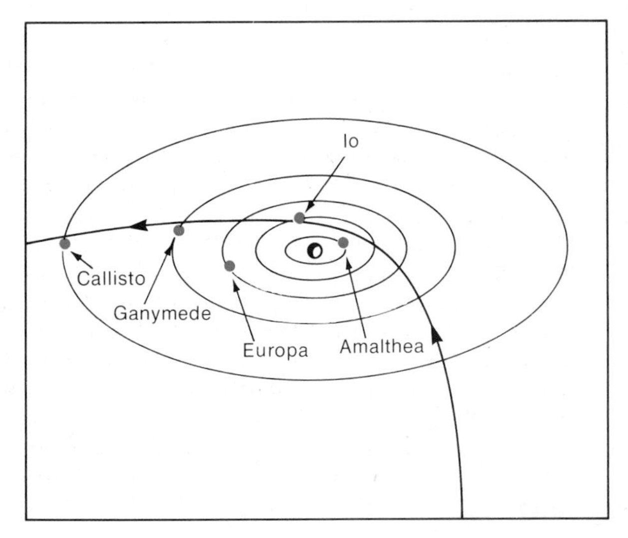

Figure 18–37 *Voyager 1's flight path allowed scientists to obtain close range photographs of five of Jupiter's 13 satellites. Each is shown at its closest point to Voyager 1's outbound flight away from Jupiter. The spacecraft approached within 280,000 km of Jupiter. (Courtesy of the National Aeronautics and Space Administration.)*

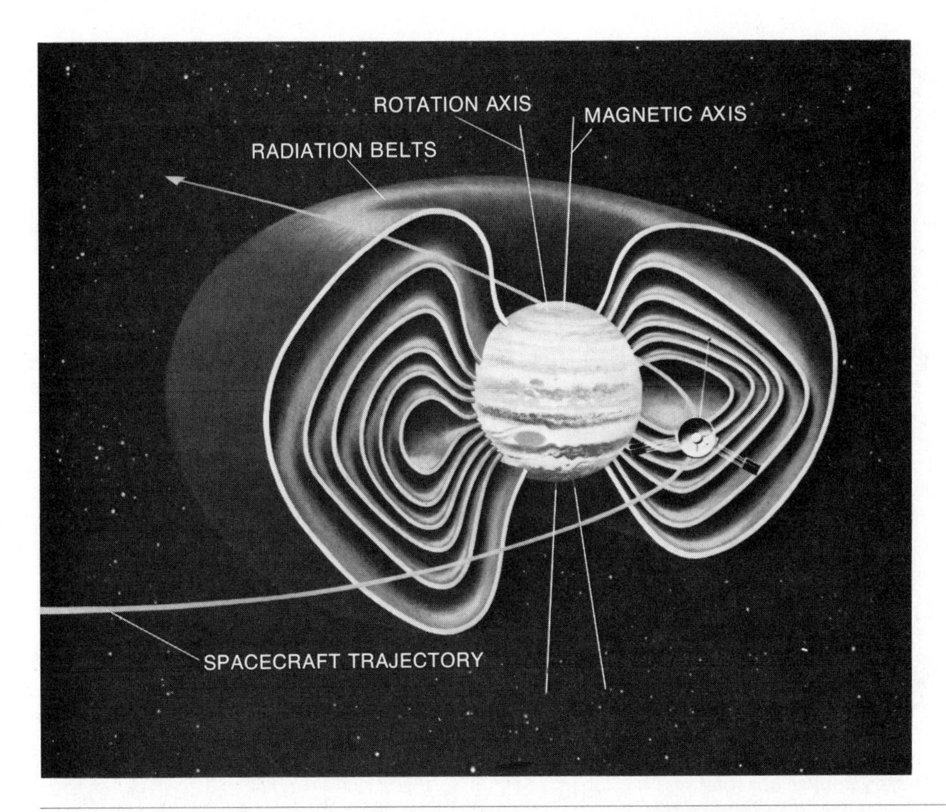

Figure 18–39 *Jupiter's magnetic field or* magnetosphere *is shaped somewhat like a giant donut surrounding the great planet. (From Pasachoff, J.M.,* Contemporary Astronomy, *2nd ed. Philadelphia, Saunders College Publishing, 1981.)*

vided even better images of Jupiter's larger moons than had been transmitted by its sister ship. These larger satellites are called the Galilean moons to commemorate their discoverer. They are named *Callisto, Ganymede, Europa,* and *Io.* Each has different characteristics. Callisto (Fig. 18–40), the outermost of the Galilean moons is riddled with craters resulting from meteorite impact. On Callisto's icy, dirt-laden surface, Voyager's cameras detected huge concentric rings that outline extraordinarily broad impact basins. Callisto's nearest neighbor is Ganymede (Fig. 18–41), the largest of Jupiter's satellites. Ganymede also shows the effects of meteorite bombardment in its dark, cratered terrain and bright streaks or rays that fan out from the rims of craters like huge splash marks. Sinuous ridges have been detected on Ganymede's surface, as well as crisscrossing fractures suggesting that fault movements have occurred on the satellite. It will take time to analyze all the data on Ganymede transmitted by Voyager, but it is already certain that the satellite shows the effects of a variety of geologic processes, including crustal movements.

Europa (see chapter opening illustration) is the brightest of the Galilean moons. It appears to have a thin crust of ice that may rest upon somewhat softer ice or water. The surface of Europa is marked by huge fractures and ridges, but few impact craters have been detected. This suggests that Europa's surface is being continuously renewed, perhaps by the development and slow migration of glacier-like masses of ice.

When images of Io first reached scientists, they immediately compared the appearance of that moon's surface to a pizza (Fig. 18–42). The distinctive mottled, yellow-orange color of the satellite is probably derived from the large amounts of frozen sulfur dioxide or amorphous silica in surface materials. Io's surface is thought to be young, for it is relatively smooth and scarred by only a few impact craters. Geologists were especially interested in the discovery of active volcanoes on Io (Fig. 18–43). Altogether eight volcanoes

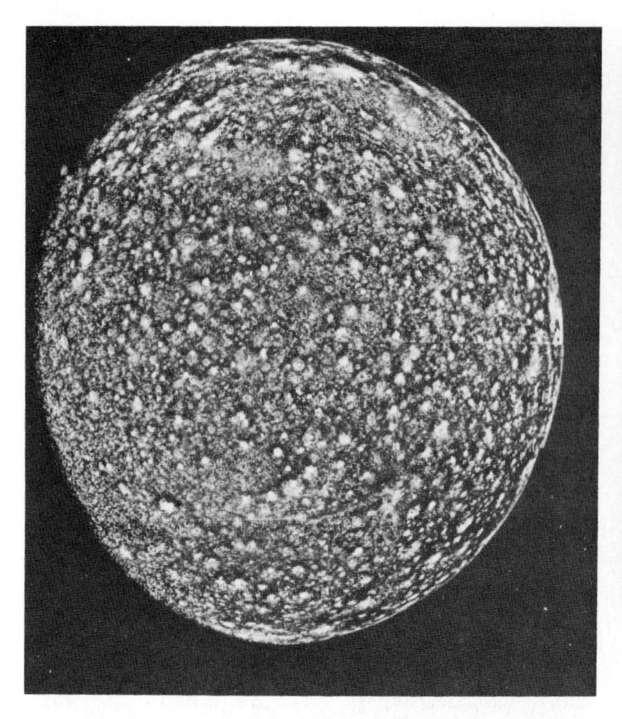

Figure 18–40 *Callisto, the outermost of the Galilean moons, and the most heavily cratered planetary body in our solar system. In this photomosaic, a special computer filter was used to provide high contrast in the surface topography. (Courtesy of the National Aeronautics and Space Administration.)*

Figure 18–41 *Ganymede, the largest of Jupiter's moons, has a radius of 2600 km (about 1.5 times as large as our moon). The satellite is probably composed of a mixture of rock and ice. (Courtesy of the National Aeronautics and Space Administration.)*

have been detected, some with plumes extending 300 km above the mean surface of the satellite. Sufficient sulfur dioxide is being emitted by the volcanoes to form a ring of ionized sulfur and oxygen atoms around Jupiter near Io's orbital path.

Voyager 2 also passed near Amalthea, Jupiter's innermost satellite. The spacecraft's camera showed that Amalthea was a reddish, nonspherical satellite about 165 km long and 150 km wide. It is without doubt the most irregularly shaped satellite known in our solar system.

In addition to the larger Galilean moons, Jupiter has two groups of small moons. The first group includes satellites designated J13, J6, J10, and J17. These are about 11,000 to 12,000 km from Jupiter, and their orbital planes are inclined 24° to 29° from the plane of Jupiter's equator. The second group of small moons circle at a distance of about 21,000 to 24,000 km. They are designated J12, J11, J8, and J9. The satellites of this outer group have orbital planes inclined 147° to 164° to Jupiter's equator.

Figure 18–42 *Io. The relatively smooth surface and the occurrence of active volcanoes suggest that Io may be Jupiter's youngest satellite. Large amounts of sulfur and sulfur dioxide frost are thought to be present in the surface material. (Courtesy of the National Aeronautics and Space Administration.)*

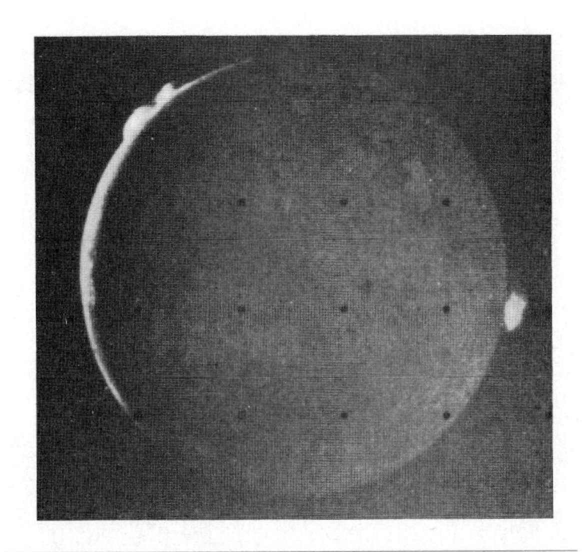

Figure 18–43 Voyager 2 photograph of Io showing three active volcanoes (black dots are calibration points on the camera). (Courtesy of the National Aeronautics and Space Administration.)

Saturn

Saturn's density is only about 70 percent that of water. Its mass, however, is equivalent to 95 earth masses. Measurements of Saturn's size and density suggest that the planet may have a central core of heavy elements (probably mostly iron), surrounded by an outer core of hot, compressed volatiles, such as methane, ammonia, and water. These two core regions, however, constitute only a small part of the total volume of the planet, in which hydrogen and helium are overwhelmingly the predominant elements.

As was true for Jupiter, our knowledge of Saturn has been greatly improved as a result of data collected by unmanned spacecraft. In particular, the electronic eyes of Voyager 2 in August 1981 provided truly spectacular views of this second largest planet, its rings, and its moons. The data transmitted to earth helped planetary scientists confirm that Saturn has a magnetic field, trapped radiation belts, and an internal source of heat.

The general appearance of Saturn's atmosphere is similar to that of Jupiter. The alternating light and dark bands and swirls of gas, however, are partially veiled by a haze layer above the denser clouds. Temperatures near the cloud tops range from 86°K (−305°F) to 92°K (−294°F) with the coolest temperatures occurring near the center of the equatorial zone. The planet is well illuminated, for even its dark face receives a substantial amount of light reflected from the rings. On this dark side, Voyager 2 detected radio emissions rather like those given off during lightning discharges.

The rings of Saturn

There is no more beautiful sight in the solar system than the rings of Saturn (Fig. 18–44). With his primitive telescope, Galileo was the first to recognize these

Figure 18–44 Voyager 1 looked back at Saturn on November 16, 1980, four days after the spacecraft flew past the planet, to observe the appearance of Saturn and its rings from this unique perspective. Voyager 1 took this image when it was 5.3 million km from the planet. Saturn's shadow falls across part of the rings. (Courtesy of the National Aeronautics and Space Administration.)

rings, but he was unable to discern them clearly. Fifty years later, in 1655, Christian Huygens was able to differentiate three concentric rings, the brightest and broadest of which was in the center. There are, however, far more than three rings around Saturn. As a result of pictures taken by Pioneer II, and the more recent and truly dazzling photographs taken by Voyager I and Voyager 2, we now know that the once neatly defined grouping of three rings is really a complex system of thousands of rings and rings within rings. Saturn's rings are composed of billions of orbiting particles of dust and ash, as well as many fragments measuring tens of meters in diameter. All revolve around Saturn in the approximate plane of the planet's equator. Saturn's 17 or more moons also revolve in this plane. The entire ring system has an outside diameter of about 275,000 km. Although the area of the rings is truly enormous, most of the material is concentrated within a thickness of only about 16 km. Thus, the entire system reminds one of a phonograph record with the rings corresponding to the record's many tiny grooves. Scientists are uncertain about the origin of Saturn's rings. Some suggest that the particles in the rings are the remains of an exploded satellite. Others argue that the rings are a remnant of the particle cloud from which satellites form.

Saturn's moons

Largely as a result of the Voyager missions, we know that Saturn ranks first among the planets in the number of its satellites. Altogether, 17 moons have been recognized. Some have been formally named; others bear only numbers. The largest of the named satellites is Titan. Titan is about 5,120 km in diameter and has a relatively low density, about twice that of water ice. The satellite's surface is obscured by thick, dense haze. Titan has a nitrogen-rich atmosphere, and lakes of liquid nitrogen may exist at the poles, where surface temperaures are an estimated 90° K (−300°F). The more important inner satellites of Saturn are named *Mimas, Enceladus, Tethys, Dione,* and *Rhea* (Fig. 18–45). Their densities and surface brightness suggest they too are composed mainly of water ice. The five satellites range in diameter from Mimas's 390 km to Rhea's 1530 km. All are densely cratered. The outer satellites of Saturn include tiny Phoebe, potato-shaped Hyperion, and Iapetus.

Voyager 2, which provided our best photographs of Saturn's moons and rings, is programmed to use the gravitational sling-shot effect provided by the pull of Saturn and Jupiter to set a course for a visit to Uranus in 1986 and Neptune in 1989. There is hope that the photographic systems, which have already shown signs of fatigue, will not fail before giving us a glimpse of these distant planets. Subsequently, the spacecraft will begin its endless voyage to the stars.

Uranus

Beyond the orbit of Saturn lies the planet Uranus. Like Saturn and Jupiter, Uranus also has rings, that have been photographed from earth with infrared film. Uranus has a low density (1.2 grams per cubic centimeter) and a frigid surface temperature of about −185°C. Five moons circle Uranus. They are named *Ariel, Miranda, Oberon, Titania,* and *Umbriel.* Thus far, earth-based instruments have been able to detect only hydrogen and methane on Uranus, although helium may also be present. The methane may be responsible for the planet's greenish hue. Perhaps the most distinctive characteristic of Uranus is the orientation of its axis of rotation, which lies nearly in the plane of its orbit (Fig. 18–46). The planet's direction of rotation is opposite that of all other planets except Venus.

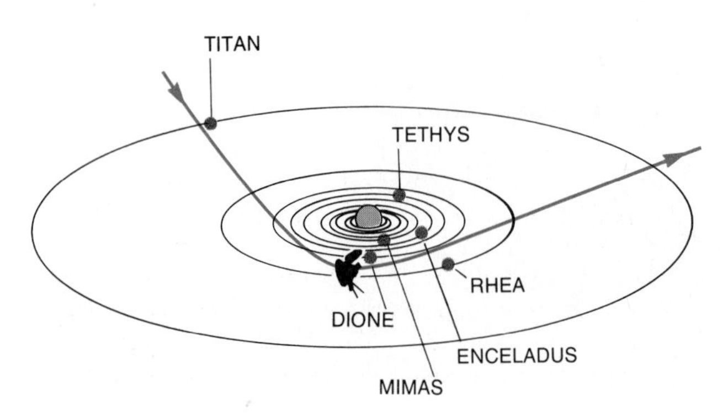

Figure 18–45 *Voyager 1 approached within 124,000 km (77,000 miles) of Saturn's cloudtops. Six of the satellites that were photographed are shown in their approximate positions at closest approach by the spacecraft. (Courtesy of the National Aeronautics and Space Administration.)*

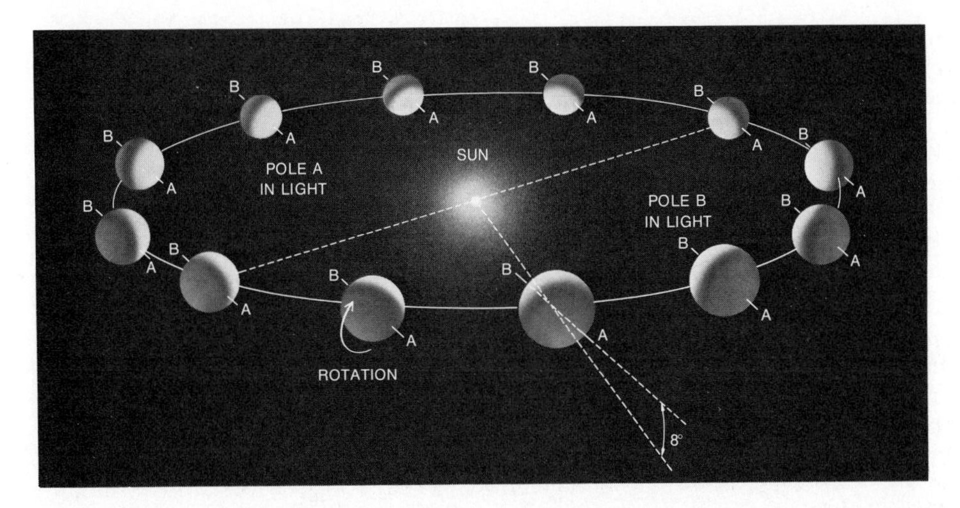

Figure 18–46 *The axis of rotation of Uranus lies approximately in the plane of its orbit, whereas other planets have their axes of rotation roughly parallel to their axes of revolution around the sun. (From Pasachoff, J.M.,* Contemporary Astronomy, *2nd ed. Philadelphia, Saunders College Publishing, 1981.)*

This retrograde, clockwise direction is also characteristic of the orbits of the five satellites around Uranus.

Neptune

Because they are similar in size and other physical properties, Neptune and Uranus are often called twin planets. Through the telescope, Neptune, like Uranus, appears as a greenish orb. Neptune's density of 1.7 grams per cubic centimeter is just slightly greater than the density of Uranus. Neptune's atmosphere may also resemble that of Uranus. Both hydrogen and helium have been detected spectroscopically, and probably methane is also present. Two moons, named *Triton* and *Nereid*, circle the planet. Triton is one of the four largest moons in the solar system (the other three being Ganymede, Titan, and Callisto). Its surface may contain ice made of frozen methane, and it may even have a thin atmosphere of methane gas. Recently, astronomers have found evidence that Neptune may have a third moon and possibly even a ring system.

Pluto

Far out on the outer limits of our solar system lies the orbit of Pluto. It is so small and so distant that less is known about it than the other planets. We do observe that Pluto's orbit is tilted an unusual 17° to the ecliptic plane and that its orbit crosses that of Neptune. As a result, Pluto is actually closer to the sun than Neptune during certain periods (as from 1979 to 1999). Because of the high inclination of Pluto's orbit, however, there is no danger of a collision with Neptune. Its small size and orbital peculiarities suggest that Pluto may not be a true planet but rather a former satellite of Neptune that escaped the gravitational pull of that planet and now occupies a separate orbit. Evidence for this theory, however, is not conclusive.

In 1978, scientists of the U.S. Naval Observatory found that Pluto has a moon. They named it *Charon*, after the mythologic boatman who ferried the souls of the dead across the River Styx to Hades. Some scientists believe the existence of Pluto's moon makes the escaped-satellite theory for Pluto's origin impossible. Others argue that Pluto and Charon may have been ejected from the Neptunian system as satellite pair.

DOOMSDAY

Anyone interested in the history of the earth is naturally led to a consideration of its future. Clearly, all that has happened, is happening, and will happen to this planet is inevitably linked to the history and destiny of the sun. The sun is a star, and like other stars, it will run its course from birth to death. By observing other stars that are at various stages in their histories, astronomers have pieced together a story of what happens when these bodies begin to deplete their store of nuclear fuel. Our star has been "burning" for about 5 billion years. In this span of time, it has consumed about half of its nuclear fuel and so has about another 5 billion years of life remaining.

What will "doomsday" for the sun and earth be like? Most astronomers believe that once the hydrogen in the sun's core has been largely converted to helium by fusion, the core will begin to contract. As the core contracts, the outer layers of the sun will expand and cool, causing the sun to puff up into an enormous sphere and to turn blood-red. Our sun will have become a red giant. As the great ball of rarefied gases continues to expand, it will engulf and vaporize Mercury and Venus. What will happen to the earth can be readily imagined. The searing heat will boil away the oceans and heat the crust even to the melting point of some surface rocks. This purgatorial state of affairs may continue for a billion years or so, until finally the sun has dissipated its energy. The dying sun may then begin to shed its outer layers, leaving behind a dense, hot, white dwarf star about the size of the earth. Such white dwarf stars have exhausted their supply of nuclear fuels and remain incandescent only because of their residual internal heat. Eventually, even this light will fade as the sun becomes a black dwarf, so-called because it is now too cool to radiate visible light. The scorched and frazzled earth will continue as a barren, rigid sphere unable to free itself from the gravitational grasp of the dead sun.

Summary

The planet earth is one of a group of nine planets revolving around a modest-sized star, our sun. Planets and sun form our solar system, which, together with myriads of other stars, planets, dust, and gas, make up the Milky Way Galaxy. Our sun provides energy to the earth, which drives geologic processes and which has ultimately determined the character of terrestrial life. The sun is composed mostly of hydrogen and helium. The helium in the core results from the fusion of hydrogen nuclei with the conversion of a part of the hydrogen into solar energy.

Theories for the origin of our solar system must conform to the known facts of the system's distribution of mass and angular momentum, as well as the known orbital and rotational characteristics of the planets. Recent theories postulate condensation and accretion of planets from a nebula of gases and dust particles. Currently, the most popular of such concepts is the protoplanet hypothesis. It suggests that planets were formed by the cold accretion of particles collected within a turbulent eddy located on a disc-shaped rotating cloud of dust and gas.

Meteorites are important to planetary geologists because they are truly extraterrestrial samples of the solar system. They can be classified as ordinary chondrites, carbonaceous chondrites, achondrites, irons, and stony-irons and are believed to be fragments resulting mostly from collisions of one or more small planetary bodies.

The moon, which has been studied by telescope for centuries, was first visited by humans in July 1969. Its general features include the relatively smooth, dark mare basins and the rugged lunar highlands. Samples of the loose regolith on the moon have been brought back to earth. They include basaltic rocks from the maria and specimens of anorthosite from the highlands. Study of the age and distribution of these rocks suggests that lunar history began about 4.6 billion years ago, when the moon had developed its skin of anorthositic igneous rock. This surface was then flooded in various regions by the basalts of the maria about 1 billion years later.

The planets of our solar system can be divided into two main groups. Those resembling the earth are the inner, terrestrial planets. The outer planets resemble Jupiter and are termed Jovian planets. The terrestrial planets tend to be small, dense, and rocky, whereas the Jovian planets are large and have many satellites and dense thick atmospheres.

The terrestrial planet closest to the sun is Mercury. It is the smallest of the inner planets and is very moonlike in appearance. Next in the progression outward from the sun is Venus. Venus is about the size of the earth. It has a dense carbon dioxide atmosphere that traps heat from the sun. The surface temperature of Venus is higher than any other planet in the solar system. Its rocky surface is enshrouded by a dense cloud cover.

Mars is of interest because of its great variety of surface features. It has a cratered topography reminiscent of the moon, and at the same time has features that appear to be the result of processes similar to those operating on earth. Dunes, stratified sediments, and sinuous rills suggest erosion by wind, water, and ice. Mars has a thin carbon dioxide atmosphere and polar caps composed of water ice and frozen CO_2.

Among the outer planets, Jupiter is known for its gargantuan size. It is larger than all other planets combined, has a massive hydrogen, helium, methane, ammonia atmosphere, and rotates so rapidly as to form beautiful arrays of colored cloud bands. The planet has an internal heat supply. Its strong magnetic field suggests the presence of a metallic core. Sixteen moons encircle Jupiter.

Saturn is rather similar to Jupiter, although smaller. Saturn's most distinctive feature is the thin, complex ring system that encircles its equator like the brim of a straw hat. At least 17 moons accompany Saturn on its revolution around the sun.

Uranus and Neptune are similar in size and general characteristics. Both have atmospheres rich in hydrogen compounds, and Uranus appears to have a small ring system. Beyond Neptune lies Pluto. Pluto is roughly the size of the earth's moon and is thus the smallest planet in the solar system.

QUESTIONS FOR REVIEW

1 What are the names of the planets in our solar system? In what galaxy is our solar system located?

2 Given the amount of solar radiation intercepted by the earth, why is the earth's surface not much hotter than it is?

3 Compare Mercury, Mars, and the moon. What do they have in common, and how do they differ?

4 What regularities or uniformities exist in the solar system that indicate it is truly a coherent system?

5 In general, how do the terrestrial planets differ from the Jovian planets?

6 The surface of Io is noted for its bright yellow and orange colors. What is the cause of this coloration?

7 What was the principal objection to the original nebular hypothesis for the origin of the solar system? How is this problem apparently resolved in the protoplanet hypothesis?

8 What are the principal kinds of meteorites that have been found on the earth? In what way are they thought to have originated? What are chondrites?

9 What evidence indicates that, unlike the earth, the moon does not have an iron-nickel core?

10 How do the rocks of the lunar maria and lunar highlands differ in composition and age?

11 What is the derivation of the reddish coloration of the planet Mars?

12 How does Mars differ from the earth in density, internal structure, and average range of equatorial temperatures?

13 What is the origin of the craters on the earth's moon?

14 How do lunar geologists determine the relative ages of craters on the moon?

15 What is lunar regolith? How is it thought to have originated?

SUPPLEMENTAL READING AND REFERENCES

Abell, G.O., 1982. *Exploration of the Universe.* 4th ed. New York, Holt, Rinehart & Winston.

Arvidson, R.E., Binder, A.B., and Jones, K.L., 1978. The surface of Mars, *Sci. Am.* 238:3, 76–91.

Beatty, J.K., O'Leary, B., and Chaikin, A. (eds.), 1982. *The New Solar System* (2d ed.) Cambridge, Mass., Sky Publishing Corp., and London, Cambridge University Press.

Cadogan, P., 1981. *The Moon: Our Sister Planet,* New York, Cambridge University Press.

Carr, M.H., 1981. *The Surface of Mars.* New Haven, Conn., Yale University Press.

Hartmann, W.K., 1983. *Moons and Planets: An Introduction to Planetary Science.* 2d ed. Belmont, Calif., Wadsworth Publishing Co.

King, E.A., 1976. *Space Geology: An Introduction.* New York, John Wiley & Sons.

Murray, B., Malin, M.C., and Greeley, R., 1981. *Earthlike Planets.* San Francisco, W.H. Freeman & Co.

Noyes, R.W., 1982. *The Sun, Our Star.* Cambridge, Mass., Harvard University Press.

Wood, J.A., 1979. *The Solar System,* Englewood Cliffs, N.J., Prentice-Hall, Inc.

19

beyond the solar system

Nature and Nature's Laws lay hid in Night.
God said, "Let Newton be!" and All was Light.
ALEXANDER POPE
Essay on Man, 1734

Prelude Human curiosity drives us to discover all that we can about our external world, not only here on earth but even into the farthest reaches of space. This drive to plumb the secrets of stars and planets began among ancient peoples who believed that the earth was the center of the universe, and it continues into our modern age of space exploration. What has been learned has shaped our thinking about our place in the cosmos. In this chapter we review some of the highlights of astronomy. The tools of astronomy are described, and the stars and galaxies are examined. Factual information about a star can be found here, but the real interest is in learning how the characteristics of light emanating from objects in space provide clues to the evolution of stars and galaxies.

The Whirlpool Galaxy in the constellation Canes Venatici. (From Pasachoff, J.M., Astronomy: From the Earth to the Universe, *2nd ed. Philadelphia, Saunders College Publishing, 1983.)*

THE GROWTH OF ASTRONOMICAL CONCEPTS

Superstition and a chronology

Ever since the first human glanced upward at the stars, the spectacle of the night sky has aroused wonder. To primitive people, the firmament was an omnipresent mystery filled with celestial manifestations of dragons, demons, and gods that controlled the destiny of every human being on the earth below. Persons thought to have an ability to interpret the stars became priests and oracles. Some of these ancient seers carefully noted the repetitive passage of stars and used this knowledge to select the most propitious time for important happenings. By observing the appearance of the moon and the position of the sun and stars, it was even possible to measure increments of time.

Time and its reckoning have always been important. By 4200 B.C., the crude astronomy of earlier ages had advanced to the stage where Egyptians were able to devise a calendar having 365 days to the year. The Egyptian calendar had 12 months of 30 days each, except for the last month, which contained 35 days. Later the Egyptians recognized that there were 365½ days in a year but did nothing to correct this discrepancy. It was left to Julius Caesar (who acted on the advice of an astronomer named Sosigenes) to correct the error by introducing a leap year. The extra half day was assimilated by having every fourth year contain 366 instead of 365 days.

The calendar established by Caesar is called the Julian Calendar. Unfortunately, like earlier calendars, it presented problems. The Julian year was 365.25 days long, compared with 365.2422 mean solar days in a true year of seasons. The difference (11 minutes and 14 seconds) builds up to one day in 128 years. Over many centuries, this accumulated difference becomes bothersome. Spring holidays like Easter would eventually be occurring in winter. In 1582, therefore, Pope Gregory XIII was compelled to reform the calendar. By proclamation, he adjusted the date of the vernal equinox to its present date of March 21. He then revised the leap year system, eliminating leap years in the century years *not* divisible by 400. (Thus 1700, 1800, and 1900 were not leap years.) The revision corrects the calendar for one day every 133 years, which is close to the aforementioned 128 years. The Gregorian calendar is still in error by about 24 seconds per year, which will accumulate to a day in 3,300 years.

The astronomers of ancient Greece

The ancient Greeks are well known for their ability to observe and make deductions. They recognized the correlation between the length of daylight and the seasons, learned that stars were self-luminous bodies that could be used in navigation, and knew that the moon did not have its own light but merely reflected light from the sun. These and other discoveries can be traced back to Thales of Miletus (624–547 B.C.), who is regarded as the founder of Greek astronomy. With remarkable insight, Thales was even able to predict the total eclipse of the sun that occurred during the battle between the Medes and Lydians on May 28 in 585 B.C.

Thales was both a remarkable scientist and an excellent teacher. Among his most notable pupils was Anaximander (611–546 B.C.), who recognized that the firmament revolves around the polar star. He is also credited with inventing the Grecian version of the sundial. Anaximander's astronomic findings were not all valid, however. He thought the earth was at the center of the universe, and this error persisted for many centuries after his death.

From about 300 to 200 B.C., an important library and observatory for Greek astronomy was established in Alexandria, Egypt. It was here, in what came to be known as the Alexandrian Museum, that Aristarchus, Eratosthenes, and Hipparchus completed significant studies in astronomy. Aristarchus (c. 300–250 B.C.), is remembered for his proposal that the earth and planets revolve around the sun. Using crude instruments but sound experimental methods, he demonstrated that the distance between the sun and the earth was far greater than the distance that separated the earth from the moon. Further, he accounted for the apparent relative immobility of the stars by correctly inferring that they were very much farther away from the sun than was the earth.

Eratosthenes, who served a term as director of the Alexandria Museum, achieved fame for having measured the circumference of the earth. As already described in Chapter 17, he accomplished this feat with the use of simple geometry and logic. In this triumph of reasoning, Eratosthenes came within 6000 km of the correct value for the average circumference of the earth.

Perhaps the greatest of the Alexandrian Museum astronomers was Hipparchus, who died about 125 B.C. He discovered that the earth's axis is not fixed in space but undergoes precession. He compiled an accurate catalogue of the positions and characteristics of more than 1000 stars. His measurements of the size of the moon and its distance from the earth were within 10 percent of present-day values. Hipparchus also accurately determined the length of the seasons and constructed a chart showing the position of the sun on the ecliptic for each day of the year.

The last of the more important Greek astronomers was Claudius Ptolemy, who studied in Alexandria between A.D. 127 and 157. His so-called Ptolemaic system of astronomy dominated theories about the universe for nearly 13 centuries. When Ptolemy formulated his system of astronomy, he was primarily interested in a mathematical model that would predict the motions of planets. He never said his system actually existed. Unfortunately, centuries later his model gradually became a physical reality in people's minds. The Ptolemaic concepts that the earth was immobile, and that the fixed stars were fastened to the ceiling of a vast dome that revolved around the earth every 24 hours, became astronomical dogma. These ideas, however, were to be challenged 1300 years later by Nicolaus Copernicus (1473 - 1543), the first of the Renaissance astronomers.

Renaissance astronomers

To his great credit, Copernicus was a discerning scientist who did not unquestionably accept the Ptolemaic theory as dogma. As a result of his own observations, he detected the many errors in earlier theories and set astronomers on the right track in his famous book *De Revolutionibus Orbium Caelestium* ("Concerning the

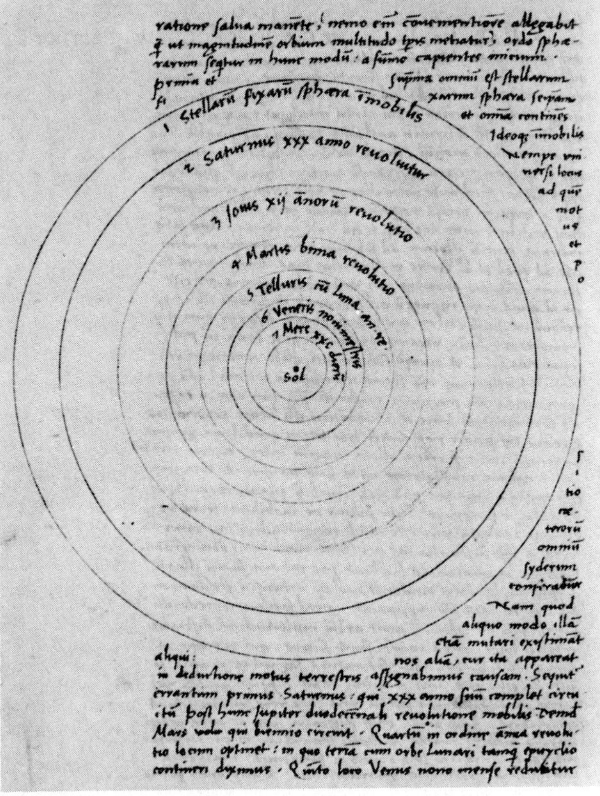

Figure 19–1 *A 16th-century woodcut depicting Copernicus and his 1543 illustration of the solar system with the sun at its center. (From Pasachoff, J.M.,* Astronomy: From the Earth to the Universe, *2nd ed. Philadelphia, Saunders College Publishing, 1983.)*

Revolutions of the Heavenly Bodies"). Copernicus proposed that the sun must be at the center of the solar system and that the earth was only one of the planets that revolve around the sun (Fig. 19-1). He argued that the earth turned from west to east on its axis, and that this rotation was the cause of day and night and the apparent motion of the stars. The moon, he said, revolved around the earth while accompanying our planet on its revolution around the sun.

Not all of these new ideas were warmly received during the Renaissance. Giordano Bruno (1548-1600) however, bravely espoused the views of Copernicus and went still further in describing the stars as scattered at varying distances through an immense and limitless outer space. They were not all located along the inside of a celestial dome or sphere. Bruno even speculated that other planets beyond the limits of observable space might be occupied by living things. Because of these views, he was considered a heretic and burned at the stake.

Tycho Brahe (1546-1601) was a contemporary of Bruno best known for his accurate measurements of the positions of stars and planets. Unfortunately, he is also known for having rejected Copernicus's ideas. He devised a scheme in which the sun revolved around a stationary earth, and the planets in turn circled the sun.

One of Brahe's most brilliant assistants was Johannes Kepler (1571-1630). Kepler returned to the Copernican concept of the sun at the center of the solar system. He discerned the elliptical orbital paths of planets and formulated three famous laws of planetary motion. The first of these is called the **law of elliptical orbits,** and states that the planets travel in elliptical paths (Fig. 19-2) with the sun at a focus nearer to one end of the ellipse. The second law, **the law of areas,** states that a planet moves so that an imaginary line drawn from it to the sun sweeps over equal areas in equal intervals of time. Thus, as a planet moves in its orbit, the area produced by the aforementioned line (radius vector) is proportional to time. In Figure 19-3, the areas and time intervals are equal. To make this possible, the orbital velocity of the planet must be greater when it passes closest to the sun.

Ten years after the publication of his first two laws, Kepler was able to state his **harmonic law.** It showed that the squares of the periods of revolution of any two planets around the sun are proportional to the cubes of their mean distances from the sun. With the use of this great empirical law, the mean distance of a planet from the sun could be calculated if its true period of revolution was known. Conversely, its period of revolution could be obtained if its mean distance from the sun was known.

While Kepler was determining the rules that describe how planets move, his Italian contemporary Galileo Galilei (1564-1642) was at work making discoveries of his own. Galileo's greatest contributions were his studies of motion and the actions of forces on bodies. His findings in this area of *mechanics* laid the foundations for the subsequent work of Sir Isaac Newton. In addition, Galileo improved the first telescope and was the first to use it for observing planets, satellites, and stars. Even with so primitive an instrument, he identified craters and mountains on the moon, discovered sun spots, some of the moons of other planets, and the ring structure around Saturn. He corrected earlier misconceptions about the perfect

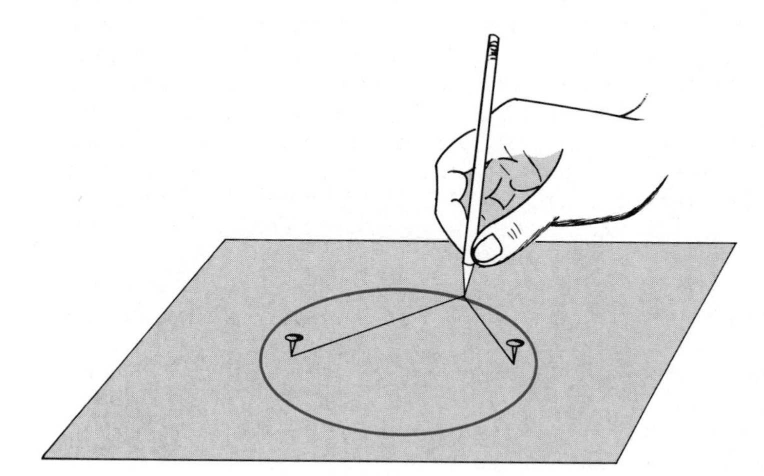

Figure 19-2 *An ellipse is a closed curve. Inside of an ellipse are two points called the* foci. *The sum of the distances from the foci to the edge of the curve is always the same for any point on the curve. To draw an ellipse, put two nails in a board and stretch a string between them as shown. Then move a pencil as shown, keeping the string taut.*

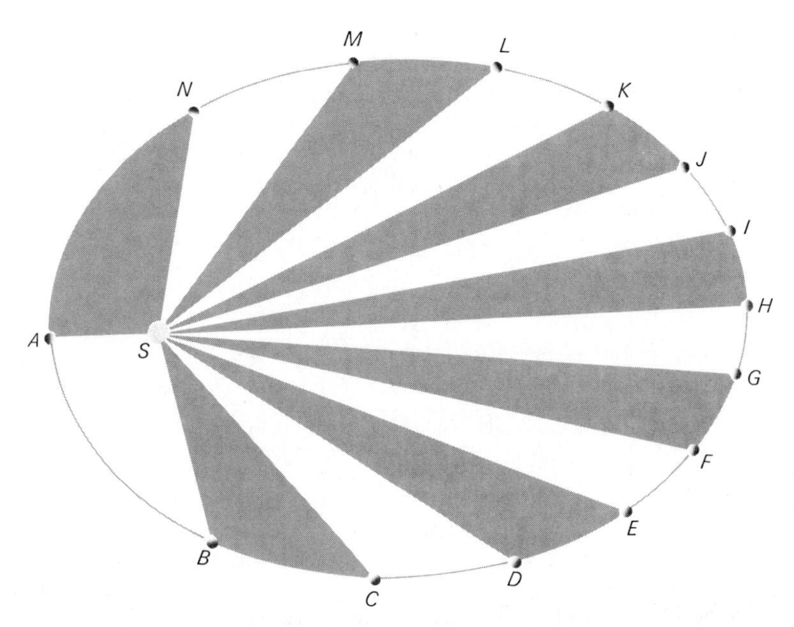

Figure 19–3 *Law of equal areas. A planet moves most rapidly on its elliptical orbit when it is at position* A, *nearest the focus of the ellipse,* S, *where the sun is. The planet's orbital speed varies in such a way that in equal intervals of time it moves distances* AB, BC, CD, *and so on, so that regions swept out by the line connecting it and the sun (shaded and clear zones) are always the same in area. (From Abell, G.O.,* Realm of the Universe, *3rd ed. Philadelphia, Saunders College Publishing, 1984.)*

sphericity of planets, and he used the phases of planets to demonstrate that they revolve around the sun (not the earth) and that they do not shine by their own light, like stars, but are illuminated by light from the sun.

Sir Isaac Newton's great laws

Preliminary concepts

Kepler's laws were based upon observation. They were empirical rules describing observable and measurable planetary movements. The reason for the observed motion of planets, however, was not known for another half century. The full explanation awaited the genius of Sir Isaac Newton (1642–1727). Newton formulated three laws of motion and a universal law of gravitation. These were based upon Kepler's earlier laws of planetary movement, as well as theories developed by Galileo. His indebtedness to these earlier investigators is acknowledged in his famous statement, "If I have seen farther than others, it is because I have stood on the shoulders of giants."

Among Newton's greatest accomplishments was his ability to clearly define the relationships among mass, force, and acceleration. Progress in modern space exploration would have been severely hampered without this knowledge, for it is needed in order to calculate the orbits of satellites and the trajectories of spacecraft. To understand Newton's discoveries, however, we must pause to consider the meaning of the terms *mass, inertia, force,* and *acceleration.*

Mass refers to the amount of matter a body contains. The mass of a body does not change whether it is on earth, in space, or on a planet. In fact, one can only change the mass of a body by adding matter to it or taking matter away from it. Here on earth, we can judge the mass of a body by its weight. Every mass is pulled to the earth's center by the force of gravity, and this provides us with the weight of the body. Provided the distance to the earth's center is kept constant, the greater the mass of a body, the greater will be its weight.

The term **inertia** describes the resistance of a material body to any change in its state of motion. Thus, a body at rest tends to resist being set in motion, and conversely, a moving body will resist being stopped because of inertia. Planets in the solar system provide excellent examples of the inertia of moving bodies. Each planet revolves around the sun in a period which has changed very little in billions of years. Force (such as a push or a pull) or friction must be applied in order to overcome the inertia of a body.

An important aspect of inertia is its relationship to mass. Both Newton and Galileo recognized that the extent to which a body has inertia is also related to its mass. A standing bowling ball would require a more vigorous kick to get it moving than a soccer ball because of the bowling ball's greater mass and, hence, greater inertia.

Once an object has been put into motion, its movement can be described in terms of average speed and velocity. **Average speed** refers to the distance traversed by a body per unit of time:

$$\text{Av. speed} = \frac{\text{distance}}{\text{time}}$$

Velocity has a somewhat different meaning; it is the average speed *in a given direction*. An automobile that has completed an 80 km trip in two hours has an average *speed* of 40 km per hour. Its *velocity*, however, would be expressed as 40 km per hour *to the east* (or some other specified direction).

The final term needed for an appreciation of Newton's discoveries is acceleration. **Acceleration** is the change in the velocity of a body per unit of time:

$$\text{Acceleration} = \frac{\text{change in velocity}}{\text{time}}$$

A freely falling body provides a good illustration of acceleration. In a vacuum (where there is no friction with air), bodies of varying weights all fall with a uniform acceleration of approximately 9.8 meters per second per second (9.8 m/sec^2). Thus, if a lead ball is dropped from a high place, its velocity will increase by 9.8 meters per second for every second that it falls. After one second, the ball will be falling 9.8 meters per second. At the end of two seconds it will be falling 19.6 meters per second, at the end of three seconds, 28.4 meters per second, and so on.

It was Galileo who first demonstrated that the weight of an object did not affect acceleration. In his famous experiment, he is described as having simultaneously released two unequal weights from the leaning tower of Pisa. The weights remained together during their fall and struck the ground at the same moment. A similar experiment was performed by an Apollo astronaut while he was standing on the moon. The astronaut dropped a feather and a heavy metal tool, and as Galileo would have predicted, both objects hit the moon's surface simultaneously.

Three laws of motion

Newton's first law. Now that we have an understanding of mass, inertia, and acceleration, we can turn to Newton's laws of motion. The first of these, known as the **law of inertia,** states that *a body at rest tends to remain at rest, and a body in motion tends to remain in motion with uniform velocity in a straight line unless acted upon by an outside force.*

Newton's second law. Newton's second law states that *when a force acts upon a body, the body accelerates in the direction of the force.* On the recent flights of the space shuttle, for example, a force in the form of a blast from the rocket engines provided acceleration whenever it was needed. The amount of acceleration is dependent on the mass of the body (the space shuttle and its occupants), and the force applied (by the shuttle's rockets). If the mass of a body is increased, the acceleration will be decreased, whereas if the force is increased, acceleration increases also. Stated more precisely, *the acceleration of a body is directly proportional to the magnitude of the force applied and inversely proportional to the mass of the object.* This is the essence of Newton's second law. It may be stated mathematically as follows:

$$\frac{\text{force}}{\text{mass}} = \text{acceleration}$$

or

$$\text{Force} = \text{mass} \times \text{acceleration}$$

Newton's third law. Simply stated, Newton's third law, **the law of reacting forces,** asserts that *to every action there is an equal and opposite reaction.* There are many familiar illustrations of this law. A rifle "kicks" when a bullet is fired. The reaction of the rifle causes it to move backward when the bullet is shot forward. The astronauts are well aware of this law as they move about in their weightless spacecraft. Only a small push in one direction would send them off in the opposite direction.

Momentum

From Newton's three laws of motion, it is possible to formulate a concept of momentum. The **momentum** of a body is defined as the product of its mass and velocity. Thus:

$$\text{momentum} = \text{mass} \times \text{velocity}$$

An object's momentum is related to its inertia. For example, from the law of inertia we know that a moving body tends to continue its motion in a straight line with undiminished speed. The extent of that tendency depends on the momentum of the body. A body having a large amount of momentum will have a strong tendency to remain in motion. It will be hard to stop. When a force acts on a body, it changes the momentum of that body, and the change is proportional to the size of the applied force and the time over which the force is applied.

The universal law of gravitation

Earlier, we noted that the *weight* of a body is due to the attraction between that body and the earth. Newton discerned that a force of attraction exists between the earth and everything on or near it, and also that every object in the entire universe attracts every other object. He called the attractive force **gravity.** Gravity is the underlying cause for a host of processes affecting the earth and planets. The earth's roughly spherical shape (like that of the sun and stars) is a consequence of gravitational forces that work to compress matter into an economical spherical shape. Gravity explains the orbits of planets and their satellites, the tides, and the erosional sculpturing of the earth's surface.

Newton's universal law of gravitation states that *the force of attraction between any two objects is proportional to the product of their masses and inversely proportional to the square of the distance between them.* Today, these relationships are expressed as follows:

$$F = G\frac{m_1m_2}{d^2}$$

In this formula m_1 and m_2 are the masses of the two objects, d^2 the square of the distance between them, and G the universal gravitational constant determined by Sir Henry Cavendish 125 years after Newton had formulated his law.

Whereas astronomers are particularly concerned with gravitational forces between planetary bodies, geologists are interested in the nature of gravitational attraction at the earth's surface. Surveys of gravity over areas of the planet's surface can be made with the aid of accurate gravity meters. Readings from these instruments reveal that gravity varies slightly and systematically with latitude. This is because the earth has an oblate shape, and the value of d in Newton's formula is therefore greater at the equator than at the poles (Fig. 19-4). The relatively greater equatorial radius of the earth results in lesser gravitational attraction at the equator. The value of d increases with altitude, and thus gravitational attraction on high mountains is slightly lower than nearer to sea level.

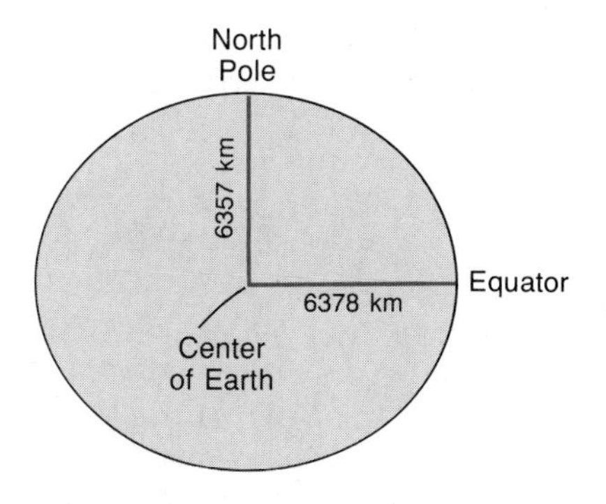

Figure 19-4 The equatorial radius of the earth exceeds its polar radius by 21 km. A given mass will weigh less at the equator than near the poles because the value of d *in Newton's equation is greater at the equator than at the poles. The values of* m_1, m_2, *and* G *are the same at both locations.*

Circular motion in the solar system

Newton's laws of motion tell us that a body in motion tends to keep moving in a straight line because of inertia. Therefore, to move a body like the earth in a circular orbit, a force must be applied that keeps pulling the body in toward the center of a circle. Such a force is called **centripetal force.** In the familiar example of a weight swung around one's head by a string, it is the string that provides the inward pull. For the earth and other planets revolving around the sun, the force of gravity provides an analogous inward pull and prevents each planet from moving off in a straight line.

In one of his volumes, Newton illustrated an imaginary experiment to help his readers understand the circular motion of the moon. The drawing depicts a great mountain on the nearly spherical earth (Fig. 19-5). A cannon, with its barrel aimed horizontally, is perched atop the mountain. A ball fired from the cannon moves directly forward under the impetus of the explosion. Because of inertia, the ball has a tendency to travel in a straight line, but gravity pulls it toward the earth's center. The combination of forward motion and downward pull results in the curved trajectory of the cannon ball. The flight distance of the ball would depend upon the size of the explosive charge. Newton asked his readers to imagine a cannon so powerful that the ball could travel completely around the earth. Again, the balance of forward impetus resulting from the cannon's explosive charge and the downward deflection caused by gravity would curve the trajectory of the cannon ball into a circular orbit around the earth. Indeed, the cannon ball would have become a small artificial satellite.

Although Newton conceived this model to explain the way the moon circled the earth, it is also useful in understanding the movement of planets and asteroids around the sun. One can imagine several possibilities if either the effect of gravity or the amount of momentum varied. An object whose forward momentum was

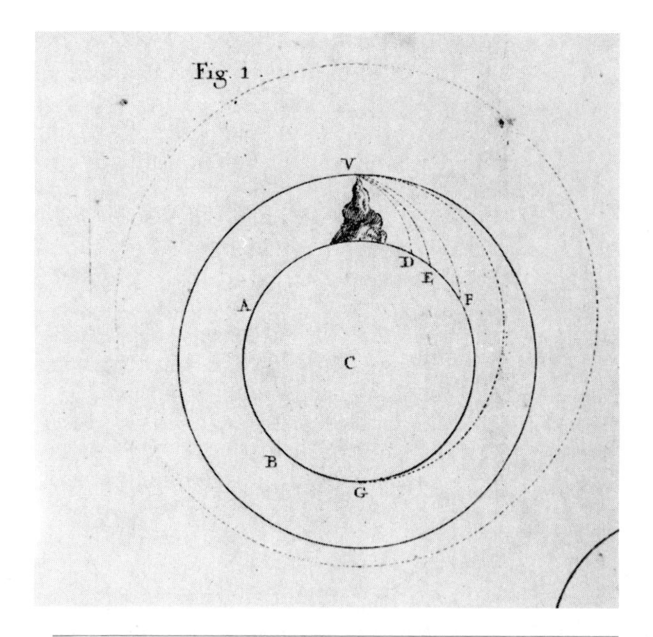

Figure 19–5 *The diagram Newton used to explain the motion of the moon. The cannon placed on the mountain top fires a shot that travels a curved path determined by its horizontal forward velocity and downward motion produced by the gravitational attraction of the earth. (Crawford Collection, Royal Observatory, Edinburgh. From Abell, G.O.,* Realm of the Universe, *3rd ed. Philadelphia, Saunders College Publishing, 1984.)*

insufficient to carry it completely around the sun must then fall into the sun (as some comets have been observed to do). On the other extreme, an object with an enormous amount of forward momentum may escape the sun's gravitational pull and leave the solar system completely. Between these two extremes there is the important third possibility, in which the forward momentum imparted to the body (planet or asteroid) is sufficient to keep it from being drawn into the sun but not so large as to cause it to fly out of the solar system. Provided the gravitational pull and forward momentum are in the proper proportion, the path of the object will be approximately circular. Should the forward momentum be somewhat too small, however, the object will approach the sun on a curved path, loop around it, and move outward again in an elliptical path. Conversely, if the forward momentum is somewhat too great, the object will move in a curved path away from the sun, but lacking enough momentum to break out of the solar system entirely, it would eventually return to complete a broadly elliptical loop.

THE TOOLS OF ASTRONOMY

Progress in obtaining knowledge about the stars and planets has always depended upon the invention and improvement of instruments used to study the firmament. Telescopes, of course, were among the earliest improvements over naked-eye observation. Until the invention of telescopes, we knew nothing at all about the fainter stars, which cannot be seen with the unaided eye. We had no concept of the Milky Way Galaxy or the multitude of other galaxies in the universe. Clearly, telescopes, of both the optical and radio kind, have vastly improved our understanding of outer space. More recently devised tools, however, have provided a bonanza of information about the planets and moons in our own solar system. These new engineering marvels include the Apollo spacecraft used to

Figure 19–6 *The Space Shuttle* Columbia *being lifted off the launch pad by its two rocket boosters on June 27, 1982. (Courtesy of the National Aeronautics and Space Administration.)*

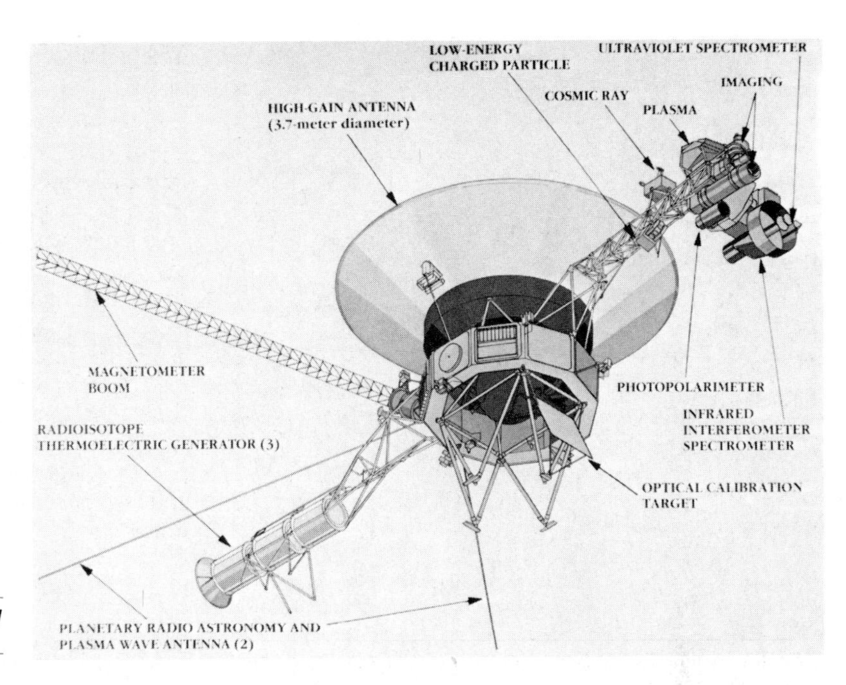

Figure 19–7 *Voyager spacecraft and its instruments.*

explore the moon, the data gathering Skylab, the Space Shuttle (Fig. 19–6), and the Pioneer, Mariner, Viking, and Voyager probes (Fig. 19–7).

Optical telescopes

In 1608 a Dutch lens grinder and maker of spectacles named Hans Lippershey invented the first telescope by experimenting with the placement of lenses at opposite ends of a length of pipe. Galileo learned of the invention, and within a short time he was able to make telescopes of his own. Like the Lippershey instrument, Galileo's telescope consisted of a piece of pipe with a convex lens at the front and a concave lens at the back. Of course, it required several trials to determine the length of pipe needed to produce a clear magnified image, but Galileo persisted. When he directed his device toward a distant church tower, he was astonished by how close it seemed to be. He experimented further, using lenses of different powers, and he made his pipes increasingly larger until he had produced a telescope that could magnify objects 30 times. When he fastened his pipe to a stand and pointed it at the moon, he was able to observe lunar craters, mountain ranges, and the moon's vast dark maria. The telescope also permitted him to see stars too faint to be seen with the unaided eye, to recognize that the Milky Way was made up of multitudes of individual stars, and to discover the four largest of Jupiter's 16 satellites. One day he observed that a small area seemed to be missing from Venus. Subsequent observations with his telescope revealed that Venus goes through phases like our moon, proving that the planet must revolve around the sun (Fig. 19–8). Never before or since had an astronomer been able to make so many major discoveries in a span of only a few years. The age of the telescope had arrived.

In the more than 300 years since Galileo, telescopes have become increasingly powerful, and yet the modern instruments have essentially the same components. An optical telescope has a lens or a mirror called the *objective.* The objective collects light from an object and forms its image (Fig. 19–9). The point at which light is brought to a focus and the image formed is called the **focal point.** The distance between the center of the objective and the focal point is termed the **focal length.** The image formed at the focal point is examined through an *eyepiece,* which is a convex lens or magnifying glass of short focal length. The size of the image depends on the ratio of the focal lengths. Image size increases as the objective's focal length increases, or as the eyepiece focal length decreases. In a familiar example of this fact, camera buffs know that increasing the focal length is the purpose of a telephoto lens. The elongate lens may have a focal length of 300 mm, and thereby provide a much larger image

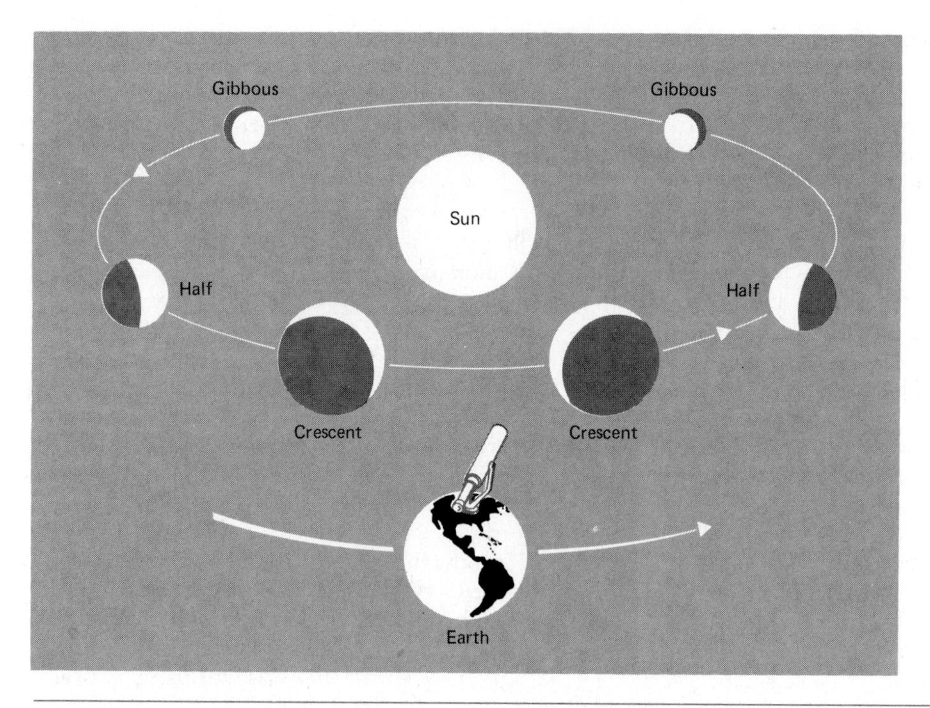

Figure 19–8 In observing Venus through his telescope, Galileo discovered that it goes through phases like the moon, an observation that supported the Copernican view of the solar system. (From Abell, G.O., The Realm of the Universe, *3rd ed. Philadelphia, Saunders College Publishing, 1984.)*

than is possible with the 50 mm focal length lens on many cameras.

In addition to magnifying objects, optical telescopes also intensify light. This is accomplished through the light-gathering power of their large objectives. Merely doubling the diameter of an objective lens or mirror will increase the apparent brightness of a star four times. For observation of very faint stars, a

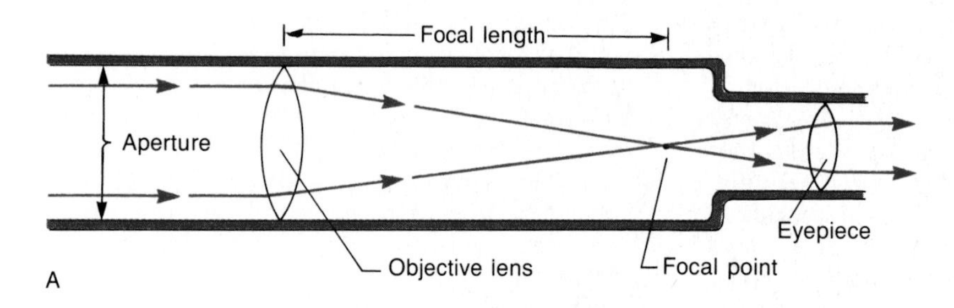

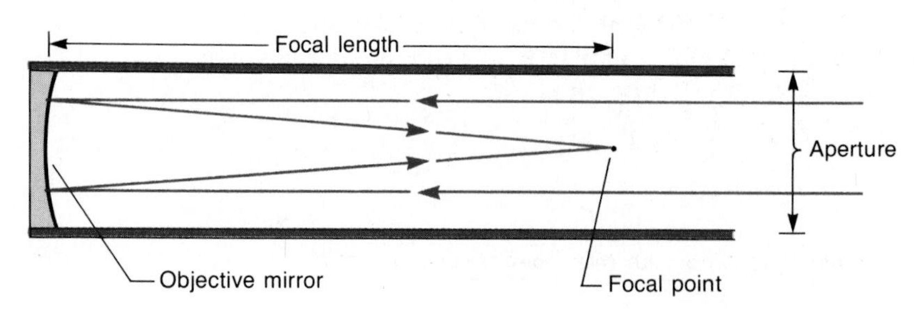

Figure 19–9 Elements of the refractor *telescope (A) and the* reflector *telescope (B).*

photograph is made through the telescope over a period of several minutes or hours. The photograph provides a permanent record for subsequent study. In addition to photographic processes, new photoelectronic devices have markedly improved our ability to record distant stars in much shorter exposure times. These devices, called image intensifiers, can be 100 times more sensitive than photography. The short exposure times they require increase the astronomer's ability to study rapid changes in celestial objects.

There are two general types of optical telescopes in use today. The most familiar type is the **refractor** with its two lenses (see Fig. 19-9A). The fixed large objective lens collects the light from the object being viewed, focuses it, and forms its image at the focal point. The eyepiece is movable. It receives light from the focal point, magnifies it, and transmits it to the eye. The other kind of telescope is a **reflector** and has a concave mirror that reflects the parallel rays of light it receives from an object and focuses them at the focal point where the image is formed (see Fig. 19 – 9B). The largest reflecting telescope in the United States is the famous 200-inch Hale instrument on Palomar Mountain in California (Fig. 19 – 10). It is surpassed in size by the 236-inch, 850-ton Soviet reflector recently constructed in the Caucasus Mountains northwest of Tiflis. In such large-objective reflectors, the observer, photographic, and spectroscopic equipment may be located directly at the focal point.

Because all optical telescopes depend upon light that passes through the earth's atmosphere, they have serious limitations. Suspended particles, charged atoms, gas molecules, and turbulence interfere with the passage of light. Thus, astronomers are eager to build future observatories on large artificial satellites, or on the moon, so as to avoid atmospheric interference. A remarkable new telescope and instrument system, designated the LST (large space telescope), is now being prepared for use beginning in 1985 (Fig. 19 – 11). The LST incorporates a telescope within a free-flying spacecraft that will be put into its own orbit by the Space Shuttle. The telescope will be able to see far out into space, producing images of unprecedented clarity. At an altitude of 600 kilometers above the earth, the telescope will be unhampered by atmospheric distortion, and it will be able to "see" objects 50 times dimmer than anything detectable by ground-based optical telescopes.

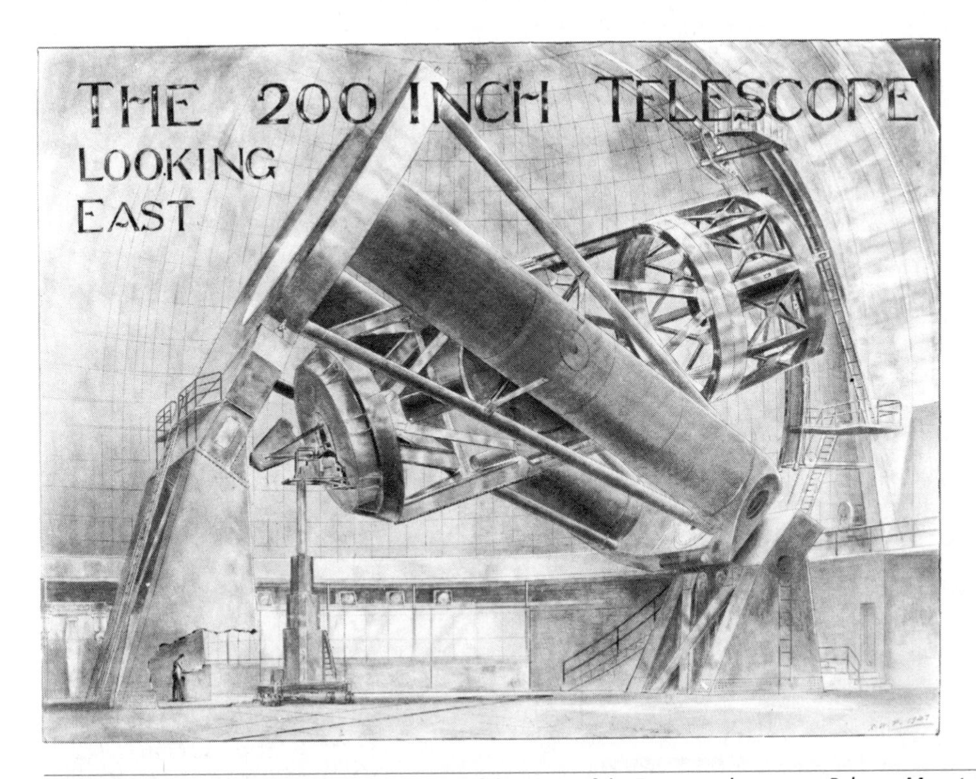

Figure 19 – 10 *One of Russell W. Porter's set of drawings of the 5-meter telescope on Palomar Mountain.*

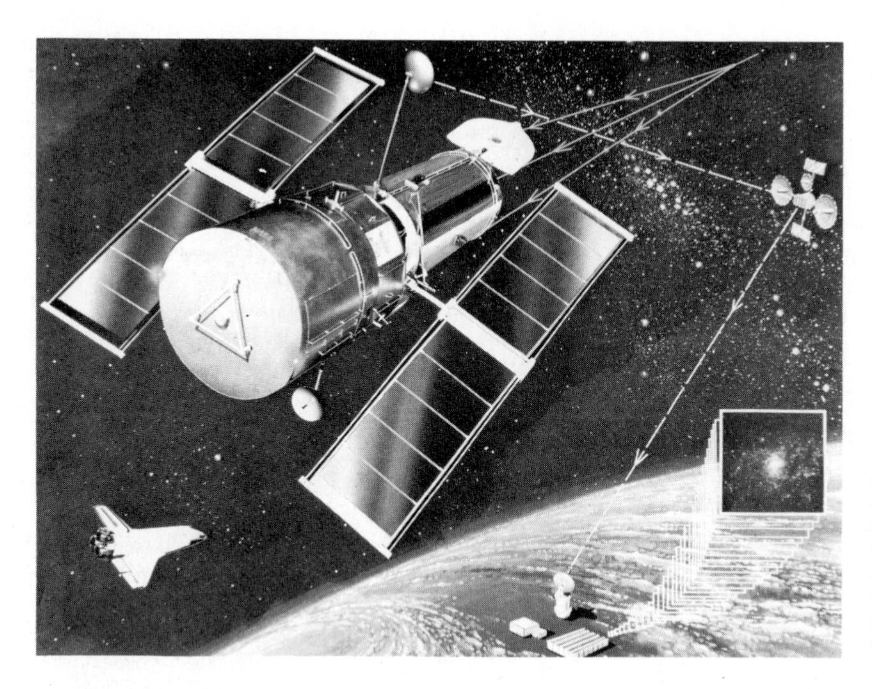

Figure 19–11 The large space telescope (LST) in orbit. The telescope is receiving light from a distant star and transmitting information to a relay satellite, which in turn transmits the data to Goddard Space Flight Center in Maryland. (This artist's concept courtesy of NASA.)

Radio telescopes

In 1931 a young engineer named Karl Jansky was experimenting with a radio antenna to track down sources of static that interfered with telephone conversations on the trans-Atlantic links of the Bell Telephone system. By accident, he discovered that some of the static and background noise actually came from outer space. He later determined that these radio waves were coming from the Milky Way Galaxy. Jansky's discovery opened a new branch of science called radio astronomy.

Rather like an optical reflecting telescope, a radio telescope has a parabolic surface or dish, but the dish is usually made of a metal mesh instead of a glass mirror (Fig. 19–12). This kind of telescope "sees" by recording the intensity of the radio waves it receives. The dish gathers and focuses incoming radio waves onto the antenna, creating an electric current. The current is then fed into a receiver, amplified, and recorded.

Figure 19–12 The twin 90-ft radio telescopes of the Radio Observatory of the California Institute of Technology. (California Institute of Technology; from Abell, G.O., Realm of the Universe, *3rd ed. Philadelphia, Saunders College Publishing, 1984.)*

Much as in an ordinary radio, the receiver can be tuned to permit study of specific wavelengths.

Radio waves are reflected and refracted in much the same way as light waves. Like light waves, radio waves may be visualized as waves that spread out from a source, as when a pebble is dropped into a pool of water. Each wave has a crest and trough, and the distance between any two successive crests is termed the *wavelength*. The wavelengths of radio waves are much longer than the wavelengths of visible light. Those larger wavelengths permit transmission of radio waves through the atmosphere with little interference — a decided advantage over the optical telescope. The larger wavelength, however, does limit the resolving power of the radio telescope. To improve on the instrument's ability to distinguish detail, the curved dish must be exceptionally large. Indeed, the large dish of the radio telescope is its most recognizable feature. The radio telescope at the Max Planck Institute in West Germany has a steerable dish more than 100 meters in diameter, and the device at Arecibo in Puerto Rico (Fig. 19-13) fills a huge natural depression and measures an astonishing 305 meters across. (According to one estimate, it could hold the annual beer consumption of the entire earth.) The Arecibo antenna is fixed in position, but some directional ability is obtained by moving the detector around in the vicinity of the focus of the bowl.

Even with the enormous reflectors of modern radio telescopes, their resolving power is still less than that of the human eye. This problem, however, has been partially solved with the use of **interferometers.** Interferometers are arrays of two or more radio telescopes which are positioned over a large area. The radio waves received by the separate instruments are combined electronically to obtain greater resolution. A noteworthy example of such an installation is the VLA (very large array) located in New Mexico. This system consists of 27 dishes, each 85 feet in diameter, mounted on rails in a Y pattern over a distance of 21 km. When in full operation, the VLA should have a resolution as good as or better than that now obtained with optical telescopes.

Next to the telescope, the most important instrument for astronomical research is the spectrograph, a device used to photograph the spectrum emitted by a body. White light (which is a mixture of all wavelengths or colors) is directed into the spectrograph through a slit placed at the telescope's focal plane (Fig. 19-14). Light rays pass through the slit, diverge, pass through a collimator lens, and emerge as parallel rays. The rays then progress through a prism that separates the components of light received into a **spectrum.** A spectrum is simply a band of colors like that in a rainbow, proceeding from red to violet, which is produced when light is dispersed by a prism. In lieu of the prism,

Figure 19-13 *The 305-meter dish of the radio telescope at Arecibo, Puerto Rico. (From Abell, G.O.,* Realm of the Universe, *3rd ed., Philadelphia, Saunders College Publishing, 1984.)*

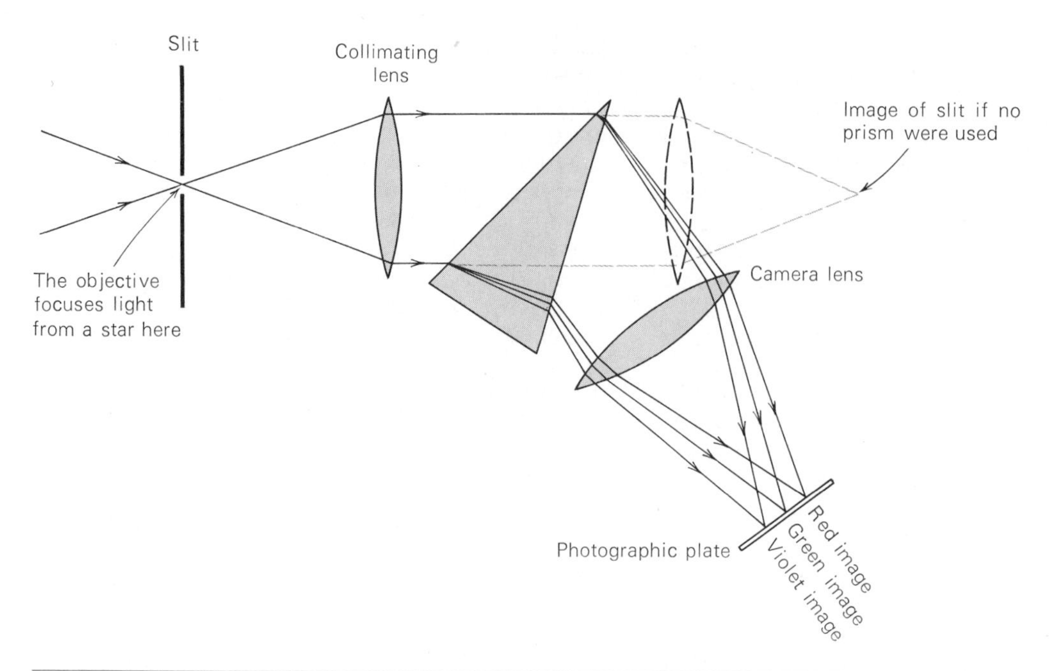

Figure 19–14 The optical elements and construction of a simple prism spectrograph. (From Abell, G.O., Realm of the Universe, *3rd ed., Philadelphia, Saunders College Publishing, 1984.)*

the spectrum can also be formed by a **diffraction grating,** which consists of a flat surface ruled with thousands of very fine parallel lines. From the diffraction grating, the rays of light, now separated into different colors or wavelengths, are transmitted to the objective lens of a camera and recorded on a photographic plate. If the photographic plate is removed and an eyepiece is used to observe the spectrum, the instrument is called a **spectroscope** rather than a spectrograph (and if used to measure the various wavelengths, it is a **spectrometer**).

From the spectrum recorded on the photographic plate, an astronomer can observe clues to temperature, chemical composition, relative movement, and many other properties of the distant light source. If, for example, the radiating source is a gas under low pressure, only specific colors will be observed and the spectrum will appear as a series of bright lines against a dark field. In contrast, a gas under high pressure, a liquid, or a solid will produce a continuous rainbow of color from red to blue. If the light from a distant star passes through a gas under low pressure, the gas will absorb only those colors it is capable of emitting, and the result will be a series of dark lines superimposed on the continuous band of color.

STARS AND GALAXIES

For millenia humans have been fascinated by stars. How far away are they? What is the source of their light? How do they originate? Do they persist forever or slowly perish? To find the answers to these questions, astronomers have assiduously studied the data derived from instruments directed toward the stars. They have learned that stars are truly great balls of gas held together by gravity and supported by opposing pressures generated deep in their stellar cores. Hot gases generate these pressures. They derive their heat from nuclear reactions, which convert light elements to heavier ones and liberate heat as a by-product. The size and brightness of a star depend upon the balance between outwardly directed pressure and gravity, as well as the balance between energy loss at the surface and energy generation in the interior.

Locating stars

Centuries ago people observed that stars maintained their positions relative to one another. With a little imagination, the patterns of these stars seemed to sug-

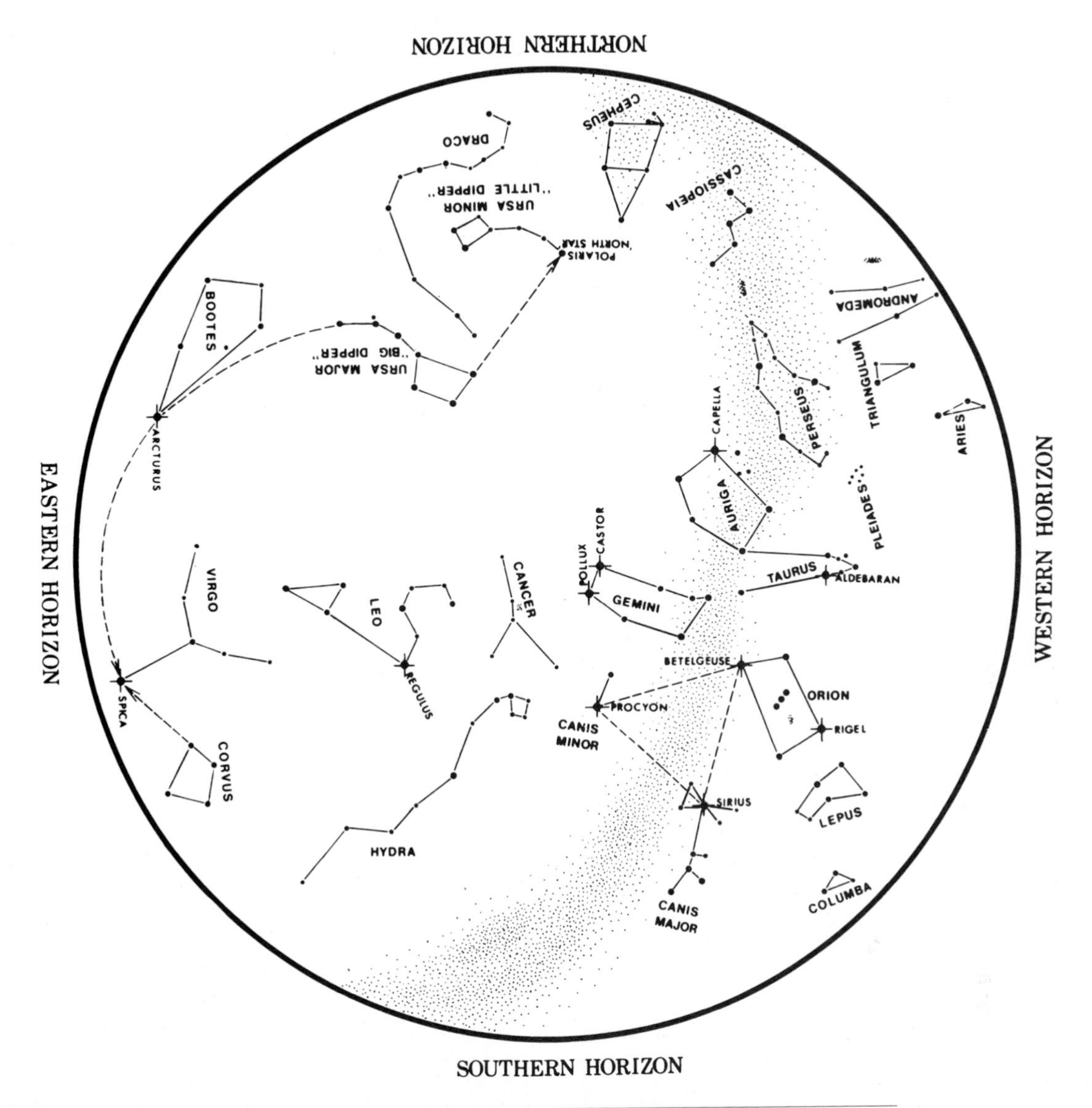

Figure 19–15 *The night sky in mid-March viewed from the United States at about 9 P.M. To use, hold the chart vertically and turn it so the direction you are facing shows at the bottom. (From Abell, G.O.,* Realm of the Universe, *3rd ed., Philadelphia, Saunders College Publishing, 1984. Courtesy of the Griffith Observatory, Los Angeles.)*

gest the shapes of animals, gods, and legendary heroes. Many star patterns, called **constellations,** are still identified by their ancient names: *Ursa Major,* the great bear; *Cygnus,* the Swan; and the mighty hero *Perseus.* Many people take pleasure in becoming familiar with the constellations (Fig. 19–15). Being able to recognize them increases our esthetic enjoyment of a clear starlit evening, and also helps us identify particular stars more readily.

The easily visible stars are usually named by their location in a constellation and by how bright they are relative to the other stars in the same constellation. The rank or level of brightness is indicated by letters of the Greek alphabet. For example, Alpha (α) Lyrae is the brightest star in the constellation of Lyra, the lyre; Beta (β) Orionis is the second-brightest star in Orion, the hunter. Some of the brightest stars have individual names in addition. The brightest three stars in the northern sky, for example are individually named Sirius, Arcturus, and Vega.

As the earth turns on its axis, the constellations seem to move across the sky as if they were fixed to the inside of a transparent globe surrounding the earth. If the axis of the earth were extended out into space it would form the axis of this imaginary globe, which is called the **celestial sphere.** Because the earth rotates from west to east, the celestial sphere appears to move in the opposite direction, from east to west. The points at which the extensions of the earth's axis intersect the celestial sphere are called the north and south celestial poles (Fig. 19-16). A star located directly on the celestial north pole would not seem to move. For example, Polaris (the so-called North Star) is very close to the celestial north pole and appears to move only very slightly. All stars appear to circle the celestial sphere. Stars near Polaris (called circumpolar stars) are visible to viewers at mid-northern latitudes through their entire circular paths; those farther to the south can only be seen for a part of their larger circular paths. These stars appear to rise in the east and then pass out of view in the west. Stars near the south celestial pole cannot be seen from the northern latitudes at all.

In addition to changes in the night sky due to the earth's rotation, there are also apparent changes caused by the earth's revolution around the sun. The earth is on different sides of the sun in winter and in summer, and so the stellar scenery changes with the seasons: in winter, for example, the constellation Orion is a conspicuous feature, whereas in summer Lyra and Cygnus are prominent.

Here on earth, geographers have long benefited from the use of a grid or coordinate system of meridians and parallels to help locate places and objects. Astronomers use an analogous system extended into the celestial sphere. The terminology is somewhat different, however. Astronomers use the names **right ascension** for the longitude coordinate, and **declination** for the latitude coordinate.

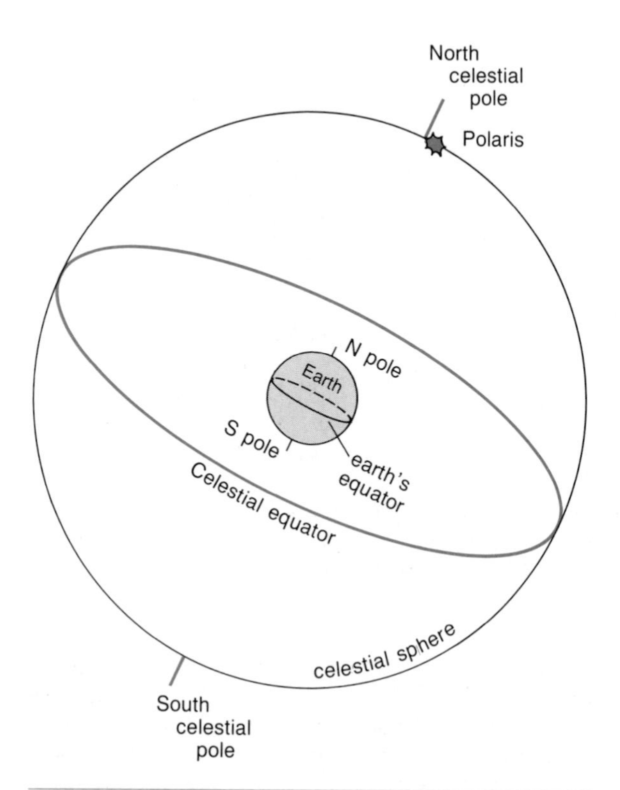

Figure 19-16 The celestial sphere, with its poles and equator relative to the earth. We observe the celestial sphere from the earth, which is at its center.

Distance to stars

The sun, of course, is the star closest to the earth. The next closest star is Proxima Centauri, which is approximately 4.3 light-years away. A light-year is a convenient way to measure the immense distances of the universe. One **light-year** is the distance light travels in one year while moving at the impressive rate of 299,792 km per second (or about 3×10^5 km/sec). If one multiplies this amount by the number of seconds in a year (roughly 31 million) then we have the actual distance that light travels in a year—9.47×10^{12} km (nearly 9½ million million kilometers). It is interesting to consider that the light we receive today from Proxima Centauri actually left there 4.3 years ago. And, if Proxima Centauri were to explode today, astronomers would know nothing of the explosion until 4.3 years from now. The light from the more distant stars observed with optical telescopes left those stars about 300 million years ago, when the most advanced vertebrates on earth were swamp-dwelling amphibians.

One of the very first things astronomers want to know about a star is its distance from the earth. For the nearer stars, a principle similar to that used in a camera's range finder is employed. To understand the method, perform this simple experiment. Hold a pencil at arm's length in front of yourself and examine it with one eye closed and then the other. Notice how the pencil seems to shift position relative to fixed background objects. This is because your eyes are a few centimeters apart, and when you look through each eye, you see the pencil from two slightly different points of view. Notice that if you repeat the experiment with the pencil closer to your eyes, the apparent shift is greater. This same effect is used to find the distance to relatively near stars, and similarly, the closer the star is to us, the farther it will appear to shift against the background of fixed stars.

To make the most of the effect, it is advantageous to observe the star in question from two points that are separated by as wide an interval as possible. For earthbound astronomers that would be the two positions of the earth at opposite sides of its orbit. Thus, the second observation would be made six months after the first. The two positions are connected by a straight line called the **baseline.** In Figure 19–17, the baseline is twice as long as the distance from the earth to the sun. Astronomers would make note of this by stating that the baseline is 2AU, where 1AU or **astronomical unit** is equal to the average distance between the earth and the sun (1.496×10^8 km). The angle that a star appears to move between two observations at the ends of the baseline 1AU (that is, half the maximum possible baseline) is termed the **parallax** of a star (see Fig. 19–17). Once parallax is known, simple trigonometry can be employed to calculate the distance to the star.

Unfortunately, the farther a star, the smaller its parallax, and for distant stars the parallax shift is simply too small for direct measurement. As a result, indirect methods dependent upon a star's brightness or luminosity are employed. The light from a star can be expressed as either **apparent brightness** or **intrinsic brightness.** The former is simply the amount of the star's light that reaches us. If we disregard intervening gas and dust, apparent brightness depends on a star's actual brightness and how far away it is. An extremely bright star, for example, may appear very faint if it is exceptionally distant. This is because the apparent brightness of a source of light varies inversely as the square of the distance from the source to the observer. This physical fact is known as the **inverse square law.** Thus, if we could move a star twice as far away from us, it would appear 4 times fainter, and if the star were moved 8 times as far from us, it would appear 64 times dimmer. Stated another way, if we know that two stars are of equal true brightness, and we observe that one is 64 times dimmer than the other, the fainter star must be 8 times farther away from us than the brighter star.

Unlike apparent brightness, intrinsic brightness is not affected by distance. Intrinsic brightness is the actual or real brightness and is a reflection of the total amount of energy being radiated by the star. If the distance to a star has been determined by the parallax method, then its intrinsic brightness can be determined by calculating how bright the star would have to be at that known distance to provide the observed apparent brightness. When the apparent brightness and distance are known, intrinsic brightness can be determined. Conversely, if apparent brightness and intrinsic brightness are known, distance can be calculated.

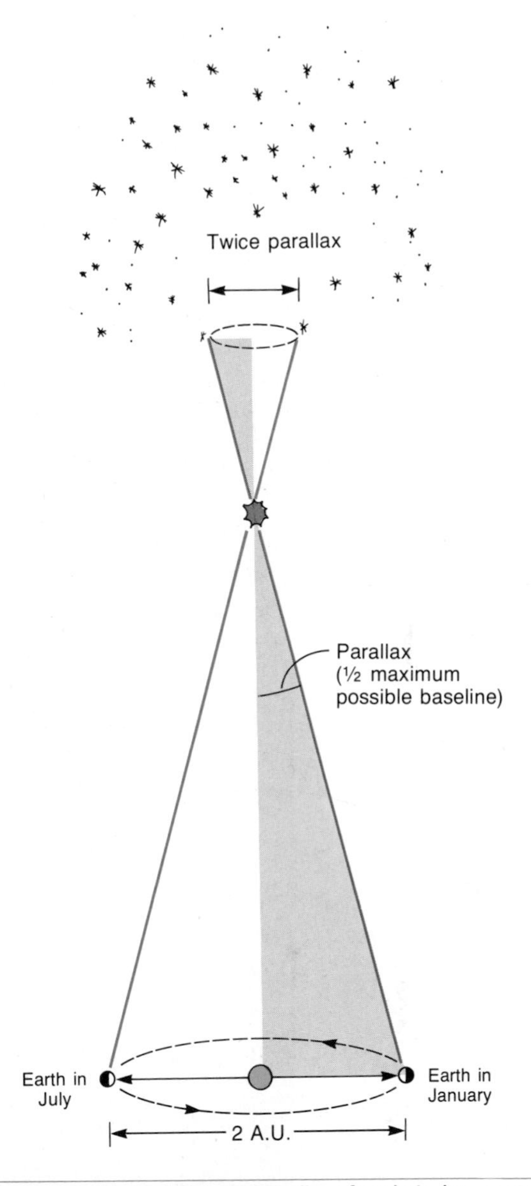

Figure 19–17 To obtain the parallax of a relatively near star, it is viewed against the background of more distant stars from nearly opposite sides of the earth's orbit.

Astronomers frequently employ a unit of distance based on parallax that is termed a **parsec.** One parsec is equal to the distance from us that a star would be if its back-and-forth parallax shift was one second of arc. (The term combines *par* in parallax with *sec* for seconds of arc.) A star having one second of arc in its parallax shift would be 3.26 light-years, or 206,265 AU, or 30 trillion kilometers away from the earth.

The brightness of stars

The most obvious way to generally describe a star is by its brightness. The brightness of a star as it appears to an observer on earth is called its **apparent magnitude.** A classification of stars according to their apparent magnitudes was developed in the second century B.C. by the Greek astronomer Hipparchus. In this scheme there were six classes of magnitude or brightness. The brightest of stars were designated first magnitude, and each successive number designated a perceptibly dimmer star. Sixth-magnitude stars are so faint as to be just barely visible. The Hipparchus classification has survived, but the units of magnitude have been given a more precise definition. Hipparchus had to estimate brightness with the naked eye, but scientists today use a photometer (essentially a very accurate exposure meter) to accurately measure the visible radiation from a given star. In the modern stellar magnitude scale (Fig. 19-18), each step is equal to about 2.5 times greater brightness (radiant energy) than the next lower step.

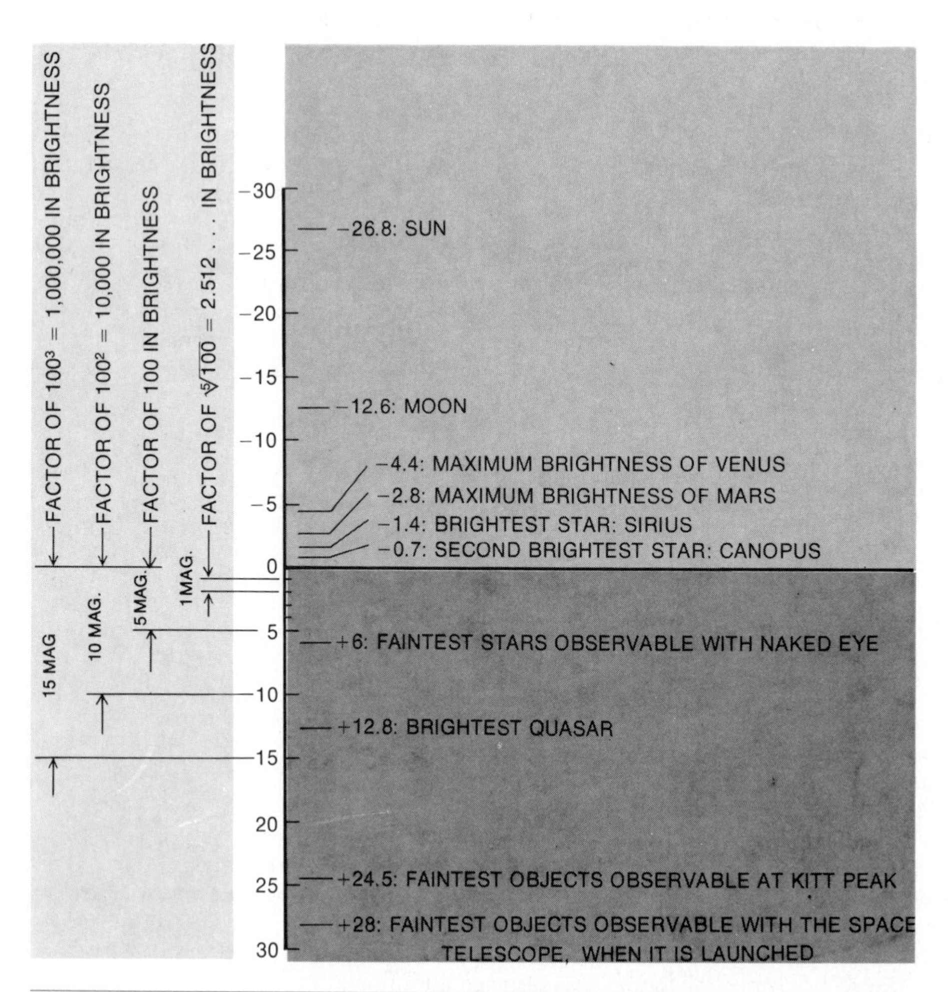

Figure 19-18 *The scale of apparent stellar magnitude. (From Pasachoff, J.M.,* Astronomy: From the Earth to the Universe, *2nd ed. Philadelphia, Saunders College Publishing, 1983.)*

Thus, a first-magnitude star radiates 2.5 times more energy than a second-magnitude star and is 2.5×2.5 or $(2.5)^2$ times brighter than a third-magnitude star. A first-magnitude star would be $(2.5)^5$ or nearly 100 times brighter than a sixth-magnitude star. Astronomers purposely defined the magnitude system so that a difference of five magnitudes would be equal to a ratio of 100.

All of the magnitudes in the modern magnitude scale are measures of how bright a star appears to us. They are, therefore **apparent magnitudes.** When stars are compared, astronomers calculate the brightness each star would have at a standard distance of 10 parsecs (32.6 light years) and designate this measurement as the star's **absolute magnitude.** Another useful term is **luminosity,** which refers to the amount of radiation (light, radio waves, infrared waves, etc.) given off by a star per unit of time.

The stellar spectra

Stars are great blazing balls of gas with surface temperatures reaching tens of thousands of degrees. They emit a continuous spectrum of light that includes all wavelengths. As this light passes through the atmosphere of a star (which consists of the same elements present in the body of the star, but at very low density), the stellar atmosphere extracts certain wavelengths of the continuous spectrum and produces a dark line or **absorption spectrum.** The pattern appears as a continuous spectrum with a series of dark lines superimposed on it. Examination of these absorption lines on photographic records helps astronomers determine the temperature and composition of the surface layers of a star. The temperature of a star's surface determines the role of its atoms in emitting and absorbing particles of light called photons. Thus, temperature has an important effect on the spectrum of light received from a star. Over the past 150 years, the spectra of several thousands of stars have been examined. The results indicate that the vast majority of stars are almost identical in chemical makeup. Stars are composed mainly of hydrogen with a little helium and less than 4 percent of the heavier elements. The surprising but very significant fact emerges that differences in the spectra of most stars result almost entirely from differences in temperature and not from differences in composition.

Over the past century and a half, astronomers have had an opportunity to examine hundreds of thousands of stellar spectra. This work has resulted in a classification of spectra into several general classes, lettered O, B, A, F, G, K, M, which reflect stars of decreasing surface temperatures. (The mnemonic device "*O*h *B*e *A* *F*ine *G*irl [or *G*uy], *K*iss *M*e," may help you remember the sequence.) In the classification, stars having the hottest surface temperatures are in the O category, whereas the coolest are M stars. For greater precision, the classes are then divided into 10 sub-

TABLE 19–1 **SURFACE TEMPERATURES AND SPECTRAL FEATURES OF DIFFERENT TYPES OF STARS**

Spectral type	*Average surface temp. (degrees Kelvin)*	*Special features*
O	30,000 (range 30,000–80,000°K)	Spectra display lines of ionized helium and weak hydrogen absorption. (Only the hottest stars contain abundant ionized helium because the energy required to strip an electron from the helium atom is greater than for any other atom.) Stars are bluish. The star Lacertae is an example.
B	20,000 (range 12,000–30,000°K)	Absorption lines from un-ionized (neutral) helium (collisions are less violent, so fewer helium atoms are ionized) and prominent hydrogen lines as well. Stars are bluish-white. Rigel is a B star.
A	10,000 (range 8,000–12,000°K)	Prominent hydrogen absorption lines. Still some absorption of un-ionized helium. Stars appear white, as in Vega and Sirius.
F	7,000 (range 6,000–8,000°K)	Hydrogen lines dominate but are diminishing. Lines for neutral metal atoms make their appearance, especially calcium, iron, and titanium. Stars have a yellowish appearance. Canopus is an F star.
G	5,500 (range 4,500–6,000°K)	Complex spectrum containing thousands of closely spaced lines. Metallic lines predominate. The sun is an example.
K	4,000 (range 3,500–4,500°K)	Increasing absorption lines from un-ionized atoms of heavy elements; no hydrogen lines are visible. Stars are orange. Aldebaran is a K star.
M	3,000 (range 2,000–3,500°K)	Absorption features of un-ionized atoms and even molecules of simple compounds. Barnard's Star is an example. Stars tend to be reddish.

classes numbered from 0 to 9. The star Aldebaran, for example, is designated class K5. The characteristics of the various spectral classes are indicated on Table 19-1.

The Hertzsprung-Russell diagram

The surface temperature of a star (as indicated by its spectrum) and its absolute magnitude are two of the most obvious ways in which stars differ from each other. One cannot help but wonder if there is a relationship between these variables. Presumably with this thought in mind, two astronomers, each working independently in his own country, found that there was indeed a remarkable relationship. In 1913, Ejnar Hertzsprung in Denmark and Henry N. Russell of the United States expressed the relationship in a diagram, which as a tribute is named the Hertzsprung-Russell diagram—or more simply, the H-R diagram (Fig. 19-19).

Hertzsprung and Russell had no prior thought of what they might find when they began to plot absolute magnitude against temperature. They hoped only to uncover some unrecognized characteristics of stars. To their great satisfaction they found that stars were not distributed over the graph at random. They do not show all combinations of absolute magnitude and temperature but rather are clustered into definite regions of the diagram.

The majority of stars in the H-R diagram are aligned along a narrow band that slants downward across the graph from upper left to lower right, covering the entire range of spectral classes. This band is called the **main sequence.** The existence of the main sequence is evidence of the close relationship between absolute magnitude and temperature for all main sequence stars. Over 90 percent of all stars plot in the main sequence. A star in this main sequence uses nuclear fuel at a uniform rate and is considered a stable star. The brightest stars on the main sequence band are bluish-white in color, and the faintest have the lowest temperatures and are reddish in color. Of course, a star's position in the main sequence or elsewhere in the H-R diagram has nothing to do with its position in space, but is only a point that represents its luminosity and temperature.

A particularly interesting relationship among stars is the correlation between a star's mass (the amount of matter in it), and its luminosity. Generally, the more massive the star, the greater its luminosity

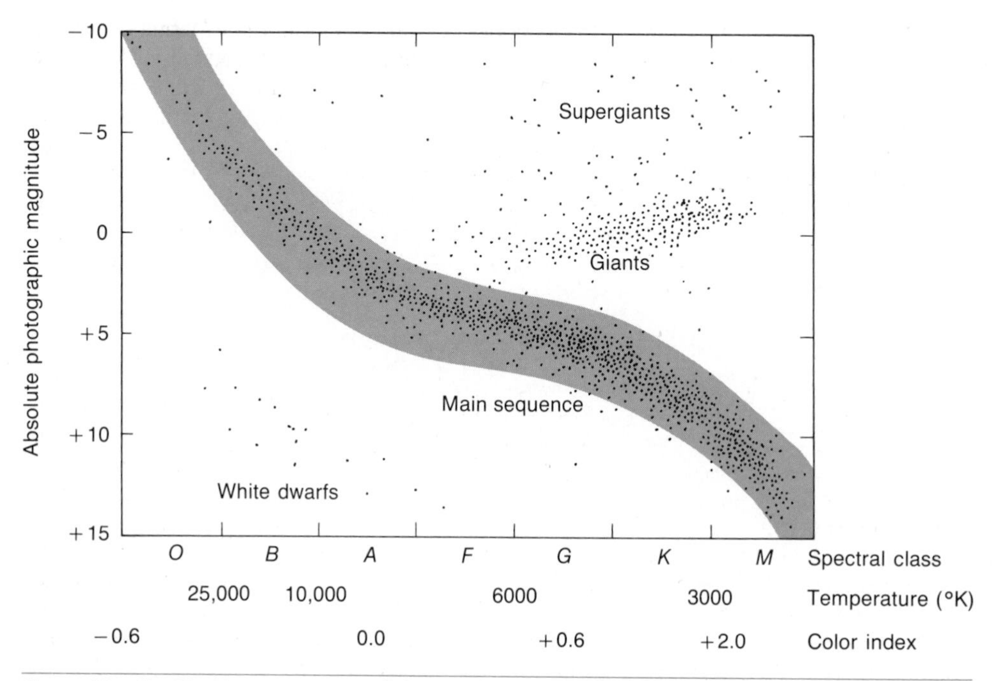

Figure 19-19 The Hertzsprung-Russell diagram for stars of known distances. (From Abell, G.O., Realm of the Universe, *3rd ed. Philadelphia, Saunders College Publishing, 1984.)*

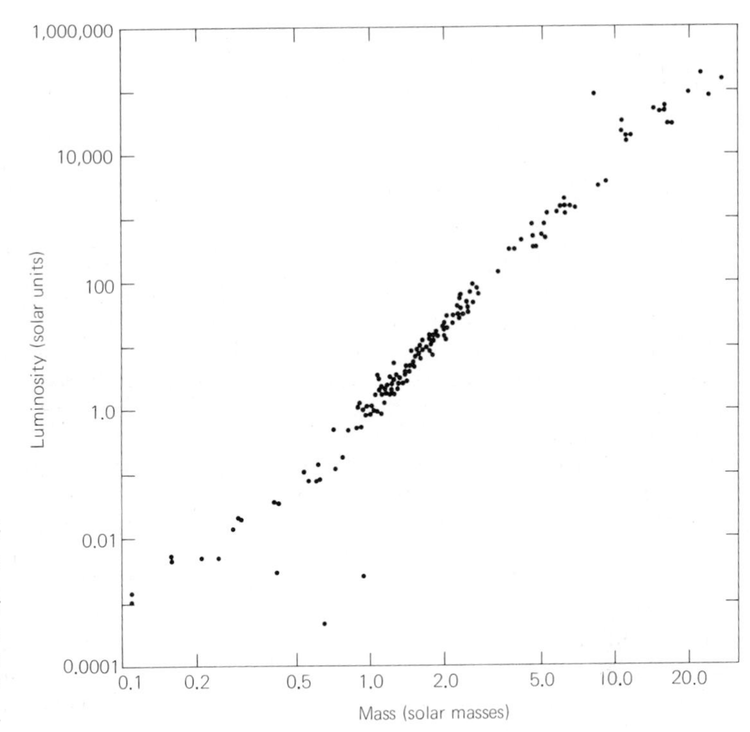

Figure 19–20 The relationship of the mass of stars to their luminosity. Each point represents a star of known mass and luminosity. Mass is indicated on the horizontal scale (abscissa) in units of the sun's mass; luminosity, on the vertical scale (ordinate) in units of the sun's luminosity. (From Abell, G.O., Realm of the Universe, *2nd ed. Philadelphia, Saunders College Publishing, 1980.)*

(Fig. 19–20). On the H–R diagram, the position of a star on the main sequence is determined by its mass. The blue-white O and B stars, for example, are both the most luminous and also the most massive.

Another prominent clustering of stars occurs above and to the right of the main sequence on the H–R diagram in the region of high luminosity but relatively lower temperatures. These are the **giant stars.** Above are stars of even higher luminosity, called **supergiants.** Supergiants may be of any spectral class (color). Because they are rather rare, only a few are found on the diagram.

Finally, there are hot, low-luminosity stars in the lower left area of the diagram, which are called **white dwarfs.** Sirius B, the small companion to the bright star Sirius, is an example of a white dwarf. These stars are faint because they are relatively small in size. In the entire universe, white dwarfs may be as numerous as main sequence stars, but they are difficult to find. This accounts for their relative rarity on the H–R diagram.

On the H–R diagram, one can find places where there are stars of the same spectral class but with radically different luminosities, and other places where stars have the same luminosity but different spectral classes. For example, both the sun and the star Capella have about the same spectral class (G), but Capella is 100 times more luminous. As both stars have the same surface temperature, Capella must be brighter because its radiating surface is 100 times larger than the sun's. By analogy, 100 candles in a cluster will be more luminous than a single candle, even though the temperature of each candle is the same. Thus, the H–R diagram indicates that a star can have the same surface temperature as another star, but it will differ in luminosity depending on its size.

Stars and nebulae

Stars are not scattered uniformly throughout the firmament, but occur mostly in gigantic arrays, which together with cosmic dust and gas are known as **galaxies** (Fig. 19–21). Even within a galaxy, stars are not uniformly spaced, and many stars appear as multiple star systems in smaller clusters. The immense interstellar clouds of gas and dust in galaxies are called **nebulae** (Fig. 19–22). About three fourths of all the matter in the nebulae is hydrogen, with helium accounting for a little less than one fourth, and all other elements only about 1 or 2 percent of the total composition. The atoms in these nebulae are widely dis-

Figure 19–21 The spiral Andromeda Galaxy with two much smaller elliptical galaxies. (From Pasachoff, J.M., Astronomy: From the Earth to the Universe. *Philadelphia, Saunders College Publishing, 1979.)*

persed. As a result, nebulae are so thin and diffuse that an average sample would represent a truer vacuum than can be achieved here on earth. Sometimes, however, a region in a gas and dust cloud may collect sufficient matter so as to begin to contract under the influence of gravity and form a sphere. Temperature and pressure build at the center of the gaseous contracting sphere until thermonuclear reactions occur. Gradually, the great gaseous sphere is converted into a star.

Figure 19–22 The Lagoon nebula in Sagittarius, photographed in red light with the 200-inch telescope. (From Abell, G.O., Realm of the Universe, *3rd ed. Philadelphia, Saunders College Publishing, 1984. Courtesy of California Institute of Technology/Palomar Observatory.)*

When we look at the night sky with the naked eye, most of the stars we see are members of our own Milky Way Galaxy. This is a tremendous, disclike spiral structure about 80,000 to 100,000 light-years in diameter and about 10,000 light-years thick. The entire system is rotating, with the central regions moving faster than the periphery. Our solar system forms a very tiny part of the Milky Way. It is unobtrusively located near the thinning margin, about 30,000 light-years away from the galactic center. When viewed from the earth, the Milky Way appears as a faint band of light, which marks the equatorial plane of the galaxy. Astronomers assume planets like those in our own solar system are probably commonplace elsewhere in our galaxy and in other galaxies as well. After all, our sun is only one of billions of stars in the Milky Way, and the universe contains billions of galaxies, each having billions of stars.

Not all galaxies are the size and spiral shape of our own. Some are discoidal, some elliptical, some globular, and some can only be described as irregular. The diameter of an average galaxy is about 40,000 light-years, and the average distance between neighboring galaxies is about 2 million light-years. Spectrographic studies reveal that, except for a few close neighbors, all galaxies are receding from us. Thus the entire universe appears to be expanding.

Next to the Milky Way, the best-known galaxy is probably Andromeda (see Fig. 19-21). Although more than 2 million light-years away, the Andromeda Galaxy is sufficiently large that it can be seen with the unaided eye. Andromeda is larger than the Milky Way, and its spiral arms are more tightly wound.

Globular clusters and galactic history

The stars within galaxies are not uniformly distributed. Some are clustered, with the stars in each cluster remaining relatively close together and moving as a unit. Open clusters of a few hundred stars are scattered throughout the galactic disc. Even larger than these clusters, however, are the great **globular clusters** (Fig. 19-23) found outside the plane of the galactic disc in the surrounding region of space known as the galactic halo. Over a million stars may reside in a globular cluster, and there are over 150 such clusters associated with our galaxy. Globular clusters are really the outposts of the Milky Way. There are few stars between them, and beyond them are even vaster regions of emptiness. Partly because of their location far out on the outside of the galactic disc, globular clusters are believed to be very old. This belief is based on our best theories for the origin of a galaxy. The Milky Way Galaxy, about which we know the most, was probably formed from a large spherical cloud of gaseous matter — mostly hydrogen — which began to condense and contract about 10,000 million years ago. As it contracted, its density increased until clusters of stars began to form. Eventually, the galaxy contracted to its present disclike shape. Some swarms of stars that were the first to appear in the galaxy were left behind as the cloud contracted. These became globular clusters. The globular clusters of our galaxy lack O and B stars, and this indicates their great age. Astronomers name such groups of old stars **population II stars.** They tend to be groups of faint, reddish stars lacking surrounding interstellar dust and gas.

During the youthful stages of the Milky Way's evolution, the earlier heavy stars gradually aged to the stage where they produced heavy elements. Upon the explosive death of these stars, the heavy elements were liberated and incorporated into the new stars that formed later. Astronomers call these younger stars with their richer store of heavy elements **population I stars.** Indeed, their bluish brilliance reflects their youthfulness. In addition, clouds of dust and gas are common in regions surrounding stars of population I. This indicates that star formation is actually taking place in these dusty regions of the galaxy. In the vicinity of older population II stars, the clouds of gas have been swept away and no new stars are developing.

Figure 19-23 *A globular cluster known as M3 and located in the constellation Canes Venatici, the hunting dogs. (From Pasachoff, J.M.,* Astronomy: From the Earth to the Universe. *Philadelphia, Saunders College Publishing, 1979.)*

Stars that vary in brilliance

Although most stars shine with a constant luminosity, a minority exhibit noticeable variations in their intrinsic brightnesses or absolute magnitudes. There are three types of variable stars. **Pulsating variables** expand and contract, pulsating in size as well as in brilliance. The second type are known as **eruptive variables.** These include the novae and supernovae yet to be discussed. Eruptive variables show sudden, usually unpredictable outpourings of light. The **eclipsing variables,** the third type, are really not true variable stars in that their intrinsic brightness is constant. The variation we witness in them is caused by one star's blocking our view of the other. Such is the case with so called binary or double stars, whose orbits of mutual revolution lie parallel to our line of sight so that they periodically eclipse one another.

Among the pulsating variable stars, a group called Cepheid variables exhibit particularly regular changes in brightness. The changes are repeated on a cycle of anywhere from two days to two months depending on the particular star being studied. Their variations in brightness can be plotted on a graph like that shown in Figure 19–24. In the early 1900s astronomers discovered that Cepheid variables exhibit a definite relationship between their absolute magnitudes and their period of light variation. (The *period* of a variable star is the time interval between successive maxima of brightness.) They were able to show that the longer the light period of a Cepheid variable, the greater is its absolute magnitude. Thus, by determining the period of such a star, its absolute magnitude is ascertained. For example, if a Cepheid is observed having a light variation period of 10 days, we know that its absolute magnitude is -1.8. The apparent magnitude may be measured by direct observation, and as we noted in our discussion of parallaxes, when both absolute and apparent magnitudes are known, the distance to a star can be calculated. Consequently, Cepheid variables enable us to obtain the distances to stars that are beyond our own galactic system.

The cause of the variation of brightness in a Cepheid star is the pulsation of the star itself. Expanding and contracting in a complex manner, it alternately becomes brighter and dimmer because of the alternate heating and cooling associated with compression and expansion of constituent gases.

Exploding stars

Sometimes a star that has been shining steadily for a long time experiences a sudden outburst of radiant energy that temporarily increases its luminosity by as much as a thousand times. During this outburst, the star blows off its outer layers. This causes it to be brighter for several days. Such exploding stars occur in binary systems containing a white dwarf (to be defined subsequently), many of which cannot be seen with the naked eye. During the outburst, however, the star is readily seen and appears to us as a new star in the sky (Fig. 19–25). For this reason, these stars are called **novae,** the Latin word for new. After a few weeks, the luminous gas from the nova has expanded into space and cooled down, and the star fades from view.

Novae are certainly impressive when they occur, but are not nearly as spectacular as supernovae. A **supernova** is a truly cataclysmic stellar explosion in which the star becomes hundreds of millions of times

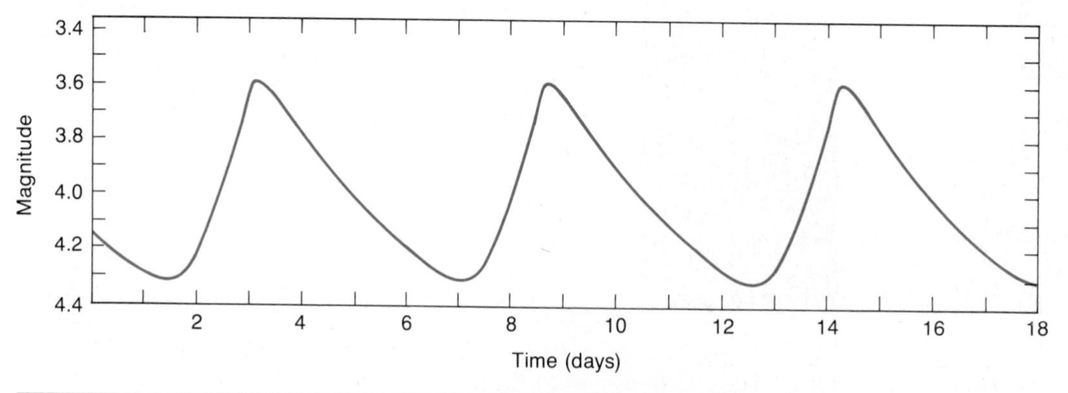

Figure 19–24 *Light curve for a typical cepheid variable. (From Abell, G.O.,* Realm of the Universe, *3rd ed. Philadelphia, Saunders College Publishing, 1984.)*

Figure 19–25 *Nova Cygni (1975):* (left) *at maximum;* (right) *2 months past maximum. (From Abell, G.O.,* Realm of the Universe, *3rd ed. Philadelphia, Saunders College Publishing, 1984. Courtesy of Ben Mayer.)*

brighter. Indeed, a supernova may temporarily rival the brilliance of the entire galaxy in which they are located. The Crab nebula (Fig. 19–26) is the remnant of a supernova that occurred in A.D. 1054.

Figure 19–26 *The Crab nebula. (From Abell, G.O.,* Realm of the Universe, *3rd ed. Philadelphia, Saunders College Publishing, 1984. Courtesy of California Institute of Technology/Palomar Observatory.)*

Novae and supernovae are both explosive in nature, but the cause of the explosions and the masses of the stars involved are distinctly different. Novae involve small white dwarfs, which have relatively modest amounts of mass. As noted above, many of these appear to be one of a pair of binary stars that circle closely around each other under the attraction of their mutual gravity. Astronomers believe that a nova may occur when matter is drawn off the partner star and deposited on the surface of the neighboring white dwarf. This would cause the white dwarf to become unstable and explode.

Supernovae occur in large stars having several times more mass than those involved in novae. They result from a catastrophic collapse of the core of the star, which triggers tremendous explosions in the overlying layers. The gases of these outer layers are thrown violently into surrounding space. This outburst provides the visible supernova display we witness on earth.

The life of a star

As an observation platform for viewing the stars, the earth has drawbacks. Its atmosphere obscures and distorts our view of astronomical objects. There is, however, also an advantage. Our planet is located about two thirds of the distance from the center of the Milky Way Galaxy. From this position, the earth is not only in the midst of the gas and dust clouds from which stars are born, but also relatively close to a multitude of stars that are in various stages of their evolution.

A convenient way to examine the life history of a star is to consider the changes as occurring in four evolutionary stages. Stars representative of each stage, and the general progression between stages, can be plotted on an H-R diagram (Fig. 19-27). Stage 1, the contraction stage, begins with a dust and gas cloud in which atoms of gas and particles of dust slowly begin to collect under the influence of their mutual gravitational attraction. As more and more matter accumulates, gravitational forces increase, and contraction accelerates until the entire mass occupies only the tiniest fraction of its original volume. The rate of contraction will vary according to the amount of matter in the parent cloud. In general, the more matter, the more rapid the rate of contraction. Eventually contraction raises central temperatures and pressures sufficiently to initiate thermonuclear reactions, and the dust and gas cloud is transformed into a star. The entire process is not considered to take much longer than about 10 million years, depending on the mass of the star.

Stage 2 may be called the equilibrium stage. At this stage, the gaseous matter in the star has reached a delicate balance between the gravitational force that causes contraction, and such factors as high temperatures and thermonuclear reactions, which promote expansion. The star has now reached a stable position in the main sequence band (see Fig. 19-27). It behaves much as does the sun, uniformly emitting the energy derived from its nuclear fuel. Such stars are important from a biological point of view, for their constancy enhances the possibility for life on associated planets. The actual luminosity of stars in this stage appears to depend upon the mass of the parent cloud. A cloud of large mass would form a large, hot, blue star of spectral type O, whereas a small mass would yield a star of spectral type M or K.

Stage 3, sometimes called the expansion stage, begins when a critical proportion of the hydrogen in the core has been converted to helium. At this stage, the core no longer has sufficient energy to balance gravitational forces, and it therefore contracts. Higher temperatures and pressures accompanying the contraction initiate nuclear reactions in a shell around the core, causing the star to expand and become a red giant. Its position would be plotted upward and to the right on the H-R diagram. Near the end of this stage, most stars stabilize in the red giant region of the H-R diagram. Temperatures in the center of the star be-

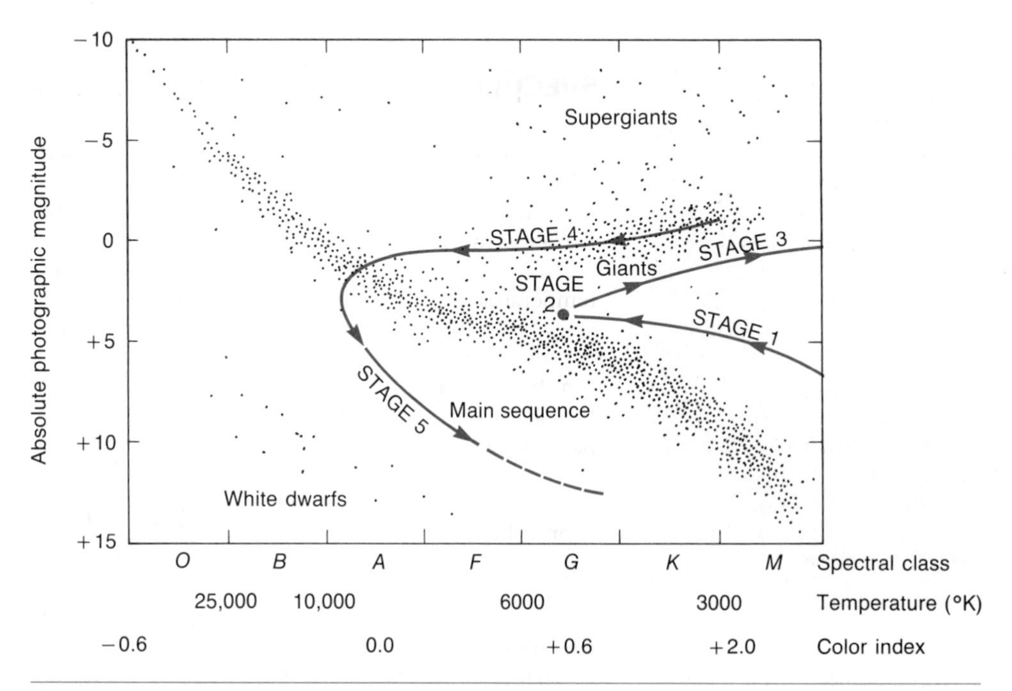

Figure 19-27 Five stages in the evolution of a star indicated on a Hertzsprung-Russell diagram.

come so great (100 million degrees Celsius) that nuclei in the core fuse into heavy elements, such as carbon and oxygen.

Ultimately, a stage is reached in the life of a red giant when it has exhausted its store of nuclear energy. Without that energy to sustain its internal temperature, the star begins to shrink, compressing its entire bulk into a volume roughly equivalent to the average for planets. The reduction in volume produces great heat, so the star assumes a blue-white glow. Such a star, representing the final stage in stellar evolution, is termed a **white dwarf.** A white dwarf is far more dense than any solid substance. In fact, a piece of such a star the size of a golf ball would weigh an incredible 50 tons if placed here on earth. So great a density is only possible when electrons are forced from their orbital shells and pushed closer to their nuclei so as to occupy less space. (Matter structured in this way is called degenerative matter.)

Having exhausted its nuclear fuel and completed its shrinking, the white dwarf's only remaining source of energy is the heat produced by motions of ions in its interior. Even this heat, however, is gradually radiated into space, and the white dwarf begins to fade away. Eventually, all that remains is a dense, cold, dark body called a **black dwarf.**

Calculations indicate that stars with cores that exceed the sun's mass by about 1.4 times are unlikely to become white dwarfs and die in the rather quiet way just described. Instead, they are blasted apart in violent supernovae explosions, characteristically accompanied by the release of tremendous amounts of radiation and emission of high-energy particles. Electrons and protons in the remaining star are forced to combine so as to form neutrons, and a so-called neutron star will become the product of the supernova. Neutron stars are notable for their small size (an average radius of only 10 km) and enormous density. In addition, many neutron stars have strong magnetic fields and rotate rapidly. As they spin, they emit radiation at radio wavelengths. This radiation sweeps across space like the light from the emergency blinker on a police car. When the rays flash across the earth, they are usually detected as regular pulsations of radio waves (although a very few have been seen optically). For this reason, they are called **pulsars.**

Black holes

Astronomical studies suggest that most stars terminate their life history as white dwarfs. The mass of an ordinary white dwarf is less than 1.4 times that of the sun. Some aging stars, however, are far more massive. When a star having a core mass greater than about three to five times the mass of the sun collapses after depletion of its nuclear fuel, it may become a black hole. A **black hole** is an agglomeration of matter that exerts such strong surface gravitational forces that nothing, not even photons of light, can escape. The collapse of a star of large mass may cause it to shrink by more than 100,000 times, in which case the distance between matter at the surface and in the center of the star decreases enormously. As this distance is squared in the gravitation equation $\left(F = G\,\frac{m_1 m_2}{d^2}\right)$, the decrease in d results in a substantial increase in gravitational force. Often this force is increased by 10 billion times. Such extreme gravitation is capable of bending light back upon itself, so that it can never escape the gravitational field. Thus, the star has become a black hole from which nothing can escape and into which approaching particles are violently drawn. With no light to detect them, astronomers must find black holes from the gravitational effects they have on other stars or from x-rays emitted by objects being drawn into them.

SPECULATIONS ABOUT THE UNIVERSE

All theories that attempt to describe the development of the universe are forced into confirmity with an important astronomic observation called the **red shift.** Earlier we noted how the study of a star's spectrum by means of the spectrograph can yield clues to its surface temperature. Spectral analysis can also reveal whether the star is moving toward or away from the earth, and at what speed. To understand how this information can be obtained, recall listening to the apparent change in pitch from a moving source of sound as it passes by. For example, if two cars pass on the highway while one is sounding its horn, persons in the other car can hear an apparent change in pitch as the vehicles approach, pass, and separate. This apparent change in pitch is a manifestation of a phenomenon known as the **Doppler shift.** As the distance between the listener and the noise-making object lessens, the rate at which sound waves enter the ear increases. It is as if the wavelength is shortened, and so the sound rises in pitch. As listener and noise source move apart, the reverse occurs. There is an *apparent* lengthening of waves and lowering of pitch. Another analogy might be imagined in which a toy boat is propelled toward the point where a rock has been thrown in a pond, causing widening circles of waves. The boat bobs up and down more frequently while moving toward the source of the waves. If, however, the boat moves away from the center of the pond, there will be longer intervals between each up-and-down motion. The waves appear to have increased in length. A similar effect occurs with light waves. An approaching star is recognized by an apparent shortening of wavelength, and spectral lines are shifted toward the *blue* end of the spectrum. If the star is moving away from us, the wavelengths are apparently increased, and the spectral lines shift toward the *red* end of the spectrum. Furthermore, the greater the velocity of separation, the larger the observed red shift. By 1914, the astronomer V.M. Slipher had determined the spectra of 14 galaxies and found that the light from 12 of those exhibited red shift. Slipher reasoned that most of the galaxies within our range of observation were moving rapidly away from us. In 1929, Edwin Hubble made the further discovery that the red shift increased with increasing distances of the galaxies: The more distant the galaxy, the higher its recessional velocity.

Assuming that the red shift does indeed tell us that the galaxies have been steadily moving apart, then it is possible to calculate that all the matter in the present galaxies began to move outward from a single location between about 15 billion and 20 billion years ago. This knowledge has led to a hypothesis for the birth of the universe that is appropriately termed the **big bang theory.** According to this theory, our universe began its evolutionary history as a gigantic, superdense, highly compressed, and completely homogeneous primeval nucleus. A violent explosion of this nucleus started the universe expanding. As it expanded and cooled, condensation produced galaxies and clusters of galaxies. If we use the analogy of a cake rising or expanding in the oven as representing the expanding universe, then raisins within the expanding cake would be rather like the galaxies receding from one another.

Another theory about the evolution of the universe stipulates that for every big bang there is a preceding "big crunch." According to this idea, after the great explosion that causes the initial expansion, gravitational forces begin to prevail and cause all the matter to be drawn back to its place of origin. The expanding universe would thus become a contracting one. In the last stages of contraction, material would be returning at enormous velocities, resulting in compression of all the returned matter into an exceedingly dense mass in which temperatures and pressures would be so extreme that a new primeval nucleus would form. From this nucleus would come the next big bang and the initiation of yet another episode of expansion.

Summary

The science of astronomy was born more than 6000 years ago. Its beginnings can be found in the writings of ancient Chinese, Egyptians, and Babylonians. The Egyptians used the apparent motions of the sun and moon to construct an annual calendar of 365 days. The astronomers of ancient Greece made accurate predictions of eclipses, demonstrated that the earth was round, and recognized precession of the earth's axis. Eratosthenes determined the circumference of the earth with surprising accuracy, and Aristarchus proposed that the sun, not the earth, was at the center of the solar system. Scientists later reverted to the concept of an earth-centered solar system, and this erroneous idea persisted until corrected by Copernicus in the sixteenth century.

Copernicus brilliantly proposed that the sun must be at the center of the solar system and outlined the basic movements within the earth-moon-sun system. In the decades that followed, Bruno developed a more valid concept of the spatial distribution of stars, Brahe gave us accurate measurements of the positions of visible stars and planets, and Kepler formulated his famous laws of planetary motion. A half century later, Kepler's work and the experiments of Galileo helped Newton formulate his laws of inertia, force and acceleration, reacting forces, and of course the universal law of gravitation expressed by the formula

$$F = G\frac{m_1 m_2}{d^2}$$

The most important tools of astronomy are optical telescopes of both the refractor and reflector type, radio telescopes, and the spectrograph. The basic function of the optical telescope is to gather light and bring it to focus so that an image is formed. The image can then be photographed, and its spectrum and brightness studied. The objective of the telescope is the main light-gathering and image-forming component and may be either a lens (as in refractor telescopes) or a curved mirror (as in reflector telescopes). In addition to optical telescopes, astronomers employ radio telescopes to collect radio waves from distant celestial sources.

Stars are self-luminous spheres of gas contained in galaxies. Galaxies are large assemblages of stars, gas, and dust. Millions of galaxies are known to exist throughout space. Nearby stars, including our sun, are members of the Milky Way Galaxy. The stars we see at night occur in identifiable configurations called constellations. Constellations seen from the Northern Hemisphere appear to circle around the north celestial pole (which is close to Polaris).

Astronomers measure the distance to stars by observing the shift in their apparent position relative to other stars as the earth moves around the sun. The angle that a star appears to move is called its parallax. The intrinsic brightness of a star is obtained by measuring its apparent brightness and then calculating how bright the star would have to be at a distance (determined from its parallax) to provide the observed apparent brightness. The calculation involves the use of the inverse square law, which states that the apparent brightness of a source varies inversely as the square of the distance from the source to the observer. The temperature of a star is obtained by analyzing its spectrum. Its radial velocity as it recedes or approaches is determined by the Doppler effect.

The absolute magnitude of a star is the brightness it would have if located at a standard distance of 10 parsecs from the observer. Absolute magnitude of stars can be plotted against their surface temperature on the Hertzsprung-Russell (H – R) diagram. Such a plot reveals that most stars lie on a narrow diagonal band called the main sequence. Average stars, like our sun, are located in about the center of the main sequence. A few very large, cooler stars called red giants are located in the upper right. Below the main sequence, in the lower left portion of the H – R diagram are very small, very hot stars named white dwarfs.

The stars in most galaxies can be grouped into two categories or populations, designated population I and population II. Population I stars are young stars thought to have been produced relatively recently from dust and gas within the galaxy. Population II stars are older and not usually found within regions of appreciable dust and gas. They are groupings of some of the earliest stars to form in the galaxy. Stars go through a definite aging process. They begin as huge clouds of gas and dust that contract, compress their gases, initiate the production of nuclear energy, and shine. If the star has a mass similar to that of our sun, it will take its place in the center of the main sequence on the H – R diagram. Typically, the star would then progress to the red giant, and subsequently, the white dwarf stage. Thus, changes in color, temperature, and size are criteria for the age of a star.

The study of light from the galaxies has revealed that spectral lines shift toward the red end of the spectrum, indicating that currently the universe is expanding. This knowledge is the basis for recent hypotheses about the origin of the universe. The big bang hypothesis, for example, suggests that the universe began with a gigantic primeval explosion. As material moved out from the center of the explosion, condensation and accretion produced the galaxies and their component stars and planets.

QUESTIONS FOR REVIEW

1 What important contributions to astronomy were made by Eratosthenes, Copernicus, Kepler, Galileo, and Newton?

2 According to the universal law of gravitation, formulated by Newton, what happens to the gravitational force between two bodies when the distance between them is decreased?

3 What determines the size of the image formed by a telescope? Why are most large research optical telescopes located on mountaintops?

4 What advantages and disadvantages does the radio telescope have compared with the optical telescope? How can the resolution of radio telescopes be improved?

5 Distinguish between the apparent brightness and intrinsic brightness of a star. Which of these can be directly observed?

6 Discuss the relation of a star's color to its surface temperature. In general, what colors characterize population I stars? population II stars? Which of these types of stars would be more characteristic of globular clusters?

7 What are the differences in the spectrum of a type A and type M star? What are the differences in the color and average surface temperatures of these two star types?

8 What is the relationship between the mass of a main sequence star and its brightness?

9 What is a black hole? What is its origin? What force draws matter into the black hole and prevents the escape of ions or even of photons of light?

10 What is the Doppler effect? Provide an example of the Doppler effect that you encounter in your everyday activities. What do we mean by red shift? Imagine that an astronomer discovered there was a blue shift instead of a red shift from a distant galaxy. How might he interpret this information?

11 Astronomers sometimes refer to the light from galaxies as "fossil light." Why is this expression appropriate?

SUPPLEMENTAL READINGS AND REFERENCES

Abell, G.O., 1984. *Realm of the Universe.* 3d ed. Philadelphia, Saunders College Publishing.

Apfel, N.H., and Hynek, J.A., 1979. *Architecture of the Universe.* 2d ed. Menlo Park, Calif., Benjamin/Cummings Publishing Co.

Pasachoff, J.M., 1985. *Contemporary Astronomy.* Philadelphia, Saunders College Publishing.

Seeds, M.A. (ed.), 1980. *Astronomy: Selected Readings.* Menlo Park, Calif., Benjamin/Cummings Publishing Co.

appendix a

topographic maps

Among the graphic tools employed by earth scientists, none are more widely used than topographic maps. Topographic maps show the size, shape, and distribution of features of the earth's surface, and from them one can learn much about the nature of rocks that lie beneath the surface. They are used as base maps for a variety of geological investigations.

The three major categories of features depicted on topographic maps are *relief* (hills, valleys, plains, cliffs, and the like), *drainage* (streams, seas, lakes, swamps, and canals), and *culture* (the works of humans including towns, cities, roads, railroads, boundaries, and names). Maps published by the U.S. Geological Survey all bear a key or legend on which symbols are defined and additional items, such as tints for forests or urban areas, are explained.

RELIEF

Contour lines connecting points of equal altitude are the standard method for portraying relief on a topographic map. These lines, printed in brown color, are an effective device for showing accurately a third dimension on a flat sheet of paper (Fig. A–1). With practice, it is possible to readily visualize hills and valleys from the pattern of contour lines. On some topographic maps, landscape features are even more easily recognized because of the use of shading. Such maps, called *shaded-relief topographic maps,* are tinted so as to simulate in color the appearance of sunlight and shadows on the landscape and thereby provide the illusion of a three-dimensional land surface.

The vertical distance between any two contour lines is termed the *contour interval.* A satisfactory contour interval is one that depicts the landscape features adequately without the need to space contour lines so closely that reading the map becomes difficult. In Figure A–1, the contour interval is 10 feet. As a further aid to readability, every fourth or fifth contour is printed somewhat heavier and labeled with the elevation. These lines constitute *index contours.* The steepness of slopes can be ascertained on a contour map by the spacing of contour lines. A simple rule is that the closer the spacing, the steeper the slope. Contour lines bend or loop upstream in valleys and can thus be used to determine direction of streamflow. In becoming acquainted with contour maps, it is also useful to keep in mind that the ground surface must rise away from stream channels.

MAP SCALE

The scale of a map depicts the relation of a given distance on a map to the actual distance on the surface of the earth. Two widely used types of scales are *fractional* and *graphic.* The fractional scale is usually printed on the bottom margin of the map as a ratio, such as 1 : 24,000 or 1 : 125,000. The meaning of a scale of 1 : 24,000 is that one unit of measurement on the map represents 24,000 of the same units on the earth's surface. A graphic scale (Fig. A–2) consists of a line that has been divided into units representing length, such as kilometers or fractions of kilometers. Graphic scales can be used directly on the map to determine distances.

DIRECTION

Compass directions on a standard U.S. Geological Survey topographic map are indicated by the meridians and parallels of latitude. By convention, the top of the map is north, and the right margin, consequently, is east. An important additional aid to orientation is the small diagram on the lower border of the map, which indicates true north, magnetic north, and the angle between the two, known as the *magnetic declination* (Fig. A–3). Magnetic declination is expressed in degrees and changes slowly. Corrections of magnetic declination can be made by referring to isogonic charts published each year by the U.S. Coast and Geodetic Survey.

LOCATION

If you noticed some interesting feature like an oxbow lake, sinkhole, cirque, or campsite on a topographic map, how would you describe its location accurately? Usually one would employ one of the two principal aids to determining location imprinted on U.S. Geological Survey topographic maps. One of these aids is the global grid of *parallels of latitude* (east-west lines parallel to the equator) and *meridians of longitude* (true north-south lines that converge at the poles). Zero degrees of latitude corresponds to the equator, and 90°N to the North Pole. By agreement, the zero circle of longitude is placed so as to cross the former site of the Royal Greenwich Observatory in England. Longitude is measured in degrees east or west of the meridian that crosses Greenwich.

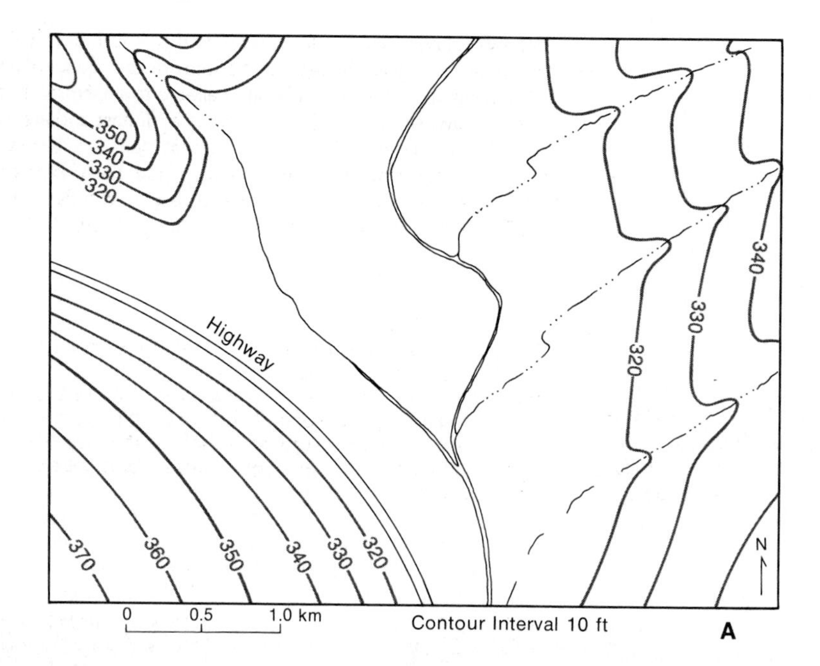

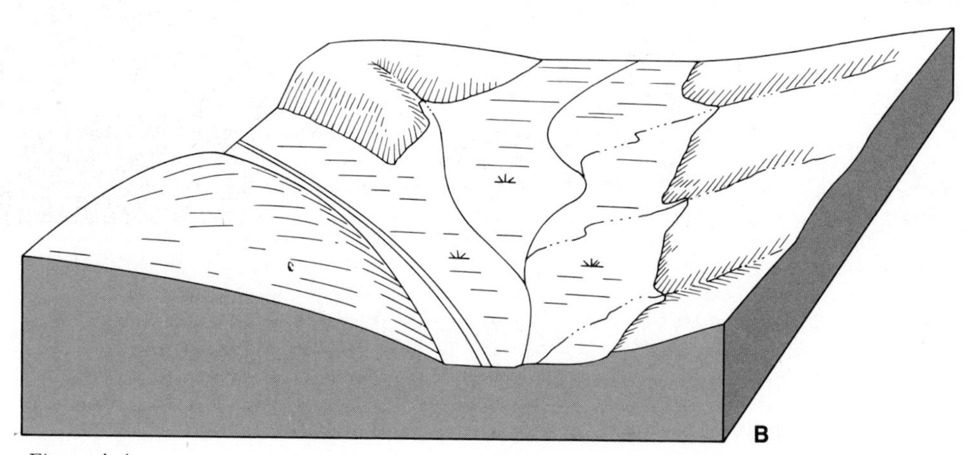

Figure A-1

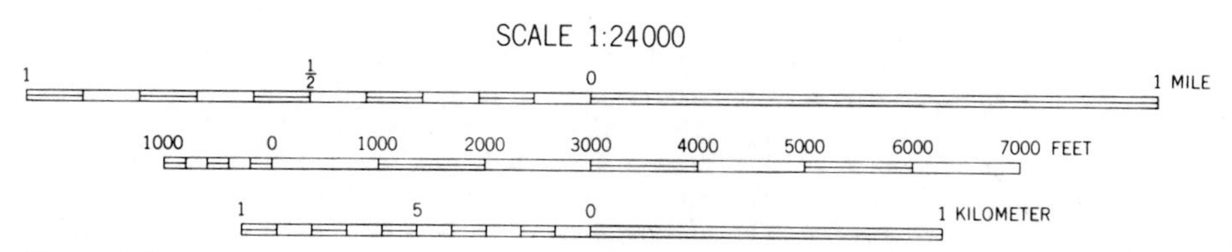

Figure A-2

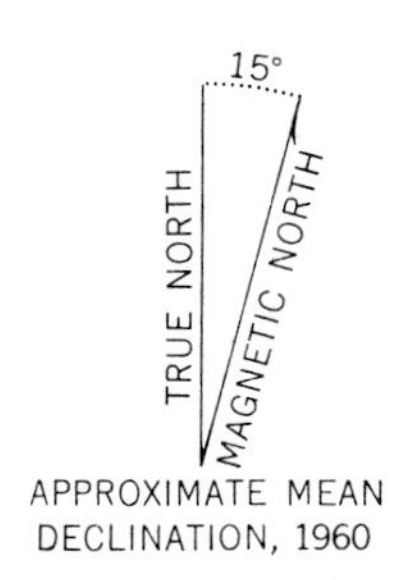

Figure A-3

Another location method widely used in the United States because of its practicality, is the *township and range system.* In this scheme, areas of the earth's surface are systematically divided according to a grid that is based on carefully surveyed east-west lines called base lines, and north-south lines called principal meridians. The basic unit of the grid is a square of land measuring six miles on a side and called a *township.* Vertical rows of townships, called ranges, are laid off east and west of the principal meridian. Each township is divided into 36 smaller squares, measuring one mile on a side and called sections. These sections contain 640 acres and may be subdivided according to fraction and geographic position as shown in Figure A–4.

REFERENCES

Dickinson, G.C., 1979. *Maps and Air Photographs.* New York, John Wiley & Sons.

Thompson, M.M., 1979. *Maps for America.* United States Geological Survey. Topographic Maps. Pamphlet obtainable without charge on request to the Map Information Office, U.S. Geological Survey, Washington, D.C. 20242.

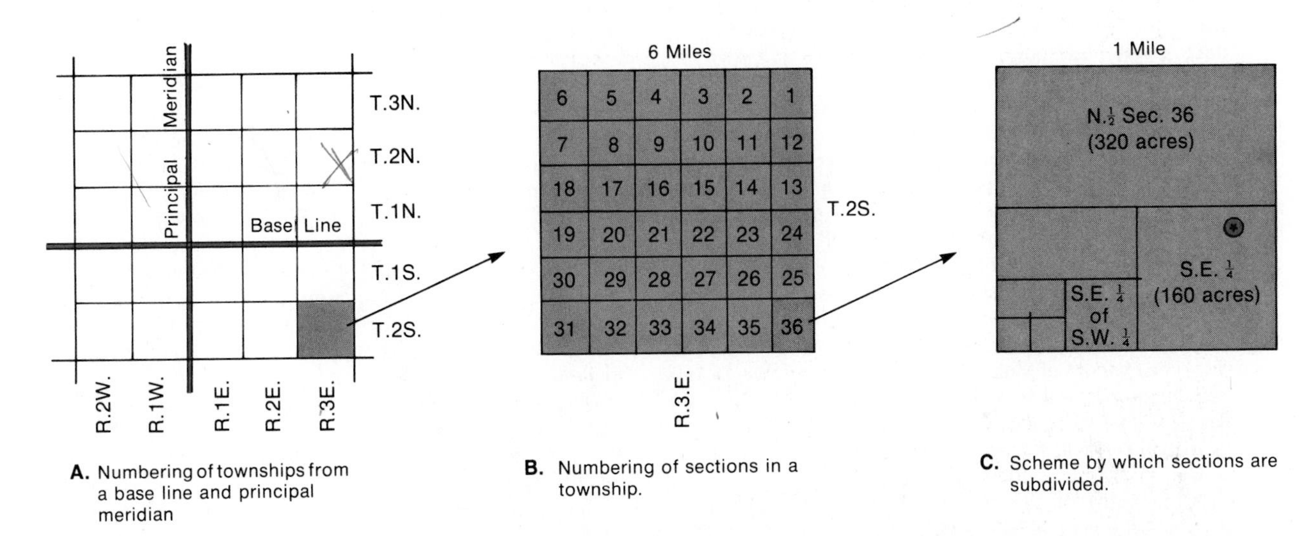

A. Numbering of townships from a base line and principal meridian

B. Numbering of sections in a township.

C. Scheme by which sections are subdivided.

Example, description of the location indicated by the circled star is N.E. ¼ of S.E. ¼ of Sec. 36, T.2S., R.3E.

Figure A-4

appendix b

physical properties of minerals

The minerals listed in the following tables have been selected on the basis of their general abundance at the earth's surface, and/or their special interest. They are arranged alphabetically to facilitate quick reference. The physical properties most often used in mineral identification include streak, color, luster, hardness, cleavage, fracture, general form, and specific gravity. In addition, some minerals can be recognized by distinctive taste, feel, odor, magnetism, or reaction to acid.

The streak of a mineral is the color of its powder, which can be obtained by firmly scraping the mineral across a hard, white porcelain streak plate. A mineral's streak is often less variable than its overall color. Luster refers to the appearance of a mineral in reflected light. The two major categories of luster are metallic and nonmetallic. Nonmetallic minerals can be further grouped into lusters described as adamantine (as in diamond), vitreous (bright, shiny), resinous, waxy, pearly, greasy or just dull. Hardness is an especially useful aid to identifying minerals in hand specimen. Hardness refers to the resistance of a mineral to scratching. One does not necessarily require the minerals of Mohs' hardness scale (see Table 2–4) to use this property. It is often more convenient to use one's fingernail (hardness of 2 to 2½), a copper penny (hardness of 3), a good knife blade (hardness of 5½) or a streak plate (hardness of 7). If, for example, you are able to scratch a mineral with your fingernail, then the mineral's hardness would be less than 2. If the mineral will scratch a copper penny but won't scratch a knife blade, its hardness would be in the range from 3 to 5½.

As described in Chapter 1, cleavage refers to the tendency of certain minerals to break in preferred directions along smooth planes. Cleavage surfaces are usually parallel to a real or possible crystal face. Minerals vary in the number and perfection of cleavage planes. Frequently this property cannot be distinguished on the many tiny grains of minerals that ordinarily occur as aggregates.

For the identification of the relatively few exceptionally common minerals, it is often useful to start by observing whether the mineral has a metallic or nonmetallic luster. If metallic, your choice may be among such minerals as galena (distinctive high specific gravity), pyrite (distinctive color and streak), magnetite (attracted to magnet), and hematite (distinctive streak). For nonmetallic minerals, consider first those that have a dark color and are harder than a steel knife blade. If good cleavage is evident, you may have a pyroxene or amphibole. If cleavage is absent or poor, the mineral may be olivine or garnet. Many other possibilities exist as well. For the softer, dark, nonmetallic specimens, one may find the identification among such minerals as biotite, sphalerite, chlorite, goethite, or again, hematite. Feldspar can often be recognized by the light coloration, hardness in excess of 5, nonmetallic luster, and prominently good cleavage. Similarly colored quartz minerals will lack this cleavage. Other light-colored minerals, such as halite, gypsum, calcite, and muscovite, can be identified by their softness and often well-developed cleavage planes.

APPENDIX B **COMMON MINERALS AND THEIR PROPERTIES**

Mineral	*Chemical composition*	*Form, cleavage, fracture*	*Usual color*	*Hard-ness*	*Streak*	*Specific gravity*	*Other properties*
Actinolite	$Ca_2(MgFe)_5Si_8O_{22}(OH)_2$	Slender crystals, radiating, fibrous	Blackish-green to light green	5–6	Pale green	3.2–3.6	Vitreous luster. Common in green schists.
Agate (onyx)	SiO_2	Always massive and banded	Banded, red, white, pink, brown, green	7	White	2.6	Variety of quartz. No cleavage, tough, translucent; waxy luster.
Alabaster (gypsum)	$CaSo_4 \cdot 2H_2O$	Granular and massive	White, gray, pink	1–2.5	White	2.2–2.4	Pearly to dull luster.
Albite	$Na(AlSi_3O_8)$ (sodic plagioclase)	Good cleavage in 2 directions, nearly 90°	White, gray	6–6.5	White	2.6	May show fine striations (twinning lines) on cleavage faces.
Almandite (garnet)	$Fe_3Al_2(SiO_4)_3$	No cleavage, crystals 12- or 24-sided	Deep red	6.5–7.5	White	4.2	Vitreous to resinous luster. Metamorphic index mineral.
Amethyst	SiO_2	No cleavage, uneven fracture	Purple	7	White	2.6	Vitreous luster. Hexagonal crystals or massive crystalline.
Andalusite	Al_2SiO_5	Usually in square prisms	White, gray, green, violet, brown	6–7.5	White	3.1	Crystals may show dark interior cross.
Anhydrite	$CaSO_4$	Granular masses, crystals with 2 good cleavage directions	White, gray, blue-gray	3–3.5	White	2.9–3	Brittle, resembles marble but acid has no effect.
Anorthite	$Ca(Al_2Si_2O_8)$ (calcic plagioclase)	Good cleavage in 2 directions at 94°	Colorless, white, gray, green	6	White	2.8	May show fine striations (twinning lines) on cleavage faces.
Apatite	$Ca_5(F,Cl)(PO_4)_3$ (calcium fluorophosphate)	Massive, granular	Green, brown, red	4.5–5	Pale red-brown	3.1	Crystals may have a partly melted appearance, glassy.
Asbestos (var. chrysotile)	$H_4Mg_3Si_2O_9$	Fibrous	White to pale olive-green	1–2.5	White	2.6–2.8	Pearly to greasy luster. Flexible, easily separated fibers.
Augite (a pyroxene)	$Ca(Mg,Fe,Al)(Al,Si_2O_6)$	Short stubby crystals have 4 or 8 sides in cross-section	Blackish-green to light green	5.6	Pale green	5–6	Vitreous, distinguished from hornblende by the 87° angle between cleavage faces.
Azurite	$Cu_3(CO_3)_2(OH)_2$ (copper carbonate)	Varied, may have fibrous crystals	Azure blue	4	Pale blue	3.8	Vitreous to earthy, effervesces with HCl. An ore of copper.
Barite	$BaSO_4$	Tabular crystals, 3 cleavages, crystalline masses	White	2.5–3.5	White	4.5	Vitreous to pearly luster. Easily determined by high specific gravity.
Bauxite	Hydrous aluminum oxides. Not a true mineral.	Earthy masses	Reddish to brown	1.5–3.5	Pale reddish-brown	2.5	Dull luster, claylike masses with small round concretions.

Appendix B (Continued)

Mineral	*Chemical composition*	*Form, cleavage, fracture*	*Usual color*	*Hardness*	*Streak*	*Specific gravity*	*Other properties*
Beryl	$Be_3Al_2(SiO_3)_6$	Uneven fracture, hexagonal crystals	Green, yellow, blue, pink	7.5–8.0	White	2.6–2.8	Vitreous luster. Gem variety is aquamarine.
Biotite	$K(Mg,Fe)_3AlSi_3O_{10}(OH)_2$	Perfect cleavage into thin sheets	Black, brown green	2.2–2.5	White, gray	2.7–3.1	Vitreous luster, divides readily into thin, flexible sheets.
Bornite	Cu_5FeS_4	Compact, massive	Purplish	3	Gray-black	4.9–5.4	Metallic luster, brittle, an ore of copper.
Calcite	$CaCO_3$	Perfect cleavage into rhombs	Usually white, but may be variously tinted	3	White	2.7	Transparent to opaque. Rapid effervescence with HCl.
Carnotite	$K_2(UO_2)_2(VO_4)_2$	Earthy, powdery	Canary yellow	Very soft	Pale yellow	4	Soft, earthy. An ore of uranium.
Cassiterite	SnO_2	Massive or as sand-size grains	Black	6–7	Dark brown	7	Submetallic, heavy, hard grains.
Chalcedony	SiO_2	Fractures uneven or conchoidal	White, gray, blue	7	White	2.6	Waxy luster, tough, translucent. A variety of quartz.
Chalcocite	Cu_2S	Usually massive	Dark lead-gray	2.5–3	Lead-gray	5.5–5.8	Metallic luster, often with bluish tarnish.
Chalcopyrite	$CuFeS_2$	Uneven fracture	Brass yellow	3.5–4.5	Greenish-black	4.2	Metallic luster, softer than pyrite.
Chlorite	Hydrous ferromagnesian aluminum silicate	Perfect cleavage as fine scales	Green	2.0–2.5	Gray, white, pale green	2.8	Pearly to vitreous luster. Low-grade metamorphic mineral.
Chromite	$FeCr_2O_4$	Massive, granular, compact	Black	5.5	Dark brown	4.4	Metallic to submetallic luster. An ore of chromite occurring in serpentine.
Cinnibar	HgS	Compact, granular masses	Scarlet red to red-brown	2.5	Scarlet red	8	Color and streak distinctive. Ore of mercury.
Clay (clay minerals)	Hydrous aluminum silicates, with Ca, Na, K, Fe, Mg	Soft, compact, earthy masses	White, but tinted by impurities	1–2	White	2.2–2.6	Greasy feel, adheres to tongue, earthy odor when moist.
Corundum	Al_2O_3	Short, six-sided barrel-shaped crystals	Gray, light blue, and other colors	9	None	3.9–4.1	Ruby and sapphire are corundum varieties. Hardness is distinctive.
Diamond	C	Octahedral crystals	Colorless or with pale tints	10	None	3.5	Adamantine luster. Hardness is distinctive.

Diopside (a pyroxene)	$CaMg(SiO_3)_2$	Usually short, thick prisms; may be granular	White to light green	5–6	White to greenish	3.2–3.6	Vitreous luster. A contact metamorphic mineral in carbonates.
Dolomite	$CaMg(CO_3)_2$	Cleaves into rhombs; granular masses	White, pink, gray, brown	3.5–4	White to pale gray	3.9–4.2	Effervesces slightly in cold dilute HCl.
Enstatite (a pyroxene)	$Mg_2Si_2O_6$	Cleavage good at 87° and 93°; usually massive	Pale green, brown, gray or yellowish	5.5	White	3.2–3.5	Vitreous luster. Common in mafic igneous rocks.
Epidote	Hydrous Ca, Al, Fe, silicate	Usually granular masses; also as slender prisms	Yellow-green, olive-green, to nearly black	6.7	Pale yellow to white	3.3	Vitreous luster, occurs as contact metamorphic mineral in carbonates.
Fluorite	CaF_2	Octahedral, and also cubic crystals	White, yellow, green, purple	4	White	3.2	Cleaves easily, vitreous, transparent to translucent.
Galena	PbS	Perfect cubic cleavage	Lead or silver gray	2.5	Gray	7.6	Metallic luster. Ore of lead.
Glauconite	Hydrous silicate of K and Fe	Occurs as grains or granular masses	Green	1–2	Greenish-white	2.2	Dull luster. Common constituent of "greensands."
Goethite	$HFeO_2$	Massive, in fibrous aggregates or foliated	Yellow-brown to dark brown	5–5.5	Yellow-brown	3.3–4.4	Adamantine to dull. An iron ore.
Graphite	C	Foliated, scaly, or earthy masses	Steel gray to black	1–2	Gray or black	2.2	Feels greasy, marks paper.
Gypsum	$CaSO_4 \cdot 2H_2O$	Tabular crystals, fibrous, or granular	White, pearly	1–2.5	White	2.2–2.4	Thin sheets (selenite), fibrous (satinspar), massive (alabaster).
Halite	NaCl	Granular masses, perfect cubic crystals	White, also pale colors and gray	2.5–3	White	2.2	Pearly luster, salty taste, soluble in water.
Hematite	Fe_2O_3	Granular, massive, or earthy	Brownish-red	2.5	Dark red	2.5–5	Often earthy, dull appearance.
Hornblende (an amphibole)	Complex Ca, Mg, Fe, Al, silicate	Elongate crystals	Dark shades of green		Pale green	3.2	Crystals 6-sided with 124° between cleavage faces.
Jadeite (a pyroxene)	$NaAl(Si_2O_6)$	Compact fibrous aggregates	Green	6.5–7	White, pale green	3.3	Vitreous luster.
Jasper	SiO_2	Fracture uneven to conchoidal	Red, yellow-brown	7	White	2.6	Dull to waxy luster.
Kyanite	Al_2SiO_5	Good cleavage in one direction; bladed aggregates	White, pale blue, gray	4–7	White	3.6	Vitreous to pearly; crystals may have blue interiors.
Labradorite	A mixture of $CaAl_2Si_2O_8$ and $NaAlSi_3O_8$	Cleavage perfect in 2 directions	Dark gray to grayish-white	6	White	2.7	Vitreous, fine striations on 1 cleavage face, play of colors.

Appendix B (Continued)

Mineral	*Chemical composition*	*Form, cleavage, fracture*	*Usual color*	*Hard-ness*	*Streak*	*Specific gravity*	*Other properties*
"Limonite" (not a true mineral)	$2Fe_2O_3 \cdot 3H_2O$	Earthy fracture	Brown or black	1.5–4	Brownish-yellow	3.6	Earthy masses that resemble clay.
Magnetite	Fe_3O_4	Uneven fracture, granular masses	Iron-black	5.5	Iron-black	5.2	Metallic luster. Strongly magnetic.
Malachite	$CuCO_3Cu(OH)_2$	Uneven, splintery fracture	Bright green, dark green	3.5–4	Emerald green	4	Effervesces with HCl. Associated with azurite.
Marcasite	FeS_2	Uneven fracture, arrow-shaped crystals	Pale brass-yellow	6	Black	4.9	Metalic luster (never in cubes as in pyrite).
Muscovite	$KAl_3Si_3O_{10}(OH)_2$	Perfect cleavage into thin sheets	Colorless if thin	2–2.5	White	2.7–3	Vitreous or pearly, flexible and elastic, splits easily.
Olivine	$(MgFe)_2SiO_4$	Uneven fracture, often in granular masses	Various shades of green	6.5–7	White	3.2–3.3	Vitreous, glassy luster. Common in basalt. Gem variety is peridote.
Opal (a mineraloid)	$SiO_2 \cdot nH_2O$	Conchoidal fracture, amorphous, massive	White and various colors	5.5–6.5	White	2.1	Vitreous, greasy, pearly luster. May show a play of colors.
Orthoclase	$K(AlSi_3O_8)$	Good cleavage in 2 directions at 90°	White, pink, red, yellow-green, gray	6	White	2.6	Vitreous to pearly. A common associate of quartz in granite.
Pyrite ("fools gold")	FeS_2	Uneven fracture cubes with striated faces, octahedrons	Pale brass-yellow (lighter than chal-copyrite)	6–6.5	Greenish-black	5	Metallic luster, brittle, very common.
Quartz	SiO_2	No cleavage, massive and as 6-sided crystals	Colorless, white, or tinted any color by impurities	7	White	2.6	Includes rock crystal, rose and milky quartz, amethyst, smoky quartz, etc.
Rutile	TiO_2	Prismatic cleavage, uneven fracture	Reddish-brown to black	6.7	Light brown to light gray	4.3	Adamantine to submetallic. An ore of titanium.
Serpentine	$Mg_3Si_2O_5(OH)_4$	Uneven, often splintery fracture	Light and dark green, yellow	2.5	White	2.5	Waxy luster, smooth feel, brittle.
Siderite	$FeCO_3$	Good rhombohedral cleavage	Brown	3.5–4	Pale yellow or yellow-brown	3.8	Vitreous luster, brittle, cleavage faces often curved.

Sillimanite	Al_2SiO_5	Good clevage in 1 direction, fibrous	Brown, light green, white	6–7	White	3.23	Vitreous luster. Index to high-grade metamorphism.
Specularite (var. hematite)	Fe_2O_3	Cleavage absent or micaceous	Dark steel-gray	2.5–6.5	Red or reddish-brown	4.4–5.3	Metallic luster, Bright sparkling scales.
Sphalerite	ZnS	Perfect cleavage in 6 directions at 120°	Shades of brown and red	3.5	Reddish-brown	4	Resinous luster. May occur with galena, pyrite. An ore of zinc.
Spinel (a mineral) group)	$MgAl_2O_4$	No cleavage, rare octahedral crystals	Black, dark green, or various colors	7.5–8.0	White	3.5–4.1	Vitreous luster. Hardness is distinctive.
Staurolite	$Fe^{+2}Al_5Si_2O_{12}(OH)$	Fair cleavage in 1 direction	Dark brown to nearly black	7	White to grayish	3.6–3.8	Subvitreous to resinous. Crystals sometimes twinned as crosses.
Stibnite	Sb_2S_3	Perfect cleavage in 1 direction	Dark lead-gray	2	Gray or black	4.5	Metallic luster, slender prismatic crystals.
Talc (soapstone)	$Mg_3Si_4O_{10}(OH)_2$	Perfect cleavage in 1 direction	Green, white gray	1–1.5	White	1–2.5	Greasy feel. Occurs in foliated masses.
Topaz	$Al_2SiO_4(F,OH)_2$	Cleavage good in 1 direction; conchoidal fracture	Colorless, white, pale tints of blue, pink	8	Colorless	3.5–3.6	Vitreous luster.
Tourmaline	Complex boron aluminum silicate with Na, Ca, F, Fe, Li	Poor cleavage, uneven fracture; striated crystals	Black, brown, green, pink	7–7.5	White to gray	4.4–4.8	Vitreous, slightly resinous. A gem stone.
Tremolite (an amphibole)	$Ca_2Mg_5(Si_8O_{22})(OH)_2$	Perfect cleavage at an angle of 124°	White, light gray, light green	5.6	White	2.9–3.1	Vitreous to silky. Crystals usually long-bladed or short and stout.
Uraninite	UO_2 with small amounts of Pb, Ra, Th, Y, N, He, and A	Massive or botyroidal; no cleavage	Brownish-black	5–6	Brownish-black	6.4–9.7	Submetallic luster to greasy or dull.
Wollastonite	$CaSiO_3$	Usually in fibrous masss of elongate crystals	White, pink, gray, colorless	4.5–5.0	White	2.8–2.9	Vitreous luster. A mineral of contact metamorphism of limestone
Zircon	$Zr(SiO_4)$	Cleavage poor, but often well-formed tetragonal crystals	Colorless, gray, green, pink, bluish	7.5	White	4.7	Adamantine luster.

appendix c

international table of atomic weights

Based on the assigned relative atomic mass of $^{12}C = 12$.

The following values apply to elements as they exist in materials of terrestrial origin and to certain artificial elements. When used with the footnotes, they are reliable to ± 1 in the last digit, or ± 3 if that digit is in small type.

	Symbol	Atomic number	Atomic weight
Actinium	Ac	89	
Aluminum	Al	13	26.98154[a]
Americium	Am	95	
Antimony	Sb	51	121.7_5
Argon	Ar	18	39.94_8[b,c,d,g]
Arsenic	As	33	74.9216[a]
Astatine	At	85	
Barium	Ba	56	137.3_4
Berkelium	Bk	97	
Beryllium	Be	4	9.01218[a]
Bismuth	Bi	83	208.9804[a]
Boron	B	5	10.81[c,d,e]
Bromine	Br	35	79.904[c]
Cadmium	Cd	48	112.40
Calcium	Ca	20	40.08[g]
Californium	Cf	98	
Carbon	C	6	12.011[b,d]
Cerium	Ce	58	140.12
Cesium	Cs	55	132.9054[a]
Chlorine	Cl	17	35.453[c]
Chromium	Cr	24	51.996[c]
Cobalt	Co	27	58.9332[a]
Copper	Cu	29	63.54_6[c,d]
Curium	Cm	96	
Dysprosium	Dy	66	162.5_0
Einsteinium	Es	99	
Erbium	Er	68	167.2_6
Europium	Eu	63	151.96
Fermium	Fm	100	
Fluorine	F	9	18.99840[a]
Francium	Fr	87	
Gadolinium	Gd	64	157.2_5
Gallium	Ga	31	69.72
Germanium	Ge	32	72.5_9
Gold	Au	79	196.9665[a]

	Symbol	Atomic number	Atomic weight
Hafnium	Hf	72	178.4_9
Helium	He	2	4.00260[b,c]
Holmium	Ho	67	164.9304[a]
Hydrogen	H	1	1.0079[b,d]
Indium	In	49	114.82
Iodine	I	53	126.9045[a]
Iridium	Ir	77	192.2_2
Iron	Fe	26	55.84_7
Krypton	Kr	36	83.80
Lanthanum	La	57	138.905_5[b]
Lawrencium	Lr	103	
Lead	Pb	82	207.2[d,g]
Lithium	Li	3	6.94_1[c,d,e,g]
Lutetium	Lu	71	174.97
Magnesium	Mg	12	24.305[c,g]
Manganese	Mn	25	54.9380[a]
Mendelevium	Md	101	
Mercury	Hg	80	200.5_9
Molybdenum	Mo	42	95.9_4
Neodymium	Nd	60	144.2_4
Neon	Ne	10	20.17_9[c]
Neptunium	Np	93	237.0482[f]
Nickel	Ni	28	58.70
Niobium	Nb	41	92.9064[a]
Nitrogen	N	7	14.0067[b,c]
Nobelium	No	102	
Osmium	Os	76	190.2
Oxygen	O	8	15.999_4[b,c,d]
Palladium	Pd	46	106.4
Phosphorus	P	15	30.97376[a]
Platinum	Pt	78	195.0_9
Plutonium	Pu	94	
Polonium	Po	84	
Potassium	K	19	39.09_8
Praseodymium	Pr	59	140.9077[a]

	Symbol	Atomic number	Atomic weight
Promethium	Pm	61	
Protactinium	Pa	91	231.0359[a,f]
Radium	Ra	88	226.0254[f,g]
Radon	Rn	86	
Rhenium	Re	75	186.207
Rhodium	Rh	45	102.9055[a]
Rubidium	Rb	37	85.467_8[c]
Ruthenium	Ru	44	101.0_7
Samarium	Sm	62	150.4
Scandium	Sc	21	44.9559[a]
Selenium	Se	34	78.9_6
Silicon	Si	14	28.08_6[d]
Silver	Ag	47	107.868[c]
Sodium	Na	11	22.98977[a]
Strontium	Sr	38	87.62[g]
Sulfur	S	16	32.06[d]
Tantalum	Ta	73	180.947_9[b]
Technetium	Tc	43	98.9062[f]
Tellurium	Te	52	127.6_0
Terbium	Tb	65	158.9254[a]
Thallium	Tl	81	204.3_7
Thorium	Th	90	232.0381[f]
Thulium	Tm	69	168.9342[a]
Tin	Sn	50	118.6_9
Titanium	Ti	22	47.9_0
Tungsten	W	74	183.8_5
Uranium	U	92	238.029[b,c,e]
Vanadium	V	23	50.941_4[b,c]
Wolfram		(see Tungsten)	
Xenon	Xe	54	131.30
Ytterbium	Yb	70	173.0_4
Yttrium	Y	30	88.9059[a]
Zinc	Zn	30	65.38
Zirconium	Zr	40	91.22

[a] Mononuclidic element.
[b] Element with one predominant isotope (about 99 to 100 percent abundance).
[c] Element for which the atomic weight is based on calibrated measurements by comparisons with synthetic mixtures of known isotopic composition.
[d] Element for which known variation in isotopic abundance in terrestrial samples limits the precision of the atomic weight given.
[e] Element for which users are cautioned against the possibility of large variations in atomic weight due to inadvertent or undisclosed artificial isotopic separation in commercially available materials.
[f] Most commonly available long-lived isotope.
[g] In some geological specimens this element has a highly anomalous isotopic composition, corresponding to an atomic weight significantly different from that given.

appendix d

measurement systems

D-1 Convenient conversion factors

To convert from	*to*	*multiply by*
Centimeters	Feet	0.0328 ft/cm
	Inches	0.394 in/cm
	Meters	0.01 m/cm
	Microns (micrometers)	10,000 μm/cm
	Miles (statute)	6.214×10^{-6} mi/cm
	Millimeters	10 mm/cm
Feet	Centimeters	30.48 cm/ft
	Inches	12 in/ft
	Meters	0.3048 m/ft
	Microns (micrometers)	304800 μm/ft
	Miles (statute)	0.000189 mi/ft
	Yards	0.3333 yd/ft
Gallons (U.S., liquid)	Cubic centimeters	3785 cm^3/gal
	Cubic feet	0.133 ft^3/gal
	Cubic inches	231 in^3/gal
	Cubic meters	0.003785 m^3/gal
	Cubic yards	0.004951 yd^3/gal
	Liters	3.785 ℓ/gal
	Quarts (U.S., liquid)	4 qt/gal
Grams	Kilograms	0.001 kg/g
	Micrograms	1×10^6 μg/g
	Ounces (avdp.)	0.03527 oz/g
	Pounds (avdp.)	0.002205 lb/g
Inches	Centimeters	2.54 cm/in
	Feet	0.0833 ft/in
	Meters	0.0254 m/in
	Yards	0.0278 yd/in
Kilograms	Ounces (avdp.)	35.27 oz/kg
	Pounds (avdp.)	2.205 lb/kg
Liters	Cubic centimeters	1000 cm^3/ℓ
	Cubic feet	0.0353 ft^3/ℓ
	Cubic inches	61.03 in^3/ℓ
	Cubic meters	0.001 m^3/ℓ
	Cubic yards	0.001308 yr^3/ℓ
	Gallons (U.S., liquid)	0.264 gal/ℓ
	Ounces (U.S., fluid)	33.81 oz/ℓ
	Quarts (U.S., liquid)	1.0567 qt/ℓ
Meters	Centimeters	100 cm/m
	Feet	3.2808 ft/m
	Inches	39.37 in/m
	Kilometers	0.001 km/m
	Miles (statute)	0.0006214 mi/m
	Millimeters	1000 mm/m
	Yards	1.0936 yd/m

(*Continued on next page*)

(*Continued*)

To convert from	*to*	*multiply by*
Miles (statute)	Centimeters	160,934 cm/mi
	Feet	5280 ft/mi
	Inches	63,360 in/mi
	Kilometers	1.609 km/mi
	Meters	1609 m/mi
	Yards	1760 yd/mi
Ounces (avdp.)	Grams	28.3 g/oz
	Pounds (avdp.)	0.0625 lb/oz
Pounds (avdp.)	Grams	453.6 g/lb
	Kilograms	0.454 kg/lb
	Ounces (avdp.)	16 oz/lb

From Turk, J., and Turk, A., *Physical Science*. Philadelphia, W.B. Saunders Company, 1977.

D-2 Metric units, prefixes, and scientific notation

Metric units

Basic units

length	meter (m)
volume	liter (l)
mass	gram (gm or g)
time	second (sec or s)

Other metric units

1 micron (μ) = 10^{-6} meter
1 Ångstrom (Å or A) = 10^{-10} meter
= 10^{-8} cm

Prefixes for use with basic units of metric system

Prefix	Symbol	Power		Equivalent
tera	T	10^{12} =	1,000,000,000,000	Trillion
giga	G	10^{9} =	1,000,000,000	Billion
mega	M	10^{6} =	1,000,000	Million
kilo	k	10^{3} =	1,000	Thousand
hecto	h	10^{2} =	100	Hundred
deca	da	10^{1} =	10	Ten
———	—	10^{0} =	1	One
deci	d	10^{-1} =	.1	Tenth
centi	c	10^{-2} =	.01	Hundredth
milli	m	10^{-3} =	.001	Thousandth
micro	μ	10^{-6} =	.000001	Millionth
nano	n	10^{-9} =	.000000001	Billionth
pico	p	10^{-12} =	.000000000001	Trillionth
femto	f	10^{-15} =	.000000000000001	
atto	a	10^{-18} =	.000000000000000001	

Examples: 1000 meters = 1 kilometer = 1 km
10^{6} hertz = 1 megahertz = 1 MHz
10^{-3} sec = 1 millisecond = 1 msec

Scientific notation

In scientific notation, a number is usually written as a figure between 1 and 9.99 multiplied by a power of 10. The power of 10 is called the exponent, and it indicates the number of places the decimal point must be moved to restore the number. The decimal point is moved to the right if the exponent is positive, and to the left if it is negative. For example, 150,000,000 is $1.5 \times 100{,}000{,}000$, which in scientific notation would be written 1.5×10^8. The figure 0.00000576 would be written 5.76×10^{-6}.

Modified from Pasachoff, J., *Contemporary Astronomy*. Philadelphia, Saunders College Publishing, 1985.

D-3 Relation between Fahrenheit, Celsius, and Kelvin temperature scales

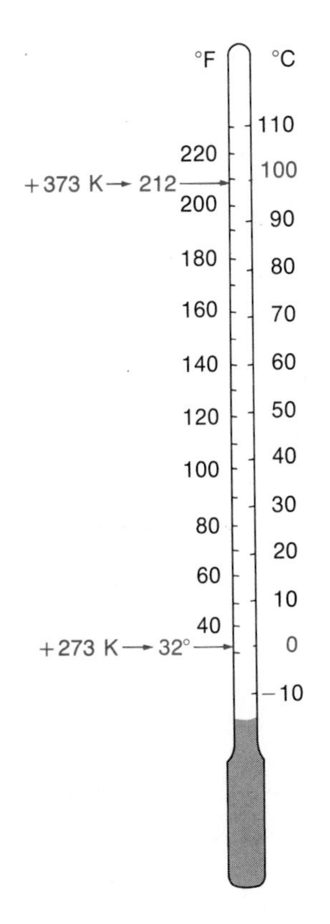

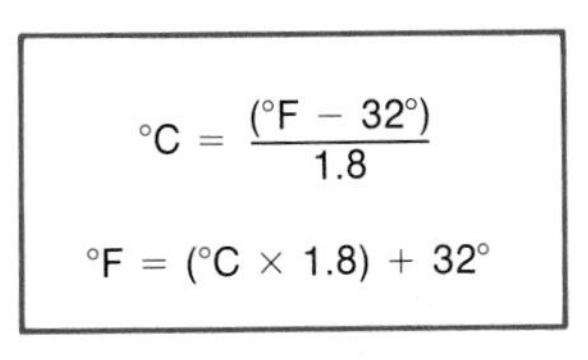

$$°C = \frac{(°F - 32°)}{1.8}$$

$$°F = (°C \times 1.8) + 32°$$

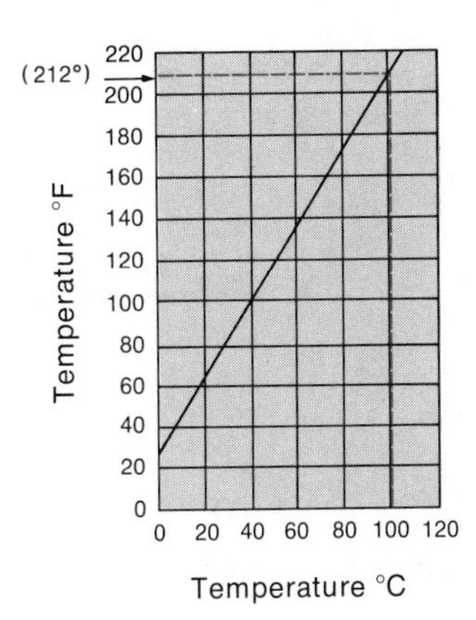

A third temperature scale, widely used in astronomy, is the *Kelvin scale*. It starts with the coldest possible temperature, −273°C and uses the same size degrees as does the Celsius scale. The Kelvin scale degrees are called *kelvins* and are abbreviated K (not °K). In converting Celsius to Kelvin, −273°C would be 0 kelvin, 0°C would be +273 K, and 100°C would be +373 K.

appendix e

earth, moon, and sun data

Volume of the earth = 1,083,230 km^3
Volume of the oceans = 1.4×10^9 km^3 (1.4 billion cubic kilometers)
Total mass of seawater = 1.4×10^{24} grams
Average depth of the ocean = 3,800 meters
Average elevation of land = 840 meters
Equatorial radius of the earth = 6,378 km
Polar radius of the earth = 6,356 km
Average density of the earth = 5.5 gm/cc
Average density of the moon = 3.3 gm/cc
Average density of the sun = 1.4 gm/cc
Mass of the earth = 5.976×10^{27} gm
Mass of the moon = 7.347×10^{25} gm
Mass of the sun = 1.971×10^{32} gm
Gravitational acceleration of the earth = 980 cm/sec^2 = 9.8 meters/sec^2
Mean distance to moon from earth = 384,393 km (238,860 mi)
Mean distance to sun from earth = 149,450,000 km (92,900,000 mi)

glossary

Aa. Basaltic lava having a jagged, rough, clinkery surface.

Ablation. The combined processes of melting, evaporation and sublimation by which a glacier wastes. The ablation zone is the lower margin of a glacier where loss of water exceeds additions.

Absolute humidity. The actual mass of water vapor contained in a given volume of moist air, expressed in grams per cubic meter.

Absolute instability. A category of stability that exists when the measured lapse rate is greater than the dry adiabatic lapse rate.

Absolute magnitude of star. The magnitude a star would appear to have if it were at a distance from us of 10 parsecs, or 32.6 light-years.

Absolute stability. A category of stability that exists when the measured lapse rate in an air column is less than the saturated lapse rate.

Absorption spectrum. Dark lines that are superimposed on a continuous spectrum.

Abyssal plains. Nearly level areas of the oceanic deeps below about 2000 meters,

Acceleration. The change in the velocity of a body per unit of time.

Achondrite. A type of stony meteorite that lacks the small rounded mineral bodies called chondrules.

Actualism. The concept that the laws of nature are invariant and responsible for changes that have occurred on earth through time.

Adiabatic lapse rate. The rate of decrease of temperature with height of a parcel of air lifted adiabatically through an atmosphere in hydrostatic equilibrium.

Adiabatic temperature change. A change in temperature that takes place without heat being gained from, or lost to, the environment.

Advective fog. Fog caused by the blowing of moist air over a cold surface and the resulting cooling of that air below the dew point.

Aerosol. A suspension of particles in air.

Air mass. An extensive body of air that is relatively homogeneous horizontally in terms of temperature and humidity.

Albedo. The ratio of the amount of incoming radiation reflected by a natural surface (ice, snow, water, clouds) to the amount incident upon that surface, commonly expressed as a percentage.

Alluvial fan. A cone-shaped deposit of unconsolidated, poorly sorted sediments made by a stream where it passes from an area of steep gradient to lower gradient.

Alluvium. Unconsolidated, poorly sorted, detrital sediments ranging from clay to gravel sizes and characteristically fluvial in origin.

Alpha particle. A particle, equivalent to the nucleus of a helium atom, that is emitted from an atomic nucleus during radioactive decay.

Altostratus. A middle-level cloud having the appearance of a gray sheet.

Ammonoid. An extinct group of cephalopods with coiled conchs divided into chambers by septa having fluted margins.

Amphiboles. Ferromagnesian minerals characterized by good prismatic cleavage in two directions intersecting at 56° and 124° and including hornblende and actinolite.

Amygdules. A gas cavity or vesicle in an igneous rock that has become filled with a secondary mineral, such as calcite or quartz.

Andesite. An extrusive igneous rock approximately intermediate in chemical composition between basalt and rhyolite. Andesites are composed of about 75 percent plagioclase and lack quartz or orthoclase.

Andesite line. A line drawn on a map of the Western Hemisphere that separates the region of basaltic extrusive rocks of the Pacific from the more andesite rocks of the circum-Pacific.

Angiosperm. Plants having flowers and seeds in a closed ovary — the "flowering plants."

Angle of repose. The maximum angle, measured in degrees, at which such material as loose rock, sand, or silt will remain stable.

Angular unconformity. An unconformity in which the older strata dip at a different (usually steeper) angle than younger strata.

Anion. An ion that carries a negative charge.

Annular drainage pattern. Streams and their tributaries that form a ringlike pattern because their development is controlled by the circular outcrop of strata associated with structural domes and basins.

Antecedent stream. A stream whose valley antedates the structures across which it flows, and which has been able to maintain its course largely unaffected by uplift, folding, or faulting.

Anthracite coal. Hard, black, lustrous coal containing a high percentage (80 to 90 percent) of fixed carbon and usually less than 5 percent volatile matter.

Anticline. A geologic structure in which strata are bent into an upfold or arch, and in which older rocks are found toward the center of curvature along the erosionally truncated surface.

Anticyclone. An atmospheric circulation pattern having a generally clockwise rotation in the Northern Hemisphere and counterclockwise rotation in the Southern Hemisphere.

Aphanitic. A textural term for rocks in which the crystalline components are too small to be recognized with the unaided eye.

Aphelion. The position of the earth's orbit farthest from the sun.

Apparent brightness. The amount of a star's light that reaches the earth, depending on its distance and its actual or intrinsic brightness.

Apparent magnitude. The measure of the observed light received from a star, determined by a standardized system.

Apparent solar day. The time that elapses between two successive noons.

Aquiclude. A rock unit that will not transmit water fast enough to provide an appreciable supply for a spring or well.
Aquifer. A formation that, because of good porosity and permeability, is able to transmit water in sufficient quantity to supply springs and wells.
Aragonite. An orthorhombic variety of calcium carbonate.
Arête. A narrow, jagged ridge developed by glacial erosion along the divide separating valley glaciers.
Arroyo. The channel of an intermittent or ephemeral stream.
Artesian well. A well in which the groundwater tapped has sufficient hydrostatic head to rise above the level of its aquifer.
Asbestos. A general term applied to minerals of the amphibole and serpentine families that are characterized by fibrous habit.
Asthenosphere. The zone within the earth between 50 and 250 km, below which seismic waves travel at much reduced speeds, possibly because of less rigidity in the rocks. Presumably, convective flow of material may occur in the asthenosphere, and plates of the lithosphere move over the asthenosphere.
Asymmetrical fold. A fold (anticline or syncline) in which the axial plane is inclined, and the two limbs dip unequally in opposite directions.
Atmosphere. The envelope of air surrounding the earth and bound to it by the earth's gravitational attraction.
Atmospheric pressure. The force exerted by the weight of the atmosphere per unit area.
Atoll. An island or circle of islands surrounding a central lagoon.
Atom. The smallest divisible unit of matter retaining the characteristics of a specific chemical element.
Atomic mass (atomic weight). A quantity essentially equivalent to the number of neutrons plus the number of protons in an atomic nucleus.
Atomic number. The number of protons in the nuclei of atoms of a particular element.
Attitude (of strata). The relation of some directional feature of a rock, such as a bedding plane or joint, to a horizontal plane. Attitude is usually recorded as a statement of strike and dip.
Autotroph. An organism that uses an external source of energy to produce organic nutrients from simple inorganic chemicals.
Autumnal equinox. An equinox is either of the two points of intersection of the sun's apparent path and the plane of the earth's equator. The autumnal equinox occurs on September 22d as the sun approaches the Southern Hemisphere.
Avalanche. A large mass of snow and ice in fast motion down a mountainside. Debris avalanches may be composed of rock waste and lack snow or ice.
Axial plane (of a fold). The plane that divides a fold as symmetrically as possible.
Bajada. A relatively flat surface formed adjacent to a mountain front by the coalescence of a series of alluvial fans.
Barograph. A self-recording barometer.
Barometer. An instrument used for measuring atmospheric pressure.
Barrier reef. A coral reef separated from the coast of the mainland or an island by a lagoon.
Base level. The limiting lower level to which stream erosion can proceed. Ultimate base level is equivalent to sea level.
Basalt. A dark-colored, finely crystalline, generally extrusive igneous rock composed predominantly of ferromagnesian minerals and plagioclase.
Basin. Structurally, a circular or elliptical downwarp of strata with younger beds at the center. In a depositional context, a basin refers to a depressed region that serves as a catchment area for sediments.
Batholith. A very large (at least 100 km^2 in area) intrusive igneous body of irregular shape.
Bathyal. Term that refers to the benthic environment from 200 to 2000 meters in depth.
Bauxite. A mixture of hydrous aluminum oxides and hydroxides commonly formed as a residual clay in tropical and subtropical regions. Bauxite is an ore of aluminum.
Bay barriers. A ridge of sand or gravel that is built up across the mouth of a bay, thus blocking entry into that bay.
Beach. The wave-washed, gently sloping accumulation of clastic sediment along a shore. The beach extends from the outermost breakers to the landward limit of wave action.
Benioff Seismic Zone. An inclined zone along which frequent earthquakes occur that marks the location of the plunging forward edge of a subducting lithospheric plate.
Benthic environment. All of the aquatic bottom environment from the shoreline to the deepest areas.
Benthos. The organisms that live at the bottom of the ocean, either stationary, attached, or able to move by crawling, burrowing, or swimming near the ocean floor.
Bergschrund. A deep crevasse located at the head of a valley glacier and separating the headwall of a cirque from the ice mass.
Berm. A terrace-like portion of a beach or backshore commonly formed by storm waves and composed of wave-transported sand and gravel.
Beta particle. A charged particle, essentially equivalent to an electron, emitted from an atomic nucleus during radioactive disintegration.
Bioclasts. Detrital particles of sediment, commonly composed of fragments of the skeletons of marine invertebrates or calcareous algae.
Biofacies. The biological aspect of a stratigraphic unit that differs discernably from that of adjacent units. Biofacies are recognized by their fossil content.
Biogenic. Pertaining to a deposit that originated as a result of physiological activities of organisms.
Bituminous coal. Compact, gray-black, brittle coal having 50 to 80 percent fixed carbon.
Black hole. A hypothetical body that has undergone gravitational collapse, from which not even light can escape because of enormous gravitational attraction.
Blowout. A basin or shallow pit excavated in the ground by wind erosion.
Bode's law. A numerical scheme that gives the radii of the orbits of the seven innermost planets and the radius of the asteroid belt.
Body seismic waves. Seismic waves that, like *P* and *S* waves, travel through the "body" of a medium rather than along a free surface.
Bolson. In arid regions, a closed basin or depression more or less rimmed by mountains and characterized by drainage directed toward the basin's center.
Bora. A cold winter wind that blows from a northerly direction over the shores of the Adriatic.
Bottomset beds. Layers of sediment deposited in an ocean or lake beyond the advancing margin of a delta and eventually covered by the delta.

Breakers. Waves that break along the shore or on encountering a shallow area, such as a reef.

Breccia. A clastic rock composed predominantly of angular fragments of granule size (2 to 4 mm) or larger.

Breeder reactor. An atomic reactor capable of producing more fissionable material than it consumes.

Caldera. A large, rather circular, steep-sided volcanic depression commonly at the summit of a volcano and containing volcanic vents.

Caliche. Calcium carbonate precipitated as surface or near-surface crusts by the evaporation of moisture in the pore spaces of soils.

Calorie. A unit of heat defined as the amount of heat required to raise the temperature of one gram of water through one degree Celsius at sea level pressure.

Capacity (of a stream). The maximum amount of solid and dissolved material that a stream can carry under a given set of conditions.

Capillary fringe. That zone immediately above the water table along which water rises in void spaces by capillarity.

Carbonation. The chemical addition of carbon dioxide to earth materials during weathering.

Carbonate spar. A variety of limestone composed of a mosaic of calcite crystals.

Cataclastic metamorphism. Metamorphism involving deformation of rocks by shattering without appreciable chemical reconstitution.

Cation. A positively charged ion.

Cave travertine. Calcium carbonate deposited from solution in limestone caverns, often in the form of stalactites and stalagmites.

Celestial sphere. The apparent sphere of the sky, on which stars and other celestial bodies appear to be located.

Celsius scale. The standard temperature scale in the metric system, in which the freezing point of water is at 0° and the boiling point at 100°.

Cenozoic Era. The most recent of the geologic eras, beginning about 65 million years ago and extending to the present.

Centrifugal force. The apparent outward force experienced by an object moving in a circular path. Centrifugal force is a manifestation of inertia—the tendency for moving things to travel in straight lines.

Centripetal force. A center-directed force that diverts a body from a straight path into a curved path.

Chalk. A white, soft, fine-grained variety of limestone composed largely of the calcium carbonate-skeletal remains of marine planktonic organisms.

Chemical element. A substance whose atoms all have the same atomic number.

Chemical precipitation. The formation of solid particles in a solution as a result of changes in composition, temperature, pressure, or evaporation.

Chemical weathering. The combination of chemical reactions that act on rocks to cause their decomposition. During chemical weathering existing minerals may be dissolved and/or converted to new minerals more stable under conditions at the earth's surface.

Chatter marks. Scars made in bedrock as debris carried along at the base of a glacier chip into the underlying surface.

Chert. A dense, hard, sedimentary rock or mineral composed of submicrocrystalline quartz. Unless colored by impurities, chert is white, unlike flint, which is gray or black.

Chinook. A hot, dry wind that develops on the eastern side of the Rocky Mountains.

Chondrites. Stony meteorites that contain rounded silicate grains called chondrules believed to have formed by solidification of droplets of liquid silicates.

Chromosphere. The portion of the sun's atmosphere that lies between the photosphere and the corona.

Cinder cone. A steeply sloping conical hill composed of pyroclastics built up around a volcanic vent.

Cirque. A steep-walled recess in the side of a mountain sculptured by glacial erosion.

Cirrostratus. A high-level cloud appearing as a whitish, sometimes fibrous veil, which may totally cover the sky.

Cirrus. A high-level cloud composed of detached elements in the form of delicate filaments, bands, or patches.

Clastic. Consisting of broken fragments of rock, mineral, or skeletal material.

Cleavage. The tendency of certain minerals to split in preferred directions along planes parallel to real or possible crystal faces.

Coal. A combustible earth material containing at least 70 percent by volume of carbonaceous matter.

Coccoliths. Tiny discoidal calcareous platelets secreted by marine, planktonic, golden-brown algae known as coccolithophorids.

Cold front occlusion. An occluded front in which the front is colder than the air ahead of it.

Columnar joints. Joints that break igneous rocks roughly into long, usually six-sided columns. Columnar jointing is particularly characteristic of basalt sills and flows.

Competence (of a stream). The largest particle that a stream can transport under a given set of conditions.

Concordant age. An age determination that has been confirmed by agreement of data from two separate isotope ratios.

Concordant pluton. An intrusive igneous body having contacts parallel to the stratification or foliation of the preexisting intruded rock.

Condensation. The process by which water vapor becomes liquid water.

Condensation level. The height at which a rising parcel of air becomes saturated, leading to cloud formation.

Condensation nuclei. Particles upon which water vapor condenses in the atmosphere.

Conditional instability. The condition that exists when the measured lapse rate is greater than the dry adiabatic lapse rate. Because the dry adiabatic lapse rate is larger, the greater measured lapse rate gives an unstable atmosphere for either saturated or unsaturated air.

Conduction (thermal). A heat-transfer process by molecular vibrations and collisions but without transport or exchange of molecules (compare *convection*).

Cone of depression. Conical depression of the water table in the area surrounding a well in which water is being withdrawn.

Conglomerate. Coarse-grained clastic sedimentary rock composed of rounded rock and mineral fragments larger than 2 mm in diameter.

Connate water. Water trapped in void spaces in a sedimentary rock at the time it was deposited.

Conodonts. Small conical or variously shaped fossils composed of calcium phosphate.

Consequent stream. Streams whose courses are a consequence of an original or preexisting slope.

Constant gases. The gases in the atmosphere that vary little in proportion through time or from place to place. The three most important constant gases are nitrogen, oxygen, and argon.

Constellation. A configuration of stars named for a particular object, person, or animal; or a region of the sky assigned to a particular configuration.

Continental rise. The ocean floor beyond the base of the continental slope, generally with lower gradient than the continental slope.

Continental shelf. The shelflike, gently sloping area extending from the shore of a continent seaward to a line marked by an abrupt increase in slope.

Continental slope. The submerged region of steep slope extending from the seaward edge of the continental shelf down to the upper margin of the continental rise.

Convection (thermal). A heat-transfer process in which energy is conveyed by transport of molecules.

Core (of the earth). The earth's innermost zone, whose outer boundary is at a depth of about 2900 km.

Coriolis effect. The tendency for a body moving on the surface of the earth to be deflected to the right in the Northern Hemisphere and the left in the Southern Hemisphere as a result of the earth's rotation.

Corona (of the sun). The outermost region of the sun, characterized by temperatures of millions of kelvins.

Corrasion. Mechanical erosion accomplished by moving ice, flowing water, and/or wind as they and the rock particles they contain strike or move against solid rock.

Correlation (stratigraphic). The matching of rock units from different areas in order to determine their age equivalence.

Cosmic rays. Extremely high energy particles, mostly protons, which move through the galaxy and frequently strike the earth's atmosphere.

Country rock. The preexisting rocks that are penetrated by, and that surround, igneous intrusive bodies.

Covalent bonding. Chemical linkage between two atoms produced by sharing electrons in the region between the atoms.

Craton. That portion of a continent that is composed of very ancient crystalline rocks, and that has been tectonically stable for several hundred million years.

Creep. The slow downhill movement of regolith that results from continuous rearrangement of particles. Creep, measured in centimeters per year, is the slowest form of mass wasting.

Cross-bedding (cross-stratification). Beds or laminations arranged at an oblique angle to the main bedding.

Crust. The outer layer of the earth extending from the solid surface to Mohorovičić discontinuity.

Cryoturbation. Mixing and sorting of regolith as a result of alternate thawing and freezing.

Cryptozoic Eon. The span of geologic time that preceded the Cambrian Period.

Cumulus. A cloud having good vertical form, somewhat resembling a cauliflower, but not as tall as cumulonimbus.

Cup anemometer. A device employing three or four hemispherical cups free to rotate about an axis as a means of measuring wind speed.

Cutoff. A relatively new stream channel formed when the stream cuts through the neck of a meander.

Cyclone. Circular wind systems that move in a counterclockwise pattern in the Northern Hemisphere and clockwise in the Southern Hemisphere.

Daily mean temperature. The average value of temperature over a period of 24 hours.

Darcy's law. A law that governs the relation between the velocity of percolation of groundwater, the permeability of the aquifer, and the hydraulic gradient. Darcy's Law can be expressed as $v = \frac{kh}{l}$ in which V is velocity, k an experimentally determined constant for the water-bearing material, h the head, and l the length or distance over which the water moves.

Declination (celestial). The angular distance north or south of the celestial equator.

Deep-sea fan. A submerged fan-shaped deposit of clastic sediment often located at the seaward margins of submarine canyons.

Deep-sea trench. A deep, narrow trough in the ocean floor believed to mark the line along which an oceanic tectonic plate is undergoing subduction.

Deflation. The removal of fine-grained sediment from a surface by wind.

Delta. An accumulation of alluvial sediments deposited at the mouth of a stream where it enters a sea or a lake.

Dendritic drainage pattern. A drainage pattern that branches irregularly, rather like the pattern of branching in trees.

Density. The mass per unit volume of a substance expressed in grams per cubic centimeter.

Desert pavement. A blanket-like residual concentration of wind-eroded, closely packed rock fragments or gravel that remains at the surface after wind has removed finer particles.

Dew point. The temperature to which air must be cooled for saturation to occur.

Diatom. A unicellular, microscopic plant commonly having a siliceous case or frustule.

Diatreme. A volcanic vent that has been filled with angular rock fragments resulting from explosive eruption.

Dike. A discordant tabular body of igneous rock that cuts across preexisting stratification or structures in the country rock.

Diorite. A usually intrusive igneous rock composed primarily of sodic plagioclase, hornblende, biotite, and/or pyroxene.

Dip. The angle formed by an inclined layer of rock or other planar feature with a horizontal plane.

Discharge (of stream). The amount of water passing through a given cross-section of a stream in a given time.

Discordant pluton. A pluton that cuts across the bedding, structure, or foliation of the country rock it intrudes.

Disintegration. (See mechanical weathering.)

Disseminated magmatic deposit. An ore deposit in which economically valuable minerals are scattered throughout the host rock.

Distributaries. The downstream branches of a stream, as seen in a river that divides upon reaching a delta.

Divergent tectonic plate boundary. The boundary along which tectonic plates move apart to permit upwelling and the formation of new lithosphere.

Dolostone. A sedimentary rock composed largely of the mineral dolomite, $CaMg(CO_3)_2$.

Dome. A rather symmetrical uparching of layered rocks so that they dip away about equally from a center point.

Doppler shift. The change in wavelength that results when the source of waves and the observer are moving relative to each other.

Drainage basin. The area or region that contributes water to a specified stream.
Drainage density. The cumulative length of all the channels in a drainage system divided by the total surface area of the drainage basin.
Drainage divide. The upland tract or ridge that separates adjacent streams or drainage basins.
Drainage system. A given stream and all of its tributaries.
Drift. Earth materials, such as clay, sand, gravel, or boulders, deposited as a result of glacial activity.
Drizzle. Precipitation consisting of very small and numerous water droplets (less than 0.5 mm in diameter).
Drumlin. A streamlined hill usually composed of till, having its long axis in the direction of glacial movement.
Dry adiabatic lapse rate. The rate of decrease of temperature with height in an air column when that rate is the same as the rate of cooling in an unsaturated adiabatic process.
Dynamothermal metamorphism. (See regional metamorphism.)
Earth flow. The sluggish, erratic flow of clayey or silty regolith down relatively gentle slopes.
Echo sounding. A geophysical method for determining depth of water by measuring the time required for a sound signal to travel to the bottom and return.
Eclipsing variable star. A binary star in which one member periodically hides the other.
Ekman spiral. The theoretical representation of the effect of a steady wind blowing across an ocean of uniform viscosity and unlimited depth and breadth, such that the surface layer of water would move at an angle of 45° to the right of the wind direction in the Northern Hemisphere and 45° to the left in the Southern Hemisphere.
Elastic limit. The maximum amount of stress to which an earth material can be subjected before it will begin to deform permanently by fracture or flow.
Elastic rebound. The abrupt release of elastic strain that has slowly accumulated in a rock mass.
Electron. A component of an atom that exists outside the nucleus, has very low mass, and carries one unit of negative charge.
Entrenched meander. A meander deeply incised into bedrock as a result of regional uplift (also called an incised meander).
Ephemeral stream. A stream or stream segment that derives its discharge almost entirely from precipitation and thus is often dry between periods of rainfall.
Epicenter. A point on the earth's surface directly above the focus (true center) of an earthquake.
Epicontinental marginal sea. A depressed, submerged area of the continental margin having greater topographic irregularity than the continental shelf.
Epoch. The chronologic subdivision of a geologic period. Rocks deposited during an epoch constitute the series for that epoch.
Erratic. A rock fragment that has been transported from a distant source, usually by glacial or floating ice, and that rests on bedrock of a different kind.
Eruptive variable star. A star whose changes in light are erratic or explosive.
Esker. A sinuous ridge composed of stratified drift.
Eugeosyncline (eugeocline). The oceanward and more unstable portion of a geosyncline, characterized by great thickness of poorly sorted clastic sediment, siliceous sediment, and volcanics.
Eustatic. Worldwide, simultaneous changes in sea level, such as may result from a change in the volume of continental glaciers.
Evaporite. Sediment precipitated from a water solution as a result of evaporation. Evaporite minerals include anhydrite gypsum ($CaSO_4$) and halite ($NaCl$).
Exfoliation. The breaking off of successive thin outer shells of a rock mass in response to weathering.
Exosphere. The extreme uppermost reaches of the earth's atmosphere.
Facies. A particular aspect of one part of a rock body that is distinct from adjacent parts.
Fahrenheit scale. A temperature scale on which 32° denotes the temperature of melting ice and 212° the temperature of boiling water.
Fault breccia. A breccia, the angular rock fragments of which were produced by crushing between the walls of a fault.
Fault plane. A generally planar fault surface.
Feldspars. A group of abundant rock-forming alumino-silicate minerals containing sodium, calcium, or potassium and having framework atomic structure. Feldspars are the most common of any mineral group in crustal rocks, constituting an estimated 60 percent of the earth's crust.
Ferromagnesian minerals. Generally dark-colored silicate minerals containing iron and magnesium, and including olivine and pyroxene.
Fetch. The continuous area of water across which wind blows in a constant direction and generates waves.
Filter pressing. The process whereby the liquid portion of a partially crystallized magma is squeezed out when the magma is compressed by earth movements.
Firn. Snow that has been partially consolidated by successive thawing and freezing but has not yet been altered into compact glacial ice.
Fissility. That property of rocks, including shale, that causes them to split into thin slabs parallel to bedding.
Fission (nuclear). The splitting of an atomic nucleus.
Focal length (of telescope). The distance from the center of a lens or mirror to the focal point.
Fjord. A narrow arm of the sea representing a former stream valley that had been further eroded by a glacier. Fjords are recognized by their steep walls, uneven bottom profile, and streams that enter as waterfalls and rapids.
Flood basalt (plateau basalt). A layered lava flow that issued from fissures and extends over an entire region as a sequence of flat or nearly flat layers.
Flood plain. The lowland that borders a stream, which is composed of sediments deposited by the stream, and which is dry except when the stream overflows its banks during flood stages.
Focus (of an earthquake). The true center of an earthquake and the point at which rupture occurs and strain energy is converted to elastic wave energy.
Fog. Numerous tiny droplets of water suspended in air close to the surface of the ground.
Fold. A bend or flexure in layered rocks.
Foliation. A textural feature especially characteristic of metamorphic rocks in which laminae develop by growth or realignment of minerals into roughly parallel orientation.
Footwall. The mass of rock that lies beneath an inclined fault.
Foraminifers. Mostly marine and microscopic protozoans that

commonly secrete skeletons composed of calcium carbonate.

Foreset bed. The inclined layers deposited along the forward slope of a body of sediment that is advancing, as in a delta or a dune.

Formation. A mappable, lithologically distinct body of rock having recognizable contacts with adjacent rock units.

Fractional crystallization. The separation of components of a cooling magma by sequential formation of particular mineral crystals at progressively lower temperatures.

Friction-layer wind. Wind that is affected by friction from the earth's solid surface, occurring within the layer of air below 1 km.

Fringing reef. A coral reef that is attached to an island, thus forming a fringe.

Frontal fog. Fog formed along a weather front.

Fumarole. A vent at the earth's surface that emits volcanic gases, steam, and hot water.

Fusion. The joining of nuclei of atoms into heavier nuclei.

Gabbro. A coarsely crystalline (phaneritic) igneous rock having the same mineralogic composition as basalt.

Galaxy. An aggregate of stars and planets separated from other such aggregates by distances greater than those between member stars.

Galilean moons. The four brightest moons of Jupiter (Io, Europa, Ganymede, and Callisto).

Geochronology. The study of time as applied to planetary history.

Geostrophic wind. A wind that blows parallel to isobars, above a height of 600 meters.

Giant star. A star of large radius and luminosity.

Glaze. A coating of ice formed on objects by the freezing of a film of supercooled rain, drizzle, or fog.

Globular cluster. A compact, spherical system of many thousands of stars.

Gradient wind. Any horizontal wind velocity in which balance is achieved between the Coriolis force, pressure force, and centrifugal force.

Graupel. A pellet formed as supercooled droplets freeze onto snow crystals.

Hail. Pellets or balls of ice that fall from cumulonimbus clouds.

Harmonic law. Kepler's third law of planetary motion, which states that the squares of times that the planets require to make complete revolutions around the sun are proportional to the cubes of their mean distance from the sun.

Heat capacity. The amount of heat that must be supplied to a cubic centimeter of a substance to warm it 1°C.

Heat equator. A line on a map joining places on the earth that have the highest average temperature.

High-pressure area. An anticyclone or area of high atmospheric pressure that has closed circulation.

Humidity. The concentration of water vapor in the atmosphere at any given time.

Hurricane. A severe tropical cyclonic storm.

Hygrometer. A meteorological device used to measure relative humidity.

Hygroscopic dust particles. Particles that are able to attract water from the atmosphere when it is unsaturated.

Inertia. The tendency of a body at rest to stay at rest and of a body that is in motion to continue its motion in a straight line.

International date line. An imaginary line at about 180° longitude, which when crossed, changes standard time by 24 hours.

Intertropical convergence zone. The region near the equator where the northeast trade winds and southeast trade winds converge.

Inverse square law. As applied to the propagation of light, this law states that light spreads out and covers an increasing area in proportion to the square of the distance it has traveled from its source.

Ionosphere. The layer of the atmosphere above the stratosphere in which molecules of ionized gas exist.

Isotherm. A line on a map connecting points having the same temperature.

Jet stream. A narrow band of very strong, high-altitude winds.

Kelvin scale. A temperature scale that assigns zero to the point where a gas reaches its absolute minimum temperature of −273°C, and which uses the same size degrees as the Celsius scale.

Knot. The unit of speed in the nautical system. A knot is equal to one nautical mile (1.1508 statute miles) per hour.

Land and sea breezes. The cycle of diurnal local winds occurring on coastlines as a result of differences in land and sea surface temperature.

Lapse rate. The change in temperature of air with altitude.

Latent heat. The heat released or absorbed with a change in state of water.

Law of areas. Kepler's second law, which states that a line joining a planet and the sun sweeps equal areas in its orbital plane in equal intervals of time.

Law of elliptical orbits. Kepler's first law of planetary motion, which states that the orbit of a planet is an ellipse with the sun at one of the foci.

Law of gravitation. The force of attraction between any two objects is proportional to the product of their masses and inversely proportional to the square of the distance between them.

Law of reacting forces. Newton's third law, which states that to every action there is an equal and opposite reaction.

Light-year. The distance traveled by light in a vacuum in one year. One light-year equals about 9.7×10^{12} km.

Lightning. The visible manifestation of electrical discharges in the atmosphere.

Low-pressure area. An area of low pressure represented on a weather map by concentric closed isobars, within which winds circulate counterclockwise in the Northern Hemisphere.

Main sequence star. A star on a narrow band on the Hertzsprung-Russell diagram along which lie the majority of stars.

Mass. The total amount of matter contained in a body.

Mean solar day. The average of all the apparent solar days through an entire year.

Measured lapse rate. The measured change in temperature with altitude.

Meridian of longitude. An imaginary line drawn from pole to pole, used with parallels of latitude to determine location on the earth.

Mesosphere. A relatively warm layer of the atmosphere above the stratosphere and below the ionosphere. In geology, the lower part of the earth's mantle.

Millibar. A standard unit of pressure used in international weather observations.

Mistral. A cold, dry north wind that blows from the Alps down the Rhone Valley in France.

Moist adiabatic lapse rate. The expansion cooling rate of a column of rising saturated air (4° to 10°C per km).

Momentum. Mass times the velocity of a body.

Monsoons. A seasonal wind blowing from the cool ocean to warmer land in the summer and from cool land to the warmer ocean in the winter.

Mountain breeze. The downslope winds generated by air cooled at night along mountain slopes.

Nebula. A cloud of interstellar gas and dust.

Nimbus. A heavy cloud from which rain falls.

Nova. A star that increases in brightness by several magnitudes as a result of a violent explosion.

Objective (of telescope). The principal image-forming component of a telescope.

Obliquity of the ecliptic. The 23.5° angle between the plane of the ecliptic and that of the celestial equator.

Occluded front. A complex front that forms when a cold front overtakes a warm front.

Orographic lifting. The lifting of air as it encounters mountainous barriers.

Ozone (O_3). Oxygen molecules composed of three atoms developed as a result of ultraviolet radiation of O_2 molecules in the stratosphere.

Parallax (of star). The apparent angular movement of a nearby star with respect to more distant stars. Parallax results from the movement of the earth around the sun.

Parallel of latitude. An imaginary line drawn on the earth's surface parallel to the equator. Together with meridians, parallels form a grid for locating points on the globe.

Parsec. The distance of an object having a stellar parallax of one second of arc. A parsec is equal to 3.26 light-years.

Partial pressure of water vapor. That part of the total pressure of air that is caused by the motion of water-vapor molecules.

Perihelion. That position along the path of a body revolving around the sun that is closest to the sun.

Plane of the ecliptic. The plane of the earth's orbit around the sun.

Polar easterly. A weak easterly planetary wind located near the poles.

Polar front. A frontal zone located between air masses that originate near the poles and those that are of tropical origin.

Population I stars. A class of relatively young stars containing rather abundant amounts of metals.

Population II stars. A class of relatively older stars having very low amounts of metals.

Precession (of earth). The slow conical motion of the earth's axis of rotation.

Precipitation. The collective name for all forms of moisture that falls from the atmosphere.

Pressure gradient. The change in atmospheric pressure per unit of horizontal distance.

Pressure-gradient force. The force exerted on air by a difference in pressure between two points.

Prevailing wind. Wind that characteristically blows in the same general direction over a given period of time.

Prime meridian. The 0° longitude line, which passes through Greenwich, England.

Principle of cross-cutting relationships. Geologic features, such as faults, dikes, and veins, must be younger than the rocks they cut across.

Principle of faunal succession. Animals have changed continuously through time, so the faunas for one part of geologic time will be different from those of earlier and later times.

Principle of superposition. In a sequence of layered rocks or strata, the oldest layers are at the bottom and sequentially younger rocks occur upward through the column.

Principle of uniformitarionism. The past history of the earth can be interpreted in terms of currently known natural laws and processes.

Psychrometer. A hygrometer that uses a wet-and-dry bulb thermometer.

Pulsar. A rapidly pulsating source of electromagnetic waves produced by rapid rotation of a neutron star.

Pulsating variable star. A variable star that periodically pulsates in size and brightness.

Radiation (thermal). The process by which heat is transmitted through space.

Radiation fog. Fog formed as objects on the ground cool by losing radiation.

Radiosonde. A device containing electronic components that record and transmit temperature, pressure, and humidity conditions at high altitudes.

Reflector telescope. A telescope in which the main optical component is a concave mirror.

Refractor telescope. A telescope in which the principal optical component is a lens or lens system.

Relative humidity. The amount of water vapor in the air divided by the the amount of water vapor at saturation expressed as a percentage.

Revolution. The orbiting of one body around another.

Right ascension. Celestial longitude, measured eastward along the celestial equator in hours of time from the vernal equinox.

Rotation, planetary. The spinning of a planetary body on its axis.

Scattering of radiation. The process whereby small particles intercept radiation and deflect it in all directions but do not alter the radiation in wavelength or amount.

Secondary air pollutants. An air pollutant produced by chemical alteration of an earlier substance.

Sidereal day. The time between two successive transits of the observer's meridian by a given star.

Sidereal time. Time measured with respect to a given star; the local hour angle of the vernal equinox.

Smog. Originally considered a mixture of smoke and fog, but now considered any air pollution.

Solar time. A system of time based on the mean solar day or the time required for the sun to make two successive transits of the observer's meridian.

Spectroscope. A device having one or more prisms or a diffraction grating, used to separate a beam of radiation into the succession of rays of different wavelengths that compose it.

Spectrum. A band of colors, proceeding from red to violet, that is produced when light is dispersed by a prism.

Stationary front. The zone between two air masses when neither is displacing the other.

Steppes. The grassy, treeless plains such as are found in Siberia and other semiarid regions.

Stratopause. The zone at the top of the stratosphere that separates the stratosphere from the overlying mesosphere.

Stratosphere. The atmospheric layer above the troposphere in which temperature is constant or rises with height.

Stratus. A low, relatively thin and continuous, uniformly gray, sheetlike cloud.

Stromatolite. A laminated body of calcium carbonate produced by marine algae.

Subtropical high-pressure area (horse latitudes). The semipermanent high-pressure regions centered over the oceans at about 30° to 35° latitude.

Supernova. A stellar explosion in which a star suddenly and tremendously increases its luminosity.

Temperature. The property of a body that determines the flow of heat from it to another body.

Temperature inversions. A condition where air temperature increases with height (rather than the more usual decrease with height).

Thermograph. A device that produces a continuous record of temperature.

Tornado. A very violent air whirl of relatively small area, which frequently gives wind velocities of over 300 km per hour.

Total pressure. The sum of all the partial pressures of the gases in air.

Trade winds. The persistent easterly winds that blow from the tropical high-pressure belts toward the equatorial region of low pressure, from northeast in the Northern Hemisphere and southeast in the Southern Hemisphere.

Tropopause. The upper limit of the troposphere.

Troposphere. The layer of the atmosphere extending from the earth's surface to the tropopause, in which temperature falls with increasing altitude.

Upslope fog. Fog formed as air flows up the slope of a topographic barrier.

Vapor pressure. The pressure exerted by water vapor alone. It is the partial pressure due to molecules of water vapor in air.

Wave cyclone. On a weather front, the cyclone occurring at the crest of a wave.

Weather front. The line of separation between warm and cold masses of air.

Westerly. A persistent west-to-east wind in the lower troposphere of middle latitudes and in the upper troposphere over most of the globe.

White dwarf. A star that has depleted nearly all of its nuclear fuel and has collapsed to a very small size.

Wind. Air in horizontal movement.

Wind rose. A diagram showing the proportion of winds blowing from each of the main points of the compass and their strength for a given locality.

index

In this index, definitions are indicated by boldface page numbers. Illustrations are indicated by an asterisk. A small *t* following a page number indicates a table. Terms in the glossary are not included.

PERIODIC TABLE OF THE ELEMENTS

IA	IIA	IIIB	IVB	VB	VIB	VIIB	VIII	VIII	VIII	IB	IIB	IIIA	IVA	VA	VIA	VIIA	0
1 H 1.0079																1 H 1.0079	2 He 4.00260
3 Li 6.941	4 Be 9.01218											5 B 10.81	6 C 12.011	7 N 14.0067	8 O 15.9994	9 F 18.998403	10 Ne 20.179
11 Na 22.98977	12 Mg 24.305											13 Al 26.98154	14 Si 28.0855	15 P 30.97376	16 S 32.06	17 Cl 35.453	18 Ar 39.948
19 K 39.0983	20 Ca 40.08	21 Sc 44.9559	22 Ti 47.90	23 V 50.9415	24 Cr 51.996	25 Mn 54.9380	26 Fe 55.847	27 Co 58.9332	28 Ni 58.70	29 Cu 63.546	30 Zn 65.38	31 Ga 69.72	32 Ge 72.59	33 As 74.9216	34 Se 78.96	35 Br 79.904	36 Kr 83.80
37 Rb 85.4678	38 Sr 87.62	39 Y 88.9059	40 Zr 91.22	41 Nb 92.9064	42 Mo 95.94	43 Tc (98)	44 Ru 101.07	45 Rh 102.9055	46 Pd 106.4	47 Ag 107.868	48 Cd 112.41	49 In 114.82	50 Sn 118.69	51 Sb 121.75	52 Te 127.60	53 I 126.9045	54 Xe 131.30
55 Cs 132.9054	56 Ba 137.33	57 ★La 138.9055	72 Hf 178.49	73 Ta 180.9479	74 W 183.85	75 Re 186.207	76 Os 190.2	77 Ir 192.22	78 Pt 195.09	79 Au 196.9665	80 Hg 200.59	81 Tl 204.37	82 Pb 207.2	83 Bi 208.9804	84 Po (209)	85 At (210)	86 Rn (222)
87 Fr (223)	88 Ra 226.0254	89 ▾Ac 227.0278	104 Unq (261)	105 Unp (262)	106 Unh (263)	107 Uns		109									

★ Lathanide Series

58 Ce 140.12	59 Pr 140.9077	60 Nd 144.24	61 Pm (145)	62 Sm 150.4	63 Eu 151.96	64 Gd 157.25	65 Tb 158.9254	66 Dy 162.50	67 Ho 164.9304	68 Er 167.26	69 Tm 168.9342	70 Yb 173.04	71 Lu 174.967

▾ Actinide Series

90 Th 232.0381	91 Pa 231.0359	92 U 238.029	93 Np 237.0482	94 Pu (244)	95 Am (243)	96 Cm (247)	97 Bk (247)	98 Cf (251)	99 Es (252)	100 Fm (257)	101 Md (258)	102 No (259)	103 Lr (260)

Note: Atomic masses shown here are 1977 IUPAC values.